Proceedings of 10th International Kimberlite Conference

D. Graham Pearson
Herman S. Grütter • Jeff W. Harris
Bruce A. Kjarsgaard
Hugh O'Brien • N. V. Chalapathi Rao
Steven Sparks
Editors

Proceedings of 10th International Kimberlite Conference

Volume 1

Special Issue of the Journal of the Geological Society of India

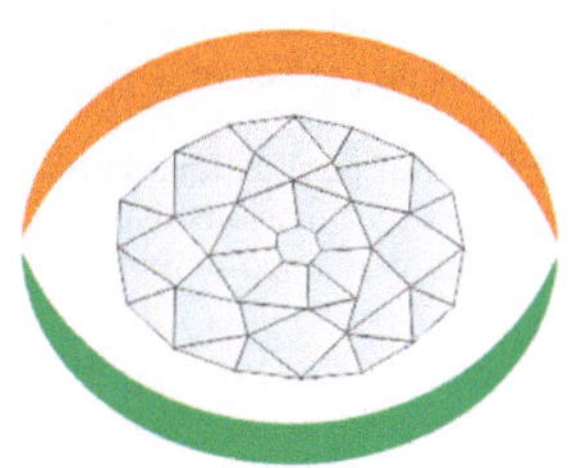

Editors
D. Graham Pearson
Herman S. Grütter
Department of Earth and Atmospheric Sciences
University of Alberta
Edmonton, AB
Canada

Jeff W. Harris
School of Geographical and Earth Sciences
University of Glasgow
Glasgow
UK

Bruce A. Kjarsgaard
Geological Survey of Canada
Ottawa, ON
Canada

Hugh O'Brien
Finland Isotope Geosciences Laboratory
Geological Survey of Finland (GTK)
Espoo
Finland

N. V. Chalapathi Rao
Department of Geology
Banaras Hindu University
Varanasi, Uttar Pradesh
India

Steven Sparks
School of Earth Sciences
Bristol University
Bristol
UK

ISBN 978-81-322-1169-3 ISBN 978-81-322-1170-9 (eBook)
DOI 10.1007/978-81-322-1170-9
Springer New Delhi Heidelberg New York Dordrecht London

Library of Congress Control Number: 2013942130

Printed on acid-free paper

Springer is part of Springer Science+Business Media (www.springer.com)

IKC Group Photograph

The 10th International Kimberlite Conference

International Kimberlite Conferences (IKCs) are special events that are held across the world once in 4–5 years. IKC is the confluence platform for academicians, scientists and industrial personnel concerned with diamond exploration and exploitation, petrology, geochemistry, geochronology, geophysics and origin of the primary diamond host rocks and their entrained xenoliths and xenocrysts (including diamond) to get together and deliberate on new advances in research made in the intervening years. Ever since the organization of the first IKC in 1973 and its tremendous success, the entire geological world eagerly looks forward to subsequent such conferences with great enthusiasm and excitement. The scientific emanations from IKCs continue to make significant impact on our understanding of the composition, nature and evolution of the planet we live on. The previous conferences were held at Cape Town (1973), Santa Fe, New Mexico (1977), Clermont-Ferrand, France (1982), Perth, Western Australia (1987), Araxa, Brazil (1991), Novosibirsk, Russia (1995), Cape Town (1998), Victoria, Canada (2003) and Frankfurt, Germany (2008).

The 10th IKC was held at Bangalore, India between 6 and 11th February 2012. The conference was organized by the Geological Society of India in association with the government organizations, academic institutions and Indian diamond mining companies. About 300 delegates from 36 countries attended the conference and 224 papers were presented. The papers include 78 oral presentations and 146 poster presentations on following topics: Kimberlite geology, origin, evolution and emplacement of kimberlites and related rocks, petrology and geochemistry of metasomatised lithospheric mantle magmas, diamond exploration, cratonic roots, diamonds, diamond mining and sustainable developments and policies and governance of diamond exploration. Pre- and post-conference field trips were organized to (i) the diamond bearing kimberlites of Dharwar Craton in South India, (ii) lamproites of Bundelkhand Craton in Northern India and (iii) diamond cutting and polishing industry of Surat, Gujarat in Western India. A series of social and cultural programmes depicting cultural diversity of India were organized during the conference. The Kimberlite fraternity enjoyed yet another socially and scientifically successful conference.

Cultural programmes organized during the 10th IKC

About the Editors

Dr. D. Graham Pearson is Canada Excellence Research Chair—Arctic Resources in the Department of Earth and Atmospheric Sciences, University of Alberta, Edmonton, Canada. Dr. Pearson obtained his B.Sc. from the Royal School of Mines, Imperial College, London and his Ph.D. from Leeds University. He taught at Durham University for 15 years, becoming Professor of Geochemistry and now holds a CERC research chair in Arctic Resources specialising in diamonds, kimberlites and cratonic roots. Pearson has been a member of the International Kimberlite Conference Advisory Committee since 2007.

Dr. Herman S. Grütter is currently associated with the Department of Earth and Atmospheric Sciences, University of Alberta, Edmonton, Canada. Dr. Grütter has over 20 years' senior-mining, junior-mining and consulting experience in kimberlite targeting, exploration and early stage resource definition on projects spanning the globe. He obtained a B.Sc. (Hons) from the University of Cape Town in 1986 and a Ph.D. in metamorphic petrology from the University of Cambridge in 1993. He maintains applied research interests in mantle mineralogy and petrology, and in craton evolution.

Dr. Jeff W. Harris retired in 2006 after a distinguished academic career at the University of Glasgow. Since retirement he has held an Honorary Research Fellowship at Glasgow. For over 30 years, he was a consultant to De Beers Consolidated Mines, managing their worldwide outside diamond research programmes conducted at Universities and equivalent institutions; a research endeavour which played a major part in furthering our understanding of geochemical processes operating in the mantle.

Dr. Bruce A. Kjarsgaard is a Senior Research Scientist in Geological Survey of Canada, Ottawa, Canada. Dr. Kjarsgaard received his Ph.D. in Geology (experimental petrology) from the University of Manchester, England in 1990. He has been with the Geological Survey of Canada (GSC) in Ottawa since 1990, working as an economic geologist. His research involves broad-based petrological studies (including mineralogy, geochemistry and volcanology) of mantle-derived volatile-rich ultramafic magmas (kimberlite, lamprophyre, nephelinite–carbonatite) and their attendant economic mineralization (diamonds, and niobium and REE, respectively).

Dr. Hugh O'Brien is a Senior Research Scientist in Finland Isotope Geosciences Laboratory (SIGL), Geological Survey of Finland (GTK), Espoo, Finland. Dr. O'Brien received his B.S. in 1982 from University of Minneapolis and his Ph.D. from UW in Seattle in 1988. After a short stint (1988–1991) with the GTK, he returned to Seattle for post-doctoral studies. Since 1997, he has been a Senior Research Scientist at GTK, covering diamond exploration and mantle and ore research using isotopic and electron beam methods.

Dr. N. V. Chalapathi Rao is currently an Associate Professor of Geology at Banaras Hindu University, Varanasi, India. Dr. Rao holds Ph.D. degrees from Osmania University, Hyderabad (India) and University of Cambridge (UK). His research specialisation concerns petrology, geochemistry and genesis of kimberlites, lamproites, lamprophyres and their entrained crustal and mantle xenoliths with focus on lithospheric evolution and diamond prospectivity.

Dr. Steven Sparks is Professor of Geology in the School of Earth Sciences at Bristol University, Bristol, UK. His research concerns volcanic and igneous processes and he has made contributions in petrology, many physical volcanology, fundamental fluid mechanics, sedimentology, and in hazard and risk assessment methods. He has been past-President of the Geological Society of London and IAVCEI President of the Volcanology, Geochemistry and Petrology section of the American Geophysical Union 2008–2012.

10th IKC Committees and Sponsors

Organising Committee

Conveners:
R. H. Sawkar, Geological Society of India, Bangalore
Fareeduddin, Geological Survey of India, Bangalore

Liaison:
R. H. Mitchell, Department of Geology, Lakehead University, Ontario, Canada

Treasurer:
S. Jitendra Kumar, Geological Society of India, Bangalore

Members:
K. Ayyasami, Geological Survey of India, Hyderabad
Abhijeet Mukherjee, National Mineral Development Corporation, Hyderabad
N. V. Chalapathi Rao, Banaras Hindu University, Varanasi
D. K. Paul, Kolkata University, Kolkata
E. V. S. S. K. Babu, NGRI, Hyderabad
J. N. Das, Geological Survey of India, Bangalore
K. Sajeev, Indian Institute of Science, Bangalore
K. S. Godhavari, Geological Society of India, Bangalore
M. Sudhakar, Ministry of Earth Science, New Delhi
M. S. Rao, Geological Society of India, Bangalore
M. S. Jairam, Geological Survey of India, Kolkata
P. Krishnamurthy, Geological Society of India, Bangalore
Rajesh K. Srivastava, BHU, Varanasi
S. S. Nayak, Geological Survey of India, Bangalore
S. K. Bhushan, MSPL, Hospet
Sivaji, Department of Science and Technology, New Delhi
Sojan Joy, De Beers
V. N. Vasudev, Geomysore Services (India) Pvt. Ltd, Bangalore
Veeranna, Department of Mines and Geology, Bangalore
Y. J. Bhaskar Rao, NGRI, Hyderabad

Programme Committee

Roger Mitchell (Chairman)
Barbara Scott Smith
E. V. S. S. K. Babu
D. Graham Pearson
Hugh O'Brien
J. Barry Dawson
Jeff Harris
N. V. Chalapathi Rao
Steve Sparks
Fareeduddin

Financial Aid Committee

Sonja Aulbach
Roger Mitchell
S. J. Patel
D. K. Paul
D. Graham Pearson
N. V. Chalapathi Rao
Thomas Stachel

International Kimberlite Conference Advisory Committee (Since 2008)

Roger Mitchell (Chairman)
Gerhard Brey (Germany)
J. Barry Dawson (UK; Deceased, 2013)
Fareeduddin (India)
Herman Grütter (Canada)
Jeff Harris (UK)
Stephan Haggerty (USA)
Peter Nixon (UK)
Hugh O'Brien (Finland)
Sue O'Reilly (Australia)
Tom Nowicki (Vancouver)
D. Graham Pearson (Canada)
Roberta Rudnick (USA)
Barbara Scott Smith (Canada)
Craig Smith (South Africa)
Nick Sobolev (Russia)
Thomas Stachel (Canada)

Sponsors

The Rio Tinto's diamond sponsorship has been utilized for the publication of the Proceeding Volume. BHP Billiton and De Beers supported the conference with platinum sponsorship. Australian Scientific Instruments, Ministry of Earth Sciences, Government of India, National Geophysical Research Institute (NGRI) and National Mineral Development Corporation (NMDC) supported the conference with silver sponsorship. The generous financial contributions made by the 7th, 8th and 9th IKC organizing committees helped in providing the financial assistance to the students and other senior scientists. The Geological Survey of India has provided logistic support during field trip to the South Indian Kimberlites. The National Mineral Development Corporation permitted the field trip delegates to visit their properties at Majhgawan. The Rio Tinto India Exploration Ltd. has permitted the delegates to visit their property, study the core samples and offered traditional camp-site hospitality at Bunder.

Conference Sponsors

Platinum sponsors

Diamond sponsors

Silver sponsors

Supporting sponsors

Contents

Roger H. Mitchell

Roger Mitchell is one of the staunchest supporters of the kimberlite/upper mantle community. He has the distinction of being one of the 9 "Old Farts" who have attended all 10 of the International Kimberlite Conferences since their inception in Cape Town in 1973, and since 1996 has been Chairman of the International Kimberlite Conference Advisory Committee. In addition he was co-convenor, with Barbara Scott Smith, of the highly-successful 8th Kimberlite Conference held in Victoria, BC in 2003.

His first foray into kimberlite geology after graduating from the University of Manchester was on strontium isotopes when he did his Ph.D. at McMaster University. Although isotope geochemistry has long been one of his specialities, Roger has been involved in the full spectrum of kimberlite geology ranging from field mapping (with relevance for kimberlite diatreme emplacement) and petrography to detailed mineralogy in which his studies on perovskite have made him a world authority. Early contacts in Norway had already excited his interests in carbonatites which, although perhaps subordinate to kimberlites and mantle xenoliths, have been another major aspect of his research career. After post-doctoral fellowships in Edmontonand Oslo, in 1972 Roger joined the faculty at Lakehead University where, in addition to his own research, he played a full role in the administration of his department and the university. His research has led him to being honoured by election to the Royal Society of Canada and to Fellowship of the Mineralogical Society of America.

Perhaps reflecting his upbringing, Roger has two Yorkshire characteristics of first, being a very hard worker and, second, of having a direct approach to both problems and people and the inability to suffer fools gladly. The first characteristic has resulted in over 230 papers and 5 monographs. Many papers have been written in conjunction with co-authors from Europe, Russia, China and India, being indicative of his approach to international cooperation. His second, no-nonsense trait, has been of great value in his editorships or associate editorships of various journals, and of the Proceedings Volumes of both the 6th and 8th International Kimberlite Conferences.

Roger retired from Lakehead, where he is now Emeritus Professor, in 2006 but, with the able support of his wife Valerie, he continues to be as active as ever. The word "retired" is used advisedly as he continues to research several aspects of kimberlite and carbonatite mineralogy, has recently taken on co-editorship of the Mineralogical Magazine, and is still widely consulted by mining companies. So much for retirement!

For his unstinting contributions to the International Kimberlite Conferences, it is only fitting that Roger is honoured by the co-dedication of the Proceedings of the Bangalore Conference in his name.

Barry Dawson
UK

Paragenesis and Oxygen Isotopic Studies of Serpentine in Kimberlite

Roger H. Mitchell

Abstract

Although serpentine is a ubiquitous mineral in kimberlites, its origin remains controversial. Some petrologists claim that all serpentine is secondary and produced by the introduction of externally derived fluids, whereas others suggest that serpentine is a primary mineral formed mainly from late-stage hydrothermal deuteric fluids. To date, the only investigations of the isotopic composition of serpentine in kimberlite have been by bulk analysis of samples that have undergone subaerial alteration, and hence undoubted interaction with meteoric water. Serpentine and chlorite in kimberlites occur as follows: (1) pseudomorphic retrograde lizardite and chlorite after primary olivine; (2) late-stage primary serpophitic serpentine; (3) prograde serpophitic serpentine replacing retrograde lizardite; and (4) serpentine–chlorite replacing magmaclast olivine and interclast phlogopite in diatreme zone rocks. Serpophitic or polygonal serpentine is a primary hydrothermal low temperature phase that common forms monomineralic segregations in the groundmass of hypabyssal kimberlites. In this study, the oxygen isotopic composition of all varieties of serpentine/chlorite was determined by ion microprobe in hypabyssal and diatreme zone kimberlites from Southern Africa, Canada and the USA. Data were acquired using the Edinburgh Cameca IMS 1270 ion microprobe and an antigorite ($\delta^{18}O = 8.8$ ‰) standard. In Iron Mountain hypabyssal kimberlites, the $\delta^{18}O$ of pseudomorphic marginal and vein serpentine is similar and ranges between 4.3 and 6.3 ‰. For compositionally zoned pseudomorphic marginal serpentines, $\delta^{18}O$ decreases from core-to-rim (cores 6.4 to 3.8 ‰ $\delta^{18}O$; rims 2.7 to 0.6 ‰ $\delta^{18}O$). Prograde pseudomorphic serpophite in the Ham West kimberlite ranges in $\delta^{18}O$ from 4.1 to -0.5 ‰, with segregationary primary serpophite $\delta^{18}O$ ranging from 1.6 to -1.8 ‰. In diatreme zone rocks at Letseng-la-terae, the earliest pseudomorphic serpentines range in $\delta^{18}O$ from 5.0 to 3.5 ‰, later-forming rim and marginal serpentines from 4.3 to -1.8 ‰, with interclast serpophite from 3.4 to 1.5 ‰. At Kao, the earliest pseudomorphic serpentines are relatively uniform in composition and range in $\delta^{18}O$ from 4.9 to 4.5 ‰, later rim and marginal serpentines from 4.9 to 4.8 ‰ and interclast serpophite from 6.3 to 3.8 ‰.These data show that the kimberlites investigated have not interacted with large volumes of light meteoric waters. Similar oxygen isotopic compositions in hypabyssal and diatreme zones rocks from different localities certainly preclude the introduction of meteoritic water as a cause of serpentinization. The trends from mantle oxygen $\delta^{18}O$ to lighter oxygen are a result of either fluid compositional evolution or very minor introduction of light water. Assuming that parental olivine has a

R. H. Mitchell (✉)
Department of Geology, Lakehead University, Thunder Bay, Ontario, P7B 5E1, Canada
e-mail: rmitchel@lakeheadu.ca

D. G. Pearson et al. (eds.), *Proceedings of 10th International Kimberlite Conference*, Volume 1, Special Issue of the Journal of the Geological Society of India,
DOI: 10.1007/978-81-322-1170-9_1,

primary $\delta^{18}O$ of about +6 ‰ and that this is decreased during serpentinization by ^{16}O enrichment by reactions with isotopically lighter water (−10 to −20 ‰), at 200–300 °C, in closed or open systems, then "water/serpentine" ratios cannot exceed 0.8. These data are interpreted to suggest that there was not an influx of significant volumes of low-temperature meteoric water as a cause of serpentinization. This conclusion is supported by the absence of extremely light oxygen (i.e. $\delta^{18}O < -5$ ‰) in all of the serpentines investigated.

Keywords
Kimberlite • Oxygen isotopes • Serpentine • Serpophite • Chlorite

Introduction

Kimberlites are rocks derived from volatile-bearing magmas as indicated by the common presence in them of primary calcite and deuteric serpentine (Mitchell 1986). Their parental magma has been proposed by Brooker et al. (2011), Kjarsgaard et al. (2009), Kopylova et al. (2008), le Roex et al. (2003) and Price et al. (2000) to be a carbonated water-bearing ultrabasic silicate melt. Although the H_2O/CO_2 ratios of the parental magmas and associated fluids are not known, Kjarsgaard et al. (2009) have emphasized that the majority of the compositions of fresh hypabyssal kimberlites are water-rich with molar $H_2O/(H_2O + CO_2)$ ratios greater than 0.5. Calculated H_2O and CO_2 contents of the primary magmas are estimated to range from 9.5 to 12.5 and 5.5 to 8.1 wt%, respectively (Brooker et al. 2011; Kjarsgaard et al. 2009). In contrast to hypabyssal kimberlites, diatreme zone kimberlites are poor in calcite and volatile-bearing phases are dominated by diverse serpentines and chlorites (Mitchell et al. 2009). These observations suggest that the volatiles associated with diatreme-forming events could be relatively much richer in water.

There have been few stable isotope studies of kimberlites, and of these, the majority have focussed upon the O and C isotopic composition of carbonates (Deines and Gold 1973; Kobelski et al. 1979; Kirkley et al. 1989; Wilson et al. 2007). Only, Sheppard and Epstein (1970), Sheppard and Dawson (1975), Ukhanov and Devirts (1983) and Ustinov et al. (1994) have provided data for the isotopic composition of H, C and O in silicates. These studies were undertaken on olivine, phlogopite megacrysts, serpentinized megacrysts, and bulk rock kimberlites. Samples were taken at random and included kimberlites (and orangeites) which have been subjected to significant subaerial alteration. The bulk rock data presented by Sheppard and Dawson (1975) for Southern African kimberlites represent the average of the isotopic compositions of all of the phases present in a particular sample; hence, these data are for mixtures of minerals rather than single phases, and interpretation is fraught with ambiguities. Regardless, Sheppard and Dawson (1975) have interpreted their data to imply that "meteoritic–hydrothermal fluids" interacted with kimberlite during low-temperature (50–200 °C) serpentinization. The origins of the "fluids" were not specifically defined, although some combination of magmatic, metamorphic or exchanged meteoric waters with a range of low water/rock ratios (<0.2) was suggested. Ukhanov and Devirts (1983) provided only H isotopic data and claimed these data supported a meteoric origin for water in serpentines in the Udachnaya and Obnazhennaya kimberlites.

One of the unresolved problems in kimberlite petrology is the origin of the common serpentine and/or chlorite found in hypabyssal and diatreme zone kimberlites which have *not* been subjected to subaerial alteration. Commonly, serpentine is regarded by many geologists as a secondary mineral, typically resulting from the in situ serpentinization of olivine, pyroxene and/or mica by fluids not necessarily genetically related to the primary minerals. Recently, Sheppard and Dawson's (1975) conclusions have been misinterpreted by some volcanologists (Stripp et al. 2006; Cas et al. 2008; Hayman et al. 2009) to support claims that all serpentine in kimberlites is the result of alteration in pre-existing material, and therefore is entirely of secondary origin, and thus genetically unrelated to fluids present during kimberlite emplacement. In contrast, Mitchell (1986, 2008) and Mitchell et al. (2009) claim that many serpentine group minerals in kimberlites are formed from low-temperature deuteric hydrothermal fluids and are in many instances primary phases.

One approach to resolving these contradictory hypotheses is to determine the hydrogen and oxygen isotopic composition of the serpentine group minerals, as these data should indicate whether or not meteoric or deuteric fluids dominate the serpentinization process. Typically, the serpentines in kimberlites are themselves mineralogically complex, very fine grained and commonly intergrown with other minerals.(Mitchell and Putnis 1988; Mitchell 2008; Mitchell et al. 2009). Bulk analysis for H and O of samples extracted by microdrilling is possible in some instances, but unfortunately, this methodology results in a loss of paragenetic information and is not applicable to complexly zoned serpentine, and moreover, the sample can include extraneous material. In contrast, compositions determined by in situ methods such as secondary ion mass spectrometry (ion microprobe) can be directly related to spatial petrographic/paragenetic information but unfortunately do not typically provide data for hydrogen isotopes. Determination of H/D ratios by ion probe methods is subject to significant instrumental contamination problems and associated with large analytical errors (Harford and Sparks 2001). This study outlines the paragenetic types of serpentine in kimberlites and presents the results of an ion microprobe investigation of the isotopic composition of oxygen in serpentines and chlorites from some hypabyssal and diatreme zone kimberlites.

Serpentine in Kimberlites

Definitions

Primary groundmass magmatic minerals in kimberlites are those precipitated as liquidus phases from kimberlite magmas at relatively high temperatures (> 500 °C; Mitchell 1995, 1986). They include olivine; perovskite; diverse Mg–Ti–Fe-rich spinels; monticellite; phlogopite–kinoshitalite; apatite; and calcite. Crystallization of the magma results in increasing volatile contents in the residual melt; this being reflected in the formation of late-stage volatile-rich calcite–serpentine segregations (Mitchell 1986). There is actually a continuum from magmatic crystallization to minerals formed from hydrothermal residua (Mitchell 2008). For example, calcite can be either a primary magmatic or a primary hydrothermal phase. Some serpentines can crystallize directly from the hydrothermal fluids, commonly in association with hydrothermal primary calcite, during the later stages of crystallization, and are here defined as *primary serpentines*, that is they are not formed by the replacement of pre-existing minerals. Serpentine formed by the latter process is commonly defined as "secondary serpentine". A variety of "secondary serpentine" can be formed in kimberlites by the reaction of hydrothermal deuteric fluids with pre-existing minerals, for example serpentinization and pseudomorphing of olivine and/or monticellite. The type of serpentine (lizardite) formed is typically structurally different to late-stage primary deuteric polygonal serpentine (see below). In this work, the replacement of pre-existing primary minerals by deuteric fluid serpentine-forming reactions is termed *retrograde serpentinization* to distinguish it from *secondary serpentine* formed as a result of the introduction of external fluids (meteoric water) genetically unrelated to kimberlite magma.

Low-temperature hydrothermal residual fluids can also permeate throughout pre-existing magmatic assemblages, resulting in the replacement of the minerals by deuteric polygonal serpentine with in some instances concomitant destruction and replacement of retrograde serpentine. This style of *deuteric replacement* is termed here *prograde serpentinization.*

Serpentine Paragenesis in Kimberlites

Serpentine group polymorphs and their polytypes occur in kimberlites in the following parageneses:

1. Pseudomorphic retrograde vein serpentine formed by the direct conversion of macrocrystal and phenocrystal forsteritic olivine to lizardite-1*T* (Jago and Mitchell 1985). Such serpentine forms anastomosing thin veins within the parental olivine and typically shows relatively coarse-grained mesh textures with pronounced (or flamboyant) anisotropy (Figs. 1 and 2). In some instances, the serpentine comprising the veins is a pale yellow Fe-poor variety containing small crystals of magnetite. In other cases, the serpentine is pale brown, Fe-rich, lacks magnetite and contains Ni-sulphides (millerite, heazlewoodite). The difference between the magnetite-bearing and sulphide-bearing serpentines depends upon whether the serpentine is produced under relatively oxidizing or reducing conditions, respectively (Mitchell 1986). The serpentinizing fluids are in hypabyssal kimberlites considered to be of deuteric origin.
2. Pseudomorphic retrograde marginal serpentine formed by the direct conversion of forsteritic olivine to compositionally diverse "serpentines", together with the replacement of earlier vein serpentine. Typically, these serpentines occur as mantles on, or as complete replacements of, olivine macrocrysts and phenocrysts (Fig. 1). Some pseudomorphed olivines exhibit several

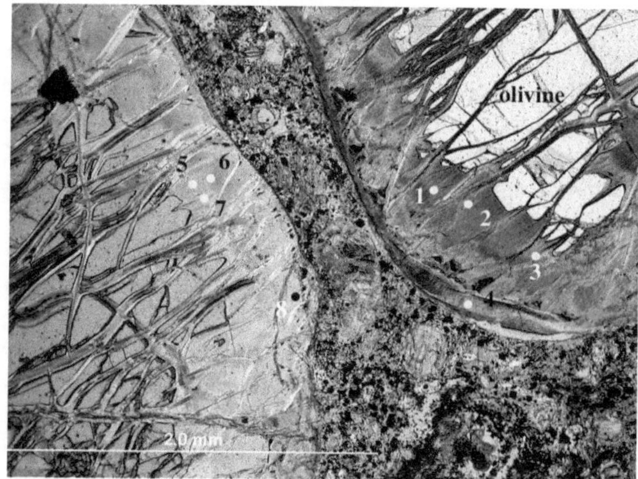

Fig. 1 Retrograde vein serpentine and pseudomorphic marginal serpentine replacing olivine macrocrysts in the Iron Mountain (USA) hypabyssal kimberlite. Plane polarized light image. Numbers indicate analytical points (see Table 1)

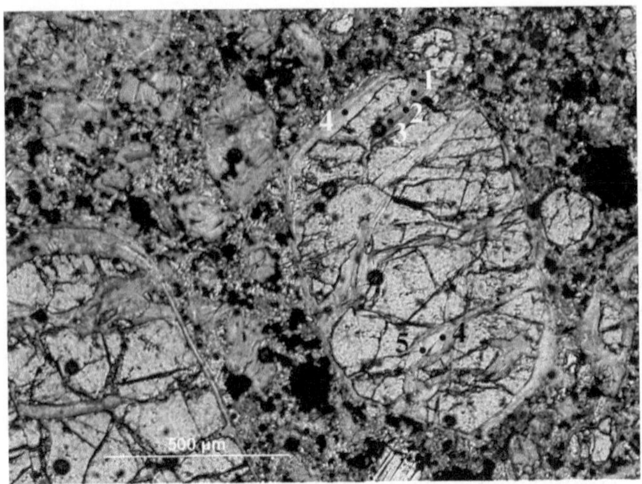

Fig. 2 Retrograde vein serpentine replacing olivine microphenocrysts in the Wesselton W7 hypabyssal kimberlite (South Africa). Plane polarized light image. Numbers indicate analytical points (see Table 1)

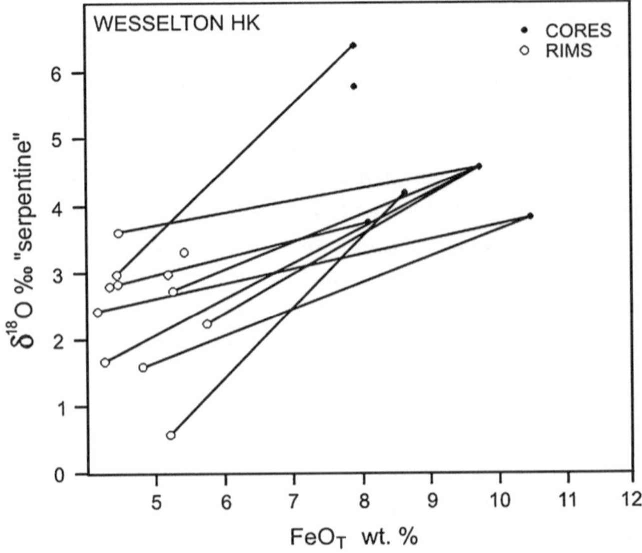

Fig. 3 Core-to-rim variations in the isotopic composition of oxygen (*^{18}O) versus FeO_T content of serpentine replacing olivine microphenocrysts in Wesselton W7 hypabyssal kimberlite (South Africa). Note that very light oxygen ($\delta^{18}O < -5$ ‰) is not present and that the cores of the crystals contain oxygen with isotopic compositions similar to those of mantle oxygen ($\delta^{18}O = +5$ to $+6$ ‰)

(1–4) discrete mantles of differing colour and composition. All are relatively coarse grained and exhibit "flamboyant" anisotropy. The "serpentines" range in character from lizardite to mixed layer serpentine–chlorites. The textures and compositions indicate that several discrete stages of serpentinization must have occurred subsequent to the formation of vein serpentine.

3. Non-pseudomorphic, segregationary serpentine (Figs. 4 and 5) is characteristic of hypabyssal kimberlite (Mitchell 1986, 2008). Typically, segregations occur as discrete amoeboid areas within the groundmass of the kimberlite and consist of serpentine alone or serpentine plus calcite. Some segregations contain marginal calcite, acicular phlogopite and/or acicular apatite. There is no textural evidence to suggest that these serpentines are pseudomorphing earlier-formed olivine or phlogopite. Calcite, apatite and mica in the segregations can be partially replaced or completely pseudomorphed by segregationary serpentine.

Segregationary serpentine can be pale yellow to light brown in colour, is characteristically extremely fine grained and, depending on grain size, can be isotropic or near-isotropic. In the latter case, it appears as an aggregate of tiny spheres with a texture rather like that of frogspawn. Such serpentines have commonly been termed "Povlen-type serpentine" or "serpophite" (Lodochnikov 1933). The latter name is a useful descriptive term but has no strict mineralogical meaning, as Wicks and Zussman (1975) initially showed the material is not amorphous and consists of fine-grained identifiable phases. Subsequent studies have shown that serpophitic serpentine consists primarily of polygonal (Middleton and Whittaker 1976; Mitchell and Putnis 1988; Baronnet et al. 1994) and polyhedral (Baronnet et al. 2007; Andreani et al. 2008) serpentine with lesser very thin chrysotile fibres; and not platy lizardite or antigorite.

Baronnet and Devourard (1996) and Devouard et al. (1997) have shown that polygonal serpentine can be synthesized from hydrous Mg–silicate gels at 300 °C and 0.07 GPa. Polygonal serpentine formation in these experiments does not require a precursor phase; consequently, polygonal serpentines in kimberlites can be regarded as *primary* phases (Mitchell 2008). Primary in this context, as noted above, implies crystallization directly from a residual deuteric hydrothermal fluid. Andreani et al. (2008) consider that polyhedral and polygonal serpentines are always the last type of serpentine to form in ultramafic rocks; their formation requiring open space, relatively low temperatures (<200–300 °C) and small amounts of Al (0.06–0.22 *apfu*). On the basis of these observations, Mitchell (2008) has regarded segregation polygonal serpentine to be a product of low-temperature transformation of a Mg–Si-bearing gel formed from residual deuteric fluids in the final stages of crystallization of hypabyssal kimberlite.

4. Pseudomorphic polygonal serpentine is produced when the residual hydrothermal sols or gel-forming fluids do not form discrete segregations (Fig. 4). In this case, the fluids migrate and permeate throughout the pre-existing kimberlite mineral assemblage, and replace both fresh and/or previously altered minerals. Thus, retrograde pseudomorphic lizardite–chlorite which initially replaced olivine can be in turn replaced by very fine-grained polygonal serpentine. Mitchell (1986) and Jago

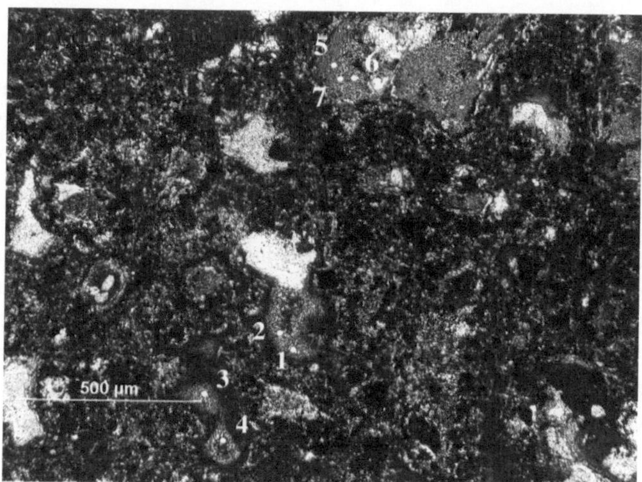

Fig. 4 Near-isotropic prograde serpophitic serpentine replacing olivine and near-isotropic primary hydrothermal serpophitic serpentine occurring as segregations in the Ham West (Canada) hypabyssal kimberlite. Crossed polarized light image. Numbers indicate analytical points (see Table 1)

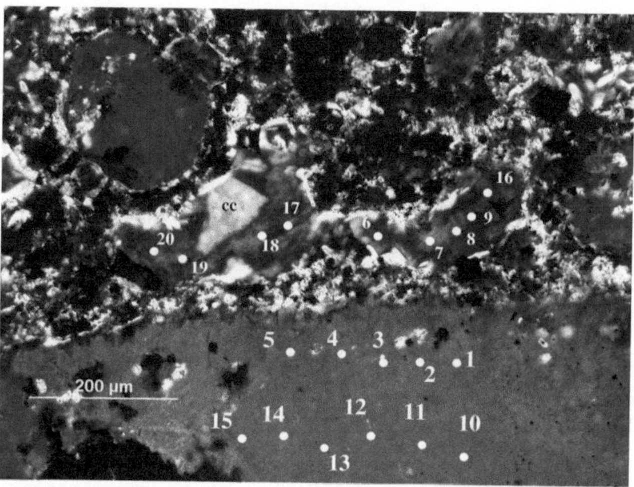

Fig. 5 Near-isotropic primary hydrothermal serpophitic serpentine occurring as segregations in the Ham West (Canada) hypabyssal kimberlite. Crossed polarized light image. Numbers indicate analytical points (see Table 1)

and Mitchell (1985) have termed this process "prograde serpentinization". Replacement of pre-existing minerals by calcite can also accompany this low-temperature serpentinization event. Kimberlites which have been extensively replaced by prograde polygonal serpentine and calcite can consist essentially of serpentine and calcite with relict spinels and rutile (after perovskite). Although at first sight, such rocks might be considered to be the products of subaerial secondary alteration by meteoric waters. However, the preserved igneous textures indicate their real origins, as secondary replacement typically results in the eradication of primary igneous textures by replacement with massive aggregates of secondary phases. It should be noted that polygonal serpentine does not form in the subaerial-weathering environment (Baronnet, pers. comm. 2006).

5. Interclast "serpentine" (Figs. 6, 7, 8) is a characteristic of diatreme zone Kimberley-type pyroclastic kimberlite (or KPK; Scott Smith et al. 2013) a.k.a. tuffisitic kimberlites. This material is commonly intimately intergrown with microlitic diopside (e.g. see Figs. 28 and 69 of Mitchell 1997); polygonal serpentine appears to be absent. Although considered to be "serpentine sensu lato", there have been no modern TEM or XRD investigations of this material. Mitchell et al. (2009) on the basis of SEM–BSE–morphological and compositional data suggest that this "serpentine" is a mixed layer serpentine–chlorite phyllosilicate, formed by the deuteric hydrothermal alteration in primary phlogopite. It should be noted that olivine in magmaclasts in KPK is commonly pseudomorphed by retrograde lizardite–chlorite that was formed in a different serpentinization event to that which produced the interclast serpentine, that is there is typically no pervasive "prograde serpentinization" of clasts and interclast matrix (Mitchell et al. 2009).

Experimental Methods

In this work, the oxygen isotopic composition of all of the above varieties of serpentine was determined in hypabyssal and diatreme zone kimberlites from Southern Africa, Canada and the USA. Data were acquired in two analytical sessions (February 2007 and January 2010) at the University of Edinburgh (Scotland, UK) with a Cameca IMS 1270 (#309) ion microprobe, using a ~5-nA primary $^{133}Cs^{+}$ beam. All data were collected on Au-coated polished thin sections on material located near the centre of the section. Secondary ions were extracted at 10 kV, and ^{16}O ($\sim 3.0 \times 10^{9}$ cps) and ^{18}O ($\sim 4.0 \times 10^{6}$ cps) were monitored simultaneously on dual Faraday cups (L′2 and H′2). Each analysis involved a pre-analysis sputtering time of 50 s, followed by automatic secondary beam and entrance slit centring, and finally data collection in two blocks of five cycles, amounting to a total count time of 40 s. The internal precision of each analysis is <0.2 ‰, and the external precision, calculated from multiple analyses of the standard, that is assumed to be homogeneous is estimated at <0.3 ‰.

The analytical standard used was a sample of antigorite (Royal Ontario Museum acquisition #M8571) whose bulk isotopic composition had been determined by standard fluorination methods at Queen's University (Kingston, Ontario), that is $\delta^{18}O = 8.8$ ‰. Secondary analytical controls included lizardite-1T (Royal Ontario Museum acquisition #M19804; $\delta^{18}O = 9.9$ ‰). Lacking a chlorite standard, it was assumed that chlorite behaves like serpentine with respect to ionization characteristics and can be analysed with the antigorite standard. Because of significant compositional differences, with respect to Al content, between chlorite and serpentine, the use of the antigorite standard for chlorite analyses means that the accuracy of these data is likely to be significantly less than that obtained for serpentine, and the values should be taken as "indicative" only. The O isotopic composition of the antigorite standard was routinely measured multiple times during each analytical session. Analytical results are presented in Tables 1 and 2 as $\delta^{18}O$ ‰ relative to SMOW.

Experimental Results

Hypabyssal Kimberlite

Figures 1 and 2 illustrate the locations of analytical points and types of serpentines analysed (Table 1) in hypabyssal kimberlites. For Iron Mountain (Coopersmith et al. 2003) and Wesselton W3 (Shee 1985; Mitchell et al. 2009) kimberlites, it is apparent that the $\delta^{18}O$ of pseudomorphic retrograde marginal and vein serpentine is similar and ranges between 4.3 and 6.3 ‰. Figure 3 illustrates core-to-rim compositions ($\delta^{18}O$ vs. FeO_T) of Wesselton W7 kimberlite (Shee 1985) compositionally zoned pseudomorphic marginal serpentines. Core serpentines range from 6.4 to 3.8 ‰ $\delta^{18}O$, and rims from 2.7 to 0.6 ‰ $\delta^{18}O$, indicating that $\delta^{18}O$ decreases from core-to-rim of these pseudomorphed olivines and is correlated with decreasing FeO_T content of the serpentine.

Prograde pseudomorphic serpophite (Fig. 4) in the Ham West HW20 and HW11 kimberlites (Jago and Mitchell 1985) ranges in $\delta^{18}O$ from 4.1 to −0.5 ‰, with segregationary primary serpophite $\delta^{18}O$ ranging from 1.6 to −1.8 ‰ (Figs. 4 and 5). These data suggest that later-stage serpophite-forming fluids are isotopically light relatively to those that form retrograde marginal pseudomorphic serpentine. The relatively heavy isotopic composition of prograde pseudomorphic serpophite in HW11 (Table 1) must represent that of a re-equilibrated mixture between that of pre-existing retrograde pseudomorphic serpentine and the residual primary serpophite-forming fluid, as all serpophite has probably crystallized from late-stage fluids. Non-pseudomorphic segregation serpophite, coexisting with calcite and apatite, in the Lac de Gras kimberlite ranges in $\delta^{18}O$ from 3.0 to 2.0 ‰ (Table 1).

Diatreme Zone Kimberlites

Figures 6, 7, 8, 9 illustrate the locations of analytical points and types of serpentines analysed (Table 2) in diatreme zone kimberlites from Letseng-la-terae (Bloomer and Nixon 1973), Kao (Clement 1973) and Lethlekane (Field et al. 2008). The pseudomorphic serpentine analysed in all cases is lizardite–chlorite after olivine macrocrysts located within magmaclasts (formerly termed "pelletal lapilli"; see Scott Smith et al. 2013; Mitchell et al. 2009). At Letseng-la-terae (Fig. 6), the earliest pseudomorphic serpentines range in $\delta^{18}O$ from 5.0 to 3.5 ‰, later-forming rim and marginal serpentines from 4.3 to −1.8 ‰, with interclast serpophite from 3.4 to 1.5 ‰. Figure 9 illustrates core-to-rim compositions ($\delta^{18}O$ vs. FeO_T) of Letseng-la-terae pseudomorphic serpentines and shows that although there is no correlation between $\delta^{18}O$ and FeO_T, marginal serpentines are all enriched in light oxygen. At Kao (Fig. 7), the earliest pseudomorphic serpentines are relatively uniform in composition and range in $\delta^{18}O$ from 4.9 to 4.5 ‰, later rim and marginal serpentines from 4.9 to 4.8 ‰ and interclast serpophite from 6.3 to 3.8 ‰. The limited data for Lethlekane (Fig. 8) indicate that pseudomorphic serpentine is relatively uniform in isotopic composition and enriched in ^{18}O relative serpentines at Letseng and Kao.

Discussion

The oxygen isotopic composition of the earliest-formed retrograde pseudomorphic serpentine after olivine is remarkably similar (4–6 ‰ $\delta^{18}O$) in all hypabyssal kimberlites from diverse localities. This cannot be fortuitous as the kimberlites investigated are of different ages and emplaced in very different geological and hydrological environments. In all examples, oxygen isotopic compositions are seen to evolve in the late-stage segregationary serpentines towards lighter oxygen ($\delta \sim 0.0$ ‰) from these initial values.

In the diatreme zone kimberlites, interclast serpophite at Letseng-la-terae is lighter than pseudomorphic serpentine and appears to continue the light isotope enrichment trend. In contrast, interclast serpophite at Kao is of variable composition but relative to pseudomorphic serpentine is slightly enriched in ^{18}O; this could imply that the fluids forming these particular serpentines were unrelated.

For later-forming zoned and mantled pseudomorphic serpentine, isotopic evolution in hypabyssal and diatreme environments is towards ^{16}O enrichment ($\delta\ ^{18}O =$ 3.5–0.5 ‰), suggesting that the light isotope enrichment

Table 1 Oxygen isotopic composition ^{18}O ‰ of serpentines in hypabyssal Kimberlites

Location	Analysis	^{18}O	Location	Analysis	^{18}O
Iron Mountain (Wyoming, USA)			*Ham West (Canada)*		
Hypabyssal Kimberlite (Jan. 2010)			HW20 Hypabyssal Kimberlite (Jan. 2010)		
Pseudomorphic serpentine	1	4.2	Pseudomorphic serpophite	1	1.9
Pseudomorphic serpentine vein	2	4.3	Pseudomorphic serpophite	2	1.2
Pseudomorphic serpentine	3	4.6	Pseudomorphic serpophite	3	1.0
Marginal serpentine	4	5.3	Pseudomorphic serpophite	4	1.0
Pseudomorphic serpentine	5	4.7	Pseudomorphic serpophite	5	1.5
Pseudomorphic serpentine	6	4.2	Pseudomorphic serpophite	10	3.2
Pseudomorphic serpentine	7	4.8	Pseudomorphic serpophite	11	2.5
Marginal serpentine	8	5.7	Pseudomorphic serpophite	12	2.4
Pseudomorphic serpentine matrix	9	6.3	Pseudomorphic serpophite	13	2.5
			Pseudomorphic serpophite	14	2.5
Wesselton Mine (South Africa)			Pseudomorphic serpophite	15	−0.5
W3 Hypabyssal kimberlite (Jan. 2010)			Segregation serpophite	6	−1.3
Pseudomorphic serpentine vein	1	4.9	Segregation serpophite	7	−1.6
Pseudomorphic serpentine vein	2	4.9	Segregation serpophite	8	−1.8
Pseudomorphic serpentine vein	3	4.7	Segregation serpophite	9	−1.8
Marginal serpentine	4	5.1	Segregation serpophite	16	−0.1
			Segregation serpophite	17	0.1
Wesselton Mine (South Africa)			Segregation serpophite	18	−0.5
W7 Hybabyssal kimberlite (Feb. 2007)			Segregation serpophite	19	−0.1
Pseudomorphic serpentine core	1	3.8			
Pseudomorphic serpentine rim	2	2.5	*Ham West (Canada)*		
Pseudomorphic serpentine matrix	3	3.4	*HW11 Hypabyssal kimberlite (Jan.2010)*		
Pseudomorphic serpentine rim	4	1.6	Segregation serpophite	1	1.5
Pseudomorphic serpentine core	5	4.2	Segregation serpophite	2	1.2
Pseudomorphic serpentine rim	6	0.6	Segregation serpophite	3	1.6
Pseudomorphic serpentine core	7	4.6	Segregation serpophite	4	1.3
Pseudomorphic serpentine rim	8	1.7	Pseudomorphic serpophite	5	4.1
Pseudomorphic serpentine rim	9	2.7	Pseudomorphic serpophite	6	3.3
Pseudomorphic serpentine rim	10	3.6	Pseudomorphic serpophite	7	3.7
Pseudomorphic serpentine rim	11	2.2			
Pseudomorphic serpentine core	12	3.7	*Lac De Gras (Canada)*		
Pseudomorphic serpentine core	13	5.8	*93T33 Hypabyssal kimberlite (Jan. 2010)*		
Pseudomorphic serpentine rim	14	2.9	Segregation serpophite	1	2.6
Pseudomorphic serpentine rim	15	2.9	Segregation serpophite	2	2.0
Pseudomorphic serpentine core	16	6.4	Segregation serpophite	3	2.9
Pseudomorphic serpentine rim	17	2.9	Segregation serpophite	4	3.0
Pseudomorphic serpentine matrix	18	1.7	Segregation serpophite	5	2.6
Pseudomorphic serpentine rim	19	2.8	Segregation serpophite	6	3.0
Pseudomorphic serpentine rim	20	3.6			

trend is either a feature of fluid compositional evolution or a influx of light meteoric water. Local alteration in situ by either present day or paleo-meteoric waters is very unlikely to produce such similarities, as the isotopic composition of meteoric water would certainly be very different at each locality. For example, it seems unreasonable to expect the

Table 2 Oxygen isotopic composition ^{18}O ‰ of serpentines in diatreme zone Kimberlites

Location	Analysis	^{18}O	Location	Analysis	^{18}O
Letseng-le-terae (Lesotho) (Jan. 2010)			*Kao (Lesotho)*		
K1 Main Pipe sample #1			K1 (Jan. 2010)		
Pseudomorphic serpentine core	1	5.0	Pseudomorphic serpentine core	1	4.5
Pseudomorphic serpentine core	2	4.9	Pseudomorphic serpentine core	2	4.9
Pseudomorphic serpentine rim	3	3.9	Pseudomorphic serpentine rim	3	4.9
Marginal serpentine	4	2.4	Marginal serpentine	4	4.8
Pseudomorphic serpentine core	5	3.5	Pseudomorphic serpentine core	5	4.9
Pseudomorphic serpentine rim	6	3.1	Pseudomorphic serpentine core	6	4.6
Marginal serpentine	7	4.3	Marginal serpentine	7	4.8
Marginal serpentine	9	4.0	Interclast serpophite	8	3.8
Pseudomorphic serpentine rim	10	4.3	Interclast serpophite	9	6.3
Pseudomorphic serpentine core	11	3.5	Interclast serpophite	10	5.3
Pseudomorphic serpentine core	12	4.2			
Interclast serpophite	13	2.1	*Lethlekane (Botswana)*		
Interclast serpophite	14	1.5	*(Jan. 2010)*		
Pseudomorphic serpentine core	15	4.4	Pseudomorphic serpentine core	1	8.1
Pseudomorphic serpentine rim	16	1.7	Pseudomorphic serpentine core	2	7.8
Interclast serpophite	17	1.7	Pseudomorphic serpentine core	3	7.5
Interclast serpophite	18	3.4	Marginal serpentine	4	9.1
Letseng-le-terae (Lesotho)					
K1 Main Pipe sample #2 (Feb. 2007)					
Pseudomorphic serpentine core	1	4.2	Pseudomorphic serpentine rim	16	5.2
Pseudomorphic serpentine core	2	3.9	Pseudomorphic serpentine core	17	4.0
Pseudomorphic serpentine core	3	4.0	Pseudomorphic serpentine core	18	3.7
Pseudomorphic serpentine rim	4	3.8	Pseudomorphic serpentine rim	19	4.8
Pseudomorphic serpentine rim	5	3.2	Pseudomorphic serpentine rim	20	4.3
Pseudomorphic serpentine rim	6	3.9	Marginal serpentine	21	2.3
Pseudomorphic serpentine vein	7	2.5	Pseudomorphic serpentine core	22	4.6
Marginal serpentine	9	1.2	Pseudomorphic serpentine rim	23	2.3
Marginal serpentine	10	1.3	Pseudomorphic serpentine rim	24	2.4
Marginal serpentine	11	1.7	Pseudomorphic serpentine core	25	4.1
Pseudomorphic serpentine core	12	4.6	Pseudomorphic serpentine rim	26	1.7
Pseudomorphic serpentine core	13	4.8	Pseudomorphic serpentine vein	27	1.4
Pseudomorphic serpentine rim	14	2.8	Pseudomorphic serpentine core	29	3.9
Marginal serpentine	15	-1.8	Marginal serpentine	30	1.8

Cretaceous local groundwater/rainfall at Letseng-la-terae (a kimberlite emplaced in basalt) to be similar to that of Devonian groundwater at the Iron Mountain kimberlite emplaced in granite.

From these data, it is also clear that there cannot be simple permeation of an externally derived fluid throughout these kimberlites, otherwise all isotopic compositions regardless of paragenesis would be similar, and trends such as those shown in Figs. 3 and 9 would not be preserved. These observations argue against any *significant* amounts of meteoric water being involved in the serpentinization events and are at variance with the hypotheses of Hayman et al. (2009) that interclast serpentine in KPK is merely a replacement of pre-existing material.

Figure 10 shows how the oxygen isotopic compositions of pseudomorphic serpentine might change as a function of water/serpentine ratios in open and closed systems (Sheppard et al. 1969; Wenner and Taylor 1973). It is assumed that parental olivine has a primary $\delta^{18}O$ of about +6 ‰, that is similar to that of fresh olivine in kimberlite

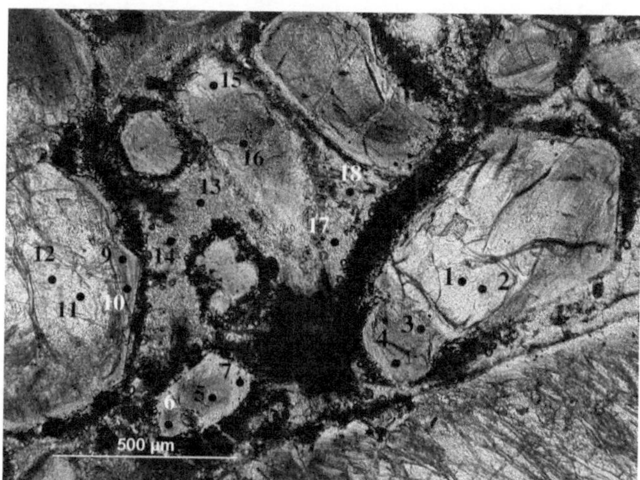

Fig. 6 Letseng-la-terae (Lesotho) diatreme zone KPK (Main Pipe) consisting of serpentine–chlorite pseudomorphs after olivine set in a chlorite–diopside interclast matrix. Back-scattered electron image. Numbers indicate analytical points (see Table 2; sample #1)

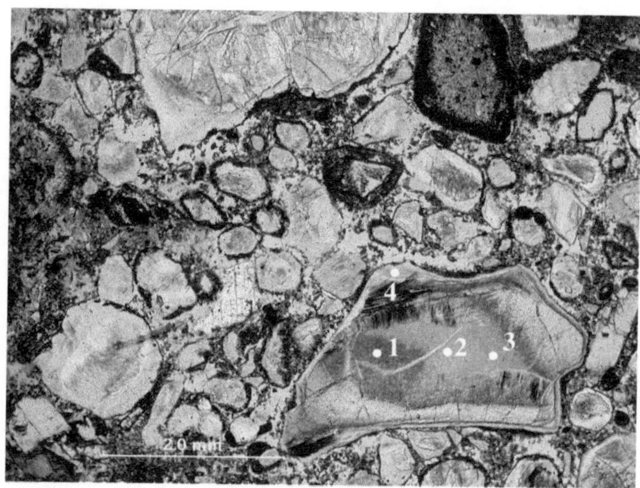

Fig. 8 Lethlekane (Botswana) diatreme zone KPK consisting of serpentine–chlorite pseudomorphs after olivine set in a chlorite–diopside interclast matrix. Back-scattered electron image. Numbers indicate analytical points (see Table 2)

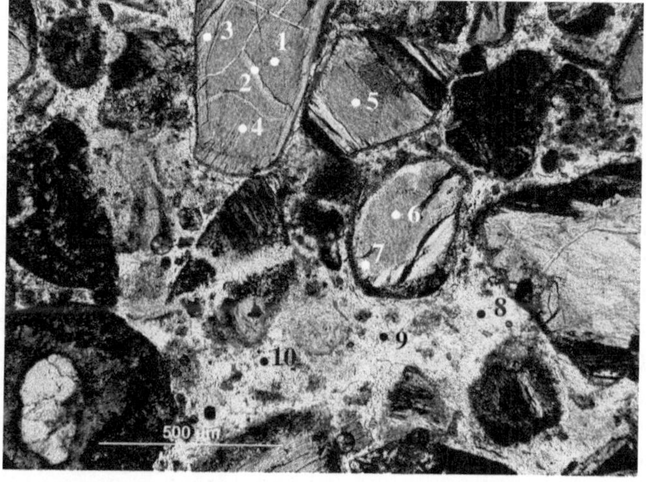

Fig. 7 Kao (Lesotho) diatreme zone KPK (K1) consisting of serpentine–chlorite pseudomorphs after olivine set in a chlorite–diopside interclast matrix. Back-scattered electron image. Numbers indicate analytical points (see Table 2)

(Ustinov et al. 1994) and mantle-derived ultramafic rocks (Taylor and Sheppard 1986), and that this is reduced by ^{16}O enrichment as a consequence of mixing and reaction with isotopically lighter water. Figure 10 is based upon the serpentine–water fractionation factor ($\Delta^{18}O$) determined by Wenner and Taylor (1973). Minor revisions to this fractionation factor by Früh-Green et al. (1996) do not change the conclusions given below. Chlorite–water fractionation factors determined by Cole and Ripley (1998) are essentially identical to those of serpentine–water, suggesting that isotopic fractionation between chlorite and water can be described by the same parameters as those for serpentine–water. Figure 10 indicates that for serpentinization at 200–300 °C in closed systems with initial water compositions ranging from −10 to −20 ‰, "water/serpentine" ratios cannot exceed 0.8. Low-temperature (<100 °C) serpentinization (Evans 2004) is considered to be an unlikely process with respect to the probable serpentinization temperatures of kimberlites (200–400 °C: Mitchell 1986; Evans 2004; Stripp et al. 2006; Andrani et al. 2007), and effectively rules out high ($\gg$1) "water/serpentine" ratios as a cause for the isotopic trend observed. Thus, these data are interpreted to suggest that there was not an influx of *significant* volumes of low-temperature meteoric water as a cause of serpentinization. This conclusion is supported by the absence of extremely light oxygen (i.e. $\delta^{18}O < -5$ ‰) in all of the serpentines occurring in the kimberlites investigated. Similar conclusions can be drawn for open systems (Fig. 10).

Figure 10 should be taken as an illustrative "end-member" case, as the $H_2O/(H_2O + CO_2)$ ratios of the fluids involved are unity and equilibrium is assumed. In reality, as noted above, fluids in kimberlites are undoubtedly mixtures of H_2O and CO_2, with dominant H_2O (Kjarsgaard et al. 2009) and equilibrium, especially in diatreme zone rocks, might not be realized. The effects of varying the composition of the fluids on the oxygen isotopic composition of carbonates in kimberlites have been discussed by Wilson et al. (2007). It was shown that $\delta^{18}O$ values can vary significantly (2–17 ‰) and depend not only on temperature, but strongly upon the assumed CO_2/H_2O ratio of the fluid. Hence, the observed isotopic trend towards enrichment in ^{16}O could simply result from changes in the fluid composition during crystallization. Calculation of such changes is beyond the scope of this preliminary investigation.

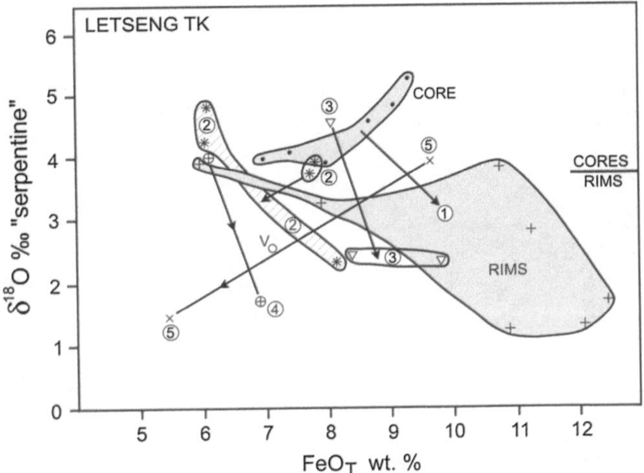

Fig. 9 Core-to-rim variations in the oxygen isotopic composition of oxygen ($\delta^{18}O$) versus the FeO_T content of pseudomorphic serpentine–chlorite replacing olivine macrocrysts in diatreme zone kimberlite from Letseng-la-terae (Lesotho)

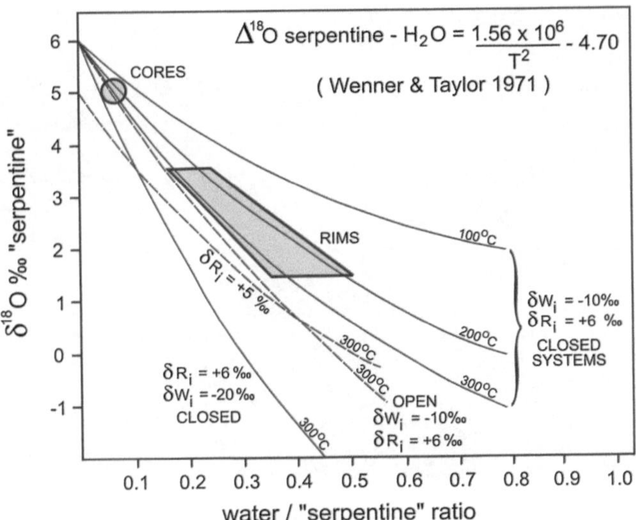

Fig. 10 Isotopic composition of "serpentine" ($\delta^{18}O$) calculated for open and closed systems as a function of (water/"serpentine") ratios, temperature and isotopic composition of external fluids (meteoric water) characterized by isotopically light oxygen. Serpentine–water fractionation factors are from Wenner and Taylor (1971). δR_i = initial $\delta^{18}O$ of olivine; δW_i = initial $\delta^{18}O$ of water

The calculated low water/serpentine ratios and ^{16}O enrichment trends, in themselves, do not rule out some influx of meteoric water, although they clearly demonstrate that this is not a volumetrically important process in the formation of serpentine in kimberlite. However, this observation does not imply that such a secondary process is the primary cause of serpentinization. This is because petrographic observations (Mitchell 1986, 2008; Mitchell and Putnis 1988) together with O and C isotopic data for carbonates (Wilson et al. 2007) have provided conclusive evidence that serpophite–carbonate segregations in kimberlites are of deuteric origin and not post-emplacement secondary replacement features as advocated, for example, by Hayman et al. (2009) and Cas et al. (2008). Such processes cannot produce the complex textural and paragenetic relationships actually observed and would result in pervasive uniform serpentinization with or without secondary calcite. It is unreasonable to expect that post-emplacement secondary serpentinization would produce similar petrographic and mineralogical features in kimberlites of diverse age and emplacement conditions in different groundwater regimes. Note that Andreani et al. (2007) have shown that serpentinization is a dynamic process and controlled by reaction kinetics rather than by P or T conditions; hence, equilibria of the type utilized by Stripp et al. (2006) are not relevant to the crystallization of serpentine from deuteric Mg–Si sols. Andreani et al. (2007) show that the sequence of serpentine formation with decreasing temperature in serpentine veins in oceanic peridotites is as follows: lizardite; polygonal serpentine; chrysotile; conical serpentine, that is essentially the same sequence as observed in kimberlites from vein and marginal pseudomorphic lizardite to serpophitic segregation serpentine. This sequence *cannot* originate from post-emplacement serpentinization. Consequently, it is concluded here that the water involved in vein and marginal serpentinization and formation of serpophite–calcite segregations is also principally of deuteric origin. On the basis of petrographic criteria, the serpentine–chlorite of the interclast matrices of PKP is considered to result from the alteration in pre-existing phyllosilicates by deuteric fluids (Mitchell et al. 2009). These fluids on the basis of the O isotopic data presented here are considered also to be of deuteric origin. However, fluids which resulted in the marginal and pseudomorphic serpentinization of olivine in magmaclasts might (e.g. Letseng-la-terae) or might not (e.g. Kao) be the same as those forming the interclast matrix serpentine chlorite.

Conclusions

The oxygen isotopic composition of serpentine and serpentine–chlorite replacing olivine in diverse hypabyssal and diatreme zone kimberlites ranges from about + 5 to −1.7 $\delta^{18}O$ ‰. These data are interpreted to indicate that early-forming retrograde serpentine replacing olivine (and/or mica) had near-mantle oxygen isotopic compositions. Subsequent isotopic evolution of these deuteric fluids to lower temperatures (200–300 °C) resulted in slight enrichment in ^{16}O. This oxygen isotopic variation can be modelled using experimentally derived water–serpentine fractionation factors and diverse externally derived water/serpentine ratios. The observed isotopic variations can be interpreted to indicate that serpentine compositions must reflect low water/serpentine ratios (< 0.8) at the time of their formation. An alternative explanation of these data is that the isotopic trends observed reflect changes in the CO_2/H_2O ratio of the serpentine-forming fluids.

These isotopic data coupled with petrographic observations suggest that pervasive replacement of kimberlites by large volumes of low-temperature externally derived (meteoric) water has not occurred. Similar trends in oxygen isotopic evolution regardless of facies or location imply that internal deuteric processes are dominant in the formation of the observed serpentine–chlorite assemblages.

Acknowledgments This research is supported by the Natural Sciences and Engineering Research of Canada, Lakehead University and Almaz Petrology. John Craven of the University of Edinburgh Ion Microprobe Facility is thanked for his expert assistance with the experimental work. The Royal Ontario Museum is thanked for providing the material used as analytical standards. Kurt Kaiser, Queen's University, is thanked for comments on the interpretation of the oxygen isotope geochemistry. Bruce Kjarsgaard, Joyashish Thakurta and Ryan Ikert are thanked for comments on the initial draft of this paper.

References

Andreani M, Mével C, Boullier AM, Escartin J (2007) Dynamic control on serpentine crystallization in veins: constraints on hydration processes in oceanic peridotites. Geochemistry, Geophysics, Geosystems G^3, 8(2)

Andreani M, Grauby O, Baronnet A, Muñoz M (2008) Occurrence, composition and growth of polyhedral serpentine. Eur J Mineral 20:159–171

Baronnet A, Devouard B (1996) Topology and crystal growth of natural chrysotile and polygonal serpentine. J Cryst Growth 122:952–960

Barronet A, Mellini M, Devouard B (1994) Sectors in polygonal serpentine, a model based on dislocations. Phys Chem Miner 21:330–343

Baronnet A, Andreani M, Grauby O, Devouard B, Nitsche S, Chaudanson D (2007) Onion morphology and microstructure of polyhedral serpentine. Am Mineral 93:687–690

Bloomer AG, Nixon PH (1973) The geology of the Letseng-la-terae kimberlite pipes. In: Nixon PH (ed) Lesotho Kimberlites, Lesotho National Development Corporation, Maseru, pp 20–36

Brooker RA, Sparks SJ, Kavanagh JL, Field M (2011) The volatile content of hypabyssal kimberlite magmas: some constraints from experiments on natural rock compositions. Bull Volcanol 73:959–981

Cas RAF, Hayman PC, Pittari A, Porritt LA (2008) Some major problems with existing models and terminology associated with kimberlites from a volcanological perspective and some suggestions. J Volcanol Geoth Res 174:209–225

Clement CR (1973) Kimberlites from the Kao pipe, Lesotho. In: Nixon PH (ed) Lesotho Kimberlites. Lesotho National Development Corporation, Maseru, pp 110–121

Cole DR, Ripley EM (1998) Oxygen isotope fractionation between chlorite and water from 170 to 350 °C: A preliminary assessment based on partial exchange and fluid/rock experiments. Geochimica et Cosmochima Acta 63:449–457

Coopersmith HG, Mitchell RH, Hausel WD (2003) Kimberlites and lamproites of Colorado and Wyoming. In: Kjarsgaard BA (ed) Field trip guidebooks VIIIth international kimberlite conference 2003. Geological Survey of Canada, Ottawa. (www.nrcan.gc.ca/gsc/bookstore)

Deines P, Gold DP (1973) The isotopic composition of carbonatite and kimberlite carbonates and their bearing on the isotopic composition of deep-seated carbon. Geochim Cosmochim Acta 37:1709–1733

Devouard B, Baronnet A, van Tendeloo G, Amelinckx S (1997) First evidence of synthetic polygonal serpentines. Eur J Mineral 9:539–546

Evans BW (2004) The serpentine multisystem revisited: chrysotile is metastable. Int Geol Rev 46:479–506

Field M, Stiefenhofer J, Robey J, Kurszluakis S (2008) Kimberlite-hosted diamond deposits of southern Africa. Ore Geol Rev 34:33–75

Früh-Green GL, Plas A, LeCouyer C (1996) Petrologic and stable isotopic constraints on hydrothermal alteration and serpentinization of the EPR shallow mantle at Hess Deep (SITE 895). In: Mevel C, Gillis KM, Allan JF, Meyer PS (eds) Proceeding ODP science results, vol 147, pp 255–291

Harford CL, Sparks RSJ (2001) Recent remobilisation of shallow-level intrusions on Montserrat revealed by hydrogen isotope composition of amphiboles. Earth Planet Sci Lett 185:285–297

Hayman PC, Cas RAF, Johnson M (2009) Characteristics and alteration origins of matrix minerals in volcaniclastic kimberlite of the Muskox pipe (Nunavut, Canada). Lithos 112S:473–487

Jago BC, Mitchell RH (1985) Mineralogy and petrology of the Ham kimberlite, Somerset Island, N.W.T., Canada. Can Mineral 23:619–634

Kirkley MB, Smith HS, Gurney JJ (1989) Kimberlite carbonates—a carbon and oxygen stable isotope study. In: Ross J (ed) Kimberlites and related rocks, vol 1. Geological Society of Australia Special Publication, Australia, pp 264–281

Kjarsgaard BA, Pearson DG, Tappe S, Nowell GM, Dowall DP (2009) Geochemistry of hypabyssal kimberlites from Lac de Gras, Canada: Comparisons to a global database and applications to the parent magma problem. Lithos 112S:236–248

Kobelski BJ, Gold DP, Deines P (1979) Variations in stable isotope composition for carbon and oxygen in some African and Lesothan kimberlites. In: Meyer HOA, Boyd FR (eds) Kimberlites, Diatremes and diamonds: their geology, petrology and geochemistry. American Geophysical Union, pp 252–271

Kopylova MG, Matveev S, Raudsepp M (2008) Searching for the parental kimberlite melt. Geochim Cosmochim Acta 71:3616–3629

le Roex AP, Bell DR, Davis P (2003) Petrogenesis of group 1 kimberlites from Kimberley, South Africa: evidence from bulk-rock geochemistry. J Petrol 44:2261–2286

Lodochnikov WN (1933) Serpentines and serpentinites and the petrological problems connected with them. Probl Sov Geol 2:145–150 (in Russian)

Middleton AP, Whittaker EJW (1976) The structure of Povlen-type chrysotile. Can Mineral 14:301–306

Mitchell RH (1986) Kimberlites: mineralogy geochemistry and petrology. Plenum Press, New York

Mitchell RH (1995) Kimberlites orangeites and related rocks. Plenum Press, New York

Mitchell RH (1997) Kimberlites, orangeites, lamproites, melilitites and minettes: a petrographic atlas. Almaz Press, Thunder Bay, Ontario. (www.almazpress.com)

Mitchell RH (2008) Petrology of hypabyssal kimberlites: relevance to primary magma compositions. J Volcanol Geoth Res 174:1–8

Mitchell RH, Putnis A (1988) Polygonal serpentine in segregation-textured kimberlites. Can Mineral 26:991–997

Mitchell RH, Skinner EMW, Scott Smith BH (2009) Tuffisitic kimberlites from the Wesselton Mine, South Africa: Mineralogical characteristics relevant to their formation. Lithos 112S:452–464

Price SE, Russell JK, Kopylova MG (2000) Primitive mantle from the Jericho Pipe, N.W.T.: constraints on primary kimberlite melt chemistry. J Petrol 41:789–808

Scott Smith BH, Nowicki TE, Russell JK, Webb KJ, Mitchell RH, Hetman CM, Harder M, Skinner EMW, Robey JV Kimberlite terminology and classification (2013)

Shee SR (1985) The petrogenesis of the Wesselton Mine kimberlite, Kimberley, Cape Province, R.S.A. Ph.D. Thesis, University of Cape Town, South Africa

Sheppard SMF, Dawson JB (1975) Hydrogen, carbon, and oxygen isotope studies of megacryst and matrix minerals from Lesothan and South African kimberlites. Phys Chem Earth 9:747–763

Sheppard SMF, Epstein S (1970) D/H and O^{18}/O^{16} ratios of minerals of possible mantle or lower crustal origin. Earth Planet Sci Lett 9:232–239

Sheppard SMF, Nielsen RL, Taylor HP (1969) Oxygen and hydrogen isotope ratios of clay minerals from porphyry copper deposits. Econ Geol 64:755–777

Stripp GR, Field M, Schumacher JC, Sparks RSJ, Cressey G (2006) Post-emplacement serpentinization and related hydrothermal metamorphism in a kimberlite from Venetia, South Africa. Metamorph Geol 24:515–534

Taylor HP, Sheppard SMF (1986) Igneous rocks: I. Processes of isotopic fractionation and isotope systematics. In: Valley JW, Taylor HP, O'Neil JR (eds) Stable isotopes in high temperature geological processes, vol 16. Mineralogical Society of America Reviews in Mineralogy, pp 227–271

Ukhanov AV, Devirts AL (1983) Meteoric origin of water serpentinizing kimberlites. Doklady Akademy Nauk SSSR 268:706–709 (in Russian)

Ustinov VI, Ukhanov AV, Gavrilov YY (1994) Oxygen isotope composition of mineral assemblages in the stages of emplacement of kimberlites. Geochem Int 31:152–156

Wenner DB, Taylor HP (1973) Oxygen and hydrogen isotope studies of the serpentinization of ultramafic rocks in oceanic environments and continental ophiolite complexes. Am J Sci 273:207–239

Wicks FJ, Zussman J (1975) Microbeam X-ray diffraction patterns of serpentine minerals. Can Mineral 13:244–258

Wilson MR, Kjarsgaard BA, Taylor B (2007) Stable isotope composition of magmatic and deuteric carbonate in hypabyssal kimberlite, Lac De Gras field, Northwest Territories, Canada. Chem Geol 242:435–454

Wyoming Craton Mantle Lithosphere: Reconstructions Based on Xenocrysts from Sloan and Kelsey Lake Kimberlites

I. V. Ashchepkov, H. Downes, R. Mitchell, N. V. Vladykin, H. Coopersmith, and S. V. Palessky

Abstract

The structure of the lithospheric mantle of the Wyoming craton beneath two Paleozoic kimberlite pipes, Sloan and Kelsey Lake 1 in Colorado, was reconstructed using single-grain thermobarometry for a large data set (>2,600 EPMA analyses of xenocrysts and mineral intergrowths). Pyrope compositions from both pipes relate to the lherzolitic field (up to 14 wt% Cr_2O_3) with a few deviations in CaO to harzburgitic field for KL-1 garnets. Clinopyroxene variations (Cr-diopsides and omphacites) from the Sloan pipe show similarities with those from Daldyn kimberlites, Yakutia, and from kimberlites in the central part of the Slave craton, while KL-1 Cpx resemble those from the Alakit kimberlites in Yakutia that sample metasomatized peridotites. LAM ICP analyses recalculated to parental melts for clinopyroxenes from Sloan resemble contaminated protokimberlite melts, while clinopyroxenes from KL-1 show metasomatism by subduction fluids. Melts calculated from garnets from both pipes show peaks for Ba, Sr and U, and HFSE troughs, typical of subduction-related melts. Parental melts calculated for ilmenites from Sloan suggest derivation from highly differentiated melts, or melting of Ilm-bearing metasomatites, while those from Kelsey Lake do not display extreme HFSE enrichment. Three P-Fe# (where Fe# = Fe/(Fe + Mg) in atomic units) trends within the mantle lithosphere beneath Sloan have been obtained using monomineral thermobarometry. At the base, the trends reveal melt metasomatized (possibly sheared) peridotites (Fe# = 13–15 %), refertilized peridotites (Fe# = 10–11 %) and primary mantle peridotites (Fe# = 7–9 %). Anomalous heating was found at depths equivalent to 4.0 and 3.0–2.0 GPa. The mantle section beneath KL-1 is widely metasomatized with several stages of refertilization with dispersed Ilm–Cpx trends. The step-like subadibatic heating in the mantle column beneath the Sloan pipe is strong in the

I. V. Ashchepkov (✉) · S. V. Palessky
Institute of Geology and Mineralogy, SD RAS, Novosibirsk, Russia
e-mail: igora57@mail.ru

H. Downes
Birkbeck College, University of London, London, UK

R. Mitchell
Lakehead University, Ontario, Canada

N. V. Vladykin
Institute of Geochemistry SD RAS, Irkutsk, Russia

H. Coopersmith
Coopersmith and Associates, Fort Collins, Colorado, P.O. Box 1916, USA

D. G. Pearson et al. (eds.), *Proceedings of 10th International Kimberlite Conference*, Volume 1, Special Issue of the Journal of the Geological Society of India, DOI: 10.1007/978-81-322-1170-9_2, © Geological Society of India 2013

base and the middle part and weaker within 2–3 GPa. Heating beneath the KL-1 pipe is more evident in the base and middle part from 7.0 to 3.0 GPa.

Keywords
Mantle lithosphere • Thermobarometry • Colorado • Wyoming craton • Pyrope • Cr-diopside • Ilmenite • Chromite

Introduction

Information about the composition and structure of the mantle beneath North America is mainly available from the Slave craton (Aulbach et al. 2004; Kopylova and Caro 2004). Less is known about the Wyoming craton, despite the rather long history of investigation of its kimberlites, mantle mineralogy, and petrology (McCallum et al. 1975; Eggler et al. 1979; Hearn 1984; Ater et al. 1984; Hausel 1998; Smith et al. 1979; Ashchepkov et al. 2001; Smith and Griffin 2005; Usui et al. 2007; De Stefano et al. 2009; Carlson et al. 2004; Schulze et al. 2008 etc.). The more productive kimberlites in the Slave and Superior cratons have attracted more attention from mantle petrologists (Menzies et al. 2004; Schmidberger et al. 2007; Smart et al. 2009; MacKenzie and Canil 1999; Kopylova et al. 1999; Kopylova and Caro 2004; Heaman et al. 2003, 2004, 2006; Kjarsgaard et al. 2004; Schmidberger and Francis 1999; Meyer et al. 1994; Eccles et al. 2010 etc.).

The Wyoming craton is a unique example in which to study the evolution of the upper mantle because kimberlites were emplaced from Late Proterozoic through Late Paleozoic to Eocene and Miocene times (Lester and Larson 1996; Griffin et al. 2004; Hausel 1998; Hearn 2004). Earlier studies of xenocrysts and xenoliths from the Wyoming craton (McCallum and Eggler 1976; Eggler et al. 1979; Ater et al. 1984; Downes et al. 2004; Davies et al. 2004; Schulze et al. 1992, 1995; Hearn 2004; Hausel 1998; Coopersmith et al. 2003; Carlson et al. 2004) revealed general variations of mineral chemistry and compositions of mantle minerals and xenoliths, as well as information about the transformation of the Wyoming craton keel. Petrologic reconstructions suggest continuous destruction of the craton by subduction tension, tectonic and plume forces with reduction in the cratonic keel (Coopersmith et al. 2003; Carlson et al. 2004; Carlson and James 2011) accompanied by extreme variations in mantle isotopic compositions (Carlson et al. 1999, 2004; Rudnick et al. 1999). We present new information for samples from the Sloan and Kelsey Lake-1 (KL-1) pipes (Ater et al. 1984; Coopersmith et al. 2003 and references therein; Carlson et al. 2004). Using mainly monomineral thermobarometry and geochemical features of minerals, we show reconstructions of the subcratonic lithospheric mantle (SCLM) beneath the Wyoming craton and its possible evolution.

Location and Geology

The Wyoming craton has experienced mantle-derived volcanism over a long period of time, including kimberlites (Vendian to Eocene), lamproites (Mitchell and

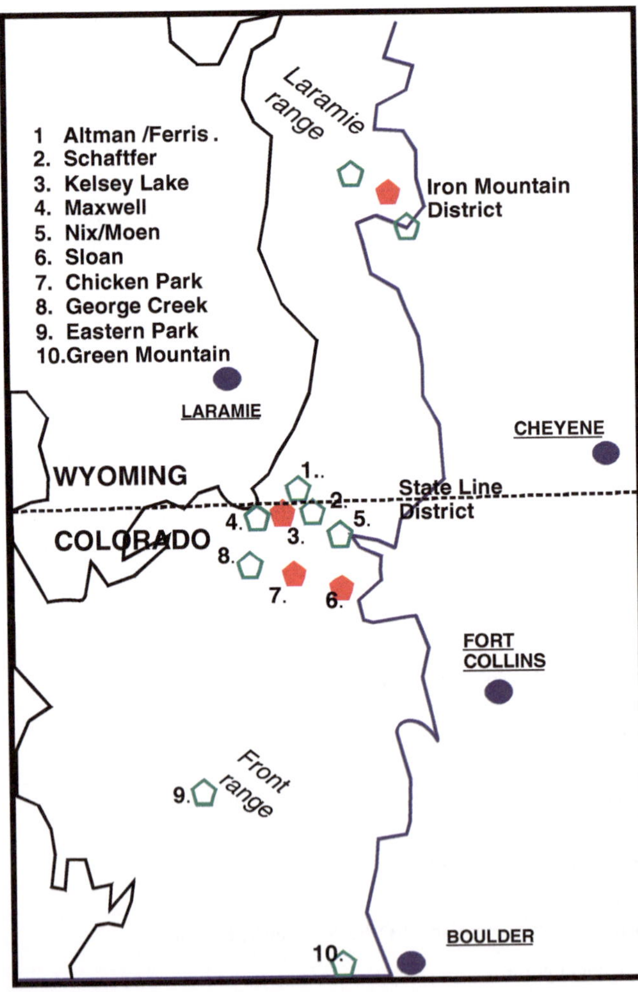

Fig. 1 Location of kimberlite pipes in Wyoming craton in state line district and surrounding area

Bergman 1991), minettes (Macdonald et al. 1992), and alkaline basalts (O'Brien et al. 1995), and most of these contain deep-seated xenoliths. The Paleozoic Sloan and Kelsey Lake kimberlite groups are located in the southern part of the Wyoming craton in the State Line kimberlite district, in northern Colorado (Fig. 1). They both consist of several kimberlites bodies (5 in Sloan and 8 in Kelsey Lake) (Hausel 1998; Coopersmith et al. 2003; Schulze et al. 2008), some of them relatively large (up to 400 m) which contain diamonds. Most of them carried abundant xenocrysts which are more frequent and fresher in the Sloan pipes. Large peridotite xenoliths are rare, and in KL-1 pipe, they are mostly altered.

Data Set and Methods

The samples from two pipes were analyzed in the Institute of Mineralogy and Petrology SD RAN Novosibirsk by Electron Microprobe (EPMA) (Lavrent'ev and Usova 1994) from KL-1 pipe (H. Coopersmith collection) and Sloan (H. Downes samples) as separate grains and mineral intergrowths (758 and 1,150 analyses, respectively). In addition, 14 mineral grains from KL-1 and 21 from Sloan were analyzed by inductively coupled mass spectrometry with laser ablation system (LAM ICP). The method and precision were described by Ashchepkov et al. (2010, 2013, 2012).

Mineralogy of Xenocrysts

Megacrysts and xenocrysts from Sloan and Kelsey Lake-1 (KL-1) pipes were analyzed by EPMA and LAM ICP-MS in the Analytic Centre IGM SB RAS (Novosibirsk). We also investigated a smaller number of xenocrysts from the Vendian Chicken Park and Ordovician Iron Mountain kimberlites.

Garnets (Gar) from Sloan (Fig. 2) mainly fall in the lherzolite field (Gr9) (Sobolev et al. 1973a, 1977; Gurney 1984) with up to 12.5 wt% Cr_2O_3. In contrast, the KL-1 garnets reach 14 wt% Cr_2O_3 and 5 % of them show

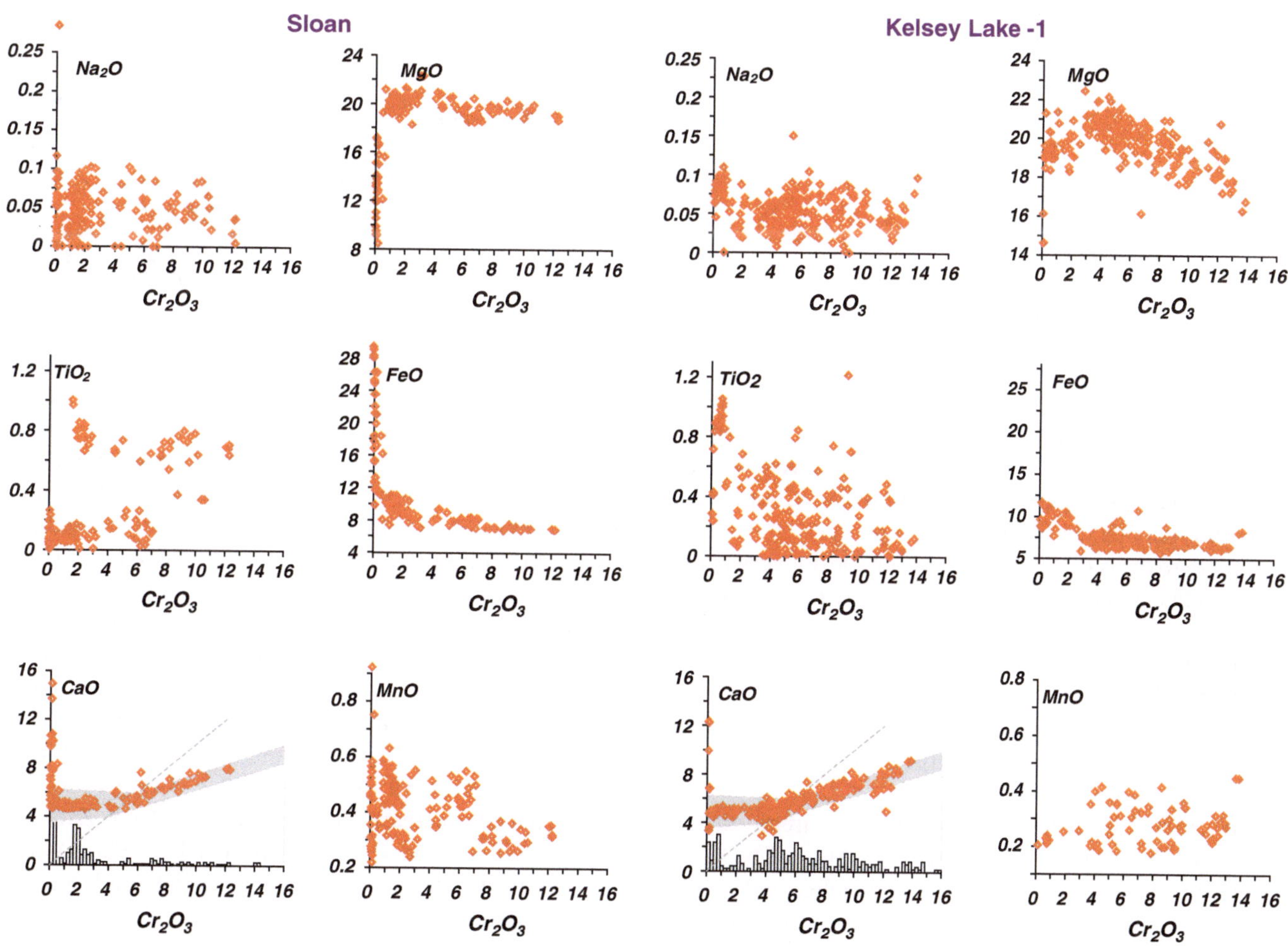

Fig. 2 Variations of major element compositions from garnets from Sloan and Kelsey Lake 1

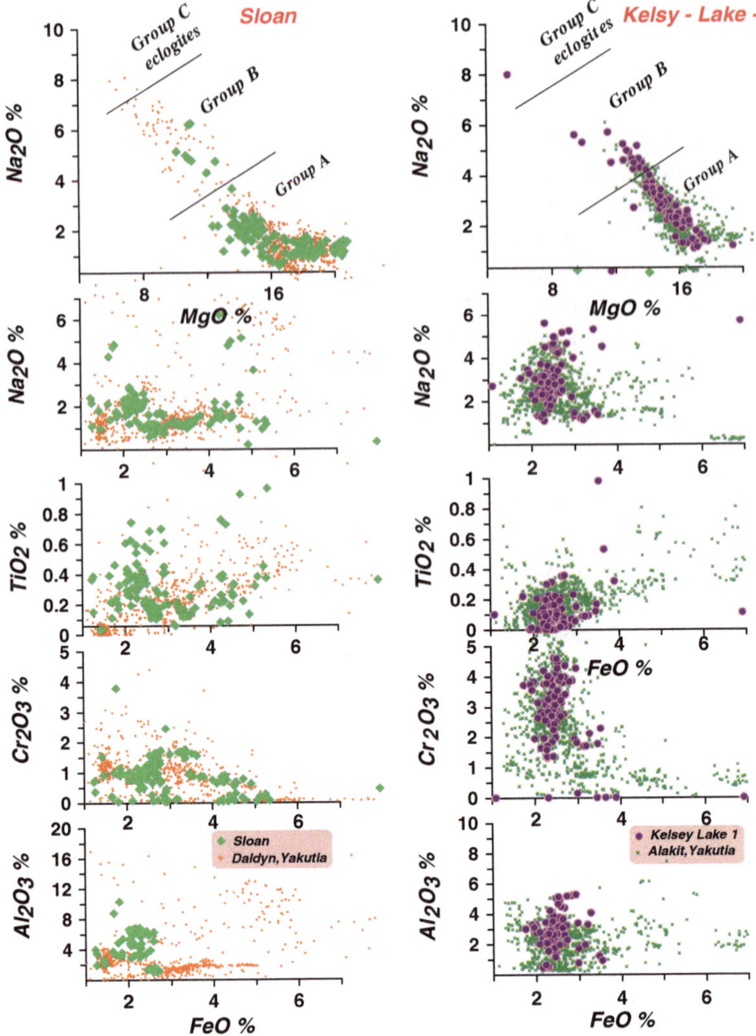

Fig. 3 Variations of major element compositions of clinopyroxenes from Sloan pipe (*A*) in comparison with published Cpx data for Daldyn field (Ashchepkov et al. 2013,2012) and for clinopyroxenes from Kelsey Lake 1 pipe (*B*) (added from Ater et al. 1984) in comparison with Alakit Cpx compositions

deviations to the harzburgite field (Gr10) (Hausel 1998) which correlates with the dominant harzburgite type of diamond inclusions in this pipe (Schulze et al. 2008). We only analyzed the +0.5 mm fraction. However, a more complete data set from prospecting work (Coopersmith et al. 2003) contained plenty of Gr10 pyropes.

Clinopyroxenes (Cpx) from Sloan are mostly very coarse grained and derived from pyroxenites. They may be divided into a common Cpx type containing 1–1.5 % FeO, corresponding mostly to the upper part of the mantle section, and a melt metasomatized type showing an increase of FeO, Na_2O, and TiO_2 which is typical for sheared peridotites (Boyd et al. 1997; Ionov et al. 2010) from the lower part of the SCLM.

The KL-1 diopsides (Fig. 3) show a trend that is restricted in FeO with a sharp increase of Al_2O_3, Cr_2O_3, and Na_2O, typical for Cpx from metasomatites and hybrid varieties from the Alakit kimberlite field, Yakutia (Ashchepkov et al. 2004, 2013). The Cpx from the Homestead pipe are similar to those from KL-1 pipe in general except for low—Al varieties.

Cr-diopsides (Cpx) from Sloan (Fig. 3) show similarities in their compositional fields to those from kimberlites in the Daldyn field, Siberia, from moderately depleted harzburgites and lherzolites and some melt-modified varieties (Boyd et al. 1997).

The eclogitic omphacites from Sloan vary in compositions from group A to C (Dawson 1980), while those for KL-1 (Ater et al. 1984) mainly refer to A and B types. Cpx from KL-1 are richer in N_2O and Cr_2O_3 and lower in TiO_2.

Chromites and Cr-spinels from Sloan (Fig. 4) form a straight peridotitic Cr–Al trend. Their TiO_2, V_2O_3, and Cr_2O_3 contents increase together, whereas NiO, ZnO, and Nb_2O_3 decrease, and Chromites from KL-1 (Fig. 4) coincide with the Cr-rich part of the Sloan spinel trend, showing a wide variation of TiO_2 values.

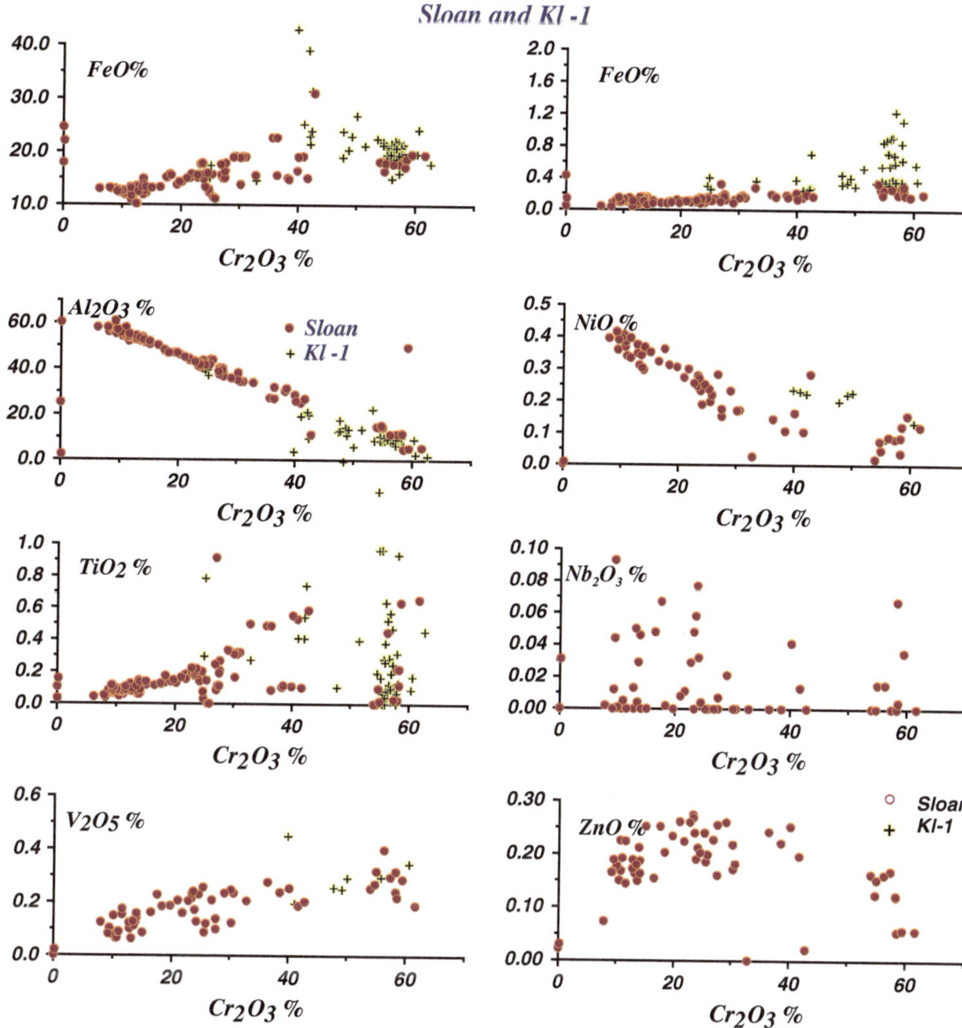

Fig. 4 Variations of major element compositions for Cr-spinels from (1) Sloan and (2) KL-1 pipes

Mg-ilmenites from Sloan pipe (Fig. 5) show a rather restricted range in TiO_2 from 56 to 44 wt%. The decrease of TiO_2 is accompanied by a continuous increase in Cr_2O_3 and decreases in Al_2O_3 and FeO. Mg-ilmenites from KL-1 pipe reveal a more extended trend (36–60 wt% TiO_2) which is more variable because many ilmenites occur in intergrowths with exsolved rutile. Nevertheless, the general trends are similar as for Sloan ilmenites, showing enrichment in Cr_2O_3 toward the low-Mg, low-Ti part of the trend. Strong enrichment in MnO to 8 wt% is common for some varieties. The compositional variations determined for ilmenites in this paper are similar to those found by Schulze et al. (1995).

Single-Grain Thermobarometry for Mantle Sections Beneath Wyoming Craton

The SCLM structure beneath the Sloan and Kelsey Lake kimberlites has been reconstructed by monomineral methods described in Ashchepkov et al. (2010, 2013) which yield PT estimates close to those obtained by orthopyroxene-based methods (Brey and Kohler 1990; McGregor 1974). PT estimates for Cr-diopsides (Ashchepkov et al. 2010) are close to those produced by Nimis and Taylor (2000) but work with a wider range of compositions including augites and omphacites.

The SCLM beneath the Sloan pipe is divided into lower and upper parts at 4.0–3.0 GPa by a layer composed largely of Opx, omphacites (6 wt% Na_2O), low-CrCpx and Cr-bearing ilmenites and diopsides from metasomatic veins. In the lower part, the PT data for Ti-rich lherzolitic garnets trace the 37 mW/m^2 geotherm to the base of the SCLM (7.5–6.0 GPa), showing an inflection (Nixon and Boyd 1973). Two advective branches rising from the base are marked by Cr-diopsides of melt metasomatized (sheared) type.

The PT plot produced by the combination of the Opx thermometer and Opx–Gar geobarometer (Brey and Kohler 1990) reproduced the PT plot for the Cpx in the upper and lower parts of the mantle section and the advective branch.

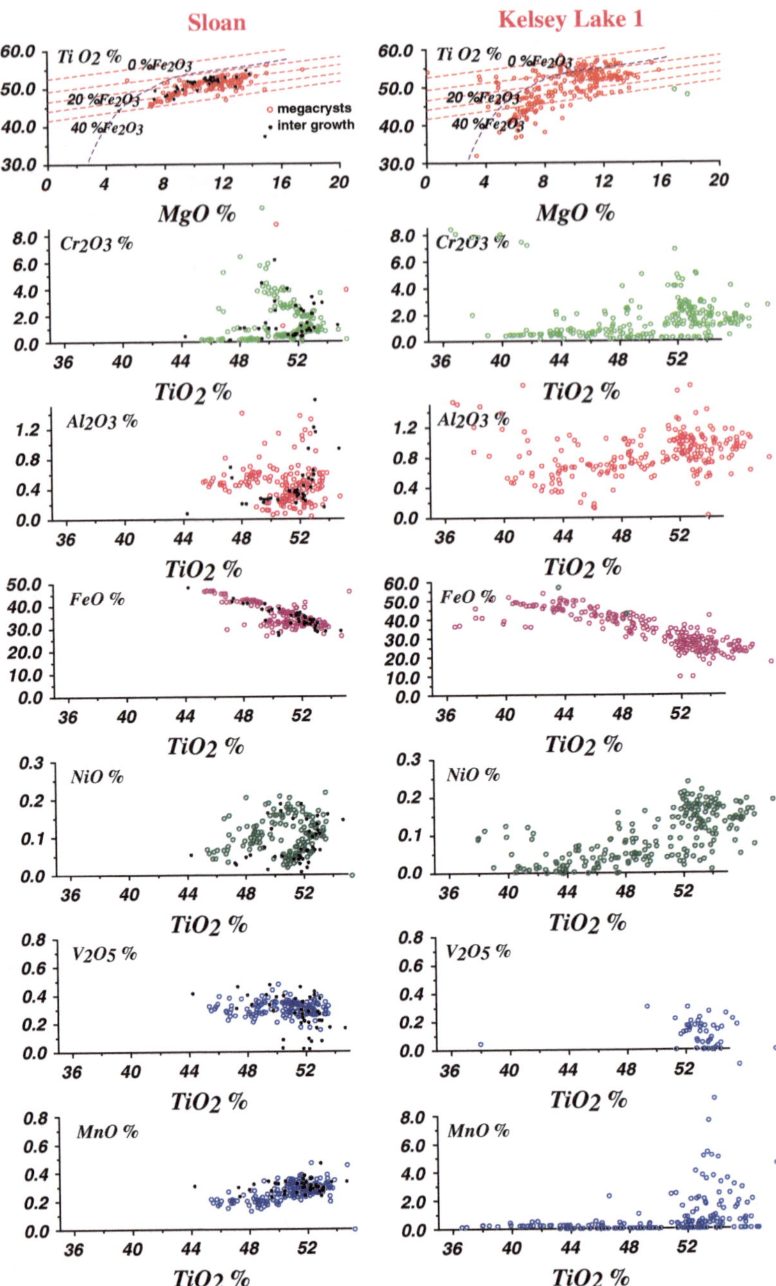

Fig. 5 Variations of major element compositions of ilmenites from Sloan (**a**) (1—xenocrysts; 2—intergrowths) and Kelsey Lake (**b**)

Cpx from Sloan are clearly divided into two large groups corresponding to the lower (more Fe rich) and upper parts of the lithospheric mantle. Stepwise increases in FeO, Na_2O, TiO_2, Al_2O_3, with three separate intervals in P = Fe# plot (Fig. 6a) and decreasing Cr_2O_3, possibly reflect several episodes of melt interaction which is similar to those for sheared peridotites. Garnets estimations are splitting into several arias. TiO_2-rich melt metasomatized garnets trace the 40–45 mW/m^2 geotherms, while common garnets are located near to the 35 mW/m^2 line. PT estimates for ilmenites and Ti-garnets reflect fractionation of protokimberlites (Fe# = 17). Ilmenites which are similar to those from metasomatic veins in xenoliths from the Alakit kimberlite region (Ashchepkov et al. 2010) show a stepped increase in Cr_2O_3 (Fe# = 12) at lower temperatures (LT), and higher Fe# than peridotitic minerals (Fe# = ~8–10). Cr-spinels from deeper levels show high TiO_2 and NiO contents, decreasing toward the top of the SCLM, also influenced by melts. The upper level <3.0 GPa shows three PT paths: the hottest for Opx and Fe-Cpx, cooler for garnets and Cpx, and the coldest for Sp, Cpx, and Gar with Fe# ~9–11 correspondingly. The latter are spinel lherzolites. Eclogitic Cpx mainly correspond to the upper mantle section and middle part, where they are more Fe rich and a few in the lower part of the SCLM.

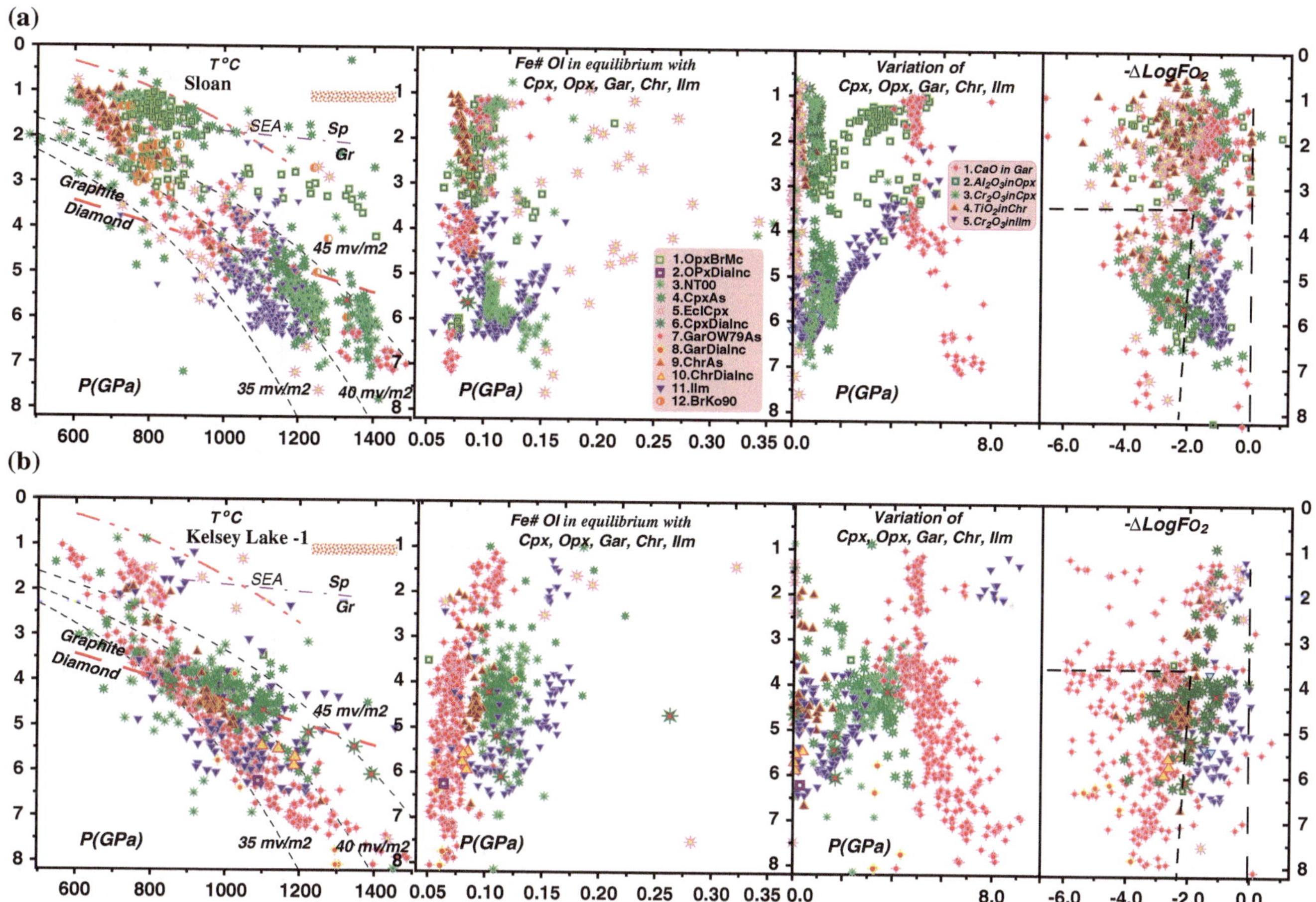

Fig. 6 PTX fO_2 plot: PT – X(Fe#, CaO, Cr_2O_3, Al_2O_3) – fO_2 (PTXF) plots using data set for Sloan (**a**) and Kelsey Lake (**b**) pipes. *Symbols* 1. Opx T °C (Brey and Kohler 1990)—P(GPa) (McGregor 1974); 2. the same for diamond inclusions; 3. *Cpx* T °C (Nimis and Taylor 2000)—P(GPa) (Ashchepkov et al. 2010); 4. T °C (Nimis and Taylor 2000)—P(GPa) (Ashchepkov et al. 2010); 5. the same for eclogites; 6. the same diamond inclusions; 7. garnet: (O'Neil and Wood 1979) monomineral—P(GPa) (Ashchepkov et al. 2010); 8. the same for diamond inclusions; *Sp* 9. T °C (O'Neil and Wall 1987)—P(GPa) (Ashchepkov et al. 2010); 10. the same for diamond inclusions; *Ilmenite* 11. T °C (Taylor et al. 1998)—P(GPa) (Ashchepkov et al. 2010); 12. Gar–Opx: (Brey and Kohler 1990). Field for diamond-bearing associations in the lower left corner after McCammon et al. (2001)

The LT mantle sequence determined from garnets and some Cpx, chromites, and ilmenites beneath KL-1 relate to a <35 mW/m^2 geotherm which is lower in general than as found previously (Carlson et al. 2004). Cr-diopsides PT estimates show heating along the graphite–diamond boundary, confirming earlier estimates (Pizzolato and Schultze 2003). High temperature (HT) branches for Cr-rich garnets from the base of the SCLM show an increase in CaO. The combined trends of ilmenites and Cr-diopside heated to 40 mW/m^2 (Fe# ~ 15–17; 11–14; 9–10) result from refertilization. PT conditions also tend to be divided into relatively cold and those overheated at the shallow level. Chromite PT estimates show split trends (Fe# ~ 10; 12) at 5.5–4.0 GPa. HT conditions with subadiabatic path are detected for some ilmenites in the middle and upper parts of the lithospheric mantle section.

Eclogitic Cpx (Ater et al. 1984) mark the upper part of the mantle section, and one point corresponds to the lower part of mantle section.

The oxygen fugacity values were calculated for ilmenites and chromites according to the single mineral version of Taylor et al. (1998) oxybarometers and for pyroxenes and garnet according to the equations published in Ashchepkov et al. (2011).

The oxidation state of peridotite minerals is higher in the lithosphere base and upper part of the SCLM beneath Sloan pipe and in the middle part of the mantle section. In KL-1, the fO_2 values generally rise upward, as in South African lithosphere (McCammon et al. 2001), and the highest heterogeneity similar to most studied mantle sections (Ashchepkov et al. 2011, 2013) including Iron Mountain SCLM. In the Eocene mantle, the peridotite minerals are more

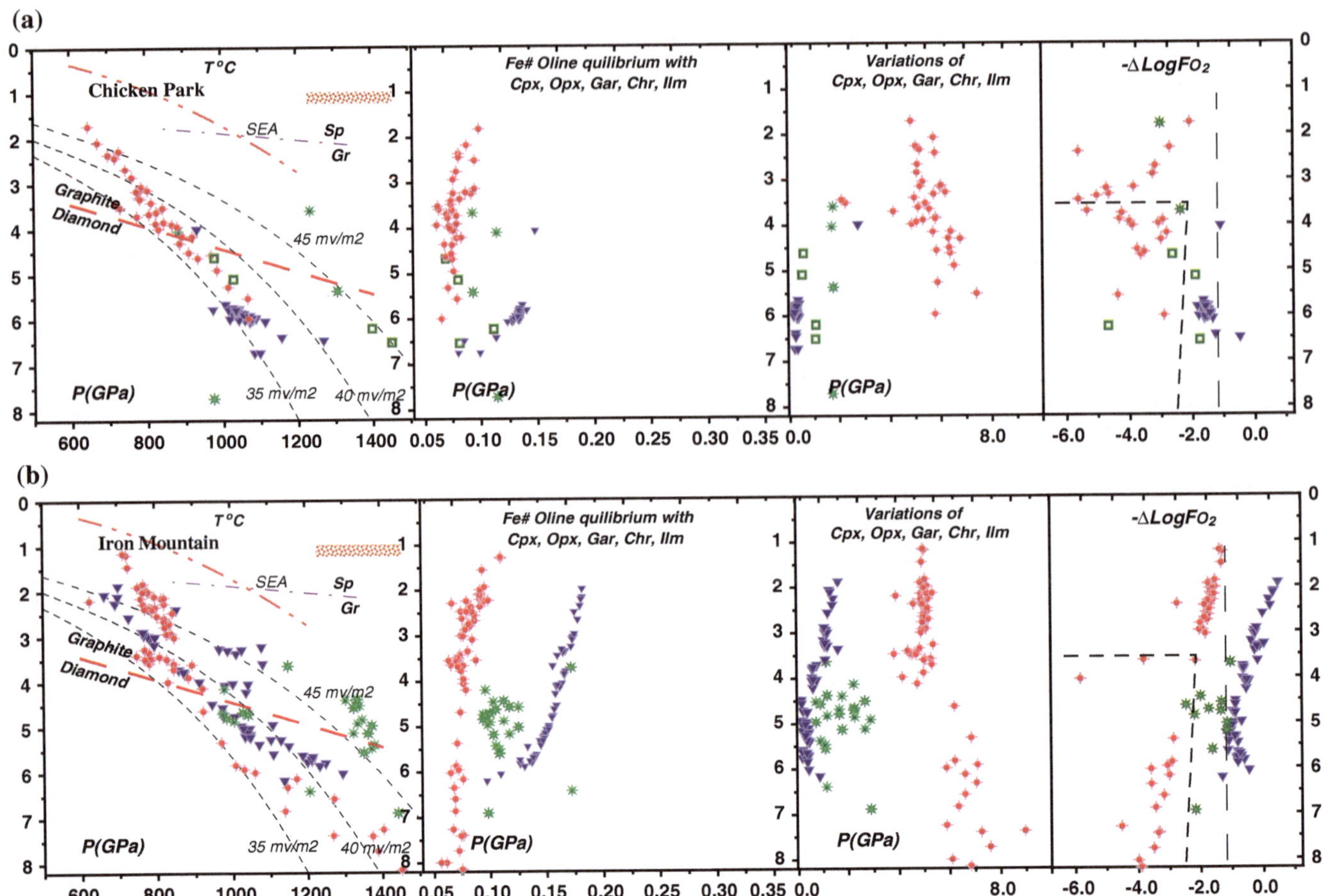

Fig. 7 PTX fO_2 diagrams for xenocrysts of the Vendian Chicken Park kimberlite pipe (**a**). Data original and from Griffin et al. (2004); Schulze et al. (1995) and for (**b**) Ordovician iron mountain kimberlite pipe. Data after Griffin et al. (2004); Schulze et al. (1995). For symbols, see Fig. 6

oxidized, probably due to reactions with melts. The fO_2 distribution in the mantle section helps to determine the details of the SCLM structure. The number of apparent layers is between 7 and 10 for KL-1 and Sloan.

Trace Element Geochemistry

Sloan peridotitic garnets show rounded REE patterns with humps to Nd to Gd and small U, Ba, Nb, and Ta peaks, with varying Zr concentrations indicating mixed subduction and metasomatic features. The Cpx patterns increase their inclinations toward the deeper parts of the mantle $50 < Gd/Yb < 80$. Small Zr troughs are typical. Enrichment in Ba is a result of the presence of fluid inclusions (Fig. 7).

Eclogitic minerals (Sobolev et al. 1973b, 1994) demonstrate different behavior in trace elements. One of the eclogites shows common rounded patterns for garnets and clinopyroxene without Eu anomalies. The next one with high amount of modal Cpx demonstrates a spoon-like pattern common for garnet-poor or garnet-free associations. Two Cpx show patterns typical for plagioclase and probably result from the inversion of this mineral without much influence of garnet partitioning. The spider diagrams for most of them show peaks in Ba, U, Sr, and Pb, which is typical for subduction-related fluids but they have no essential negative anomalies in HFSE.

The trace elements for ilmenites are very uniform with strong peaks in Ta–Nb, weaker ones in Hf–Zr, and slight ones for Pb. Some ilmenites have higher Rb and Ba, probably due to metasomatism. One ulvöspinel shows an inclined REE pattern and moderate enrichment in HFSE.

Garnets from KL-I with increasing Cr show lower REE and progressive S-shaping, with increasing pressure. REE patterns of Cpx show rotation at Gd probably due to variations in modal garnet in their source. Zr–Hf depletion is rare but Ta–Nb troughs are common, increasing with U peaks which are common for subduction processes. Deep-seated Cpx LREE and Th enrichment reflects interaction with carbonatite. Cpx from splherzolites display increases in LREE, U, and Ba, relating to ancient subduction. Ilmenites reveal inclined REE patterns with varying

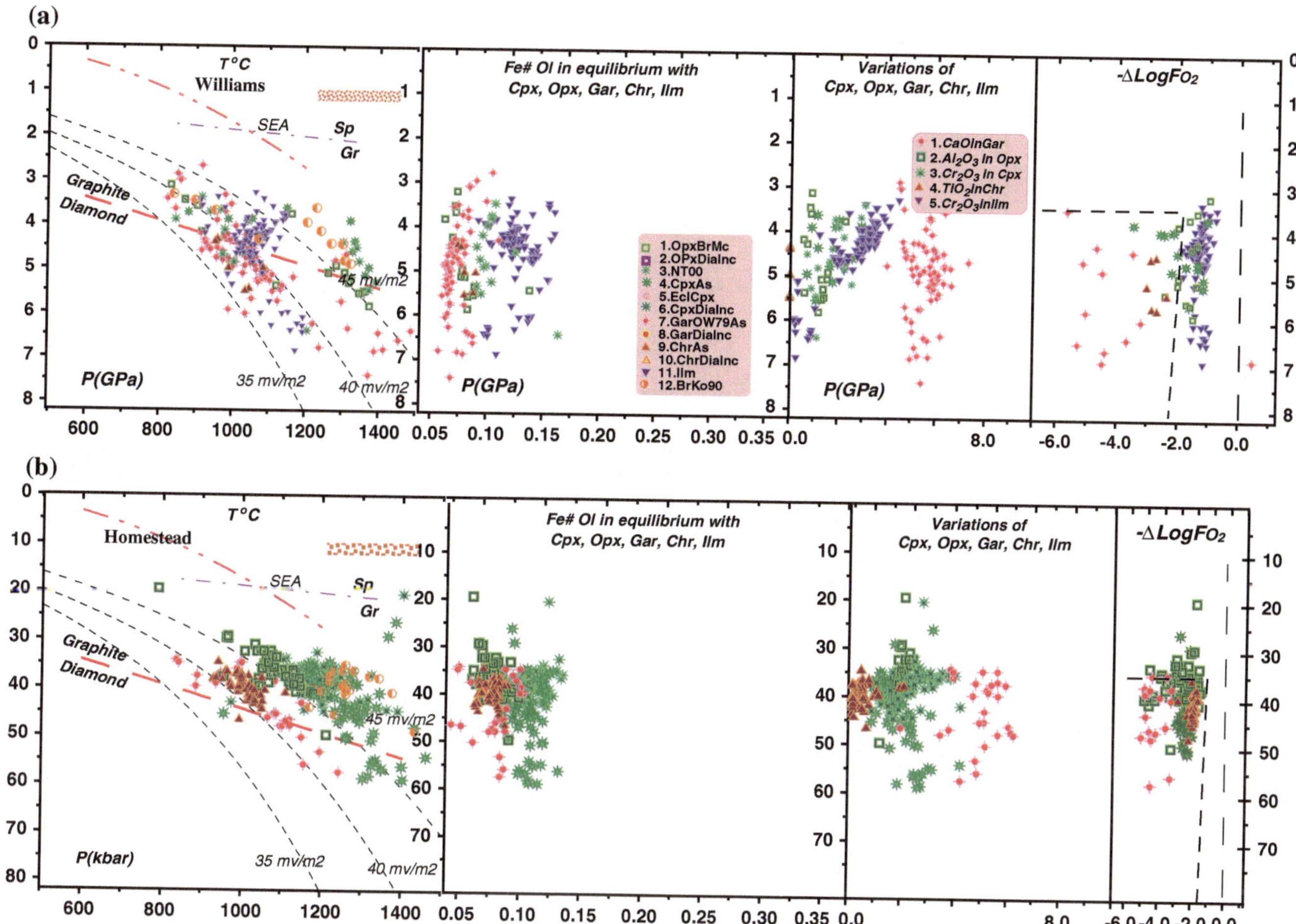

Fig. 8 PTX fO_2 diagrams for minerals from concentrates of (**a**) Eocene Williams kimberlite pipe data after Eggler et al. (1979); Eggler et al. (1987); Hearn (2004); Hearn and McGee (1984); Griffin et al. (2004); Schulze et al. (1995) and for (**b**) Eocene Homestead kimberlite pipe (Montana). Data after Hearn (2004). For symbols, see Fig. 6

HREE and show lower U, Ta–Nb, and Zr–Hf peaks common for metasomatites.

Discussion

Structure of the SCLM Beneath Sloan and Kelsey Lake 1 and Other Pipes from Wyoming Craton

The sharp division of the SCLM beneath Sloan pipe (Fig. 6a) into three parts is possibly the result of sharp primary discontinuities in the SCLM marked by the presence of eclogites which are most Fe rich in the boundary at ~3.5–4 GPa. The strong interaction occurred in two stages (judging by difference in Fe# including shearing possibly means high concentration of melts). The high Fe# of Opx in the middle part means high permeability of the mantle. Some parts of the SCLM at 6.0–5.0 and 4.5–3.0 GPa are practically devoid of Gar and Cpx, and lithologies here are possibly close to dunites which concentrate the melts. Such melts produced high degrees of interaction not only in lower part but also in the upper part of the mantle which is shown by rather homogeneous trace element patterns.

The more continuous mantle column in the lower part of the SCLM beneath KL-1 (Fig. 6b) was possibly originally more heterogeneous, including rare Fe-rich eclogites and Mn-rich rocks (possibly deep water sediments). The geochemical signature, specifically the U, Pb, Sr, Ba enrichment, indicates a possible subduction-related metasomatism of the SCLM which was accompanied by melting. The similarity with the Alakit metasomatic trend for Cpx supports this suggestion. The lower part of the mantle column was subjected to several refertilization events. A higher concentration of Cpx-rich rocks, possibly pyroxenites, occurs near 4.0 GPa in SCLM beneath Udachnaya (Pokhilenko et al. 1999).

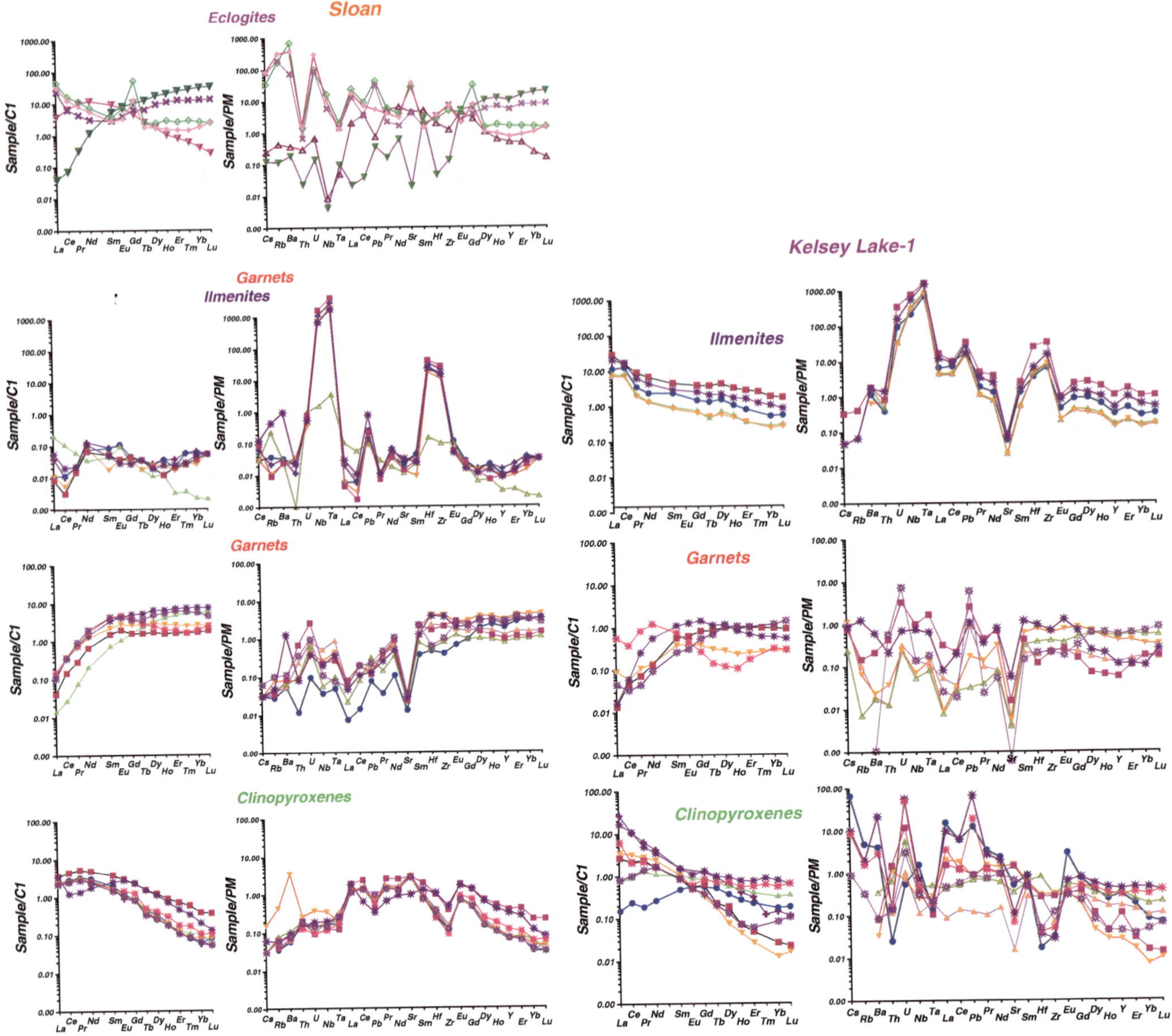

Fig. 9 Trace element patterns for minerals from xenoliths from Sloan (**a**) and Kelsey Lake-1 (**b**) pipes. Normalization to primitive mantle after McDonough and Sun (1995)

Comparison with SCLM in Other Pipes of the Wyoming Craton

Examples from two other pipes from the State Line district (Fig. 8) give some indications about the development of the mantle sections. Chicken Park (Fig. 7a) is probably one of the earliest pipes in the Wyoming craton (Heaman et al. 2004). The SCLM obtained with the restricted analyses shows a subdivision into two parts at 4.0 GPa where the garnet PT trend reveals an inflection. The base of the lithosphere judging by PT conditions for two Opx is heated to 1,500 °C. The Fe# enrichment is restricted to 0.15 for peridotite minerals, and high Fe# values are rare in the upper levels. The next heated horizon is at 3.0–4.0 GPa, and hot associations are also apparent at 2.0 GPa.

The SCLM beneath Iron Mountain (Wyoming) (Fig. 7b) is similar with a sharp division at 4.0 GPa. Garnets in the upper section demonstrate an increase in FeO and equal amounts of CaO, while in the lower part, there is dispersion in CaO and an increase in Fe within the 5.0–7.0 GPa pressure interval. The chromites show the enrichment in Ti, Nb in Cr—rich indicating the greater influence of a HFSE-rich melt at the base of the SCLM and its differentiation accompanied by peridotite assimilation (AFC) toward the top of the mantle column. The HT convective branch is pronounced there for PT estimates from both garnets and

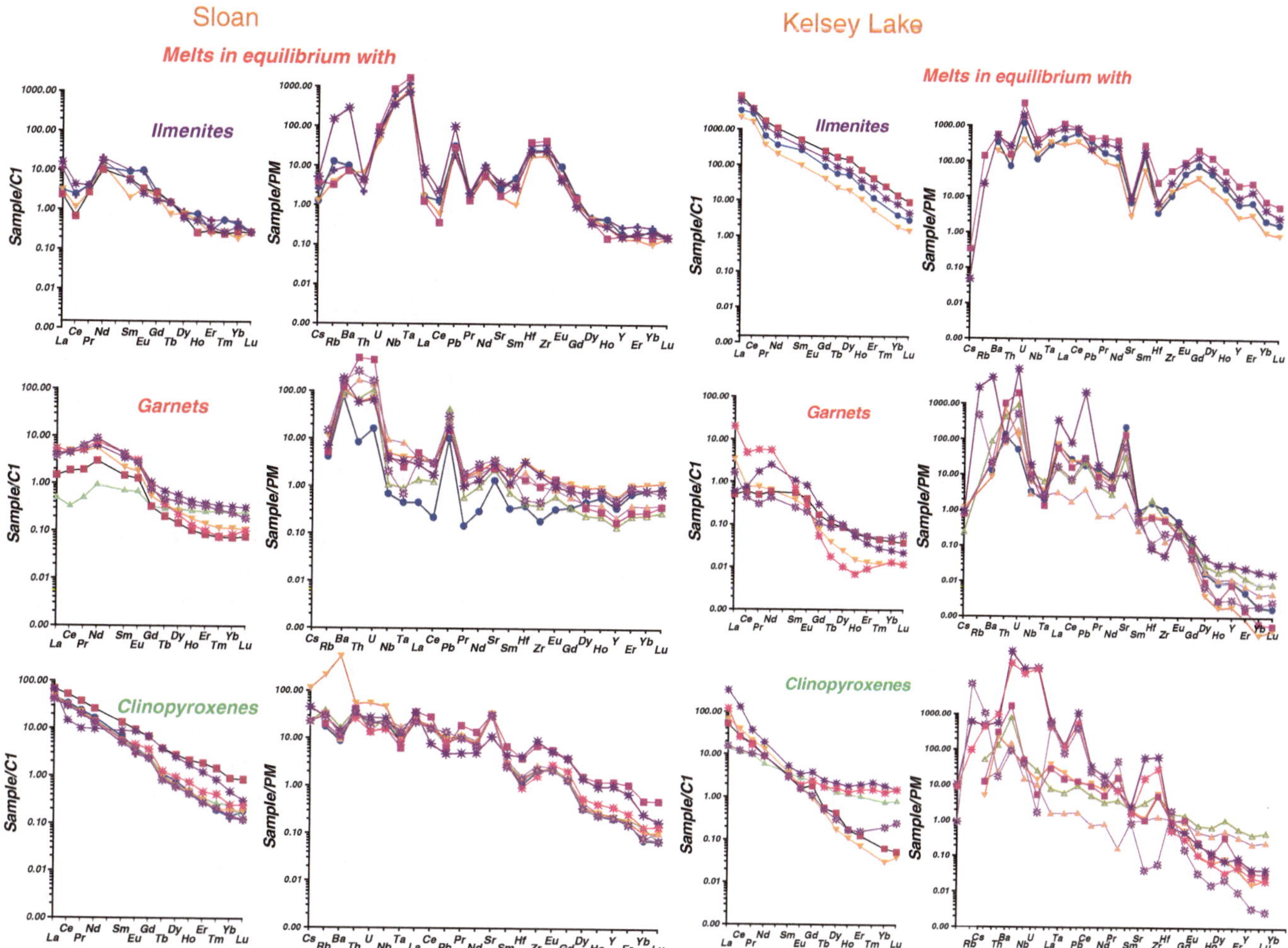

Fig. 10 Primitive mantle-normalized REE and trace element patterns for the melts parental for xenocryst minerals from (**a**) Sloan and (**b**) Kelsey Lake I pipes. Partition coefficients for the minerals are from Hart and Dunn (1993) for clinopyroxene; Hauri et al. (1994) for garnet, and Zack and Brumm (1998) for ilmenite

ilmenites. All analyzed Cpx have Fe# intermediate between those for garnets and ilmenites. The latter show a polybaric trend of continuous increase in Fe# and Cr with decreasing pressure which is also common for some pipes in Siberia, for example Mir (Ashchepkov et al. 2010).

The Paleocene (48 Ma) pipes from the northern part of the craton (Williams and Homestead in Montana) show thinner SCLM (Fig. 9). The mantle sequence beneath the Williams pipe (Fig. 8b) reveals rather reduced SCLM, high heating and splitting of the geothermal gradients into two branches of 47 and 40 mW/m^2, according to PT estimates for the data of Hearn and McGee (1984) determined according to Brey and Kohler (1990). But it still keeps its lithospheric root at about 6.5 GPa as calculated for garnets containing >10 wt% of Cr_2O_3 (Hearn and McGee 1984) and ilmenites Schulze et al. (1995) and it is proved also by Cpx estimates, but most xenoliths were brought from the heated upper part of the mantle section. The northern Homestead kimberlite (Hearn 2004) shows a very similar configuration of the PT plot but relates mainly to the mantle section above the graphite–diamond transition and 45 mW/m^2 geotherm from 4.0 to 2.0 GPa. The xenoliths demonstrate disequilibrium between garnets and pyroxenes, and the Cpx are higher than Opx and garnets in Fe# and definitely were formed as a result of fertilization.

Geochemistry of Parental Melts for the Mantle Minerals from Wyoming Craton

The melts reconstructed using KD values for Cpx (Hart and Dunn 1993) reveal rather regular patterns which are closer to the plume melts, but are lower in REE than typical kimberlites (Fig. 10). Small dips in Ba, Ta, and Hf, as well as peaks in Sr, suggest that such melts are products of interaction of protokimberlites and partial melts from slightly metasomatized peridotites. Parental melts for garnets, reconstructed using KD values from Hauri et al. (1994), reveal patterns which possibly reflect the assimilation of minerals which have humped patterns such as Cpx. The enrichment in Ba, Th, U, and Sr possibly reflects the influence of subduction-related fluids. The parental melts calculated from ilmenites (KDs from Zack and Brumm 1998; Klemme et al. 2006) show that they were enriched in HFSE and may be produced by the fusion of metasomatic assemblages which already contained ilmenite.

The parental melts calculated for Cpx from KL-1 show two types of trace element patterns. The first have very high La/Yb_n ratios and peaks at U, Ba, Th, Sr, Zr, and Pb and should result from melting of garnet-rich rocks, metasomatic rocks with ilmenite, or possibly phlogopites or under the influence of enriched fluids. The other type of parental melts shows rather primitive mantle-normalized patterns, which may be formed by melting of lherzolites. The garnet parental melts show different REE patterns. Those humped in MLREE are possibly the result of contamination in Cpx. The spider diagrams of melts parental to the garnets have peaks in Pb, Sr, U and troughs in Nb–Ta (sometimes also in Zr–Hf) which resemble subduction-related melts.

Parental melts calculated from ilmenites show very strong enrichments in trace elements mainly in REE. The inclined REE patterns indicate high Gar/Cpx ratios in the coexisting solid. The peaks in U, Sr, and Hf indicate the AFC (assimilation + fractionation crystallization), evolution of ilmenite parental melts and dissolution of metasomatites containing specific minerals like phlogopite, amphibole and Cpx with the precipitation of olivine, Cpx and ilmenite.

Thus, reconstruction of the geochemical features of mantle melts reveals their high heterogeneity, possibly reflecting different stages of mineral formation.

Integrated Knowledge About Evolution of the SCLM Beneath Wyoming

The thick (200–270 km) and layered structure of the Wyoming craton SCLM (Dueker et al. 2001) determined by geophysical methods show flat banding consisting of 10–12 units which is close to our thermobarometric estimates and corresponds to high Cr_2O_3 content of the garnets. These observations may support the hypothesis of low-angle subduction in the SE part of the craton. But a rather highly inclined SCLM structure in SE–NW direction may be evidence for heterogeneity of mantle lithosphere as found by a comparison of mantle columns beneath Sloan and KL-1 pipes.

Previous calculations suggested a thinner craton keel (Coopersmith et al. 2003), as shallow as 180 km (5 GPa; Pizzolato and Schultze 2003) for KL-1 while for Sloan the highest PT estimates referred to 6 GPa and 1,320 °C correspond to 42 mW/m^2 (Carlson et al. 2004) which is supported by our estimates, but most HT conditions determined for Cpx and garnets relate to 45 42 mW/m^2 typical for sheared peridotites.

Differences in the structure of the lithospheric mantle may result from the evolution of the lithosphere mainly by the repeated plume melt intrusions which produced the heating that resulted in melting and melt percolation to the melt traps near 4.0–4.5 and 2.0–3.0 GPa.

Carlson et al. (2004) determined an initial formation in the Proterozoic for Sloan mantle and an Archean age of formation for Homestead peridotites, with later Mesozoic reactivation. Conclusions about rather heated geothermal gradient at the base of SCLM beneath Sloan (Carlson et al. 2004) and irregular heating under KL-1 are supported by our calculations. The geochemistry of garnets from the SCLM of the Wyoming craton and other regions of North America (Griffin et al. 2004) gives a preliminary version of the craton's structure, showing a large number of PT estimates near the graphite–diamond boundary, which is not supported by PT calculations for pyroxenes.

Comparison of the combined data for radiogenic isotopes, trace elements and PT for garnets, clinopyroxenes and other minerals show that different minerals possibly grew in different stages in SCLM which is supported by the disequilibrium in Nd–Sm, Lu–Hf, and Re/Os systems (Carlson et al. 2004; Carlson and james 2011; Rudnick et al. 1999). Most isotopic systems show ancient depletion events. The ancient metasomatism and incompatible element enrichment and metasomatism took place in upper part of SCLM in Achaean time (Carlson and Irving 1994). They initially show mainly subduction signatures which formed low Ti-garnets and Opx and few pyroxenes. But later Cpx obtain mainly plume signatures which formed coarse Cpx possibly growing in pyroxenites, veinlets, and porphyritic peridotites. In initially highly metasomatised columns like that beneath KL-1, the heating produced by plumes produced partial melting and melt percolation from the lithosphere base to middle part where they were crystallizing. This process was completed in the Late Mesozoic–Early Cenozoic time which is seen in the SCLM beneath Williams (Hearn and McGee 1984; Carlson et al. 1999) and Homestead kimberlites (Hearn 2004).

Conclusions

1. Xenocrysts from the Sloan and KL-1 pipes reveal different processes of extensive reaction of the mantle protolith with ancient melts and protokimberlites. Throughout its entire section, the Sloan SCLM interacted with protokimberlite melts, as revealed by the uniform relatively smooth Cpx trace element patterns

from all parts of the mantle section and continuous compositional trends of chromites.

2. The KL-1 SCLM is more diverse in composition and contains abundant subducted material (high Mn, Ba, and Sr). It was metasomatized and homogenized by subduction-related melts, plumes and their alkaline derivatives (O'Brien et al. 1995) and then by protokimberlites.
3. The high heterogeneity of the mantle beneath the Wyoming craton was probably an original feature of the craton keel, composed of high-angle slabs merged in craton margins. Position and compositions of eclogites vary in mantle sections across northern Colorado.

Acknowledgments This study was supported by grant RBRF 05-11-00060a, 11-05-91060-PICSa. Thanks to Prof. G.D.Pearson for corrections of the text.

References

Ashchepkov IV, Vladykin NV, Mitchell RH, Howard C, Garanin VG, Saprykin AI, Khmelnikova OS, Anoshin GN (2001) Mineralogy of the mantle xenocrysts from KL-1 (Kelsey Lake) kimberlite pipe, Colorado. Rev Bras Geocienc 31:645–652

Ashchepkov IV, Vladykin NV, Nikolaeva IV, Palessky SV, Logvinova AM, Saprykin AI, Khmel'Nikova OS, Anoshin GN (2004) Mineralogy and geochemistry of mantle inclusions and mantle column structure of the yubileinaya kimberlite pipe, alakit field, Yakutia. Doklady Russ Acad Sci Earth Sci Sect 395:517–523

Ashchepkov IV, Pokhilenko NP, Vladykin NV, Logvinova AM, Kostrovitsky SI, Afanasiev VP, Pokhilenko LN, Kuligin SS, Malygina LV, Alymova NV, Khmelnikova OS, Palessky SV, Nikolaeva IV, Karpenko MA, Stagnitsky YB (2010) Structure and evolution of the lithospheric mantle beneath Siberian craton, thermobarometric study. Tectonophysics 485:17–41

Ashchepkov IV, Vladykin NV, Ntaflos T, Downes H, Mitchel R, Smelov AP, Alymova NV, Kostrovitsky SI, RotmanAYa, Smarov GP, Makovchuk IV, StegnitskyYuB, Nigmatulina EN, Khmelnikova OS (2013) Regularities of the mantle lithosphere structure and formation beneath Siberian craton in comparison with other cratons. Gondwana Res 23:4–24

Ashchepkov IV, Rotman AY, Somov SV, Afanasiev VP, Downes H, Logvinova AM, Nossyko S, Shimupi J, Palessky SV, Khmelnikova OS, Vladykin NV (2012) Composition and thermal structure of the lithospheric mantle beneath kimberlite pipes from the Catoca cluster, Angola. Tectonophysics 530–531:128–151

Ater PC, Eggler DH, McCallum ME (1984) Petrology and geochemistry of mantle eclogites xenoliths from Colorado-Wyoming kimberlites: recycled oceanic crust? In: Kornprobst J (ed) Kimberlites II: The mantle and crust

Aulbach S, Griffin WL, Pearson NJ, O'Reilly SY, Kivi K, Doyle BJ (2004) Mantle formation and evolution, Slave Craton: constraints from HSE abundances and Re-Os isotope systematics of sulfide inclusions in mantle xenocrysts. Chem Geol 208:61–88

Boyd FR, Pokhilenko NP, Pearson DG, Mertzman SA, Sobolev NV, Finger LW (1997) Composition of the Siberian cratonic mantle: evidence from Udachnaya peridotite xenoliths. Contrib Miner Petrol 128:228–246

Brey GP, Kohler T (1990) Geothermobarometry in four-phase lherzolites II. New thermo-barometers, and practical assessment of existing thermobarometers. J Petrol 31:1353–1378

Carlson RW, Irving AJ (1994) Depletion and enrichment history of subcontinental lithospheric mantle: an Os, Sr, Nd, and Pb isotopic study of ultramafic xenoliths from the northwestern Wyoming craton. Earth Planet Sci Lett 126:457–472

Carlson RW, James DE (2011) Craton destruction: an ongoing example in Western North America. International conference on craton formation and destruction abstracts. Beijing, pp 14–15

Carlson RW, Irving AJ, Hearn Jr BC (1999) Chemical and isotopic systematics of peridotite xenoliths from the Williams kimberlite, Montana: clues to processes of lithospheric formation, modification and destruction. In: Gurney JJ, Gurney JL, Pascoe MD, Richardson SH (eds) Proceedings of 7th international kimberlite conference, Vol 1. Red Roof Design, Cape Town, pp 90–98

Carlson RW, Irving AJ, Schulze DJ, Carter B, Hearn J (2004) Timing of precambrian melt depletion and phanerozoic refertilization events in the lithospheric mantle of the Wyoming craton and adjacent central plains orogen. Lithos 77:453–472

Coopersmith HG, Mitchell RH, Hausel WD (2003) Kimberlites and lamproites of Colorado and Wyoming, USA. Guidebook for the 8th international kimberlite conference, Colorado and Wyoming Field Trip, 32 p

Davies RM, Griffin WL, O'Reilly SY. Doyle BJ (2004). Mineral inclusions and geochemical characteristics of microdiamonds from the DO27, A154, A21, A418, DO18, DD17 and Ranch Lake kimberlites at Lac de Gras, Slave Craton, Canada. Lithos 77:39–55

Dawson B (1980) Kimberlites and their xenoliths. Springer, Berlin 200 p

De Stefano A, Kopylova MG, Cartigny P, Afanasiev VP (2009) Diamonds and eclogites of the Jericho kimberlite (Northern Canada). Contrib Miner Petrol 158:295–315

Downes H, Macdonald R, Upton BGJ, Cox KG, Bodinier J-L, Mason PRD, James D, Hill PG, Hearn BC Jr (2004) Ultramafic xenoliths from the Bearpaw mountains, Montana, USA: evidence for multiple metasomatic events in the lithospheric mantle beneath the Wyoming Craton. J Petrol 45:1631–1662

Dueker K, Yuan H, Zurek B (2001) Thick-structured Proterozoic lithosphere of the Rocky Mountain region. GSA Today 11:4–9

Eccles DR, Simonetti SS, Cox R (2010) Garnet pyroxenite and granulite xenoliths from Northeastern Alberta: evidence of ~ 1.5 Ga lower crust and mantle in Western Laurentia. Precambr Res 177:339–354

Eggler DH, McCallum ME, Smith CB (1979) Megacryst assemblages in kimberlite from northern Colorado and southern Wyoming: petrology, geothermometry–barometry and areal distribution. In: Boyd FR, Meyer HOA (eds) Kimberlites, diatremes, and diamonds: their geology, petrology, and geochemistry. American Geophysical Union, Washington, DC, pp 213–226

Eggler DH, McCallum ME, Kirkley MB (1987) Kimberlite-transported nodules from Colorado-Wyoming: a record of enrichment of shallow portions of an infertile lithosphere. In: Morris EM and Pasteris JD (eds) Mantle metasomatism and alkaline magmatism. Geol Soc Am Spec Paper 215:77–90

Griffin WL, O'Reilly SY, Doyle BJ, Pearson NJ, Coopersmith H, Kivi K, Malkovets V, Pokhilenko N (2004) Lithosphere mapping beneath the North American plate. Lithos 77:873–922

Gurney JJ (1984) A correlation between garnets and diamonds. In: Glover JE, Harris PG (eds) Kimberlite occurrence and origin: a basis for conceptual models in exploration Geology Department and University Extension Univ WA Publ, vol 8, pp 143–166

Hart SR, Dunn T (1993) Experimental cpx/melt partitioning of 24 trace elements. Contrib Miner Petrol 113:1–8

Hauri EH, Wagner TP, Grove TL (1994) Experimental and natural partitioning of Th, U, Pb and other trace elements between garnet, clinopyroxene and basaltic melts. Chem Geol 117:149–166

Hausel WD (1998) Diamonds and mantle source rocks in the Wyoming Craton, with a discussion of other US occurrences. Wyoming State geological survey report of investigations 53, 93 p

Heaman LM, Kjarsgaard BA, Creaser RA (2003) The timing of kimberlite magmatism in North America: implications for global kimberlite genesis and diamond exploration. Lithos 71:153–184

Heaman LM, Kjarsgaard BA, Creaser RA (2004) The temporal evolution of North American kimberlites. Lithos 76:377–397

Heaman LM, Creaser RA, Cookenboo HO, Chacko T (2006) Multistage modification of the northern slave mantle lithosphere evidence from zircon- and diamond-bearing eclogite xenoliths entrained in Jericho Kimberlite, Canada. J Petrol 47:821–858

Hearn BC (2004) The Homestead kimberlite, central Montana, USA: mineralogy, xenocrysts, and upper-mantle xenoliths. Lithos 77:473–491

Hearn Jr BC, McGee ES (1984) Garnet peridotites from Williams kimberlites, north-central Montana, USA. In: Kornprobst J (ed) Kimberlites II: the mantle and crust-mantle relationships. Elsevier, Amsterdam, pp 57–70

Ionov DA, Doucet LS, Ashchepkov IV (2010) Composition of the lithospheric mantle in the Siberian Craton: new constraints from fresh peridotites in the udachnaya-east kimberlite. J Petrol 51:2177–2210

Kjarsgaard IM, McClenaghan MB, Kjarsgaard BA, Heaman LM (2004) Indicator mineralogy of kimberlite boulders from eskers in the Kirkland Lake and Lake Timiskaming areas, Ontario, Canada. Lithos 77:705–731

Klemme S, Gunther D, Hametner K, Prowatke S, Zack T (2006) The partitioning of trace elements between ilmenite, ulvospinel, armalcolite and silicate melts with implications for the early differentiation of the moon. Chem Geol 234(3):251–263

Kopylova MG, Caro G (2004) Mantle xenoliths from the Southeastern slave craton: evidence for chemical zonation in a thick, cold lithosphere. J Petrol 45:1045–1067

Kopylova MG, Russell JK, Cookenboo H (1999) Petrology of peridotite and pyroxenite xenoliths from the Jericho kimberlite: implications for the thermal state of the mantle beneath the slave craton, northern Canada. J Petrol 40:79–104

Lavrent'ev YUG, Usova LV (1994) New version of KARAT program for quantitative X-ray spectral microanalysis. ZhurnalAnaliticheskoiKhimii 5:462–468

Lester A, Larson E (1996) New geochronological evidence for the late proterozoic emplacement of Colorado Wyoming kimberlites belt. EOS Trans Am Geoph Union 77:821

Macdonald R, Upton BGJ, Collerson KD, Jr Hearn BC, James D (1992) Potassic mafic lavas of the Bearpaw Mountains, Montana: mineralogy, chemistry, and origin. J Petrol 38:305–346

MacGregor ID (1974) The system $MgO-Al_2O_3-SiO_2$: solubility of Al_2O_3 in enstatite for spinel and garnet peridotite compositions. Am Miner 59:110–119

MacKenzie JM, Canil D (1999) Composition and thermal evolution of cratonic mantle beneath the central Archean Slave Province, NWT, Canada. Contrib Miner Petrol 134:313–324

McCallum ME, Eggler DH, Burns LK (1975) Kimberlitic diatremes in northern Colorado and southern Wyoming. Phys Chem Earth 9:149–161

McCallum ME, Eggler DH (1976) Diamonds in an upper mantle peridotite nodule from kimberlite in southern Wyoming. Science 192:253–256

McCammon CA, Griffin WL, Shee SR, O'Neill HSC (2001) Oxidation during metasomatism in ultramafic xenoliths from the Wesselton kimberlite, South Africa: implications for the survival of diamond. Contrib Miner Petrol 141:287–296

McDonough WF, Sun SS (1995) The Composition of the Earth. Chem Geol 120:223–253

Menzies A, Westerlund K, Grutter H, Gurney J, Carlson J, Fung A, Nowicki T (2004) Peridotitic mantle xenoliths from kimberlites on the Ekati Diamond Mine property, NWT, Canada: major element compositions and implications for the lithosphere beneath the central Slave craton. Lithos 77:395–412

Meyer HOA, Waldman MA, Garwood BL (1994) Mantle xenoliths from kimberlite near Kirkland Lake, Ontario. Can Miner 32:295–306

Mitchell RH, Bergman SC (1991) Petrology of lamproites. Plenum Press, New York 447 p

Nimis P, Taylor W (2000) Single clinopyroxene thermobarometry for garnet peridotites. Part I calibration and testing of a Cr-in-Cpx barometer and an enstatite-in-Cpx thermometer. Contrib Miner Petrol 139:541–554

Nixon PH, Boyd FR (1973) Petrogenesis of the granular and sheared ultrabasic nodule suite in kimberlites. In: Nixon PH (ed) Lesotho kimberlites. Lesotho Nat Dev Corp Maseru, Lesotho, pp 48–56

O'Brien HE, Irving AJ, McCallum IS, Thirlwall MF (1995) Sr, Nd and Pb isotopic evidence for the interaction of post-subduction asthenospheric potassic mafic magmas of the Highwood Montana, Montana, USA, with ancient Wyoming craton lithospheric mantle. Geochimicaet Cosmochima Acta 59:4539–4556

O'Neill HSTC, Wall VJ (1987) The olivine orthopyroxene-spinel oxygen geobarometer, the nickel precipitation curve, and the oxygen fugacity of the Earth's upper mantle. J Petrol 28:1169–1191

O'Neill HSTC, Wood BJ (1979) An experimental study of Fe-Mg-partitioning between garnet and olivine and its calibration as a geothermometer. Contributions to Mineralogy and Petrology 70:59–70

Pizzolato L, Schultze DJ (2003) Geochemistry of peridotites and eclogites from the Kelsey Lake kimberlite, Colorado–Wyoming, USA. In: Proceedings of the 8th International Kimberlite Conference, Victoria, BC, Canada, June 2003:FLA–0300

Pokhilenko NP, Sobolev NV, Kuligin SS, Shimizu N (1999) Peculiarities of distribution of pyroxenite paragenesis garnets in Yakutian kimberlites and some aspects of the evolution of the Siberian craton lithospheric mantle. In: Proceedings of the VII international kimberlite conference. The P.H. Nixon volume, pp 690–707

Rudnick RL, Ireland TR, Gehrels G, Irving AJ, Chesley JT, Hanchar JM (1999) Dating mantle metasomatism: U–Pb geochronology of zircons in cratonic mantle xenoliths from Montana and Tanzania. In: Gurney JJ, Gurney JL, Pascoe MD, Richardson SH (eds) Proceedings of 7th international kimberlite conference, vol 2. Red Roof Design, Cape Town, pp 728–735

Schmidberger SS, Francis D (1999) Nature of the mantle roots beneath the North American craton: mantle xenolith evidence from Somerset Island kimberlites. Lithos 48:195–216

Schmidberger SS, Simonetti A, Heaman LM, Creaser RA, Whiteford S (2007) Lu-Hf, in situ Sr and Pb isotope and trace element systematics for mantle eclogites from the Diavik diamond mine: Evidence for Paleoproterozoic subduction beneath the Slave craton, Canada. Earth Planet Sci Lett 254:55–68

Schulze DJ (1992) Diamond eclogite from Sloan Ranch, Colorado and its bearing on the diamond grade of the Sloan kimberlite. Econ Geol 87:2175–2179

Schulze DJ, Anderson PFN, Hearn Jr BC, Hetman CM (1995) Origin and significance of ilmenite megacrysts and macrocrysts from kimberlite. Int Geol Rev 37:780–812

Schulze DJ, Coopersmith HG, Harte B, Pizzolato L-A (2008) Mineral inclusions in diamonds from the Kelsey Lake Mine, Colorado, USA: depleted Archean mantle beneath the Proterozoic Yavapai province. Geochimicaet Cosmochimica Acta 72:1685–1695

Smart KA, Heaman LM, Chacko T, Simonetti A, Maya Kopylova, Mah D, Daniels D (2009) The origin of high-MgO diamond eclogites from the Jericho Kimberlite, Canada. Earth Planet Sci Lett 284:527–537

Smith D, Griffin WL (2005) Garnetite xenoliths and mantle–water interactions below the Colorado plateau, Southwestern United States. J Petrol 46:1901–1924

Smith CB, McCallum ME, Coopersmith HG, Eggler DH (1979) Petrochemistry and structure of kimberlites in the Front Range and Laramie Range, Colorado-Wyoming. In: Boyd FR, Meyer HOA (eds) Kimberlites, diatremes, and diamonds; their geology, petrology, and geochemistry. American Geophysical Union, Washington, DC, pp 178–189

Sobolev NV (1977) Deep-seated inclusions in kimberlites and the problem of the composition of the mantle. American Geophysical Union, Washington, DC, 279 p

Sobolev NV, Lavrent'ev YUG, Pokhilenko NP, Usova LV (1973a) Chrome-rich garnets from the kimberlites of Yakutia and their parageneses. Contrib Miner Petrol 40:39–52

Sobolev NV, Sobolev VN, Snyder GA, Yefimova ES, Taylor LA (1973b) Significance of eclogitic and related parageneses of natural diamonds. Int Geol Rev 41:129–140

Sobolev VN, Taylor LA, Snyder GA, Sobolev NV (1994) Diamondiferous eclogites from the Udachnaya pipe, Yakutia. Int Geol Rev 36:42–64

Taylor WL, Kamperman M, Hamilton R (1998) New thermometer and oxygen fugacity sensor calibration for ilmenite and Cr-spinel-bearing peridotite assemblage. In: Gurney JJ, Gurney JL, Pascoe MD, Richardson SH (eds) 7th international kimberlite conference. Red Roof Design, Capetown, pp 986–988

Usui T, Kobayashi K, Nakamura E, Helmstaedt H (2007) Trace element fractionation in deep subduction zones inferred from a lawsonite-eclogite xenolith from the Colorado Plateau. Chem Geol 239:336–351

Zack T, Brumm R (1998) Ilmenite/liquid partition coefficients of 26 trace elements determined through ilmenite/clinopyroxene partitioning in garnet pyroxenite. In: Gurney JJ, Gurney JL, Pascoe MD, Richardson SH (eds) 7th international kimberlite conference. Red Roof Design, Cape Town, pp 986–988

Contrasting Garnet Lherzolite Xenolith Suites from the Letšeng Kimberlite Pipes: Inferences for the Northern Lesotho Geotherm

N. P. Lock and J. B. Dawson

Abstract

The Main and Satellite Pipe kimberlites at Letseng-la-Terae, Lesotho, both contain abundant crustal and mantle xenoliths. In garnet lherzolite xenoliths that are the subject of this paper, textures range from coarse to highly deformed in samples from both pipes, but the most extreme deformation is seen only in Main Pipe samples. Development of reaction coronas around the garnets in all samples indicates retrogression to spinel peridotite facies conditions prior to entrainment in the host kimberlites. Bulk rock compositions of most samples are depleted relative to pristine mantle compositions. Most Main Pipe samples equilibrated at 950–1,000 °C and pressures of 40–45 Kbar, with others extending the temperature range to 820 °C and 1,350 °C. In the temperature range 950–1,000 °C, there is no overlap with Satellite Pipe xenoliths which equilibrated at <800 and >1,305 °C. High-T samples from both pipes lie on the high-T side of the Cretaceous Kalahari geotherm of Rudnick and Nyblade (1999), indicating a perturbed geotherm, similar to that deduced from Thaba Putsoa xenoliths but unlike the non-perturbed geotherm inferred from mantle xenoliths in other Cretaceous Lesotho kimberlites. Hence, xenolith suites from individual kimberlite intrusions each have distinctive thermal characteristics, and there is no consistently perturbed Cretaceous geothermal gradient beneath Lesotho.

Keywords

Garnet lherzolites • PTs of equilibration • Lesotho geothermal gradient

Introduction

Kimberlite pipes and dykes intrude the Karoo (Carboniferous–Triassic) sandstones and basalts in the Maluti Mountains of northern Lesotho, to the east of the major Cretaceous kimberlite province of South Africa (Nixon 1973). Zircon from the Mothae pipe has been dated at 87.1 Ma (Davis 1977), and perovskite from other pipes (Kao, Liqhobong, Lipelaneng, Mothae, Ngopetseu) has yielded ages ranging from 88.8 ± 1.4 to 94.7 ± 1.1 Ma (1 sigma errors—Hoese 2009).

The Main and Satellite kimberlite pipes at Letšeng-la-Terae (29°00′S 28°43′E) (for brevity hereafter referred to as Letšeng) contain abundant crustal and mantle xenoliths and afford the opportunity to compare two peridotite xenolith suites which can reasonably be assumed to have been transported up the same or closely adjacent upper mantle conduit(s). The pipes, which are only ~350 m apart (Fig. 1), crop out at ~3,100 m above sea level, forming the world's highest diamond mine. Features of the bulk diamond production are

N. P. Lock
Coffey Mining, 20 Meteor Drive, Etobicoke, Ontario, M9W 1A4, Canada

J. B. Dawson (✉)
Grant Institute of Geosciences, University of Edinburgh, Edinburgh, EH9 3JW, UK
e-mail: jbdawson@glg.ed.ac.uk

D. G. Pearson et al. (eds.), *Proceedings of 10th International Kimberlite Conference*, Volume 1, Special Issue of the Journal of the Geological Society of India, DOI: 10.1007/978-81-322-1170-9_3,

the overall low diamond grade but the occasional occurrence of exceptionally large, high-quality (type IIA) diamonds. Harris et al. (1979) and Bowen et al. (2009) report significant differences in diamond morphology and colour, but not of size–frequency distribution, between the diamond populations from the two pipes. An objective of this study is to compare the Letseng xenolith suite with that from Thaba Putsoa, Lesotho, from which a perturbed geothermal gradient has been inferred with implications for major frictional heating at the base of the Kalahari craton (Boyd 1973a).

General Geology

The two pipes (Fig. 1) are separated by basalt wall rock that forms abundant xenoliths in both pipes. In the Lesotho province, kimberlites intruding the Drakensberg basalts have a maximum stratigraphic age of ca 180 Ma (Duncan and Marsh 2006), but the Letseng pipes have not been dated radiometrically. Other xenoliths are of Karoo sediments; gneisses and granulites that have their analogues in the basement of southern Africa; and upper mantle peridotites (*s.l*).

The Main Pipe (surface area 17.2 ha) comprises four petrographically distinct types of kimberlite, the contacts between which are gradational over a few metres (Lock, 1980). The kimberlites are broadly volcanoclastic (terminology of Scott Smith et al. 2008), but with distinctive characteristics, and the pipe is mainly infilled by autolithic kimberlite. This is penetrated by a smaller body of highly garnetiferous kimberlite which is zoned, with increasing abundance and/or size of peridotite xenoliths and garnet megacrysts towards its centre. It is cut by three dykes, which are interpreted as the hypabyssal facies of the garnetiferous kimberlite. Penetrating the garnetiferous kimberlite are two smaller bodies of highly micaceous kimberlite, the younger of which contains fresh olivines. The intrusive relationships between the different kimberlite types suggest a younging towards the centre of the diatreme.

The Satellite Pipe (surface area 5.2 ha) mainly contains kimberlite that is similar to the Main Pipe autolithic kimberlite, though Palmer et al. (2008) recognise five texturally different varieties and, in the south-west part of the pipe, reworked kimberlite that surrounds a major basalt raft. Sedimentary xenoliths are more abundant than in the Main

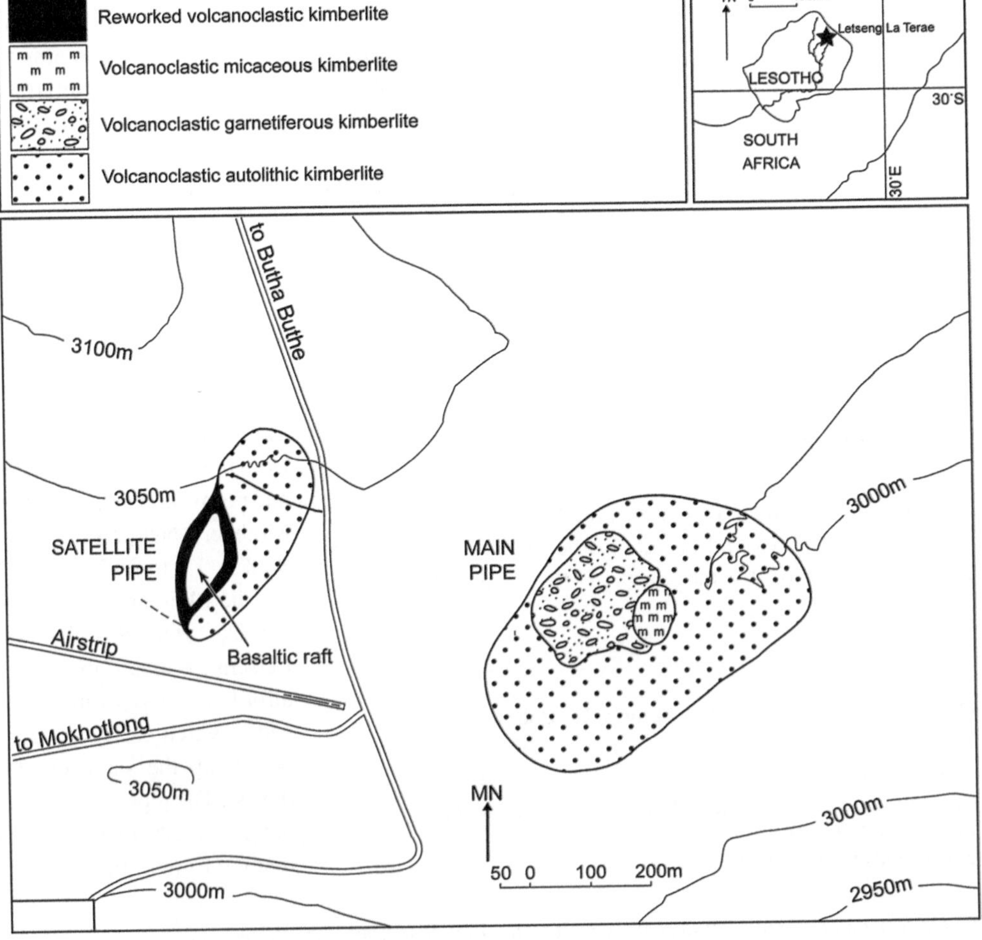

Fig. 1 The Letšeng kimberlite pipes, showing the distribution of kimberlite types in the Main Pipe at the 55 m level (Lock 1980) and the Satellite Pipe at the 70 m level (Palmer et al. 2008)

Pipe, whereas peridotite xenoliths are much rarer. An early dyke of micaceous kimberlite is truncated by the pipe. The emplacement of the two pipes is assumed to be pene-contemporaneous, but there are no dates on either intrusion.

Ultramafic Xenoliths

The xenoliths reported here were collected in situ underground from the Main Pipe and from the surface diggings on the Satellite Pipe. However, kimberlite from both pipes is processed at the plant and the xenoliths piled in a common dump; so, when collected from the dump, the source of xenoliths is unclear. In the case of samples studied by Simon et al. (2003), neither texture nor location (?Main or ?Satellite Pipe) is reported so they are here referred to as being from "Pipe not known".

The most common ultramafic xenoliths are garnet or chromite lherzolites and harzburgites, but rarer types are lherzolites containing both garnet and Cr-spinel, garnet dunites, garnet websterites, garnet pyroxenites, Al-spinel lherzolites (some of which also contain garnet) and barren harzburgites (i.e. containing low-Ca–Al enstatites—Hervig et al. 1980). Primary phlogopite is present in a few lherzolites, and secondary phlogopite, generally adjacent to garnet, is present in samples that also show incipient serpentinisation. Most rock types are found in both the Main and Satellite pipes, but garnet lherzolites are approximately three times more abundant in the Main Pipe. Barren harzburgites, in which the orthopyroxenes contain very low amounts of CaO and Al_2O_3 (Hervig et al. 1980), have been recorded only from the Main Pipe.

Petrography

Garnet lherzolites are the focus of this contribution. Measured modes and maximum grain sizes in coarse specimens are as follows: olivine 54–78 modal %, up to 6 mm; enstatite 11–41 %, 8 mm; garnet 3–14 %, 5 mm (including the reaction coronas); and green diopside 0.3–10.3 %, 4 mm. Diopside is often spatially close to garnet. A small number of samples also contain rare chrome-rich spinel or Al-rich spinel (up to 0.3 mm). Grain sizes in deformed rocks are less than in coarse samples.

Due to varying deformation and recrystallisation, the peridotites exhibit a range of textures from coarse texture (not recrystallised), through increasingly recrystallised porphyroclastic and mosaic-structured to laminated and disrupted (LAD) textures (terminology of Harte 1977). In the LAD rocks, garnet and its surrounding reaction coronas are disrupted and strung out into fine-grained chains, interspersed, with laminae of fine-grained olivine and pyroxene neoblasts. Rocks showing recrystallisation are collectively referred to hereafter as "deformed". A range of textures are seen in samples from both pipes (Table 1); however, the Main Pipe suite is dominated by coarse-textured rocks and the most extreme LAD deformation has not been recognised in Satellite Pipe xenoliths.

Whole Rock Chemistry

Due to small xenolith size in some cases, not all samples listed in Table 1 were analysed. Compositions of the analysed samples are given in Table 2, with comparable analyses of other pertinent garnet lherzolites and a reference analysis for primitive mantle. With one high-Fe exception, the Letšeng rocks, whether coarse or deformed, are depleted in TiO_2, Al_2O_3, FeO and CaO relative to the primitive mantle composition. With the exception of the high-Fe sample, the Mg# values (a depletion indicator) of the Letseng lherzolites have a narrow range (91.0–93.2) with a slight tendency for lower values in deformed samples (Fig. 2). Coarse "sterile" (i.e. depleted) and sheared "fertile" (i.e. undepleted) garnet lherzolites from Thaba Putsoa show greater differences. Cr/(Cr + Al) ratios (another indicator of depletion) show a range from 0.149 to 0.367, with an overlap between coarse and deformed samples; however, there is no overall correlation between the two depletion parameters. Also, in the Letseng samples, there is no correlation between degree of depletion and deformation, unlike Thaba Putsoa where high deformation is linked to high "fertility". The Fe-rich sample 3 is also relatively rich in Ti and Ca, and although differing in containing little Na_2O, in both its Mg# and Cr/(Cr + Al) ratio, it strongly resembles Thaba Putsoa fertile lherzolite which has been interpreted as having been metasomatised during the kimberlite event. The bulk compositions of most Letšeng lherzolites are similar to many other Kalahari craton lherzolite suites in Cretaceous kimberlites and also show little compositional difference, in terms of SiO_2, FeO, Al_2O_3 and CaO, to off-craton peridotites (Janney et al. 2010). However, they are more refractory than garnet lherzolites in the Proterozoic Premier kimberlite (Table 2), the undepleted composition of which is attributed to sampling before the early Mesozoic major melting/depletion event that produced the Mesozoic Karoo volcanics (Danchin 1979).

Phase Chemistry

Primary Phases

Analyses of the phases in twenty-four garnet lherzolites were carried out by WDS on electron microprobes at the universities of Edinburgh and Cambridge. At Edinburgh,

Table 1 Textures and equilibration conditions of Letšeng garnet lherzolites

Samples	Field Number	Texture	NG85[a]	TA98[b]
Main Pipe—coarse			kb	°C
1	NL001	Coarse	45.3	989
2	NL006	Coarse	40.8	976
3	NL012	Coarse, Fe-rich	41.0	1,017
4	NL103	Coarse	38.8	981
5	NL104	Coarse	39.6	963
6	NL128	Coarse + Cr-spinel	25.8	757
7	NL144	Coarse + phlogopite	40.9	991
8	NL145	Coarse + Cr-spinel	40.1	864
9	NL153	Coarse	36.6	928
10	NL197	Coarse	39.7	973
11	NL492	Coarse	39.9	982
Deformed				
12	NL007	Porphyroclastic	42.0	963
13	NL141	Porphyroclastic	51.6	1,211
14	NL169	Porphyroclastic	49.2	1,190
15	NL426	LAD	42.2	1,029
16	NL427	LAD	40.8	1,015
17	NL441	Porphyroclastic	40.2	1,001
18	NL494	Mosaic porphyroclastic	55.0	1,356
19	NL495	Mosaic porphyroclastic	46.3	1,181
20	PHN2573[c]	"Sheared"	51.1	1,337
21	PHN2575/2[c]	"Sheared"	52.0	1,357
22	PHN2575/3[c]	"Sheared"	50.8	1,347
Satellite Pipe—coarse				
23	BD1870-1	Coarse,	19.3	697
24	BD1870-2[d]	Coarse	30.4	778
25	BD1894	Coarse Fe-rich	18.6	693
Deformed				
26	NL021	Mosaic porphyroclastic	43.5	1,305
27	NL503	Mosaic porphyroclastic	43.5	1,313
28	BD1899	Porphyroclastic	49.2	1337
Pipe not known				
29	LET29[e]	Texture not stated + Cr-spinel	39.0	992
30	LET38[e]	Texture not stated	40.6	987
31	LET64[e]	Texture not stated	41.4	1,020

[a] Pressure by Nickel and Green (1985). [b] Temperature by Taylor (1998)
[c] From Boyd (1973b); [d] From Hervig et al. (1980); [e] From Simon et al. (2003)

where most analyses were performed, the probe was a Cameca SX50, and standards used were the following: for Si and Ca—wollastonite; Ti—rutile; Al—corundum; Fe, Mn—metals; Mg—periclase; Na—jadeite; K—orthoclase; Na and K were analysed early in the routine to avoid migration/volatility effects. Counting times were 30 s on peaks and 15 s on backgrounds. Analyses were performed with a spot beam of ~2 μm at 20 kV and a probe current of 20 nA. Data were reduced using the PAP routine procedure (Pouchou and Pichoir 1991). Our data are supplemented by data on seven additional lherzolites from Boyd (1973b), Hervig et al. (1980) and Simon et al. (2003) (Table 1).

Olivines (Table 3) lie mainly in the compositional range Mg#90–94 with two exceptions, one from each pipe, that

Table 2 Whole-rock analyses of Letseng lherzolites[a] (numbers from Table 1) and other reference samples

		Letseng	Coarse			Letseng	Deformed		
Sample	**1**	**3**	**9**	**10**	**15**	**16**	**17**	**18**	**19**
SiO_2	44.79	45.83	44.93	42.46	47.49	45.31	46.11	44.19	42.60
TiO_2	<0.10	0.11	<0.10	<0.10	0.06	0.07	0.03	0.01	0.09
Al_2O_3	0.89	3.57	0.84	0.61	1.31	0.93	2.44	0.89	0.83
Cr_2O_3	0.40	0.49	0.48	0.23	0.42	0.24	0.88	0.55	0.35
FeO^t	6.35	9.70	6.00	6.99	6.66	7.79	5.96	8.64	8.28
MnO	0.09	0.18	0.91	0.09	0.12	0.14	0.13	0.15	0.11
MgO	46.34	39.12	45.97	48.62	43.04	44.46	43.32	44.28	46.55
NiO	0.30	0.32	0.25	0.29	0.29	0.37	0.28	0.35	0.35
CaO	0.56	0.98	0.38	0.51	0.50	0.50	0.77	0.73	0.72
Na_2O	0.26	0.09	0.22	0.17	0.10	0.21	0.06	0.21	0.11
Mg#	92.8	87.8	93.2	92.5	92.0	91.0	92.8	90.1	90.9
Cr/(Cr + Al)	0.367	0.083	0.280	0.202	0.174	0.149	0.195	0.292	0.221
Sample	**A**	**B**	**C**	**D**					
SiO_2	48.54	44.46	44.31	44.95					
TiO_2	0.00	0.20	0.33	0.158					
Al_2O_3	0.66	2.95	3.40	3.52					
Cr_2O_3	0.26	0.36	0.34	0.385					
FeO^t	5.70	9.59	9.46	7.97					
MnO	0.11	0.15	0.15	0.131					
MgO	44.14	38.19	38.14	39.5					
NiO			0.22	0.252					
CaO	0.50	3.44	2.59	2.79					
Na_2O	0.04	0.33	0.18	0.298					
Mg#	93.2	87.6	88.1	89.6					
Cr/(Cr + Al)	0.207	0.077	0.063	0.067					

A. "Sterile" coarse garnet lherzolite PHN1569, Thaba Putsoa (Boyd and McCallister 1976)
B. "Fertile" sheared garnet lherzolite PHN1611, Thaba Putsoa (Boyd and McCallister 1976)
C. Fertile garnet lherzolite from Proterozoic Premier kimberlite (Danchin 1979)
D. Primitive mantle (Lyubetskaya and Korenaga 2007)
[a] Recalculated on an H_2O^- , CO_2- and K_2O- free basis, and all iron as FeO

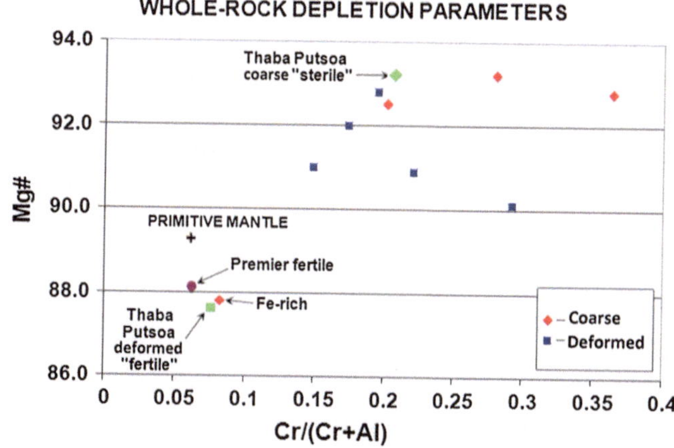

Fig. 2 Bulk rock depletion parameters for Letšeng garnet lherzolites and other mantle compositions. Data from Table 2

are more Fe-rich (Mg# < 90). There are no significant differences in the olivine compositions in coarse and deformed xenoliths, and there is a reasonable correlation between their Mg# and the Mg# of the coexisting pyroxenes and garnet.

Orthopyroxenes (Table 4) show a narrow range of Mg# (91.6–93.7) with the exception of the Fe-rich samples (Mg# 89.8 and 90.7). There is a range in CaO (0.20–1.59 wt%) and, with the exception of the Fe-rich specimens, there is a tendency towards slightly higher Ca and Fe but lower Mg in enstatites from deformed xenoliths in both pipes. All enstatites contain appreciable Al_2O_3 (0.69–1.29 wt %), and those with $Al_2O_3 > 1.2$ wt % are consistently from deformed samples and coexist with sub-calcic diopsides (Ca/(Ca + Mg) < 0.37).

Table 3 Compositions of olivines in Letšeng garnet lherzolites

Samples	1	2	3	4	5	6	7	8
SiO_2	41.18	41.09	40.77	41.33	41.14	41.22	42.40	41.36
TiO_2	0.03	0.00	0.03	0.01	0.00	0.00	0.03	0.00
Al_2O_3	0.03	0.05	0.06	0.05	0.10	0.05	0.06	0.06
Cr_2O_3	0.01	0.07	0.00	0.02	0.04	0.00	0.02	0.01
FeOt	7.12	7.72	11.36	7.41	7.70	7.45	7.21	7.32
MnO	0.10	0.10	0.13	0.10	0.12	0.09	0.08	0.10
NiO	0.42	0.39	0.52	0.50	0.40	0.44	0.43	0.43
MgO	51.62	51.24	48.10	51.28	50.79	50.99	51.65	52.02
CaO	0.03	0.04	0.05	0.01	0.04	0.00	0.03	0.03
Na_2O	0.02	0.03	0.04	0.01	0.04	0.01	0.00	0.01
Total	100.56	100.73	101.06	100.72	100.37	100.25	101.91	101.34
Mg#	92.8	92.2	88.3	92.5	92.2	92.4	92.7	92.7
Samples	**9**	**10**	**11**	**12**	**13**	**14**	**15**	**16**
SiO_2	41.42	41.49	41.05	41.17	41.24	41.19	41.09	40.84
TiO_2	0.02	0.02	0.05	0.01	0.09	0.01	0.04	0.04
Al_2O_3	0.02	0.04	0.04	0.03	0.00	0.08	0.06	0.03
Cr_2O_3	0.02	0.02	0.04	0.03	0.05	0.06	0.03	0.02
FeOt	6.99	7.28	7.76	7.96	7.98	7.64	8.02	7.44
MnO	0.08	0.06	0.08	0.11	0.07	0.10	0.10	0.10
NiO	0.38	0.44	0.40	0.38	0.41	0.37	0.44	0.44
MgO	51.88	51.06	50.88	50.55	50.57	51.07	50.85	50.86
CaO	0.02	0.01	0.01	0.01	0.07	0.06	0.04	0.04
Na_2O	0.04	0.03	0.01	0.02	0.04	0.08	0.05	0.03
Total	100.87	100.45	100.32	100.27	100.52	100.66	100.72	99.84
Mg#	93.0	92.6	92.1	91.9	91.8	92.3	91.9	92.4
Samples	**17**	**18**	**19**	**20**	**21**	**22**	**23**	**24**
SiO_2	41.25	40.68	40.35	40.30	40.41	40.00	40.95	40.90
TiO_2	0.03	0.03	0.03	0.03	0.02	0.03	0.00	0.02
Al_2O_3	0.06	0.05	0.05	0.05	0.06	0.09	0.00	0.00
Cr_2O_3	0.01	0.10	0.05	0.04	0.06	0.05	0.00	0.10
FeOt	6.81	8.71	8.81	9.38	9.09	8.84	8.93	8.80
MnO	0.07	0.12	0.10	0.10	0.13	0.12	0.14	0.10
NiO	0.44	0.39	0.40	0.00	0.39	0.38	0.34	0.41
MgO	51.64	50.07	49.97	50.23	50.30	50.66	50.12	50.70
CaO	0.04	0.07	0.05	0.07	0.08	0.09	0.00	0.01
Na_2O	0.03	0.02	0.04	0.04	0.00	0.00	0.00	0.00
Total	100.38	100.24	99.85	100.24	100.54	100.26	100.48	101.04
Mg#	93.1	91.1	91.0	90.5	90.8	91.1	90.9	91.1
Samples	**25**	**26**	**27**	**28**	**29**	**30**	**31**	
SiO_2	41.13	41.19	41.90	41.07	42.0	42.0	41.4	
TiO_2	0.00	0.00	0.00	0.00				
Al_2O_3	0.00	0.00	0.00	0.00	0.01	0.01	0.02	
Cr_2O_3	0.00	0.00	0.00	0.00	0.02	0.02	0.03	

(continued)

Table 3 (continued)

Samples	25	26	27	28	29	30	31
FeOt	10.16	9.05	8.83	9.25	7.25	7.03	7.10
MnO	0.12	0.10	0.10	0.12	0.09	0.08	0.09
NiO	0.34	0.36	0.29	0.37	0.42	0.48	0.44
MgO	49.33	49.91	50.61	49.81	51.4	51.5	51.5
CaO	0.00	0.09	0.11	0.10	0.02	0.03	0.02
Na_2O	0.00	0.00	0.00	0.00			
Total	101.08	100.70	101.84	100.72	101.21	101.15	100.60
Mg#	89.7	91.0	91.1	90.9	92.7	92.9	92.8

Sample numbers from Table 1

Clinopyroxenes (Table 5) range from ureyitic and chrome diopsides to diopsides and subcalcic diopsides (terminology of Stephens and Dawson 1977), with a wide range in Cr_2O_3 (0.75–3.66 wt %). The wide range of Ca/(Ca + Mg) (0.338–0.487) indicates a range of equilibrium temperatures (Davis and Boyd 1966). Subcalcic diopsides (Ca/[Ca + Mg] < 0.40) are present in deformed rocks from both pipes. Intermediate values (0.40–0.46) are common in the clinopyroxenes in Main Pipe samples, but not in those from the Satellite Pipe which have a bimodal grouping of Ca/(Ca + Mg). In the Main Pipe samples, there is no overall correlation between the Ca/(Ca + Mg) ratio and texture, but in coarse samples from both pipes, clinopyroxenes have relatively high Ca/(Ca + Mg) ratios and those from highly deformed samples show relatively low values, with implications for equilibration temperatures.

Garnet compositions (Table 6) show a considerable range in both Mg# (range 78.0–86.5) and Cr content (Cr_2O_3 = 2.77–11.1 wt %); Cr_2O_3 varies directly with CaO and inversely with Al_2O_3. There are no systematic chemical differences between garnets from coarse and deformed rocks in the Main Pipe samples, but those in coarse samples from the Satellite Pipe are more iron-rich than in the deformed rocks and are most similar to the single Fe-rich coarse sample from the Main Pipe. Most, including those co-existing with relatively Fe-rich olivines and pyroxenes, are chrome pyropes (Group 9 of Dawson and Stephens 1975), but one high Ca–Cr garnet in sample 8 is a knorringitic uvarovite pyrope. None is compositionally similar to low-Ca, high-Cr inclusions in diamond.

Spinel A small number of the garnet lherzolites contain magnesian chromite in addition to garnet. Compositionally, in having higher Cr but lower Mg, they are unlike the spinels occurring in both granuloblastic spinel lherzolites and spinel–garnet lherzolites from Letseng (Lock 1980).

Neoblast compositions. Compared with primary grains, both olivine and enstatite neoblasts show no changes in Mg# but consistent, small increases in CaO, for example 0.08 wt % CaO in an olivine neoblast compared with 0.04 wt % in primary olivine. Although this Ca increase might appear small, it is relevant to the whole rock Ca budget due to the modal dominance of olivine in the neoblast population. In the case of deformed clinopyroxenes, the neoblasts contain less CaO, resulting in a slight decrease in the Ca/(Ca + Mg) ratio. Overall lower Ca in the clinopyroxene neoblasts coincides with a concomitant increase in the olivine and enstatite neoblasts. Based on the enstatite solubility in diopside criterion, this can be interpreted as reflecting slightly increased temperatures during recrystallisation.

Conditions of Equilibration of the Letseng Garnet Lherzolites: Inferences for the Lesotho Geotherm

Garnet lherzolites permit calculation of their temperatures and pressures of equilibration, as determined in experimental systems. We have used the two-pyroxene thermometer of Taylor (1998) combined with the Al-in-enstatite in equilibrium with garnet (Nickel and Green 1985) for pressure estimates, as recommended by Nimis and Grütter (2010) in a rigorous review of the various methods of PT calculation. The same methods have also been used to recalculate, for comparison, the data of Boyd (1973a) and Nixon and Boyd (1973), used in their much-discussed model for a perturbed Cretaceous geothermal gradient below northern Lesotho.

The equilibration PTs of the primary assemblages in the Letšeng garnet lherzolites are given in Table 1 and plotted on Fig. 3. Also shown is the Kalahari craton geotherm of Rudnick and Nyblade (1999), which is based on both surface heat-flow and xenolith equilibrium data calculated by the methods of Brey and Köhler (1990); Brey et al. (1990) which, for a given pressure (depth), in most cases give similar temperatures to those obtained by the method used in this paper. Also shown is the graphite–diamond transition of Bundy (1980), preferred to that of Kennedy and Kennedy (1976)

Table 4 Compositions of orthopyroxenes in Letšeng garnet lherzolites

Samples	1	2	3	4	5	6	7	8
SiO_2	57.63	58.19	57.47	57.92	57.64	57.73	58.82	57.25
TiO_2	0.06	0.01	0.13	0.00	0.02	0.01	0.01	0.02
Al_2O_3	0.80	0.72	0.82	0.80	0.77	1.02	0.81	0.76
Cr_2O_3	0.42	0.33	0.22	0.32	0.37	0.35	0.40	0.29
FeOt	4.36	4.67	6.85	4.49	4.78	4.67	4.44	4.39
MnO	0.12	0.12	0.14	0.10	0.10	0.13	0.11	0.12
NiO	0.11	0.13	0.14	0.11	0.15	0.12	0.00	0.16
MgO	35.85	35.77	33.94	35.63	35.62	35.83	36.30	36.70
CaO	0.45	0.48	0.55	0.51	0.49	0.26	0.43	0.39
Na_2O	0.04	0.12	0.18	0.13	0.13	0.08	0.12	0.06
Total	99.84	100.54	100.44	100.01	100.07	100.20	101.44	100.14
Mg#*	93.6	93.2	89.8	93.0	93.2	93.6	93.7	93.8
Ca	1.3	1.0	1.1	0.8	1.0	0.5	0.8	0.7
Mg	92.0	92.3	88.9	92.7	92.1	91.8	92.8	93.0
Fe	6.7	6.7	10.0	6.5	6.9	6.7	6.4	6.3
Samples	**9**	**10**	**11**	**12**	**13**	**14**	**15**	**16**
SiO_2	56.79	58.35	57.55	57.58	57.36	57.60	57.52	57.66
TiO_2	0.05	0.03	0.05	0.13	0.25	0.14	0.09	0.07
Al_2O_3	0.88	0.69	0.79	0.81	0.95	0.96	0.84	0.83
Cr_2O_3	0.50	0.26	0.23	0.46	0.44	0.35	0.36	0.32
FeOt	4.26	4.38	4.67	4.90	4.83	4.59	4.83	4.66
MnO	0.12	0.10	0.09	0.13	0.10	0.12	0.12	0.11
NiO	0.14	0.09	0.10	0.13	0.12	0.13	0.12	0.14
MgO	36.23	35.71	35.60	35.26	34.82	35.14	35.24	35.43
CaO	0.44	0.50	0.48	0.50	0.86	0.93	0.54	0.58
Na_2O	0.19	0.16	0.11	0.16	0.27	0.20	0.16	0.16
Total	99.60	100.27	99.67	100.06	100.00	100.16	99.82	99.96
Mg#	93.8	93.6	93.1	92.8	92.8	93.2	92.9	93.1
Ca	0.9	0.9	0.9	0.9	1.1	1.8	1.0	1.1
Mg	93.0	92.7	92.3	91.9	91.8	91.5	91.9	92.1
Fe	6.1	6.4	6.8	7.2	7.1	6.7	7.1	6.8

Mg# = 100 Mg/(Mg + Fe)

Samples	17	18	19	20	21	22	23	24
SiO_2	57.84	57.25	56.36	56.98	56.90	56.24	57.50	58.30
TiO_2	0.01	0.01	0.20	0.19	0.09	0.16	0.00	0.09
Al_2O_3	0.83	0.94	0.99	1.26	1.16	1.29	1.14	0.87
Cr_2O_3	0.39	0.46	0.40	0.24	0.27	0.26	0.24	0.25
FeOt	4.20	5.22	5.55	5.56	5.18	5.36	5.48	5.38
MnO	0.12	0.15	0.12	0.13	0.15	0.14	0.13	0.11
NiO	0.09	0.17	0.12	0.00	0.00	0.00	0.15	0.10
MgO	35.81	34.32	34.22	34.45	33.75	34.57	35.10	36.00
CaO	0.49	1.59	0.74	1.42	1.34	1.41	0.20	0.26
Na_2O	0.14	0.09	0.26	0.32	0.29	0.35	0.16	0.05

(continued)

Table 4 (continued)

Samples	17	18	19	20	21	22	23	24
Total	99.92	100.20	98.96	100.55	99.13	99.78	100.10	101.41
Mg#	93.8	92.1	91.7	91.7	92.1	92.0	91.9	92.3
Ca	0.9	3.1	1.5	2.6	3.0	2.6	0.4	0.4
Mg	93.0	89.3	90.3	89.3	89.3	89.6	91.6	91.9
Fe	6.1	7.6	8.2	8.1	7.7	7.8	8.0	7.7
Samples	**25**	**26**	**27**	**28**	**29**	**30**	**31**	
SiO_2	57.79	57.79	58.30	57.59	58.70	58.70	58.30	
TiO_2	0.00	0.00	0.00	0.00	0.01	0.01	0.02	
Al_2O_3	1.33	1.44	1.58	1.28	0.84	0.77	0.77	
Cr_2O_3	0.27	0.34	0.30	0.25	0.36	0.29	0.30	
FeOt	6.28	5.50	5.33	5.52	4.47	4.28	4.27	
MnO	0.12	0.11	0.17	0.16	0.10	0.10	0.10	
NiO	0.12	0.17	0.00	0.14	0.11	0.10	0.12	
MgO	34.50	33.75	34.09	33.69	35.60	35.70	35.50	
CaO	0.22	1.40	1.38	1.46	0.46	0.48	0.48	
Na_2O	0.10	0.36	0.34	0.14	0.15	0.10	0.14	
Total	100.73	100.86	101.49	100.23	100.80	100.53	100.00	
Mg#	90.7	91.6	91.9	91.6	93.4	93.7	93.7	
Ca	0.5	2.7	2.7	2.8	0.9	0.9	0.9	
Mg	90.3	89.2	89.6	89.0	92.6	92.8	92.8	
Fe	9.2	8.1	7.8	8.2	6.5	6.3	6.3	

Sample numbers from Table 1

because a diamond-bearing garnet lherzolite xenolith from the Mothae pipe, 7 km north-west of Letseng (PT conditions 43.0 kb and 1035 °C—Dawson and Smith 1975), plots closer to the diamond stability field of Bundy (1980) although still plotting within the graphite stability field.

From Fig. 3, several facts are apparent:

1. Xenoliths with coarse textures have apparently equilibrated under relatively low PT conditions, though low-temperature estimates (<900 °C) should be treated with caution (Nimis and Grutter 2010).
2. The highest PT conditions are found for deformed xenoliths from both pipes; in this respect, they are similar to high-T, sheared "fertile" xenoliths in the Thaba Putsoa kimberlite (Boyd 1973a). However, a major difference, as noted earlier, is that the Letšeng deformed xenoliths are depleted.
3. The Main Pipe xenoliths (with one exception) have equilibrated along a relatively continuous linear trend, whereas the PT conditions for Satellite Pipe xenoliths are bimodal—one group being low PT, the other at much higher PT.
4. In the Main Pipe xenolith suite, at around 1,000 °C and 40 kb, there is overlap between coarse and deformed xenoliths (including those with the most extreme LAD textures), in this respect being like garnet lherzolite xenoliths from the Kimberley area (Dawson et al. 1975). The most highly deformed xenoliths (i.e. with LAD textures) did not equilibrate at the highest temperatures (see data in Table 1), so there is no overall correlation between the temperature of equilibration and degree of deformation as found at Thaba Putsoa.
5. Many Main Pipe xenoliths, coarse and deformed, derive from an intermediate PT regime apparently not sampled by the Satellite Pipe kimberlite. (The "Pipe not known" samples of Simon et al. (2003) also fall within this PT area).
6. For a given temperature, the Satellite Pipe xenoliths have equilibrated at lower pressures than the Main Pipe xenoliths, thus showing substantial differences in the mantle PT regimes sampled by the two pipes.
7. Compared with the Kalahari craton geotherm, the Main Pipe xenolith linear array is parallel to, but cooler than, the geotherm down to a depth of ~45 kb. The lowest PT xenoliths from the Satellite Pipe plot close to the geotherm, but they are exceptions, and as noted above (a), their temperature estimates should be treated with caution. However, at $P > 45$ kb, xenoliths from both pipes plot on the high-T side of the geotherm. This indicates a perturbed geotherm as proposed for northern Lesotho by Boyd (1973a).

Table 5 Compositions of clinopyroxenes in Letšeng garnet lherzolites

Samples	1	2	3	4	5	6	7	8
SiO_2	54.57	54.67	54.99	54.89	54.44	54.44	55.12	53.91
TiO_2	0.10	0.05	0.31	0.00	0.08	0.03	0.03	0.04
Al_2O_3	2.54	1.75	2.75	2.27	2.42	2.05	2.35	1.38
Cr_2O_3	2.65	1.54	1.32	1.80	1.87	1.33	2.88	1.29
FeOt	2.12	2.24	3.61	2.27	2.33	1.66	2.12	1.66
MnO	0.10	0.07	0.10	0.08	0.10	0.08	0.10	0.11
NiO	0.03	0.01	0.08	0.09	0.07	0.04	0.07	0.09
MgO	16.23	17.15	15.85	16.79	16.55	16.62	16.48	17.76
CaO	19.00	20.34	18.49	19.70	19.46	21.98	19.11	22.31
Na_2O	2.38	1.69	2.41	2.03	2.16	1.42	2.37	1.02
Total	99.72	99.51	99.91	99.92	99.48	99.65	100.63	99.57
Mg#	93.2	93.2	88.7	93.0	92.7	94.7	93.3	95.0
Ca	43.9	44.3	42.7	43.9	45.6	47.4	43.7	46.1
Mg	52.2	51.9	50.9	52.1	54.0	49.8	52.5	51.1
Fe	3.9	3.8	6.4	4.0	4.4	2.8	3.8	2.8
Ca/(Ca + Mg)	0.458	0.460	0.456	0.457	0.458	0.487	0.454	0.475
Samples	**9**	**10**	**11**	**12**	13	**14**	**15**	**16**
SiO_2	53.81	55.22	54.36	54.27	54.19	54.72	54.49	54.47
TiO_2	0.07	0.11	0.12	0.31	0.46	0.36	0.29	0.26
Al_2O_3	3.09	2.14	2.12	2.42	2.37	2.15	2.65	2.58
Cr_2O_3	3.66	1.70	1.37	3.13	2.00	1.48	2.08	1.98
FeOt	2.13	2.26	2.18	2.30	2.90	2.71	2.45	2.42
MnO	0.13	0.09	0.11	0.12	0.12	0.09	0.09	0.09
NiO	0.06	0.09	0.01	0.04	0.05	0.05	0.00	0.11
MgO	15.44	16.42	16.98	15.68	17.98	18.34	16.34	16.51
CaO	17.64	19.43	20.13	18.65	17.41	17.98	18.84	18.89
Na_2O	3.13	2.20	1.71	2.55	1.87	1.79	2.18	2.22
Total	99.16	99.66	99.09	99.47	99.35	99.67	99.41	99.53
Mg#	92.8	92.8	93.3	92.4	91.7	92.3	92.2	92.4
Ca	43.3	44.1	44.3	44.1	38.9	39.3	43.3	43.2
Mg	52.7	51.9	52.0	51.6	56.0	55.9	52.3	52.5
Fe	4.0	4.0	3.7	4.3	5.1	4.8	4.4	4.3
Ca/(Ca + Mg)	0.451	0.459	0.460	0.461	0.410	0.413	0.453	0.451
Samples	**17**	**18**	**19**	**20**	**21**	**22**	**23**	**24**
SiO_2	54.46	54.71	54.29	54.43	55.53	55.24	54.24	53.80
TiO_2	0.04	0.03	0.36	0.29	0.15	0.28	0.22	0.26
Al_2O_3	2.42	0.85	2.29	2.46	2.23	2.66	2.62	2.86
Cr_2O_3	2.48	0.75	1.68	0.68	0.74	0.75	1.66	1.56
FeOt	2.11	3.25	3.18	4.04	3.71	3.77	2.11	2.11
MnO	0.11	0.14	0.11	0.14	0.13	0.14	0.00	0.06
NiO	0.06	0.05	0.11	0.00	0.00	0.00	0.00	0.03
MgO	16.39	21.05	18.00	21.16	20.62	20.41	15.68	16.00

(continued)

Table 5 (continued)

Samples	17	18	19	20	21	22	23	24
CaO	18.78	18.25	17.19	15.03	15.20	14.26	20.70	20.60
Na_2O	2.33	0.34	2.06	1.59	1.41	1.77	2.15	2.04
Total	99.18	99.42	99.27	99.82	99.72	99.28	99.38	99.32
Mg#	93.3	92.0	91.0	90.3	90.8	90.6	93.0	93.1
Ca	43.4	36.5	38.4	31.5	32.5	31.3	46.8	46.3
Mg	52.7	58.5	55.9	61.8	61.3	62.3	49.4	50.0
Fe	3.9	5.0	5.7	6.7	6.2	6.4	3.8	3.7
Ca/(Ca + Mg)	0.452	0.384	0.407	0.338	0.346	0.344	0.486	0.480
Samples	**25**	**26**	**27**	**28**	**29**	**30**	**31**	
SiO_2	54.51	55.30	55.82	55.51	55.60	55.20	55.50	
TiO_2	0.23	0.17	0.14	0.00	0.01	0.01	0.11	
Al_2O_3	2.68	1.86	1.79	1.10	2.55	2.00	2.60	
Cr_2O_3	2.11	1.07	1.06	0.53	2.30	1.62	1.71	
FeOt	2.43	3.74	3.51	3.60	2.25	1.90	2.23	
MnO	0.00	0.15	0.00	0.07	0.08	0.09	0.09	
NiO	0.00	0.00	0.00	0.00	0.03	0.04	0.05	
MgO	15.40	20.25	20.66	21.05	16.20	16.90	16.50	
CaO	20.50	15.33	16.16	17.26	18.90	20.40	19.00	
Na_2O	2.34	1.63	1.32	0.75	2.46	1.69	2.28	
Total	100.20	99.50	100.46	99.87	100.38	99.85	100.07	
Mg#	91.9	90.6	91.3	91.2	92.8	94.1	93.0	
Ca	46.6	33.0	33.9	34.9	43.7	44.9	43.5	
Mg	49.1	60.7	60.3	59.3	52.0	51.8	52.5	
Fe	4.3	6.3	5.8	5.8	4.3	3.3	4.0	
Ca/(Ca + Mg)	0.487	0.352	0.360	0.371	0.457	0.464	0.453	

Sample numbers from Table 1

8. The Letšeng PT array ranges from low- to high-PT conditions, in this respect resembling the lherzolites from Thaba Putsoa, but unlike a relatively narrow PT range found in xenolith suites in some other nearby Lesotho kimberlites—Matsoku (Gurney et al. 1975) and Pipe 200 (Carswell et al. 1979)—and from the Kimberley area (Dawson et al. 1975; Boyd and Nixon 1978).
9. Most Main Pipe and all Satellite Pipe samples lie within the graphite stability field.

Conclusions

1. Most Letšeng samples derive from depleted mantle, in this respect being like most lherzolites in other Cretaceous kimberlites. Only a very few samples are more Fe-rich.
2. The garnet lherzolite suites in the two pipes apparently derive from different sections of the upper mantle. This conclusion is in accord with observations that the two pipes have different diamond populations.
3. Some, but not all, deformed xenoliths from both pipes have equilibrated on the high-temperature side of the Kalahari craton geotherm, indicating they were thermally perturbed, but to different extents.
4. Although the thermal perturbation is like that found in deformed rocks from Thaba Putsoa, other xenolith suites from nearby kimberlite pipes do not have this high-temperature overprint. Hence, there is not a consistently perturbed geothermal gradient Lesotho-wide in the upper mantle in the mid-Cretaceous. The reason for these localised thermal differences is debatable, but a possible explanation is differing thermal input into the mantle aureoles surrounding either individual ascending kimberlites or around coarsely crystalline mantle intrusions that have been proposed as the source of the mantle megacryst suites found in most kimberlites (Gurney et al. 1975)

Table 6 Compositions of garnets in Letšeng garnet lherzolites

Samples	1	2	3	4	5	6	7	8
SiO_2	41.52	41.65	42.03	42.03	41.45	42.04	42.29	41.62
TiO_2	0.07	0.08	0.34	0.00	0.07	0.07	0.01	0.02
Al_2O_3	18.91	19.25	21.18	19.75	19.15	21.68	16.60	15.79
Cr_2O_3	6.71	6.52	2.77	5.90	6.43	3.17	8.77	10.10
FeOt	6.40	6.83	9.78	6.63	6.93	7.61	6.80	6.65
MnO	0.39	0.41	0.41	0.38	0.41	0.47	0.45	0.54
NiO	0.00	0.01	0.04	0.00	0.02	0.01	0.00	0.03
MgO	20.42	19.63	19.40	20.10	19.67	20.17	20.33	16.68
CaO	5.70	6.45	4.73	6.08	6.01	5.43	5.85	8.73
Na_2O	0.04	0.03	0.07	0.01	0.05	0.01	0.02	0.05
Total	100.16	100.86	100.75	100.88	100.19	100.66	101.12	100.21
Ca	14.6	16.6	11.9	15.5	15.4	13.7	14.8	25.4
Mg	72.5	69.8	68.8	71.2	70.6	71.2	71.8	61.0
Fe	12.9	13.6	19.3	13.3	14.0	15.1	13.4	13.6

Samples	9	10	11	12	13	14	15	16
SiO_2	41.37	41.73	41.98	41.41	42.01	42.25	42.05	41.70
TiO_2	0.07	0.07	0.24	0.32	0.65	0.36	0.10	0.11
Al_2O_3	20.36	19.03	20.88	18.79	19.49	20.39	19.79	19.76
Cr_2O_3	5.83	5.03	3.84	6.39	4.88	4.11	5.47	5.48
FeOt	6.47	6.49	6.98	7.12	6.63	6.14	6.81	6.69
MnO	0.40	0.33	0.36	0.47	0.31	0.30	0.40	0.39
NiO	0.03	0.00	0.02	0.00	0.03	0.00	0.02	0.00
MgO	21.47	20.66	20.85	20.10	21.51	21.98	20.45	20.40
CaO	4.72	5.39	5.14	5.51	4.91	4.85	5.49	5.58
Na_2O	0.06	0.04	0.04	0.08	0.10	0.08	0.07	0.05
Total	100.78	98.77	100.33	100.19	100.52	100.46	100.65	100.16
Ca	12.0	13.8	12.9	14.2	12.2	12.1	14.0	14.3
Mg	75.3	73.4	73.2	71.5	75.0	72.1	72.0	72.4
Fe	12.7	12.8	13.9	14.3	12.8	11.8	14.0	13.3

Samples	17	18	19	20	21	22	23	24
SiO_2	41.63	41.90	41.28	41.95	41.91	42.37	41.90	41.40
TiO_2	0.09	0.07	0.93	0.58	0.55	0.73	0.14	0.11
Al_2O_3	19.14	18.19	17.92	20.70	20.44	21.02	21.36	21.80
Cr_2O_3	6.47	8.57	6.27	2.34	2.50	2.11	2.94	2.90
FeOt	6.36	5.76	7.15	7.14	6.80	6.45	8.76	8.80
MnO	0.37	0.37	0.34	0.26	0.26	0.25	0.38	0.40
NiO	0.00	0.00	0.03	0.00	0.00	0.00	0.00	0.00
MgO	20.97	17.59	20.46	22.50	22.27	22.14	19.38	19.80
CaO	5.32	6.69	5.64	4.38	4.48	4.16	5.12	4.90
Na_2O	0.03	0.04	0.10	0.04	0.04	0.04	0.00	0.03
Total	100.38	99.18	100.12	99.89	99.25	99.27	99.98	100.14
Ca	13.5	16.8	14.3	10.6	11.0	10.4	13.1	12.4

(continued)

Table 6 (continued)

Samples	17	18	19	20	21	22	23	24
Mg	73.9	70.3	71.7	75.9	76.0	77.0	69.2	70.1
Fe	12.6	12.9	14.0	13.5	13.0	12.6	17.7	17.5

Samples	25	26	27	28	29	30	31
SiO_2	42.10	42.58	43.37	42.80	42.10	42.30	42.10
TiO_2	0.15	0.65	0.35	0.15	0.03	0.02	0.03
Al_2O_3	21.00	19.85	21.05	20.25	19.70	20.40	20.10
Cr_2O_3	3.51	3.59	2.80	3.59	5.85	5.07	4.61
FeOt	9.71	6.91	6.62	7.01	6.63	6.31	6.33
MnO	0.40	0.22	0.22	0.23	0.42	0.37	0.35
NiO	0.00	0.00	0.00	0.00	0.00	0.00	0.00
MgO	18.95	21.75	22.29	21.41	20.30	20.30	20.80
CaO	5.27	4.82	4.70	5.30	5.24	5.49	5.32
Na_2O	0.00	0.00	0.04	0.00	0.02	0.01	0.03
Total	101.09	100.37	101.44	100.74	100.29	100.27	99.67
Ca	13.3	12.0	11.5	13.0	13.5	14.3	13.6
Mg	67.3	74.6	75.8	73.4	73.1	73.0	73.8
Fe	19.4	13.4	12.7	13.6	13.4	12.7	12.6

Sample numbers from Table 1

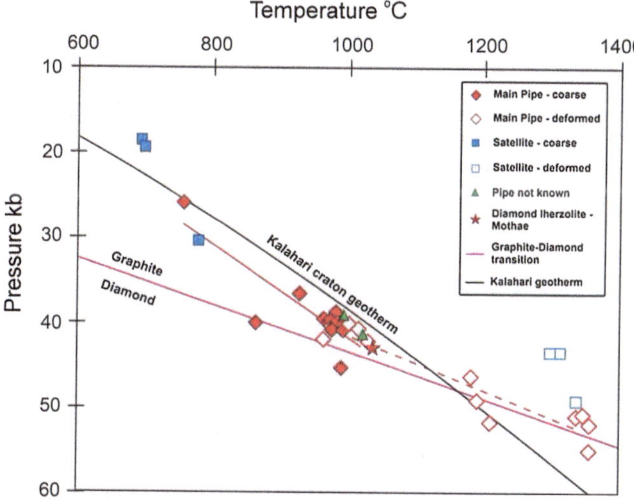

Fig. 3 PT equilibration of Letseng garnet lherzolites (data from Table 1). Other data sources: Kalahari geotherm—Rudnick and Nyblade (1999); graphite–diamond transition—Bundy (1980); Matsoku diamondiferous lherzolite—Dawson and Smith (1975). The solid and broken lines in red are the trends for Main Pipe coarse and deformed xenoliths, respectively

Acknowledgments NPL thanks N.E.R.C for financial support under a Research Scholarship held at the universities of St Andrews and Sheffield, and the Anglo-American Corporation for permission to undertake the research. Whilst NPL was employed with De Beers in Lesotho, Barry Hawthorne and Roger Clement provided encouragement and advice, and Keith Whitelock, the mine manager at that time, laid the foundation for this study. Numerous people helped with the analytical work in St Andrews and Sheffield, and also at Edinburgh, Manchester and Cambridge electron probe laboratories. JBD thanks NERC and the Carnegie Trust for the Universities of Scotland for research funding, and the Organizing Committee of the 10th International Kimberlite Conference for funding that enabled attendance at the Conference. We thank Bill Griffin for drawing our attention to the kimberlite perovskite dates of E. Hoese. Darren Wilkinson drafted the line diagrams. Guest editor H. O'Brien and N.V.C. Rao provided helpful comments which improved the original manuscript.

References

Bowen DC, Ferraris RD, Palmer C, Ward JD (2009) On the unusual characteristics of the diamonds from Letšeng-la-Terae kimberlites, Lesotho. Lithos 112S:767–774

Boyd FR (1973a) A pyroxene geotherm. Geochim Cosmochim Acta 37:2533–2546

Boyd FR (1973b) Appendix of mineral analyses Letšeng-la Terae. In: Nixon PH (ed) Lesotho kimberlites. Lesotho National Development Corporation, Maseru, pp 33–36

Boyd FR, McCallister RH (1976) Densities of fertile and sterile garnet peridotites. Geophys Res Lett 3:509–512

Boyd FR, Nixon PH (1978) Ultramafic nodules from the Kimberley pipes, South Africa. Geochim Cosmochim Acta 42:1367–1382

Brey GP, Köhler T (1990) Geothermobarometry in four-phase lherzolites II. New thermobarometers, and practical assessment of existing thermobarometers. J Pet 31:1353–1378

Brey GP, Köhler T, Nickel KG (1990) Geothermobarometry in four-phase lherzolites I. Experimental results from 10 to 60 kb. J Pet 31:1313–1352

Bundy FP (1980) The P, T phase and reaction diagram for elemental carbon, 1979. J Geophys Res B 85:6930–6936

Carswell DA, Clarke DB, Mitchell RH (1979) The petrology and geochemistry of ultramafic nodules from Pipe 200, northern Lesotho. In: Boyd FR, Meyer HOA (eds) The mantle sample: Inclusions in kimberlites and other volcanics. Proceedings of the second international kimberlite conference, Washington, DC, vol 2. Am Geophys Union, pp 127–144

Danchin RV (1979) Mineral and bulk chemistry of garnet lherzolite and garnet harzburgite xenoliths from the premier mine, South Africa. In: Boyd FR, Meyer HOA (eds) The mantle sample: inclusions in kimberlites and other volcanics. Proceedings of the second international kimberlite conference, Washington, DC, vol 2. Am Geophys Union, pp 104–126

Davis GL (1977) The ages and uranium contents of zircons from kimberlites and associated rocks. Carnegie Inst 76:631–654

Davis BTC, Boyd FR (1966) The join $MgSiO_3$ and $CaMgSi_2O_6$ at 30 Kb and its application to pyroxenes from kimberlite. J Geophys Res 71:3567–3576

Dawson JB, Gurney JJ, Lawless PJ (1975) Palaeogeothermal gradients derived from xenoliths in kimberlite. Nature 257:299–300

Dawson JB, Smith JV (1975) Occurrence of diamond in a mica-garnet lherzolite xenolith from kimberlite. Nature 254:580–581

Dawson JB, Stephens WE (1975) Statistical classification of garnets from kimberlite and associated xenoliths. J Geol 83:589–607

Duncan AR, Marsh JS (2006) The Karoo igneous province. In: Johnson MR, Anhausser CR, Thomas RJ (eds) The geology of South Africa, Johannesburg. pp 501–520

Gurney JJ, Harte B, Cox KG (1975) Mantle xenoliths in the Matsoku kimberlite pipe. Phys Chem Earth 9:507–523

Harris JW, Hawthorne JB, Oosterveld MM (1979) Regional and local variations in the characteristics of diamonds from some southern African kimberlites. In: Boyd FR, Meyer HOA (eds) The mantle sample: Inclusions in kimberlites and other volcanics. Proceedings of the second international kimberlite conference, Washington, DC, vol 1. Am Geophys Union, pp 27–41

Harte B (1977) Rock nomenclature with particular relation to deformation and recrystallisation textures in olivine-bearing xenoliths. J Geol 85:279–288

Hervig RL, Smith JV, Dawson JB (1980) Fertile and barren Al-Cr-spinel harzburgites from the upper mantle: ion and electron probe analyses of trace elements in olivine and orthopyroxene: relation to lherzolites. Earth Planet Sci Lett 50:41–58

Hoese E (2009) The use of perovskite to explore characteristics of kimberlite. Unpubl Honours thesis, Macquarie University, p 52

Janney PE, Shirey SB, Carlson RW, Pearson DG, Bell DR, le Roex AP, Ishikawa A, Nixon PH, Boyd FR (2010) Age, composition and thermal characteristics of South Africa off-craton mantle lithosphere: evidence for a multi-stage history. J Pet 51:1849–1890

Kennedy CS, Kennedy GC (1976) The equilibrium boundary between graphite and diamond. Jour Geophy Res B 81:2467–2470

Lock NP (1980) The geology of the Letšeng kimberlites, Lesotho, Unpublished Ph.D. thesis, University of Sheffield

Lyubetskaya T, Korenaga J (2007) Chemical composition of the Earth's primitive mantle and its variance; 1. Method and results. J Geophys Res B 112:03211

Nickel KG, Green DH (1985) Empirical geothermobarometry for garnet peridotites and implications for the nature of the lithosphere, kimberlites and diamonds. Earth Planet Sci Lett 73:158–170

Nimis P, Grütter H (2010) Internally consistent geothermometers for garnet peridotites and pyroxenites. Contrib Mineral Pet 159:411–427

Nixon PH (1973) (ed) Lesotho kimberlites. Lesotho National Development Corporation, Maseru, p 350

Nixon PH, Boyd FR (1973) Petrogenesis of the granular and sheared ultrabasic nodule suite in kimberlites. In: Nixon PH (ed) Lesotho kimberlites. Lesotho National Development Corporation, Maseru, pp 48–56

Palmer CE, Ward JD, Stiefenhofer J, Whitelock TK (2008) Volcanological processes and their effect on diamond distribution in the Letšeng Satellite Pipe, Lesotho. 9th international kimberlite conference extended abstract no. 9IKC-A-00096

Pochou JL, Pichoir F (1991) Quantitative analysis of homogeneous or stratified microvolumes applying the model "PAP". In: Heinrich KJF, Newbury DE (eds), Electron Probe Quantitation. Plenum Press, New York, pp 31–75

Rudnick RL, Nyblade AA (1999) The thickness and heat production of Archaean lithosphere: constraints from xenolith thermobarometry and surface heat flow. In: Yingwei F, Berika CM, Mysen BO (eds) Mantle petrology: field observations and high pressure experimentation: a tribute to Francis R. (Joe) Boyd. Geochemical Society, Special Publication 6, Houston, pp 3–12

Scott smith BH, Nowicki TE, Russell JK, Webb KJ, Hetman CM, Harder M Mitchell RH (2008) Kimberlites: descriptive geological nomenclature and classification. 9th international kimberlite conference extended abstract no. 9IKC-A-00124

Simon NSC, Irvine GJ, Davies GR, Pearson DG, Carlson RW (2003) The origin of garnet and clinopyroxene in "depleted" Kaapvaal peridotites. Lithos 71:289–322

Stephens WE, Dawson JB (1977) Statistical comparison between pyroxenes from kimberlites and their associated xenoliths. J Geol 85:433–449

Taylor WR (1998) An experimental test of some geothermometer and geobarometer formulations for upper mantle peridotites with application to fertile lherzolite and garnet websterite. Neues Jahrb Mineral. Abhandl 172:381–408John Barry Dawson 1932–2013

John Barry Dawson 1932–2013

Barry Dawson died suddenly at his home in Edinburgh in late January 2013. He will be especially missed by all of his family, colleagues and diverse friends including those of the kimberlite carbonatite and upper mantle community. Barry was known and loved for his wit, charm and conviviality together with his extensive knowledge of kimberlite

geology, poetry, ornithology, and African history. He was a raconteur and a great after-dinner speaker, with a penchant for singing rather rude and amusing songs. Barry was widely regarded as the patriarch of the "kimberlite family". He was well-respected by university faculty, students and industrial geoscientists. Typically, he never held a grudge against persons who held opposing views. Many of us will really miss staying at the "Braid Farm Road Hotel" enjoying the hospitality and G&Ts so generously provided by our host.

Barry was born in Leeds (UK) and attended Leeds University from which he graduated with a B.Sc. in Geology in 1957. His interest in all things African was developed as graduate student with the Research Institute of African Geology. His work there took him to Basutoland (now Lesotho) to study kimberlite intrusions and culminated in his Ph.D (1960). Parts of his thesis appeared in the seminal paper Basutoland Kimberlites (Geol. Soc. America Bull., 73, 1962). This work was the first modern study of kimberlites and was notable in that Barry introduced the term "fluidization" to kimberlite geology; there are still arguments as to the role of this process in diatreme emplacement! Subsequent, to his graduation Barry, and his late wife Christine, embarked again for Africa to join the Tanganyika Geology Survey. It was here he discovered the amazing sodium carbonate volcanism of Oldoinyo Lengai and earned himself a permanent place in petrological history with the publication of his paper in Nature (1962) on these extraordinary rocks. At the time of his death, some 50 years later, Barry was still studying Oldoinyo Lengai and other Tanzanian volcanoes in the Gregory Rift.

In his academic career, Barry was initially employed at the University of St. Andrews (1964–1978), followed by appointment as Sorby Professor of Geology University of Sheffield, and ultimately at the University of Edinburgh 1988–1997. He retired from active teaching in 1997 and continued an active program of research in his two favourite topics; upper mantle-derived xenoliths, and Recent volcanic rocks of northern Tanzania. In recognition of his research he was awarded the Collins medal of the Mineralogical Society of Great Britain in 2012 and the Clough Medal of the Geological Society of Edinburgh (1999). He was a Fellow of the Royal Society of Edinburgh and the German Academy of Scientists.

Barry's involvement with kimberlites continued throughout his life. In particular, he was one of the convenors of the 1st International Kimberlite Conference held in Cape Town (1973) with field excursions led by him and Peter Nixon to Lesotho. This conference resulted in worldwide attention being given to kimberlites and upper mantle-derived xenoliths, and to Barry as "the authority" on all things "kimberlitic". Subsequently, Barry was a convenor of the 2nd Kimberlite Conference held in Santa Fe (1978); an event at which major advances in kimberlite geology were reported; these resulting from studies of the material collected in Lesotho and South Africa. As a consequence of the success of these conferences Barry was instrumental in the formation of the International Kimberlite Conference Advisory Committee and was Chairman of this committee from 1986–1998.

During the last years of his life Barry undertook several expeditions to the Oldoinyo Lengai area, even making the difficult trek to the top of the volcano Kerimasi. As late as December 2012 he was back in Lesotho doing field work. One thing he regretted was that he had not yet climbed all of the Scottish Munros—something he was working on still. His passing is a great loss to us all. To some of us involved in kimberlite and carbonatite research he can never be replaced as he was a mentor and guide throughout our careers. Perhaps Barry was one of the last of the great "gentlemen scientists"?

Roger H. Mitchell
Emeritus Professor of Geology
Lakehead University, Thunder Bay, Ontario
Canada

Tectonic Relationships Between E-Type Cratonic and Ultra-High-Pressure (UHP) Diamond: Implications for Craton Formation and Stabilization

H. H. Helmstaedt

Abstract

Many eclogite xenoliths from kimberlites show geochemical and isotopic signatures compatible with an origin by subduction of oceanic crustal rocks, and it has been proposed earlier that progressively metamorphosed coesite- and diamond-bearing eclogitic assemblages and coesite inclusions in E-type diamonds may be viewed as an expression of Archean and Proterozoic UHP metamorphic events. Yet the observation that kimberlite-born cratonic eclogitic diamonds and diamonds in UHP metamorphic terranes lie on opposite ends of the geological age spectrum, is still commonly used to infer that UHP metamorphism is a Neoproterozoic or younger process that was not possible in a much hotter Archean Earth. The present paper re-evaluates the UHP model for the Archean eclogitic upper mantle sample by discussing (1) examples for igneous exhumation of subducted Phanerozoic and Proterozoic UHP eclogites and diamonds; (2) seismic images showing that tectonic accretion of oceanic lithosphere was a major factor during assembly of Paleoproterozoic and Archean cratons, and (3) recently discovered Archean metabasic eclogites representing the oldest surface record for subduction zone metamorphism. It is concluded that crustal UHP melanges and the diamondiferous eclogite upper mantle sample are complementary end products of subduction zone metamorphism. Lower density crustal melanges are tectonically exhumed soon after continental or microcontinental collision. The mantle eclogites and their E-type cratonic diamonds are parts of the higher density oceanic slab subducted prior to collision and accreted to the roots of microcontinental nuclei. They had long mantle residence times and could be exhumed only when picked up by younger igneous transport media (e.g. kimberlites). As shown by ~2.9 Ga ages of the oldest known eclogitic diamonds and xenoliths with subduction signatures, deep subduction along UHP-gradients began latest in the Mesoarchean, producing lithospheric roots sufficiently thick and cool to extend into the diamond stability field. The range of Proterozoic E-type diamond ages on several diamondiferous cratons, even within individual kimberlites, suggests that craton roots were repeatedly modified during the Proterozoic.

Keywords

Eclogitic (E-type) diamond • Archean ultra-high-pressure (UHP) metamorphism • Onset of subduction • Lithosphere stabilization

H. H. Helmstaedt (✉)
Department of Geological Sciences and Geological Engineering, Queen's University, Kingston, ON K7L 3N6, Canada
e-mail: helmstaedt@geol.queensu.ca

D. G. Pearson et al. (eds.), *Proceedings of 10th International Kimberlite Conference*, Volume 1, Special Issue of the Journal of the Geological Society of India,
DOI: 10.1007/978-81-322-1170-9_4,

Introduction

Economic kimberlites and lamproites are located almost exclusively on or near Archean cratons (Janse 1994), and thus their diamonds are commonly referred to as cratonic (e.g., Stachel and Harris 2008). Close to 99 % of such diamonds are derived from peridotitic and eclogitic source rocks in the subcontinental lithospheric mantle (SCLM), and Archean to Proterozoic mineral inclusion ages (reviewed by Gurney et al. 2010) suggest that they are xenocrysts within their normally much younger igneous transport media. It is generally assumed that the growth events of cratonic diamonds are closely related to the Precambrian processes of formation and early modifications of the cratonic roots in which they are stored. Yet although a subduction origin for many of the eclogitic diamond sources and coesite-bearing eclogites (Fig. 1) has been well established (e.g., Schulze and Helmstaedt 1988; Taylor and Neal 1989; Schulze et al. 2000, 2003, 2004; Jacob 2004), relatively little attention has been paid to the tectonics of the emplacement process and its general implications for Early Precambrian cratonic root formation and preservation.

Diamond discoveries in surface outcrops of non-volcanic and non-xenolithic rocks were reported in the literature since the beginning of the twentieth century, initially mainly from ultramafic complexes and derived placers, but most of these early reports were dismissed as anomalous and as based on insufficient documentation. In a review of these 'anomalous' diamond occurrences, Nixon (1995) commented that geologists rejected such reports because

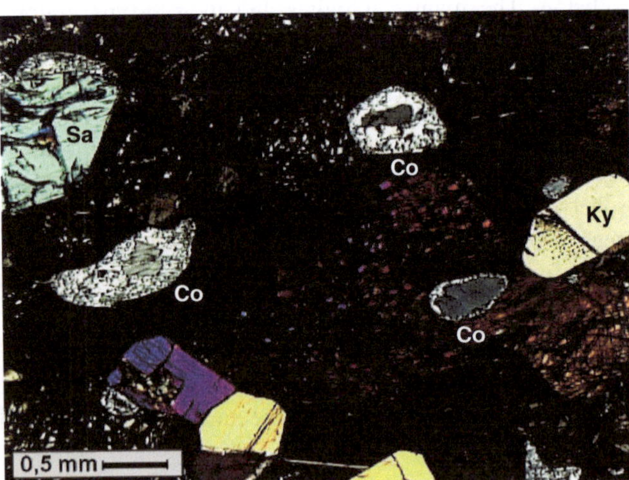

Fig. 1 Photomicrograph (crossed nicols) of coesite- and sanidine-bearing grospydite (calcic garnet- and kyanite-bearing eclogite) xenoliths from the Roberts Victor mine, South Africa. This xenolith was first described by Smyth and Hatton (1977), though its subduction origin was not recognized until later. Coesite grains (Co) are surrounded by retrogressive polycrystalline quartz rims. Coesite, sanidine (Sa), and kyanite (Ky) are set in a matrix of altered clinopyroxene and garnet

(1) they could not envisage that mantle rocks could have been exhumed from depths corresponding to the diamond stability field (ca. 120–150 km), or (2) even if exhumed by obduction, diamonds would not have survived graphitization during the relatively slow tectonic exhumation, or (3) the diamonds may have been the result of sample contamination, or (4) other minerals may have been misidentified as diamonds. Indeed, as most of the early 'anomalous' diamond reports were from ultramafic rocks thought to have originated in the upper mantle, only a tectonic one-way street (obduction) was contemplated as possible emplacement mechanism. Most geologists could not have imagined that some of the ultramafic diamondiferous rocks might have traveled a tectonic two-way street, that is, was first subducted into the diamond stability field and then returned to the Earth's surface by tectonic exhumation. For this to become accepted, diamonds would have to be found in clearly recognizable recycled crustal rocks.

The first ultra-high-pressure (UHP) mineral discovered in metasedimentary rocks was coesite and found almost simultaneously in quartzites and schists of the Dora Maira Massif in the Western Alps (Chopin 1984) and in eclogites of the Western Gneiss Region of Norway (Smith 1984) (Fig. 2). The occurrence of coesite inclusions in pyropes from metamorphic rocks with clearly supracrustal protoliths from Dora Maira confirmed the existence of a tectonic two-way street from the crust into the upper mantle and back (Chopin 1984) (Fig. 3). A UHP metamorphic interpretation of microdiamond inclusions in garnets and zircons of metasedimentary garnet–biotite gneisses and schists in the Kokchetav Massif of northern Kazakhstan (Sobolev and Shatsky 1990) suggested that this two-way street may continue through the stability field of coesite (minimum depth of ~ 90 km) into that of diamond (> 120 km) (Fig. 3).

UHP terranes with microdiamond inclusions in various crustal minerals (garnet, zircon, kyanite, carbonates, etc.) have now been described from at least ten localities (Ogasawara 2005; Liou et al. 2009) (Fig. 2). They occur in fossil subduction scars, but are confined to domains in which subduction of ocean floor (Pacific-type subduction) was followed by Alpine-type subduction of continental crust during continental or microcontinental collision (Maruyama et al. 1996). From the observation that ages derived from UHP metamorphic mineral assemblages are consistently much younger than the supracrustal protoliths from which they are derived (Table 1), it has been suggested that the involvement of old, cold, and relatively dry continental crust is an essential requirement for UHP metamorphism (Coleman and Wang 1995). Ages of metamorphism in surface occurrences of UHP terranes are generally younger than ~ 1 Ga (e.g., Ernst 2006, 2009; Liou et al. 2009; Kylander-Clark et al. 2012) and, where dated, tectonic

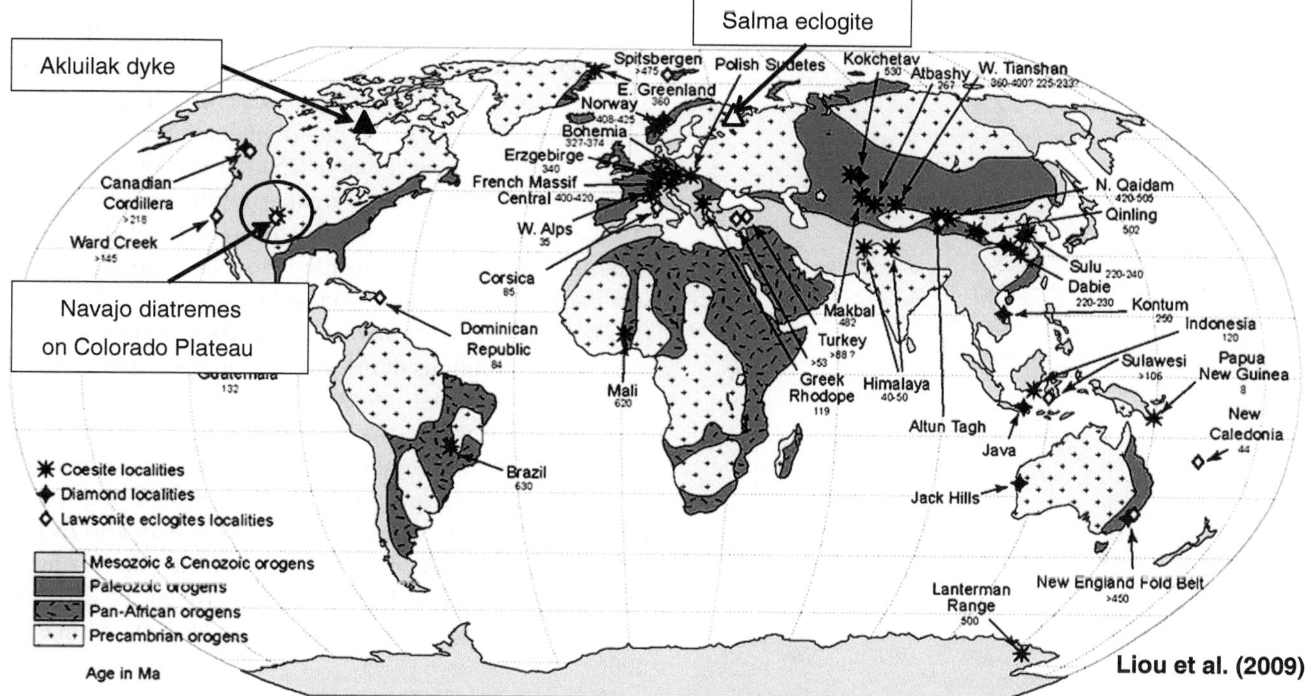

Fig. 2 Global distribution of UHP metamorphic terranes and lawsonite eclogites (from Liou et al. 2009). Location of lawsonite–coesite eclogite xenoliths from Navajo diatremes on Colorado Plateau is indicated by black circle. Ward Creek eclogite locality (to the west) is part of Franciscan accretionary melange. Solid black triangle is approximate location of 1,832 Ma Akluilak dyke in Canadian Shield. Open triangle denotes location of Salma eclogite at the margin of the Kola province of the Fennoscandian Shield. The Bingara and Copeton area subduction diamonds are located in the New England Fold Belt of eastern Australia

exhumation of the subduction melanges follows peak metamorphic conditions at depths of up to 150 km within the relatively brief time spans of 10–20 million years (Coleman and Wang 1995; Ernst 2006; Kylander-Clark et al. 2012) (Table 1). Although both Precambrian mantle eclogite xenoliths and eclogites from UHP metamorphic assemblages may contain coesite and/or diamond, the major index minerals for what has been defined as UHP metamorphism (Coleman and Wang 1995), the observation that cratonic diamonds and UHP diamonds lie on opposite ends of the geological age spectrum (Tables 1 and 2) is still being used to infer that UHP metamorphism is a Neoproterozoic or younger process that was not possible in a much hotter Archean Earth (e.g., Stern 2007; Ernst 2009). On the other hand, it has been conjectured that diamond- and coesite-bearing eclogite and grospydite xenoliths from kimberlites may be expressions of earlier UHP metamorphic events, dating as far back as the Archean (e.g., Schulze and Helmstaedt 1988; Schreyer 1995). As UHP metamorphic terranes are the only tectonic setting in which lithospheric diamond can be observed in situ, the present paper re-evaluates the viability of a UHP metamorphic model for the kimberlitic eclogitic upper mantle sample in the light of more recent evidence, consisting of (1) examples of Phanerozoic and Precambrian UHP products exhumed by igneous rocks, (2) seismic images showing that tectonic underplating was a major factor during the constructive phase of Paleoproterozoic and Archean cratons, and (3) recently discovered Archean metabasic eclogites recording the oldest evidence for subduction zone metamorphism.

Igneous Exhumation of UHP Metamorphic Products

Coesite Eclogite Xenoliths from Navajo Diatremes

Lawsonite- and coesite-bearing eclogite xenoliths from ca. 30 Ma kimberlite-like igneous breccias of the Navajo diatremes on the Colorado Plateau (Fig. 2) have been interpreted as fragments of subducted oceanic lithosphere (e.g., Helmstaedt and Doig 1975; Usui et al. 2003, 2006), the emplacement of which can be correlated with the subduction of the Farallon plate under western North America in the Mesozoic and Cenozoic. Pacific-style subduction of the oceanic Farallon plate began in the Jurassic (Figs. 4 and 5), with the slab descending at a steep angle, causing formation of the Sierra Nevada magmatic arc, together with the Great

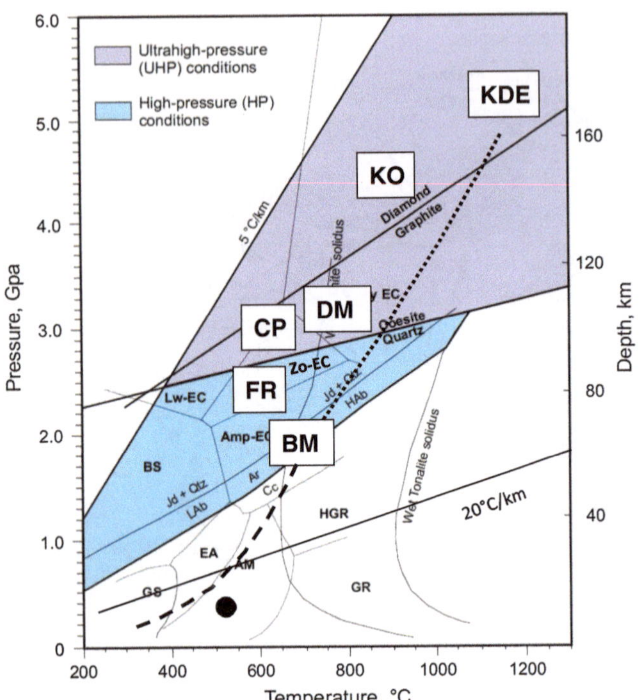

Fig. 3 Petrogenetic grid for high- and ultra-high-pressure metamorphic P–T environments as used by Liou et al. (2009). P–T conditions are indicated for DM—Dora Maira (Western Alps, see also Fig. 2) (Chopin 1984); KO—Kokchetav Massif, Kazakhstan (see also Fig. 2) Sobolev and Shatsky 1990); *CP*—Coesite–lawsonite eclogite xenoliths from Navajo diatremes on Colorado Plateau (Usui et al. 2003); *FR*—Franciscan eclogites (Tsujimori et al. 2006); *BM*—Salma eclogite, Belomorian accretionary orogen (Mints et al. 2010). *Solid black circle* is aluminosilicate triple point for reference. *Dashed line* to BM is prograde metamorphic P–T path of Salma eclogites; *dotted line* is inferred continuation of Salma eclogite P–T path toward approximate stability field of kimberlite-borne diamond eclogite xenoliths (KDE)

Valley forearc and the outboard Franciscan accretionary prism in which lenses of ca. 150–155 Ma low-temperature eclogites were exhumed (Anczkiewicz et al. 2004). Arc magmatism in the Sierra Nevada ceased at ca. 85 Ma, and calc-alkaline volcanism migrated eastward, far into the continental interior (Lipman et al. 1972). The eastward shift in magmatism coincided with a period of rapid convergence between the North American and Farallon plates following the opening of the North Atlantic (Fig. 4), and the corresponding decrease in dip of the descending slab of the Farallon Plate (e.g., Hyndman 1972; Coney and Reynolds 1977; Dickinson and Snyder 1978) caused subhorizontal eastward directed underplating of oceanic lithosphere for as far as 900 km (Fig. 5). The rate of underplating between 85 and 30 Ma was about 2 cm/yr (Usui et al. 2003; Liu et al. 2008). Transport of the xenoliths to the surface by the 30 Ma Navajo diatremes was related to the upwelling of hot material from the upper mantle below the slab that in turn was caused by the tearing and sinking of the Farallon plate (Dickinson 1997). Phanerozoic Re–Os mantle extraction ages for the protoliths of the prograde metamorphic eclogite xenoliths (Ruiz et al. 1999) and 80–30 Ma U–Pb zircon ages for eclogite metamorphism (Usui et al. 2003) are consistent with this emplacement model, but the possibility exists that the diverse eclogite xenolith suite of the Navajo diatremes (Helmstaedt and Schulze 1988) may also include older eclogites plucked from a Proterozoic suture zone in the Colorado Plateau basement (e.g., Roden et al. 1990; Helmstaedt and Schulze 1991; Smith et al. 2004; Usui et al. 2006).

The inferred P–T trajectory for the coesite-bearing lawsonite eclogite xenoliths (Fig. 6) follows a relatively cool subduction path into the UHP part of the lawsonite eclogite stability field (Okamoto and Maruyama 1999). In contrast, eclogites of the Franciscan accretionary prism contain neither coesite nor prograde metamorphic lawsonite and show peak metamorphic conditions in the high-pressure part of the zoisite/epidote eclogite field (Tsujimori et al. 2006) (Fig. 3). Key to the tectonic emplacement of the coesite–lawsonite eclogites under the Colorado Plateau was the change in subduction geometry from a relatively steep slab descent to shallow-angle subduction, allowing the subducting slab to maintain contact with the subcontinental lithosphere of the overriding plate (Fig. 5). It thus appears that a normal Pacific-style subduction mode causes oceanic slabs to descend into the upper mantle leading to eclogite megalith formation at the transition zone (Ringwood 1991), whereas tectonic underplating of oceanic under subcontinental lithosphere is possible only after a change from a

Table 1 Comparison of ages and metamorphic conditions of selected UHP metamorphic terranes [Data from Liou et al. (2009) and Kylander-Clark et al. (2012)]

UHP terrane	Kokchetav Dia, Coe	W. Gneiss Region Dia, Coe	Dabie Shan Dia, Coe	Erzgebirge Dia, Coe	Dora Maira Coe
Protolith formation age	2.2–2.3 Ga	1.6–1.8 Ga	650–800 Ma	>400 Ma	~300 Ma
Age of UHP metamorphism	~533 Ma	405–400 Ma	245–222 Ma	336.8 ± 2.8 Ma	35.4 ± 2.7 Ma
Peak temperatures	900 ± 75 °C	775 ± 75 °C	750 ± 75 °C	650–800 °C	725 ± 50 °C
Maximum depth	~140 km	90–130 km	90–125 km	95 to > 130 km	90–110 km
Rise time to mid-crust	~6 m.y.	>15 m.y.	>20 m.y.	<7 m.y.	~4 m.y.

Table 2 Summary of isotopic ages derived from diamond inclusions of cratonic diamonds from kimberlites and lamproites (L) on five cratons [modified from Gurney et al. (2010)]

Diamond age/Paragenesis	Kaapvaal/Kalahari	Siberia	Slave	Australia	Congo
P lherz ~ 3.57–3.2 Ga	Kimberley	Udachnaya	Panda		
	Finsch				
E-type	Kimberley	Udachnaya			
~2.9 Ga	Koffiefontein				
~2.6 Ga	Jwaneng				
	Orapa				
	Klipspringer				
P lherz ~ 2.0 Ga	Premier				
	Venetia				
E-type	Premier	Udachnaya	Diavik	Argyle (L)	
~2.2–1.8 Ga	Venetia		Jericho		
~1.7	Jagersfontein				
~1.5	Finsch				
~1.2	Jwaneng				
~1.1–1.0	Premier				
	Jagersfontein				
	Koffiefontein				
	Orapa				
P lherz ~ 1.5 Ga				Ellendale (L)	
E-type ~ 0.6 Ga					M'buji Mayi

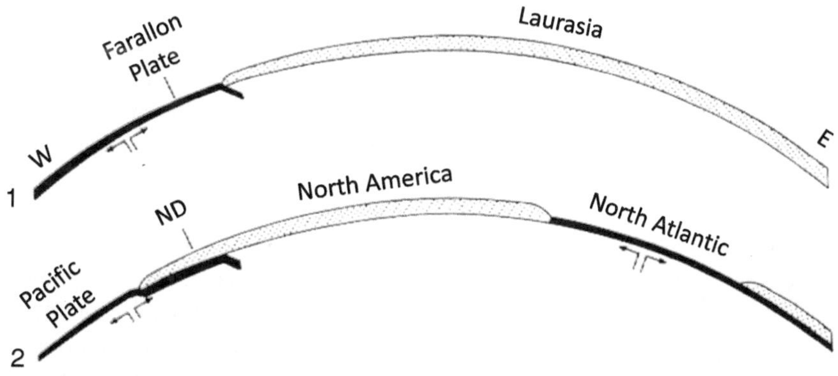

Fig. 4 Schematic cross sections depicting tectonic setting of change from normal subduction to shallow subduction of Farallon Plate under the western margin of North America (modified from Helmstaedt and Schulze 1988) between 80 Ma (*1*) and 30 Ma (*2*). ND in (*2*) stands for Navajo diatremes

steep to a shallower subduction mode (see also Helmstaedt and Schulze 1988; Abbott 1991).

Phanerozoic 'Subduction' Diamonds in Eastern Australia

Comprehensive studies of diamonds and inclusion minerals from alluvial diamond fields in New South Wales, Australia, including stones from the Copeton and Bingara areas (Barron et al. 2008) (Fig. 2), suggest the possibility of a multiple origin for the diamonds of these enigmatic fields. Davies et al. (2002) distinguished two diamond groups: one with low to moderate N and high %B aggregation, resembling diamonds from kimberlites and lamproites, but showing signs of a long recycling history, and the other with high N, low %B aggregation and other unusual characteristics (Table 3) consistent with an origin in a Paleozoic

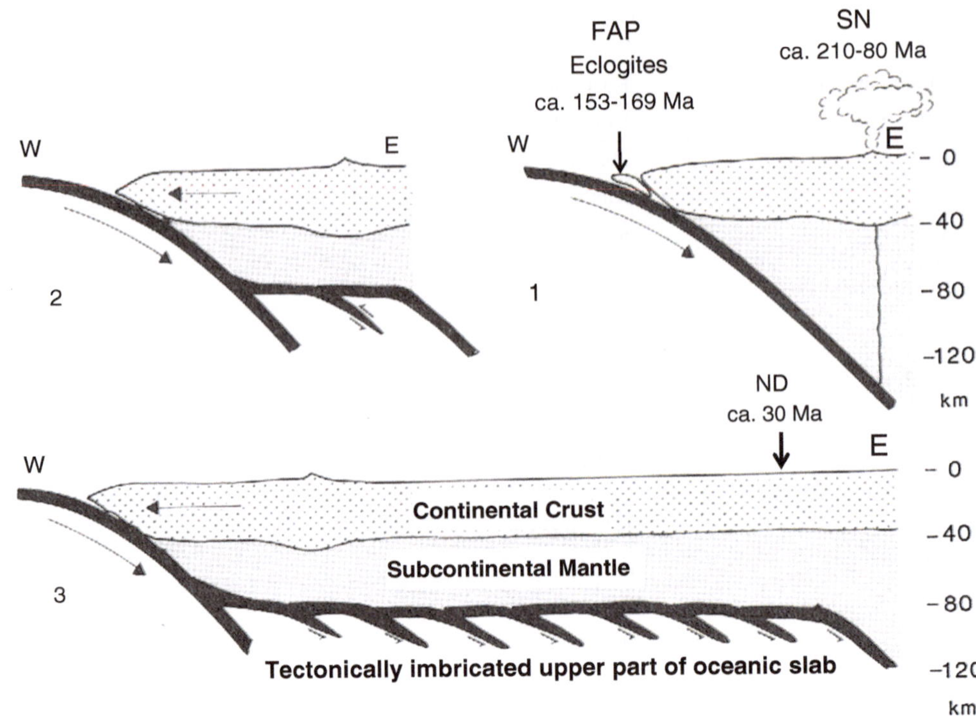

Fig. 5 Model depicting possible geometry of change from normal subduction (*1*) to shallow subduction (*2* and *3*) following extinction of magmatism in Sierra Nevada arc (SN) at ca. 80 Ma. Ages of Franciscan eclogites (FAP) after Anczkiewicz et al. (2004). *ND*—Navajo diatremes. The upper part of subducted oceanic slab is thought to be imbricated as Farallon plate is overridden by North American plate (modified from Helmstaedt and Schulze 1988). Distance in section (3) between the Navajo diatremes and the continental margin at ca. 30 Ma was about 900 km

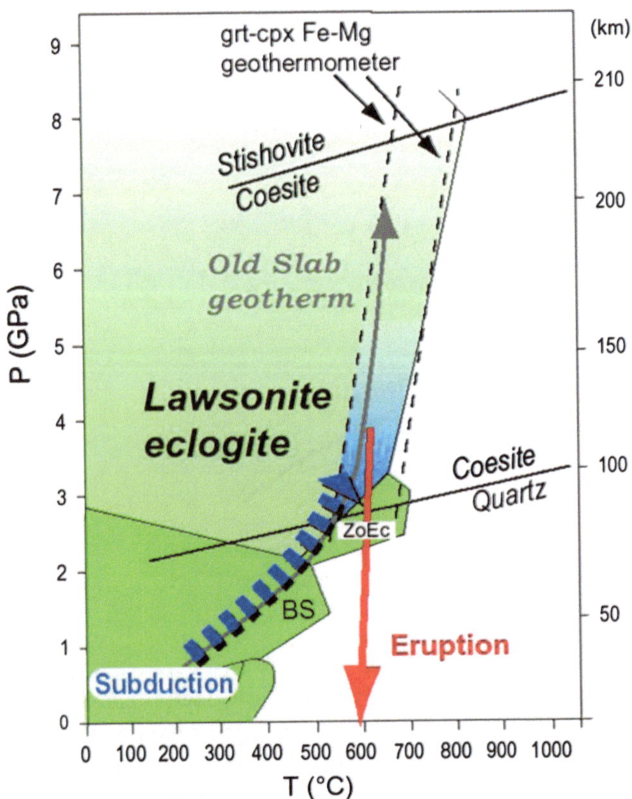

Fig. 6 Inferred P–T trajectory of coesite–lawsonite eclogite xenoliths from Moses Rock (Navajo diatremes) showing prograde metamorphic path during subduction (broken blue arrow) and exhumation path in diatreme (*solid red arrow*) (after Usui et al. 2003). ZoEc is stability field of zoisite eclogite. Old slab geotherm is after Okamoto and Maruyama (1999)

subduction environment, as previously proposed for these alluvials by Barron et al. (1994, 1996). In addition to 'subduction' diamond, Barron et al. (2005) studied eclogitic and UHP crustal garnets with decompression microstructures, such as crystallographic exsolutions of rutile, apatite, and ilmenite, suggesting partial exhumation from mantle depths. Some of the UHP garnets and a resorbed diamond were recovered near a composite ca. 181 Ma basanite body. The diamonds and garnets are inferred to have formed in Carboniferous/Triassic subducted slabs related to the formation of the Devonian-to-Permian New England Fold Belt. They were exhumed in a two-stage process: first by partial tectonic exhumation of the slab which remained buried at intermediate depth, and then by igneous exhumation in shallow-sourced basaltic magmas. The fact that tectonic exhumation was only partial prior to igneous transport to the surface is thought to be the reason for the unusual preservation of macrocrystals of subduction (UHP) diamond.

Paleoproterozoic Subduction Diamonds in Minette Dyke in Canadian Shield

The oldest documented UHP diamond-bearing igneous rock to date is the 1.832 Ga Akluilak lamprophyre dyke (MacRae et al. 1995, 1996) which has intruded Archean rocks of the Gibson-MacQuoid Lake Greenstone Belt in the Baker Lake area of Nunavut, Canada. It overlaps in age with the 1.84–1.83 Ga Christopher Island Formation of the

Table 3 Comparison of properties of UHP and cratonic diamonds

	UHP diamonds	Cratonic diamonds
Major host rocks	Metamorphosed continental crustal rocks, including metasedimentary rocks	Mantle rocks of Ga-peridotitic (Pharz and Plherz) and eclogitic composition (E)
	Subordinate mafic and ultramafic rocks	
Geotectonic setting	Collision zones involving old continental crust	Stored in lithospheric roots of stable continental cratons
Mode of occurrence and grain size	Microdiamonds included in other minerals (garnet, kyanite, zircon, carbonates, etc.)	Micro- to macrodiamonds with P-type and E-type inclusion minerals
Ages of diamond formation	Mainly phanerozoic	Pharz-type Paleo- to Mesoarchean
	UHP diamonds included in Paleoproterozoic lamprophyre dyke	Plherz-type Proterozoic E-type Mesoarchean to Neoproterozoic fibrous diamonds are younger, but predate kimberlites
Ages of host rocks	Always older than age of UHP metamorphism	Equal to or predating diamond formation
Crystal forms	S-type star shaped	Octahedra, cubes, cubo-octahedra macles, resorbed forms and crystal fragments fibrous diamond and fibrous diamond coatings
	R-type rugged surfaces	
	T-type fine-grained crystals	
Carbon isotopes ($\delta^{13}C$ in ‰)	−3 to −30 ‰	P-type near −5 ‰
		E-type −38.5 to +2.7 ‰
Nitrogen	Up to 9,000 ppm	From traces to 3,500 ppm
		P-type average 200 ppm
		E-type average 300 ppm
Nitrogen aggregation	Poorly aggregated	Well aggregated, except for fibrous diamond

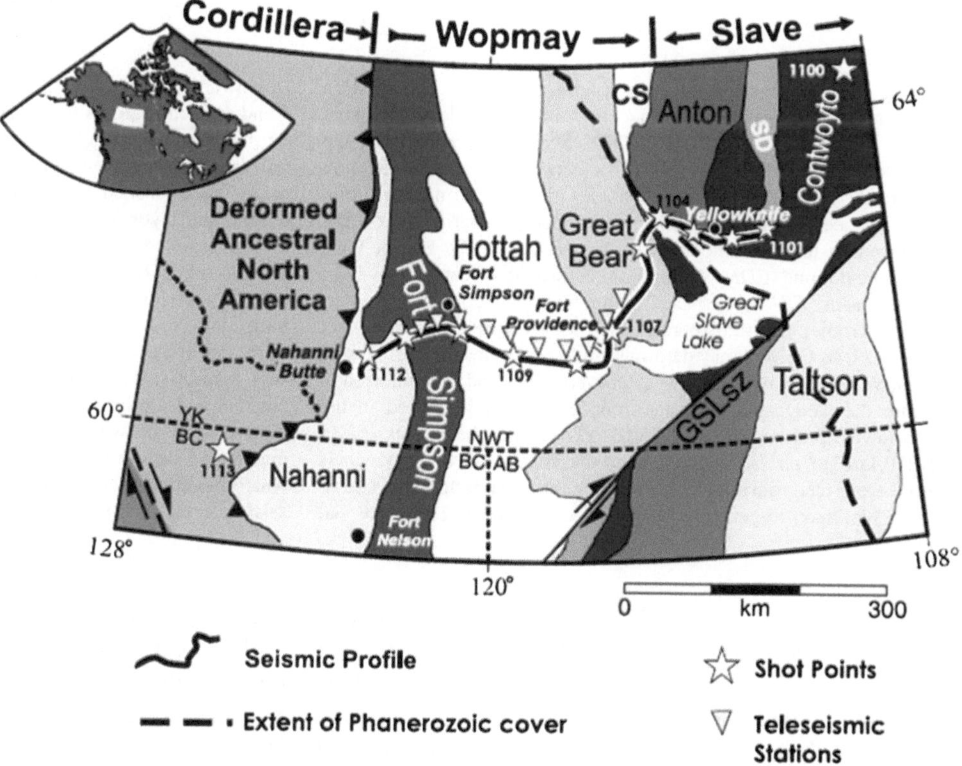

Fig. 7 Tectonic map showing location of seismic profile of LITHOPROBE Slave—Northern Cordillera lithospheric evolution (SNORCLE) transect 1 from southwestern part of the Archean slave province through the Paleoproterozoic Wopmay orogen to the Cordilleran margin, west of Paleoproterozoic Fort Simpson terrane (from Oueity and Clowes 2010). Black line shows continuous seismic reflection line. White stars are shot locations for the wide-angle experiment, and the inverted triangles are teleseismic station locations

Baker Lake Group, part of a large ultrapotassic rock province that straddles the Snowbird tectonic zone, bounding the Archean Rae and Hearne Provinces of the Canadian Shield (Peterson and LeCheminant 1996). The dyke is a metamorphosed minette which is unusually rich in microdiamonds, with > 1,700 diamonds reported from a 22 kg

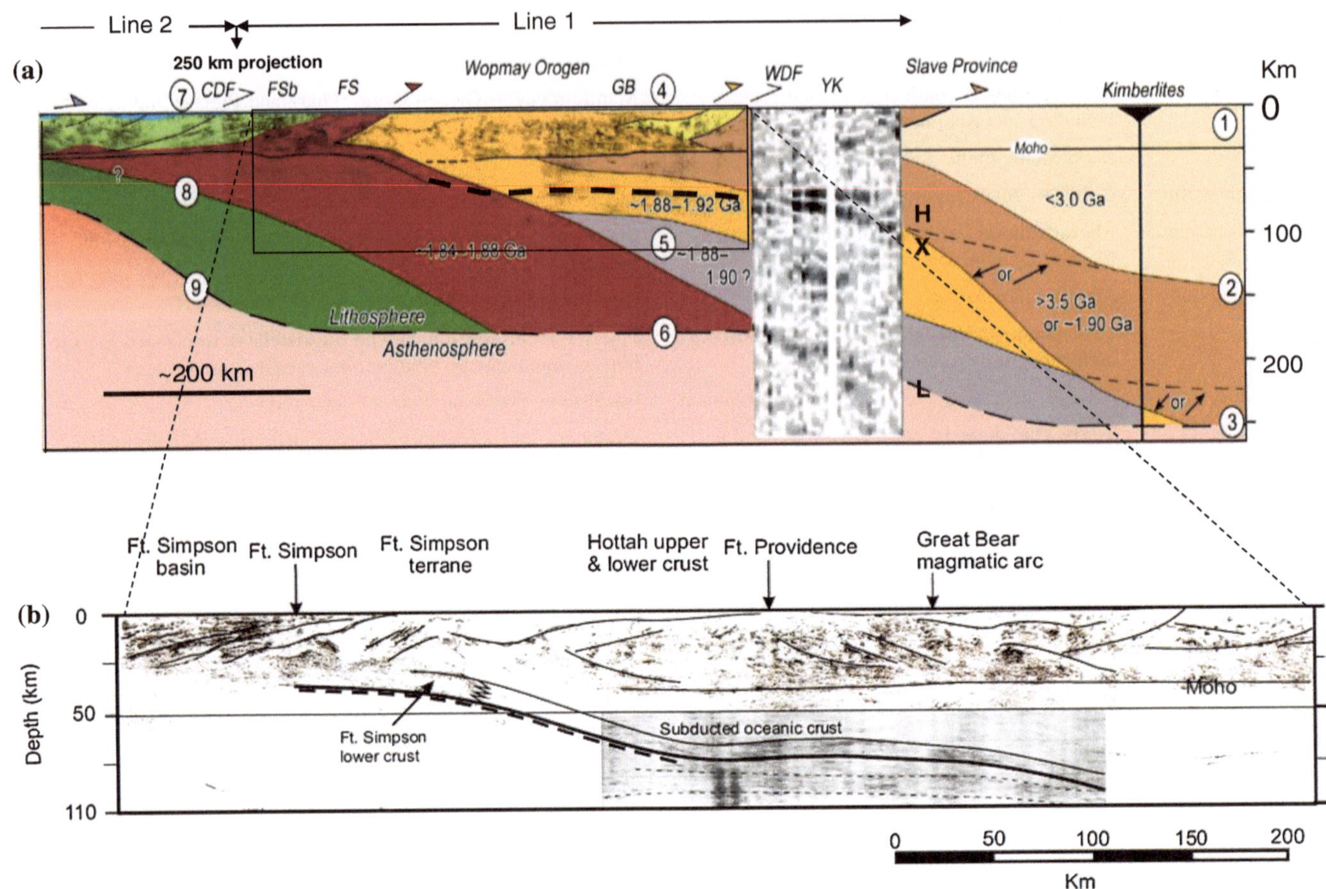

Fig. 8 **a** Lithosphere-scale interpretive cross section of the western and central Slave craton and adjacent younging Paleoproterozoic orogenic belts based on LITHOPROBE SNORCLE and follow-up geophysical and xenolith data (from Cook and Erdmer 2005). Section incorporates teleseismic data of Bostock (1997, 1998) near Yellowknife (YK) (see layers H, X, and L) and projects eastwards into central Slave Province to the diamondiferous kimberlites of the Lac de Gras field. Abbreviations: *CDF*—Cordilleran deformation front; *FSB*—Fort Simpson Basin; *FS*—ca. 1.71 Ga suture with Paleoproterozoic Fort Simpson terrane; *GB*—Paleoproterozoic Great Bear magmatic arc; *WDF*—ca. 1.88 Ga Wopmay deformation front. Lithosphere of central and eastern Slave Province (between *circled nos. 1* and *2*) is inferred to thin into the wedge-shaped intra-Archean indenter identified in the mid-to-lower crust under Yellowknife (YK). Lithosphere thickness of 230–250 km (to *circled no. 3*) under the central Slave province is inferred from electromagnetic, teleseismic, and xenolith data. Age of crust in Wopmay orogen (*circled no. 4*) ranges from 2.5 to 1.9 Ga. Downward younging of imbricate Paleoproterozoic slices (*circled nos. 5* and *6*) is inferred from upward projections of seismic reflectors into crust. Lithosphere thickness of ~ 180 km under the Wopmay orogen (*circled no. 6*) is from wide-angle seismic reflection data. Note that heavy black dashed line connecting reflections under Fort Simpson terrane and layer X under Yellowknife teleseismic array corresponds to reinterpreted section depicted in Fig. 8b. **b** Reinterpretation of seismic reflection profile of SNORCLE line 1 by Oueity and Clowes (2010) showing the shallow-dipping geometry of subducted Fort Simpson terrane lower crust and oceanic crust under Wopmay orogen. Thin and thick solid lines represent the top and bottom of the crustal slab, respectively. The flatness and slight arch shown in the section are not real and are due to acquisition geometry; the slab dips shallowly (Oueity and Clowes 2010). Thick broken line coincides with the mantle lid inferred from teleseismic data by Mercier et al. (2008) which is inferred to connect with layer X under the Yellowknife seismic array (see *broken line* in Fig. 8a)

sample (Kaminsky et al. 1998). The diamonds are unlike normal cratonic diamonds (Table 3). Most of them are intensely colored, exhibiting shades of green, yellow, burgundy, and brown. The $\delta^{13}C$ values range from −12.2 to −3.3 ‰ (Kaminsky et al. 1998; Chinn et al. 2000), and nitrogen contents are as high as 8,000 ppm. However, in spite of these high N contents, they are type Ib–IaA diamonds without any 4-nitrogen IaB centers, meaning that even at low temperatures of ~900 °C, these crystals are unlikely to have resided in the mantle for more than 1 million years (Chinn et al. 2000). The light carbon isotope ratios, high N content, and low N aggregation are suggestive of a collision-related eclogitic origin, and Cartigny et al. (2004) have interpreted them as UHP diamonds related to continental collisions during the formation of the Trans-Hudson Orogen about 1.8–2 billion years ago. If so, they would predate the oldest known surface occurrences of UHP diamonds by more than 1 billion years (Table 1).

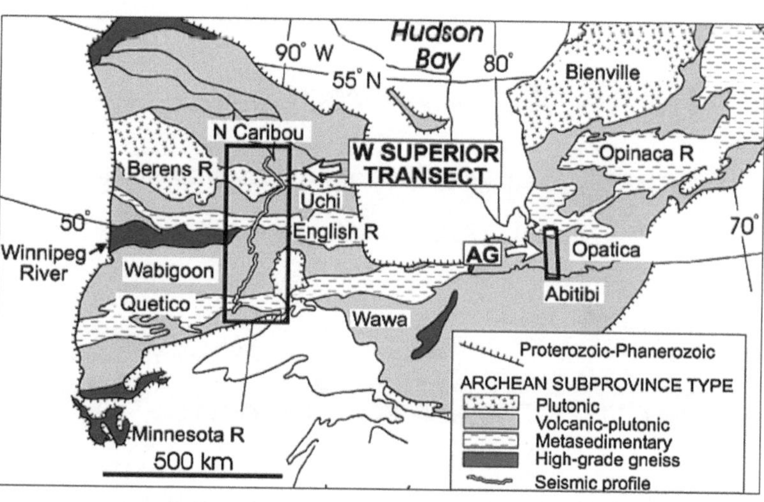

Fig. 9 Tectonic map of Superior Province showing locations of part of LITHOPROBE Abitibi–Grenville transect (*small rectangle* AG), crossing the boundary between the Abitibi and Opatica subprovinces, and the location of the Western Superior transect (*large rectangle* with seismic profile) (after White et al. 2003)

Seismic Images Suggesting Proterozoic and Archean Subduction

Slave Province and Its Western Margin: Evidence for Paleoproterozoic Underplating

Several seismic transects of the Canadian LITHOPROBE project provide compelling evidence that the constructive phase of the Precambrian lithosphere of the Canadian Shield involved horizontal plate interactions with subduction of intervening seafloor. Of particular, relevance for the topic of the present paper is SNORCLE Corridor 1 crossing the western margin of the Archean Slave Province and the Proterozoic terranes between the Slave Province and the eastern margin of the Northern Cordillera (Bostock 1998; Cook et al. 1999; Cook and Erdmer 2005) (Fig. 7). The seismic reflection profile shows evidence for east-dipping imbrication of the Moho under the western Slave Province as well as numerous east-dipping reflectors to a depth of ca. 110 km consistent with westward accretion of imbricate slices of Paleoproterozoic lithosphere to an Archean keel of the Slave Province (Fig. 8a). By reprocessing the original nearly vertical incidence (NVI) and refraction—wide-angle reflection (R/WAR) seismic data along part of Corridor 1, using 2-D and 3-D procedures made possible by crooked-line acquisition geometry, Oueity and Clowes (2010) were able to show that the eastward-dipping slab under the Paleoproterozoic Fort Simpson terrane, first identified by Cook et al. (1999), lies directly above the teleseismically identified upper mantle lid of Mercier et al. (2008) and extends nearly horizontally for a distance of about 300 km under Wopmay Orogen before dipping under the Slave Province (Fig. 8b). The eastern part of the slab coincides with a pair of reflectors previously considered as separate features (Fig. 8a) that project eastwards to lie above layer X identified under the Yellowknife seismic array by Bostock (1998). Layer X may thus be considered as the lid of Paleoproterozoic mantle dipping under the western part of the Archean Slave Province.

Seismic Evidence for Intra-Archean Subduction in Superior Province

Although intra-Archean subduction has been invoked in models for the formation of the oldest identified E-type diamonds on the Kaapvaal craton (e.g., Shirey and Richardson 2004; Gurney et al. 2005), such models are based on surface geological evidence and the petrologic and isotopic characteristic of the kimberlitic upper mantle sample, but they have not yet been confirmed by detailed seismic reflection sections. However, seismic evidence for Archean paleo-subduction has been identified in two LITHOPROBE transects through the Superior Province of the Canadian Shield. In the Grenville–Abitibi transect section (Fig. 9), Calvert and Ludden (1999) described a zone of north-dipping seismic reflectors beneath the Abitibi–Opatica subprovince boundary that extends to a depth of about 70 km and has been interpreted as the remnant of Archean subducted oceanic crust. In addition, the northward imbrication of the Abitibi crust is compatible with southward propagating tectonic accretion of this juvenile Archean subprovince. The Western Superior Province (Fig. 9) is the type area for one of the first Archean plate tectonics models proposed to explain its characteristic east–west striking subprovince structure by lateral tectonic accretion from south to north (Langford and Morin 1976). The Western Superior LITHOPROBE transect, designed to test this model, is discussed below.

Seismic images indeed corroborate the earlier accretion model by displaying two sutures consistent with northward subduction (Fig. 10b). The southern suture (S2) bounds a subhorizontal high-velocity slab (Fig. 10 a) inferred to be

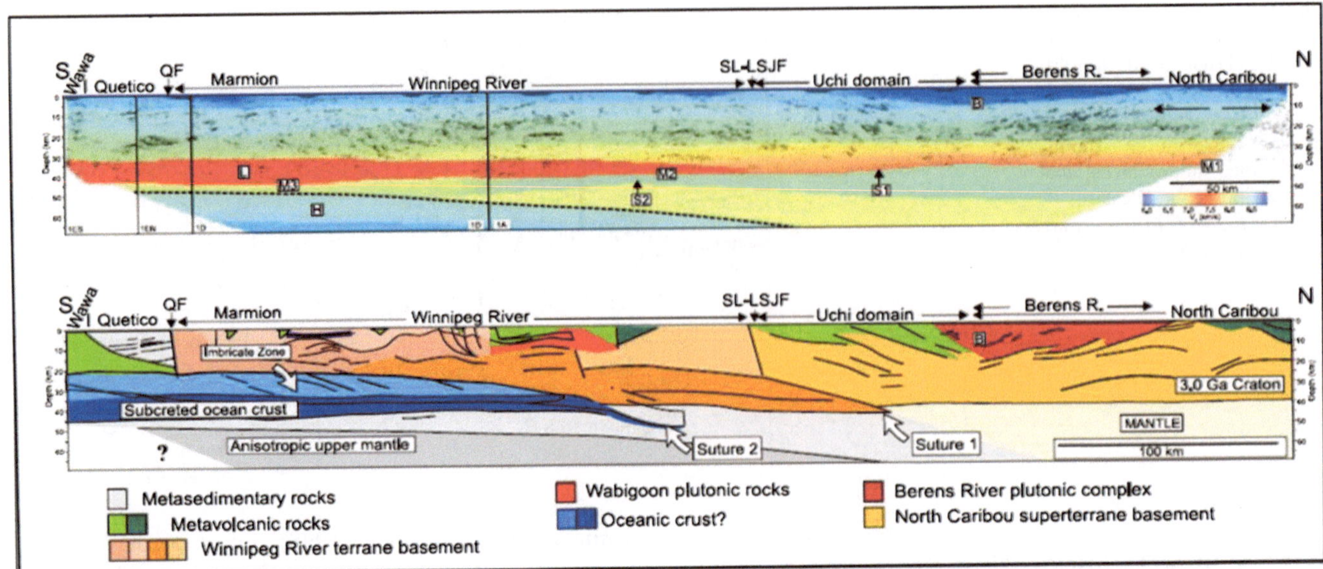

Fig. 10 **a** Seismic reflection data and seismic refraction velocity model of LITHOPROBE Western Superior transect (for location see Fig. 9) (from Percival et al. 2006). Note high-velocity lower crustal region (L) in southern part of section that in part corresponds to subcreted oceanic crust (see Fig. 10b). S1 and S2 correspond to two seismic sutures identified in Fig. 10b. **b** Tectonic interpretation of seismic data in Fig. 8a [from Percival et al. (2006) with interpretation after White et al. (2003)]. Note imbricate sutures S1 and S2 discussed in text. Note also imbricate nature in top of high-velocity slab interpreted as oceanic crust

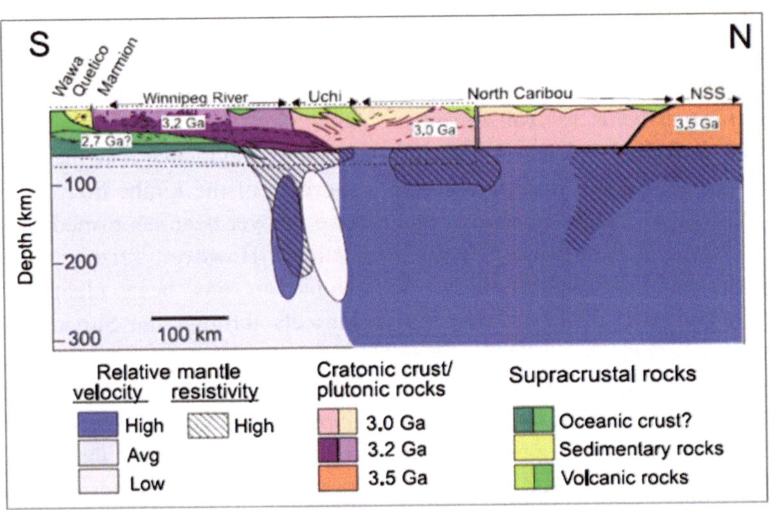

Fig. 11 Combined crustal and upper mantle section of Western Superior transect using data in Fig. 10 and seismic data of Kendall et al. (2002) and magnetotelluric data of Craven et al. (2001). Section is from Percival et al. (2006). Note that the area below the two crustal sutures is marked by steeply dipping to vertical velocity and resistivity anomalies, separating a generally higher-velocity upper mantle in the north from a lower-velocity upper mantle, in the south

composed of garnet amphibolite and to represent a remnant of oceanic crust subducted for more than 200 km beneath the ca. 3.2 Ga Winnipeg River terrane (White et al. 2003). The upper boundary of the slab is an imbricate zone consistent with near-horizontal, northward tectonic underplating by what is interpreted as Neoarchean oceanic crust. The northern suture (S1) is thought to mark the boundary of the Mesoarchean Winnipeg River terrane with the lower crust of the ca. 3 Ga Uchi domain. The sutures appear to mark a major boundary within the subcontinental upper mantle (Fig. 11), with a steeply north-dipping high-resistivity zone and a coincident subvertical high-velocity zone in the region between the two sutures separating a relatively high-velocity (and probably older) upper mantle domain, in the north, from an average velocity, but anisotropic mantle domain to the south (Percival et al. 2006). Both the seismic and electrical upper mantle structures beneath the crustal suture zones are interpreted steep slab-like features consistent with Archean subduction (Craven et al. 2001; Kendall et al. 2002).

Mesoarchean Metabasic Eclogites from the Fennoscandian Shield

Recently discovered 2.87 Ga metabasic eclogite bodies in the Keret tectonic nappe at the southwestern margin of the Kola craton, located just above its suture with the Belomorian accretionary orogen in the Fennoscandian Shield (Mints et al. 2010) (Fig. 2), negate the argument that an apparent lack of eclogites and other high-pressure rocks in the Archean surface record (e.g., Hamilton 1998) should be interpreted as evidence that subduction along plate margins did not occur in the Archean. The prograde part of the clockwise pressure–temperature–time (P–T–t) path inferred for the Salma eclogite (Fig. 3) follows a gradient of 12–15 °C/km to peak metamorphic conditions of ~700 °C at slightly above 14–15 Kbar (1.5 Gpa). The fact that this gradient is somewhat higher than those observed in younger accretionary prisms, such as the Franciscan (Fig. 3), may reflect warmer conditions in the Archean mantle (Mints et al. 2010), or it may be a consequence of the subduction of younger seafloor. Nevertheless, the gradient is consistent with subduction zone metamorphism, and mafic rocks subducted to greater depth along this gradient would still be considered UHP metamorphic rocks, that is, they may reach the coesite and diamond eclogite fields without melting (Fig. 3). Judging from seismic images, the tectonic slices within the suture region between the Belomorian accretionary orogen and the Kola craton, dip northeastward beneath the Kola craton (Mints et al. 2009), suggesting that the seafloor represented by the Salma eclogite followed a similar, northeastward subduction path. It is possible that some of the eclogite xenoliths occurring in diamondiferous kimberlites of the Arkhangelsk Province, located on the buried southeastern extension of the Kola–Kuloi craton, just east of the White Sea (Beard, et al. 2000), represent fragments of this subducted Archean seafloor, though this has to be tested by geochronological studies. Pressure–temperature conditions for eclogite xenoliths from the Grib pipe at the time of entrainment by the kimberlite have been estimated at ca. 950 °C and 40 kbar, corresponding to a depth of about 140–145 km (Skublov et al. 2011).

Discussion

As the concept of UHP metamorphism carries the important geodynamic implication of 'ultradeep' subduction of metasedimentary crustal rocks (Schreyer 1995), the UHP metamorphic literature is dominated by studies of continental crustal rocks and mechanisms of their return to the Earth's surface by tectonic exhumation as the result of continental or microcontinental collisions. Although continuous subduction of oceanic crust does not appear to provide the necessary tectonic setting for UHP metamorphism as defined above (Coleman and Wang 1995), the immense volumes of oceanic crust subducted prior to collisions are also expected to undergo UHP metamorphism after descending to the appropriate depths. All present evidence is consistent with a model that diamond- and coesite-bearing UHP melanges and diamond/coesite mantle eclogite xenoliths are complementary end products of subduction zone metamorphism. The former represent low-density crustal assemblages that are not permanently subductable and thus are tectonically exhumed soon after continental or microcontinental collision. The latter are parts of the oceanic slabs that subducted prior to collision and were accreted to the lithospheric roots of preexisting cratons or cratonic nuclei. Metamorphosed to eclogite, they reside long enough in the diamond stability field for carbon to crystallize as cratonic diamond with well-aggregated nitrogen before exhumation by younger igneous rocks, such as kimberlites or lamproites.

Tectonic accretion of permanently subducted oceanic UHP assemblages requires special tectonic circumstances, as demonstrated by the subduction of the Farallon plate under western North America. Here, it was the transition from steep subduction, with a long-lived arc-trench system and an accretionary prism with high-pressure Franciscan eclogites, to shallow-angle subduction which allowed the large-scale tectonic underplating of oceanic crust under western North America. Fragments of this crust had reached the UHP lawsonite–coesite eclogite facies by the time they were exhumed as xenoliths in the Navajo diatremes, more than 100 million years after tectonic exhumation of the Franciscan eclogites and several hundred kilometers inland from the original trench. The change in subduction geometry from steep to shallow-angle facilitated tectonic accretion of the eclogites under the North American lithosphere.

As seen by the LITHOPROBE SNORCLE corridor 1 seismic section, a shallow subduction geometry was responsible also for the tectonic underplating of the Slave Province during the Paleoproterozoic. The reflections of the subduction scar resulting from the 1.85 to 1.84 Ga accretion of the Fort Simpson terrane (Cook and Erdmer 2005) (Fig. 8) can be traced eastwards under Wopmay orogen toward the Slave Province, where a depth of ~100 km they match correlative reflections in teleseismic data of Bostock (1998) that continue to dip under the Slave Province. Eastward tectonic underplating of Paleoproterozoic oceanic lithosphere under the Slave craton is corroborated by the discovery of Paleoproterozoic diamond-bearing eclogite xenoliths with subduction signatures in the Jericho and Diavik kimberlites (Heaman et al. 2002, 2006; Schmidberger et al. 2005, 2007) and of Paleoproterozoic E-type diamonds from Diavik (Aulbach et al. 2009). As Archean E-type diamonds and eclogites have not been identified to date in Slave Province kimberlites (Helmstaedt 2009), it is possible that part of the Archean lithosphere under the Slave Province was tectonically eroded by the Paleoproterozoic underplating, and that lenses of eclogite within the underplated Paleoproterozoic oceanic lithosphere were the main contributor to the Slave Province E-type diamond budget. Furthermore, the presence of underplated Paleoproterozoic lithosphere under the Slave Province corroborates the models of Bostock (1997, 1998) and Oueity and Clowes (2010) that relatively shallow Paleoproterozoic subduction was the principal agent of stabilization and preservation for the Archean Slave craton root. It also allows, for the first time, to link a kimberlitic diamondiferous eclogitic upper mantle sample from the center of an Archean craton to a seismically identified Paleoproterozoic subduction zone along the craton margin. The timing of Paleoproterozoic subduction under the Slave Province more or less correlates with the minimum age of the Akluilak dyke subduction diamonds (as given by a dyke intrusion age of ca. 1,832 Ma), and the formation of which was inferred to be related to subduction during the Trans-Hudson Orogeny (Cartigny et al. 2004), both ages coinciding with tectonic assembly of Laurentia, the Paleoproterozoic nucleus of the North American continent (Hoffman 1989).

With Mesoarchean subduction zone metamorphism demonstrated in the geological surface record by the occurrence of prograde metamorphic eclogite lenses in the suture zone between the Belomorian accretionary orogen and the Kola craton (Mints et al. 2010), earlier assertions that the $\sim$2.9 Ga ages of the oldest known eclogitic diamonds and eclogite xenoliths with subduction signatures in South African and Siberian kimberlites (e.g., Pearson et al. 1995; Jacob 2004; Gurney et al. 2005, 2010; Shirey and Richardson 2004, 2011; Helmstaedt et al. 2010) (Table 2) represent a minimum age for the onset of continental root formation by tectonic subcretion and are corroborated. Does this mark the beginning of plate tectonics?

Ernst (2009) pointed out that the question of how and when plate-tectonic processes began on Earth is essentially one of terminology, for mantle circulation beneath a mobile outer rind has been in progress ever since the beginning of solidification of the latest magma ocean. The argument can thus be rephrased as to when lithosphere–asthenosphere interactions began to resemble more modern plate-margin processes. The appearance of eclogitic UHP metamorphic assemblages at ca. 3 Ga implies that >3 Ga microcontinental nuclei (small cratons) where coupled to lithospheric roots sufficiently thick and cool to reach into the diamond stability field, an inference supported by the occurrence of even older P-type diamonds (Table 2). The lowering of geothermal gradients required for UHP metamorphism of the underplated mafic rocks was likely achieved by large-scale tectonic imbrication of oceanic lithospheric slabs (Fig. 5), as envisaged earlier by Helmstaedt and Schulze (1988, 1989) and de Wit et al. (1992) and now inferred from LITHOPROBE reflection seismic profiles of Archean and Proterozoic cratons (e.g., Calvert and Ludden 1999; White et al. 2003; Cook and Erdmer 2005) (Figs. 8 and 10). Nevertheless, the survival of ca. 3 Ga UHP metamorphic assemblages resulting from early subduction processes should not be construed as the beginning of plate tectonics in the modern sense, including the onset of Wilson cycles as recently implied by Shirey and Richardson (2011). Continental plates with the requisite freeboard for developing passive margin sequences did not evolve before the end of the Archean (Hynes 2008), and it likely took longer before the entire Earth surface was covered by rigid plates as we see them throughout much of the Phanerozoic.

Considering the high mobility of the Archean lithosphere, it is remarkable that many of the ca. 3 Ga microcontinental nuclei retained remnants of their >3 Ga roots throughout their post-Mesoarchean tectonic history. Indeed, the worldwide harzburgitic P-type diamond budget is derived entirely from these roots (e.g., Gurney et al. 2005). Reasons commonly invoked for the survival of the roots are the refractory nature and the slightly lower density of the harzburgitic component with respect to the surrounding, more fertile lherzolitic mantle. However, until revealed recently through reflection seismic sections under the Archean Slave Province, the role of Paleoproterozoic tectonic underplating in protecting and stabilizing the earlier roots was not fully appreciated. Judging from the range of Proterozoic ages for E-type diamonds on several cratons, even within individual kimberlites, it appears that these cratons were repeatedly underplated by slices of Proterozoic oceanic lithosphere with lenses of eclogite.

References

Abbott D (1991) The case for accretion of the tectosphere by buoyant subduction. Geophys Res Lett 18:585–588

Anczkiewicz B, Platt JP, Thirwall MF, Wakabayashi J (2004) Franciscan subduction off to a slow start: evidence from high-precision Lu-Hf garnet ages on high-grade blocks. Earth Planet Sci Lett 225:147–161

Aulbach S, Stachel T, Creaser RA, Heaman LM, Shirey SB, Muehlenbachs K, Eichenberg D, Harris JW (2009) Sulphide survival and diamond genesis during formation and evolution of Archaean subcontinental lithosphere: a comparison between the slave and Kaapvaal cratons. Lithos 112S:747–775

Barron LM, Lishmund SR, Oakes GM, Barron BJ (1994) Subduction diamonds in New South Wales: implications for exploration in eastern Australia. Q Geol Surv NSW 46:4–7

Barron LM, Lishmund SR, Oakes GM, Barron BJ, Sutherland FL (1996) Subduction model for the origin of some diamonds in the phanerozoic of eastern New South Wales. Aust J Earth Sci 43:257–267

Barron BJ, Barron LM, Duncan G (2005) Eclogitic and ultrahigh-pressure crustal garnets and their relationships to phanerozoic subduction diamonds, Bingara area, New England fold belt, eastern Australia. Econ Geol 100:1565–1582

Barron LM, Barron BJ, Mernagh TP, Birch WD (2008) Ultrahigh pressure macro diamonds from Copeton (New South Wales, Australia), based on Raman spectroscopy of inclusions. Ore Geol Rev 34:76–86

Beard AD, Downes H, Hegner E, Sablukov SM (2000) Geochemistry and mineralogy of kimberlites from the Arkhangelsk Region, NW Russia: evidence for transitional kimberlite magma types. Lithos 51:47–73

Bostock MG (1997) Anisotropic upper-mantle stratigraphy and architecture of the Slave craton. Nature 390:392–395

Bostock MG (1998) Mantle stratigraphy and evolution of the Slave province. J Geophys Res B103:21183–21200

Calvert AJ, Ludden JN (1999) Archean continental assembly in the southeastern Superior province of Canada. Tectonics 18:412–429

Cartigny P, Chinn I, Viljoen F, Robinson DN (2004) Early Proterozoic ultrahigh pressure metamorphism: evidence from microdiamonds. Science 304:853–855

Chinn I, Kyser K, Viljoen F (2000) Microdiamonds from the thirsty lake (Akluilak) dykes, northwest territories, Canada. J Conf Abstr 5(2):307–308

Chopin C (1984) Coesite and pure pyrope in high-grade blueschists of the western Alps: a first record and some consequences. Contrib Mineral Petrol 86:107–118

Coleman RG, Wang X (1995) Ultrahigh pressure metamorphism. Cambridge University Press, Cambridge, p 510

Coney PJ, Reynolds SJ (1977) Cordilleran Benioff zones. Nature 270:403–405

Cook FA, Erdmer P (2005) An 1800 km cross section of the lithosphere through the northwestern North American plate: lessons from 4.0 billion years of Earth's history. Can J Earth Sci 42:1295–1311

Cook FA, van der Velden AJ, Hall KW, Roberts BJ (1999) Frozen subduction in Canada's northwest territories: lithoprobe deep lithospheric reflection profiling of the western Canadian shield. Tectonics 18:1–24

Craven JA, Kurtz RD, Boerner DE, Skulski T, Spratt J, Ferguson IJ, Wu X, Bailey RC (2001) Conductivity of western Superior province upper mantle in northwestern Ontario. Geological Survey of Canada, Current Research, Canada, p E6

Davies RM, O'Reilly SY, Griffin WL (2002) Multiple origins of alluvial diamonds from New South Wales, Australia. Econ Geol 97:109–123

de Wit MJ, de Ronde CEJ, Tredoux M, Roering C, Hart RJ, Armstrong RA, Green RWE, Peberdy E, Hart RA (1992) Formation of an Archean continent. Nature 357:553–562

Dickinson WR (1997) Tectonic implications of cenozoic volcanism in coastal California. Geol Soc Am Bull 109:936–954

Dickinson WR, Snyder WS (1978) Plate tectonics of the Laramide orogeny. Geol Soc Am Mem 151:355–366

Ernst WG (2006) Preservation/exhumation of ultrahigh-pressure subduction complexes. Lithos 92:321–335

Ernst WG (2009) Archean plate tectonics, rise of proterozoic supercontinentality and onset of regional, episodic stagnant-lid behavior. Gondwana Res 15:243–253

Gurney JJ, Helmstaedt HH, Le Roex AP, Nowicki TE, Richardson SH, Westerlund KJ (2005) Diamonds: crustal distribution and formation processes in time and space and an integrated deposit model. In: Economic geology 100th anniversary volume, pp 143–177

Gurney JJ, Helmstaedt H, Richardson SH, Shirey SB (2010) Diamonds through time. Econ Geol 105:689–712

Hamilton WB (1998) Archean tectonics and magmatism. Int Geol Rev 40:1–39

Heaman LM, Creaser RA, Cookenboo HO (2002) Extreme enrichment of high field strength elements in Jericho eclogite xenoliths: a cryptic record of Paleoproterozoic subduction, partial melting, and metasomatism beneath the Slave craton, Canada. Geology 30:507–510

Heaman LM, Creaser RA, Cookenboo HO, Chacko T (2006) Multi-stage modification of the northern Slave mantle lithosphere: evidence from zircon- and diamond-bearing eclogite xenoliths entrained in Jericho kimberlite, Canada. J Petrol 47:821–858

Helmstaedt H (2009) Crust-mantle coupling revisited: the Archean Slave craton, NWT, Canada. Lithos 112S:1055–1068

Helmstaedt H, Doig R (1975) Eclogite nodules from kimberlite pipes of the Colorado Plateau - samples of subducted Franciscan-type oceanic lithosphere. Phys Chem Earth 9:95–111

Helmstaedt HH, Schulze DJ (1988) Eclogite-facies ultramafic xenoliths from Colorado Plateau diatreme breccias: comparison with eclogites in crustal environments, evaluation of the subduction hypothesis, and implications for eclogite xenoliths from diamondiferous kimberlites. In: Smith DC (ed) Eclogites and eclogite-facies rocks. Elsevier, New York, pp 387–450

Helmstaedt HH, Schulze DJ (1989) Southern African kimberlites and their mantle sample: implications for Archean tectonics and lithosphere evolution. In: Ross J (ed) Kimberlites and related rocks. Proceedings of 4th kimberlite Conference, Perth 1986, vol. 1, Geological Society of Australia, special publication 14, pp 358–368

Helmstaedt HH, Schulze DJ (1991) Early to mid-Tertiary inverted metamorphic gradient under the Colorado Plateau: evidence from eclogite xenoliths in ultramafic microbreccias, Navajo volcanic field. J Geophys Res 96:13225–13235

Helmstaedt HH, Gurney JJ, Richardson SH (2010) Ages of cratonic diamond and lithosphere evolution: constraints on Precambrian tectonics and diamond exploration. Can Mineral 48:1385–1408

Hoffman PF (1989) Precambrian geology and tectonic history of North America. In: The Geology of North America—An Overview. Geological Society of America, Boulder, Colorado, pp 447–512

Hyndman RD (1972) Plate motions relative to the deep mantle and the development of subduction zones. Nature 238:264–265

Hynes A (2008) Effects of a warmer mantle on the characteristics of Archean passive margins. Geol Soc Am Spec Pap 440:149–156

Jacob DE (2004) Nature and origin of eclogite xenoliths from kimberlites. Lithos 77:295–316

Janse AJA (1994) Is Clifford's rule still valid? Affirmative examples from around the world. In: Meyer HOA, Leonardos O (eds) Diamonds: characterization, genesis and exploration. Proceedings of the 5th international kimberlite conference 1991, Araxa, Brazil, vol 2, CPRM special publication 1B, pp 215–235

Kaminsky FV, Sablukov SM, Sablukova LI, Shpanov VE, Zhuravlev DZ (1998) Diamondiferous minette dykes from the Parker Lake area, N.W.T., Canada. In: 7th international kimberlite conference, Cape Town, S.A., Extended Abstracts, pp 392–394

Kendall JM, Sol S, Thomson CJ, White DJ, Asudeh L, Snell CS, Sutherland FH (2002) Seismic heterogeneity and anisotropy in the western Superior province, Canada: insights into the evolution of an Archean craton. In: Fowler CMR, Ebinger CJ, Hawkesworth CJ (eds) The early Earth: physical, chemical and biological development. Geological Society of London, special publications, vol 199, pp 27–44

Kylander-Clark ARC, Hacker BR, Mattinson CG (2012) Size and exhumation rate of ultrahigh-pressure terranes linked to orogenic stage. Earth Planet Sci Lett 321–322:115–120

Langford EF, Morin JA (1976) The development of the Superior province of northwestern Ontario by merging island arcs. Am J Sci 276:1023–1034

Liou JG, Ernst WG, Zhang RY, Tsujimori T, Jahn BM (2009) Ultrahigh-pressure minerals and metamorphic terranes - The view from China. J Asian Earth Sci 35:199–231

Lipman PW, Protska HJ, Christiansen RL (1972) Cenozoic volcanism and plate-tectonic evolution of the western United States, I. Early and middle Cenozoic. Philos Trans R Soc Ser A 271:217–248

Liu L, Spasojevic S, Gurnis M (2008) Reconstructing Farallon plate subduction beneath North America back to the Late Cretaceous. Science 322:934–938

Macrae ND, Armitage AE, Jones AL (1995) A diamondiferous lamprophyre dike, Gibson Lake area, northwest territories. Int Geol Rev 37:212–229

Macrae ND, Armitage AE, Miller AR, Roddick JC, Jones AL, Mudry MP (1996) The diamondiferous Akluilak lamprophyre dyke, Gibson Lake area, N.W.T. Geological Survey of Canada, Open File 3228, pp 101–107

Maruyama S, Liou JG, Terabayashi M (1996) Blueschists and eclogites of the world and their exhumation. Int Geol Rev 38:485–594

Mercier J-P, Bostock MG, Audet P, Gaherty JB, Garnero EJ, Revenaugh J (2008) The teleseismic signature of fossil subduction: northwestern Canada. J Geophys Res 113:B04308. doi: 04310.01029/02007JB005127

Mints M, Suleimanov A, Zamozhniaya N, Stupak V (2009) A three-dimensional model of the Early Precambrian crust under the southeastern Fennoscandian Shield: Karelia craton and Belomorian tectonic province. Tectonophysics 472:323–339

Mints MV, Belousova EA, Konilov AN, Natapov LM, Shchipansky AA, Griffin WL, O'Reilly SY, Dokukina KA, Kaulina TV (2010) Mesoarchean subduction processes: 2.87 Ga eclogites from the Kola Peninsula, Russia. Geology 38:739–742

Nixon PH (1995) The morphology and nature of primary diamondiferous occurrences. J Geochem Explor 53:41–71

Ogasawara Y (2005) Microdiamonds in ultrahigh-pressure metamorphic rocks. Elements 1:91–96

Okamoto K, Maruyama S (1999) The high-pressure synthesis of lawsonite in the MORB + H_2O system. Am Mineral 84:362–373

Oueity J, Clowes RM (2010) Paleoproterozoic subduction in northwestern Canada from near-vertical and wide-angle reflection data. Can J Earth Sci 47:35–52

Pearson DG, Snyder GA, Shirey SB, Taylor LA, Carlson RW, Sobolev NV (1995) Archean Re-Os age for Siberian eclogites and constraints on Archean tectonics. Nature (London) 374:711–713

Percival JA, Sanborn-Barrie M, Skulski T, Stott GM, Helmstaedt H, White DJ (2006) Tectonic evolution of the western Superior province from NATMAP and LITHOPROBE studies. Can J Earth Sci 43:1085–1117

Peterson TD, Lecheminant AN (1996) Ultrapotassic rocks of the Dubawnt Supergroup, District of Keewatin, N.W.T. In: Lecheminant AN, Richardson DG, Dilabio RNW, Richardson KA (eds) Searching for diamonds in Canada. Geological Survey of Canada, Open File 3228, pp 97–100

Ringwood AE (1991) Phase transformations and their bearing on the constitution and dynamics of the mantle. Geochimica Cosmochimica Acta 55:2083–2110

Roden MF, Smith D, Murthy VR (1990) Geochemical constraints on lithosphere composition and evolution beneath the Colorado Plateau. J Geophys Res 95:2811–2831

Ruiz L, McCandless TE, Helmstaedt HH (1998) Re-Os model ages for Eclogite Xenoliths from the colorado plateau, USA. In: Gurney JJ, Gurney JL, Pascoe MD, Richardson SH (eds) Proceedings of the 7th international kimberlite conference, vol 2, P.H. Nixon Volume, Cape Town, pp 736–740

Schmidberger SS, Heaman LM, Simonetti A, Creaser RA, Cookenboo HO (2005) Formation of Paleoproterozoic eclogitic mantle, Slave province (Canada): insights from in situ Hf and U-Pb isotope analyses of mantle zircons. Earth Planet Sci Lett 240:621–633

Schmidberger SS, Simonetti A, Heaman LM, Creaser RA, Whiteford S (2007) Lu-Hf, in situ Sr and Pb isotope and trace element systematics for mantle eclogites from the Diavik diamond mine: evidence for Paleoproterozoic subduction beneath the Slave craton, Canada. Earth Planet Sci Lett 254:55–68

Schreyer W (1995) Ultradeep metamorphic rocks: a retrospective viewpoint. J Geophys Res 100:8353–8366

Schulze DJ, Helmstaedt H (1988) Coesite-sanidine eclogites from kimberlite: products of mantle fractionation or subduction. J Geol 96:435–443

Schulze DJ, Valley JW, Spicuzza MJ (2000) Coesite eclogites from the Roberts Victor kimberlite, South Africa. Lithos 54:23–32

Schulze DJ, Harte B, Valley JW, Brenan JM, Channer DMD (2003) Extreme crustal oxygen isotope signature preserved in coesite in diamond. Nature 423:68–70

Schulze DJ, Harte B, Valley JW, Channer DMD (2004) Evidence of subduction and crust-mantle mixing from a single diamond. Lithos 77:349–358

Shirey SB, Richardson SH (2004) Integrated models of diamond formation and craton evolution. Lithos 77:923–944

Shirey SB, Richardson SH (2011) Start of the Wilson cycle at 3 Ga shown by diamonds from subcontinental mantle. Science 333:434–436

Skublov SG, Shchulina EV, Guseva NS, Malkovets VG, Golovin NN (2011) Geochemical characteristics of zircons from xenoliths in the V. Grib kimberlite pipe, Archangelsk diamondiferous province. Geochem Int 49:415–421

Smith DC (1984) Coesite in clinopyroxene in the Caledonides and its implications for geodynamics. Nature 310:641–644

Smith D, Connelly JN, Manser K, Moser DE, Housh TB, McDowell FW, Mack LE (2004) Evolution of Navajo eclogites and hydration of the mantle wedge below the Colorado Plateau, southwestern United States. Geochem Geophys Geosyst 5:GC000675

Smyth JR, Hatton CJ (1977) A coesite-sanidine grospydite from the Roberts Victor kimberlite. Earth Planet Sci Lett 34:284–290

Sobolev NV, Shatsky VS (1990) Diamond inclusions in garnet from metamorphic rocks. Nature 343:742–746

Stachel T, Harris JW (2008) The origin of cratonic diamonds—constraints from mineral inclusions. Ore Geol Rev 34:5–32

Stern RJ (2007) When and how did plate tectonics begin? Theoretical and empirical considerations. Chin Sci Bull 52:578–591

Taylor LA, Neal CR (1989) Eclogites with oceanic crustal and mantle signatures from the Bellsbank kimberlite, South Africa, Part I: mineralogy, petrography, and whole rock chemistry. J Geol 97:551–567

Tsujimori T, Matsumoto K, Wakabayashi J, Liou JG (2006) Franciscan eclogite revisited: revaluation of the P-T evolution of tectonic blocks from Tiburon Peninsula, California, U.S.A. Mineral Petrol 88:243–267

Usui T, Nakamura E, Kobayashi K, Maruyama S, Helmstaedt H (2003) Fate of the subducted Farallon plate inferred from eclogite xenoliths in the Colorado Plateau. Geology 31:589–592

Usui T, Nakamura E, Helmstaedt H (2006) Petrology and geochemistry of eclogite xenoliths from the Colorado Plateau: implications for the evolution of subducted crusts. J Petrol 47:929–964

White DJ, Musacchio G, Helmstaedt HH, Harrap RM, Thurston PC, van der Velden AJ, Hall K (2003) Images of a lower-crustal oceanic slab: direct evidence for tectonic accretion in the Archean western Superior province. Geology 31:997–1000

Deep-Seated Xenoliths from the Brown Breccia of the Udachnaya Pipe, Siberia

I. V. Ashchepkov, T. Ntaflos, S. S. Kuligin, E. V. Malygina, A. M. Agashev, A. M. Logvinova, S. I. Mityukhin, N. V. Alymova, N. V. Vladykin, S. V. Palessky, and O. S. Khmelnikova

Abstract

A detailed study in thin section of a set of 330 fresh small xenoliths from the brown breccia within a deep level of the Udachnaya kimberlite pipe allowed us reconstruction of the mantle section from 8.0 to 1.0 GPa, which represents 6 main capturing intervals. Interaction of the protokimberlite melt with the deeper level was responsible for peridotite heating, shearing, metasomatism, and remelting of eclogites. The most compositionally and thermally diverse mantle horizon is detected from 7.0 to 5.0 GPa. The pyroxenite-rich levels are found in the lithosphere base, near 4.0 GPa and in the upper part at 30–2.0 GPa.

High precision data for Cr, Al, Ti, Ca, and Ni in olivines from 64 xenoliths, obtained using Cameca SX 100 electron microprobe, reveal a high-Ca group that corresponds to porphyroclastic peridotites and a high-Ca–Ti group that corresponds to sheared peridotites. The low Ca but high Ni values are typical for coarse garnet peridotites, while the lower Ni values are derived from the Sp varieties. Phlogopite metasomatism throughout the whole section of the mantle column is due to interaction with deep-seated melts resembling Group II kimberlites. Larger phlogopite grains relate to ancient metasomatism. The deep-seated phlogopite breccias were formed at a deep level of the mantle column due to the H_2O-rich fluid interaction with mantle peridotites and crystallized in the 2.5–1.0 GPa interval.

Keywords

Mantle • Olivine • High precision geochemistry • Thermobarometry • Peridotites • Eclogites • Layering • Trace elements

I. V. Ashchepkov (✉) · S. S. Kuligin · E. V. Malygina
A. M. Agashev · A. M. Logvinova · S. V. Palessky
O. S. Khmelnikova
Institute of Geology and Mineralogy, SD RAS Novosibirsk, 630090, Russia
e-mail: igora57@mail.ru

T. Ntaflos
Department of Lithospheric Research, University of Vienna, A-1090, 366403, Vienna, Austria

S. I. Mityukhin
The Stock Company "ALROSA", Chernyshevsky Street, 7, Mirny, 678170, Russian Federation

N. V. Alymova · N. V. Vladykin
Institute of Geochemistry SD RAS, Favorsky Street 1a, Irkutsk, Russia

Introduction

Deep-seated xenoliths from Udachnaya were described in a number of works (Boyd et al. 1997; Ionov et al. 2010; Kuligin 1997; Malygina 2000; Pokhilenko et al. 1993, 1999; Pokhilenko 2006; Ashchepkov et al. 2010a, 2012b; Shimizu et al. 1997; Snyder et al. 1997; Sobolev et al. 1973; Sobolev et al. 1984, 1994; Solov'eva et al. 2008; Agashev et al. 2010; Yaxley et al. 2012). Peridotites were divided generally into three petrographic types: granular (coarse and medium grained) sheared, and intermediate or porphyroclastic (Ionov et al. 2010; Yaxley et al. 2012). In addition,

D. G. Pearson et al. (eds.), *Proceedings of 10th International Kimberlite Conference*, Volume 1, Special Issue of the Journal of the Geological Society of India, DOI: 10.1007/978-81-322-1170-9_5, © Geological Society of India 2013

Fig. 1 Photomicrographs of the deep-seated xenoliths from Udachnaya pipe. *1* Wehrlite with green garnet, *2* sheared peridotite with zoned garnet, *3* sheared peridotite with two directions of shearing, *4* Phl breccia, *5* Garnets with Phl rims, *6* granular garnet peridotite, *7* garnet peridotite with veined garnet, *8* granular garnet; harzburgite, *9* fine-grained garnet peridotite

eclogites (Jacob et al. 1994; Sobolev et al. 1994; Misra et al. 2004; Shatsky et al. 2008) and pyroxenites (Kuligin et al. 2003; Pokhilenko et al. 1999) were analyzed. The collection of extremely fresh small xenoliths from the brown breccias located in deepest level (550 m) of the Udachnaya quarry containing complete set of petrographic varieties was analyzed by EPMA (110 in thin section) and LAM-ICP-MS (>65) for detail reconstructions of mantle sections. Several intergranular assemblages possibly relate to intermediate stops in mantle magma chambers during upwelling of the material to the surface.

Petrographic Description

The commonest sheared peridotites, ellipsoidal in shape, contain large red or orange garnets, sometimes zoned with cores being pink or red and surrounded by 1–2-mm rims of orange garnets (Figs. 1 and 2). Fine-grained mosaic olivine aggregates surround elongated or pen-like grains of Cpx and Opx. The detail of the thin sections shows that there were two or more stages of deformation because several populations of already deformed aggregates are cut by the more fine-grained material tracing another direction of shearing (Figs. 1, 2 and 3). Sometimes in the intergranular space, it is possible to determine grains of phlogopites, which appear to be equilibrated with the olivines. (Figs. 1, 2, 3, 4 and 5). They are common in the rims on garnets which appear equilibrated and have a mantle origin in kelyphitic aggregates.

Porphyroclastic varieties are formed by coarse-grained Ol and partly Opx grains, and mosaic Ol intergranular aggregates in different proportions. One is represented by the common recrystallized fine-grained olivine aggregates and the second one by veins or veinlets of Cpx, Opx, and sometimes garnets showing signs of fertilization (Figs. 1, 2,

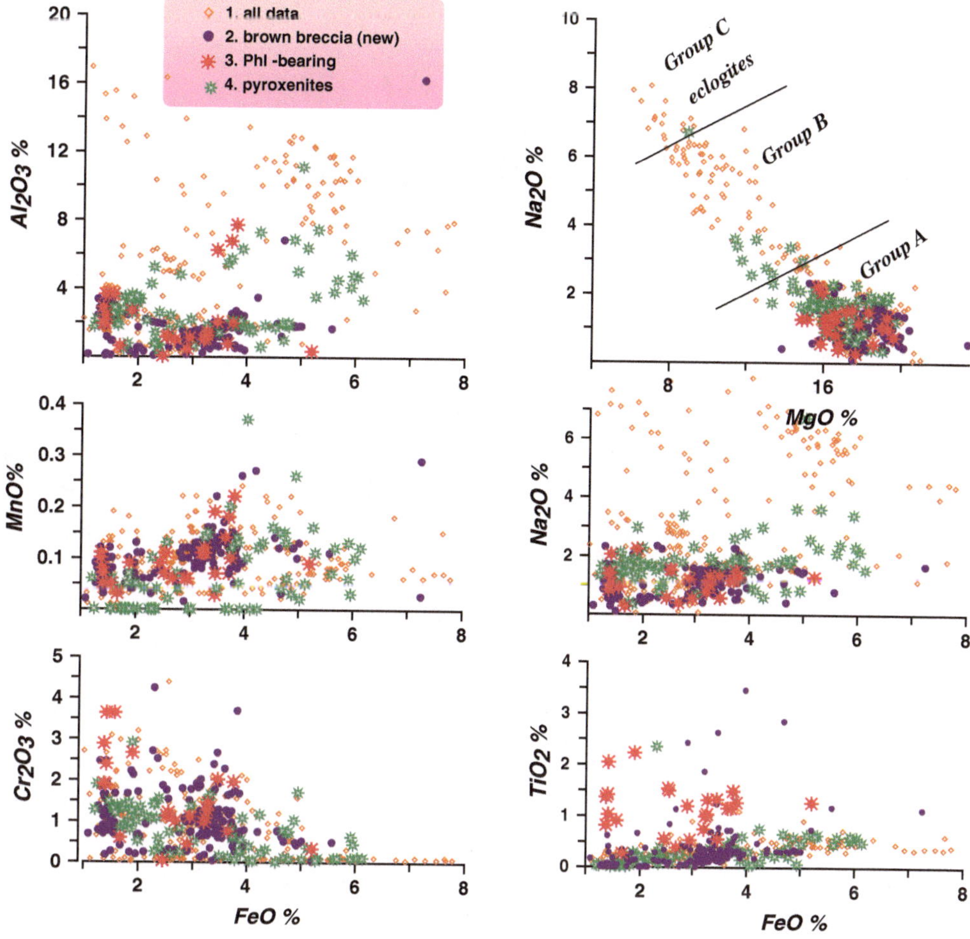

Fig. 2 Variations of major element compositions in clinopyroxenes from xenoliths in the brown kimberlite breccia in comparison with other published associations from Udachnaya pipe. *1* previous data, *2* new data, *3* Phl-bearing associations, *4* pyroxenites

3, 4, 5, 6 and 7). Rare garnet wehrlites with green garnets and harzburgites also reveal porphyroclastic structures and several generations of minerals.

Common granular garnet peridotites also vary in structure. Some show large grains of Gar and Opx and finer Cpx and Ol, but in another, coarse-grained olivines are surrounded by subordinate grains of pyroxenes and garnets. The nearly equal grain structures also occur in the rare garnet peridotites. Spinel harzburgites that are extremely depleted more then peridotites from recent continental arcs (Ntaflos et al. 2007) in pyroxenes are mostly very coarse grained. The fine- or medium-grained peridotites contain more pyroxenes. Rims of metasomatic richterite are common on orthopyroxenes in the harzburgites. Symplectites of Cpx, Opx, Sp, and very often Phl are common in low-temperature spinel peridotites.

The largest grain size is shown by dunites, which are of two types: garnet- or chromite-bearing that occasionally contain pale Opx (Pokhilenko et al. 1993). The Fe-rich varieties with olive colored olivine and intergranular amphiboles and rare phlogopites (Solovieva et al. 2008) also occur. Veins with pargasite, ilmenite Ti–spinels, and phlogopite are found in spinel- and coarse-grained garnet peridotites. The amount of Phl-bearing rocks is about 1/3 or more among these xenoliths which is higher than 1/20 reported by Boyd et al. (1997). Most phlogopites are intergranular, occur as thin films, and are associated with secondary Cpx and sometimes Sp. A rare and unusual inclusion in the deep-seated breccia is composed of fine-grained material of the debris or microxenoliths, essentially olivine aggregates, cemented by phlogopites and clinopyroxene, amphibole (richterites) chromite, and magmatic olivine aggregates.

The secondary minerals in the intergranular space are alkali feldspar, Na-Cpx, sulfides, Sr-carbonates, etc. The breccias contain olivine aggregates typical of sheared peridotites or coarser grains. While spinel is only occasionally present, all breccias are lacking of pyroxenes and garnets. Chromites as well as olivines show two morphological varieties. The first has xenocrystic shapes and zonation visible in thin section in color and the second is elongate or idiomorphic cut grains.

Another type of breccias contains abundant Phl, picroilmenites (but no chromites), and various Cpx grains ranging from xenocrystic Cr-diopsides to dominated augites in the matrix.

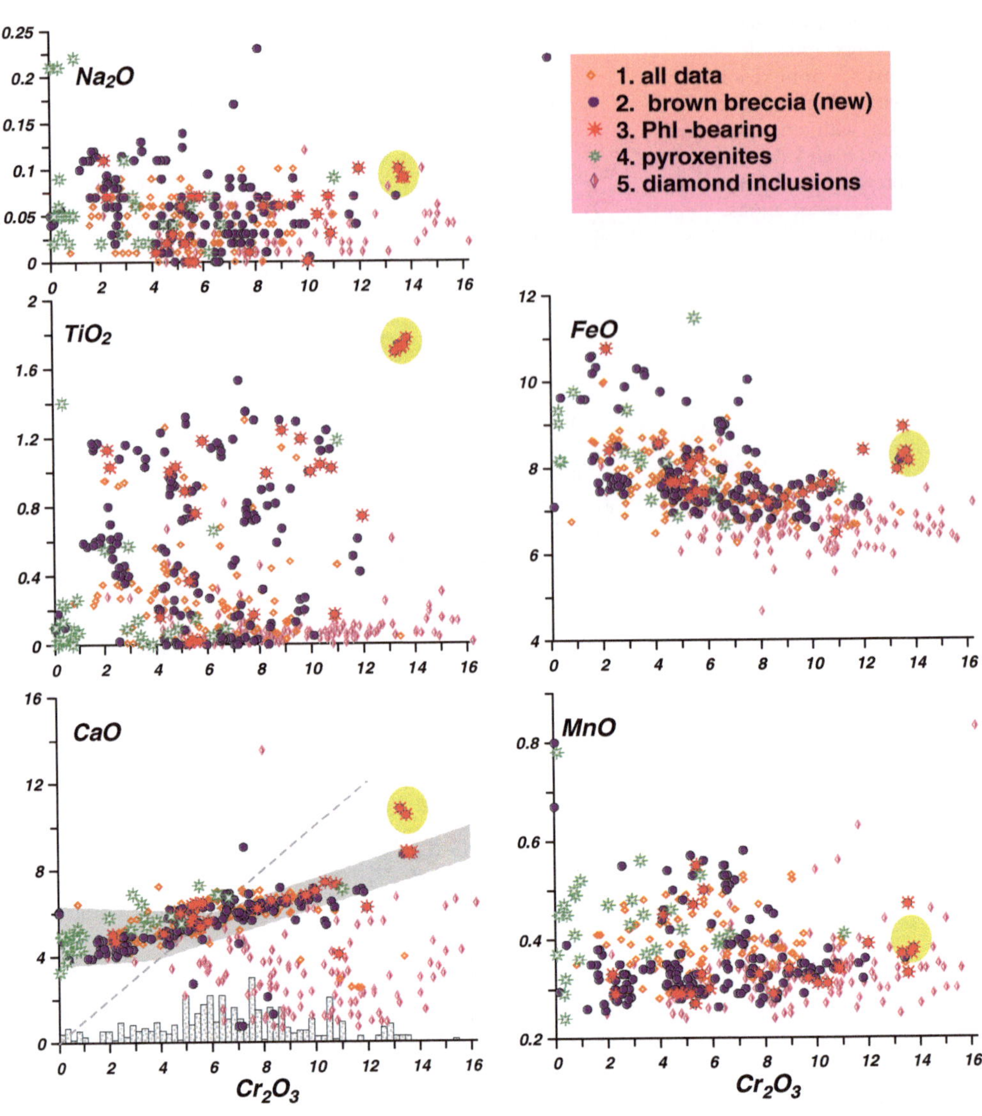

Fig. 3 Variations of major element compositions of garnets from xenoliths in the brown kimberlites breccia in comparison with other published associations from Udachnaya pipe. Symbols are the same as in Fig. 2 and, *5* diamond inclusions

Analytic Methods

The sample collection was analyzed by EPMA and ICP-MS. Several analyses were made using a Jeol Superprobe in IGM SD RAS. For comparison, one mount with 60 associations was analyzed using Camebax Micro machine and methods established by Lavrentiev and colleagues (1994) in UIGM SB RASc. The detection limit is about 2 % for most elements for the measurements made on the Jeol Superprobe. A slightly higher detection limit (~3 %) was found for the Camebax Micro. Most samples were studied in thin section in Vienna University Department of Lithospheric Research on Cameca SX 100. All analyses were made against mineral standards with wavelength-dispersive spectrometers; acceleration voltage and beam current were 15 kV and 20 nA, respectively, and standard correction procedures were applied.

For the high precision analyses of the olivine the acceleration voltage of 20 kV, a slightly defocused beam current of 60 nA was applied. In order to increase the precision and reduce the effect of noisy background on very low elemental concentrations, a 120-s counting time on peak position and on each, low and high background positions, was used. Precision varies from 10 to 25 ppm for several trace elements like Ni, Cr, Al, Mn, Ca, and Zn. Abundances of approximately 30 trace elements were determined in situ in selected minerals by laser ablation inductively coupled plasma mass spectrometry (LA-ICP-MS) at the Analytic Centre of IGM on a Finnigan Element ICP-MS equipped with a UV laser probe using the international reference standards NIST 612 SRM and NIST 614

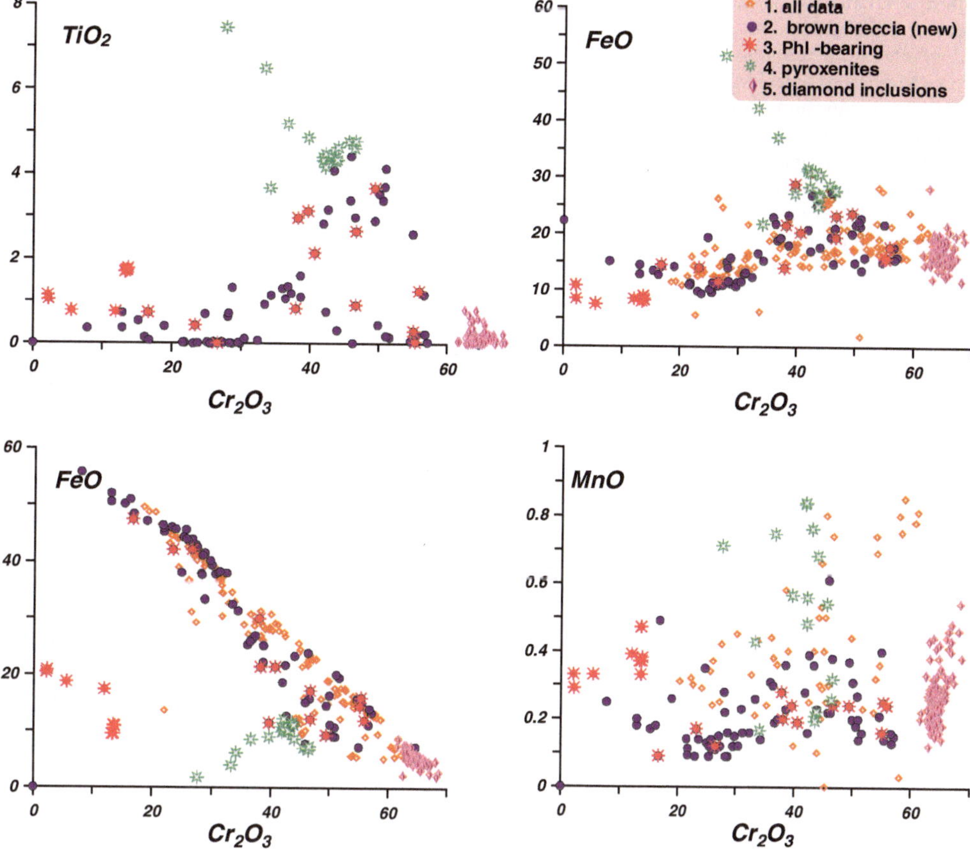

Fig. 4 Variations of major element compositions of chromites from xenoliths in the brown kimberlites breccia in comparison with other published associations from Udachnaya pipe. Symbols are the same as in Fig. 3

SRM and internal standard Ca for silicates and Ti for ilmenites.

Variations of the Major Components

Clinopyroxenes. Major element variations of clinopyroxenes from the studied collection are shown on Fig. 2 in comparison with the previous published data. The peridotites show strong variation in Fe (1–5 wt%) and other components. The low Fe type relates to coarse spinel harzburgites and rare porphyroclastic peridotites with large olivine grains. Variations of the other components are also rather wide: 1–4 wt% for Al_2O_3 and 0.5–2.0 wt% for Cr_2O_3 and Na_2O. Common garnet lherzolites (Ionov et al. 2010) show 1.5–2.5 wt% FeO and smaller fluctuations in Na, Al, and Ti. An increase in these components was found for Cpx from sheared peridotites (2.7–4.0 wt% FeO) (Agashev et al. 2010). Enrichment in TiO_2 was found for most of the Phl-bearing associations. Eclogites mostly refer to groups B–C (Dawson 1980) and are characterized by broad variations in FeO and high Al, Na contents.

Garnets. Garnets from studied peridotites mostly fall within the lherzolite field (Sobolev et al. 1973) and show wider variations in Cr_2O_3 than previously determined (Malygina 2000; Ionov et al. 2010; Yaxley et al. 2012) to 14 wt% for green garnets from porphyroclastic harzburgites (Ionov et al. 2010) and wehrlites (Fig. 3). The later are enriched in TiO_2 and CaO resembling pyroxenitic garnets (Sobolev et al. 1973). Rare giant-grained dunites and harzburgites have sub-calcic garnet compositions (Pokhilenko et al. 1993; Burgess and Harte 2004) close to those from diamond inclusions (Logvinova et al. 2005). Garnets from the Phl-bearing associations also enriched in TiO_2.

Chromites. This mineral shows variations from 10 to 60 wt% Cr_2O_3. The varieties having >40 wt% Cr_2O_3 are enriched in TiO_2 and FeO. This is more typical for chromites in association with phlogopites (Fig. 4).

Phlogopites. Compositions of phlogopites range from 2 to 6.5 wt% FeO (Fig. 5). The amount of TiO_2 is not highly dependent on Fe content. The highest values relate to secondary phlogopites in rims around minerals, while larger grains are commonly low in TiO_2. In Phl from the polymict breccia, FeO varies from 5.0 to 8.5 wt%. The latter are low in Cr_2O_3 but generally higher in Na_2O.

High Precision Data for Olivines

High precision data for olivines were analyzed from 64 associations. They can be divided according to the petrographic features. The highest temperature Ca-rich

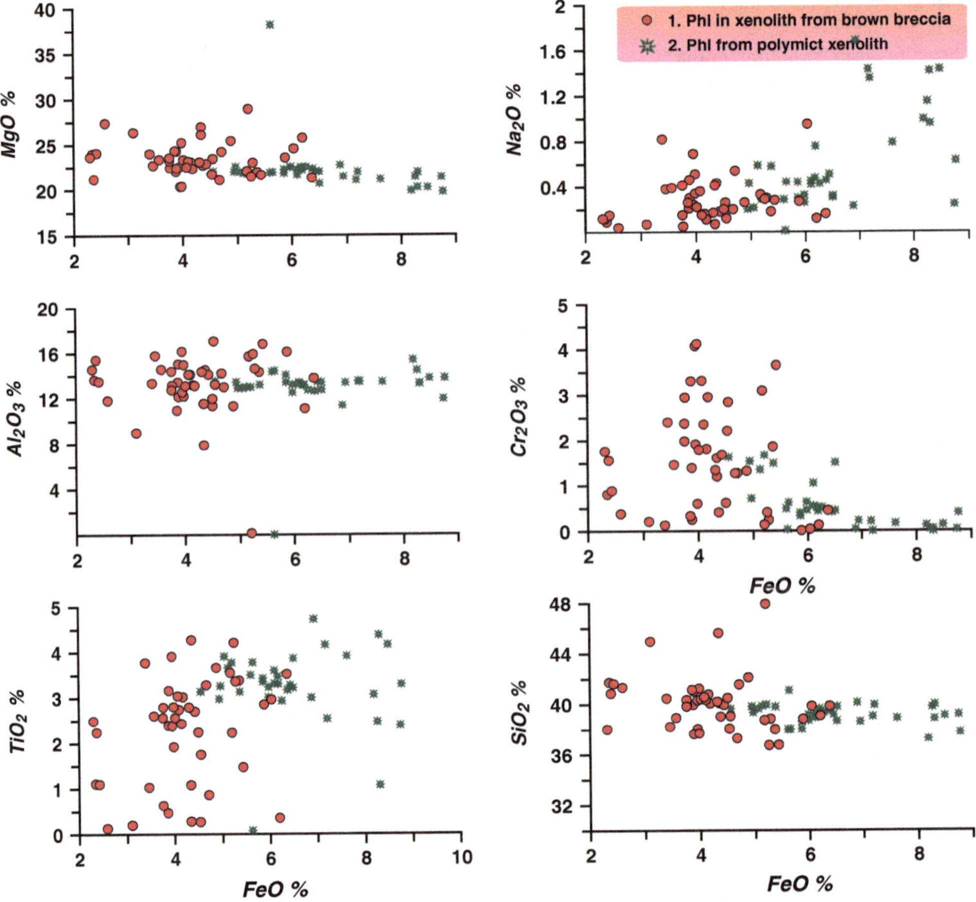

Fig. 5 Variations of major element compositions of phlogopites from xenoliths in the brown kimberlite breccia in comparison with those from deep-seated polymict breccia from Udachnaya pipe. *1* Phlogopites from xenoliths, *2* Phl from deep-seated Phl-bearing breccia

associations are some porphyroclastic peridotites (Fig. 6). They are low in Ti but rather high in Cr. Olivines from sheared peridotites show Ca contents near 400 ppm. All of them are enriched in Ti, Cr, and Al. In the coarse-grained spinel peridotites, olivines have low Ca content, which in general correlates with the low temperatures, low Cr, and high Ni content. Olivines from coarse garnet peridotites are higher in Ni but lower in Cr. Data for the xenoliths coincide in the diagrams with the clouds of the data points obtained for single olivine grains from Udachnaya pipe (Sobolev et al. 2009).

Thermobarometry

Thermobarometry of Olivine

High precision trace element analyses of the olivines have been used to compile the new thermometers (De Hoog et al. 2010). We checked most of them and found that the best one is that using Cr–Al exchange, although those using Ca and Al thermometry are also close (Fig. 7b). Using this thermometry and common Opx barometry, it possible to obtain nearly the same geotherm (Ashchepkov et al. 2010a, 2012b) as obtained by common single Opx methods (Brey and Koehler 1990; McGregor 1974) (Fig. 7a) or polymineral methods (Brey and Kohler 1990).

De Hoog et al. (2010) reported that these thermometric equations in inverse form may be used as barometers. Nevertheless in general, they give distinctly lower pressures than Opx-based methods (Boyd et al. 1997). Ca-Ol-Cpx barometry (Kohler and Brey 1990) for olivine also does not give a regular geotherm when using this equation together with the Ol thermometers (De Hoog et al. 2010). Nevertheless, the single olivine trace element thermobarometer is promising and may be developed using the combination of several components in the equation.

Common Orthopyroxene-Garnet Thermobarometry

Opx-based thermobarometry applied to our new data set (Fig. 8) (Brey et al. 1990; McGregor 1974) gives a PT plot very similar to those obtained with the previous data together for peridotites (Boyd et al. 1997; Ionov et al. 2010; Agashev et al. 2010; Yaxley et al. 2012) (Fig. 9). The PT estimates for extended data sets including pyroxenites

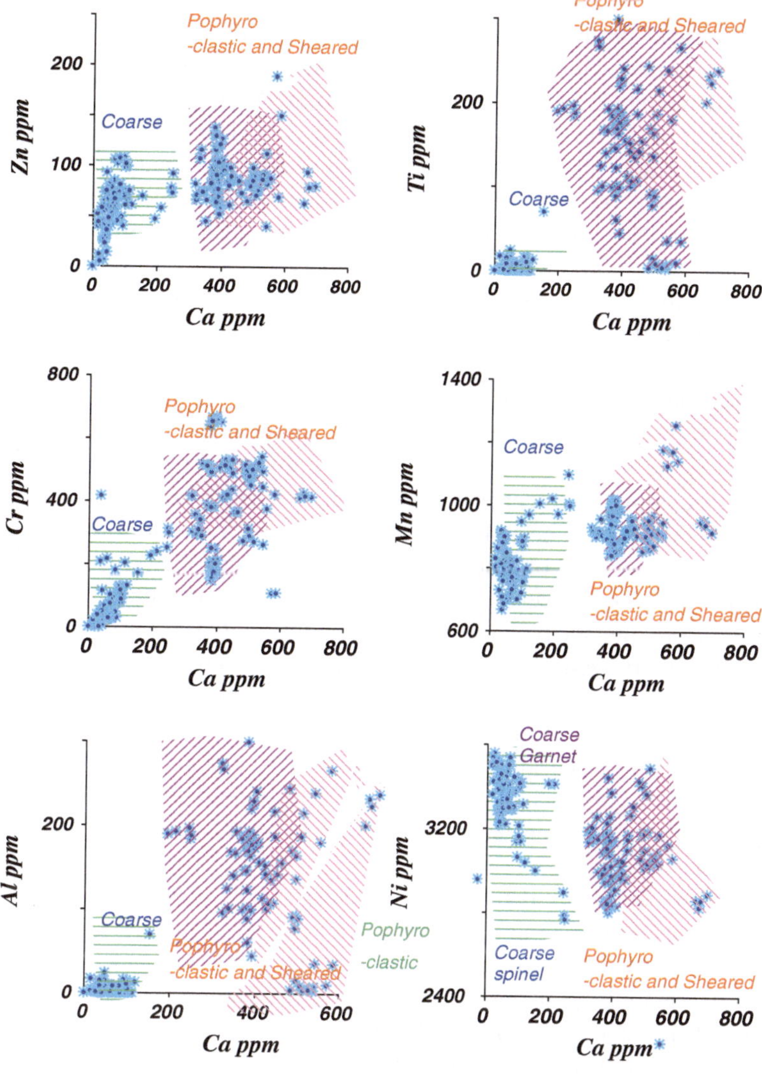

Fig. 6 Minor element variations in olivine obtained with high precision measurements

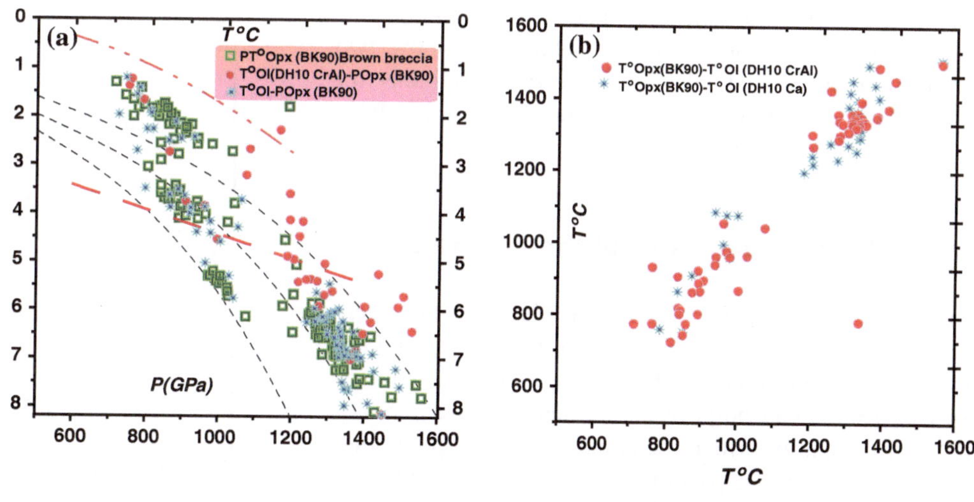

Fig. 7 PT conditions for the new data for peridotites from the brown kimberlite breccias from Udachnaya pipe. **a** PT Correlations of the Ol-based T °C according to De Hoog et al. (2010). Symbols see Fig. 9. **b** High precision analyses and Opx T °C correlations of the temperatures

(Pokhilenko et al. 1999; Kuligin 1997), diamond-bearing rocks, and inclusions (Logvinova et al. 2005; Simizu et al. 1997; Sobolev et al. 1984, 2004) show a wide range of PT parameter, especially in the lower part of the mantle section. The difference is that the convective branch is now more extensive reaching 8.0 GPa and possibly deeper, which corresponds to a depth of about 270 km and a 35 mw/m^2 conductive geotherm (Artemieva 2009). It divides into several arrays with different PT gradients. There is a relic low-temperature (LT) gradient (34–36 mw/m^2) mainly found from 5.0 to 3.0 GPa and below 6.5 GPa. These data are supported also by Cpx and garnet thermobarometry (Ashchepkov et al. 2012a, b). The common and most abundant PT points for the spinel peridotites also correspond to the 40 mw/m^2 geotherm which was published previously (Boyd et al. 1997) obtained with the Brey and Kohler (1990) and McGregor (1974) methods.

The convective branch, traced also by points produced using the Brey et al. (1990) method, starts from 8.0 to 7.5 GPa. It raises upward showing sub-adiabatic inclination and crosses the 45 mw/m^2 conductive geotherm near 6.0 GPa and becoming less steeper in upper part. Two more groups of HT PT points near 4–3.5 GPa and 3–2 GPa possibly mark the position of intermediate magmatic chamber(s). This result now receives support from the more detailed data from the analyses in thin sections. The Cpx thermobarometry (Ashchepkov et al. 2010a; Nimis and Taylor 2000) also yields abundant HT PT points corresponding to the base of the sub-cratonic lithospheric mantle (SCLM). But the cold branch still exists to 8.0 GPa marked mainly by the Cr-rich garnets and cold eclogitic PT estimates. This level is heterogeneous and is represented by rocks showing irregular heating.

Oxygen conditions determined for the for ilmenites and chromites according to the single grain versions of mineral (Taylor et al. 1998) oxybarometers and for pyroxenes and garnet and clinopyroxenes according to the equations published in (Ashchepkov et al. 2011) are shown in diagrams (Figs. 8 and 9). The common trend of the reducing of fO_2 to –4–5 (Yaxley et al. 2012), as in South African lithosphere (McCammon et al. 2001), is pronounced for clinopyroxenes, garnets, and chromites. The upper part show two trends of reduced and relatively oxidized conditions for garnets and clinopyroxenes found for the samples from brown breccia (Fig. 8).

Trace Element Chemistry of the Minerals

The trace element patterns for minerals from the deep-seated xenoliths are highly inclined with small humps which are shifted to the LREE side (Fig. 10). TRE patterns for Cpx from small porphyroclastic xenoliths are complex and vary comparing with those found for sheared peridotites which show enriched patterns without strong fluctuations and inflections (Ionov et al. 2010; Agashev et al. 2010).

The Cpx from common garnet lherzolites and harzburgites reveal patterns with LREE enrichment at 10x to chondrite C1, a the small hump near Pr, and high and varying La/Yb$_n$, ratios while pyroxenitic Cpx show smaller

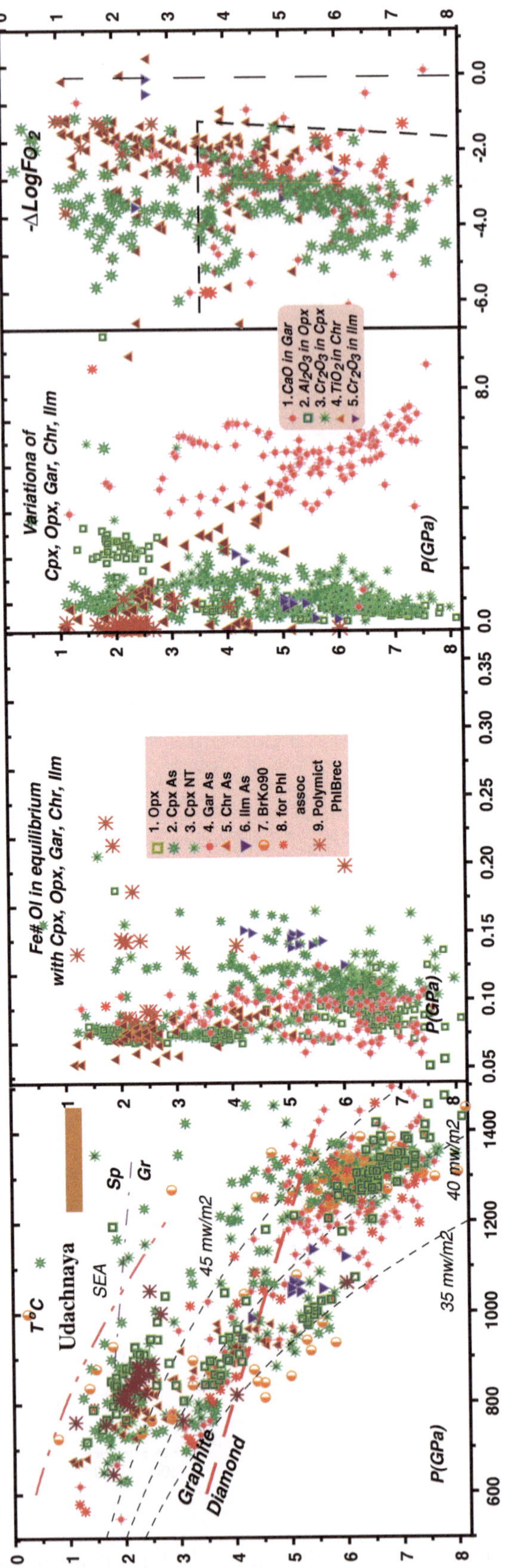

Fig. 8 PT conditions for the new data for peridotites from the brown breccia and Phl-bearing associations (*semi-filled squares*). Symbols see Fig. 9. In addition 9 T °C (Nimis and Taylor 2000)—P(GPa) (Ashchepkov et al. 2011) for Cpx from Phl polymict breccia

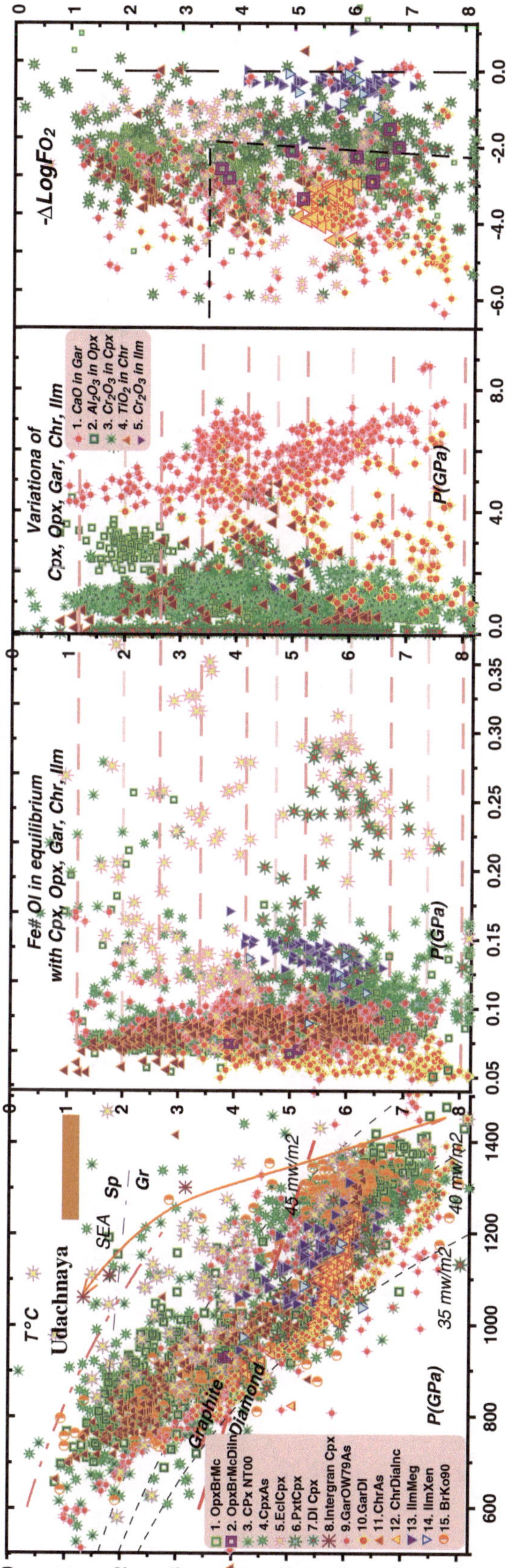

◀ **Fig. 9** PT conditions for the new data and for all previous data set of deep-seated xenoliths from Udachnaya pipe. *1* Orthopyroxene thermobarometry: T °C (Brey and Kohler 1990)-P(GPa) (McGregor 1974) for peridotites, *2* The same for diamond inclusions, *3* Clinopyroxene thermobarometry T °C and P(GPa) (Nimis and Taylor 2000), *4* T °C (Nimis and Taylor 2000)—P(GPa) (Ashchepkov et al. 2010a), *5* The same for eclogites, *6* The same for pyroxenites (Pokhilenko et al. 1999; Kuligin 1997), *7* The same for diamond inclusions (Sobolev et al. 1973, 2009; Logvinova et al. 2005) and xenoliths (Sobolev et al. 2004; Jacob et al. 1999), *8* The same for secondary Cpx in xenoliths; Garnet thermobarometry: *9* O'Neil Wood 1979 monomineral- P(GPa) (Ashchepkov 2010a), *10* the same the same for diamond inclusions; Cr-spinel thermobarometry, *11* ToC (O'Neil and Wall 1987)—P(GPa) (Ashchepkov et al. 2010a), *12* the same for diamond inclusions;. Ilmenite thermobarometry, *13* ToC (Taylor et al. 1998)—P(GPa) (Ashchepkov et al. 2010a), *14* the same for xenoliths from (Alymova et al. 2006), *15* ToC and—P(GPa) (Brey and Kohler 1990), *2* The field for the diamond-bearing associations in the lower left corner after (McCammon et al. 2001). Data original (Ashchepkov et al. 2012b) and after (Boyd et al. 1997; Ionov et al. 2010; Kuligin 1997; Malygina 2000; Pokhilenko 2006; Alymova 2006)

inclinations, and a more gentle hump which reflects the higher degree of melting and lower Gar/Cpx ratio in the melting source. The small dip in Zr is typical for all garnet lherzolites and harzburgites. Some of them reveal minor Sr enrichment.

The most complex REE patterns are shown by Cpx from highly depleted harzburgites with small amount of garnets. The LREE enrichment is higher for typical garnet-bearing varieties. They reveal also Sr dips but strong Pb peaks. Most of the diagrams show Zr as well as Y and moderate Ta- and Nb-depletion, and Ba dips although Rb is higher especially in Phl-bearing associations. Small peaks for U exist for most of samples. Essential varying peak in Pb in TRE for Cpx from small xenoliths may be the result of interaction with melts enriched in volatiles, some HFSE (Ti) and sulfides. On the other hand, the Cpx from the large xenoliths without phlogopite reveal small dips in Pb (Ionov et al. 2010).

Cpx from pyroxenites have Pb troughs as well as Zr, Nb, Ta, and Ba, suggesting small degree crystallization differentiation with precipitation of sulfides and oxides (ilmenites) as well as phlogopites.

Eclogitic Cpx have very different TRE patterns. Some of them with rather regular patterns suggest magmatic crystallization or deep interaction with the protokimberlites (hot type or remelted). Others reveal typical subduction features –dips or peaks in Eu, HFSE depletion and enrichment in U.

The REE and TRE patterns for peridotite garnets from are more regular in xenoliths from our collection, though some of them reveal S-shaped or even U-shaped REE patterns, which are typical for depleted compositions with lower TRE content. They all reveal deep Sr minima and Zr and Hf dips as well as for Ba, Th but fluctuating Nb and Ta (Fig. 9). Orthopyroxenes show mostly flat patterns with

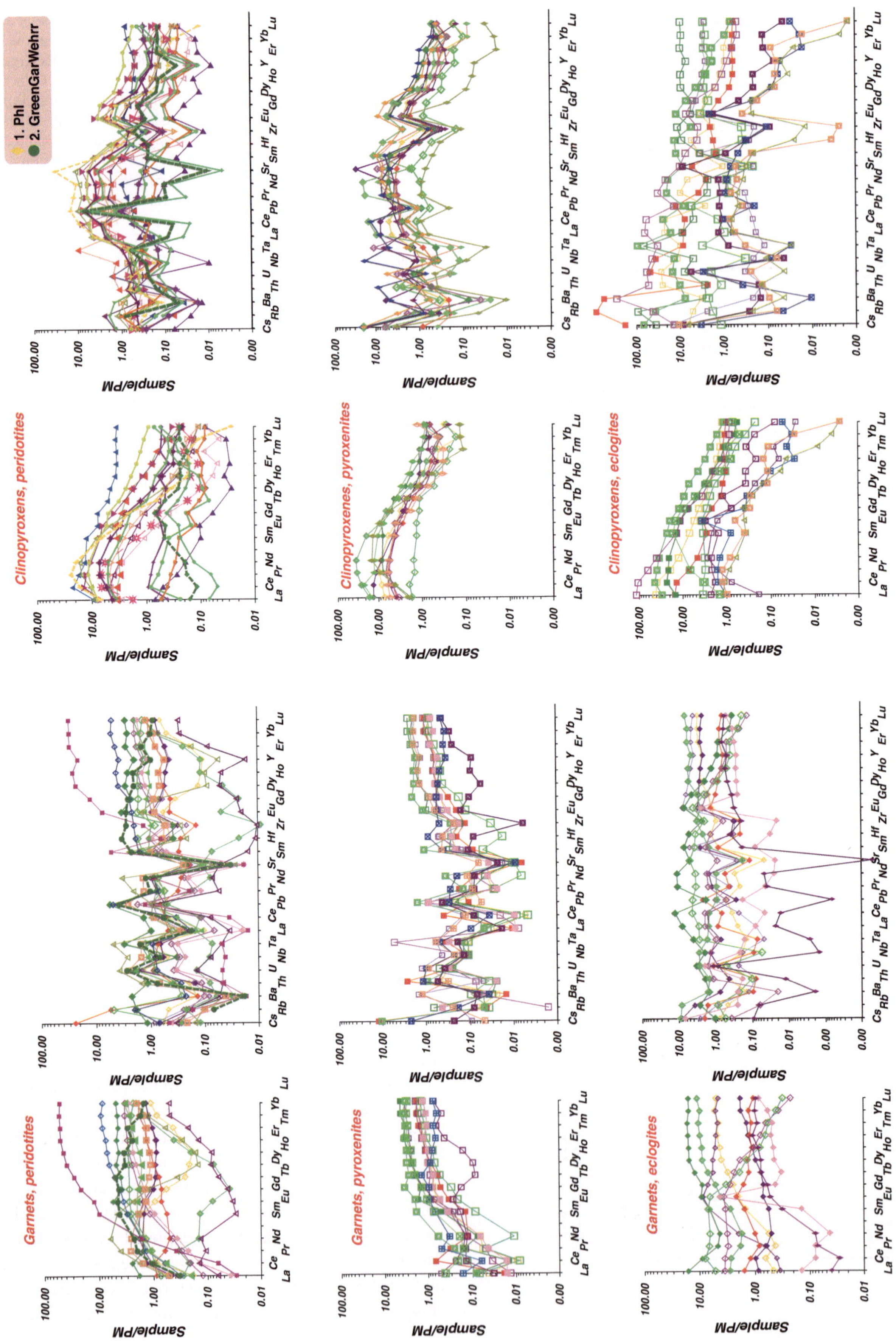

Fig. 10 TRE patterns for minerals from deep-seated xenoliths from Udachnaya pipe. Normalization after (McDonough and Sun 1995)

small negative inclination but those with lower concentrations of TRE have La/Yb > 1.

The Ti-rich green garnets from the porphyroclastic wehrlite have a very enriched rounded REE pattern, but the content of incompatible elements is very low. This agrees with the very depleted nature of the source rock. The TRE element content of the coexisting Cpx shows a low La/Yb ratio corresponding to the low-Gar source and overall depletion in incompatible elements. This suggests disequilibrium and later growth of the garnets from the Ti-rich melts.

The TRE patterns for phlogopites reveal high inclinations in general. They are jagged with the peaks in LILE components and Sr, Pb, and depression in Th and some HFSE.

Discussion

Structure of the Mantle

Detailed thermobarometry for the mantle section beneath Udachnaya and comparisons with petrographic features visible in thin section allows reconstruction of the set of rocks in each level in the mantle column. The uppermost horizon from 1.0 to 2.0 GPa is composed of coarse-grained relatively depleted spinel lherzolites and harzburgites. Pyroxenites are rare, but they are abundant just near the garnet-spinel boundary close to 2.0 GPa. They belong to both the Cr-diopside suite and the Fe–Ti-rich cumulates.

Xenoliths from 2.0 to 3.0 GPa are peridotites with relatively scarce and fine garnet grains which are dispersed in the mantle column. The level near 3.0–3.5 GPa is enriched in pyroxenites and some Fe-enriched eclogites (Jacob et al. 1994; Sobolev and Sobolev 1993; Sobolev et al. 1994; Snyder et al. 1997). The low-temperature coarse-grained harzburgites and dunites are related to the lower part and bottom of layers in paleo-subduction slices. Pyroxenites (Pokhilenko et al. 1999) and lherzolites (Ionov et al. 2010; Boyd et al. 1997) marking mainly the 4.0 GPa level contain pyroxenes enriched in Al, but the Cr content is not high.

The level between 6.0 and 4.0 GPa is very heterogeneous. The sequence of Cr-pyroxenites (Kuligin 1997; Kuligin and Pokhilenko 1998; Pokhienko et al. 1999) located close to 4.0 GPa shows an increase in Cr upward and Ca downward in the mantle section. Relative Ti enrichment is demonstrated by Cr-spinels from this interval; peridotites are very heterogeneous and are mostly lherzolites which are sometimes rich in pyroxenes. Phlogopite metasomatism is typical in this level and form veins that probably relate to crystallization of percolating carbonatite melts (Tappe et al. 2007). Phl is found in diamond-bearing associations (Sobolev et al. 2009). The boundary from 3.7 to 4.2 GPa is probably the sharpest in the mantle section. Dunites and harzburgites coincide here with eclogites and pyroxenites showing very wide variations in Fe in pyroxenes and Ca in garnets. The level between 4.5 and 5.5 GPa is essentially peridotitic but shows 4 separate trends of Ca and Fe increase in peridotitic garnets with the calculated pressures, suggesting presence of marble type peridotite structures which consist of pyroxenites and eclogites at the top and then lherzolites, harzburgites, and dunites. In the P–Fe# diagram, there are also 4 trends of Fe# decrease with decreasing pressures in this direction for the clinopyroxenes. Two most Fe# rich trends of Cpx coincide with the trends for ilmenites and may thus reflect two stages of protokimberlite melt percolation. Among them, the higher in Fe# trend corresponds to the HT diamond-bearing eclogites (pyroxenites) (Sobolev et al. 1994; Snyder et al. 1997). Some studies suggest that protokimberlite melts near 4.0 GPa are essentially carbonatitic (Tappe et al. 2007). More Fe-rich eclogite compositions also containing diamonds, but more LT are probably of primary subduction origin. Cr-spinel-bearing relatively fertile lherzolites with chains or veinlets of Gar, Cpx, and Chr probably mark melt migration paths and refertilization judging by the relatively high Ti in Chr. The high Cr content of some Cpx probably fixes the levels of the depletion and specific Na–Cr metasomatism which is more typical for the mantle columns beneath the Alakit pipes. (Ashchepkov et al. 2004). This level coincides with the presence of Gar harzburgites.

The boundary at 4.8–5.0 GPa is characterized by the appearance of abundant rather LT eclogites. They are all diamondiferous and differ in Fe contents (Sobolev et al. 1994; Snyder et al. 1997) separating into four groups. Two groups of ilmenites bearing Gar-clinopyroxenites are probably cumulate from protokimberlites. The group of eclogites with Fe# 20–30 is probably of subduction origin—the intermediate range may correspond to submelted varieties (Misra et al. 2004; Shatsky et al. 2005, 2008). The horizon between 5.0 and 6.5 GPa marks the so-called kink of the geotherm (Boyd et al. 1997). The range of 1,200–1,000 °C coincides for eclogites (Sobolev et al. 1994), for Cr-spinels from diamond inclusions, and for garnet harzburgites as well as for some ilmenite pyroxenites (Alymova 2006) and sheared peridotites. Most of the peridotitic rocks including harzburgites have Fe# that range from 12 to 5 and correlate positive with pressure. A similar range in Fe# 12-8 has been observed in the so-called sheared peridotites (Agashev et al. 2010; Boyd et al. 1997). Ca content of the garnets increases downward, but for the Cr-rich sub-calcic garnets, it decreases with depth. Probably, sub-calcic garnets reflect an ancient depletion event formed by the specific fluids that formed megacrystalline dunites (Pokhilenko et al. 1993), which were also studied in our collection. This is the level of the appearance of megacrystic Ilm pyroxenites (Alymova 2006), which sometime exhibit contacts with deformed peridotites.

A similar lens between 6.5 and 7.2 GPa is characterized by a higher abundance of Ilm pyroxenites. Gar dunites are less abundant but are more depleted. The sheared peridotites are more HT and show veins of magmatic Cpx, Opx, and even garnets. They coincide with the porphyroclastic varieties. The newly formed Ti-enriched green garnets in the Ti–rich wehrlitic and harzburgitic peridotites reflect the latest melt metasomatism, which coincides with the Phl formation. Metasomatic diamond-bearing Ilm lherzolite (Pokhilenko et al. 1976) with very high Cr content in its Cpx also come from this level. Some subduction-type eclogites (Jacob et al. 1994) relate to the LT conditions.

The deepest level (8 GPa) recorded by the xenoliths and xenocrysts corresponds to the most Cr-rich garnets with the high range of Ca decreasing with pressure. More Fe-rich compositions relate to HT cumulates probably left by the protokimberlites and to the hot diamond-bearing eclogites as well as some porphyroclastic peridotites.

Evolution of Mantle Column

Several pipes like those in Yubileinaya Ashchepkov et al. (2004) demonstrate a difference in the structure of the mantle columns reconstructed by use of xenocrysts and xenoliths. Since the Udachnaya West has not been studied in detail due to the low abundance of fresh xenoliths, a complete reconstruction of the mantle stratigraphy in the area is not yet available. The xenoliths from the brown breccias which are in general freshest show slightly more restricted intervals but also deeper levels. The brown breccias seems to be related to the latest stage of the pipe formation when the magmatic roots were affected by HFSE-rich hydrous melt as can be seen by the Ti enrichments (Figs. 2, 3 and 4).

Indirect evidence of the wide variations of mineral compositions and the high positions of the geotherms in the deeper horizons even to depths of 2.0 GPa shows that there were abundant reactions and melt interactions at several levels within the mantle columns. The deeper levels were definitely subjected to the influence of protokimberlite melts which created two magmatic chamber-vein systems near 7.0 and 6.0 GPa and caused deformation of the peridotites and high-scale interactions possibly due to hydraulic fracturing (Ave Lallement et al. 1980; Mei et al. 2002). It seems that the coexistence of the relatively easy melted material like eclogites and melt conduits like dunites is not causal. The upper 5.5–4.0 GPa levels correspond mainly to interactions around veins formed by carbonatitic melts which may be created by splitting of primary protokimberlites during the rising and cooling. Such systems containing abundant carbonates are favorable for the creation of diamonds especially during interactions with reduced peridotites. Pervasive melt percolation ceased near 4.0 GPa where the large dense layer of eclogite-pyroxenite layer exists.. Possibly, it was created during the subduction in hot conditions of early Earth (van Hunen and van den Berg 2008) when the Fe-rich eclogites cannot pass through the 4 GPa boundary without melting The large units in the mantle columns were primary created by subduction processes (Pearson et al. 1999; Ashchepkov et al. 2012b), and eclogites should correspond to the upper parts of the subducted slabs. The sequence of such slabs is responsible for rhythmic marble-like SCLM layering.

The physical–chemical boundaries correspond to the temperature minima of oxidized peridotite melts at 7–8 GPa, carbonated peridotites at 5–4 GPa, and hydrous peridotite at 2- GPa (Tappe et al. 2007). These boundaries should cause melt concentration and creation of the pyroxenite lenses. The separate high temperature level in the upper part near the Gar-Sp boundary also concentrates the melts. Possibly, a separate most evolved H_2O-bearing protokimberlite melt portion reached this level.

Phlogopite Metasomatism

The nature of phlogopite metasomatism is debatable. It occurs in the deepest levels in diamond inclusions (Sobolev et al. 2009). Phlogopites with FeO >4.5 wt% and also high in Na_2O may be related to melt metasomatism. They are close to the varieties found in the Phl-bearing deep-seated breccia. There are two types of Phl according to Ti content which can mark the interaction with melts percolating through the mantle. Phlogopite microveins or well-equilibrated grains with other minerals have low TiO_2 content and may be related to introduction of melts or fluids that correspond earlier Precambrian stage or close to Rodinia break up (Pokhilenko et al. 2012) as it is the case in Alakit (Ashchepkov et al. 2010b) field. Nevertheless, many phlogopites in Udachnaya peridotites may relate to later events corresponding to the protokimberlite melts.

Phlogopites from the breccia have enriched Na and Ti contents but do not contain Ba and Cl as well as F. One can suggest that there is no relation to subduction-related rocks or melts. But intergranular material in this breccia contains barite and strontianite as well as abundant apatites and rare sulfides. Amphiboles are Na-richterites and kaersutites. The presence of picroilmenites (MgO ~10 wt%) as well as the debris of sheared peridotites indicates a very deep origin of these rocks and polybaric crystallization of melts. Thus, the channels of the protomagmatic systems before rising to the surface were subjected to interaction with H_2O-rich fluid rising from levels of >8.0 GPa. So, the K-rich fluids found also in diamonds (Zedgenizov et al. 2009) were widespread and concentrated in the mantle just before the eruption. Many porphyroclastic and even sheared peridotites contain intergranular phlogopites especially in rims around garnets.

One can suggest that they are related to the kimberlite II melts which appeared in the Daldyn field from the deep 410-km boundary. Another possibility is they developed from subducted material reactivated by the kimberlites.

Melt Rising and Capturing

The adiabatic branches traced by the secondary Cpx PT conditions show that there were several intermediate levels (chambers) where xenoliths interacted with kimberlites during their rise to the surface. Several separate pressure intervals (6) as shown in Figs. 6 and 7, marked by xenoliths from the brown breccia possibly refer to such chambers.

Trace Element Evidences

TRE content of rocks and minerals is usually is used for recognition of the genetic signatures and links of the studied rocks as well as the degree of melting or differentiation, contamination, and characteristics of the melting sources. Trace elements in mantle peridotites from Udachnaya (Shimizu et al. 1997; Ionov et al. 2010; Agashev et al. 2010) indicate the preference of Cpx input in the bulk rock composition. Sheared peridotites more enriched in HREE possibly refer to interaction with melts (Ionov et al. 2010; Agashev et al. 2010; Boyd et al. 1997; Griffin et al. 1989) with less inclined TRE which were protokimberlites.

Cpx and Gar are the major repositories of TRE in peridotites. Cpx show different levels of REE which are higher together with the La/Yb_n ratios reflecting both increasing Gar/Cpx ratios and decreasing melting degree in the mantle columns with depth. The melting degree is less than 0.1 % for most harzburgitic rocks according to the La/Ce_n ratios. This suggests a low degree melting usually corresponding to infiltration of fluids. Presence of garnets with S-shaped patterns (Shimizu et al. 1997) possibly reflects various stages of garnet-melt interaction.. The complex shapes of the REE and TRE patterns and inflection and local dips or elevations in MREE may mean that there were several stages of melt fluid percolations through the mantle rocks, especially in the lower part of the mantle column and changes of the volatile content from CO_2 or H_2O or H_2-rich. It was also subject to pervasive HFSE metasomatism (Grifiin et al. 1989, 1993). The permanent dips in Zr may also suggest that the rocks initially were melted in a high activity of P, creating apatites or zircons.

Conclusions

There are 6 large unities in the structure of the lithospheric mantle beneath Udachnaya pipe, probably related to ancient subduction(s). The structure of mantle columns was essentially transformed by the different melt percolation events. The occurring temperatures in the separate levels from 2 to 7.5 GPa could be attributed to the upwelling of protokimberlite melts, which produced intermediate magmatic bodies.

1. The brown breccia from the deepest level of the Udachnaya kimberlite quarry contains the deepest and freshest mantle xenoliths in the pipes, captured down to about 8.0 GPa and representing separate capturing intervals.
2. Phlogopite metasomatism is developed in the whole section of the mantle column beneath the pipe. Well-equilibrated phlogopites are probably related to ancient stages of melt percolation. The deeper levels probably correspond to deep-seated melts close to kimberlite II which appeared from the deep level of the mantle.
3. The deep-seated phlogopite breccias came from a deep level of the mantle column but were crystallized near the 2–1 GPa level.
4. Interaction of protokimberlite melt with this suggests a low degree melting usually corresponding to infiltration of fluids was responsible for peridotite shearing and wide scale interaction and possibly creation of some population of the diamonds. Phlogopite metasomatism is accompanied these interactions.

Acknowledgments To Austrian Academy of Sciences for the invitation of I.V. Ashchepkov for the study of xenoliths. To prof. H. Downes for the correction of the text. To Dr L.Pokhilenko for the analyses and eclogite samples. To company ALROSA for the permission to collect the samples. Supported by grant RBRF 05-11-00060a, 11-05-91060-PICSa.

References

Agashev AM, Pokhilenko NP, Cherepanova YuV, Golovin AV (2010) Geochemical evolution of rocks at the base of the lithospheric mantle: evidence from study of xenoliths of deformed peridotites from kimberlite of the Udachnaya pipe. Doklady Earth Sci 432(4):510–513

Alymova NA (2006) Peculiarities of ilmenites and ilmenite—bearing association from kimberlites of Yakutian province. PhD dissertation thesis. Irkutsk, Institute of Geochmistry SB RAS, p 175

Artemieva IM (2009) The continental lithosphere: reconciling thermal, seismic, and petrologic data. Lithos 109:23–46

Ashchepkov IV, Pokhilenko NP, Vladykin NV, Logvinova AM, Kostrovitsky SI, Afanasiev VP, Pokhilenko LN, Kuligin SS, Malygina LV, Alymova NV, Khmelnikova OS, Palessky SV, Nikolaeva IV, Karpenko MA, Stagnitsky YB (2010a) Structure and evolution of the lithospheric mantle beneath Siberian craton, thermobarometric study. Tectonophysics 485:17–41

Ashchepkov I, Travin A, Ntaflos T, Ionov D, Logvinova A, Vladykin N, Palessky S (2010b) Metasomatism in peridotites beneath the Daldyn-Alakit region Yakutia. Geochimica et Cosmochimica Acta 74(12 Supplement 1):A35

Ashchepkov IV, André L, Downes H, Belyatsky BA (2011) Pyroxenites and megacrysts from Vitim picrite-basalts (Russia): Polybaric fractionation of rising melts in the mantle? J Asian Earth Sci 42:14–37

Ashchepkov IV, Rotman AY, Somov SV, Afanasiev VP, Downes H, Logvinova AM, Nossyko S, Shimupi J, Palessky SV, Khmelnikova OS, Vladykin NV (2012a) Composition and thermal structure of the lithospheric mantle beneath kimberlite pipes from the Catoca cluster, Angola. Tectonophysics 530–531:128–151

Ashchepkov IV, Vladykin NV, Ntaflos T, Downes H, Mitchel R, Smelov AP Rotman AYa, Stegnitsky Yu, Smarov GP, Makovchuk I V, Nigmatulina EN, Khmelnikova OS (2012b) Regularities of the mantle lithosphere structure and formation beneath Siberian craton in comparison with other cratons. Gondwana Research. 23:4–24

Ave Lallement HL, Carter NL, Mercier JCC, Ross JV (1980) Rheology of the uppermost mantle: inferences from peridotite xenoliths. Tectonophysics 70:221–234

Boyd FR, Pokhilenko NP, Pearson DG, Mertzman SA, Sobolev NV, Finger LW (1997) Composition of the Siberian cratonic mantle: evidence from Udachnaya peridotite xenoliths. Contrib Miner Petrol 128:228–246

Brey GP, Kohler T (1990) Geothermobarometry in four-phase lherzolites. II. New thermobarometers, and practical assessment of existing thermobarometers. J Petrol 31:1353–1378

Burgess SR, Harte B (2004) Tracing lithosphere evolution through the analysis of heterogeneous G9–G10 garnets in peridotite xenoliths, II: REE chemistry. J Petrol 45:609–634

Dawson B (1980) Kimberlites and their xenoliths. Springer, Berlin Heidelberg, New York, p 200

De Hoog JCM, Gall G, Cornell DH (2010) Trace-element geochemistry of mantle olivine and application to mantle petrogenesis and geothermobarometry. Chem Geol 270:196–215

Griffin WL, Smith D, Boyd FR, Cousens DR, Ryan CG, Se SH, Suter GF (1989) Trace element zoning in garnets from sheared mantle xenoliths. Geochim Cosmochim Acta 53:561–567

Griffin WL, Sobolev NV, Ryan CG, Pokhilenko NP, Win TT, Yefimova ES (1993) Trace elements in garnets and chromites: diamond formation in the Siberian lithosphere. Lithos 29:235–256

Ionov DA, Doucet LS, Ashchepkov IV (2010) Composition of the lithospheric mantle in the Siberian craton: new constraints from fresh peridotites in the Udachnaya-East kimberlite. J Petrol 51:2177–2210

Jacob D, Jagoutz E, Lowry D, Mattey D, Kudrjavtseva G (1994) Diamondiferous eclogites from Siberia: remnants of Archean oceanic crust. Geochim Cosmochim Acta 58:5191–5207

Kohler T, Brey GP (1990) Calcium exchange between olivine and clinopyroxene calibrated as a geothermobarometer for natural peridotites from 2 to 60 kb with applications. Geochim Cosmochim Acta 54(2375):2388

Kuligin SS (1997) Complex of pyroxenite xenoliths in kimberlites from different regions of Siberian platform. PhD thesis, United Institute of Geology Geophysics and Mineralogy, Novosibirsk, p 220

Kuligin SS, Pokhilenko NP (1998) Mineralogy of xenoliths of garnet pyroxenites from kimberlite pipes of Siberian platform. Extended Abstracts 7IKC. Cape Town, pp 480–482

Kuligin SS, Malkovets VG, Pokhilenko NP, Vavilov MA, Griffin WL, O'Reilly SY (2003) Mineralogical and geochemical characteristics of a unique mantle xenolith from the Udachnaya kimberlite pipe. Extended abstracts of the 8th international kimberlite conference. FLA_0114

Lavrent'ev YUG, Usova LV (1994) New version of KARAT program for quantitative X-ray spectral microanalysis. Zh Anal Khim 5:462–468

Logvinova AM, Taylor LA, Floss C, Sobolev NV (2005) Geochemistry of multiple diamond inclusions of harzburgitic garnets as examined in situ. Int Geol Rev 47:1223–1233

MacGregor ID (1974) The system MgO-Al_2O_3-SiO_2: solubility of Al_2O_3 in enstatite for spinel and garnet peridotite compositions. Am Miner 59:110–119

Malygina EV (2000) Xenoliths of granular mantle peridotites in Udachnaya pipe. PhD thesis, United Institute of Geology Geophysics and Mineralogy, Novosibirsk, p 187

McCammon CA, Griffin WL, Shee SR, O'Neill HSC (2001) Oxidation during metasomatism in ultramafic xenoliths from the Wesselton kimberlite, South Africa: implications for ival of diamond. Contrib Miner Petrol 141:287–296

McDonough WF, Sun SS (1995) The composition of the Earth. Chem Geol 120:223–253

Mei S, Bai W, Hiraga T, Kohlstedt DL (2002) Influence of melt on the creep behavior of olivine basalt aggregates under hydrous conditions. Earth Planet Sci Lett 201:491–507

Misra KC, Anand M, Taylor LA, Sobolev NV (2004) Multi-stage metasomatism of diamondiferous eclogite xenoliths from the Udachnaya kimberlite pipe, Yakutia. Siberia Contrib Miner Petrol 146(6):696–714

Nimis P, Taylor W (2000) Single clinopyroxene thermobarometry for garnet peridotites. Part I. Calibration and testing of a Cr-in-Cpx barometer and an enstatite-in-Cpx thermometer. Contrib Miner Petrol 139:541–554

O'Neill HSTC, Wall VJ (1987) The olivine orthopyroxene-spinel oxygen geobarometer, the nickel precipitation curve, and the oxygen fugacity of the Earth's upper mantle. J Petrol 28:1169–1191

O'Neill HSTC, Wood BJ (1979) An experimental study of Fe-Mg-partitioning between garnet and olivine and its calibration as a geothermometer. Contrib Miner Petrol 70:59–70

Pearson DG (1999) The age of continental roots. Lithos 48:171–194

Pokhilenko LN (2006) Volatile composition and oxidation state of mantle xenoliths from Siberian kimberlites. PhD thesis, United Institute of Geology Geophysics and Mineralogy, Novosibirsk, pp 225

Pokhilenko NP, Sobolev NV, Sobolev VS, Lavrentiev YuG (1976) Xenolith of diamond bearing ilmenite-pyrope lherzolite from the Udachnaya kimberlite pipe, Yakutia. Doklady AN SSSR 231:149–151

Pokhilenko NP, Sobolev NV, Boyd FR, Pearson DG, Shimizu N (1993) Megacrystalline pyrope peridotites in the lithosphere of the Siberian platform: mineralogy, geochemical peculiarities and the problem of their origin. Russ Geol Geophys 34:56–67

Pokhilenko NP, Sobolev NV, Kuligin SS, Shimizu N (1999) Peculiarities of distribution of pyroxenite paragenesis garnets in Yakutian kimberlites and some aspects of the evolution of the Siberian craton lithospheric mantle. In: Proceedings of the 7th international kimberlite conference. The P.H. Nixon volume, pp 690–707

Pokhilenko LN, Alifirova TA., Yudin DS (2012) $^{40}AR/^{39}AR$ Dating of phlogopite in mantle xenoliths from kimberlite pipes of Yakutia: evidence for deep ancient metasomatism of the Siberian platform. In: 10th international kimberlite conference long abstarcts 10th IKC-057

Shatsky VS, Buzlukova LV, Jagoutz E, Koz'menko OA, Mityukhin SI (2005) Structure and evolution of the lower crust of the Daldyn-Alakit district in the Yakutian Diamond Province (from data on xenoliths). Russ Geol Geophys 46:1252–1270

Shatsky V, Ragozin A, Zedgenizov D, Mityukhin S (2008) Evidence for multistage evolution in a xenolith of diamond-bearing eclogite from the Udachnaya kimberlite pipe. Lithos 105:289–300

Shimizu N, Pokhilenko NP, Boyd FR, Pearson DG (1997) Geochemical characteristics of mantle xenoliths from the Udachnaya kimberlite pipe. Russ Geol Geophys 38:205–217

Snyder GA, Taylor LA, Crozaz G, Halliday AN, Beard BL, Sobolev VN (1997) The origins of Yakunan eclogite xenoliths. J Petrol 8:85–113

Sobolev NV (1977) Deep-seated inclusions in kimberlites and the problem of the composition of the mantle. Amer Geophys Union, Washington, DC, p 279

Sobolev VN, Sobolev NV (1993) Nd and Sr isotopes from diamondiferous eclogites, Udachnaya kimberlite pipe, Yakutia, Siberia: evidence of differentiation in the early Earth? Earth Planet Sci Lett 118:91–100

Sobolev NV, Lavrentev YG, Pokhilenko NP, Usova LV (1973) Chrome-rich garnets from the Kimberlites of Yakutia and their parageneses. Contrib Miner Petrol 40:39–52

Sobolev NV, Pokhilenko NV, Efimova ES (1984) Xenoliths of diamond bearing peridotites in kimberlites and problem of the diamond origin. Russ Geol Geophys 25(12):63–80

Sobolev VN, Taylor LA, Snyder GA, Sobolev NV (1994) Diamondiferous eclogites from the Udachnaya pipe, Yakutia. Int Geol Rev 36:42–64

Sobolev NV, Logvinova M, Zedgenizov DA, Seryotkin YV, Yefimova ES, Floss C, Taylor LA (2004) Mineral inclusions in microdiamonds and macrodiamonds from kimberlites of Yakutia: a comparative study. Lithos 77:225–242

Sobolev NV, Logvinova AM, Zedgenizov DA, Pokhilenko NP, Malygina EV, Kuzmin DV, Sobolev AV (2009) Petrogenetic significance of minor elements in olivines from diamonds and peridotite xenoliths from kimberlites of Yakutia. Lithos 112(S1):701–713

Solov'eva LV, Lavrent'ev YuG, Egorov KN, Kostrovitskii SI, Korolyuk VN, Suvorova LF (2008) The genetic relationship of the deformed peridotites and garnet megacrysts from kimberlites with asthenospheric melts. Russ Geol Geophys 49:207–224

Tappe S, Foley SF, Stracke A, Romer RL, Kjarsgaard BA, Heaman LM, Joyce N (2007) Craton reactivation on the Labrador Sea margins: $^{40}Ar/^{39}Ar$ age and Sr-Nd-Hf-Pb isotope constraints from alkaline and carbonatite intrusives. Earth Planet Sci Lett 256:433–454

Taylor WL, Kamperman M, Hamilton R (1998) New thermometer and oxygen fugacity sensor calibration for ilmenite and Cr-spinel- bearing peridotite assemblage. In: Gurney JJ, Gurney JL, Pascoe MD, Richardson SH (eds) Proceedings of 7th international kimberlite conference, Red Roof Design, Capetown, pp 986–988

van Hunen J, van den Berg AP (2008) Plate tectonics on the early Earth: limitations imposed by strength and buoyancy of subducted lithosphere. Lithos 103:217–235

Yaxley GM, Berry AJ, Kamenetsky VS, Woodland AB, Golovin AV (2012) An oxygen fugacity profile through the Siberian Craton—Fe K-edge XANES determinations of Fe3 +/ΣFe in garnets in peridotite xenoliths from the Udachnaya East kimberlite. Lithos 140–141:142–151

Zedgenizov DA, Ragozin AL, Shatsky VS, Araujo D, Griffin WL, Kagi H (2009) Mg and Fe-rich carbonate–silicate high-density fluids in cuboid diamonds from the Internationalnaya kimberlite pipe (Yakutia). Lithos 112(S2):638–647

The "Exceptionally Fresh" Udachnaya-East Kimberlite: Evidence for Brine and Evaporite Contamination

S. I. Kostrovitsky, M. G. Kopylova, K. N. Egorov, and D. A. Yakovlev

Abstract

The composition of the serpentine-free Udachnaya-East kimberlite containing alkali carbonate, gypsum, halite, and other Na-, Cl-, and S-rich minerals has been the basis for a model of alkali-rich primary kimberlite melt. The interpretation of these minerals as mantle-derived, however, contradicts geology and hydrogeology of the Yakutian kimberlite province, as well as petrographic, geochemical, and isotopic evidence. The Udachnaya-East pipe is similar to many other southern Yakutian kimberlites, which emplace through 2-km-thick evaporite-bearing terrigenous carbonate sediments saturated with brines. A secondary origin of Na-, Cl-, and S-rich minerals in the southern Yakutian kimberlites is supported by (1) a regional correlation between the geology and hydrogeology of the local country rocks and the kimberlite mineralogy, in particular the difference between southern and northern Yakutian kimberlites; (2) a restriction of halite or gypsum mineralization to certain depth horizons where pipes intersect country rock strata with similar mineralogy; (3) the localization of the highest abundances of Na–Cl–S-bearing minerals at a depth interval that correlates across three magmatic phases of kimberlites and coincides with the roof of the aquifer carrying Na brines; (4) the presence of evaporite xenoliths and veins of halite, gypsum, and carbonate cutting through kimberlite and xenoliths; (5) crystallization of halite and alkali carbonate after serpentine and other groundmass minerals as evidenced by the rock textures; (6) geochemical evidence for crustal contamination, including high bulk CO_2 and CaO content, the absence of correlation between bulk Na_2O and any geochemical parameters, as well as initial Sr ratios, $\delta^{13}C$, $\delta^{18}O$, $\delta^{37}Cl$, and $\delta^{34}S$ intermediate between crustal and mantle values. We propose that the Udachnaya-East kimberlites acquired high Na, S, and Cl contents by interaction with buried Cambrian Na–Ca–Cl brines or assimilating evaporite xenoliths from the 500-m-thick Chukuck suite formed in the Daldyn-Markha carbonate bank.

Revised for the 10th IKC Proceedings Volume, November 2012.

S. I. Kostrovitsky (✉) · D. A. Yakovlev
Institute of Geochemistry SB RAS, Irkutsk, Russian Federation
e-mail: serkost@igc.irk.ru

M. G. Kopylova
Department of Earth & Ocean Sciences, University of British Columbia, Vancouver, Canada

K. N. Egorov
Institute of the Earth's Crust SB RAS, Irkutsk, Russian Federation

D. G. Pearson et al. (eds.), *Proceedings of 10th International Kimberlite Conference*, Volume 1, Special Issue of the Journal of the Geological Society of India,
DOI: 10.1007/978-81-322-1170-9_6,

Keywords
Yakutian province • Udachnaya-East kimberlite pipe • Halite • Mantle • Brine • Evaporite xenolith • Primary kimberlite melt

Introduction

Kimberlites are difficult rocks to study as they have a hybrid nature and combine several unequilibrated mineral parageneses. Kimberlite research is further complicated by late deuteric and secondary hydrothermal alteration. This alteration may lead to intense serpentinization of primary kimberlitic olivine. In Yakutian kimberlites, all olivine, microphenocrystal, macrocrystal, and megacrystal, is completely serpentinized, to the extent that the kimberlites are composed of 80–90 % serpentine and carbonates (e.g., Khar'kiv et al. 1991). In contrast to this, some kimberlites of the Udachnaya-East pipe show fresh, unserpentinized olivine and a lack of interstitial, late magmatic serpentine in the groundmass. The absence of all kinds of serpentine in the Udachnaya-East pipe gave rise to a theory that these kimberlites are "exceptionally fresh" in the global context (e.g., Kamenetsky et al. 2004; Maas et al. 2005). The composition of serpentine-free Udachnaya-East (SFUE) kimberlite was proposed as the best approximation of the deep kimberlite melt, which, unlike melt in other kimberlites, was not altered by late addition of crustal water. The authors further claimed that all serpentine in kimberlites is secondary and all hypabyssal kimberlites with interstitial groundmass serpentine, that is, all hypabyssal kimberlites in the world except the Udachnaya-East, are altered. As the composition of the SFUE kimberlite is poor in H_2O and rich in Cl, K, and Na, these characteristics were assigned to the composition of the primary kimberlite melt equilibrated with the mantle (e.g., Maas et al. 2005; Kamenetsky et al. 2007a). This model provided an attractive explanation for compositions of fluid inclusions in fibrous diamonds, which are also rich in halogens and alkalis (e.g., Klein-BenDavid et al. 2009). The new model of the alkali-rich kimberlite melt has provided a new standard for mantle chlorine isotopes (Sharp et al. 2007) and has been used to constrain physical properties of the primary kimberlite melt (Kamenetsky et al. 2007a) and in numerous experimental studies of mantle melting.

To test this model, we collected geological data on the Udachnaya kimberlite and its country rocks. This geological background provides the basis for a petrographic and geochemical characterization of the SFUE kimberlite. The data suggest that the high Na, S, and Cl contents are the result of interaction with buried Cambrian Na–Ca–Cl brines or assimilation of evaporite xenoliths.

Samples and Methods

SFUE kimberlite was sampled (34 specimens) in drill cores (holes 218 and 222) and in the open pit mine in 2000–2010 with a goal of building a representative collection of all kimberlite phases, including petrographic and mineralogical variations within the phases. As it is not possible to visually estimate a mode of Na-, S-, and Cl-rich minerals in the kimberlite, the selected specimens are not restricted to the extreme modes of these minerals.

Concentrations of ten major elements were analyzed by X-ray fluorescence (XRF) spectrometry on the SRM-25 Orelnauchpribor machine in the Geochemistry Institute Sib RAN (Irkutsk, Russia). The loss on ignition was determined at 950 °C in an induction furnace. Samples were then homogenized into fused beads from a 1:2 mixture of sample to lithium tetraborate at 1,100 °C in carbon glass containers in the induction furnace. The analysis was performed at 30 kV and current of 40 mA using an Rh anode X-ray tube. The standards were natural samples of carbonatite CI-2, basalt B-1, and peridotite P-1. The amount of FeO and CO_2 was determined for 0.2–0.3 g samples using $HF–HNO_3–HClO_4$ acid digestion and ammonium metavanadate titration. Chlorine was analyzed by XRF on S4 Pioneer spectrometer (AXS, Bruker), with a minimum detection limit of 0.01 % and a relative error <1 %. Water was determined gravimetrically using the Brush-Penfield method.

Trace elements were analyzed on an Elan 6100 DRC inductively coupled plasma mass spectrometer (Geochemistry Institute Sib RAN, Irkutsk, Russia) using $HF–HNO_3–HClO_4$ acid digestion. The precision of repeat trace metal measurements on standards was below 5 rel. % for Sr, La, Ce, Nd, Sm, Tb, Ho, Er, and Yb and 5–15 rel. % for Y, Zr, Pr, Eu, Gd, Dy, Tm, Lu, and Hf. Minimum detection limits for this analysis varied from 0.01 to 0.03 ppm.

The isotopic analysis of Sr was performed at the Geochemistry Institute Sib RAN (Irkutsk, Russia) with a Finnigan MAT-262 mass spectrometer. The analysis was carried out on the single band activation regime using Ta_2O_5 and the Re band. The measurements were normalized for the NBS

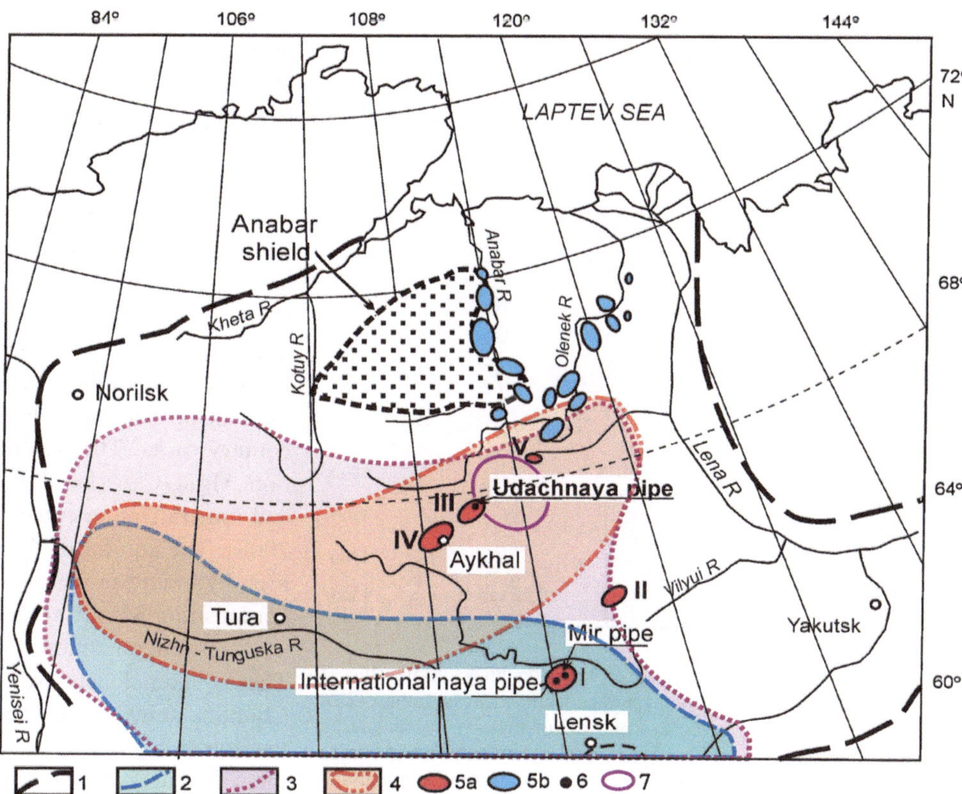

Fig. 1 Kimberlite fields of the Siberian platform (Khar'kiv et al. 1998) with occurrences of salt-bearing deposits and brines (after Alekseev et al. 2007). *1–3* Boundaries of Siberian platform (*1*), the lower Cambrian evaporite-bearing sedimentary rocks (*2*), and brine occurrences (*3*); *4* zone of complete saturation of sedimentary cover in metamorphosed brine; *5* kimberlite fields: **a** southern diamondiferous (*I* Malobotuobiya, *II* Nakyn, *III* Daldyn, *IV* Alakit-Markha, *V* Verhnemuna) and **b** northern; *6* pipes; *7* Daldyn-Markha carbonate bank (Sukhov 2001; Polozov et al. 2008a, b; Drozdov et al. 2008)

SRM 987 standard. The value of $^{87}Sr/^{86}Sr$ during the analysis was determined as 0.710254 ± 11 ($n = 28$), with the relative error of 0.002 %.

Geology of the Udachnaya Kimberlite

The Udachnaya kimberlite is emplaced into the Siberian craton and is part of the Yakutian kimberlite province. It is comprised of 20 fields (Fig. 1). Some of the 5 southern fields contain economic pipes, whereas the 15 northern fields are non-economic. The kimberlites erupted through Paleozoic terrigenous carbonate sediments, with a thickness of 1–3 km in the southern part of the kimberlite province and 0–1 km in the northern part (Brakhfogel 1984). The sediments are Cambrian and Ordovician limestone, dolomite with minor marls, evaporites, mudstone, and sandstone (Bobrievich et al. 1959).

The 353–367 Ma Udachnaya pipe from the Daldyn field was emplaced through the Lower to Upper Cambrian sedimentary rocks (limestones, dolomites, argillites, sandstones, and conglomerates) over 2 km thick (Fig. 2). The 400-m-thick Upper Cambrian clays and carbonates are underlain by clastic dolomite of the Chukuck suite at depths of 420–900 m. It, in turn, lies above the Early Cambrian limestones and sandstones. Although the kimberlite does not intersect massive evaporites, the clastic dolomite of the Chukuck suite contains halite and gypsum cement, and the total mode of halite in some Chukuck dolomites may reach 20–30 %. For example, the drill core 2 km to the northeast of the pipe cuts through halite-rich dolomite layers 1–3 m thick. The dolomite contains numerous karst cavities infilled with halite, gypsum, oil, and bitumen. The Chukuck suite in proximity to Udachnaya varies in thickness from 20–50 to 400 m based on drilling. The Chukuck suite and the coeval 520 Ma sedimentary rocks are thought to have accumulated in the sabkha environment (Polozov et al. 2008a, b) in a supratidal arid coastline and lagoon setting on a shallow bank with reefs (Fig. 3). This Daldyn-Markha carbonate bank (Sukhov 2001; Drozdov et al. 2008) is mapped in the area 200 by 50 km, encompassing the Udachnaya pipe (Fig. 1).

Hydrogeology of the Udachnaya Kimberlite

The southern kimberlite fields of the Yakutian province that host economic diamond deposits occur in platform areas where thick sediments contain buried ancient groundwater (Fig. 1) that varies in composition from ultrafresh to Na–Ca brines. Brines are sealed in continuous, 1,200-m-thick unit of gypsum-bearing Cambrian limestones, dolomites, and halite (Alekseev et al. 2007). Two of the southern kimberlite fields—Daldyn and Alakit-Markha (III, IV on Fig. 1)

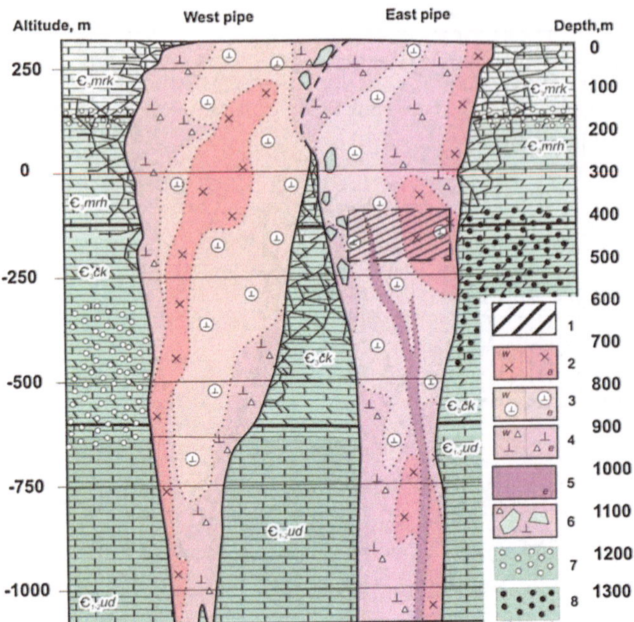

Fig. 2 A geological cross section of the Udachnaya kimberlite. *1* SFUE kimberlite; *2* phase 1 of hypabyssal kimberlites; *3* phase 2 of volcaniclastic kimberlite breccia; *4* phase 3 of massive volcaniclastic kimberlite breccia; *5* phase 4 of hypabyssal dyke kimberlites; *6* large blocks of country rocks in kimberlite; *7* Ca–Cl waters in the *upper* and *lower aquifers*; *8* Na–Cl waters in the *lower aquifer*. *Country rocks* $Є_{1-2}$ud (the Udachnaya suite, 500–600 m)—limestones, carbonate sandstones, gravelites, $Є_3$ck—the Chukuck suite, 260–340 m, coarse clastic dolomites and limestones cemented by cryptocrystalline calcite and gypsum. The middle part of the suite comprises karst dolomite with leached cavities infilled with gypsum, halite, bitumen, and oil. $Є_3$mrh (the Markha suite, 370 m thick)—limestones and dolomites with occasional beds of carbonate siltstone and argillite. $Є_{13}$mrk (the Morkoka suite, 220 m thick)—interstratified dolomitized limestones, dolomites, carbonate siltstones, argillites, carbonate conglomerates, and sandstones

(containing Udachnaya and Aykhal pipes, among others)—are situated within the zone of complete saturation of the sedimentary cover in metamorphosed highly mineralized brines (>0.2–0.4 g/cm^3) (Alekseev et al. 2007). In contrast to the southern kimberlite fields, the northern kimberlites erupt through buried low mineralized waters (0.03–0.05 g/cm^3) (Khar'kiv et al. 1991).

The Udachnaya kimberlite cuts through 2 aquifers (Drozdov et al. 1989 and Fig. 2) of the Olenek artesian basin. The upper aquifer was restricted to the fractured kimberlite close to the eroding surface and had disappeared after mining deepened the open pit. The roof of this aquifer was situated at 150–220 m into the mine pit. The thickness of the aquifer was no more than 5–10 m. The waters ($Mg_{40}Ca_{44}Na_{14}K_2$ $Cl_{98.2}Br_{0.7}$ $SO_{41.0}$ $HCO_{3\ 0.1}$) were Cl and Ca brines, with traces of Br, B, Rb, Li, and Sr and mineralization equal to 91.4 g/dm^3 (Drozdov et al. 2008). The lower aquifer of the Olenek artesian basin is mapped below 350 m to the east of the pipe and below 630 m to the west of the pipe. The aquifer has a very pronounced arched relief of the roof; the highest point of this convex roof is 200–250 m higher than that of the aquifer roof in the Udachnaya West and in the sedimentary country rocks (Fig. 4). This arched aquifer is hosted by clastic dolomites of the Chukuck suite and porous, sponge-like kimberlites with high permeability, comparable to that of the dolomites. Hydrogeological studies confirm that the clastic dolomites have a hydraulic connection with the porous kimberlites (Drosdov et al. 1989, 2008). The aquifer changes its hydrological properties and the composition of the water from the east to the west of the pipe. The latter has a low internal pressure and flow rate, similar to the sedimentary country rocks. The waters are Ca–Cl brines with composition $Mg_{19,5}Ca_{56.8}Na_{18.6}K_{5.1}$ $Cl_{99}Br_{0.85}$ $SO_{4\ 0.05}HCO_{3\ 0,0.1}$ and mineralization equal to 300–380 g/dm^3 (Drozdov et al. 2008). The aquifer of the Udachnaya-East kimberlite has a high hydraulic pressure. The waters are Na–Cl brines with composition $Mg_{4.4}Ca_{5.4}Na_{82.2}K_8$ $Cl_{99.1}Br_{0.47}$ $SO_{4\ 0.42}HCO_{3\ 0.01}$ and mineralization of 322 g/dm^3 (Drozdov et al. 2008). These anomalously sodic brines are restricted to the Chukuck suite of the Daldyn-Markha bank. The waters of the lower aquifer flow into the underground Udachnaya mine and deposit abundant halite and gypsum in the Udachnaya open pit (Fig. 4).

Geology and Mineralogy of the Udachnaya-East Kimberlite

The Udachnaya kimberlite comprises two separate pipes, the older West and the younger East pipe, that merge at 250 m below the present surface. Each of the Udachnaya pipes formed through 3–5 eruptive phases (Khar'kiv et al. 1991; Zinchuk et al. 1993) (Fig. 2). The East pipe is infilled with early hypabyssal kimberlite, commonly occurring on pipe margins (Fig. 2) and several phases of volcaniclastic kimberlite breccia cut by 10–50-cm-thick dykes of hypabyssal monticellite kimberlite (Kornilova et al. 1998). The breccia commonly contains clasts of early olivine- and calcite-rich hypabyssal kimberlite and occasionally clasts of olivine-, diopside-, and phlogopite-rich hypabyssal kimberlite (Egorov et al. 1986). The Udachnaya West kimberlite hosts serpentinized olivine through the entire explored depth (1,200 m), whereas the profile of olivine serpentinization of the Udachnaya-East kimberlite is complex. In the upper 400 m of the kimberlite, olivine is 80–95 % serpentinized; at 400–500 m, it is fresh; and at 700–1,200 m, the degree of serpentinization gradually increases to 100 %.

The SFUE kimberlite at 400–500 m depth (Fig. 2) was first described in 1976 (Marshintsev et al. 1976). The serpentine-free kimberlite comprises all 3 magmatic phases of the pipe at this depth. The contact between SFUE kimberlite

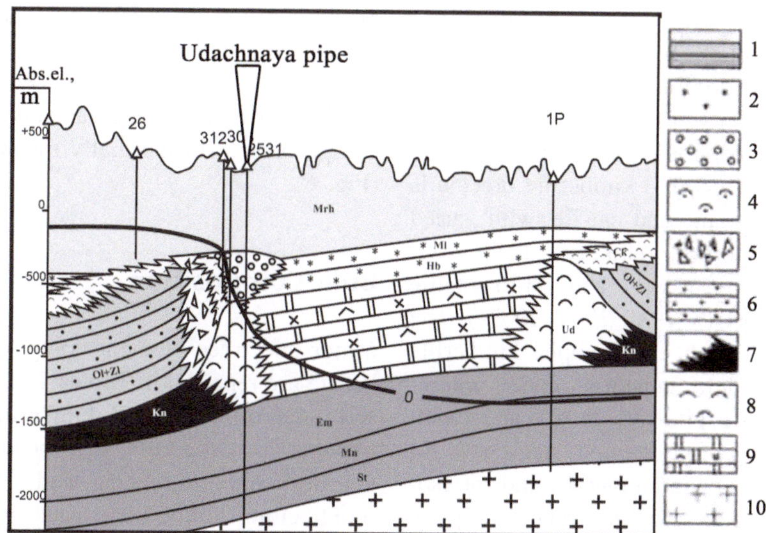

Fig. 3 A cross section through the Daldyn-Markha carbonate bank (Polozov et al. 2008a, b; Sukhov 2001; Kononov 2009) showing localization of the Udachnaya pipe. *1* Cambrian carbonates, *2* sabha deposits, *3* carbonate bar deposits, *4* terrigenous carbonate rock marls and dolomite breccias, *5* carbonate terrigenous rocks, *7* black shales, *8* reef rocks, *9* lagoon sulfate–salt–carbonate rocks, *10* granite–gneiss basement

Fig. 4 **a**, **b** Deposition of modern halite in the Udachnaya open pit at the 450–480 m level in 2008; **c**, **d** geodes in a leached kimberlite partly (**c**) or completely (**d**) infilled with gypsum

and more common kimberlites with serpentinized olivine is very sharp even on the scale of a single thin section (Fig. 5). The SFUE kimberlite is predominantly fragmental, containing 1–30 % xenoliths of country rocks and clasts of early massive hypabyssal phlogopite kimberlite (Egorov et al. 1986). The matrix of the SFUE kimberlite breccia is composed of phenocrystal olivine and calcite, with varied abundances of interstitial shortite, zemkorite, halite, sylvine, phlogopite, apatite, perovskite, and Ti-magnetite (Egorov et al. 1986; Kamenetsky et al. 2004, 2007a). The groundmass of hypabyssal diopside–phlogopite kimberlite found as clasts is composed of euhedral olivine microphenocrysts (20–25 %), phlogopite (30–35 %), and clinopyroxene (15 %) set in a fine-grained aggregate of carbonate, sodalite, shortite, opaque minerals, and interstitial shortite (Egorov et al. 1986). Textural relationships of S-, Na-, and Cl-rich minerals in thin sections demonstrate their crystallization in sequence serpentine – anhydrite + carbonate – halite (Pavlov and Ilupin 1973). Similar conclusions on the late crystallization of halite, shortite, and zemkorite with respect to other matrix and groundmass minerals have been made by Kornilova et al. (1998) and Egorov et al. (1988). The SFUE kimberlite breccia is cut by dykes of the SFUE coarse macrocrystal monticellite kimberlite (Kornilova et al. 1998). In contrast to the breccias, the groundmass of the late hypabyssal macrocrystal kimberlite contains up to 15–20 % monticellite and lower modes of carbonate minerals (Kornilova et al. 1998).

All varieties of SFUE kimberlite show development of secondary carbonate. Recrystallization of calcite makes it more coarse-grained and forms monomineralic calcite veins. SFUE kimberlite is also cut by late fine-grained gypsum veinlets and coarse veins of gypsum occasionally reaching 20 m in thickness. The kimberlite often hosts geodes with carbonate–chloride–gypsum minerals precipitated in late leached cavities (Fig. 4c–d). The SFUE kimberlite carries large angular xenoliths of carbonate–halide composition with thermally metamorphosed blue selvages (Fig. 6).

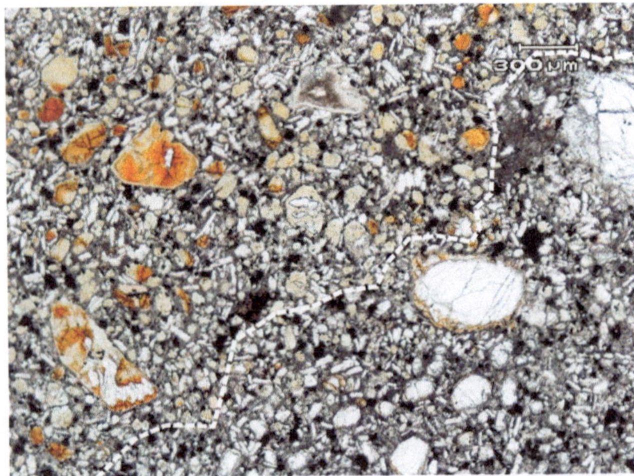

Fig. 5 A photomicrograph of hypabyssal SFUE kimberlite with calcite microlites. Note the sharp boundary between kimberlite with serpentinized olivine (SFUE, *upper left corner*) and kimberlite with fresh olivine (*low right corner*)

Bulk Composition

Major and trace element compositions of the SFUE kimberlite are reported in Tables 1, 2, and 3. The compositions were determined for 3 kimberlite phases: (1) the early hypabyssal kimberlite that occurs as clasts in breccia; (2) the kimberlite breccia; and (3) the coarse macrocrystal monticellite kimberlite occurring in late dykes. Oxide contents averaged by the kimberlite phase (Table 1) show

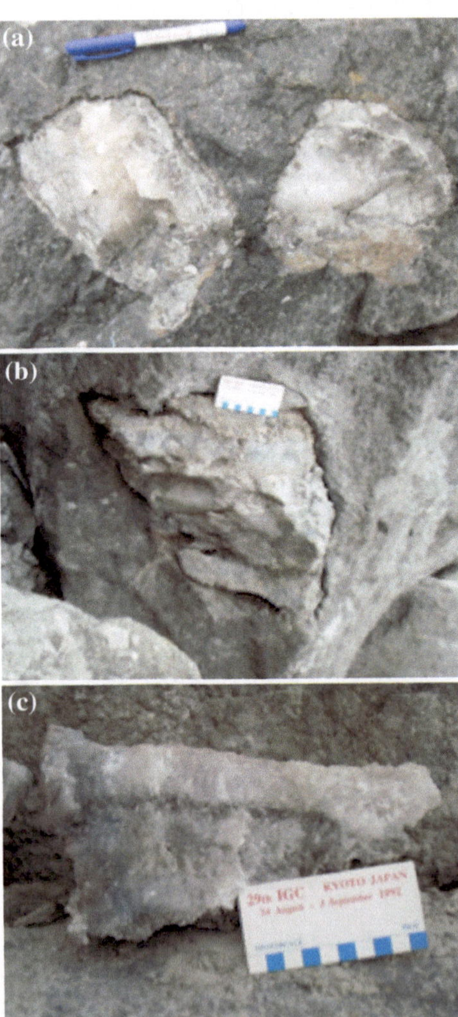

Fig. 6 Xenoliths of halite in the Udachnaya-East kimberlite. Note angular shapes of the xenoliths and a blue thermally metamorphosed margin of the xenolith on photograph C (Polozov et al. 2008a, b)

Table 1 Statistical parameters for major element compositions (in wt.%) for SFUE kimberlites of different phases

Number of analyses	Phase 1 hypabyssal kimberlite	Kimberlite breccia	Dyke hypabyssal kimberlite
	19	11	3
SiO_2, wt.%	**29.0 ± 2.4**	**28.2 ± 2.19**	**26.8 ± 4.2**
	24.1 – 33.3	*24.7 – 31.5*	*23.5 – 31.5*
TiO_2	**1.64 ± 0.48**	**1.37 ± 0.38**	**1.8 ± 0.51**
	0.96 – 2.91	*0.53 – 1.83*	*1.5 – 2.38*
Al_2O_3	**2.22 ± 0.5**	**2.13 ± 0.46**	**2.2 ± 0.43**
	1.51 – 3.21	*1.66 – 3.3*	*1.85 – 2.68*
Fe_2O_3	**3.3 ± 1.17**	**3.47 ± 0.52**	**4.6 ± 1.36**
	0.84 – 5.66	*2.56 – 4.05*	*3.14 – 5.86*
FeO	**5.84 ± 1.0**	**5.31 ± 1.08**	**5.8 ± 1.47**
	4.08 – 7.9	*3.33 – 7.25*	*4.77 – 7.5*
MnO	**0.16 ± 0.03**	**0.15 ± 0.03**	**0.19 ± 0.04**
	0.12 – 0.21	*0.08 – 0.19*	*0.16 – 0.23*
MgO	**31.4 ± 2.93**	**31.0 ± 2.38**	**30.5 ± 2.97**
	24.9 – 36.4	*26.7 – 35.5*	*27.7 – 33.6*
CaO	**13.3 ± 3.15**	**13.7 ± 2.44**	**14.2 ± 1.51**
	7.47 – 19.1	*9.25 – 16.8*	*12.4 – 15.14*
Na_2O	**0.70 ± 0.85**	**0.79 ± 0.94**	**0.26 ± 0.06**
	0.07 – 2.82	*0.09 – 3.1*	*0.19 – 0.3*
K_2O	**0.94 ± 0.35**	**1.29 ± 0.53**	**1.16 ± 0.27**
	0.42 – 1.63	*0.57 – 2.07*	*0.92 – 1.45*
H_2O	**2.02 ± 1.36**	**1.98 ± 1.41**	**1.71 ± 1.09**
	0.24 – 5.49	*0.55 – 4.19*	*0.87 – 2.94*
CO_2	**8.17 ± 3.27**	**9.58 ± 2.91**	**9.2 ± 3.25**
	3.19 – 17.1	*5.83 – 14.5*	*5.43 – 11.2*
P_2O_5	**0.47 ± 0.12**	**0.48 ± 0.16**	**0.53 ± 0.15**
	0.30 – 0.79	*0.30 – 0.85*	*0.36 – 0.66*
F	**0.16 ± 0.06**	**0.16 ± 0.06**	**0.14 ± 0.05**
	0.09 – 0.3	*0.06 – 0.22*	*0.08 – 0.17*
Cl	**0.67 ± 0.44**	**0.73 ± 0.67**	
	0.33 – 1.53	*0.34 – 2.23*	*0.46 – 0.52*

Comment The *bold values* in each cell refer to mean contents ± standard deviation of two sigma; the *numbers in italics* refer to the minimum and maximum values of oxide contents

that the early hypabyssal kimberlite and hypabyssal kimberlite of late dykes are richer in FeO_{tot} and TiO_2 than the breccia. Phlogopite kimberlite (sample Eg-Phl in Table 2) is characterized by extremely high contents of TiO_2, FeO_{tot}, and K_2O at very low concentrations of MgO and CO_2. The main trend in the major element composition for all 3 phases of the SFUE kimberlite shows an inverse correlation of CaO with SiO_2 and MgO, typical of all worldwide kimberlites.

The content of Na_2O (0.07–3.1 wt.%, 0.74 wt.% on average) in the Udachnaya-East kimberlite is higher than that of a typical kimberlite (0.16 wt.% Na_2O in Group I southern African kimberlite, Becker and Le Roex 2006). In our samples, sodium oxide is the only major element oxide that does not correlate with any other oxides, including H_2O.

An interesting pattern is observed in the covariation of Na_2O and CO_2. There is a threshold concentration of 9 wt.% CO_2, above which Na_2O content in kimberlite does not exceed 0.5 wt.%. Since much of CO_2 reside in secondary carbonate, this pattern could be interpreted as sealing of carbonated kimberlite to Na–Cl brines.

Concentrations of Na_2O and Cl in the Udachnaya-East vary widely. In our analyzed samples, the chlorine content is 0.3–2.2 wt.%. Depth-dependent Cl and Na_2O concentrations have been found by previous studies. The subsurface levels above 160 m in Udachnaya-East are characterized by low (<0.2 wt.%) Na_2O and Cl contents, which increase below 160 m (Pavlov and Ilupin 1973). The highest contents of Na_2O and Cl (up to 9 wt.%) are reached in SFUE kimberlites at 400–500 m (Kamenetsky et al. 2007a). Elevated Cl levels (2.2–4.2 wt.% Cl) are also noted at 760–860 m (Pavlov and Ilupin 1973). Superimposed on these depth patterns are spatially localized heterogeneities in the Na_2O and Cl contents within a depth horizon. It is interesting that the kimberlites of the three magmatic phases locally demonstrate identical contents of Na_2O and H_2O, although the contents vary widely depending on the sample location. This is shown for kimberlites of 3 phases (the hypabyssal kimberlite clasts in breccia, the breccia, and the late hypabyssal dykes) sampled in 6 locations in the 400–500-m depth interval (Fig. 7). The contents of H_2O, Na_2O, and, by inference, the mode of halite are thus controlled only by the spatial position of the kimberlite specimen, rather than by the composition of the kimberlite melt.

Trace element patterns of all 3 phases of the SFUE kimberlite demonstrate identical shapes at different absolute concentrations (Fig. 8, Table 3). The early hypabyssal kimberlite is richer in incompatible elements than in the breccia, while the highest concentrations of trace elements are found in the late hypabyssal kimberlite dykes.

Strontium Isotope Composition of Halite in the Udachnaya Kimberlite

We analyzed $^{87}Sr/^{86}Sr_i$ ratios for different textural varieties of halite, including salts deposited by brines in the open pit (Fig. 4), angular xenoliths of halite (Fig. 6), and a coarse halite–sylvine geode in a secondary leached cavity in the kimberlite. Salts deposited in the Udachnaya-East open pit have a $^{87}Sr/^{86}Sr_i$ of 0.708348, whereas extreme $^{87}Sr/^{86}Sr_i$

Table 2 Bulk compositions (in wt.%) of the Udachnaya-East SFUE kimberlite from mine level 400–500 m

Sample number	1	2	3	4	5	6	7	8	9
	07-6d	07-6b	01-213k	01-213b	05-75k	05-75b	00-137k	00-137b	10-4d
SiO_2, wt.%	25.34	24.75	28.08	30.32	30.20	29.00	24.06	25.68	23.49
TiO_2	1.50	1.06	2.48	1.83	1.45	1.83	1.09	0.53	2.38
Al_2O_3	1.85	1.96	2.41	1.82	1.85	1.98	2.19	1.66	2.68
Fe_2O_3	4.68	3.99	5.66	3.88	3.71	3.99	0.84	3.48	5.86
FeO	4.77	3.33	5.19	5.49	5.39	5.63	5.4	3.73	5.17
MnO	0.17	0.12	0.21	0.15	0.14	0.18	0.13	0.08	0.23
MgO	30.13	26.66	29.95	34.08	32.81	30.67	24.93	30.71	27.67
CaO	14.97	16.75	16.98	11.57	11.74	15.07	19.12	16.77	15.14
Na_2O	0.19	0.24	0.49	0.66	1.21	1.25	0.08	0.16	0.30
K_2O	1.11	2.03	0.42	1.03	0.55	0.57	0.70	1.47	1.45
H_2O	1.32	1.61	0.24	0.55	2.49	1.30	3.43	3.86	2.94
CO_2	10.97	14.54	6.25	7.49	7.08	6.79	17.1	13.15	11.16
P_2O_5	0.56	0.30	0.79	0.36	0.37	0.64	0.30	0.85	0.66
F	0.17	0.20	0.15	0.20	0.09	0.09	0.15	0.11	0.16
Cl	0.46	0.47	0.34	0.62	0.81	0.69	0.33	0.34	0.52
Total	98.19	98.01	99.64	100.05	99.89	99.68	99.85	102.58	99.81

	10	11	12	13	14	15	16	17	18
	10-4b	222/420k	222/420b	222/403k	218/415k	222/437b	03-180/450b	03-179/450b	222/467k
SiO_2	28.32	30.8	31.5	31.25	28.4	29.51	29.79	27.02	30.71
TiO_2	1.34	1.65	1.58	1.44	1.93	1.35	1.49	1.08	1.89
Al_2O_3	1.97	2.5	2.13	2.40	2.7	3.30	2.39	1.70	3.21
Fe_2O_3	3.42	2.74	2.56	2.64	3.64	4.05	2.87	2.97	3.88
FeO	5.64	7.45	7.25	6.70	6.00	6.20	5.63	4.81	6.35
MnO	0.17	0.17	0.15	0.15	0.19	0.19	0.14	0.14	0.20
MgO	30.98	33.56	35.48	34.64	31.85	30.38	29.36	31.56	30.49
CaO	13.46	11.22	9.25	11.77	13.64	14.11	10.56	14.34	13.02
Na_2O	0.28	2.63	1.83	0.38	0.88	0.62	3.10	0.19	2.82
K_2O	1.16	0.78	0.81	0.57	1.03	1.14	2.07	0.72	0.75
H_2O	2.11	0.94	0.59	1.85	0.76	0.65	4.19	3.97	0.54
CO_2	9.89	4.73	5.83	5.72	8.36	8.34	7.04	9.52	5.38
P_2O_5	0.56	0.53	0.33	0.47	0.51	0.51	0.39	0.47	0.63
F	0.16	0.09	0.06	0.09	0.23	–	0.22	0.22	–
Cl	0.38	–	–	–	–	–	–	–	–
Total	99.84	99.79	99.35	100.07	100.12	100.35	99.23	98.71	99.87

	19	20	21	22	23	24	25	26	27
	222/471k	222/476k	222/489k	222/504b	08-11b	07-5k	03\33k	03\91k	03\101k
SiO_2	30.19	33.3	26	25.48	28.89	28.56	27.79	28.74	28.10
TiO_2	1.78	1.9	1.34	1.45	1.569	0.957	1.53	1.29	2.91
Al_2O_3	2.46	2.88	2.43	2.43	2.11	2.39	1.52	1.77	1.78
Fe_2O_3	3.11	1.9	1.8	3.15	3.77	3.41	3.64	4.94	5.22
FeO	7.44	7.9	5.1	5.45	5.28	4.54	5.50	4.08	5.39
MnO	0.19	0.16	0.12	0.17	0.142	0.129	0.17	0.18	0.18
MgO	31.69	35.9	27.04	28.96	31.71	29.09	31.76	31.90	30.63

(continued)

Table 2 (continued)

	19	20	21	22	23	24	25	26	27
	222/471k	222/476k	222/489k	222/504b	08-11b	07-5k	03\33k	03\91k	03\101k
CaO	13.43	8.86	18.06	15.67	13.41	15.41	13.75	11.20	11.70
Na_2O	0.39	0.4	0.07	0.09	0.27	0.49	0.12	0.08	0.09
K_2O	0.95	0.49	1.63	1.93	1.273	1.289	0.92	0.82	1.06
H_2O	0.58	2.55	2.88	2.01	0.93	0.81	1.82	5.49	3.69
CO_2	7.39	3.19	12.54	13.20	9.6	10.25	9.57	7.59	7.59
P_2O_5	0.42	0.40	0.31	0.50	0.397	0.334	0.66	0.55	0.45
F	–	0.13	0.10	–	0.19	0.34	0.27	0.17	0.15
Cl	–	–	–	–	0.39	–	–	–	0.47
Total	100.02	99.96	99.42	100.49	99.93	98.19	99.02	98.80	99.41

	28	29	30	31	32	33	34	35	36
	03\142k	03\164k	04-114 k	222/A-1	222/A-8	Eg-Phl	Eg-j	Middle 34an	10-10
SiO_2	27.80	32.18	26.04	26.60	31.75	35.59	31.50	28.73	27.32
TiO_2	1.84	1.29	1.08	1.83	1.45	4.47	1.50	1.65	1.23
Al_2O_3	1.66	1.54	1.51	2.32	2.72	3.2	2.05	2.22	1.95
Fe_2O_3	3.83	3.11	3.52	2.47	3.46	4.24	3.14	3.52	4.56
FeO	6.12	6.17	4.90	5.55	5.75	7.65	7.50	5.72	2.96
MnO	0.16	0.17	0.14	0.14	0.17	0.19	0.16	0.16	0.13
MgO	30.52	36.37	31.50	28.22	34.68	24.45	33.58	31.00	30.56
CaO	12.37	7.47	16.15	16.4	9.39	10.01	12.44	13.39	10.36
Na_2O	1.91	0.67	0.26	0.17	0.19	2.53	0.28	0.74	0.72
K_2O	1.16	1.40	0.68	1.42	1.25	2.87	0.92	1.13	0.54
H_2O	2.44	2.96	0.70	2.72	1.54	2.88	0.87	2.01	10.48
CO_2	8.22	4.88	11.41	11.00	7.00	1.83	5.43	8.53	7.90
P_2O_5	0.54	0.42	0.43	0.45	0.43	0.07	0.36	0.47	0.38
F	0.17	0.30	0.18	0.10	0.10	0.04	0.08	0.15	0.14
Cl	1.53	1.11	0.44	–	–	–	–		2.23
Total	98.74	98.93	98.51	99.39	99.88	100.02	99.81	99.42	99.23

Comment In sample numbers, the last letter refers to kimberlite breccia (*b*), phase 1 hypabyssal kimberlite (*k*), hypabyssal kimberlite of late dykes (*d*), – not analyzed

values (0.712052) are measured in coarse halite in geodes (Table 4).

Discussion

Contrasts in Southern and Northern Yakutian Kimberlites

The mineralogy and genesis of the serpentine-free Udachnaya kimberlite cannot be understood without data on geology and country rocks of Yakutian kimberlites in general. As kimberlite is a hybrid rock, its composition is affected by lithology of the host country rocks (Mitchell 1986). Classical southern African kimberlites erupt through dolerites, quartz sandstones, and crystalline basement (Field and Scott Smith 1999) and show relatively low contents of CaO and CO_2 (Fig. 9a). The Archangelsk kimberlites that erupt through sandstones contain no more than 5 wt.% $CaCO_3$ (Fig. 9a). In this context, the average content of 23.6 wt.% CaO in Yakutian kimberlites is exceptional and relates to emplacement through and contamination by terrigenous carbonate sediments (Khar'kiv et al. 1991; Kostrovitsky 1986). Compared to other economic kimberlites, Yakutian kimberlites have incorporated distinctly more carbonate-rich country rock xenoliths (assessed as 15–20 vol %; Khar'kiv et al. 1991). In addition to high and variable CaO and CO_2 bulk contents in Yakutian kimberlites, the carbonate-rich country rocks bring about a very broad range of $\delta^{13}C$ values and higher positive $\delta^{18}O$ (Fig. 9b). When separated by the thickness of the sedimentary cover into northern and southern Yakutian

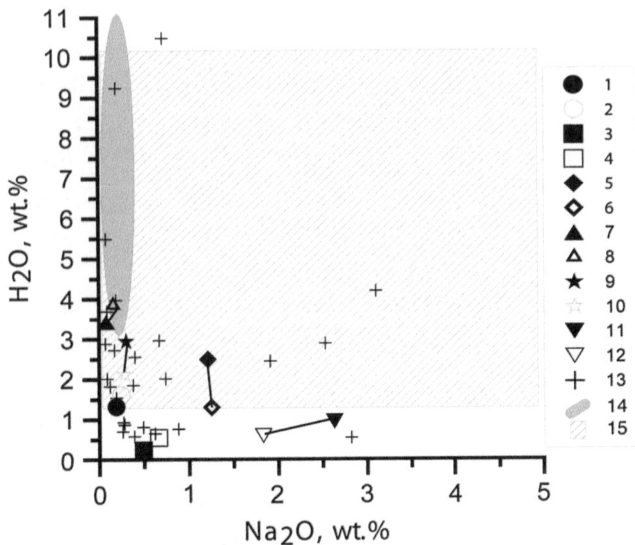

Fig. 7 A plot of Na_2O versus H_2O for SFUE kimberlites. Symbols connected by *tie-lines* show compositions of distinct kimberlite phases in composite specimens represented by either clasts of hypabyssal kimberlite in breccia, or breccia cut by late hypabyssal dykes. *1, 3, 5, 7, 9, 11* hypabyssal kimberlite of the clasts and the dykes; *2, 4, 6, 8, 10, 12* kimberlite breccia (see Table 1); *13* non-composite SFUE kimberlites; *14* Slave kimberlites (Kopylova and Hayman 2008); *15* Yakutian kimberlites (Khar'kiv et al. 1991)

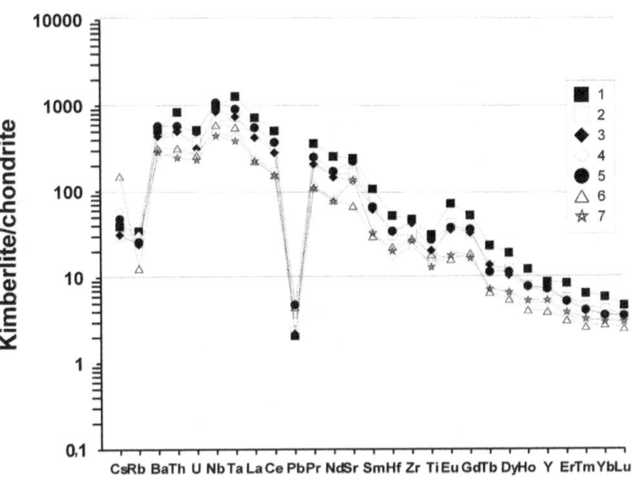

Fig. 8 Trace element concentrations in the Udachnaya-East kimberlite normalized to chondrite abundances (McDonough and Sun 1995). *Symbols 1* sample 10-4 HK, *2* sample 10-4 breccia, *3* sample 07-6 HK, *4* sample 07-6 breccia, *5* sample 03-142 PK, *6* sample 08-3 breccia, *7* sample 07-5 breccia

kimberlites, these two divisions demonstrate contrasting C and O isotope ratios. Northern kimberlites plot around the worldwide Group I kimberlites and incorporate values for magmatic carbonate, whereas southern kimberlites shift to higher values of $\delta^{18}O$ (15–23 ‰) and a broader range in $\delta^{13}C$. The systematics of the O–C isotopes can be explained by mixing of mantle melts with modern and ancient sedimentary carbonates (Fig. 9b).

Sedimentary carbonates in the local Paleozoic stratigraphy are accompanied by evaporite beds and strata of various lithologies saturated with ancient waters and brines. Mixing of country rock evaporites and groundwaters into kimberlites can be traced geochemically by elevated Cl, Na, and S. Southern Yakutian pipes (fields 1–4 in Fig. 1) are uniquely high in Cl and S contents. For example, sulfur is elevated in Mir, Udachnaya, and International'naya kimberlites (0.8–5.3 wt.%), but is low (0–0.3 wt.%) in northern Yakutian pipes (Khar'kiv et al. 1991), as it is in typical kimberlites worldwide. The average Cl content in Indian and African kimberlites is 0.03–0.06 wt.% (Ilupin et al. 1978), which is 10–100 times lower than that in the SFUE kimberlite (Tables 1, 2). Southern Yakutian kimberlites also demonstrate elevated Na_2O content, up to 5 wt.%, much higher than other worldwide kimberlites, exemplified by the Slave Craton kimberlites on Fig. 7. The halite abundance and spatial distribution match the lithology of country rocks at the corresponding levels. For example, the highest (8 vol %) abundance of halite is restricted to depth levels of southern diatremes (Mir, Udachnaya, and International'naya) where they cut through halite-rich evaporite. This is also true for gypsum and anhydrite (Khar'kiv et al. 1991).

Petrographic and Geochemical Evidence for Crustal Contamination of the SFUE Kimberlite

Mineralogical mapping of the Udachnaya-East pipe shows that the highest abundances of Na–Cl–S-bearing minerals occur at a depth interval that transcends 3 magmatic phases of kimberlites formed at different times from different magma batches (Fig. 2). This unequivocally suggests a secondary origin for the Na–Cl–S-bearing minerals. A secondary origin is also supported by the textural relationships of halite and alkali carbonate to serpentine and other groundmass minerals as observed in thin sections (Pavlov and Ilupin 1973; Egorov et al. 1988; Kornilova et al. 1998). Furthermore, halite, gypsum, and carbonate are present not only in veins cutting through kimberlite and xenoliths (Egorov et al. 1988; Sharygin et al. 2007a, b; Polozov et al. 2008a), but also in modern brine deposits in the pipe and in the secondary geodes (Fig. 4).

The secondary nature of the Na–Cl–S mineralization is also suggested by our observation that contents of Na_2O and H_2O are a function of the spatial position of the kimberlite specimen in the pipe (Fig. 9) rather than the melt compositions of distinct magma batches. Moreover, we found that the inverse correlation between H_2O and Na_2O demonstrated on Fig. 1 of Kamenetsky et al. (2009b) disappears if the data are plotted on a linear rather than on a logarithmic scale. Our analyses of the Udachnaya-East kimberlite show no inverse correlation between H_2O and Na_2O (Fig. 9). The absence of

Table 3 Rare element composition (in ppm) of SFUE kimberlites

	03-142 k	07-6d	07-6b	10-4d	10-4b	08-3b
Li, ppm	–	–	–	22	14.3	
Be	1.54	1.37	1.08	2.25	1.81	1.21
Sc	12.0	11.5	6.5	18	11.4	8.6
V	143	91	88	137	104	98
Cr	1,244	1,266	706	2,186	1,261	968
Co	67	67	52	76	81	28.5
Ni	885	939	817	801	1154	844
Cu	64.9	50.6	36.5	60	60	0.95
Zn	68	67	51	72	59	8.7
Ga	3.54	3.92	2.56	7.0	3.97	2.92
Ge	0.8	0.72	0.63	0.69	0.71	0.83
Rb	59.6	55.8	71.3	79	59	29.5
Sr	1,617	964	979	1,745	1,194	486
Y	11.5	11.5	7.9	13.7	9.8	6.2
Zr	169	160	85	179	119	107
Nb	258	203	118	232	153	143
Mo	2.22	0.43	0.55	1.47	1.10	1.58
Sn	1.78	1.45	1.22	1.56	1.08	1.01
Cs	0.91	0.6	1.11	0.75	0.56	2.88
Ba	1,391	1,059	619	1,258	983	768
La	131	100	57	171	109	53
Ce	228	173	105	310	185	95
Pr	23.2	19.2	10.7	33	20	10.3
Nd	77.8	66.5	36	116	70	35.1
Sm	9.7	9.0	4.8	16	9.3	4.4
Eu	2.1	1.98	1.06	3.94	2.36	0.90
Gd	7.1	6.5	3.7	12.0	7.0	3.7
Tb	0.41	0.5	0.26	0.98	0.79	0.24
Dy	2.83	2.58	1.62	4.70	3.08	1.35
Ho	0.43	0.42	0.27	0.67	0.44	0.22
Er	0.84	0.83	0.56	1.35	0.91	0.50
Tm	0.1	0.1	0.07	0.16	0.11	0.07
Yb	0.58	0.58	0.52	0.94	0.67	0.45
Lu	0.09	0.09	0.07	0.11	0.085	0.06
Hf	3.51	3.44	1.78	5.3	3.1	2.31
Ta	12.4	10.1	5.82	17	10.0	7.60
W	6.65	1.99	0.81	2.43	1.47	1.57
Pb	11.8	5.41	7.5	5.2	5.9	11.2
Th	16.7	14.5	8.07	24	13.6	9.3
U	3.7	2.36	2.07	3.83	2.8	1.95

Comment In sample numbers, the last letter refers to kimberlite breccia (*b*), phase 1 hypabyssal kimberlite (*k*), hypabyssal kimberlite of late dykes (*d*)

any correlation between Na_2O and H_2O in Udachnaya-East is typical of kimberlites in general, including kimberlites that contain unserpentinized macrocrysts and phenocrysts of olivine. Such samples are rare in Siberia, but common worldwide (Scott Smith 1996; Kopylova and Hayman 2008). These fresh kimberlites have no more than 0.2 wt.% Na_2O,

Table 4 Major, trace element and Sr–Nd isotopic composition of salts in the SFUE kimberlite

Sample no.	Salt recently deposited in the open pit	Halite xenoliths	Geode with coarse halite and sylvine in SFUE
	08-21	10-10 Na	10-22
Mg, wt.%	0.287	0.36	0.215
Al	0.0073	0.039	0.033
Si	0.12	0.241	0.171
NaCl	99.25	98.18	95.66
K	0.0459	0.03	3.41
Ca	0.246	0.616	0.224
Fe	0.0154	0.0578	0.0753
Br	0.0283	0.0342	0.041
S	–	0.386	0.0011
Sr, ppm	21	140	11
Rb, ppm	0.39	0.64	6.8
$^{87}Rb/^{86}Sr$	0.05374	0.01323	1.79268
$(^{87}Sr/^{86}Sr)_i$	0.708348	0.708227	0.72205
εSr_i	54.6	59	255.3
Age (Ma)	0	360	360

while H_2O contents vary independently from 3.8 to 10.7 wt.% (Fig. 9). The absence of Na_2O–H_2O correlation is supported by findings of kimberlite with serpentinized olivine that contains halite in the groundmass and halite xenoliths. For example, in sample 10-10 (Table 2), high H_2O content (10.48 wt.%) is accompanied by high (0.74 % wt.%) Na_2O, implying a significant mode of halite in the matrix. This evidence demonstrates that Na-rich kimberlite compositions are not solely restricted to unserpentinized kimberlites and that groundmass serpentine does not replace primary alkali- and chlorine-bearing minerals.

The strongest evidence for the secondary origin of Na-, Cl-, and S-rich minerals in the Udachnaya-East kimberlite and in all southern Yakutian kimberlites, in our opinion, is the regional correlation between the geology and hydrogeology of the local country rocks and the mineralogy of Yakutian kimberlites, in particular the difference between southern and northern kimberlites. Na–Cl–S mineralization is more significant for southern kimberlites that erupt through the thicker, evaporite-bearing sequence of carbonate sediments (Fig. 2b).

Isotopic Evidence for Crustal Contamination of the SFUE Kimberlite

Values of C and O isotopic ratios for the Udachnaya-East kimberlite were reported by Egorov et al. (1986), Kostrovitsky (1986), and Khar'kiv et al. (1991, 1998). Carbon isotopic ratios in SFUE kimberlite plot among the highest values in the field of contaminated Yakutian kimberlites (Fig. 9b). Isotopic compositions of C (−3.4 to −0.8 ‰ $\delta^{13}C$) are heavier than those of the mantle and require mixing with sedimentary marine carbonates (Fig. 9b) that may have been sourced from country rock carbonate xenoliths. Isotopic compositions of O vary from 15.36 to 18.59 ‰ $\delta^{18}O$, which is between the mantle (+8 ‰) and the marine sedimentary (+22 ‰) values. Both C and O isotopes of the Udachnaya-East kimberlite thus reflect mixing of a mantle melt with crustal carbonates.

The same conclusion was reached by Khar'kiv et al. (1991). The Udachnaya-East kimberlite has $\delta^{13}C$ equal to (−3.4)–(−2.7) ‰ at 20–400 m, (−1.6)–(−1.3) ‰ at 400–600 m, and (−1.5)–(−2.1 ‰) at 600–900 m (Khar'kiv et al. 1991). The highest $\delta^{13}C$ values are observed at the depth where SFUE kimberlite occurs, suggesting the most significant contamination by heavy crustal carbon at 400–600 m.

Sulfur isotopes in southern Yakutian kimberlites are characterized by high $\delta^{34}S$ (15–53 ‰) (Fig. 10). Mantle samples and magmas, including kimberlites, generally plot around $\delta^{34}S = 0$ ‰, but the Udachnaya $\delta^{34}S$ values are closer to that of the local sedimentary rocks. Another study of sulfur isotopes in sulfate minerals in the southern Yakutian kimberlites (Ilupin et al. 1978) also indicates a sedimentary source of sulfur.

Chlorine isotopic ratios of halite in SFUE kimberlite (−0.25 to 0.4 ‰ $\delta^{37}Cl$, Sharp et al. 2007) are consistent with evaporites (−0.5 to 0.5 ‰ $\delta^{37}Cl$) and Udachnaya brines (−0.33 to 0.52 ‰, Alekseev et al. 2007).

Strontium isotopic ratios measured for different kimberlite minerals and rock components constrain the history of kimberlite crystallization. Sr isotopic ratios ($^{87}Sr/^{86}Sr_i$) of the SFUE megacrystic clinopyroxene and groundmass perovskite are 0.70292–0.70308 (Kamenetsky et al. 2009b). These values are much lower than Sr isotopic ratios of the bulk Udachnaya kimberlite, water, and acid leachates (Maas et al. 2005; Kamenetsky et al. 2009b). The difference $\Delta^{87}Sr/^{86}Sr_i$ of up to 0.006 between the bulk kimberlite Sr ratios (0.7033–0.709; Kostrovitsky 1986; Kornilova et al. 1998; Kostrovitsky et al. 2007; Maas et al. 2005; Fig. 11) and those of the kimberlite minerals exceeds that for other kimberlites, reflecting a particularly complex history of crustal contamination of the Udachnaya kimberlite. Typically, Sr ratios of the bulk kimberlite are 0.001–0.0053 higher than those of perovskite and megacrysts (Paton et al. 2007; Kostrovitsky et al. 2007; Kopylova et al. 2009; Woodhead et al. 2009).

The $^{87}Sr/^{86}Sr_i$ values in bulk SFUE kimberlite are 0.70438–0.709, which are higher than those in Group I kimberlites (Fig. 11). These values do not include analyses

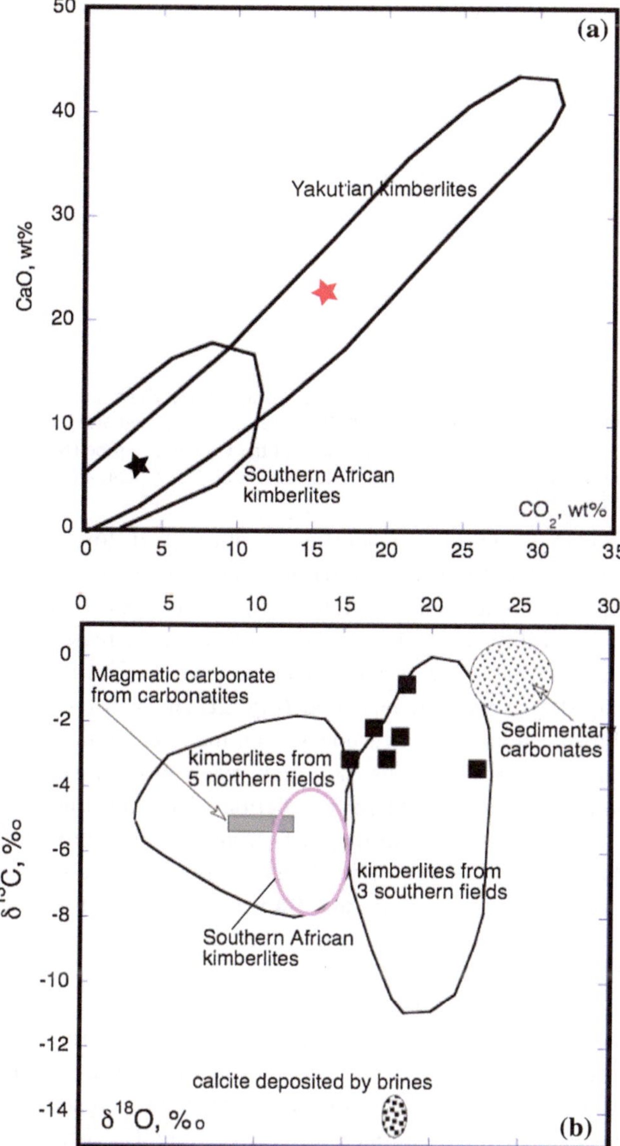

Fig. 9 Compositions of Yakutian kimberlites, modified after Khar'kiv et al. (1991). **a** A CaO–CO_2 (wt.%) plot. The average contents of these oxides in Yakutian kimberlites (*red star*) are calculated for 171 analyses. The average for the Archangelsk kimberlites (*black star*) is from Bogatikov et al. (1999). **b** A plot of O and C isotopes in the SFUE kimberlite in comparison with data on Yakutian and southern African kimberlites and carbonates in various rock types. Isotopic ratios for SFUE kimberlite (Egorov et al. 1986) are denoted as *black squares*

of salts, water-soluble fraction of the SFUE kimberlite, and acid leachates that yielded "unrealistically" low initial $^{87}Sr/^{86}Sr$ ratio (<0.700) (Maas et al. 2005), in the authors' words. The highest $(^{87}Sr/^{86}Sr)_i$ ratio of 0.709 is observed in the bulk sample with the most intense development of secondary carbonate. The Sr isotopes reflect the mixture of the kimberlite with Udachnaya brines ($^{87}Sr/^{86}Sr_i = 0.7088 - 0.7092$, Alekseev et al. 2007) or halite xenoliths (0.707871,

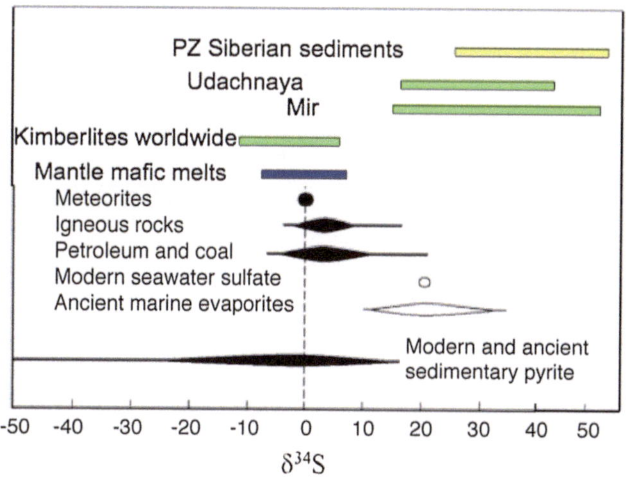

Fig. 10 Stable isotope ratio $\delta^{34}S$ for various terrestrial rocks and minerals (Seal 2006), southern Yakutian kimberlites, and their country rock sediments (Vinogradov and Ilupin 1972)

Table 4). εNd of the SFUE kimberlite is 2–4 (Maas et al. 2005), typical of the asthenospheric magmas, including Group I kimberlites. As Nd isotopes are not affected by crustal carbonate contamination (Kostrovitsky et al. 2007), mantle values of Nd ratios in the SFUE kimberlite cannot be used as an argument for a pristine mantle origin.

The Origin of Na–Cl–S Minerals in the SFUE Kimberlite

The contamination of the SFUE kimberlite and other southern Yakutian kimberlites by Na–Cl–S crustal material occurred, in our opinion, in two different ways. The kimberlites may have been contaminated by buried brines. The upper contact of the SFUE kimberlite coincides with the roof of the deep Udachnaya aquifer, with a high hydraulic pressure and anomalously Na-rich brines. The Udachnaya pipe is on the boundary between strata saturated with Na–Cl and Ca–Cl brines. The western part of the pipe was penetrated by Ca–Cl brines, whereas the eastern part of the pipe at the same depth is inundated by Na–Cl brines. It is possible that this hydrogeological feature controlled formation of the SFUE kimberlite with high modes of Na-, Cl-, and S-bearing minerals. It is these brines that precipitate halite in the Udachnaya-East open pit and leach solid, fully crystallized kimberlite producing cavities and geodes with Na–Cl–S minerals (Fig. 4). It was shown experimentally that even short (2–6 days) contact of kimberlite with brines changes its composition (Khar'kiv et al. 1991).

The second possible origin of extraneous Cl, S, and alkalies may be assimilation of carbonate-, sulfate-, and chloride-rich country rock xenoliths. Large xenoliths of halite found in the kimberlite were likely sourced from the

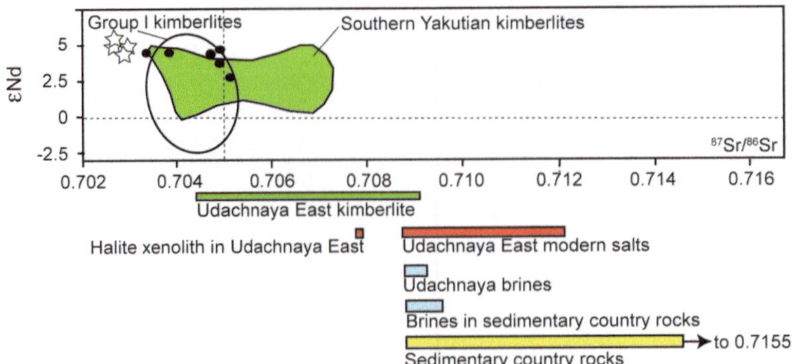

Fig. 11 Initial ratios of Sr and Nd isotopes for Group I and southern Yakutian kimberlites. Open stars mark compositions of perovskite and a clinopyroxene megacryst from the SFUE kimberlite (Kamenetsky et al. 2009b). *Bars below* the Sr–Nd plot indicate bulk Sr compositions of the Udachnaya kimberlite (Maas et al. 2005; Kamenetsky et al. 2009b; Kostrovitsky et al. 2007), halite xenoliths and modern salts (Table 4), sedimentary country rocks of the Udachnaya pipe (Kononov 2009), and Udachnaya brines (Alekseev et al. 2007; Kononov 2009)

Middle Cambrian Chukuck suite of the Daldyn-Markha bank. The xenoliths cannot be late chloride-rich segregations from a residual kimberlite melt (Kamenetsky et al. 2007b) as they have thermally metamorphosed margins (Polozov et al. 2008a, b) and angular shapes (Fig. 6). The contrasting compositions of evaporites and kimberlite melt should cause skarn-like exchange of components on contacts between xenoliths and kimberlite. Similar zones of mini-skarns are observed on contacts of felsic xenoliths with kimberlites, introducing elements enriched in xenoliths to kimberlite and leading to pervasive crystallization of hybrid minerals (Caro et al. 2004). Assimilation of evaporite xenoliths occurs at relatively high, magmatic temperatures producing hybrid melt with elevated contents of Na, K, Cl, and S. This melt is trapped as Cl-rich fluid inclusions in secondary fractures of olivine (Golovin et al. 2003). Alkali-, sulfur-, and chlorine-rich minerals may have crystallized from this late hybrid melt and may be "comagmatic" with kimberlite, as found by Maas et al. (2005).

The Relationship Between Serpentinization and Na–Cl–S Mineralization

The reason for a serpentine-free character of some Udachnaya-East kimberlite rich in Na, Cl and S minerals is not fully understood. Serpentine in kimberlites has a variety of origins, depending on its textural position. Interstitial serpentine of the groundmass may be magmatic (Scott Smith 1996) crystallizing at T > 600 °C (Kopylova et al. 2007). Serpentine replacing monticellite and olivine rims may be deuteric (T > 300 °C, Mitchell 2006), and serpentine replacing large olivine may have formed after the emplacement from meteoric groundwaters.

The absence of all textural types of serpentine in SFUE kimberlite is easily explained by the model of the evaporite country rock contamination. The hybrid residual kimberlite melt that digested evaporite xenoliths had lower H_2O activity due to increased halogen and alkali abundances (e.g., Aranovich and Newton 1996). It has been found that the solubility of H_2O in haplogranite melt decreases from 5 to 3.5 wt.% with increasing concentrations of alkali chlorides in the aqueous fluid (Fig. 2B of Webster 1992). Chlorine solubility in carbonate-rich melts is unknown, but when Cl concentrations in a melt reach the critical minimum amount required for exsolution of hydrosaline chloride fluid, the fluid removes H_2O from the melt completely (Aiuppa et al. 2009). High-salinity fluids reduce the stability field of serpentine (Sharp and Barnes 2004) so that the melt is expected to form late interstitial alkali carbonates and chlorides instead of serpentine. The absence of serpentinization of olivine macrocrysts is also controlled by the less hydrous composition of residual melts and deuteric fluids. Low-P crystallization experiments on kimberlites demonstrated that at 895 °C, forsterite peritectically reacts with the residual melt saturated with H_2O and CO_2 (Franz and Wyllie 1967). The absence of carbonatitic hydrous reactive melt would hinder the peritectic reaction and keep olivine intact. Experiments on crystallization of K-rich melts also show that in the absence of water, olivine is stable to lower temperatures (by 50–150 °C, Fig. 11 of Yoder 1986) than in the respective wet composition. Moreover, it has been observed that olivine is fresher in calcite-rich kimberlites (Mitchell, personal communication), again emphasizing the critical role of deuteric fluid composition on the kimberlite mineralogy.

A model of the SFUE kimberlite origin due to a reaction with buried brines explains the fresh character of olivine by the impeded serpentinization. Experiments on interaction between a solid kimberlite and meteoric water with elevated Na and Cl contents at room temperatures demonstrated that such water cannot serpentinize olivine (Lashkevich and Egorov 1988). This effect was also modeled for abyssal peridotites (Novoselov 2010). This model, however, cannot explain the absence of magmatic serpentine that crystallizes as late groundmass mineral in fresh kimberlites. For this phenomenon, we propose a uniquely low H_2O content (<0.5–1.0 %) for the Udachnaya-East primary kimberlite melt. The observed H_2O contents of up to 5.5 % in the SFUE kimberlite (Table 2) may be secondary, similar to the Na_2O contents, as evidenced by the spatial control on the H_2O and Na_2O concentrations (Fig. 7).

Do Melt and Fluid Inclusions in Olivine and Fibrous Diamond Prove the Mantle Origin of Chlorides in Kimberlites?

Melt inclusions in kimberlitic olivines with chloride- and carbonate-rich composition (Golovin et al. 2003, 2007; Sharygin et al. 2007a, b) and alkali- and Cl-rich fluid inclusions in fibrous diamonds were cited (Kamenetsky et al. 2004) as evidence for the mantle origin of chlorides in kimberlites. Below we present a critique of these statements.

Chloride-and alkali-rich melt inclusions in kimberlitic olivines may not imply the mantle source of Cl, Na, and K, as the inclusions are found only in olivine microfractures and are clearly secondary (Golovin et al. 2003, 2007). The inclusions trapped by thermally and explosively cracked olivines supply information on late residual melts and fluids equilibrated with kimberlite melts in the subsurface. Compositions of these inclusions contrast to primary deep-seated CO_2-rich inclusions in Udachnaya olivine (Sobolev et al. 1989). Furthermore, secondary composite melt/fluid inclusions in Udachnaya olivines differ significantly from those in the Slave and Greenland kimberlitic olivines. The former are distinctly higher in (Na + K)/(Na + Ca + K) (Fig. 8A of Kamenetsky et al. 2009a). Only the Udachnaya olivine inclusions contained anhydrous carbonates or sulfates, whereas inclusions in all other studied kimberlites worldwide hosted hydrated carbonates and sulfates indicating a greater amount of water in residual magmas (Mernagh et al. 2012). The distinct mineralogy and composition of the secondary fluid inclusions in Udachnaya-East imply a unique process of late, subsurface H_2O depletion and Na enrichment.

The similarity between compositions of fluid inclusions in fibrous diamonds and constrained alkali-rich composition of primary kimberlite melt (Fig. 3 of Kamenetsky et al. 2004) is fictitious. The conventional representation of the compositional data in the (K + Na)–Ca–Si triangle plots all alkalis together and masks significant differences between Na-rich Udachnaya compositions (Table 2) and K-rich diamond inclusions (e.g., Klein-BenDavid et al. 2009). Moreover, the sulfur-rich character of the Udachnaya kimberlite (Khar'kiv et al. 1991) does not have parallels with mantle fluids or diamond fluid inclusions.

Conclusions

SFUE kimberlite and other kimberlites of southern Yakutian fields rich in halite, alkali carbonate, gypsum, and other salts acquired the high Na, S, and Cl contents by interaction with buried Cambrian Na–Ca–Cl brines or assimilating evaporite xenoliths. The classic view of kimberlites as Na_2O-poor rocks with high K/Na ratios (Dawson 1980; Mitchell 1986) remains unchanged.

Acknowledgments The studies were supported by the integration projects of the Russian Academy of Sciences № 59, 115, and 27.1 and by an NSERC Discovery Grant to MGK. The authors thank G. Sandimirova and Yu. Pakhol'chenko for isotopic analyses, A. Polozov for helpful discussions, and E. Smith for the editorial help.

References

Aiuppa A, Baker DR, Webster JD (2009) Halogens in volcanic systems. Chem Geol 263:1–18

Alekseev SV, Alekseeva LP, Borisov VN, Shoukar-Stash O, Frape SK, Chabaux F, Kononov AM (2007) Isotopic composition (H, O, Cl, Sr) of ground brines of the Siberian platform. Russ Geol Geophys 48:225–236

Aranovich LY, Newton RC (1996) H_2O activity in concentrated NaCl solutions at high pressures and temperatures measured by the brucite-periclase equilibrium. Contrib Mineral Petr 125(2–3): 200–212

Bobrievich AP, Bondarenko MN, Gnevushev MA, Krasov LM, Smirnov GI, Yurkevich RK (1959) Diamondiferous deposits of Yakutia. State scientific-techniques publishing house, Moskow, p 527 (in Russian)

Bogatikov OA, Garanin VK, Kononova VA, Kudryavtceva GP, Vasil'eva EP, Verjak VV, Verichev EM, Posuhova TV (1999) The Arkhangelsk diamondiferous province (geology, petrography, geochemistry and mineralogy). MGU, Moskow, p 524

Brakhfogel FF (1984) Geological aspects of kimberlite magmatism in the north-east of the Siberian platform. Yakutsk, USSR, p 128 (in Russian)

Caro G, Kopylova MG, Creaser RA (2004) The hypabyssal 5034 kimberlite of the Gahcho Kue cluster, southeastern slave craton, Northwest Territories, Canada: a granite-contaminated Group-I kimberlite. Can Mineral 42:183–207

Dawson JB (1980) Kimberlites and their xenoliths. Springer, New York, p 252

Drozdov AV, Egorov KN, Gotovcev SP, Klimovsky IV (1989) Hydrogeology and hydrochemical zoning of the Udachnaya kimberlite. In: Complex cryo-hydrogeological investigations. Yakutian Branch of Russian Academy Sciences, Yakutsk, pp 145–146 (in Russian)

Drozdov AV, Iost NA, Lobanov VV (2008) Criohydrogeology of diamond mines of Western Yakutia. Irkutsk State Technical University, Irkutsk, p 508 (in Russian)

Egorov KN, Kornilova VP, Safronov AF, Filippov ND (1986) Micaceous kimberlite from the Udachanaya-East pipe. Trans Russ Acad Sci 291(1):199–202 (in Russian)

Egorov KN, Ushzhapovskaya ZF, Kashaev AA (1988) Zemkorite, the new carbonate from Yakutian kimberlites. Dokl Acad Sci USSR 301(1):188–193 (In Russian)

Field M, Scott Smith BH (1999) Contrasting geology and near-surface emplacement of kimberlite pipes in Southern Africa and Canada. In: Gurney JJ, Gurney JL, Pascoe MD, Richardson SH (eds) 7th International kimberlite conference, vol 1. Red Roof Design, Cape Town, RSA, pp 214–237

Franz GW, Wyllie PJ (1967) Experimental studies in the system CaO-MgO-SiO_2-CO_2-H_2O. In: Wyllie PJ (ed) Ultramafic and related rocks. Wiley, New York, pp 323–326

Golovin AV, Sharygin VV, Pokhilenko NP, Malkovets VG, Kolesov BA, Sobolev NV (2003) Secondary melt inclusions in olivine of unaltered kimberlites of the Udachnaya-East. Trans Russ Acad Sci 388(3):199–202 (in Russian)

Golovin AV, Sharygin VV, Pokhilenko NP (2007) Melt inclusions in olivine from unaltered kimberlites of the Udachnaya-East: some aspects of kimberlite melt evolution on late stages of crystallization. Petrologiya 15(2):178–195 (In Russian)

Ilupin IP, Kaminsky FV, Frantceson EV (1978) Geochemistry of kimberlites. Moskow, Nedra, p 352 (in Russian)

Kamenetsky MB, Sobolev AV, Kamenetsky VS, Maas R, Danyushevsky LV, Thomas R, Pokhilenko NP, Sobolev NV (2004) Kimberlite melts rich in alkali chlorides and carbonates: a potent metasomatic agent in the mantle. Geology 32(10):845–848

Kamenetsky VS, Kamenetsky MB, Sharygin VV, Golovin AV (2007a) Carbonate-chloride enrichment in fresh kimberlites of the Udachnaya-East pipe, Siberia: a clue to physical properties of kimberlite magmas? Geophys Res Lett 34(9):L09316

Kamenetsky VS, Kamenetsky MB, Sharygin VV, Faure K, Golovin AV (2007b) Chloride and carbonate immiscible liquids at the closure of the kimberlite magma evolution (Udachnaya-East kimberlite, Siberia). Chem Geol 237(3–4):384–400

Kamenetsky VS, Kamenetsky MB, Weiss Y, Navon O, Nielsen TFD, Mernagh TP (2009a) How unique is the Udachnaya-East kimberlite? Comparison with kimberlites from the Slave Craton (Canada) and SW Greenland. Lithos 112:334–346

Kamenetsky VS, Maas R, Kamenetsky MB, Paton C, Phillips D, Golovin AV, Gornova MA (2009b) Chlorine from the mantle: magmatic halides in the Udachnaya-East kimberlite, Siberia. Earth Planet Sci Lett 285(1–2):96–104

Khar'kiv AD, Zuenko VV, Zinchuk NN, Kruchkov AI, Ukhanov VA, Bogatykh MM (1991) Kimberlite petrochemistry. TCNIGRI, Yakutsk, p 302 (in Russian)

Khar'kiv AD, Zinchuk NN, Kruchkov AI (1998) Primary diamond deposits of the world. Moscow, NEDRA, p 556 (in Russian)

Klein-BenDavid O, Logvinova AM, Schrauder M, Spetius ZV, Weiss Y, Hauri EH, Kaminsky FV, Sobolev NV, Navon O (2009) High-Mg carbonatitic microinclusions in some Yakutian diamonds—a new type of diamond-forming fluid. Lithos 112S:648–659

Kononov AM (2009) Salty waters and brines of the Olenek aquifer. PhD Thesis, The Institute of the Earth's Crust, Irkutsk, pp 350 (in Russian)

Kopylova MG, Hayman P (2008) Petrology and textural classification of the Jericho kimberlite, Northern Slave Province, Nunavut, Canada. Can J Earth Sci 45:701–723

Kopylova MG, Matveev S, Raudsepp M (2007) Searching for parental kimberlite melt. Geochim Cosmochim Acta 71(14):3616–3629

Kopylova MG, Nowell GM, Pearson DG, Markovic G (2009) Crystallization of megacrysts from protokimberlitic fluids: Geochemical evidence from high-Cr megacrysts in the Jericho kimberlite. Lithos 112 (S1):284–295

Kornilova VP, Egorov KN, Safronov AF, Filippov ND, Zaytsev AI (1998) Mointichellite kimberlite of the Udachnaya pipe and some aspects of the kimberlite melt evolution. Russ Geol 6:48–51 (In Russian)

Kostrovitsky SI (1986) Geochemistry of kimberlite minerals. Novosibirsk, Nauka, p 263 (in Russian)

Kostrovitsky SI, Morikio T, Serov IV, Yakovlev DA, Amirzhanov AA (2007) Isotope and geochemical systematics of the Siberian platform kimberlites. Geol Geophys 48(3):350–371 (in Russian)

Lashkevich VV, Egorov KN (1988) Theoretical modeling of hydrothermal-metasomatic processes in the kimberlites. In: Thermodynamics in geology. Abstracts of the second all-union symposium, vol 1, Miass, pp 139–141 (in Russian)

Maas R, Kamenetsky MB, Sobolev NV, Kamenetsky VS, Sobolev AV (2005) Sr, Nd, and Pb isotope evidence for a mantle origin of alkali chlorides and carbonates in the Udachnaya kimberlite, Siberia. Geology 33(7):549–552

Marshintsev VK, Migalkin KN, Nikolaev NS, Barashkov YuP (1976) Unaltered kimberlite of the Udachnaya-East kimberlite. Dokl Acad Sci USSR, Earth Sci Ser 231(4):961–964 (in Russian)

McDonough WF, Sun SS (1995) The composition of the Earth. Chem Geol 120:223–253

Mernagh TP, Kamenetsky VS, Kamenetsky MB (2012) A Raman microprobe study of melt inclusions in kimberlites from Siberia, Canada, SW Greenland and South Africa. Spectrosc Acta Pt A-Molec Biomolec Spectr 80(1):82–87

Mitchell RH (1986) Kimberlites: mineralogy, geochemistry and petrology. Plenum Press, New York, p 442

Mitchell RH (2006) Petrology of hypabyssal kimberlites. In: Abstract of the kimberlite emplacement workshop, 8th international kimberlite conference, Saskatoon, Sept 2006. http://www.venuewest.com/8IKC/files/00%20zContents.pdf

Novoselov AA (2010) Geochemical modeling of hydrothermal transformation of peridotites of slowly-spreading mid-ocean ridges. PhD Thesis Abstract, Institute of Geochemistry, Russian Academy of Science, Moscow, p 16

Paton C, Hergt JM, Phillips D, Woodhead JD, Shee SR (2007) New insights into the genesis of Indian kimberlites from the Dharwar Craton via in situ Sr isotope analysis of groundmass perovskite. Geology 35(11):1011–1014

Pavlov DI, Ilupin IP (1973) Halite in Yakutian kimberlite, its relations to serpentine and the source of its parent solutions. Dokl Acad Sci USSR, Earth Sci Ser Engl Transl 213:178–180

Polozov AG, Sukhov SS, Gornova MA, Grishina SN (2008a) Salts from Udachnaya-East kimberlite pipe (Yakutia, Russia): occurrences and mineral composition. Extended abstracts of 9IKC-A-00247

Polozov AG, Svensen H, Planke S (2008b) Chlorine isotopes of salts xenoliths from Udachnaya-East kimberlite pipe (Russia). Extended abstracts of 9IKC-A-00249

Scott Smith BH (1996) Kimberlites. Chapter 10, Mineralogical association of Canada short course series. In: Mitchell RH (ed) Undersaturated alkaline rocks: mineralogy, petrogenesis, and economic potential, short course, vol 24, pp 217–243

Seal RR (2006) Sulfur isotope geochemistry of sulfide minerals. In: Sulfide Mineralogy and Geochemistry 61:633–677

Sharp ZD, Barnes JD (2004) Water-soluble chlorides in massive seafloor serpentinites: a source of chloride in subduction zones. Earth Planet Sci Lett 226(1–2):243–254

Sharp ZD, Barnes JD, Brearley AJ, Chaussidon M, Fischer TP, Kamenetsky VS (2007) Chlorine isotope homogeneity of the mantle, crust and carbonaceous chondrites. Nature 446(5): 1062–1065

Sharygin VV, Faure K, Golovin AV (2007a) Chloride and carbonate immiscible liquids at the closure of the kimberlite magma evolution (Udachnaya-East kimberlite, Siberia). Chem Geol 237:384–400

Sharygin VV, Kamenetsky VS, Kamenetsky MB, Golovin AB (2007b) Mineralogy and genesis of chloride-bearing nodules from the Udachnaya-East kimberlites. In: Abstracts, international conference on alkaline terrestrial magmatism and its ores. http://geo.web.ru/conf/alkaline/2007/80.pdf, pp 267–271 (in Russian)

Sobolev AV, Sobolev NV, Smith CB, Dubessy J (1989) Fluid and melt compositions in lamproites and kimberlites based on the study of inclusions in olivine. In: Ross J et al (eds) Kimberlites and related rocks: their composition, occurrence, origin and emplacement. Proceedings of the 4th IKC, vol 1. Blackwell Scientific Publications, Sydney, pp 220–241

Sukhov SS (2001) Facies and stratigraphic model of Daldyn-Markha bank: illustration to accumulation mechanism of carbonate deposits on the Siberian craton. In: Lithology and oil-and-gas potential of carbonate deposits. Proceedings of the second all-Russian lithologic conference and 8th all-Russian symposium on fossilized corals and reefs. Syktuvkar, pp 237–239 (in Russian)

Vinogradov VI, Ilupin IP (1972) Isotope compositions of sulfur in kimberlites of the Siberian Platform. Dokl USSR Acad Sci 204(6):1452–1456 (in Russian)

Webster JD (1992) Water solubility and chlorine partitioning in Cl-rich granitic systems—effects of melt composition at 2 kbar and 800 °C. Geoch Cosmoch Acta 56(2):679–687

Woodhead J, Hergt J, Phillips D, Paton C (2009) African kimberlites revisited: in situ Sr-isotope analysis of groundmass perovskite. Lithos 112:311–317

Yoder HS (1986) Potassium-rich rocks—phase-analysis and heteromorphic relations. J Petrol 27(5):1215–1228

Zinchuk NN, Spetsius ZV, Zuenko VV, Zuev VM (1993) The Udachnaya kimberlite pipe. Publishing House of the Novosibirsk University, Novosibirsk, p 147 (in Russian)

Petrology, Bulk-Rock Geochemistry, Indicator Mineral Composition and Zircon U–Pb Geochronology of the End-Cretaceous Diamondiferous Mainpur Orangeites, Bastar Craton, Central India

N. V. Chalapathi Rao, B. Lehmann, E. Belousova, D. Frei, and D. Mainkar

Abstract

The end-Cretaceous diamondiferous Mainpur orangeite field comprises six pipes (Behradih, Kodomali, Payalikhand, Jangara, Kosambura and Bajaghati) located at the NE margin of the Bastar craton, central India. The preservation of both diatreme (Behradih) and hypabyssal facies (Kodomali) in this domain implies differential erosion. The Behradih samples are pelletal and tuffisitic in their textural habit, whereas those of the Kodomali pipe have inequigranular texture and comprise aggregates of two generations of relatively fresh olivines. The Kosambura pipe displays high degrees of alteration and contamination with silicified macrocrysts and carbonated groundmass. Olivine, spinel and clinopyroxene in the Behradih and the Kodomali pipes share overlapping compositions, whereas the groundmass phlogopite and perovskite show conspicuous compositional differences. The bulk-rock geochemistry of both the Behradih and Kodomali pipes has a more fractionated nature compared to southern African orangeites. Incompatible trace elements and their ratios readily

Electronic supplementary material The online version of this article (doi:10.1007/978-81-322-1170-9_7) contains supplementary material, which is available to authorized users.

N. V. Chalapathi Rao (✉)
Centre of Advanced Study in Geology, Banaras Hindu University, Varanasi, 221005, India
e-mail: nvcr100@gmail.com

B. Lehmann
Mineral Resources, Technical University of Clausthal, 38678, Clausthal-Zellerfeld, Germany
e-mail: Lehmann@min.tu-clausthal.de

E. Belousova
GEMOC, Department of Earth and Planetary Sciences, Macquarie University, Sydney, NSW 2109, Australia
e-mail: ebelouso@els.mq.edu.au

D. Frei
Department of Earth Sciences, Stellenbosch University, Private Bag X1, Matieland, 7602, South Africa
e-mail: dirkfrei@sun.ac.za

D. Mainkar
Directorate of Mines and Geology, Chhattisgarh, Raipur, 492007, India
e-mail: dmainkar@gmail.com

D. G. Pearson et al. (eds.), *Proceedings of 10th International Kimberlite Conference*, Volume 1, Special Issue of the Journal of the Geological Society of India,
DOI: 10.1007/978-81-322-1170-9_7,

distinguish them from the Mesoproterozoic Wajrakarur (WKF) and the Narayanpet kimberlites (NKF) from the eastern Dharwar craton, southern India, and bring out their similarity in petrogenesis to southern African orangeites. The pyrope population in the Mainpur orangeites is dominated by the calcic-lherzolitic variety, with sub-calcic harzburgitic and eclogitic garnets in far lesser proportion. Garnet REE distribution patterns from the Behradih and Payalikhand pipes display "smooth" as well as "sinusoidal" chondrite-normalised patterns. They provide evidence for the presence of a compositionally layered end-Cretaceous sub-Bastar craton mantle, similar to that reported from many other cratons worldwide. The high logfO_2 of the Mainpur orangeite magma (ΔNNO (nickel-nickel oxide) of +0.48 to +4.46 indicates that the redox state of the lithospheric mantle cannot be of first-order control for diamond potential and highlights the dominant role of other factors such as rapid magma transport. The highly diamondiferous nature, the abundance of calcic-lherzolitic garnets and highly oxidising conditions prevailing at the time of eruption make the Mainpur orangeites clearly "anomalous" compared to several other kimberlite pipes worldwide. U–Pb dating of zircon xenocrysts from the Behradih pipe yielded distinct Palaeoproterozoic ages with a predominant age around 2,450 Ma. The lack of Archean-aged zircons, in spite of the fact that the Bastar craton is the oldest continental nuclei in the Indian shield with an Eoarchaean crust of 3.5–3.6 Ga, could either be a reflection of the sampling process or of the modification of the sub-Bastar lithosphere by the invading Deccan plume-derived melts during the Late Cretaceous.

Keywords
Petrology • Geochemistry • Indicator minerals • U–Pb age • Zircon • Orangeite • Mainpur • Bastar craton • India

Introduction

For more than a century, Group II kimberlites (also termed as orangeites) are believed to have been restricted only to the Kaapvaal craton of southern Africa (Skinner 1989; Mitchell 1995). However, recent discovery of diamondiferous orangeites, synchronous with the eruption of Deccan flood basalts, from the Bastar craton of central India (Lehmann et al. 2010; Chalapathi Rao et al. 2011a) provides a rare opportunity to investigate the controls on genesis of such magmas. In this study, we report new data involving petrology, bulk-rock geochemistry, indicator mineral chemistry (garnet, Cr-diopside and spinel xenocrysts) and U–Pb zircon geochronology of the Behradih, Kodomali, Payalikhand and Kosambura pipes from the Mainpur orangeite field, Bastar craton, Central India. We also estimate oxygen fugacity (fO_2) of the Mainpur orangeite magmas by applying Fe-Nb oxybarometry to their perovskite chemistry and attempt to assess the role of oxidation state in their diamond grade. The findings of this study provide additional insights into the (1) origin of these pipes, (2) diamond prospectivity, (3) nature and composition of the sub-Bastar lithosphere, and (4) in furthering our understanding on the petrological and geochemical differences between the Mainpur orangeites and well-studied Mesoproterozoic kimberlites from the Wajrakarur (WKF) and Narayanpet (NKF) fields, Dharwar craton, southern India as well as those from the Kaapvaal craton of southern Africa.

Geology of the Bastar Craton and the Mainpur Orangeite Field

The Bastar craton is one of the oldest nuclei in the Central Indian shield and is regarded to be a component of the early Archaean supercontinent "Ur" (Rogers and Santosh 2003). The basement consists largely of granitic rocks and mafic dyke swarms which are of Meso-Neoarchaean age (Crookshank 1963; Ramakrishnan and Vaidyanadhan 2008). A thickened continental crust (35–40 km) in the Bastar craton since at least 3.6 Ga to the present day has been documented from geochronological and geophysical studies (e.g., Rajesh et al. 2009; Jagadeesh and Rai 2008). The granitic basement is unconformably overlain by several supracrustal, intracratonic Meso- to Neoproterozoic sedimentary sequences, such as the Sabri Group (Sukma basin), the Indrāvati Group (Indrāvati basin), Pairi Group (Khariar basin) and Chhattisgarh Supergroup (Chhattisgarh basin) from south to north (see Ramakrishnan and Vaidyanadhan 2008).

Three kimberlite/orangeite fields are so far known in the Bastar craton: (1) the southern Indrāvati kimberlite field (IKF) represented by the ≥620-Ma non-diamondiferous kimberlite pipes intruded into the sedimentary rocks of the Indravāti basin at Tokapal, Bejripadhar and Dunganpal (e.g., Mainkar et al. 2004; Lehmann et al. 2006, 2007); (2) the Dharambanda lamproite field in the Nawapara area at the NE part of the Bastar craton (Patnaik et al. 2002); and (3) the Mainpur orangeite field comprising six diamondiferous pipes at Behradih, Kodomali, Payalikhand, Jangra, Bajaghati and Kosambura at the NE part of the Bastar craton (e.g., Newlay and Pashine 1993; Chatterjee et al. 1995; Mainkar and Lehmann 2007; Fig. 1). The last two mentioned fields are located very close to the contact between the Bastar craton and the Eastern Ghats mobile belt (EGMB). This contact is characterised by ultra-high-temperature metamorphosed rocks of EGMB over-thrusted on the Bastar cratonic footwall (Gupta et al. 2000).

The Archaean basement is not exposed in the Mainpur orangeite field. Bundeli granitoids (equivalents of nearby 2,300 Ma Dongargarh granites) together with gabbroic rocks and dolerites constitute the oldest outcropping litho-units (Mishra et al. 1988). The Bundeli granitoids are overlain by Meso-Neoproterozoic platformal sedimentary rocks of the Khariar and Pairi basins in close western vicinity of the thrust fault between the Bastar craton and the Eastern Ghats Mobile belt (EGMB) (Fig. 1). The orangeite pipes of the Mainpur area, together with minor basalts of Deccan Trap (Mainkar and Lehmann 2007; Chalapathi Rao et al. 2011b), constitute the youngest intrusives in the domain and intrude the Bundeli granites in a spread of about 19 × 6 km extending WNW-ESE (Fig. 1). The pipes have circular to oval shape in outcrop with variable length in the longest direction from ~45 m (Bajaghati) to as much as ~300 m (Behradih). The prominent NW–SE trending Sondhur fault zone dissects them into two clusters: (1) the eastern Payalikhand cluster (comprising Payalikhand and Jangara pipes) and (2) the western Behradih cluster (comprising Behradih, Kodomali Bajaghati and Kosambura pipes). A brief description of each of the orangeite pipes is provided below:

Behradih orangeite (82°12′6.3″; 20°12′54.5″): This is the first discovered diamondiferous primary host from the Bastar craton (Newlay and Pashine 1993) located ~15 km south of Mainpur town. A thin veneer of bluish-black gritty eluvial soil horizon is present at places and rests over deeply weathered, smectite-rich greenish clayey "yellow ground". Dimensions of the pipe are ~300 × 300 m, making it the largest pipe in the Mainpur field. Drilling data by Directorate of Mines and Geology (DGM), Chhattisgarh, reveals that the Behradih pipe is carrot-shaped, filled by tuffisitic breccia and contains both pyroclastic as well as diatreme facies rocks. Macrodiamonds up to 200 ct (Newlay and Pashine 1993) as well as ~486 microdiamonds have been recovered from treatment of pit material (Verma and Saxena 1997). Diamonds are mostly in the form of discrete octahedrons and dodecahedrons with eclogitic and peridotitic inclusions reported (Jha et al. 1995). Drilling of the pipe was carried out to depths of 188 m via five bore holes by DGM and least weathered to fresh bluish-grey to black, hard and compact pipe rock has been generally encountered beyond 65 m in all of them. Compositions of xenocrystal pyrope, clinopyroxene and chromite from the Behradih pipe are presented in this paper.

Kodomali orangeite (82°14′25.6″; 20°11′18.2″): Located ~4.8 km S-SE of the Behradih pipe. It is the only "harde bank" outcrop in the Mainpur field with dimensions of 80 × 30 m and extending in E-W direction. The outcrop is dark greyish green, hard, compact and massive and forms a gently elevated (>3 m) topography. An older gabbroic rock of ~500 m length and with NNW-SSE trend is exposed in eastern close vicinity (Fig. 1). One surface sample weighing 45 kg was collected during prospecting and yielded 6 macro- and 1 microdiamond (ORAPA 2000). Compositions of xenocrystal pyrope, clinopyroxene and chromite from Kodomali orangeite are provided in this paper.

Payalikhand orangeite (82°20′33.7″; 20°10′8″): Unexposed, and located ~15.4 km east of Behradih pipe. Diamond, pyrope, chromite, chrome diopside and phlogopite containing smectite-rich greenish soil comprising brownish and greyish shale, sandstone, grit and granite fragments were recovered from pits made by DGM, Chhattisgarh, during 1995–1997. Ground magnetic survey demarcated an E-W trending body occupying an area of 240 × 60 m with an irregular outline and concealed below a weathered zone of 15–20 m thickness (Singh et al. 2000). No suitable samples are available for petrographic and geochemical studies. Gravel collected over the orangeite yielded colourless, clear and octahedral/dodecahedral diamonds weighing 0.29 and 0.08 ct (Newlay and Pashine 1995). Compositions of only xenocrystal pyrope, which are available to us, are provided in this paper.

Jangara orangeite (82°19′33″; 20°10′3″): Located ~14 km ESE of Behradih pipe and ~2 km west of Payalikhand pipe. It is also a concealed pipe occupying an area of 50 m in diameter with a well-developed typical yellow ground (Chatterjee and Jha 1994). No drilling has been done for this pipe, and no samples are available for study.

Kosambura orangeite (82°12′06.5″; 20°10′42.6″): Located 3.7 km south of Behradih pipe and occurs as a highly weathered and deeply altered 80-m long dyke having a width of 0.5–1.0 m as delineated from ground electromagnetic survey (Small and Vaidya 2002). No diamonds are reported from this pipe. Surface exposures are unsuitable for petrography and geochemistry owing to their highly silicified and weathered nature. However, compositions of

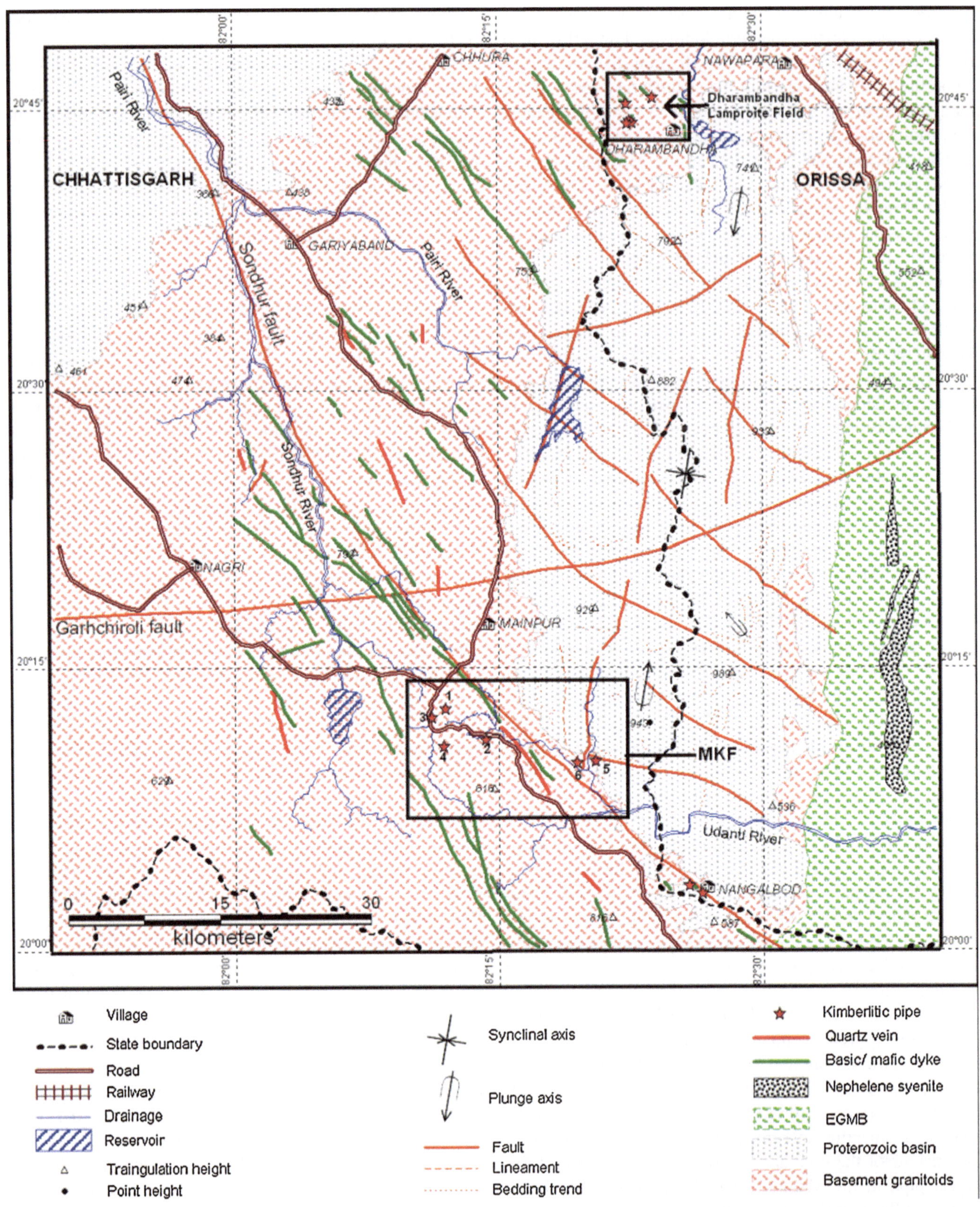

Fig. 1 Regional geological map of kimberlite/lamproite fields of the north-eastern Bastar craton which includes a part of Raipur district, Chhattisgarh district and Nuapada district, Orissa (geology after Mishra et al. 1988; Ramakrishnan and Vaidyanadhan 2008; Mainkar 2010). MKF = Mainpur kimberlite field; 1 = Behradih; 2 = Kodomali; 3 = Bajaghati; 4 = Kosambura; 5 = Payalikhand; 6 = Jangara

xenocrystal pyrope and clinopyroxene are presented in this paper.

Bajaghati (Temple) orangeite (82°11′13.3″; 20°12′ 37.5″): Located 1.6 km WSW of Behradih pipe with a surface area of 45 × 25 m in dimension as inferred from ground electromagnetic survey (Small and Vaidya 2002). The pipe is represented at surface by bluish-black alluvial soil of unknown thickness. No samples are available for petrography and geochemistry. A solitary gem-variety macrodiamond of ~4 mm diameter has been recorded from this pipe by the DGM, Chhattisgarh.

Earlier Work

The discovery of "yellow ground" at the Behradih pipe by Newlay and Pashine (1993) paved the way for the finding of other pipes in the Mainpur area (Chatterjee and Jha 1994; Newlay and Pashine 1995). The earliest publications are merely extended abstracts that deal with the geology, mineralogy and geochemistry of the pipes and mantle-derived xenocrysts including the nature of diamonds (Chatterjee et al. 1995; Jha et al. 1995). Reconnaissance petrography and geochemistry of the Behradih pipe was given by Mainkar and Lehmann (2007), and its detailed petrological and petrogenetic aspects were provided in Chalapathi Rao et al. (2011a). Petrology of the Kodomali pipe has been discussed by Fareeduddin et al. (2006) wherein the pipe rock is inferred to be an orangeite based solely on petrography and mineral chemistry. However, it should be pointed out here that in a subsequent publication, Mitchell and Fareeduddin (2009) re-interpreted Kodomali pipe to represent a lamproite. Geochemical aspects of Kodomali pipe are also dealt by Paul et al. (2006) and Marathe (2010) wherein its kimberlite status has been deduced. On the contrary, only a few studies reporting the composition of mantle-derived xenoliths/xenocrysts are available (e.g., Mukherjee et al. 2000).

$^{39}Ar/^{40}Ar$ whole-rock dating of drill cores from the Behradih orangeite gave a weighted mean plateau age of 66.7 ± 0.8 Ma (Lehmann et al. 2010). These ages are indistinguishable from in situ $^{206}Pb/^{238}U$ perovskite ages of 65.10 ± 0.80 Ma and 65.09 ± 0.76 Ma (2 σ) obtained on two samples of the Behradih orangeite (Lehmann et al. 2010). On the other hand, the Kodomali orangeite gave slightly younger $^{39}Ar/^{40}Ar$ (whole-rock) ages of 62.1 ± 1.4 and 62.3 ± 0.76 Ma and $^{206}Pb/^{238}U$ (perovskite) ages of 62.3 ± 0.8 Ma (Lehmann et al. 2010). It should be pointed out here that an older Pan-African age of 491 ± 11 Ma reported earlier for Kodomali pipe (Chalapathi Rao et al. 2007) is due to inherited excess ^{40}Ar in the sampled chloritised phlogopite megacryst (see Lehmann et al. 2010). These ages are much younger than the previously known episodes of potassic alkaline magmatism at ca. 1.1–1.5 Ga involving diamondiferous Mesoproterozoic kimberlites and lamproites from the Dharwar (Gopalan and Kumar 2008; Osborne et al. 2011) and Bundelkhand (Gregory et al. 2006; Masun et al. 2009) cratons and the ca. 117 Ma event encompassing potassium-rich ultramafics from the Damodar valley, Chotanagpur Mobile belt, off Singhbhum craton, eastern India (Kent et al. 1998) and constitute the youngest yet known diamondiferous event in the Indian shield.

Analytical Techniques

Mineral compositions of groundmass phases from the Kodomali and Behradih pipes reported in this study (Tables 1, 2, 3, 4) were determined by using a CAMECA SX100 Electron Probe Micro Analyzer (EPMA) at the Technical University of Clausthal, Germany. An accelerating voltage of 15 kV, a beam current of 20 nA and a beam diameter of 2 μm was used. The analyses were carried out using wavelength-dispersive spectrometers employing TAP, PET and LLIF crystals and a PAP online correction programme. Several in-house natural standards were used for calibration. After repeated analyses, it was found that the error on major element concentrations is <1 %. For the perovskite analyses (Table 5), an acceleration voltage of 20 kV, beam current of 40 nA, beam diameter of 2 μm and a counting time of 60 s and a number of synthetic standards such as REE6-1, REE6-2, REE6-3, REE6-4, pure Nb and Th-G were used.

Whole-rock major elements of nine Behradih and one Kosambura samples (Table 6) were analysed by X-ray fluorescence spectrometry (XRF) at Bundesanstalt für Geowissenschaften und Rohstoffe, Hannover, Germany. Relative uncertainties of the XRF analyses are ≤3 %. Whole-rock trace elements were measured with a Perkin-Elmer Sciex ELAN 6000 ICP-MS instrument with Meinhardt-nebulizer at TU Clausthal, and the relative uncertainties range from 10 to 30 %. Whole-rock analyses of the three Kodomali samples were carried out at the laboratories of the National Geophysical Research Institute (NGRI), Hyderabad, India. Concentrations of major elements were determined by X-ray fluorescence spectrometry (XRF) using a Philips MAGIX PRO Model 2440. Typical uncertainties of the XRF analyses are <5 %. Concentrations of trace, REE and HFSE were determined by ICP-MS using a PerkinElmer SCIEX ELAN DRC II. UB-N (French standard) along with BHVO-1 and JB-2 were used as reference materials following Balaram and Gnaneshwar Rao (2003). Overall, an accuracy of better than ±5 % was obtained for most determinations with a precision of better than ±6 % RSD. Data on rare earth elements (REE) on garnet concentrates from the Behradih and Payalikhand pipes were acquired using a laser-ablation PerkinElmer Sciex ELAN 5100 ICPMS (LAM) at the School of Earth Sciences, Macquarie University, Australia, and are given in Table 7. The external standard was the NIST 610 multi-element glass, accuracy of the measurements is expected to be within 1-sigma errors and the analytical conditions are as described by Norman et al. (1996) and Girffin et al. (1999).

Zircon grains were set in epoxy resin mounts, sectioned and polished to approximately half of their thickness. Prior to U–Pb isotopic analysis, cathodoluminescence images were obtained for all grains using a scanning electron microscope (JEOL JXA 8900 RL instrument) at the Geological Survey of Denmark and Greenland (GEUS), Copenhagen, Denmark (Fig. 9a). U–Pb geochronology of zircons was performed using a Thermo-Finnigan Element II sector-field ICPMS system coupled to a Merchantek/New-Wave 213-nm Nd-YAG laser system at the Geological Survey of Denmark and Greenland (GEUS), Copenhagen, Denmark, and the results are given in Table 8. The method applied essentially followed that described by Frei and Gerdes (2008). For the interpretation of the zircon data, analyses with 95–105 % concordance [calculated from 100 x ($^{206}Pb/^{238}U$ age)/($^{207}Pb/^{235}U$ age)] are considered to be concordant and those with a discordance >10 % were rejected and consequently not considered.

Mineral chemistry of the indicator minerals (xenocrysts) is given in Electronic Supplementary Tables 1 to 6 and is from Mainkar (2010). Data of indicator minerals have been provided to one of us (DM, working with Directorate of Mines and Geology, Chhattisgarh) as proprietary information by exploration companies such as De Beers, ORAPA, B.Vijaya Kumar Chhattisgarh Exploration (BVCE) and Rio Tinto, and their analytical conditions could not be retrieved. For such data generated at Technical University of Clausthal, the instrument and analytical conditions remain the same as given earlier in this section.

Petrography and Mineral Chemistry

The Kodomali occurrence is the best exposed and least altered pipe of the Mainpur field. The studied samples have a distinct inequigranular texture dominated by macrocrysts (0.5–1 mm) and microphenocrysts of fresh and unaltered olivine set in a fine-grained groundmass dominated by phlogopite, clinopyroxene, spinel, perovskite and apatite (Fig. 2a). Microlites of clinopyroxene, present as acicular laths, are a characteristic feature of the pipe along with clusters of spinel and perovskite (Fig. 2b). Serpentinisation of olivine and chloritisation of phlogopite is minimal. The groundmass displays magmatic flow layering, and the olivine macrocrysts display occasional evidence of corrosion. Autoliths of earlier erupted pipe rock are also reported (Chatterjee et al. 1995).

Rounded to sub-rounded olivine macrocrysts and pelletal lapilli set in a very fine-grained glassy to cryptocrystalline serpentine-chlorite matrix dominated by olivine microphenocrysts are a characteristic feature of the Behradih pipe (Fig. 2c). The mesostasis is dominated by phlogopite, clinopyroxene, perovskite, spinel and apatite (Fig. 2d). The macrocrysts as well as the matrix phases are affected by pervasive carbonate–talc–serpentine alteration. Fresh and unaltered olivine is not observed and has been reported only from autoliths (Chalapathi Rao et al. 2011a). The mode of occurrence of clinopyroxene as microlitic laths is identical to that observed in the Kodomali pipe. Petrography of the Behradih pipe has been reported by Mainkar and Lehmann (2007) and Chalapathi Rao et al. (2011a).

The Kosambura pipe is highly silicified with fragmental texture wherein completely altered macrocrysts are set in fine-grained altered and indistinguishable groundmass which constitutes a major proportion. Relict textures of the serpentinised, talcose and carbonated remnants are prominent along with secondary silica, chlorite and clay minerals.

New microprobe data for minerals from the Kodomali pipe and for perovskite from Behradih are given in Tables 1, 2, 3, 4, 5 and discussed below.

Olivine

Phenocrystal olivine data from the Kodomali pipe are given in Table 1. The olivine compositions (Fo_{86-90}) of this study are similar to those (Fo_{87-91}) reported earlier from the Kodomali pipe (Fareeduddin et al. 2006) and overlap with the olivine phenocryst compositions available for the Behradih pipe (Fo_{83-90}; Chalapathi Rao et al. 2011a). However, NiO contents of the Kodomali olivines vary from 0.36 to 0.45 wt% and are higher than those from the Behradih pipe and also from the NKF and WKF pipes in southern India shown for comparison (Fig. 3a).

Phlogopite

Compositional data for groundmass phlogopite from the Kodomali pipe are given in Table 2. In the TiO_2 versus Al_2O_3 (Fig. 3b) plot, its composition is similar to that reported for lamproites from Smoky Butte and Leucite Hills, USA, and contrasts with that of the Behradih pipe which has distinctly lower TiO_2 and a far wider range in Al_2O_3 contents (Fig. 3b). In the FeO_T versus Al_2O_3 (Fig. 3c) plot for discriminating between groundmass phlogopites from various potassic–ultrapotassic alkaline rocks, the Behradih phlogopites show two distinct clusters—macrocrystal and

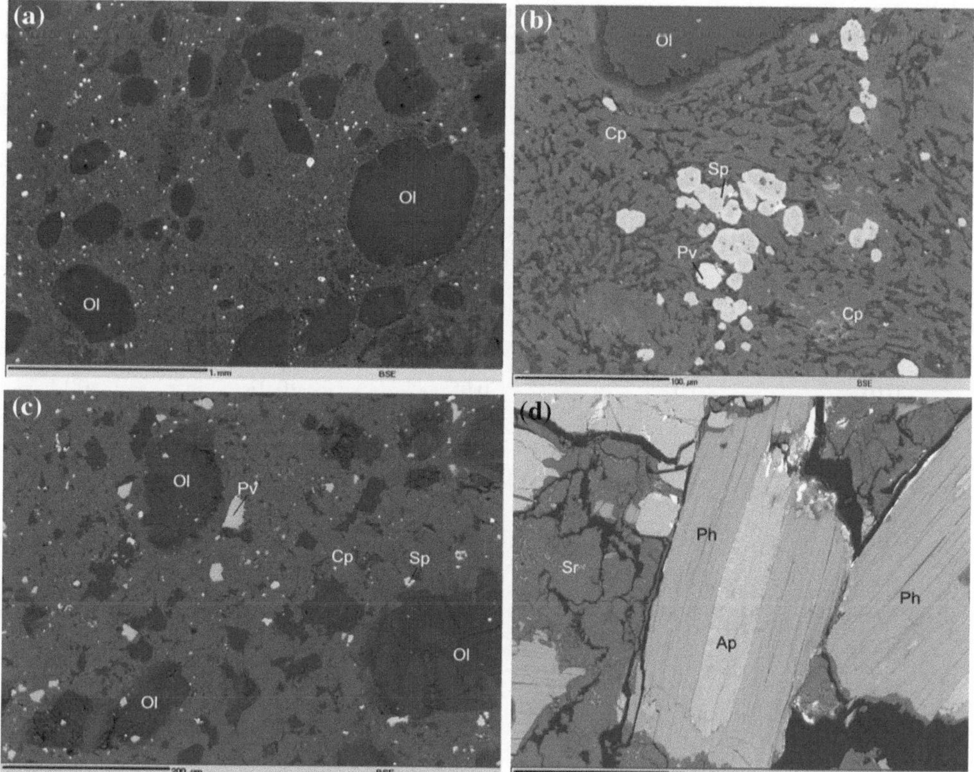

Fig. 2 Backscattered electron (BSE) images depicting the petrographic features of the Mainpur pipes of this study. **a** Macrocrystal texture of the Kodomali pipe showing two generations of fresh and unaltered subhedral to euhderal olivines set in a very fine-grained groundmass. **b** High magnification image showing the microlites of clinopyroxene which is a dominant phase in the groundmass of the Kodomali pipe; also seen are clusters of perovskite and spinel in close association with each other. **c** Rounded to sub-rounded olivine macrocrysts set in a finer groundmass dominated by clinopyroxene, perovskite and spinel are a characteristic feature of the Behradih pipe; and **d** Phlogopite microcrysts with apatite blebs are a characteristic feature of some of the Behradih drill core samples. Note that serpentine is also an important component of the mesostatis (Ol = olivine; Cp = clinopyroxene; Ph = phlogopite; Sr = serpentine; Ap = apatite; Pv = perovskite; Sp = spinel)

microphenocrystal—with the latter compositionally closer to the phlogopite from Kodomali samples and another plotting away from them. The mica compositions from WKF and NKF display a far greater range in composition in either of the plots (Fig. 3b, c).

Spinel

Groundmass spinel composition of the Kodomali samples is given in Table 3. A distinct compositional difference exists in MgO-Al_2O_3 space amongst the groundmass spinels from Kodomali and Behradih (Fig. 3d). The Kodomali spinel is relatively Mg–depleted and shows far greater variation in the Al_2O_3 content than the spinel from Behradih and is similar to the spinel composition in WKF and NKF kimberlites (Fig. 3d). Macrocryst spinel from the Tokapal kimberlite located in the Indravati field in the Bastar craton has an altogether different composition. Groundmass spinel from kimberlites has been shown to define two distinct trends (Mitchell 1986): (1) a magnesio–ulvöspinel–magnetite trend (trend 1; characteristic of Group I kimberlites) and (2) a Ti-magnetite trend (trend 2 which is characteristic of orangeites and is also similar to zoning of spinel in basalts; Roeder and Schulze 2008). Spinel from the Kodomali as well as from Behradih pipes shows a strong affinity towards trend 2 (Fig. 3e).

Clinopyroxene

Clinopyroxene in the Kodomali samples is diopsidic (Table 4) with a considerable range in composition of $Wo_{34–44}$ $En_{35–46}$ $Fs_{11–13}$ $Ac_{7–15}$. The data reported in this study are distinct from the Fe-poor clinopyroxene ($Wo_{50–52}En_{44–47}Fs_{0.9–4.3}$) reported earlier by Fareeduddin et al. (2006). In Al-Ti (atomic) space (Fig. 3f), clinopyroxene from the Mainpur pipes overlaps with the compositional fields of diopside from the NKF, Krishna lamproites and orangeites from southern Africa as well as those of worldwide lamproites (Mitchell 1995).

Table 1 Mineral chemistry (wt%) of olivine from the Kodomali orangeite

Oxide wt%	KDK-2	KDK-2C	KDK-3	KDK-3	KDK-4	KDK-4
SiO_2	41.38	41.79	40.92	40.66	41.77	41.95
TiO_2	0.01	0.00	0.04	0.02	0.04	0.01
Al_2O_3	0.00	0.03	0.04	0.02	0.02	0.05
Cr_2O_3	0.02	0.03	0.00	0.02	0.06	0.08
FeO	7.46	7.35	12.71	12.60	9.22	9.34
MnO	0.09	0.10	0.09	0.08	0.10	0.10
MgO	49.65	49.33	45.64	45.34	48.12	47.75
NiO	0.42	0.40	0.41	0.43	0.36	0.45
CaO	0.03	0.02	0.05	0.07	0.13	0.16
Total	99.04	99.07	99.91	99.24	99.82	99.91
Cations for 4 oxygens						
Si	1.013	1.021	1.016	1.017	1.022	1.026
Ti	0.000	0.000	0.001	0.000	0.001	0.000
Al	0.000	0.001	0.001	0.001	0.001	0.001
Cr	0.000	0.000	0.000	0.000	0.001	0.002
$Fe^{(ii)}$	0.153	0.150	0.264	0.263	0.189	0.191
Mn	0.002	0.002	0.002	0.002	0.002	0.002
Mg	1.811	1.796	1.690	1.690	1.755	1.741
Ni	0.008	0.008	0.008	0.009	0.007	0.009
Ca	0.001	0.001	0.001	0.002	0.003	0.004
Total	2.988	2.979	2.984	2.984	2.980	2.976
End-members						
Fo	92.15	92.18	86.41	86.44	90.20	90.02
Fa	7.76	7.71	13.50	13.47	9.69	9.87
Tp	0.09	0.11	0.10	0.09	0.11	0.11

Perovskite

The composition of perovskite from the Behradih and Kodomali pipes is presented in Table 5. Their CaO (Behradih: 36.5–37.3 wt%; Kodomali: 35.3–37.2 wt%) and TiO_2 (Behradih: 51.8–52.9 wt% Kodomali: 52.8–54.0 wt%) contents show little variation. However, their Fe_2O_3* (Behradih: 2.51–2.90 wt% and Kodomali: 1.24–2.53 wt%; both re-calculated from FeO) and Nb_2O_5 contents (Behradih: 0.70–0.73 wt%; Kodomali: 1.04–1.64 wt%) display consistent differences. On the other hand, their $LREE_2O_3$ (4.24–5.29 wt%) and SrO (0.44–1.17 wt%) show a tight compositional range. Na_2O (0.22–1.45 wt%) is present in substantial concentrations. $LREE_2O_3$ contents of the perovskites are higher than those reported for NKF (0.88–0.97 wt%) and WKF (2.57–5.37 wt%) kimberlites. Likewise, Nb_2O_5 contents are also higher than those from NKF pipes (which range from 0.22–0.33 wt%) but are similar to those from the WKF pipes (0.34–2.33 wt%; data sources: Chalapathi Rao et al. 2004; Chalapathi Rao and Dongre 2009; Chalapathi Rao et al. 2012). The Fe and Nb contents of perovskite from the Mainpur orangeites are further used for evaluating the redox conditions of the magma (below).

Bulk-Rock Geochemistry

Whole-rock major, trace (including REE) and radiogenic isotope (Sr and Nd) data for drill core samples of Behradih (Mainkar and Lehmann 2007; Lehmann et al. 2010; Chalapathi Rao et al. 2011a) and Kodomali (Paul et al. 2006; Fareeduddin et al. 2006; Marathe 2010; Lehmann et al. 2010) are available. There are no published geochemical data for the other pipes from the Mainpur field. In this study, we generated new bulk-rock geochemistry data on nine drill core samples from the Behradih pipe, three surface samples from the Kodomali pipe and one drill core sample from the Kosambura pipe (Table 6). These new data together with earlier data sets are utilised in this study.

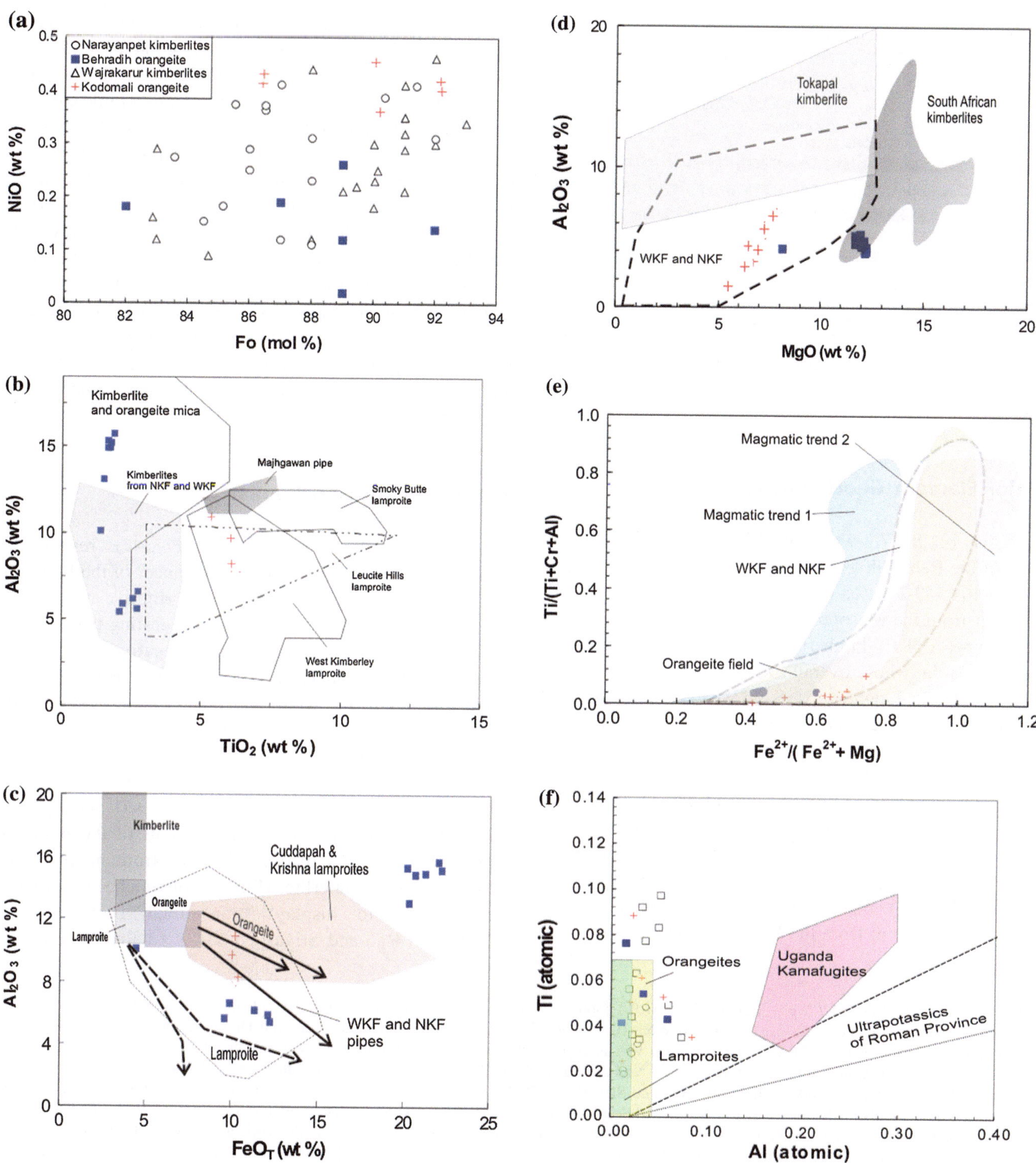
(a)
Narayanpet kimberlites
Behradih orangeite
Wajrakarur kimberlites
Kodomali orangeite
NiO (wt %)
Fo (mol %)
(b)
Kimberlite and orangeite mica
Majhgawan pipe
Kimberlites from NKF and WKF
Smoky Butte lamproite
Leucite Hills lamproite
West Kimberley lamproite
Al2O3 (wt %)
TiO2 (wt %)
(c)
Kimberlite
Cuddapah & Krishna lamproites
Orangeite
Lamproite
Orangeite
WKF and NKF pipes
Lamproite
Al2O3 (wt %)
FeOT (wt %)
(d)
Tokapal kimberlite
South African kimberlites
WKF and NKF
Al2O3 (wt %)
MgO (wt %)
(e)
Magmatic trend 2
Magmatic trend 1
WKF and NKF
Orangeite field
Ti/(Ti+Cr+Al)
Fe2+/(Fe2++ Mg)
(f)
Orangeites
Uganda Kamafugites
Ultrapotassics of Roman Province
Lamproites
Ti (atomic)
Al (atomic)

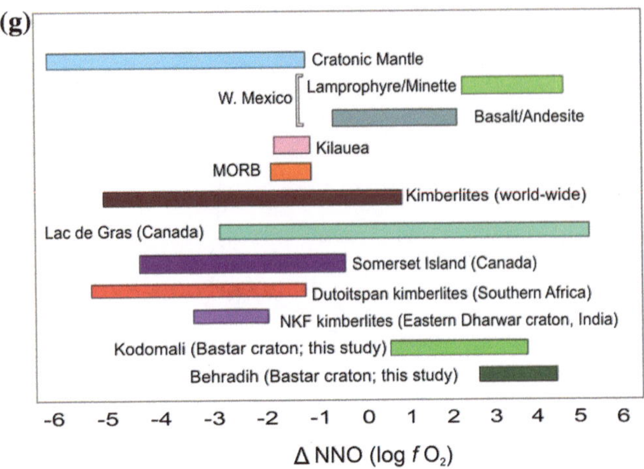

Fig. 3 continued

K_2O contents (1.42–2.92 wt%) of the Mainpur pipes are higher than the corresponding Na_2O contents (0.57–1.18 wt%; Table 6) and display the potassic nature of the pipes. As K is the most mobile of the large-ion lithophile elements and Ti is highly immobile as a high-field strength element, a K_2O-TiO_2 plot is widely used to discriminate kimberlites and orangeites (Smith et al. 1985; Mitchell 1995). Such a plot (Fig. 4b) clearly brings out the higher K_2O and lower TiO_2 contents of the Behradih and Kodomali pipes compared to those from the WKF and NKF. The solitary drill core sample from the Kosambura pipe shows very high degrees of contamination as is evident from its elevated SiO_2 (74.2 wt%) and depleted MgO (3.08 wt%), and therefore, its geochemistry serves little purpose in making any meaningful geochemical inference and hence is not considered in the plots.

Major Element Geochemistry

The SiO_2 (41.8–47.2 wt%) and MgO (22–27.2 wt%) contents of the Behradih as well as those of the Kodomali pipe (SiO_2: 46.5–49.2 wt%; MgO: 22.1–26.3 wt%) are clearly different from the Mesoproterozoic Wajrakarur (WKF) and Narayanpet (NKF) kimberlites, Eastern Dharwar craton, southern India, whose overall SiO_2 contents (<43 wt%) are lower and MgO contents (18–33 wt%) display a much wider range. The extent of negative correlation between SiO_2 and MgO implies magmatic fractionation processes. When compared to the archetypal kimberlites and orangeites of southern Africa, the Mainpur pipes as well as kimberlites from NKF and WKF clearly have a more fractionated nature (Fig. 4a). On the other hand, the $Fe_2O_3^*$ contents of the Behradih (6.47–8.28 wt%) and Kodomali (7.54–9.27 wt%) pipes show strong affinities towards orangeites, different from the WKF and NKF pipes.

Trace Element Geochemistry

Normalised multi-element plots (Fig. 4c) reveal the strongly enriched trace element abundances of the Behradih and Kodomali samples compared to primitive mantle. All patterns are parallel to sub-parallel suggesting petrogenetic similarity (Fig. 4c). Conspicuous negative spikes are observed at K, Sr and Ti for all samples and negative Ta for some samples of the Kodomali pipe, whereas positive Ba spikes are noticed in many of the samples (Fig. 4c). Recent studies have shown that incompatible element abundances (Ba, Nb, La and Rb) can be used to discriminate between southern African kimberlites and orangeites (Donnelly et al. 2011). Lower Nb and higher Ba contents of the Behradih and Kodomali pipes distinguish them from many of the WKF and NKF kimberlites (Fig. 4d) of which some are also demonstrated to display "transitional" characteristics (Chalapathi Rao and Srivastava 2009; Chalapathi Rao and

◀ **Fig. 3** **a** Variation of Fo [Mg/(Mg + Fe)] and NiO (wt%) contents in olivine from the Mainpur orangeites. Data *sources*: Kodomali (this work); Behradih (Chalapathi Rao et al. 2011a); NKF and WKF (Chalapathi Rao et al. 2011b; Chalapathi Rao and Dongre 2009; Chalapathi Rao et al. 2004). **b** TiO_2 (wt%) versus Al_2O_3 (wt%) and **c** FeO_T versus Al_2O_3 (wt%) of groundmass phlogopite from this study. *Data sources*: Kodomali (this work); Behradih (Chalapathi Rao et al. 2011a); NKF, WKF and Cuddapah and Krishna lamproites from Chalapathi Rao et al. (2011b), Chalapathi Rao and Dongre (2009), Chalapathi Rao and Srivastava (2009), Chalapathi Rao et al. (2004, 2010) and Paul et al. (2007); Fields of kimberlite, Leucite Hills madupite, West Kimberley and Smoky Butte lamproites are taken from Dawson and Smith (1977), Gibson et al. (1995). *Arrows* (evolutionary trends of mica composition) in (**b**) are taken from Mitchell (1995). Symbols are the same as in Fig. 3. **d** MgO (wt%) versus Al_2O_3 (wt%) of the groundmass Kodomali spinel. *Data sources*: Kodomali (this study); Behradih (Chalapathi Rao et al. 2011a); WKF and NKF (Chalapathi Rao et al. 2011b; Chalapathi Rao and Dongre 2009; Chalapathi Rao and Srivastava 2009; Chalapathi Rao et al. 2004); southern African kimberlites (Scott-Smith and Skinner 1984). Data for spinel from crater facies Tokapal kimberlite, Central India, are from the unpublished data set of the authors **e** $Fe^{2+}/(Fe^{2+}+Mg^{2+})$ versus Ti/(Ti + Cr + Al) (mol fraction) for groundmass Kodomali spinel projected onto the front face of the "reduced" spinel prism. Magmatic trends 1 and 2 exhibited by southern African spinel and spinel trend from orangeites are from Mitchell (1995). Data sources for Behradih, WKF and NKF are same as in 3A. Symbols are the same as in Fig. 3a. **f** Al_{total} versus Ti_{total} expressed as atoms per formula unit for clinopyroxene for the Kodomali (this work) and Behradih (Chalapathi Rao et al. 2011a). Also shown are the data for clinopyroxene from the Krishna lamproites (*open squares*; Paul et al. 2007; Chalapathi Rao et al. 2010) and Narayanpet kimberlites (*open circles*; after Chalapathi Rao et al.). Other fields are taken Mitchell (1995). **g** Oxygen fugacity (fO_2) (ΔNNO) conditions of the Behradih and Kodomali orangeites (this study) compared with those recorded by cratonic mantle lithosphere, mantle-derived magmas and global kimberlites (adapted from Canil and Bellis 2007). Data for Dutoitspan kimberlites are from Ogilvie-Harris et al. (2009); NKF kimberlites (Chalapathi Rao et al.)

Table 2 Mineral chemistry (oxide wt%) of clinopyroxene from the Kodomali orangeite

	KDK-2	KDK-2	KDK-3	KDK-3	KDK-4	KDK-4
SiO_2	52.23	53.20	53.15	52.39	51.71	51.71
TiO_2	1.94	0.87	1.80	1.27	2.20	3.18
Al_2O_3	1.27	0.27	0.43	1.95	0.71	0.50
Cr_2O_3	0.28	0.02	0.01	0.04	0.00	0.02
FeO	8.39	6.93	8.05	7.29	6.48	7.45
MnO	0.09	0.14	0.16	0.11	0.13	0.13
MgO	13.31	13.52	12.70	16.54	16.65	13.75
CaO	18.18	22.34	20.07	16.96	19.13	20.94
Na_2O	4.20	2.04	3.07	2.12	1.70	2.31
K_2O	0.83	0.04	0.04	0.37	0.22	0.25
Total	100.74	99.40	99.57	99.06	98.94	100.25
6 oxygens						
Si	1.882	1.967	1.962	1.918	1.904	1.901
Ti	0.053	0.024	0.050	0.035	0.061	0.088
Al	0.054	0.012	0.019	0.084	0.031	0.022
Cr	0.008	0.001	0.000	0.001	0.000	0.001
Fe(ii)	0.253	0.214	0.249	0.223	0.199	0.229
Mn	0.003	0.004	0.005	0.003	0.004	0.004
Mg	0.715	0.745	0.699	0.903	0.914	0.754
Ca	0.702	0.885	0.794	0.665	0.755	0.825
Na	0.293	0.146	0.220	0.150	0.121	0.165
K	0.038	0.002	0.002	0.017	0.010	0.012
Total	4.000	4.000	4.000	4.000	4.000	4.000
Wo	35.86	44.42	40.44	34.271	37.90	41.812
En	36.53	37.42	35.63	46.496	45.92	38.196
Fs	12.62	10.83	12.73	11.487	10.09	11.642
Ac	14.99	7.34	11.19	7.7467	6.092	8.3496

Dongre 2009; Chalapathi Rao et al. 2012). Lower La contents are characteristic of the Behradih pipe, whereas Kodomali has relatively higher La contents with both sharing indistinguishable Rb abundances (Fig. 4e).

Indicator Mineral Chemistry

Various indicator minerals, including diamond, were obtained from the Mainpur orangeites by dense media separator (DMS) processing by the Directorate of Mines and Geology, Chhattisgarh (Mainkar 2010). Whereas detailed studies on diamond are addressed separately in this volume, the present paper discusses the composition of pyrope garnet (Electronic Supplementary Tables 1 to 3), chrome diopside (Electronic Supplementary Tables 4) and chromite (Electronic Supplementary Tables 5 and 6) from Behradih, Payalikhand and Kosumbura orangeites.

Pyrope crystals are mostly <3 mm in size and rarely larger (up to 5 mm), and most of them are characterised by conchoidal fractures. The composition of the analysed pyropes from Behradih, Payalikhand, Kodomali and Kosambura pipes is plotted (Fig. 5a–d) in the garnet classification scheme for Cr_2O_3 (wt%) and CaO (wt%) given by Grutter et al. (2004) which is widely followed for inferring different paragenesis. A predominant proportion of the Behradih garnets corresponds to the calcic-lherzolitic variety (i.e., G9 type of Dawson and Stephens 1975) and <5 % belong to the high-interest sub-calcic harzburgitic category (i.e., G10 type of Dawson and Stephens 1975) and the remainder, to other fields (Fig. 5a). The Payalikhand data are limited (n = 7) and most garnets plot in the G5 and G9 fields (Fig. 5b). Likewise, in the case of Kodomali, most of the pyropes are classified into the G9 variety and some in G5 with only one solitary grain plotting in the harzburgitic G10 category (Fig. 5c). Pyropes of the silicified Kosambura

Table 3 Mineral chemistry (wt%) of phlogopite from the Kodomali orangeite

Oxide wt%	KDK-2	KDK-2	KDK-3	KDK-3	KDK-4
SiO_2	40.12	38.43	37.62	39.868	39.96
TiO_2	6.29	6.08	5.37	6.11	6.09
Al_2O_3	7.76	9.73	10.92	7.98	8.24
FeO	10.34	10.08	10.26	10.30	10.39
MnO	0.06	0.08	0.12	0.07	0.08
MgO	18.83	18.09	19.29	18.64	18.97
CaO	0.01	0.14	0.04	0.00	0.03
Na_2O	0.55	0.32	0.32	0.55	0.44
K_2O	9.53	9.29	8.52	9.55	9.12
Total	93.49	92.23	92.46	93.07	93.32
Cations for 22 oxygens					
Si	6.012	5.829	5.675	6.003	5.982
Ti	0.709	0.693	0.610	0.691	0.686
Al	1.371	1.739	1.940	1.416	1.453
Fe(ii)	1.296	1.279	1.294	1.297	1.301
Mn	0.007	0.010	0.016	0.009	0.011
Mg	4.206	4.090	4.337	4.185	4.233
Ca	0.002	0.023	0.007	0.000	0.005
Na	0.160	0.093	0.092	0.160	0.127
K	1.822	1.798	1.640	1.834	1.741
Total	15.59	15.554	15.55	15.554	15.5537

pipe are similar to those of the Behradih pipe in most aspects and plot significantly into the G9 and with ~5 % in the G10 category.

REE abundances of garnet concentrates from Behradih and Payalikhand could only be analysed in this study (Table 7), and their chondrite-normalised patterns are presented in Fig. 6. Data from low-Cr lherzolite pyropes from the Finsch mine, Kaapvaal craton, southern Africa (Gibson et al. 2008) are also provided in Fig. 6 for comparison. Two broad types of garnet REE distribution patterns are recognised from studies from the Kaapval craton, southern Africa and elsewhere (e.g., Shimizu 1975; Stachel et al. 1998; Burgess and Harte 2004; Lehtonen 2005; Gibson et al. 2008): (1) "smooth" chondrite-normalised patterns without any kinks with a strong LREE depletion relative to MREE and HREE and a gradual enrichment from Sm_N to Yb_N which is considered typical of Ca-saturated lherzolitic garnets and (2) "sinusoidal" chondrite-normalised REE patterns showing peaks at Nd which are characteristic of sub-calcic (harzburgitic) garnets and also of the rare Ti-poor lherzolitic class. The Behradih pyropes display both types of REE patterns with a clear predominance of garnet with "smooth" patterns, whereas the Payalikhand garnets lack "sinusoidal" patterns.

Emerald green to bright green macrocrystic clinopyroxene (essentially diopside) is of ~2 mm size in the Kodomali and Kosambura pipes. No information is available on the macrocrystic clinopyroxene from other orangeites of the Mainpur field. In terms of composition in Al_2O_3-Cr_2O_3 wt% space, both localities are similar and predominantly have been derived from garnet peridotite with very few of them transgressing into the spinel peridotite field (Fig. 7). Jha et al. (2002) also reported chrome diopsides of similar composition from the Payalikhand pipe.

Chromites collected from the Behradih and Kodomali pipes mostly occur as euhedral and octahedral crystals. Behradih spinels have Cr_2O_3 varying between 27–66 wt% with most of the data having >52 wt% Cr_2O_3. About 55 % chromites are high chromium (58–66 wt% Cr_2O_3), and out of these 26 chromites, most have between 10.5 to 13.5 wt% MgO and low TiO_2 (< 1 wt%) and fall in the diamond inclusion and intergrowth fields (Fig. 8a, b). This is also duplicated by about 30 % of the Kodomali chromites whose Cr_2O_3 (35–65 wt%), MgO (7–18 wt%) and TiO_2 (7.5 wt%; but mostly <3 wt%) contents extend into the diamond stability field (Fig. 8c, d). The chemistry of ~75 % chromites from the Kosambura pipe belongs to the high-chrome variety Cr_2O_3 (55–67 wt%), and MgO (mostly 10.0–13.5 wt%) and TiO_2 (mostly <1.15 wt%) fall into the diamond inclusion and intergrowth fields suggesting encouraging diamond potential (Fig. 8e, f).

U–Pb Zircon Geochronology

A number of zircon grains have been recovered whilst processing samples from the Behradih pipe. They are essentially of euhedral to subhedral shape, and their grain size mostly varies from 100–150 μm (Fig. 9a–l) and is essentially <500 μm. Sub-rounded to rounded zircons are not observed suggesting a lack of magmatic resorption. It is well known that zircon in kimberlites can be of crustal as well of mantle (megacrystic) origin and the chemical composition plays an important role in their discrimination (e.g., Konzett et al. 2000; Belousova et al. 2002). The elevated contents of U (113–1096 ppm) in the studied samples (Table 8) exclude their mantle derivation since megacrystic mantle-derived zircons are known to have U contents <60 ppm (see Page et al. 2007).

U–Pb dating (see Table 8 for results and Fig. 9m for graphical representation) demonstrates that the zircon population is essentially of Palaeoproterozoic age (~2450 Ma) and provides compelling evidence for their crustal-derivation.

Table 4 Mineral chemistry (wt%) of groundmass spinel from the Kodomali orangeite. Fe^{2+} and Fe^{3+} redistribution done by spinel stoichiometry

Oxide (wt%)	KDK-2	KDK-2	KDK-2	KDK-3	KDK-3	KDK-3	KDK-3	KDK-4	KDK-4	KDK-4	KDK-4	KDK-4	KDK-4
SiO_2	0.08	0.11	0.16	0.08	0.02	0.07	0.08	0.12	0.02	0.03	0.05	0.07	0.05
TiO_2	1.95	1.81	1.96	2.07	4.70	2.21	2.71	1.95	1.72	0.50	1.86	1.88	1.89
Al_2O_3	6.70	5.76	4.06	6.50	1.53	3.29	2.93	5.21	28.34	37.43	4.40	4.14	5.59
Cr_2O_3	55.05	54.17	51.09	53.37	38.26	51.76	48.14	55.89	22.84	27.67	54.95	55.58	55.71
Fe_2O_3	5.33	7.05	12.30	6.57	23.53	11.36	14.47	5.72	14.28	2.90	7.60	7.22	5.86
FeO	22.71	23.26	24.42	22.89	28.22	23.90	25.01	23.2	21.31	17.52	24.12	23.43	23.29
MnO	0.00	0.00	0.00	0.00	0.00	0.00	0.00	0.00	0.00	0.00	0.00	0.00	0.00
MgO	7.83	7.21	6.63	7.64	5.47	6.70	6.27	7.29	11.42	13.55	6.46	6.95	7.26
CaO	0.14	0.19	0.17	0.13	0.22	0.17	0.17	0.15	0.07	0.03	0.26	0.14	0.16
Total	99.80	99.55	100.81	99.25	101.95	99.47	99.79	99.60	100.00	99.63	99.72	99.41	99.81
Cations for 32 oxygens													
Si	0.022	0.031	0.045	0.021	0.006	0.020	0.023	0.032	0.005	0.006	0.015	0.021	0.014
Ti	0.401	0.375	0.409	0.430	0.994	0.469	0.577	0.406	0.308	0.088	0.391	0.395	0.392
Al	2.159	1.876	1.328	2.112	0.507	1.093	0.975	1.693	7.951	10.232	1.450	1.363	1.817
Cr	11.896	11.843	11.198	11.625	8.510	11.521	10.750	12.191	4.864	5.075	12.139	12.285	12.153
V	0.000	0.000	0.000	0.000	0.000	0.000	0.000	1.267	0.000	0.000	0.000	0.000	0.000
$Fe^{(iii)}$	1.097	1.469	2.567	1.362	4.982	2.409	3.076	0.000	2.559	0.507	1.599	1.521	1.217
$Fe^{(ii)}$	5.193	5.379	5.662	5.274	6.639	5.626	5.907	5.368	4.242	3.399	5.636	5.479	5.373
Mn	0.000	0.000	0.000	0.000	0.000	0.000	0.000	0.000	0.000	0.000	0.000	0.000	0.000
Mg	3.189	2.972	2.740	3.138	2.294	2.811	2.641	2.998	4.052	4.687	2.693	2.895	2.986
Ca	0.041	0.055	0.051	0.039	0.067	0.051	0.052	0.046	0.019	0.008	0.077	0.041	0.047
Total	24	24	24	24	24	24	24	24	24	24	24	24	24
Ti/(Ti + Cr + Al)	0.03	0.03	0.03	0.03	0.10	0.04	0.05	0.03	0.02	0.01	0.03	0.03	0.03
$Fe^2/(Fe^2 + Mg)$	0.62	0.64	0.67	0.63	0.74	0.67	0.69	0.64	0.51	0.42	0.68	0.65	0.64

Table 5 Major oxide (wt%) of perovskites from Behradih and Kodomali orangeites. Note that total iron is expressed as Fe_2O_3 in all the analyses. ΔNNO_3 is log *f*O2 relative to NNO (nickel–nickel oxide) buffer

Behradih pipe									Kodomali pipe						
Oxide wt%	BHS2-4	BHS 10-12	BHS 15-18	BHS 36-37	BHS 36-37	BHS36-37	BHS 52	BHS 61	KDK-2	KDK-2	KDK-2	KDK-3	KDK-3	KDK-3	KDK-4
Na_2O	0.24	0.22	0.29	0.24	0.27	0.25	0.27	0.28	0.64	0.55	0.6	0.97	0.59	1.15	0.73
Al_2O_3	0.28	0.29	0.27	0.26	0.28	0.31	0.28	0.29	0.10	0.09	0.13	0.07	0.13	0.05	0.07
SiO_2	0.15	0.44	0.12	0.25	0.10	0.25	0.09	0.09	0.22	0.16	0.17	0.28	0.13	0.15	0.11
CaO	37.23	36.95	37.24	37.06	37.23	36.46	37.25	37.09	37.17	36.71	36.92	36.69	36.44	35.29	37.20
TiO_2	51.80	52.85	52.33	52.32	51.99	52.44	52.53	52.70	53.41	53.36	53.03	53.48	52.86	53.74	54.02
Fe_2O_3	2.84	2.57	2.51	2.59	2.90	2.68	2.78	2.65	2.18	2.27	2.53	2.23	2.20	1.48	1.91
SrO	0.44	0.47	0.45	0.45	0.48	0.44	0.42	0.46	0.32	0.58	0.79	0.48	0.41	1.03	0.57
Nb_2O_5	0.70	0.73	0.70	0.78	0.75	0.70	0.68	0.73	1.08	1.04	1.38	1.04	1.07	1.64	1.05
La_2O_3	1.06	1.00	1.07	0.96	1.09	1.02	1.01	1.07	0.97	1.11	1.07	0.88	1.14	1.21	0.96
Ce_2O_3	3.08	2.62	2.94	2.50	2.95	2.78	2.97	2.83	2.35	2.73	2.53	2.46	2.87	2.80	2.51
Pr_2O_3	0.29	0.26	0.31	0.18	0.28	0.26	0.29	0.24	0.17	0.24	0.25	0.21	0.29	0.20	0.24
Nd_2O_3	1.01	0.86	0.95	0.84	0.97	0.97	0.99	0.92	0.75	0.81	0.76	0.81	0.88	0.77	0.76
Ta_2O_5	0.10	0.04	0.13	0.13	0.09	0.10	0.09	0.09	0.02	0.18	0.02	0.01	0.15	0.07	0.05
ThO_2	0.39	0.25	0.33	0.25	0.31	0.30	0.37	0.27	0.22	0.17	0.12	0.22	0.24	0.26	0.21
Total	99.60	99.53	99.63	98.79	99.70	98.94	99.99	99.71	99.60	100.01	100.28	99.83	99.39	99.85	100.38
ΔNNO_3	4.28	2.93	2.76	3.03	4.46	3.57	3.97	3.30	0.71	1.20	1.88	1.01	0.93	3.08	0.48

Table 6 Bulk-chemistry (wt%) of Behradih, Kodomali and Kosambura orangeites from the Mainpur field

Behradih										Kodomali			Kosambura
Oxide wt%	BHS-2-4	BHS-10-12	BHS-15-18	BHS-23-25	BHS-36-37	BHS-39	BHS-52	BHS-54	BHS-61	KDK-2	KDK-3	KDK-4	MKO
SiO_2	45.3	44	47.2	45.8	42	41.8	43.3	46.5	44	47.34	46.47	46.53	74.24
TiO_2	1.25	1.07	1.08	1.02	1.21	1.14	1.13	1.02	1.23	1.29	1.42	1.28	0.50
Al_2O_3	5.53	4.58	4.92	4.83	4.57	4.48	5.34	5.38	4.92	6.57	6.57	6.4	2.44
Fe_2O_3*	7.59	7.06	7.22	7.39	8.28	7.54	7.71	7.35	8.15	7.58	7.75	7.54	5.67
MnO	0.12	0.13	0.1	0.11	0.12	0.13	0.13	0.11	0.13	0.11	0.11	0.11	0.23
MgO	22	23	24.7	24.6	27.2	25.5	24.2	23.7	23.9	22.13	22.14	22.66	3.08
CaO	6.66	7.48	4.63	5.33	4.85	7.01	6.3	4.97	6.47	8	7.68	7.86	2.67
Na_2O	0.6	0.78	0.73	0.78	0.57	0.74	0.66	0.95	1.18	0.89	0.87	1.09	0.01
K_2O	2.92	2.15	1.42	1.63	2.19	2.18	2.41	2.15	2.92	1.6	2.11	1.75	0.13
P_2O_5	0.701	0.597	0.319	0.234	0.235	0.424	0.275	0.274	0.263	1.6	2.11	1.75	1.78
LOI	6.51	8.46	6.97	7.51	8.1	8.49	7.7	6.9	5.75	3.81	3.69	3.99	8.26
Total	99.675	99.801	99.837	99.785	99.738	100.14	99.672	99.878	99.76	100.92	100.92	100.96	99.09
Mg#	85.2	86.6	87.1	86.8	86.7	87.0	86.1	86.5	85.3	84.5	84.8	84.3	
Sr	926	909	643	739	801	1010	895	774	1040	683.58	644.78	600.32	384
Rb	132	101	54.8	66.8	181	133	161	68.8	162	81.47	87.2	87.22	7
Cu	26	33	29	31	34	35	34	29	36	48.39	44.64	42.684	15
Ni	696	875	859	950	1070	945	829	966	1040	1131.2	1131.5	1114	860
Cr	495	410	470	440	515	490	550	450	530	2487.6	1556.6	1532.3	2165
Co	44	51	49	56	60	56	48	54	59	72.88	72.88	73.19	50
V	64	80	90	70	58	54	62	116	76	138.75	137.89	138.66	80
Sc	12.5	10.5	10	10	11	11	12	9.5	11.5	16.89	16.93	16.17	20
Y	18.1	14.3	13.5	13.8	15.9	14.9	16.1	13.6	15.7	15.39	15.31	15.15	24
Zr	206	128	156	160	226	207	195	154	217	264.81	262.17	262.66	369
Nb	84	55	48.5	48	63.5	65.5	72	51.5	59.5	94.1	107	122	319
Ba	1510	1380	940	1560	1430	2260	1540	2290	4130	2325.3	2088.1	2278.1	967
Hf										4.67	4.68	4.64	
Ta	5.7	3.1	3.3	3.2	4.2	4.4	4	3.5	4	1.11	1.21	1.23	11
Pb	15	9	9	11	10	13	16	13	12	7.66	6.99	7.48	13
Th	27.3	15.6	18.3	15.8	18.7	17.9	23	16.2	22.8	8.25	8.53	8.3	54
Ga										12.97	12.79	13.14	
U	4.55	3.25	3.15	2.1	3.35	3.4	3.2	2.7	4.7	1.35	1.33	1.32	6

(continued)

Table 6 (continued)

Behradih										Kodomali			Kosambura
Oxide wt%	BHS-2-4	BHS-10-12	BHS-15-18	BHS-23-25	BHS-36-37	BHS-39	BHS-52	BHS-54	BHS-61	KDK-2	KDK-3	KDK-4	MKO
Zn	62	48	44	55	57	51	61	48	56	110.99	116.06	116.72	54
La	153	102	109	103	122	117	153	108	126	94.22	96.14	93.82	375
Ce	261	180	184	175	209	201	247	179	213	159.46	163.65	159.5	563
Pr										18.8	19.35	18.91	
Nd	91.8	64.1	69.4	65	78.4	74.6	85	65.9	77.7	66.77	69.04	67.17	213
Sm	12.2	8.95	9.3	8.85	10.7	10.3	11.3	8.8	10.9	9.77	10.1	9.65	32
Eu	2.85	2.2	2.2	2.1	2.5	2.4	2.65	2.1	2.55	3.42	3.43	3.09	
Gd	6.78	5.66	5.62	5.58	6.3	6.12	6.89	5.42	6.44	10.47	10.6	10.34	
Tb										0.97	0.99	0.966	
Dy										3.26	3.2	3.16	
Ho										0.6	0.6	0.58	
Er										1.63	1.65	1.63	
Tm										0.19	0.19	0.19	
Yb	1.3	1.05	1	0.95	1.1	1.05	1.2	0.95	1.05	0.96	1.01	0.99	
Lu	0.2	0.16	0.16	0.14	0.14	0.14	0.18	0.14	0.16	0.15	0.15	0.14	

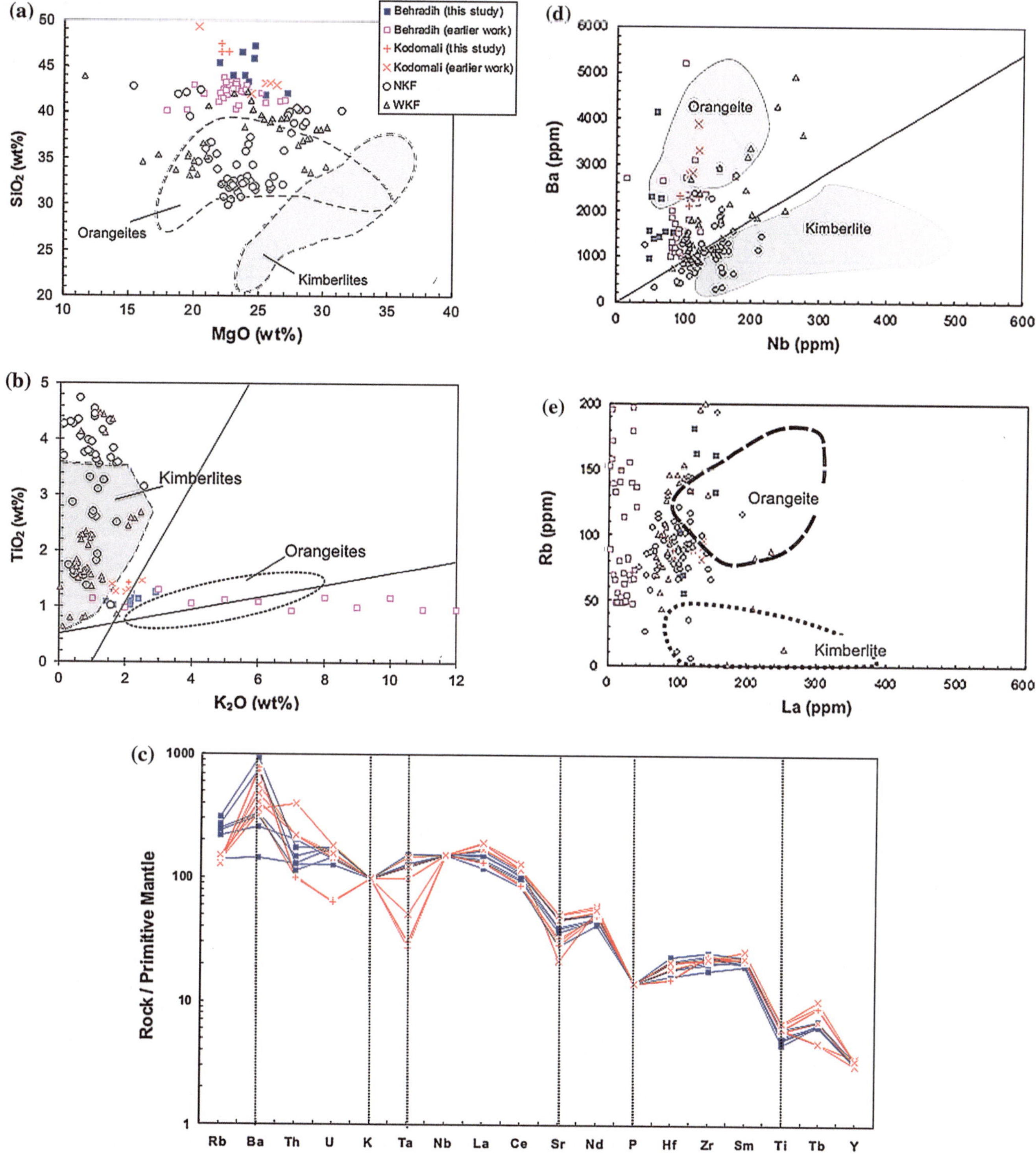

Fig. 4 Whole-rock variations of **a** MgO (wt%) versus SiO_2 (wt%) and **b** K_2O (wt%) versus TiO_2 (wt%) for the Mainpur pipes. *Data sources*: Behradih (earlier work; Chalapathi Rao et al. 2011a; Lehmann et al. 2007); Kodomali (earlier work; data from Fareeduddin et al. 2006; Paul et al. 2006; Lehmann et al. 2006); WKF and NKF (Chalapathi Rao and Dongre 2009; Chalapathi Rao and Srivastava 2009; Chalapathi Rao et al. 2004). Various kimberlite and orangeite fields are from Donnelly et al. 2011 and the references therein. **c** Primitive-mantle-normalised (Sun and McDonough 1989) multi-element spidergram for the Behradih and Kodomali orangeite samples. (x = Kodomali earlier work; Fareeduddin et al. 2006; Paul et al. 2006; Lehmann et al. 2006). Variation of **d** Nb (ppm) versus Ba (ppm) and **e** La (ppm) versus Rb (ppm). The symbols and the data sources are the same as in Fig. 4a. Various fields are taken from Donelly et al. (2011)

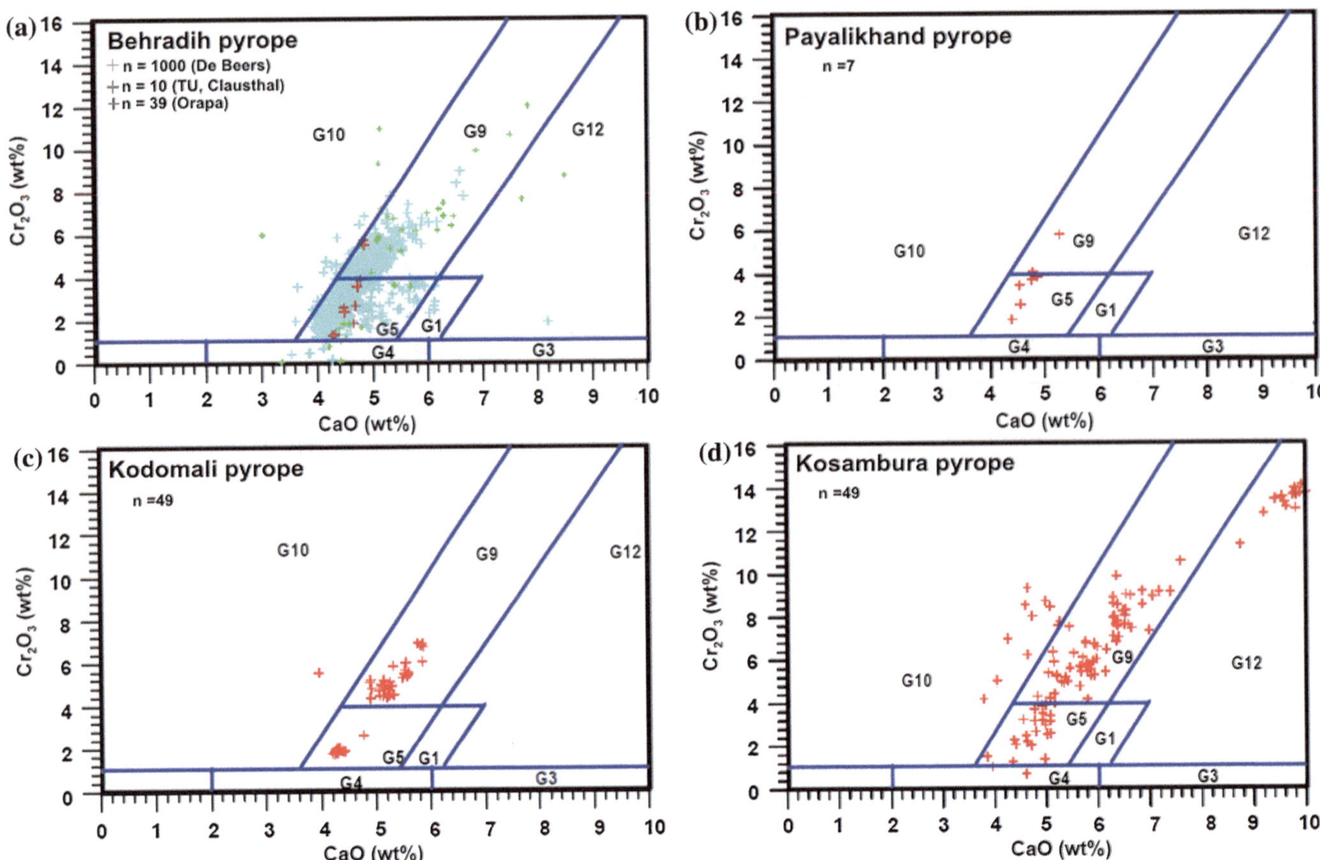

Fig. 5 Cr_2O_3 versus CaO (wt%) variation plots of pyrope garnet separates from orangeites (Mainkar 2010). The fields are adopted from Grutter et al. (2004). Source of analyses is also given

Discussion

The geology of the Mainpur orangeites reveals that they preserve diatreme (Behradih) as well as hypabyssal facies (Kodomali) and implies differential erosion in this domain. Except for the Behradih and Kodomali pipes, all other occurrences in the Mainpur field are highly altered, contaminated and silicified. Radiometric age data reveal that the Behradih and Kodomali pipes constitute the youngest yet recorded diamondiferous magmatic event in the Indian shield, straddling the Cretaceous-Tertiary (K-T) boundary, synchronous with the eruption of Deccan flood basalts (see Lehmann et al. 2010). The Mainpur orangeites have also recently been inferred to be a part of the kimberlite–lamprophyre–carbonatite–alkaline rock spectrum in the Deccan Large Igneous Province with variable lithospheric thickness controlling the geographic distribution of the different rock variants (Chalapathi Rao and Lehmann 2011).

In terms of mineral chemistry, similarities as well as differences have been noticed between the Kodomali and Behradih pipes. Olivine, spinel and clinopyroxene in both pipes have overlapping compositions, whereas the groundmass phlogopite (TiO_2) and perovskite (Nb_2O_5 and Fe_2O_3) show marked differences. Distinctness in the composition of clinopyroxene of the Kodomali samples of this study and those reported by earlier workers (Fareeduddin et al. 2006) raises the possibility of multiple eruptions within the same pipe as autoliths of earlier kimberlite intrusions were also recorded (Chatterjee and Jha 1994). As multiple kimberlite eruptions (up to 8) within a short span of a few million years have been recorded from the Prieska region, southern Africa (Smith et al. 1994) and the Fort á La Corne field, Canada (Kjarsgaard et al. 2009), radiometric dating on mineralogically distinct samples is in progress to ascertain such episodes within the Kodomali pipe. Nevertheless, mineral compositions of the Behradih and Kodomali pipes together show considerable differences to the Mesoproterozoic kimberlites of WKF and NKF, southern India, and resemble the southern African orangeites in this regard.

The major element contents viz., SiO_2, MgO, Fe_2O_3*, K_2O and TiO_2 of the Behradih and Kodomali pipes are clearly different from bulk rocks of the WKF and NKF which have relatively lower SiO_2 and a much wider range in MgO. Furthermore, when compared to the archetypal

Table 7 Rare earth element (REE) contents of garnet separates from the Behradih (BEH series) and Payalikhand (PAY series) orangeites

ppm	BEH-A-07	BEH-A-08	BEH-A-13	BEH-A-18	BEH-A-19	BEH-A-26	BEH3-01	BEH3-02	BEH3-04	BEH3-05	BEH3-07	BEH3-08	BEH3-10	BEH3-12
La	0.03	0.03	0.03	0.03	0.04	0.06	0.05	0.04	0.10	0.12	0.11	0.07	0.06	0.04
Ce	0.12	0.16	0.21	0.16	0.10	0.26	0.76	0.30	1.06	1.05	0.98	0.71	0.82	0.25
Pr	0.04	0.07	0.09	0.07	0.04	0.08	0.30	0.12	0.30	0.29	0.28	0.34	0.31	0.06
Nd	0.42	0.60	0.78	0.83	0.52	0.41	3.02	1.06	1.96	2.04	1.81	2.8	2.4	1.18
Sm	0.37	0.65	0.72	0.65	0.58	0.25	1.98	0.77	0.70	0.62	0.62	1.89	1.79	0.62
Eu	0.25	0.35	0.36	0.39	0.27	0.07	0.95	0.35	0.11	0.09	0.23	0.98	0.99	0.19
Gd	0.89	1.28	1.47	1.23	1.42	0.33	4.32	1.04	0.55	0.23	0.37	4.06	3.96	0.92
Dy	1.38	3.28	1.98	2.25	2.8	0.55	6.32	1.39	0.70	0.30	0.57	6.57	6.36	2.04
Ho	0.39	0.73	0.40	0.50	0.66	0.21	1.46	0.45	0.07	0.13	0.10	1.45	1.40	0.46
Er	1.51	2.15	1.29	1.8	1.89	0.60	4.47	1.28	0.62	0.45	0.48	4.84	4.56	1.79
Yb	1.73	2.67	1.74	1.79	2.35	1.20	6.32	2.3	0.61	0.61	0.43	6.37	6.03	2.21
Lu	0.30	0.41	0.29	0.31	0.36	0.23	1.01	0.31	0.08	0.09	0.11	1.07	0.98	0.3
ppm	BEH3-15	BEH3-16	BEH3-21	BEH4-07	BEH4-09	PAY-A-02	PAY-A-13	PAY-A-15	PAY2-05	PAY1-03	PAY1-04	PAY1-07		
La	0.05	0.07	0.12	0.03	0.04	0.04	0.05	0.05	0.04	0.08	0.04	0.03		
Ce	0.24	0.88	0.98	0.09	0.31	0.27	0.12	0.12	0.27	0.77	0.17	0.26		
Pr	0.14	0.31	0.29	0.03	0.19	0.10	0.04	0.04	0.10	0.13	0.07	0.09		
Nd	1.31	3.06	1.95	0.37	1.37	0.83	0.46	0.46	1.33	0.93	0.69	1.14		
Sm	0.51	2.2	0.66	0.56	1.2	0.52	0.51	0.51	0.82	0.77	0.97	0.92		
Eu	0.18	0.94	0.14	0.31	0.63	0.24	0.30	0.30	0.41	0.33	0.39	0.56		
Gd	0.66	3.89	0.48	1.45	1.94	0.93	1.18	1.18	1.57	1.12	1.71	1.85		
Dy	0.46	6.13	0.54	3.26	2.18	1.82	2.53	2.53	2.57	1.85	1.66	2.76		
Ho	0.06	1.37	0.07	0.75	0.51	0.44	0.51	0.51	0.65	0.49	0.39	0.69		
Er	0.41	4.54	0.26	2.47	1.31	1.33	1.73	1.73	1.72	1.69	1.17	1.78		
Yb	0.59	6.27	0.43	2.72	1.41	1.41	1.88	1.88	2.03	1.7	1.23	1.76		
Lu	0.08	1.04	0.08	0.42	0.21	0.24	0.27	0.27	0.32	0.21	0.17	0.31		

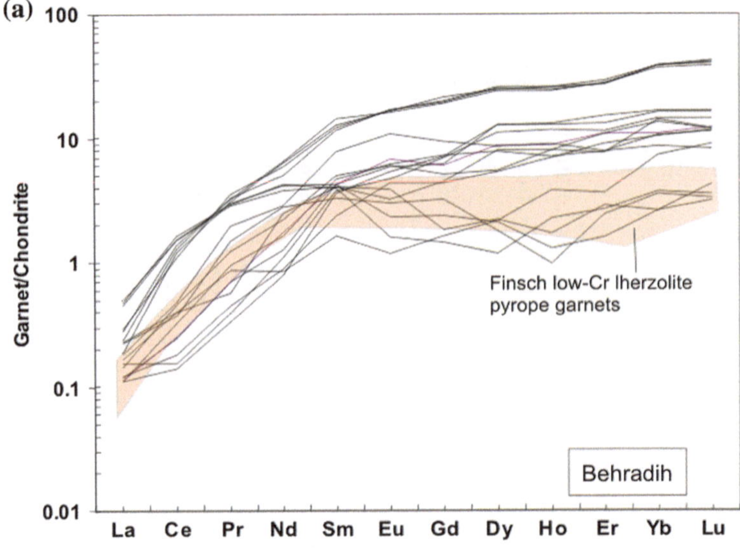

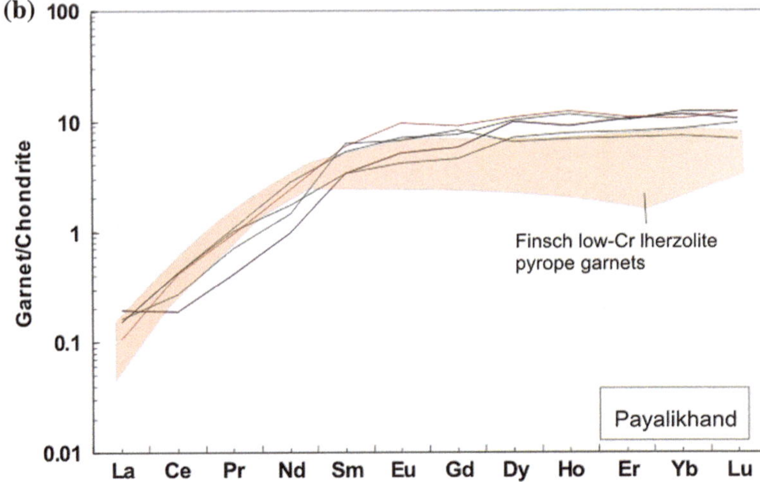

Fig. 6 Chondrite-normalised (Sun and McDonough 1995) rare earth element distribution patterns for pyrope garnet separates from Behradih (**a**) and Payalikhand (**b**). The field of Finsch low-Cr lherzolite pyropes, shown for comparison, is from Gibson et al. (2007)

kimberlites and orangeites of southern Africa, the Mainpur pipes as well as kimberlites from NKF and WKF clearly display a more fractionated nature and show strong affinities towards orangeites. Prominent negative spikes at K, Sr, P, Ta (for some samples from Kodomali) and Ti and positive Ba spikes are noticed in the primitive-mantle-normalised plots (Fig. 4c), and the samples under study can be utilised to infer the nature of their mantle source regions. Spikes in the multi-element plots are variably interpreted to reflect residual phases, fractional crystallisation or even hydrothermal alteration. Negative K-Ta-Sr-P-Ti spikes in southern African kimberlites have been attributed to be a primary feature of the source rock (Becker and Le Roex 2006), whereas negative K-Ti in southern African orangeites have been linked to the fractionation of phlogopite (Coe et al. 2008). Positive Ba spikes are also a characteristic feature of southern African orangeites and have been explained by concentration of Ba in phlogopite (Howarth et al. 2011). Similar explanations can be invoked in the case of samples under this study.

The depth of generation of kimberlite magmas is a matter of debate with sources ranging from sub-continental lithospheric mantle (SCLM) (e.g., Le Roex et al. 2003; Chalapathi Rao et al. 2004; Becker et al. 2007; Donnelly et al. 2011), convecting (asthenospheric) upper mantle (e.g., Mitchell 2006; Woodhead et al. 2009), recycled oceanic crust at the transition zone or lower mantle (e.g., Ringwood et al. 1992; Paton et al. 2009), core-mantle boundary (e.g., Haggerty 1999; Torsvick et al. 2010; Collerson et al. 2010) and even multiple reservoirs (Tappe et al. 2011). On the other hand, there is general agreement for SCLM to be the source for orangeites (e.g., Skinner 1989; Mitchell 2006; Coe et al. 2008; Chalapathi Rao et al. 2011a). Incompatible trace element ratios of Ce/Y, La/Yb and Zr/Nb can be exploited to infer the degree of melting involved in the generation of kimberlites and orangeites, and increasing

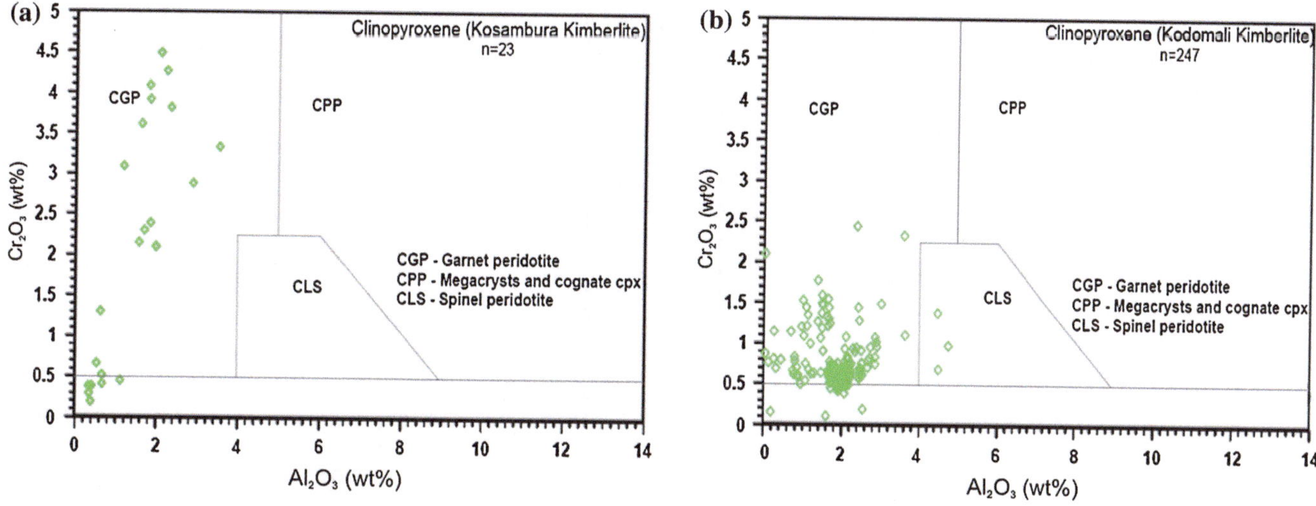

Fig. 7 Cr_2O_3 (wt%) versus Al_2O_3 (wt%) plots of Cr-diopside xenocrysts from the Kosambura (**a**) and Kodomali (**b**) pipes (Mainkar 2010). The fields are adapted from Ramsay (1992)

La/Yb is interpreted to represent lower degrees of partial melting of a peridotitic source (Mitchell 1995; Howarth et al. 2011). Figure 10a, b, involving these ratios, suggests that (1) the Behradih and Kodomali pipes have undergone degrees of partial melting similar to the unevolved Kroonstad orangeites from the Kaapvaal craton, southern Africa, (2) are typified by higher Zr/Nb ratios than those from WKF and NKF kimberlites, Dharwar craton and (3) their compositions cannot be derived by single-stage partial melting of a peridotite mantle source. We have compared the observed REE ratios of Behradih and Kodomali as well as those of others from the WKF and NKF (the latter data were taken from Chalapathi Rao et al. 2004; Chalapathi Rao and Srivastava. 2009; Chalapathi Rao and Dongre 2009), with the melting trajectories of inferred southern African Group I and orangeite source regions presented by Becker and Le Roex (2006), in an effort to test whether the southern African model is applicable for the Indian samples. Results presented in Fig. 10c show that a simple melting trajectory, although slightly away from the assumed southern African orangeite, can account for the observed REE compositions of the Behradih and Kodomali samples. On the other hand, the composition of both the NKF and WKF samples can be better explained by a kimberlite source or a combination of kimberlitic and orangeitic sources.

The bulk-rock compositions of the Kodomali and Behradih samples are also compared (Fig. 11) with the experimental data obtained from a multi-component (natural) bulk system at 10 GPa to constrain the generation and differentiation of orangeite magma (Ulmer and Sweeney 2002). Their compositions are consistent with the differentiation of an orangeite composition by olivine–garnet–orthopyroxene fractionation (Fig. 11).

The pyrope garnet population in the Behradih, Kodomali, Payalikhand and Kosambura pipes is dominated by the calcic-lherzolitic variety, with <5 % belonging to the high-interest sub-calcic harzburgitic category and the remainder to other fields such as eclogitic types (Fig. 5). The lherzolitic trend is suggestive of garnet in equilibrium with clinopyroxene (see Gibson et al. 2008) which is consistent with the garnet-peridotite affinities of the Mainpur diopside xenocrysts (Fig. 7). The findings of our study are also consistent with those reported earlier by (1) Jha et al. (2002) wherein the pyrope population in the Mainpur pipes has been inferred to be dominated by the calcic-lherzolitic variety corresponding to G9 and some eclogitic (G3, G4 and G5) and iron–titanium low calcium (G1 and G2) garnets with the sub-calcic harzburgitic category (G10) pyrope garnet population between 1 and 5 % and (2) Mukherjee et al. (2000) who reported garnets of eclogitic paragenesis from the Behradih pipe.

The presence of "sinusoidal" REE_N patterns in Behradih garnets of this study (Fig. 6) is significant since such patterns are regarded to result from a two stage process involving an (1) initial extensive komatiite melt extraction event which results in extreme LREE and HREE depletion leading to a depleted lithosphere followed by (2) a second stage involving fluid metasomatism wherein repeated pulses of fractionated melt with low HREE and variable LREE/MREE react with the depleted lithosphere (Stachel et al. 2004; Creighton et al. 2010). The sub-calcic harzburgitic garnets displaying sinusoidal REE_N are only confined to depths <175 km and temperatures of 1150 °C, whereas lherzolitic garnets and Ti–rich pyrope megacrysts originate from depths in excess of 175 km (e.g., Stachel et al. 1998, 2004; Lehtonen 2005). Therefore, the present study

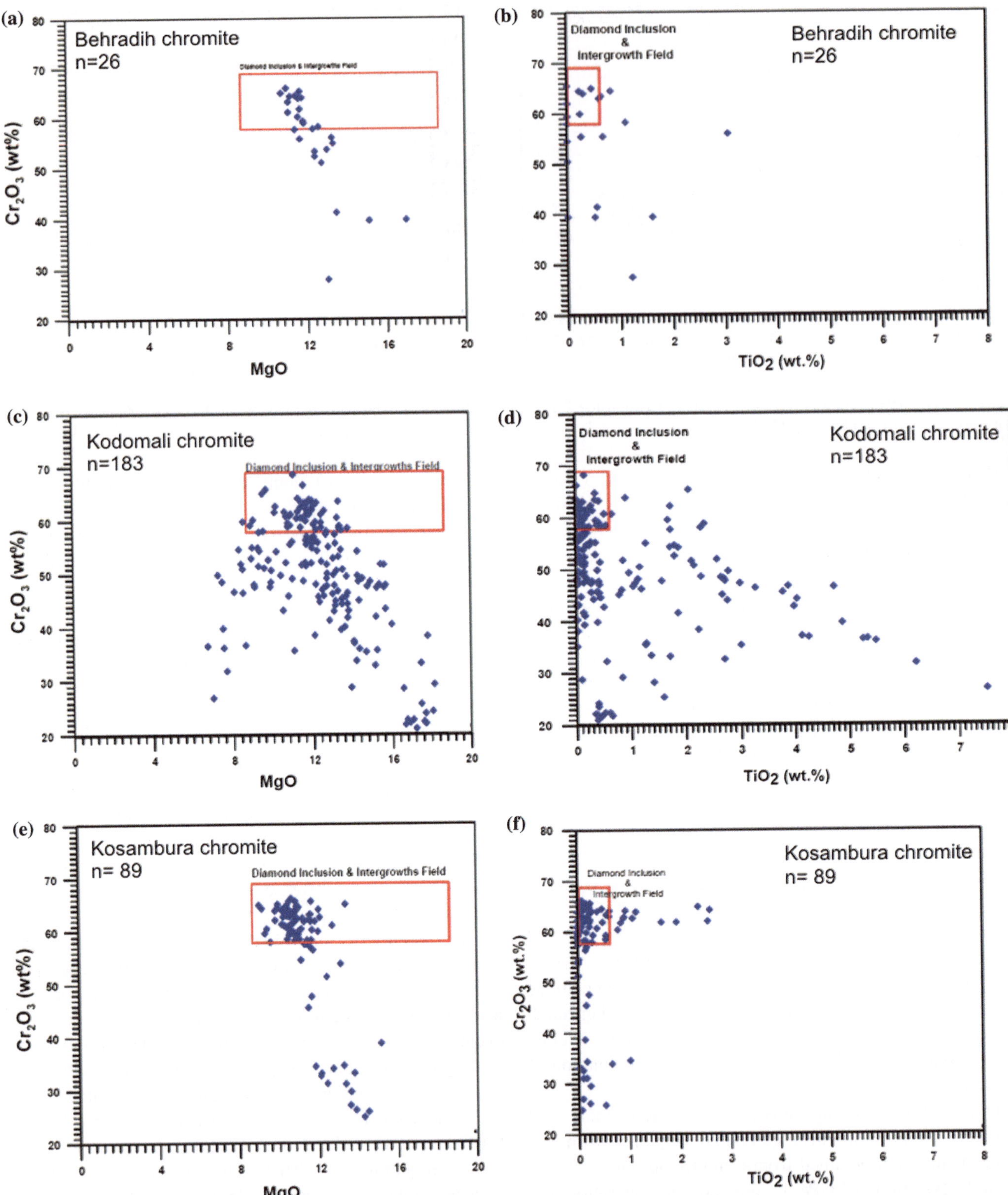

Fig. 8 Cr_2O_3 (wt%) versus MgO (wt%) and Cr_2O_3 (wt%) versus TiO_2 (wt%) variation plots of Behradih (A and B), Kodomali (**c**, **d**) and Kosambura (**e** and **f**) chromite xenocrysts (Mainkar 2010). Diamond inclusion and intergrowth field is after Fipke et al. (1995). The plots suggest that the source region of most of the indicator chromite crystals from the Mainpur pipes is from the diamond stability field

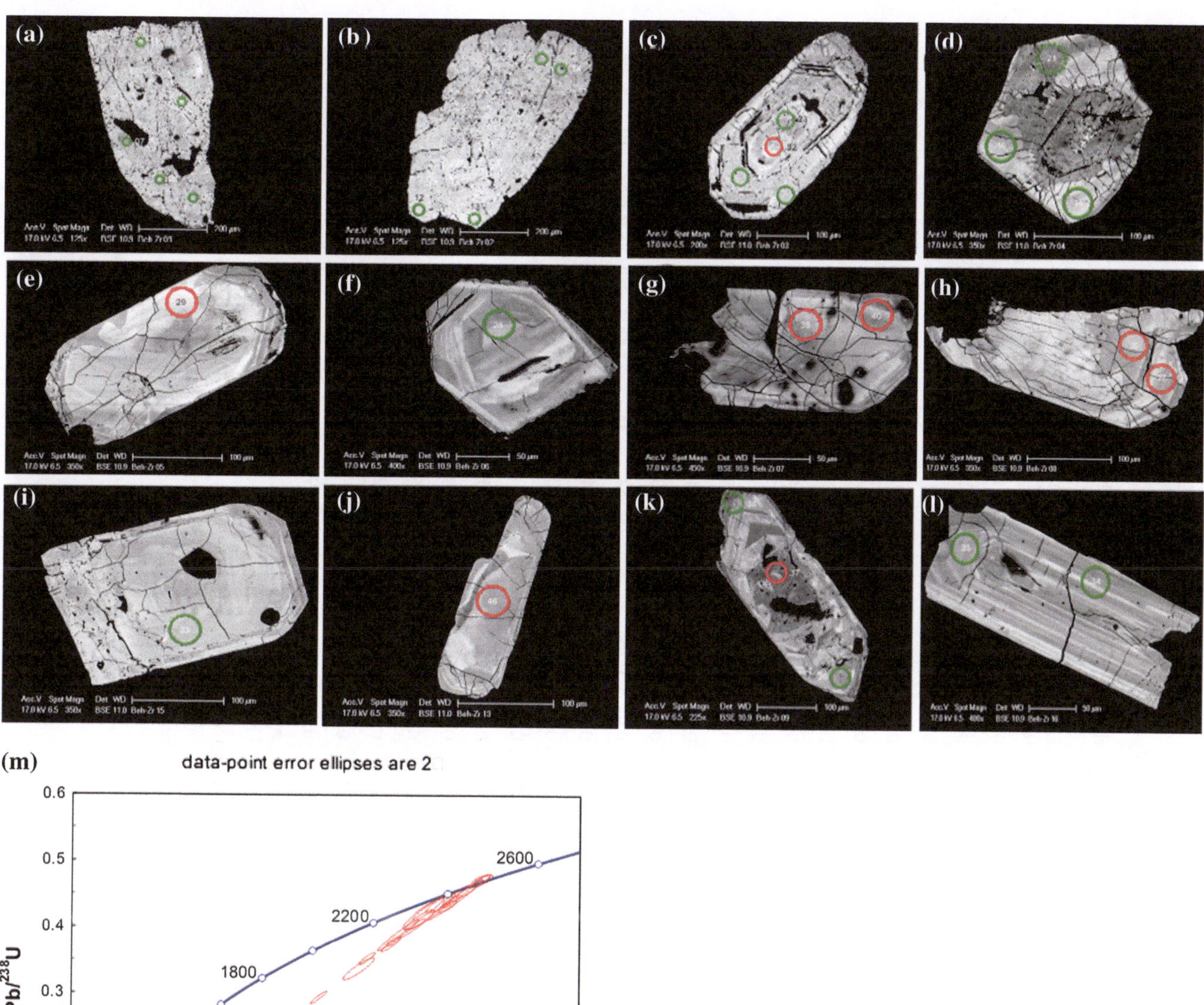

Fig. 9 Backscattered electron (BSE) images of zircon xenocrysts from the Behradih pipe depicting their varied morphology. The circles and their inscribed numbers correspond to their U–Pb isotopic analysis provided in Table 8. Figure 9m U–Pb concordia diagram of zircon analyses in (**a–l**). Data point ellipses are given as 2-sigma uncertainties

identifies, for the first time, the presence of a compositionally layered mantle in the end-Cretaceous sub-Bastar craton similar to that reported from other cratons elsewhere such as Kaapvaal craton (e.g., Gregoire et al. 2003; Gibson et al. 2008), western Guyana shield (e.g., Schulze et al. 2006), Slave craton (e.g., Griffin et al. 1999; Kopylova and Caro 2004), Siberian craton (e.g., Ashchepkov et al. 2010), North Atlantic craton (Sand et al. 2009) and Karelian craton (Lehtonen et al. 2004).

Some of the chromites from the Behradih, Kodomali and Kosambura orangeites are compositionally similar to those found as inclusions in diamonds (Fig. 8), implying their derivation from the diamond stability field, which also finds support in the diamondiferous nature of these pipes. It is

Table 8 U–Pb isotopic data of zircon grains from the Behradih pipe. Morphology of most of the analysed grains is presented in Fig. 9a to L

		^{207}Pb	U[a]	Pb[a]	Th[a]	^{207}Pb[b]	2 σ[d]	^{206}Pb[b]	2 σ[d]	rho[c]	^{207}Pb[e]	2 σ[d]	^{207}Pb	2 σ	^{206}Pb	2 σ	^{207}Pb	2 σ	Conc[f]
Crystal #	Analysis #	(cps)	(ppm)	(ppm)	U	^{235}U	%	^{238}U	%		^{206}Pb	%	^{235}U	(Ma)	^{238}U	(Ma)	^{206}Pb	(Ma)	(%)
Zr-01	zircon 07	180281	452	198	0.34	8.7352	3.1	0.3996	3.0	0.95	0.1585	1.0	2311	72	2167	65	2440	8	89
Zr-01	zircon 08	118098	262	136	0.73	8.9235	3.4	0.4121	3.0	0.88	0.1570	1.6	2330	78	2225	66	2424	13	92
Zr-01	zircon 09	90658	167	94	0.95	9.8652	3.0	0.4496	2.8	0.92	0.1592	1.2	2422	72	2393	66	2447	10	98
Zr-01	zircon 10	156653	461	199	0.30	8.3623	5.2	0.3883	5.1	0.98	0.1562	1.0	2271	117	2115	107	2415	8	88
Zr-01	zircon 11	91263	1096	104	0.26	0.9181	4.4	0.0733	1.9	0.44	0.0909	3.9	661	29	456	9	1444	37	32
Zr-02	zircon 12	69018	140	74	0.73	9.7218	2.0	0.4398	1.8	0.89	0.1603	0.9	2409	48	2350	41	2459	8	96
Zr-02	zircon 13	114272	296	139	0.77	8.1815	2.4	0.3741	2.2	0.92	0.1586	1.0	2251	55	2048	46	2441	8	84
Zr-02	zircon 14	137284	421	154	0.66	6.3353	2.8	0.2930	2.6	0.96	0.1568	0.8	2023	56	1657	44	2421	7	68
Zr-02	zircon 15	123420	582	144	0.69	4.0046	8.0	0.1872	7.8	0.98	0.1551	1.8	1635	130	1106	86	2403	15	46
Zr-03	zircon 23	59834	139	65	1.65	7.3606	4.4	0.3370	4.1	0.92	0.1584	1.8	2156	96	1872	76	2439	15	77
Zr-03	zircon 24	183960	1089	245	0.56	3.1068	2.4	0.1797	2.1	0.90	0.1254	1.1	1434	34	1065	23	2034	9	52
Zr-03	zircon 25	107667	232	121	0.87	9.4203	2.9	0.4356	2.5	0.85	0.1568	1.6	2380	70	2331	58	2422	13	96
Zr-04	zircon 16	74596	149	79	1.16	9.1437	3.6	0.4288	3.4	0.94	0.1547	1.2	2352	85	2300	78	2398	10	96
Zr-04	zircon 20	104647	188	100	0.91	9.1573	5.9	0.4250	5.3	0.91	0.1563	2.5	2354	139	2283	122	2416	21	94
Zr-04	zircon 21	127200	520	232	1.23	8.2583	1.8	0.3794	1.6	0.86	0.1579	0.9	2260	41	2073	32	2433	8	85
Zr-06	zircon 28	111559	215	124	0.97	9.6236	2.0	0.4321	1.7	0.85	0.1615	1.1	2399	49	2315	40	2472	9	94
Zr-09	zircon 36	91793	169	97	0.95	10.1780	2.6	0.4584	2.4	0.94	0.1610	0.9	2451	64	2432	60	2467	7	99
Zr-09	zircon 38	123189	596	126	0.35	3.2918	3.1	0.1734	2.7	0.88	0.1377	1.5	1479	46	1031	28	2198	13	47
Zr-10	zircon 41	90473	120	88	1.30	10.4589	1.7	0.4721	1.2	0.71	0.1607	1.2	2476	43	2493	31	2463	10	101
Zr-10	zircon 42	102138	267	124	1.53	7.5730	2.3	0.3524	2.1	0.92	0.1558	0.9	2182	50	1946	41	2411	7	81
Zr-15	zircon 33	93535	156	102	1.04	10.4786	1.9	0.4708	1.7	0.88	0.1614	0.9	2478	48	2487	43	2470	8	101
Zr-16	zircon 34	121186	283	178	1.78	9.8323	5.8	0.4430	5.7	0.98	0.1610	1.1	2419	140	2364	134	2466	9	96
Zr-16	zircon 35	75637	113	77	1.17	10.5258	2.0	0.4719	1.5	0.77	0.1618	1.3	2482	49	2492	38	2474	11	101

[a] U and Pb concentrations and Th/U ratios are calculated relative to GJ-1 reference zircon

[b] Corrected for background and within-run Pb/U fractionation and normalised to reference zircon GJ-1 (ID-TIMS values/measured value); ^{207} Pb/^{235} U calculated using (^{207} Pb/^{206} Pb)/(^{238} U/^{206} Pb * 1/137.88)

[c] Rho is the error correlation defined as the quotient of the propagated errors of the ^{206} Pb/^{238} U and the ^{207} /^{235} U ratio

[d] Quadratic addition of within-run errors (2 SD) and daily reproducibility of GJ-1 (2 SD)

[e] Corrected for mass-bias by normalising to GJ-1 reference zircon (~ 0.6 per atomic mass unit) and common Pb using the model Pb composition of Stacey and Kramers (1975)

[f] degree of concordance = (^{206} Pb/^{238} U age * 100/^{207} Pb/^{235} U)

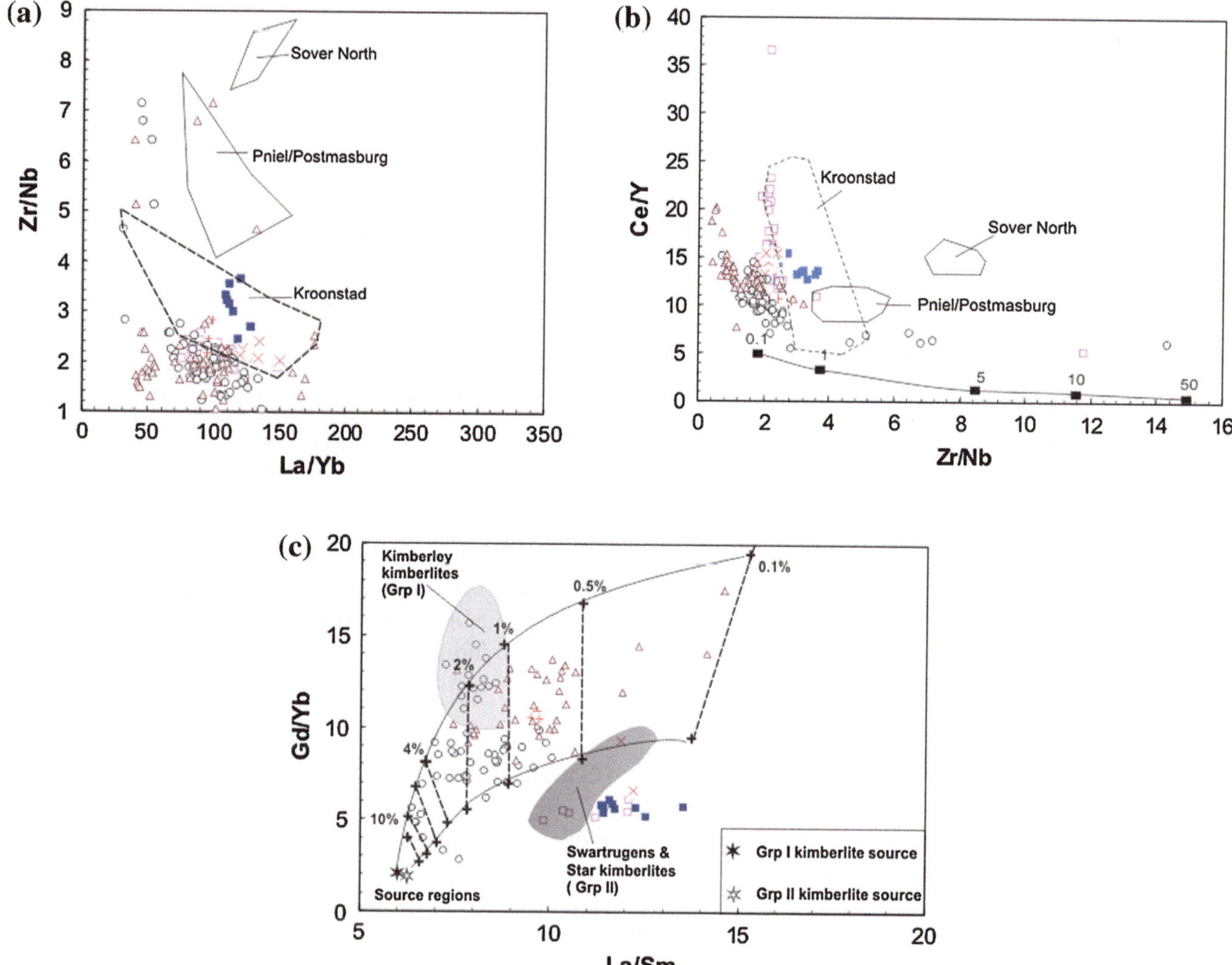

Fig. 10 **a** La/Yb versus Zr/Nb for the samples under study. Other kimberlite and orangeite fields are after Becker and Le Roex (2006). Sover North and Pniel/Postmasburg represent fields for evolved orangeites (Mitchell 1995) from southern Africa, whereas Kroonstad represents recently documented data for an orangeite field involving the Lace, Voorspoed and Besterskraal pipes from southern Africa (Howarth et al. 2011). **b** Zr/Nb versus Ce/Y and **c** La/Sm versus Gd/Yb of the Mainpur orangeites of this study. The symbols and the other data sources are the same as in Fig. 4. (**b**) Solid curved line indicates compositions of melts formed by various degree (%) of equilibrium partial melting of a peridotite containing 1.4 ppm Ce, 3.45 ppm Y, 8.51 ppm Zr and 0.54 ppm Nb and is taken from Tainton (1992). La/Sm versus Gd/Yb for Behradih and Kodomali as well as for the other kimberlites from the WKF and NKF. The symbols and the data sources are the same as in Fig. 4. **c** Illustrated curves (from Becker and Le Roex 2006) represent melting trajectories of inferred Group I and II kimberlite source regions having a residual mineralogy as follows: Grp I, ol:opx:cpx:gt = 0.67:0.26:0.04:0.03; Grp II, ol:opx:cpx:gt = 0.67:0.26:0.06:0.01. Numbers shown represent the degree of melting. Fields for Kimberley (Grp I) and Swartrugens and Star (Grp II) kimberlites are from Becker and Le Roex (2006).The symbols and the data sources are the same as in Fig. 4

now well established that the occurrence of diamonds in their primary source rock is a resultant of three major factors: (1) presence of diamondiferous zones in the sub-continental lithospheric mantle sampled by a kimberlitic magma, (2) a magmatic process that can disrupt lithospheric wall rock (various peridotites and eclogites) and incorporate some portion into the magma as xenoliths and (3) relatively non-destructive transport of diamonds to the surface by rapid ascent (e.g., Pearson et al. 2003; Scott-Smith and Smith 2009; Gurney et al. 2010). It has also been well established that with increasing fO_2, diamond in kimberlite magma oxidises to CO_2 and that prolonged residence times in high fO_2 conditions increase diamond resorption (e.g., Fedortchouk et al. 2005), resulting in a decrease in diamond grade of kimberlites. From experimental studies, Bellis and Canil (2007) proposed that diamond stability depends mainly on the fO_2 of the kimberlite magma and calibrated an empirical oxygen barometer based on Fe^{3+} content of perovskite to estimate oxygen fugacity (fO_2) during the crystallisation and emplacement of kimberlites.

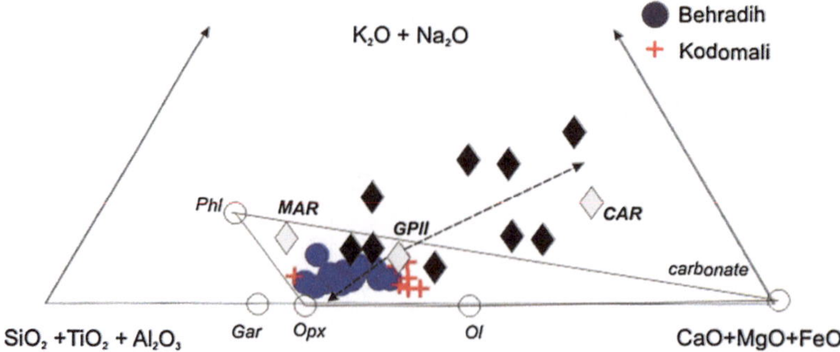

Fig. 11 Composition of the Behradih and Kodomali orangeites in comparison with summary of liquid compositions (solid diamonds) obtained by Ulmer and Sweeney (2002) in comparison with mantle carbonatite composition obtained by Sweeney (1994) and pseudoternary (wt%) diagram after Freestone and Hamilton (1980) projected from $CO_2 + H_2O$. Open circles = compositions of principal mineral phases used in the system; Gar = Garnet, Opx = Orthopyroxene; Ol = Olivine and Phl = Phlogopite. Grey-filled diamonds = location of starting orangeite (GPII) composition, MAR = MARID and CAR = carbonatite. Dashed lines = direction of differentiation of Group II composition by olivine–garnet–opx fractionation to produce alkali-dolomitic carbonatitic liquids. Note that the Mainpur samples exclude differentiation towards a MARID composition

The oxybarometry of Bellis and Canil (2007) has been applied to perovskite from the Kodomali and Behradih pipes of this study, and their ΔNNO estimates (see Table 5) exhibit a range from +0.71 (Kodomali) to +4.28 (Behradih) with Kodomali showing a much greater variation amongst the two (Fig. 3g). Our results highlight that (1) the ΔNNO conditions of the Mainpur pipes (Table 5) are higher than those from non-prospective NKF pipes as well as other prospective diamondiferous kimberlites located elsewhere such as Dutoitspan, southern Africa, Somerset Island, Canada, and (2) are indistinguishable in the redox conditions of the highly diamondiferous Lac de Gras kimberlites, Canada (Fig. 3g). We therefore conclude that oxidation state cannot explain the high incidence of diamonds in the Mainpur field and other factors very likely have played a significant role. The Mainpur orangeites are thus indeed "anomalous", compared to many other kimberlites (Fig. 3g), in terms of their high diamond incidence considering the preponderance of calcic-lherzolitic garnets and the relatively oxidising conditions at the time of their eruption.

All zircons recovered from the Behradih pipe are crustal-derived xenocrysts and not mantle-derived megacrysts, as revealed by their composition and U–Pb ages (Table 8). However, the absence of Archaean age data is surprising since the Bastar craton is regarded as the oldest continental nuclei in the Indian shield with an Eoarchaean crust as evidenced by the 3.50–3.6 Ga zircons from tonalites and granites (Sarkar et al. 1993; Ghosh 2004; Rajesh et al. 2009) and a thickened modern-day crust of 35–40 km (Jagdeesh and Rai 2008). Therefore, the lack of Archaean-aged zircons could well be a reflection of the sampling. Alternately, it may also represent modification of the sub-Bastar lithosphere by the invading Deccan plume-derived melts during the end-Cretaceous synchronous with the eruption of the Mainpur orangeites (see Chalapathi Rao and Lehmann 2011 for a detailed discussion). Further studies involving the U–Pb dating of zircons from the Mainpur field are expected to provide further clues in this direction.

Conclusions

Following major conclusions can be drawn from this study:

- Contrasting textures in case of Behradih (pelletal and tuffisitic habit) and Kodomali (inequigranular texture with two generations of fresh olivine) orangeites imply a differential erosion in this domain.
- Olivine, spinel and clinopyroxene in the Behradih and the Kodomali orangeites share overlapping compositions, whereas the groundmass phlogopite and perovskite show conspicuous compositional differences.
- Incompatible trace elements and their ratios readily distinguish Mainpur orangeites from the Mesoproterozoic Wajrakarur (WKF) and Narayanpet kimberlites (NKF) from the eastern Dharwar craton, southern India. However, there is an overall similarity in petrogenesis between the Mainpur and the southern African orangeites.
- The Mainpur orangeites are dominated by calcic-lherzolitic variety of pyrope population in comparison with that of sub-calcic harzburgitic and eclogitic variety of garnets.
- "Smooth" as well as "sinusoidal" chondrite-normalised REE patterns are displayed by garnets from the Behradih and Payalikhand orangeites. Such contrasting patterns provide evidence for the presence of a compositionally layered end-Cretaceous sub-Bastar craton mantle.
- Perovskite oxybarometry indicates that the redox state of the lithospheric mantle cannot be of first-order control for

diamond potential and implies the dominant role of other factors viz., rapid magma transport.

- Mainpur orangeites are clearly "anomalous" compared to several other kimberlite/orangeite pipes worldwide in view of their (1) highly diamondiferous nature, (2) abundance of calcic-lherzolitic garnets, and (3) highly oxidising conditions prevailing at the time of eruption.
- U–Pb dating of zircon xenocrysts from the Behradih pipe yielded distinct Palaeoproterozoic ages. The lack of Archean-aged zircons could either be a reflection of the sampling process or of the modification of the sub-Bastar lithosphere by the invading Deccan plume-derived melts at ca. 65 Ma.

Acknowledgments NVCR thanks the Head, Department of Geology, Centre of Advanced Study, Banaras Hindu University for providing logistics for undertaking field trip to Mainpur area and to the Humboldt Foundation for support. DM thanks Directorate of Mines and Geology, Chhattisgarh for support, permission and encouragement for carrying out studies on the Mainpur pipes and their xenocrysts. Chiranjeeb Sarkar and Graham Pearson are thanked for their helpful reviews.

References

Ashchepkov IV, Pokhilenko NP, Vladykin NV, Logvinova AM, Afanasiev VP, Pokhilenko LN, Kuligin S, Malygina EV, Alymova NA, Kostrovitsky SI, Rotman AY, Mityukhin SI, Karpenko MA, Stegnitsky YB, Khemelnikova OS (2010) Structure and evolution of the lithospheric mantle beneath Siberian craton, thermobarometric study. Tectonophys 485:17–41

Balaram V, Gnaneswar Rao T (2003) Rapid determination of REE's and other trace elements in geological samples by microwave acid digestion and ICP-MS. Atom Spectros 24:206–212

Becker M, Le Roex AP (2006) Geochemistry of South African on- and off-craton Group I and II kimberlites: petrogenesis and source region evaluation. J Petrol 47:673–703

Becker M, Le Roex AP, Class C (2007) Geochemistry and petrogenesis of South African transition kimberlites located on and off the Kaapvaal Craton. S. Afr. J Geol 110:631–646

Bellis AJ, Canil D (2007) Ferric iron in $CaTiO_3$ perovskite as an oxygen barometer for kimberltic magmas I: experimental calibration. J Petrol 48:219–230

Belousova EA, Griffin WL, O'Reilly SY, Fisher NI (2002) Igneous Zircon: trace element composition as an indicator of host rock type. Contrib Mineral Petrol 143:602–622

Burgess SR, Harte B (2004) Tracing Lithosphere Evolution through the Analysis of Heterogeneous G9, and G10 Garnets in Peridotite Xenoliths, II: REE Chemistry. J Petrol 45:609–633

Canil D, Bellis AJ (2007) Ferric iron in $CaTiO_3$ perovskite as an oxygen barometer for kimberlite magmas II. Applications. J Petrol 48:231–252

Chalapathi Rao NV, Dongre A (2009) Mineralogy and geochemistry of kimberlites NK-2 and KK-6, Narayanpet kimberlite field, Eastern Dharwar craton, southern India: evidence for a transitional (South African) kimberlite signature. Can Mineral 47:855–873

Chalapathi Rao NV, Lehmann B (2011) Kimberlites, flood basalts and mantle plumes: new insights from the Deccan Large Igneous Province. Earth Sci Rev 107:315–324

Chalapathi Rao NV, Srivastava RK (2009) Petrology and Geochemistry of diamondiferous mesoproterozoic kimberlites from Wajrakarur kimberlite field, eastern Dharwar Craton, Southern India: genesis and constraints on mantle source regions. Contrib Mineral Petrol 157:245–265

Chalapathi Rao NV, Gibson SA, Pyle DM, Dickin AP (2004) Petrogenesis of proterozoic lamproites and kimberlites from the Cuddapah basin and Dharwar craton, southern India. J Petrol 45(5):907–948

Chalapathi Rao NV, Burgess R, Anand M, Mainkar D (2007) 40Ar–39Ar dating of the Kodomali pipe, Bastar craton, India: a Pan-African (491 ± 11 Ma) age of diamondiferous kimberlite emplacement. J Geol Soc India 69:539–546

Chalapathi Rao NV, Kamde G, Kale HG, Dongre A (2010) Mesoproterozoic lamproites from the Krishna valley, eastern Dharwar craton, southern India: petrogenesis and diamond prospectivity. Precambr Res 177:103–130

Chalapathi Rao NV, Lehmann B, Mainkar D, Belyatsky B (2011a) Petrogenesis of the end-cretaceous diamondiferous Behradih orangeite pipe: implication for mantle plume—lithosphere interaction in the Bastar craton, central India. Contrib Mineral Petrol 161:721–742

Chalapathi Rao NV, Burgess R, Lehmann B, Mainkar D, Pande SK, Hari KR, Bodhankar N (2011b) $^{40}Ar/^{39}Ar$ ages of mafic dykes from the mesoproterozoic Chhattisgarh basin, Bastar craton, central India: implication for the origin and spatial extent of the Deccan Large Igneous Province. Lithos 125:994–1005

Chalapathi Rao NV, Paton C, Lehmann B (2012) Origin and diamond prospectivity of the Mesoproterozoic kimberlites from the Narayanpet field, Eastern Dharwar craton, southern India: insights from groundmass mineralogy, bulk-chemistry and perovskite oxybarometry. Geological J 47:186–212

Chatterjee B, Jha N (1994) Diamondiferous kimberlitic diatremes of Payalikhand, Behradih and Jangra, Raipur district, Madhya Pradesh. Rec Geol Surv India 127(6):240–243

Chatterjee B, Smith CB, Jha N, Khan MWY (1995) Kimberlites of the Southeastern Raipur kimberlitic field, Raipur district, Madhya Pradesh, Central India. In: Extended abstracts sixth international kimberlite conference, Novosibirsk, Russia, pp 106–108

Coe N, Le Roex AP, Gurney J, Pearson G, Nowell G (2008) Petrogenesis of the Swartruggens and Star Group II kimberlite dyke swarms, South Africa: constraints from whole rock geochemistry. Contrib Mineral Petrol 156:627–652

Collerson KD, Williams Q, Ewart AE, Murphy DT (2010) Origin of HIMU and EM-1 domains sampled by ocean island basalts, kimberlites and carbonatites: The role of CO_2-fluxed lower mantle melting in thermochemical upwellings. Phys Earth Planet Interiors 181:112–131

Creighton S, Stachel T, Eichenberg D, Luth RW (2010) Oxidation state of the lithospheric mantle beneath Diavik diamond mine, central Slave craton, NWT, Canada. Contrib Mineral Petrol 159:645–657

Crookshank H (1963) Geology of southern Bastar and Jeypore from the Bailadila range to the eastern Ghats. Geol Surv India Mem 87:150

Dawson JB, Smith JV (1977) The MARID suite of (mica-amphibole-rutile- ilmenite-diopside) suite of xenoliths in kimberlite. Geochim Cosmochim Acta 41:309–323

Dawson JB, Stephens WE (1975) Statistical analysis of garnets from kimberlites and associated xenoliths. J Geol 83:589–607

Donnelly CL, Griffin WL, O'reilly SY, Pearson NJ, Shee SR (2011) The kimberlites and related rocks of the Kuruman kimberlite province, Kaapvaal craton, south Africa. Contrib Mineral Petrol 161:351–371

Fareeduddin, Pant NC, Neogi S, (2006) Petrology of the Kodomali diatreme, Mainpur area, Chhattisgarh, Central India: implications for a Palaeozoic orangeite field. J. Geol.Soc. India 68:19-34

Fedortchouk Y, Canil D, Carlson JA (2005) Dissolution forms in Lac de Gras diamonds and their relationship to the temperature and redox state of kimberlite magma. Contrib Mineral Petrol 150:1725–1745

Fipke CE, Gurney JJ, Moore RO (1995) Diamond exploration techniques emphasising indicator mineral geochemistry and Canadian examples. Geol Surv Canada Bull 423:86

Freestone IC, Hamilton DL (1980) The role of liquid immiscibility in the genesis of carbonatites- an experimental study. Contrib Mineral Petrol 73:105–117

Frei D, Gerdes A (2008) Precise and accurate in situ U–Pb dating of zircon with high sample throughput by automated LA-SF-ICP-MS. Chem Geol 261:261–270

Ghosh JG (2004) 3.56 Ga Tonalite in the central part of the Bastar craton, India: oldest Indian date. J Asian Earth Sci 23:359–364

Gibson SA, Thompson RN, Leonardos OH, Dickin AP, Mitchell JG (1995) The Late Cretaceous impact of the Trindade mantle plume: evidence from large-volume, mafic, potassic magmatism in SE Brazil. J Petrol 36:189–229

Gibson SA, Malarkey J, Day J (2008) Melt depletion and enrichment beneath the western Kaapvaal Craton: evidence from Finsch Peridotite xenoliths. J Petrol 49:1817–1852

Gopalan K, Kumar A (2008) Phlogopite K–Ca dating of Narayanpet kimberlites, South India: Implications to the discordance between their Rb–Sr, Ar/Ar ages. Precambrian Res 167:377–382

Gregoire M, Bell DR, le Roex AP (2003) Garnet lherzolites from the Kaapvaal Craton (South Africa): trace element evidence for a metasomatic history. J Petrol 44:629–657

Gregory LC, Meert JG, Pradhan V, Pandit MK, Tamrat E, Malone SJ (2006) A palaeomagnetic and geochronologic study of the Majhgawan kimberlite, India: implications for the age of the Upper Vindhyan SuperGroup. Precamb Res 149:65–75

Griffin WL, Shee SR, Ryan CG, Win TT, Wyatt BA (1999) Harzburgite to lherzolite and back again: Metasomatic processes in ultramafic xenoliths from the Wesselton kimberlite, Kimberley, South Africa. Contrib Mineral Petrol 134:232–250

Grutter HS, Gurney JJ, Menzies AH, Winter F (2004) An updated classification scheme for mantle-derived garnet, for use by diamond explorers. Lithos 77:841–857

Gupta S, Bhattacharya A, Raith M, Nanda JK (2000) Contrasting pressure-temperature-deformation history across a vestigial craton-mobile belt boundary: the western margin of the Eastern Ghats belt at Deobhog (India). J Metm Geol 18:683–697

Gurney JJ, Helmstaedt HH, Richardson SH, Shirey SB (2010) Diamonds through time. Econ Geol 105:689–712

Haggerty SE (1999) Diamond formation and kimberlite clan magmatism. Geochem Soc Spec Publ 6:105–123

Howarth GH, Skinner EMW, Prevec SA (2011) Petrology of the hypabyssal kimberlite of the Kroonstad Group II kimberlite (orangeite) cluster, South Africa: evolution of the magma within the cluster. Lithos 125:795-808

Jagadeesh S, Rai SS (2008) Thickness, composition and evolution of the Indian Precambrian crust inferred from broadband seismological measurements. Precamb Res 16:4–15

Jha N, Smith CB, Griffin BJ, Chatterjee B, Pooley GD (1995) Diamonds from the kimberlites of Southeastern Raipur kimberlitic field, Raipur district, Madhya Pradesh, Central India. Extended Abstracts Sixth International Kimberlite Conference Novosibirsk, Russia, 266–268

Jha N, Chatterjee B, Mishra BK, Sarkar SK (2002) Kimberlites of Mainpur kimberlite field, Raipur district, Chhattisgarh, central India. In: Abstracts volume of the international conference on Diamonds and Gemstones, Raipur

Kent RW, Kelley SP, Pringle MS (1998) Mineralogy and 40Ar/39Ar geochronology of orangeites (Group II kimberlites) from the Damodar Valley, Eastern India. Mineral Magz 62:313–323

Kjarsgaard BA, Harvey S, Mcclintock M, Zonneveld JP, Du Pleissis P, Mcneil D, Heaman L (2009) Geology of the Orion South kimberlite, Fort à la Corne, Canada. Lithos 112:600–617

Konzett J, Armstrong RA, Gunther D (2000) Modal metasomatism in the Kaapvaal craton lithosphere: constraints on timing and genesis from U–Pb zircon dating of metasomatized peridotites and MARID-type xenoliths. Contrib Mineral Petrol 139:704–719

Kopylova MG, Caro G (2004) Mantle xenoliths from the Southeastern Slave craton: the evidence for a thick cold stratified lithosphere. J Petrol 45:1045–1067

Le Roex AP, Bell DR, Davis P (2003) Petrogenesis of Group I kimberlites from Kimberley, South Africa: evidence from bulk-rock geochemistry. J Petrol 44:2261–2286

Lehmann B, Mainkar D, Belyatsy B (2006) The Tokapal Crater-facies kimberlite system, Chattisgarh, India: Reconnaissance petrography and geochemistry. J Geol Soc India 68:9–18

Lehmann B, Storey C, Mainkar D, Jeffries T (2007) In-situ U–Pb dating of titanite in the Tokapal-Bejripadar kimberlite system, Central India. J Geol Soc India 69:553–556

Lehmann B, Burgess R, Frei D, Belyatsky B, Mainkar D, Chalapathi Rao NV, Heaman LM (2010) Diamondiferous kimberlites in central India synchronous with Deccan flood basalts. Earth Planet Sci Lett 290:142–149

Lehtonen ML (2005) Rare earth element characteristics of pyrope garnets from the Kaavi-Kuopio kimberlites- implications for mantle metasomatism. Bull Geol Soc Finland 77:31–47

Lehtonen ML, O'brien HE, Peltonen P, Johnson BS, Pakkanen LK (2004) Layered mantle at the Karelian craton margin: P-T of mantle xenocrysts from the Kaavi-Kuopio kimberlites, Finland. Lithos 77:593–608

Mainkar D (2010) Petrological and geochemical investigation of the Behradih kimberlite from the Bastar craton, Central India, with special reference to its diamond potential. Unpublished PhD thesis, Pt. Ravishankar Shukla University, Raipur, 171)

Mainkar D, Lehmann B (2007) The diamondiferous Behradih kimberlite pipe, Mainpur kimberlite field, Chhattisgarh, India: reconnaissance petrography and geochemistry. J Geol Soc India 69:547–552

Mainkar D, Lehmann B, Haggerty SE (2004) The crater facies kimberlite system of Tokapal, Bastar district, Chattisgarh, India. Lithos 76:210–217

Marathe T (2010) Significance of Petrographical and Geochemical Characteristics of Kodomali Kimberlite Clan Rock, Mainpur Kimberlite Field, Raipur District, Chhattisgarh, India. In: Karmalkar NR, Duraiswami RA, Pawar NJ, Sivaji Ch (eds) Origin and evolution of the deep continental crust. Narosa Publishers, New Delhi

Masun K, Sthapak AV, Singh A, Vaidya A, Krishna C (2009) Exploration history and geology of the diamondiferous ultramafic Saptarshi intrusions, Madhya Pradesh. India Lithos 112:142–154

Mishra VP, Singh P, Dutta NK (1988) Stratigraphy, structure and metamorphic history of Bastar craton. Rec Geol Surv India 117:1–26

Mitchell RH (1986) Kimberlites: mineralogy, geochemistry and petrology. Plenum Press, New York, p 442

Mitchell RH (1995) Kimberlites, orangeites and related rocks. Plenum press, New York, p 406

Mitchell RH (2006) Potassic magmas derived from metasomatised lithospheric mantle: nomenclature and relevance to exploration for diamond-bearing rocks. J Geol Soc India 67(3):317–327

Mitchell RH, Fareeduddin (2009) Mineralogy of peralkaline lamproites from the Raniganj Coalfield, India. Mineral Magz 73:457–477
Mukherjee A, Tripathi A, Singh PK, BABU EVSSK (2000) Chemistry of eclogitic garnets from Behradih kimberlite, Raipur district, Madhya Pradesh. J Geol Soc India 56:425–430
Newlay SK, Pashine JK (1993) New find of diamond-bearing kimberlite in Raipur district, Madhya Pradesh, India. Curr Sci 65:292–293
Newlay SK, Pashine JK (1995) New find of diamond bearing gravel horizon in Payalikhand area of Raipur district, Madhya Pradesh. J Geol Soc India 46:309–312
Norman MD, Pearson NJ, Sharma A, Griffin WL (1996) Quantitative analysis of trace elements in geological materials by laser ablation ICPMS: instrumental operating conditions and calibration values of NIST glasses. Geostand Newslett 20:247–261
Ogilve-Harris RC, Field M, Sparks RSJ, Walter MJ (2009) Perovskite from Dutoitspan kimberlite, Kimberley, South Africa: implications for magmatic processes. Mineral Magz 73:915–928
ORAPA (2000) Project Report, Raipur diamond project. Block-7, Madhya Pradesh. July 2000. website: http://www.orapa.com.au
Osborne I, Sherlock S, Anand M, Argles T (2011) New Ar–Ar ages of southern Indian kimberlites and a lamproite and their geochemical evolution. Precambrian Res 189:91–103
Page Z, Fu B, Kita NT, Fournelle J, Spicuzza MJ, Schulze DJ, Viljoen F, Basei MAA, Valley JW (2007) Zircons from kimberlite: new insights from oxygen isotopes, trace elements and Ti in zircon thermometry. Geochmi Cosmochim Acta 71:3887–3903
Patnaik BC, Mishra BP, Maharana RC (2002) A new discovery of diamond bearing pipe rocks in orissa. Proceedings international conference on diamonds & gemstones, (organized by SAEEG & DGM, Chattisgarh), Raipur, 90
Paton C, Hergt JM, Woodhead JD, Phillips D, Shee SR (2009) Identifying the asthenospheric component of kimberlitic magmas from the Dharwar craton, India. Lithos 112:296–310
Paul DK, Nayak SS, Pant NC (2006) Indian kimberlites and related rocks: petrology and geochemistry. J Geol Soc India 67:328–355
Paul DK, Crocket JH, Reddy TAK, Pant NC (2007) Petrology and geochemistry including Platinum Group element abundances of the Mesoproterozoic ultramafic (lamproite) rocks of Krishna district, southern India: implications for source rock characteristics and ptrogenesis. J Geol Soc India 69:577–596
Pearson DG, Canil D, Shirey SB (2003) Mantle samples included in volcanic rocks: xenoliths and diamonds. Treatise on Geochemistry 2:171–276 Holland HD, Turekian KK (eds)
Rajesh HM, Mukhopadhyay J, Beukes NJ, Gutzmer J, Belyanin GA, Armstrong RA (2009) Evidence for an early Archaean granite from Bastar craton, India. J Geol Soc London 166:193–196
Ramakrishnan M, Vaidyanadhan R (2008) Geology of India, Geological society of India p 1550
Ramsay RR (1992) Geochemistry of diamond indicator minerals. Unpublished PhD thesis, University of Western Australia p 246
Ringwood AR, Kesson SE, Hibberson W, Ware N (1992) Origin of kimberlites and their related magmas. Earth Planet Sci Lett 113:521–538
Roeder PL, Schulze DJ (2008) Crystallization of groundmass spinel in kimberlite. J Petrol 49:1473–1495
Rogers JJW, Santosh M (2003) Supercontinents in earth history. Gondwana Res 6:357–368
Sarkar G, Corfu F, Paul DK, McNaughton NJ, Gupta SN, Bishui PK (1993) Early Archean crust in Bastar Craton, Central India-a geochemical and isotopic study. Precamb Res 62:127–137
Sand KK, Waight TE, Pearson DG, Nielsen TFD, Makovicky E, Hutchison MT (2009) The lithospheric mantle below southern West Greenland: A geothermobarometric approach to diamond potential and mantle stratigraphy. Lithos 112s:1155-1166
Schulze DJ, Canil D, Channer DMDR, Kaminsky F (2006) Layered mantle structure beneath the western Guyana Shield, Venezuela: Evidence from diamonds and xenocrysts in Guaniamo kimberlites. Geochim Cosmochim Acta 70:192–205
Scott-Smith BH, Skinner EMW (1984) A new look at the Prairie Creek Arkansas. In: Kornprobst J (ed) Kimberlites 1: kimberlites and related rocks. Proceedings 3rd international kimberlite conference developments in petrology, vol 1. Elsevier, Amsterdam, pp 255–283
Scott-Smith BH, Smith SCS (2009) The economic implications of kimberlite emplacement. Lithos 112S:10–22
Shimizu N (1975) Rare earth elements in garnets and clinopyroxenes from garnet lherzolite nodules in kimberlites. Earth Planet Sci Lett 25:26–32
Singh SB, Chandrakar NK, Verma D, Saxena V (2000) Report on the prospecting for diamond in Behradih-Kodomali area, district Raipur (M.P) Field Season 1996–1998. Unpublished Report of DGM, M.P
Skinner EMW (1989) Contrasting Group I and II kimberlite petrology: towards a genetic model for kimberlites. Geol Soc Austral Spec Publ 14(1):528–544
Small M, Vaidya A (2002) Field exploration for diamond, Raipur area, Chhattisgarh. In: Extended Abstracts volume International Conference on Diamonds and Gemstones, Raipur
Smith CB, Clark TC, Barton ES and Bristow J (1994) Emplacement ages of kimberlite occurrences in the Prieska region, southwest border of the Kaapvaalc raton, South Africa. Chemical Geology, 113:149-169
Smith CB, Gurney JJ, Skinner EMW, Clement CR, Ebrahim N (1985) Geochemical character of southern African kimberlites: a new approach based on isotopic constraints. Trans Geol Soc South Africa 88:267–280
Stachel T, Viljoen KS, Brey G, Harris JW (1998) Metasomatic processes in lherzolitic and harzburgitic domains of diamondiferous lithospheric mantle—REE in garnets from xenoliths and inclusions in diamonds. Earth Planet Sci Letts 159:1–12
Stachel T, Aulbach S, Brey GP, Harris JW, Leost I, Tappert R, Viljoen KS (2004) The trace element composition of silicate inclusions in diamonds: a review. Lithos 77:1–19
Sun SS, McDonough WF (1989) Chemical and isotopic systematics of oceanic basalts: implications for mantle composition and processes. Geol Soc Lond Spec Publ 42:313–345 Saunders AD, Norry MJ (eds) Magmatism in ocean basins
Sweeney RJ (1994) Carbonatite melt compositions in the Earth's mantle. Earth Planet Sci Lett 28:259–270
Tainton KM (1992) The petrogenesis of group II kimberlites and lamproites from the northern Cape Province, South Africa. Unpublished PhD thesis, University of Cambridge, UK
Tappe S, Pearson DG, Nowell G, Nielson T, Milstead P, Muehlenbachs K (2011) A fresh isotopic look at Greenland kimberlites: cratonic mantle lithosphere imprint on deep source signal. Earth Planet Sci Lett 305:235–246
Torsvik TH, Burke K, Steinberger B, Webb SJ, Ashwal LD (2010) Diamonds sampled by plumes from the core-mantle boundary. Nature 466:352–355
Ulmer P, Sweeney RJ (2002) Generation and differentiation of Group II kimberlites: constraints from a high-pressure experimental study to 10 GPa. Geochim Cosmochim Acta 66:2139–2215
Verma D, Saxena VK (1997) A report on the investigation for diamond and other precious and semi-precious stones in south-eastern parts of district Raipur, M.P. Field Season 1992–1994, Unpublished Report of DGM, M.P
Woodhead J, Hergt J, Phillips D, Paton C (2009) African kimberlites re-visited: in situ Sr-isotope analysis of groundmass perovskite. Lithos 112S:311-317

Nd–Hf Isotope Systematics of Megacrysts from the Mbuji-Mayi Kimberlites, D. R. Congo: Evidence for a Metasomatic Origin Related to Kimberlite Interaction with the Cratonic Lithospheric Mantle

M. Pivin, V. Debaille, N. Mattielli, and D. Demaiffe

Abstract

Garnet and clinopyroxene megacrysts from the Cretaceous (70 Ma) Mbuji-Mayi kimberlites and one garnet megacryst from the lower Oligocene (32 Ma) Kundelungu kimberlites in Democratic Republic of Congo have been investigated for combined Nd and Hf isotope compositions. These megacrysts are thought to result from the metasomatic (re)crystallization of lithospheric mantle peridotites during the infiltration of a proto-kimberlitic melt/fluid. In addition, zircon and baddeleyite megacrysts from the Mbuji-Mayi kimberlites have been investigated for Hf isotope composition. Although baddeleyites are uncommon in kimberlite megacryst suites, their origin is most probably related to that of zircon megacrysts from Mbuji-Mayi. Mbuji-Mayi garnet megacrysts display ranges of $\varepsilon Nd_{(t)}$ from −0.6 to +6.1 and $\varepsilon Hf_{(t)}$ from +6.6 to +12.1; the Kundelungu garnet megacryst has overlapping isotopic compositions (+0.8 and +6.0, respectively). By contrast, Mbuji-Mayi clinopyroxene megacrysts display a more restricted range of $\varepsilon Nd_{(t)}$ values (+2.7 to +4.6) and extend toward lower $\varepsilon Hf_{(t)}$ compositions (+3.0 to +9.1). Mbuji-Mayi zircon and baddeleyite megacrysts have similar $\varepsilon Hf_{(t)}$ values (+6.5 to +7.1 and +6.0 to +8.4, respectively). Differences in initial isotopic composition between garnets on the one hand and clinopyroxenes, zircons, and baddeleyites on the other have been confirmed by various Hf and Nd model ages calculations. Clinopyroxene, zircon, and baddeleyite megacrysts plot close to the worldwide kimberlite field in a combined $\varepsilon Hf_{(t)}$-$\varepsilon Nd_{(t)}$ plot, which favors recent (i.e. at or shortly before the time of kimberlite eruption) crystallization through interaction between the infiltrating proto-kimberlite melt/fluid and peridotite wall rocks. On the other hand, the wide range of $\varepsilon Nd_{(t)}$ values and the higher $\varepsilon Hf_{(t)}$ values for garnet megacrysts suggest that they have been formed through recrystallization of old garnet-bearing peridotitic protoliths in the subcontinental lithospheric mantle. Calculated rare earth element patterns of liquids in equilibrium with clinopyroxene megacrysts confirm the direct relationship to Group I kimberlites, while those in equilibrium with garnet megacrysts show more variability, which could also reflect that their formation results from more complex processes.

Keywords

Zircon • Baddeleyite • Megacrysts • Mbuji-Mayi kimberlites • Nd–Hf isotopes • Mantle metasomatism

M. Pivin (✉) · V. Debaille · N. Mattielli · D. Demaiffe
Département des Sciences de la Terre et de l'Environnement (DSTE) (CP 160/02), Université Libre de Bruxelles (ULB), 50 Avenue F.D. Roosevelt, 1050 Brussels, Belgium
e-mail: Marjorie.Pivin@ulb.ac.be

D. G. Pearson et al. (eds.), *Proceedings of 10th International Kimberlite Conference*, Volume 1, Special Issue of the Journal of the Geological Society of India,
DOI: 10.1007/978-81-322-1170-9_8,

Introduction

Cr-poor megacrysts from worldwide Group I kimberlites form a suite of minerals referred to as the "megacryst suite" or the "discrete nodule suite" (Nixon and Boyd 1973), which is mostly composed of Cr-poor garnet, clinopyroxene, magnesian ilmenite, and zircon megacrysts. Olivine, orthopyroxene, and phlogopite occur sometimes but are less common. There is a general agreement about their deep magmatic origin through fractional crystallization from a silicate parental melt, close to the lithosphere–asthenosphere boundary (Moore et al. 1992). A Cr-rich suite has first been reported from the Colorado-Wyoming kimberlites (Eggler et al. 1979) and is now recognized in many other localities worldwide (Kostrovitsky et al. 1997; Moore and Belousova 2005; Kopylova et al. 2009), including Democratic Republic of Congo (DRC) (Pivin et al. 2009). The chemical distinction between those two suites is not always clear (Garrison and Taylor 1980), but a genetic link is traditionally suggested.

However, direct relationship with the kimberlite host magma has been challenged by Sr, Nd, and Pb radiogenic isotope studies (Weis and Demaiffe 1985; Jones 1987; Hops et al. 1992; Davies et al. 2001). Megacrysts usually have a slightly more depleted mantle source signature than their host. Yet, the difficulty of analyzing fresh whole-rock kimberlites is well known and recent improvements have been made in the determination of isotope ratios (Sr and Nd) in kimberlite groundmass or phenocryst phases, which are thought to better reflect the primary parental melt signature (Woodhead et al. 2009; Malarkey et al. 2010; Tappe et al. 2012). These results increasingly favor a cognate origin between kimberlites and megacrysts.

Recently, combined Nd and Hf isotope studies have also reemphasized the direct genetic relationship between megacrysts and their hosts. Group I kimberlites and associated megacrysts from South Africa have similar negative $\Delta\varepsilon$Hf values (Nowell et al. 1999, 2004) and plot below the "Mantle Array" defined by almost all terrestrial magmas [$\Delta\varepsilon$Hf is the difference between the measured εHf and the theoretical value calculated from the oceanic basalt Nd–Hf regression of Vervoort et al. (1999)]. This suggests a decoupling of the Lu–Hf and Sm–Nd isotope systems, which is unexpected during mantle melting. This peculiar isotopic signature has also been recognized in kimberlites from the Dharwar craton, South India (Paton et al. 2009), and in megacrysts from kimberlites in Africa, Siberia, and Australia through Hf isotope analyses of zircon megacrysts combined with Nd isotope ratios of associated—and assumed cogenetic—low-Cr silicate megacrysts (Griffin et al. 2000). The origin of this decoupling, which has been encountered in megacrysts and kimberlites of various ages, located on different continents (on- and off-craton localities), has been prone to several interpretations and fueled the debate regarding the potential sources of kimberlite magmas and megacrysts (Pearson et al. 2008). While Sr, Nd, and Pb isotope compositions of Group I kimberlites and megacrysts favor a mantle source similar to that of OIB, the negative $\Delta\varepsilon$Hf values are only found in some HIMU-type OIB (Ballentine et al. 1997). These negative compositions have been related to an old subducted oceanic crust stored for a long time (>1 Ga) in a deep layer (transition zone or core–mantle boundary) and isolated from convection (Blichert-Toft and Albarède 1997; Nowell et al. 1999, 2004), while others favored an origin in ancient depleted and metasomatized harzburgitic subcontinental lithospheric mantle (SCLM) with high Hf/Nd (Griffin et al. 2000). However, Group I kimberlites and megacrysts from the North Atlantic craton (Tappe et al. 2011) and some kimberlites and megacrysts from the Slave craton (Schmidberger et al. 2002; Pearson et al. 2008; Kopylova et al. 2009) have Hf–Nd isotope compositions that plot within the mantle array. Tappe et al. (2011) recently suggested that there is no negative $\Delta\varepsilon$Hf component in the source of kimberlite melts and that the apparent displacement below the mantle array observed in some localities results from the mixing of sublithospheric and lithospheric components, both lying within the mantle array (coupled Lu–Hf and Sm–Nd fractionation), but having contrasted Hf/Nd ratios.

In order to better understand the relationship between Hf and Nd isotope systematics in kimberlite megacrysts, garnet, clinopyroxene, zircon, and baddeleyite megacrysts from the Mbuji-Mayi kimberlite province and one garnet megacryst from Kundelungu (DRC) have been selected. Contrary to garnet and clinopyroxene megacrysts from most localities, those from DRC do not show evidence of an origin by fractional crystallization. They have been interpreted as fragments of mantle peridotites having suffered a "proto-kimberlitic" metasomatic event accompanied by recrystallization in the deep cratonic lithospheric mantle (Pivin et al. 2009). The isotopic signature of their source should thus reflect the petrogenetic relationships between their host magma and the cratonic lithospheric mantle in DRC. In addition, the relationship(s) between the different megacryst minerals is explored. While zircons are common in kimberlite megacryst suites, baddeleyite has never been reported, except from Mbuji-Mayi.

Geological Setting

The Cretaceous (69.8 ± 0.5 Ma; Schärer et al. 1997) Mbuji-Mayi kimberlites, East Kasai (Demaiffe et al. 1991), cut across the Archean Congo-Kasai craton (>2.9 Ga) that crops out about 100 km S-SE to the Mbuji-Mayi mining center (Fig. 1). These diamond-rich Group I kimberlites comprise a large suite of megacrysts, including garnet, clinopyroxene, zircon, and baddeleyite, but also Mg-ilmenite, Nb- or Cr-rich rutile, corundum, kyanite, chlorite, and rutile-silicate intergrowths. Eclogite nodules dominate the xenolith population (El Fadili and Demaiffe 1999), while only few pyroxenites and peridotites have been found and one exceptional Cr-rich kyanite-bearing clinopyroxenite (Pivin et al. 2011).

The Kundelungu kimberlite province (Katanga) is located ~600 km to the southeast of Mbuji-Mayi in the Neoproterozoic Katangan belt (Fig. 1). These diamond-poor Group I kimberlites are of lower Oligocene age (32.3 ± 2.2 Ma; Batumike et al. 2008) and most probably intrude old basements at depth, the Mesoproterozoic Kibaran belt (~1.3 Ga) and the Paleoproterozoic Bangweulu block (~1.9 Ga) (Batumike et al. 2007). The megacryst suite is composed of garnet, clinopyroxene, olivine, phlogopite, orthopyroxene, and ilmenite; xenoliths are mostly peridotites, with few eclogites and one clinopyroxenite (Kampata 1993).

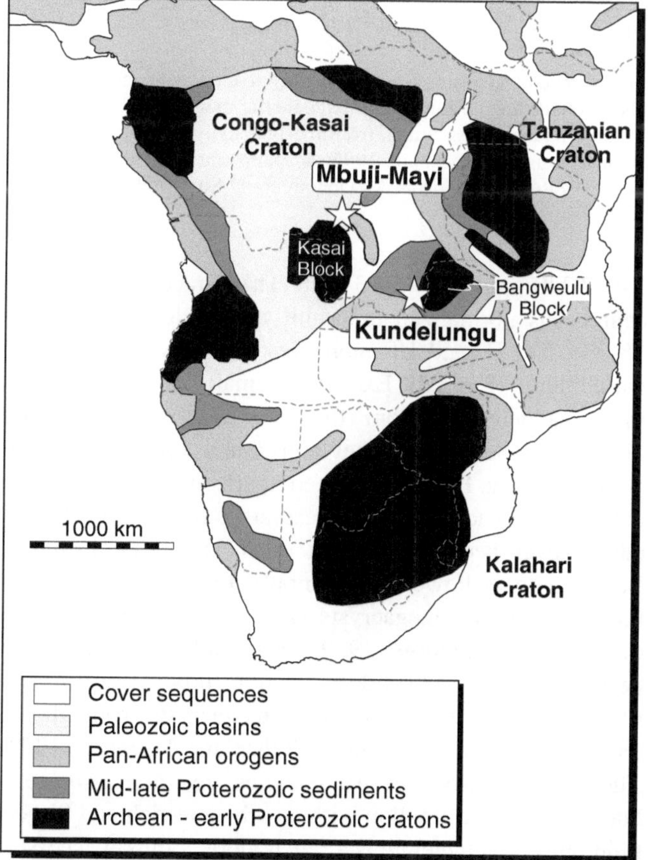

Fig. 1 Location of the Mbuji-Mayi and Kundelungu kimberlite provinces (Democratic Republic of Congo) in a simplified geological map of Central and South Africa (modified after Foster et al. 2001)

Megacryst Samples

Mbuji-Mayi garnet and clinopyroxene megacrysts and Kundelungu garnet megacrysts were previously investigated for major and rare earth element (REE) contents (Pivin et al. 2009). Pyrope megacrysts from both localities share many textural and geochemical similarities and were subdivided into three groups based on Cr content: low-, medium-, and high-Cr (0.00–1.79; 1.93–5.16; 5.42–7.10 wt% Cr_2O_3, respectively). Mbuji-Mayi diopside megacrysts were classified according to their Ca# [Ca/(Ca + Mg)] and Cr content into low-Ca (Ca#: 39.5–42.6; 0.61–0.92 wt% Cr_2O_3), medium-Ca (Ca#: 43.9–49.5; 0.30–1.94 wt% Cr_2O_3), and high-Ca/high-Cr (Ca#: 47.0–50.2; 1.13–2.77 wt% Cr_2O_3) groups. Major element compositions of DRC megacrysts differ from those of megacrysts from many localities worldwide: (1) DRC garnets are poorer in Ti, Fe, and Al and extend toward high-Cr contents; (2) DRC clinopyroxenes are depleted in Fe and Ti but richer in Mg; (3) both phases do not display the classical bi-element correlations assigned to fractional crystallization processes observed for many other megacryst suites [see Fig. 6 in Pivin et al. (2009)]. DRC garnet and clinopyroxene megacrysts are actually compositionally intermediate between kimberlite megacrysts from other occurrences and peridotite minerals. They are notably enriched in Fe and Ti and depleted in Cr by comparison with the latest. DRC garnet megacrysts display "normal" REE patterns (LREE depleted relative to M-HREE and flat HREE), while clinopyroxenes show a relative LREE enrichment and a smooth decrease toward HREE. The REE patterns and contents are similar to those from garnets and clinopyroxenes in metasomatized (LREE-enriched) peridotites. DRC megacrysts were thus interpreted as resulting from a proto-kimberlitic metasomatic event that notably enriched peridotite mantle minerals in Fe, Ti, and LREE and led to their simultaneous recrystallization and overgrowth (Pivin et al. 2009); similar processes have also been invoked for the formation of Jericho kimberlite megacrysts (Kopylova et al. 2009) or polymict breccias (Pokhilenko 2009).

In Mbuji-Mayi, large megacrysts of zircon (up to 2 cm) and baddeleyite (up to 1.3 cm) were first investigated 30 years ago by Fieremans and Ottenburgs (1979) who notably observed mutual coatings of zircon and baddeleyite on each other. Few zircon and baddeleyite megacrysts were also investigated for U-Pb and Lu–Hf isotopes (Schärer et al. 1997) and, recently, for trace element concentrations (Schärer et al. 2011). The LA-ICP-MS analyses of trace element abundances presented in our study for Mbuji-Mayi

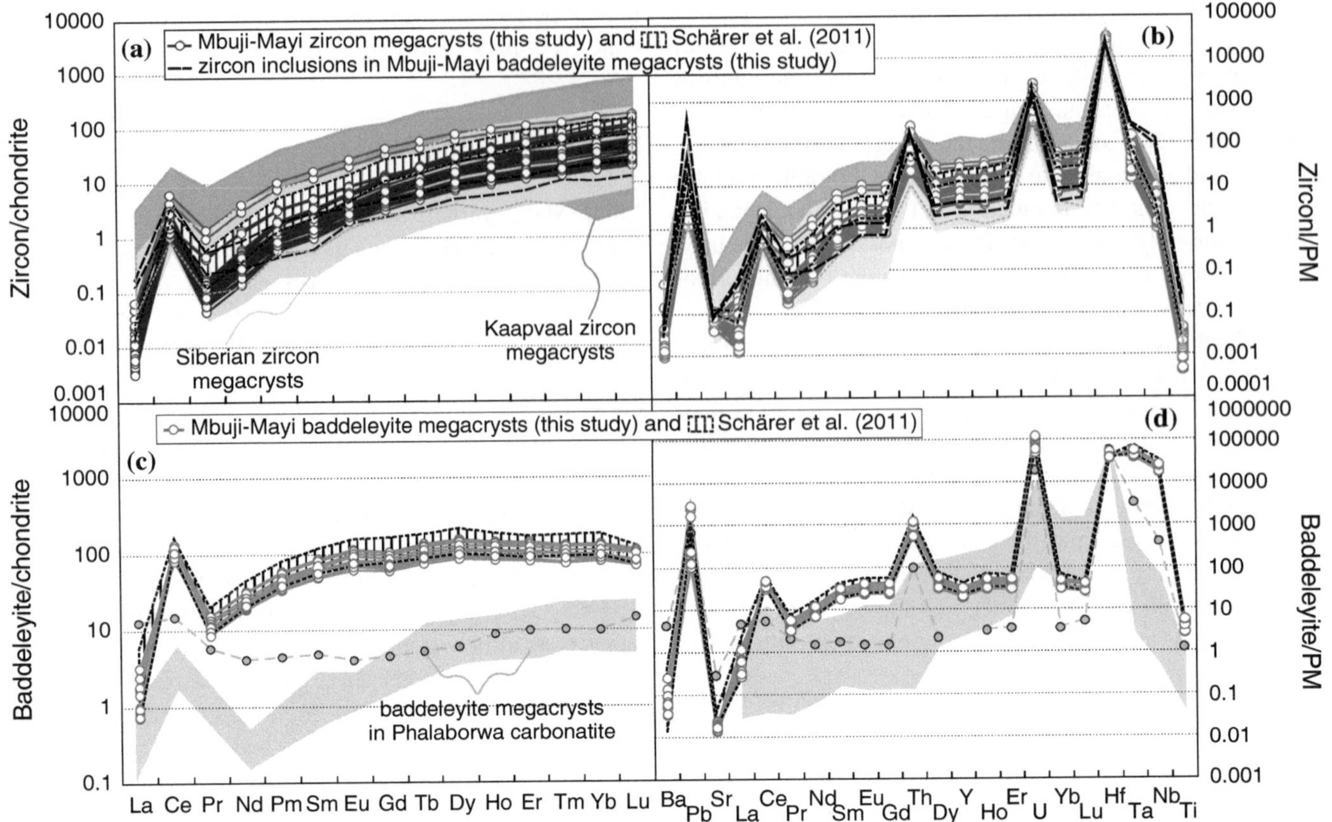

Fig. 2 Chondrite normalized rare earth element (**a–c**), primitive mantle normalized incompatible trace element (**b–d**), patterns for Mbuji-Mayi zircon (**a–b**), and baddeleyite megacrysts (**c–d**). Mbuji-Mayi zircons are compared to kimberlite zircon megacrysts from Siberia (Spetsius et al. 2002; Page et al. 2007) and South Africa (Belousova et al. 1998; Page et al. 2007). Trace element compositions of two zircon inclusions in Mbuji-Mayi baddeleyite megacrysts are also plotted (**a–b**). Mbuji-Mayi baddeleyite megacrysts are compared to large baddeleyite crystals from the Phalaborwa carbonatite (South Africa; Reischmann et al. 1995; Rubatto and Scambelluri 2003). Normalization values from McDonough and Sun (1995)

zircon (Fig. 2a, b) and baddeleyite (Fig. 2c, d) megacrysts overlap those of Schärer et al. (2011).

The Mbuji-Mayi zircon megacryst population has homogeneous major element concentrations, with <2.0 wt% HfO_2. They have low trace element contents (e.g., 10–99 ppm ΣREE, 16–143 ppm Y, 6–50 ppm U, 2–21 ppm Th) and positive Ce, Pb, Th, U, and Hf anomalies that are typical of zircon megacrysts in kimberlites (Fig. 2a, b) but have low Ti content (0.5–5.4 ppm). Zircon inclusions in two Mbuji-Mayi baddeleyite megacrysts display trace element concentrations (e.g., 8–16 ppm ΣREE) and patterns (Fig. 2a, b) similar to those of zircon megacrysts, except for the slight Nb enrichment that might be explained by host baddeleyite contamination during laser analysis.

Mbuji-Mayi baddeleyite megacrysts are mainly composed of ZrO_2 (94.7–97.0 wt%), with low HfO_2 (0.78–1.44 wt%), TiO_2 (0.41–1.11 wt%), FeO (0.17–0.66 wt%), Nb_2O_5 (0.57–1.97 wt%), and Ta_2O_5 (<0.3 wt%) contents. The REE patterns are quite similar to zircon megacryst patterns, with positive Ce anomalies (Fig. 2c) but much higher trace element contents (e.g., 140–237 ppm ΣREE; 81–138 ppm Y, 940–2,340 ppm U, 41–93 ppm Th). Trace element concentrations normalized to primitive mantle show positive Pb, Ce, Th, U, and Hf anomalies, along with Ta and Nb enrichments (Fig. 2d). Except for small (<50 μm) groundmass baddeleyites from the carbonate-rich Benfontein kimberlite (Scatena-Wachel and Jones 1984), which are enriched in Ca, Fe, and Hf by comparison with Mbuji-Mayi megacrysts, baddeleyite megacrysts have, to our knowledge, never been reported in other kimberlite localities. Baddeleyite is however commonly found as coatings on kimberlite zircon megacrysts (Kresten et al. 1975; Fieremans and Ottenburgs 1979; this study) and was also described as a subsolidus reaction product at zircon–ilmenite interface in bimineralic megacrysts from Monastery and Mothae kimberlites (Raber and Haggerty 1979). These zircon-associated baddeleyites are often assigned to late metasomatic reaction with a carbonate-rich and Si-poor fluid phase. Baddeleyite is a much more common accessory mineral in carbonatites (Tappe et al. 2009), where it is sometimes described as large crystals (up to 6 cm; Reischmann et al. 1995). Carbonatite- or phoscorite-derived

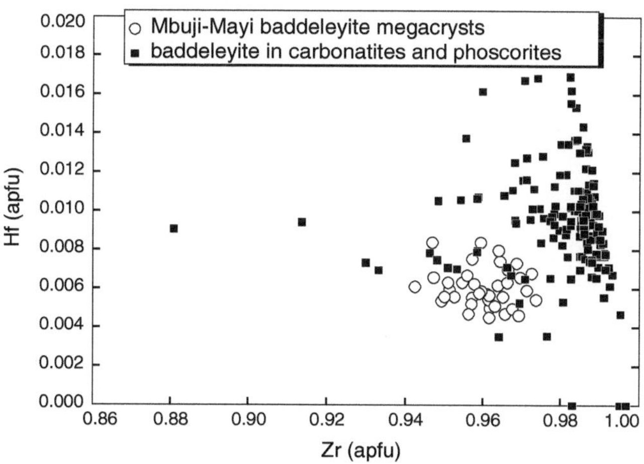

Fig. 3 Hf versus Zr (apfu) binary plot of Mbuji-Mayi baddeleyite megacrysts by comparison with baddeleyites in carbonatites and phoscorites (Kapustin 1980; Lumpkin 1999; Williams 1996; Chakhmouradian and Williams 2004)

baddeleyites are, however, on average richer in Zr and Hf (Fig. 3) and extend toward higher Nb contents (up to 7.5 wt% Nb_2O_5; Chakhmouradian and Williams 2004) than Mbuji-Mayi megacrysts that are enriched in Fe and Ti (not shown). Moreover, baddeleyite megacrysts from the Phalaborwa carbonatite (Fig. 2c, d) have lower trace element contents, which is consistent with the usually low U (50–150 ppm) and Th (<50 ppm) contents reported for baddeleyites in carbonatites (Heaman and LeCheminant 1993).

Analytical Procedure

Carefully selected fragments of the megacrysts were crushed using an agate mortar and pestle. Under a class 100 laminar flow hood in a class 1,000 cleaned room, garnet and clinopyroxene megacrysts were dissolved in Teflon Savilex® beakers using sub-boiled concentrated acids following 3 steps: (1) 5 ml HF-HNO_3 (9:1) for 2 days at 150 °C; (2) 5 ml HCl for 3 days at 150 °C; and (3) 2 ml HNO_3 on a hot plate at 120 °C for 24 h followed by 5 ml HCl in the same conditions. In order to get reasonable amounts of Hf and Nd for high-precision analyses ($\sim$50 ng Hf and $\sim$50 ng Nd), 200 mg of powder was dissolved for most samples. After re-dissolution in 2 NHCl + traces of HF, a 5 % aliquot of each sample was spiked with mixed ^{179}Hf–^{176}Lu and ^{148}Nd–^{149}Sm spikes, for concentration analyses by isotopic dilution. The Lu–Hf and Sm–Nd chromatographic separation methods were adapted from Debaille et al. (2008).

Zircon and baddeleyite ($\sim$30 mg) megacrysts were dissolved and treated following the method described in Goolaerts et al. (2004) for Lu–Hf systematic.

All analyses were performed on a Nu Plasma multicollector ICP-MS instrument (MC-ICP-MS) at Université Libre de Bruxelles (ULB), Belgium. Measured Hf and Nd isotope ratios are corrected for mass fractionation using $^{179}\mathrm{Hf}/^{177}\mathrm{Hf} = 0.7325$ and $^{146}\mathrm{Nd}/^{144}\mathrm{Nd} = 0.7219$, respectively. Analytical details and results are given in Table 1.

Results: Sm–Nd and Lu–Hf Systematics

DRC megacryst isotope compositions (Table 1) are presented on a $\varepsilon\mathrm{Hf}_{(t)}$ versus $\varepsilon\mathrm{Nd}_{(t)}$ plot (Fig. 4a) along with the data of other megacrysts and kimberlites from the literature, which have been recalculated (their initial isotopic ratios) using their respective emplacement age and the same parameters as those reported in Table 1.

Mbuji-Mayi garnets display a large range of initial $\varepsilon\mathrm{Hf}_{(t)}$ values from +6.6 to +12.1. Except for the low-Cr pyrope G3 with an unradiogenic $\varepsilon\mathrm{Nd}_{(t)}$ value of −0.6, the corresponding initial $\varepsilon\mathrm{Nd}_{(t)}$ values vary from +3.6 to +6.1 (Fig. 4a). Mbuji-Mayi diopside megacrysts also show a wide range of $\varepsilon\mathrm{Hf}_{(t)}$ from +3.0 to +9.1, which is less radiogenic than garnets (Fig. 4a). The corresponding $\varepsilon\mathrm{Nd}_{(t)}$ range (+2.7 to +4.6) overlaps that of garnets (except for the outlier) and, though slightly less radiogenic, is within error in agreement with the previously published $\varepsilon\mathrm{Nd}_{(t)}$ data for two Mbuji-Mayi diopside megacrysts (+6.5 and +7.0; Weis and Demaiffe 1985).

Three Mbuji-Mayi zircon megacrysts have radiogenic initial $\varepsilon\mathrm{Hf}_{(t)}$ values ranging from +6.5 to +7.1 that are close to those of baddeleyite megacrysts from the same locality: +6.0 to +8.4 (Fig. 4a). Schärer et al. (1997) have analyzed similar megacrysts and found similar values within error, which nevertheless extend the Mbuji-Mayi zircon–baddeleyite field toward more radiogenic $\varepsilon\mathrm{Hf}_{(t)}$ values: +8.7 to +9.0 for zircons and +5.7 to +10.7 for baddeleyites. Hf isotope compositions of baddeleyite and zircon megacrysts are included within the range displayed by Mbuji-Mayi garnet or clinopyroxene megacrysts.

The Kundelungu garnet megacryst has a low $\varepsilon\mathrm{Nd}_{(t)}$ value of +0.8 and a radiogenic $\varepsilon\mathrm{Hf}_{(t)}$ of +6.0, which are comparable to those of Mbuji-Mayi garnets (Fig. 4a). Previously published $\varepsilon\mathrm{Nd}_{(t)}$ data on one Kundelungu garnet are slightly more radiogenic (+2.5) and correspond to $\varepsilon\mathrm{Nd}_{(t)}$ values of diopside megacrysts from the same locality (+2.2 to +2.5) (Kampata et al. 1995).

DRC pyrope megacrysts plot within the mantle array (Fig. 4a), with either slightly negative or positive $\Delta\varepsilon\mathrm{Hf}_{(t)}$ values: −1.5 to +4.4 for Mbuji-Mayi and +1.7 for Kundelungu (Table 1). Most Mbuji-Mayi diopside megacrysts have slightly negative $\Delta\varepsilon\mathrm{Hf}_{(t)}$ from −6.4 to +0.3 but still plot within the mantle array ($-7 < \Delta\varepsilon\mathrm{Hf}_{(t)} < +7$; Janney et al. 2002). The ranges of $\varepsilon\mathrm{Hf}_{(t)}$ data obtained on zircon and

Table1 Hf–Nd isotope data for DRC garnet and clinopyroxene megacrysts and Hf isotope data for zircon and baddeleyite megacrysts

Locality/mineral	Sample	Lu	Hf	$^{176}Lu/^{177}Hf_m$	$^{176}Hf/^{177}Hf_m$	εHf(0)	εHf(t)	ΔεHf(t)	Sm	Nd	$^{147}Sm/^{144}Nd$	$^{143}Nd/^{144}Nd_m$	εNd(0)	εNd(t)
Mbuji-Mayi		(ppm)	(ppm)		2σ	2σ			(ppm)	(ppm)		2σ	2σ	
(70 Ma)[a]														
Garnet	G3	0.528	2.40	0.031317	0.282974 ± 11	6.7 ± 0.4	6.8	4.4	0.882	3.88	0.1374	0.512571 ± 09	−1.1 ± 0.2	−0.6
Garnet	G5[c]	0.413	1.70	0.034445	0.282978 ± 11	6.8 ± 0.4	6.8	−1.4	0.925	1.12	0.4976	0.512960 ± 12	6.4 ± 0.2	3.7
Garnet		0.422	1.72	0.034776	0.282973 ± 08	6.6 ± 0.3	6.6	−1.4	0.917	*	0.4934	0.512950 ± 08	6.2 ± 0.2	3.6
Garnet	G6	0.299	1.39	0.030449	0.283095 ± 10	11.0 ± 0.3	11.1	1.4	0.549	0.76	0.4366	0.512990 ± 08	7.0 ± 0.2	4.9
Garnet	G12[c]	0.289	1.05	0.039011	0.283121 ± 10	11.9 ± 0.4	11.6	0.8	0.496	0.76	0.3951	0.513018 ± 10	7.6 ± 0.2	5.8
Garnet		0.297	1.06	0.039725	0.283136 ± 08	12.4 ± 0.3	12.1	0.8	0.448	0.76	0.3563	0.513015 ± 08	7.5 ± 0.2	6.1
Garnet	GRB7	0.252	1.13	0.031638	0.283008 ± 12	7.9 ± 0.4	8.0	−1.5	0.702	0.80	0.5308	0.513026 ± 09	7.7 ± 0.2	4.7
CPX	CPX1 CPX3E[d]	0.010	0.53	0.002743	0.282830 ± 18	1.6 ± 0.6	3.0	−6.4	1.30	5.78	0.1362	0.512840 ± 13	4.1 ± 0.3	4.6
CPX		0.013	1.27	0.001431	0.282931 ± 11	5.2 ± 0.4	6.6	−0.2	1.86	8.57	0.1314	0.512741 ± 10	2.2 ± 0.2	2.7
CPX					0.282930 ± 07	5.1 ± 0.2	6.6	−2.0				0.512809 ± 16	3.5 ± 0.3	4.1
CPX	CPX3F	0.013	0.30	0.006276	0.282932 ± 14	5.2 ± 0.5	6.5	−2.8	1.15	4.96	0.1401	0.512839 ± 08	4.1 ± 0.2	4.6
CPX	CPX3G	0.013	0.18	0.010079	0.282881 ± 22	3.4 ± 0.8	4.5	−4.5	1.08	4.74	0.1373	0.512828 ± 12	3.9 ± 0.2	4.4
CPX	CPX3H	0.013	1.86	0.001026	0.282954 ± 09	6.0 ± 0.3	7.5	−0.1	1.95	8.96	0.1318	0.512771 ± 10	2.7 ± 0.2	3.3
CPX	Di26[c]	0.007	0.62	0.001545	0.283000 ± 10	7.6 ± 0.4	9.1	0.3	1.51	7.18	0.1270	0.512816 ± 12	3.6 ± 0.2	4.2
CPX		0.007	0.62	0.001481	0.282977 ± 10	6.8 ± 0.3	8.3	−0.9	1.51	7.21	0.1262	0.512828 ± 09	3.9 ± 0.2	4.5
CPX[e]	Diopside	A							1.3	6.3	0.1248	0.512929 ± 46	5.8 ± 0.9	6.5
CPX[e]	Diopside	B							1.4	5.9	0.1435	0.512967 ± 45	6.6 ± 0.9	7.0
Zircon	Zr 2[d]				0.282938 ± 05	5.4 ± 0.2	7.0							
Zircon					0.282926 ± 10	5.0 ± 0.3	6.5							
Zircon	Zr 4				0.282941 ± 07	5.5 ± 0.3	7.1							
Zircon	Zr 91				0.282941 ± 05	5.5 ± 0.2	7.1							
Zircon[g]	M I/1	4	7631	0.000074	0.282993 ± 17	7.4 ± 0.6	8.9							
Zircon[g]	M I/2	3	7631	0.000062	0.282995 ± 16	7.4 ± 0.6	9.0							
Zircon[g]	M II	3	7631	0.000061	0.282987 ± 18	7.1 ± 0.6	8.7							
Badd	Bd 17				0.282938 ± 09	5.4 ± 0.3	7.0							
Badd	Bd 11				0.282978 ± 06	6.8 ± 0.2	8.4							
Badd	Bd 14[d]				0.282910 ± 07	4.4 ± 0.2	6.0							
Badd					0.282919 ± 07	4.8 ± 0.3	6.3							
Badd[g]	M I	4	11024	0.000054	0.282902 ± 13	4.1 ± 0.5	5.7							
Badd[g]	M II	2	11024	0.000027	0.282931 ± 15	5.2 ± 0.5	6.7							
Badd[g]	M III	2	11024	0.000026	0.282927 ± 11	5.0 ± 0.4	6.6							
Badd[g]	M IV	2	11024	0.000022	0.283045 ± 15	9.2 ± 0.5	10.7							
Badd[g]	M V	5	11024	0.000067	0.282940 ± 11	5.5 ± 0.4	7.0							

(continued)

Table1 (continued)

Locality/mineral	Sample	Lu	Hf	$^{176}Lu/^{177}Hf_m$	$^{176}Hf/^{177}Hf_m$	εHf(0)	εHf(t)	ΔεHf(t)	Sm	Nd	$^{147}Sm/^{144}Nd$	$^{143}Nd/^{144}Nd_m$	εNd(0)	εNd(t)
Mbuji-Mayi		(ppm)	(ppm)		2σ	2σ			(ppm)	(ppm)		2σ	2σ	
(70 Ma)[a]														
Kundelungu														
(30 Ma)[b]														
Garnet	K14	0.298	0.385	0.109841	0.282999 ± 08	7.6 ± 0.3	6.0	1.7	0.545	0.91	0.3606	0.512706 ± 13	1.5 ± 0.2	0.8
Garnet[f]	MsGR								1.4	2.02	0.4190	0.512806 ± 07	3.4 ± 0.1	2.5
CPX[f]	ZeCPX2								1.44	6.2	0.1404	0.512747 ± 15	2.3 ± 0.3	2.5
CPX[f]	MsCPX2								1.56	7.1	0.1328	0.512731 ± 10	2.0 ± 0.2	2.2

Kimberlite eruption ages (t): [a] Schärer et al. (1997); [b] Batumike et al. (2008); [c] whole procedure duplicate (except for dissolution step); [d] run duplicate; [e] data from Weis and Demaiffe (1985); [f] data from Kampata et al. (1995); [g] data from Schärer et al. (1997). All literature values were recalculated following the parameters given below. Errors on Lu, Hf, Sm, and Nd concentrations and on $^{176}Lu/^{177}Hf$ and $^{147}Sm/^{144}Nd$ ratios are <0.5 % (2σ). During the analysis sessions, average $^{176}Hf/^{177}Hf$ of the JMC-475 standard was 0.282148 ± 16 (2σ, n = 24) for garnet and clinopyroxene analyses and 0.282144 ± 13 (2σ, n = 16) for zircon and baddeleyite analyses. All isotope ratios were subsequently corrected using the JMC-475 standard value of 0.282160 recommended by Vervoort and Blichert-Toft (1999). Initial εHf values calculated at age (t) using $\lambda^{176}Lu = 1.865 \times 10^{-11}\ yr^{-1}$ (Scherer et al. 2001), $(^{176}Hf/^{177}Hf)_{CHUR} = 0.282785$, and $(^{176}Lu/^{177}Hf)_{CHUR} = 0.0336$ (Bouvier et al. 2008). Initial εHf values for zircon and baddeleyite megacrysts from this study were calculated using average $^{176}Lu/^{177}Hf$ ratios of 0.000065 for zircons and 0.000039 for baddeleyites. These values, however, have no influence in the final result that only changes because of the evolution of the CHUR reservoir with time. Average $^{143}Nd/^{144}Nd$ of the Rennes standard was 0.511947 ± 16 (2σ, n = 20) for garnet and clinopyroxene analyses. Nd isotope ratios were corrected using $^{143}Nd/^{144}Nd$ of 0.511961 (Chauvel and Blichert-Toft 2001). Initial εNd values calculated at age (t) using $\lambda^{147}Sm = 6.539 \times 10^{-12}\ yr^{-1}$ (Begemann et al. 2001), $(^{143}Nd/^{144}Nd)_{CHUR} = 0.512630$, and $(^{147}Sm/^{144}Nd)_{CHUR} = 0.1960$ (Bouvier et al. 2008). ΔεHf is the difference between the measured εHf value and theoretical value calculated from the oceanic basalt Nd–Hf regression of Vervoort et al. (1999); that is, the mantle array: εHf (OIB) = 1.33 εNd + 3.19. *Sample lost, value (1.12 ppm Nd) taken from the other analysis of the duplicate sample

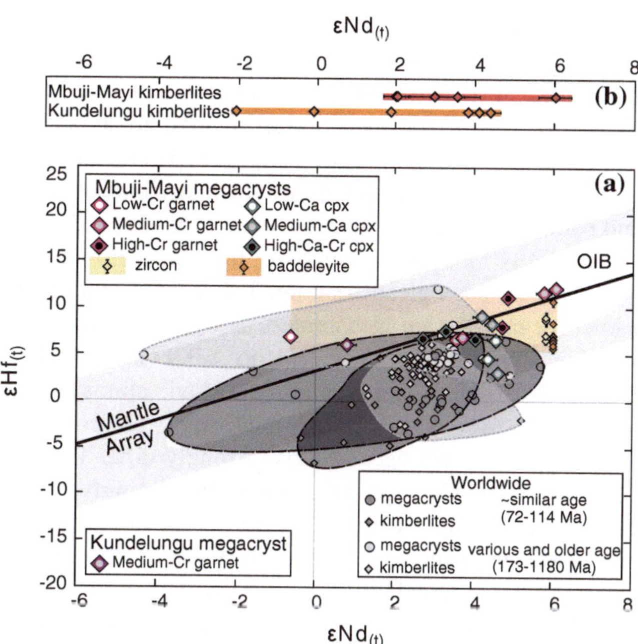

Fig. 4 **a** Initial $\varepsilon Nd_{(t)}$ versus $\varepsilon Hf_{(t)}$ values of DRC megacrysts (2σ error bars range from ±0.2 to ±0.3 for $\varepsilon Nd_{(t)}$ and from ±0.4 to ±0.8 for $\varepsilon Hf_{(t)}$, which are smaller than symbols) compared to available data for on- and off-craton Group I kimberlites and megacrysts worldwide. Different colors make the distinction between those of roughly similar age to DRC kimberlites (*dark gray*; South Africa, Lesotho, Botswana, and Somerset Island) and those of older age (*light gray*) that extend over a wider range of age and geographic distribution (South Africa, India, Canada, and Greenland) (Nowell et al. 1999, 2004; Schmidberger et al. 2002; Kopylova et al. 2009; Paton et al. 2009; Tappe et al. 2011). Present day OIB field is also plotted (Salters and White 1998). Fields for Mbuji-Mayi baddeleyite and zircon megacrysts comprise $\varepsilon Hf_{(t)}$ data from this study (2σ error bars from ±0.2 to ±0.3) and those from Schärer et al. (1997) and range over $Nd_{(t)}$ data of garnet and cpx megacrysts. **b** Ranges of $\varepsilon Nd_{(t)}$ values for the 70 Ma old Mbuji-Mayi and 32 Ma old Kundelungu kimberlite whole-rocks (Weis and Demaiffe 1985; Kampata 1993)

baddeleyite megacrysts from Mbuji-Mayi are comparable to those of garnet and clinopyroxene megacrysts from the same locality and would also plot within the mantle array if Nd isotope compositions are assumed to be similar (Fig. 4a).

Discussion

Timing Constraints

If megacrysts are cogenetic to their host kimberlitic magmas, they would be potentially good indicators of the kimberlite eruption age (Nowell et al. 2004; Kopylova et al. 2009). However, whatever the combinations of minerals tested (e.g., all of them or only a selection of two or three species), no reliable geochronological constraint has been obtained, neither from Lu–Hf nor from Sm–Nd systems for the Mbuji-Mayi megacrysts. This reemphasizes our earlier suggestion that these megacrysts are not related to their kimberlitic hosts by simple processes (Pivin et al. 2009).

The ranges of Hf model ages relative to a depleted mantle (DM) source [T_{DM}, using the parameters from Griffin et al. (2000)] for Mbuji-Mayi clinopyroxene, zircon, and baddeleyite megacrysts are similar (362–694, 355–451, and 286–484 Ma, respectively), but these ages do not have any apparent geological meaning. T_{DM} ages calculated for garnet megacrysts are meaningless as well and much more variable (from highly negative to >3.9 Ga). Depleted mantle Nd model ages of Mbuji-Mayi clinopyroxene megacrysts (460–837 Ma) are relatively well correlated with their Hf model ages and also strongly differ from the negative Nd model ages of garnets. Interestingly, differences in model ages between garnet on the one hand, and clinopyroxene–zircon–baddeleyite megacrysts on the other have also been found, for both Nd and Hf, using various $^{176}Lu/^{177}Hf$ and $^{147}Sm/^{144}Nd$ ratios (e.g., relative to an average kimberlitic source composition with $^{176}Lu/^{177}Hf = 0.00316$; $^{176}Hf/^{177}Hf = 0.282725$; $^{147}Sm/^{144}Nd = 0.0807$; and $^{143}Nd/^{144}Nd = 0.512700$—calculated using published values on kimberlites with age similar to those of DRC; Nowell et al. 1999, 2004; Schmidberger et al. 2002—or a CHUR source). Even though the different model ages are difficult to interpret, probably because we are lacking the right model to use, and will thus not be detailed further, it is clear that garnet on the one hand and zircon–baddeleyite–clinopyroxene megacrysts on the other could reflect distinct geological processes and/or histories.

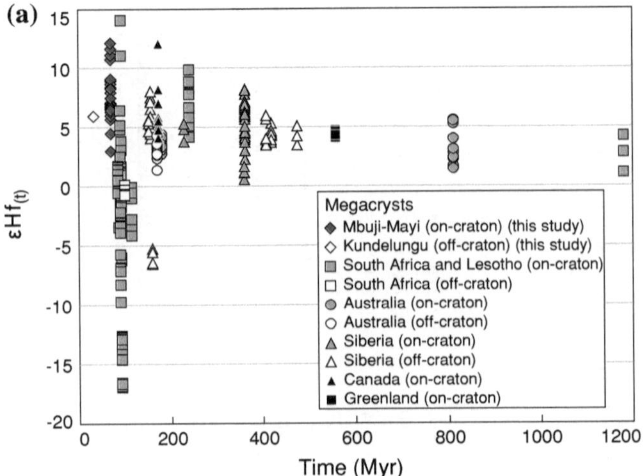

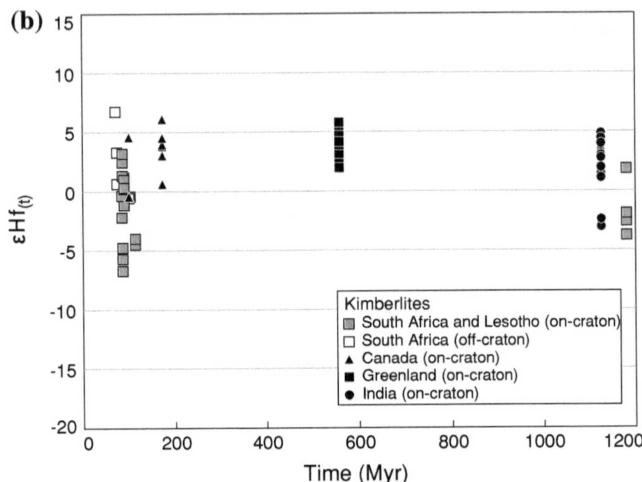

Fig. 5 Initial $\varepsilon Hf_{(t)}$ values of Group I megacrysts (**a**) and kimberlite whole-rocks (**b**) from worldwide localities plotted as a function of their emplacement age. The distinction is made between megacrysts and kimberlites from different continents and on- versus off-craton localities. In contrast to Fig. 4, even megacrysts for which only Hf isotope composition has been measured (i.e., zircon) are represented, which extends the range of data. Literature values from Nowell et al. (1999, 2004), Griffin et al. (2000), Schmidberger et al. (2002), Spetsius et al. (2002), Kopylova et al. (2009), Paton et al. (2009), and Tappe et al. (2011)

Comparison to Worldwide Megacrysts and Kimberlites

Because interpretations of $\varepsilon Hf_{(t)}$-$\varepsilon Nd_{(t)}$ diagrams for kimberlitic rocks (Fig. 4a) need to be age-specific, the distinction is made between kimberlites and megacrysts with emplacement ages similar to DRC kimberlites (~ <100 Ma; dark gray symbols and fields) and older occurrences (up to 1.2 Ga; light gray symbols and fields).

Megacrysts from many localities and different ages show a similar range of $\varepsilon Nd_{(t)}$ values from −4 to +6, most of them being positive. By contrast, Hf isotope compositions of megacrysts allow discrimination and are slightly more radiogenic for older occurrences (most $\varepsilon Hf_{(t)}$ values between 0 and +10) relative to younger ones (most $\varepsilon Hf_{(t)}$ are between −5 and +5). However, this distinction may derive from their distinct cratonic setting since all available data for Cretaceous megacrysts are from the Kaapvaal craton area. On a worldwide basis (Fig. 5a), the apparent correlation between age and Hf isotopic composition is mainly driven by the large variability of the Cretaceous data set from South Africa and Lesotho that includes many samples from various kimberlite provinces. There is no strong evidence for a direct correlation between the age of the megacrysts (i.e., the age of the kimberlite host rock) and their $\varepsilon Hf_{(t)}$ value (Fig. 5a). Megacrysts from each kimberlite province most probably have their own, distinct, isotopic signature. Nevertheless and interestingly, by comparison with the worldwide megacryst database, the Mbuji-Mayi megacryst population (especially garnets) significantly extends the isotopic field toward more radiogenic $\varepsilon Hf_{(t)}$ and $\varepsilon Nd_{(t)}$ values, that is, to the upper right and toward the mantle array (Fig. 4a).

Like megacrysts, kimberlite whole-rocks from older provinces also define a trend to more radiogenic Hf isotope compositions in the $\varepsilon Hf_{(t)}$ versus $\varepsilon Nd_{(t)}$ plot (Fig. 4a), but there is again no clear correlation between age and $\varepsilon Hf_{(t)}$ values (Fig. 5b). Most kimberlites, whatever their age or locality, display a rather restricted isotopic field (Fig. 4a), characterized by weighted means in $\varepsilon Nd_{(t)}$ of $+2.5 \pm 2.7$ and $\varepsilon Hf_{(t)}$ of $+1.9 \pm 6.0$, suggesting that their sources all evolve with relatively similar $^{147}Sm/^{144}Nd$ and $^{176}Lu/^{177}Hf$ ratios. Nd isotope compositions of Mbuji-Mayi and Kundelungu kimberlite whole-rocks (Weis and Demaiffe 1985; Kampata 1993; Fig. 4b) fall in the range shown by DRC megacrysts (Fig. 4a) and match the worldwide kimberlite data set (Fig. 4a).

Comparison to the Cratonic Subcontinental Lithospheric Mantle

The cratonic SCLM is known to be highly heterogeneous in isotopic composition, because of its long history and the numerous interactions it underwent with various melts (Simon et al. 2007). However, only few localities have been investigated for combined Nd–Hf isotope composition and analyses have been done either on whole rocks or on mineral separates (Fig. 6).

In εHf versus εNd diagram, low-temperature cratonic peridotites display a subvertical trend with highly variable and radiogenic $\varepsilon Hf_{(t)}$ compositions (up to +52) and more restricted $\varepsilon Nd_{(t)}$ range (−5.1 to +3.9), which might reflect that Nd isotopes are more sensitive than Hf isotopes to melt/fluid (often LREE-rich) interactions in the mantle. Hf isotopes most probably record time-integrated long-term depletion of the cratonic SCLM (Schmidberger et al. 2002). When garnets and clinopyroxenes are considered, a much wider range of isotope compositions is observed, with mostly radiogenic to highly radiogenic $\varepsilon Hf_{(t)}$ compositions (up to few hundred for garnets; not shown in Fig. 6) and variable $\varepsilon Nd_{(t)}$ values (from −13 to +20) that sometimes mirror a LREE enrichment expected in old metasomatized SCLM (Menzies and Hawkesworth 1987).

Interestingly, the high-temperature peridotite xenoliths, which are commonly inferred to represent the base of the SCLM and to have suffered extensive kimberlite-related metasomatic interaction (Hops et al. 1992; Moore and Belousova 2005), display a more restricted isotopic composition ($\varepsilon Hf_{(t)}$: −2 to +21; $\varepsilon Nd_{(t)}$: 0 to +5) that overlap the kimberlite field, as do their constitutive garnets (Fig. 6). Combined Hf–Nd isotope data of high-temperature peridotites, even if scarce, appear displaced toward the kimberlite compositional field, that is, toward less radiogenic $\varepsilon Hf_{(t)}$ values, by comparison with low-temperature peridotites. In addition, there is increasing geochemical evidence

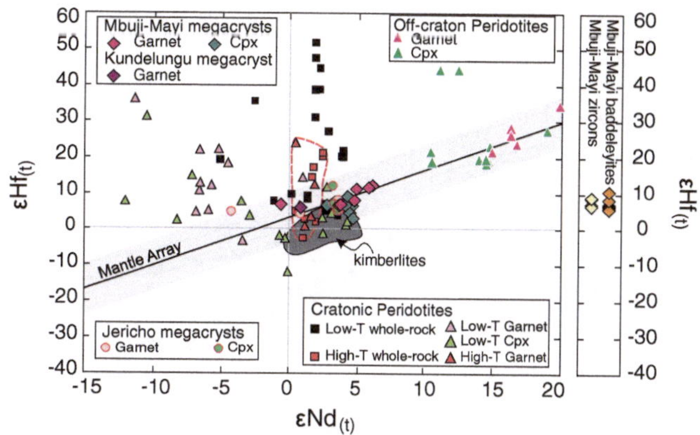

Fig. 6 Initial $\varepsilon Nd_{(t)}$ versus $\varepsilon Hf_{(t)}$ values of DRC garnet and clinopyroxene megacrysts compared to those of (1) peridotite xenoliths and constituting minerals from cratonic SCLM from the Kaapvaal and northern Canadian cratons (Schmidberger et al. 2001, 2002; Bedini et al. 2004; Simon et al. 2007) and (2) fertile off-craton peridotite xenoliths from Vitim volcanic field, Siberia (Ionov et al. 2005). High-Cr megacrysts from the Jericho kimberlite field are also plotted for comparison (Kopylova et al. 2009)

in the literature suggesting that clinopyroxenes in low-temperature cratonic peridotites might have formed recently through interaction with the erupting kimberlite melt or a precursor (Van Achterbergh et al. 2001; Grégoire et al. 2003; Simon et al. 2003, 2007; Coussaert 2005). Nd–Hf isotope systematic also supports such an origin since some low-temperature peridotitic diopsides effectively plot in, or close to, the kimberlite field (Fig. 6).

The similarity of isotopic compositions between clinopyroxenes of inferred metasomatic origin in SCLM low-temperature peridotites and Mbuji-Mayi diopside megacrysts (Fig. 6) might reflect similar petrogenetic processes, that is, recent (re)crystallization in metasomatized mantle peridotites. A metasomatic origin has also been proposed for bimineralic megacrysts from the Jericho kimberlites (Slave craton; Kopylova et al. 2009), and Nd–Hf isotope compositions of the latter compare well to those from Mbuji-Mayi (Fig. 6). In garnet–clinopyroxene pairs from Jericho, clinopyroxenes have more homogeneous $\varepsilon Nd_{(t)}$ isotope compositions (+3.0 to +3.4) than garnets (−4.5 to +3.4), whereas $\varepsilon Hf_{(t)}$ values are more variable for both clinopyroxenes (+6.1 to +13.0) and garnets (+5.1 to +9.2) (Kopylova et al. 2009).

Trace Element Modeling

As already pointed out, DRC garnet and clinopyroxene megacrysts do not show any evidence for an origin by fractional crystallization from a parental kimberlitic melt but display major and trace element geochemical similarities with the SCLM peridotite minerals. They are rather

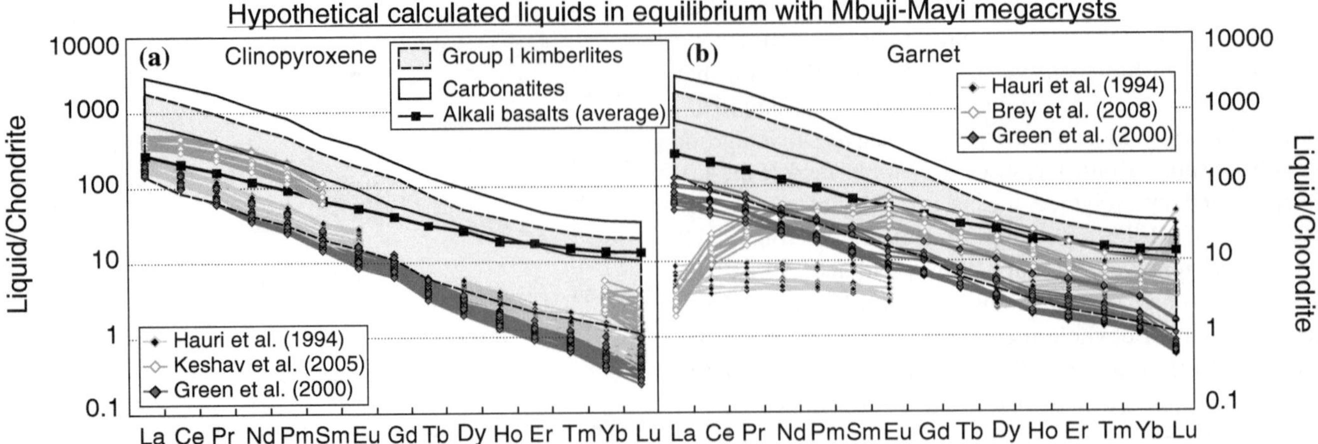

Fig. 7 Modeled REE patterns for hypothetical melts in equilibrium with Mbuji-Mayi clinopyroxene (**a**) and garnet megacrysts (**b**) using various experimental partition coefficients for silicate and/or carbonate liquids (references in Fig.; see text for details). The comparison is made with carbonatite and alkali basalt melt compositions (GEOROC database for lava from the East African rift) and worldwide Group 1 kimberlite compositions (DRC: Fieremans et al. 1984; Kampata 1993; South Africa: le Roex et al. 2003; Becker and le Roex 2006; Siberia: Kostrovitsky et al. 2007). Normalization values from McDonough and Sun (1995)

interpreted as resulting from metasomatic transformation and recrystallization of peridotite wall rocks through interaction with an infiltrating proto-kimberlite melt and/or fluid (Pivin et al. 2009).

In order to better constrain the suggested relationship, REE patterns of hypothetical liquids in equilibrium with Mbuji-Mayi clinopyroxene and garnet megacrysts have been modeled using experimentally derived partition coefficients (D) for various liquid compositions (basaltic to carbonatitic) at high pressures and temperatures (Fig. 7). Representative REE compositions of DRC megacrysts have been published in Pivin et al. (2009). Clinopyroxene and garnet D values in basaltic systems are from Hauri et al. (1994) (1,430 °C and 2.5 GPa) and Green et al. (2000) (1,080–1,200 °C and 3–4 GPa; hydrated conditions). The D values from the latter study have notably been used to model liquids in equilibrium with minerals in LT mantle peridotites from the Kaapvaal craton (Simon et al. 2007; Rehfeldt et al. 2008). Partition coefficient values are also available for clinopyroxenes in a kimberlitic system (CMAS-CO_2; ~1,400 °C and 6 GPa; Keshav et al. 2005) and for garnets in a carbonated-silicated system (1,500 °C and 6 GPa; Brey et al. 2008).

Whatever the D values used, calculated REE patterns of liquids in equilibrium with Mbuji-Mayi clinopyroxene megacrysts are comparable. They are all in good agreement with Group I kimberlite compositions, especially those modeled with the experimental D values for a kimberlite system (Keshav et al. 2005), which show a similar REE fractionation $[(La/Yb)_N = 185–394]$ to Group I kimberlites $[(La/Yb)_N = 42–253]$; references in Fig. 7). On the other hand, REE patterns of melt buffered by Mbuji-Mayi garnet megacrysts vary according to the experimental partition coefficients used (Fig. 7b). The calculated REE patterns for melts in equilibrium with garnet in a basaltic system (Hauri et al. 1994) show a flat $[(La/Yb)_N = 0.2–3.5]$, REE-depleted ($La_N = 3–9$) pattern. Using D values obtained for higher P-T conditions and more carbonated-rich system (Brey et al. 2008), the calculated liquids have MREE-HREE abundances in the kimberlite compositional range, but they have abnormal low LREE abundances ($La_N = 2–5$). Nevertheless, melts calculated with D values from a hydrated basaltic system (Green et al. 2000) show a fractionation $[(La/Yb)_N = 8–148]$ close to Group I kimberlites and REE concentrations that are close to the lower part of the kimberlite range.

Calculated liquids in equilibrium with clinopyroxene megacrysts are globally similar to Group I kimberlite melts, while those in equilibrium with garnet megacrysts appear much more variable. All modeled liquids, for garnets and clinopyroxenes, are poorer in REEs than carbonatite melts, and they differ from typical alkaline basalts (Fig. 7).

Concealing the Relationships Between Megacrysts, Kimberlites, and the SCLM

Megacrysts (garnet, clinopyroxene, zircon, and baddeleyite) from DRC kimberlites do not show significant downward displacement relative to the mantle array in the $\varepsilon Hf_{(t)}$ versus $\varepsilon Nd_{(t)}$ plot. They are all characterized by positive $\varepsilon Hf_{(t)}$, as observed in some other localities.

Except for one unradiogenic low-Cr pyrope from Mbuji-Mayi, DRC megacrysts from both Mbuji-Mayi and

Kundelungu provinces have positive $\varepsilon Nd_{(t)}$ values that compare well to those of their kimberlite host rocks but seem to have slightly more radiogenic $\varepsilon Hf_{(t)}$ values than worldwide kimberlites.

Although calculated model ages have no significant geological meaning, the systematic difference between Hf–Nd initial ratios and model ages of garnet megacrysts on the one hand and of clinopyroxene–zircon–baddeleyite megacrysts on the other suggests a different origin and/or evolution. A contrasting behavior is also apparent with the calculated trace element contents of liquids buffered by Mbuji-Mayi clinopyroxene and garnet megacrysts: the former all have a Group I kimberlite composition, whereas the latter show more variability, depending on the D values chosen. Some of them nevertheless have a composition that is very close to that of Group I kimberlites, suggesting that the variation observed might reflect more complex formation processes for garnet megacrysts.

The large range of $\varepsilon Nd_{(t)}$ data (−0.6 to +6.1) and high $\varepsilon Hf_{(t)}$ values (+6.6 to +12.1) of garnets may reflect that they are inherited from old cratonic garnet-bearing peridotites and subsequently recrystallized through interaction with the proto-kimberlite melt/fluid. The original Hf–Nd isotopic signature of protolith garnets is difficult to infer as the isotopic field of cratonic garnet peridotitic minerals is extremely large: garnets from low-temperature SCLM peridotites display ranges of $\varepsilon Hf_{(t)}$ values from −3.2 to +432.5 and $\varepsilon Nd_{(t)}$ values from −36.1 to +493.4; most between −14.5 and +9.7 [note that the range of values in Fig. 6 is more restricted because only the combined $\varepsilon Hf_{(t)}$-$\varepsilon Nd_{(t)}$ data were reported; additional data from Richardson et al. (1985), Walker et al. (1989), Pearson et al. (1995)]. Whatever their initial isotopic composition, subsequent interaction with a proto-kimberlite melt/fluid and simultaneous recrystallization would then force their isotopic composition toward the narrower kimberlite field.

By contrast, the more restricted initial Nd isotope compositions for Mbuji-Mayi clinopyroxene megacrysts (+2.7 to +4.6) and the less radiogenic $\varepsilon Hf_{(t)}$ values for clinopyroxene (+3.0 to +9.1), zircon (+6.5 to +7.1), and baddeleyite (+6.0 to +7.4) megacrysts suggest a closer link to the Group I kimberlite isotopic field, characterized by $\varepsilon Nd_{(t)}$ values from −2.1 to +6.0 and $\varepsilon Hf_{(t)}$ values from −6.7 to +6.7 [the range of values reported in Fig. 6 is slightly smaller because only the combined $\varepsilon Hf_{(t)}$-$\varepsilon Nd_{(t)}$ data were plotted; additional data from Weis and Demaiffe (1985), Kampata (1993), Pearson et al. (1995), Davies et al. (2001)]. Therefore, as suggested for clinopyroxenes in low-temperature mantle peridotites, the Mbuji-Mayi diopside megacrysts could have recently crystallized during the interaction of a proto-kimberlite melt/fluid infiltrating cratonic peridotites. As clinopyroxene megacrysts from DRC kimberlites do not show evidence for an origin by fractional crystallization (Pivin et al. 2009), they have been likely formed through metasomatic processes such as "percolative fractional crystallization" (Harte et al. 1993). The slight positive correlation between the initial Hf isotope composition of Mbuji-Mayi clinopyroxene megacrysts and their Cr content ($\varepsilon Hf_{(t)}$ = +3.0 for a medium-Ca diopside with 0.53 wt% Cr_2O_3; +4.5 for a low-Ca diopside with 0.84 wt% Cr_2O_3, and up to +9.1 for a medium-Ca diopside with 1.33 wt% Cr_2O_3) might reflect a lower melt/fluid-wall rock ratio for Cr-rich megacrysts by comparison with those with lower Cr content that plot closer to the kimberlite field ($\varepsilon Hf_{(t)}$: −6.7 to +6.7). The apparent lack of such a correlation for garnet megacrysts from the same locality ($\varepsilon Hf_{(t)}$ = +6.8 for a low-Cr pyrope with 0.29 wt% Cr_2O_3; +12.1 for a medium-Cr pyrope with 2.50 wt% Cr_2O_3 or +6.8 for another one with 5.16 wt% Cr_2O_3 and +8.0 for a high-Cr pyrope with 6.12 wt% Cr_2O_3) is most probably related to the variable compositions of their protolith garnet.

Zircon and baddeleyite megacrysts have very similar $\varepsilon Hf_{(t)}$ values and are also thought to originate through metasomatic interaction between SCLM peridotites and infiltrating proto-kimberlite melt/fluid. Such an origin has also been proposed for zircon megacrysts from the Mir kimberlite (Siberia, Spetsius et al. 2002) that have trace element composition similar to Mbuji-Mayi zircon megacrysts (Fig. 2) and in which inclusions of peridotite-derived minerals (e.g., Cr-diopside, chromite, Mg-olivine, phlogopite and sulfide) have been observed. In addition, metasomatic zircons have been found in metasomatized mantle peridotites (Konzett et al. 2000; Dawson et al. 2001) and some are effectively similar to kimberlite megacrysts: zircon in phlogopite K-richterite peridotites (PKP) from Kimberley have low trace element abundances ($\Sigma REE = 9$ and 134 ppm; 6–97 ppm U; 2–21 ppm Th; Konzett et al. 2000). Odling (1995) has also shown experimentally that the reaction between a natural Group I kimberlite and a refractory garnet-bearing harzburgite in a thermal gradient (1,225–1,000 °C) at high pressure (3 GPa) progressively transforms the harzburgite into various mineralogical assemblages, the coolest (that is also the more distal) being a PKP lherzolite with accessory zircon and Nb-Cr-rutile that most probably results from reaction between refractory harzburgite and a water-rich fluid derived from the kimberlite.

A similar metasomatic origin might be suggested for Mbuji-Mayi baddeleyite megacrysts, and their exceptional presence in these kimberlites might be related to the CO_2-rich nature (carbonate-rich) of the Mbuji-Mayi kimberlites (Demaiffe and Fieremans 1981). Indeed, the only kimberlite locality where baddeleyite has been reported as a groundmass phase is the carbonate-rich Benfontein sill. In a model of percolative proto-kimberlite melt/fluid, local change of Si saturation (or silica activity) caused by interaction with surrounding peridotites might explain the local crystallization of baddeleyite on the one hand or zircon on the other hand. Zircon inclusions in Mbuji-Mayi baddeleyite megacrysts favor such local changes in Si saturation.

Conclusion

Several megacrysts (garnet, clinopyroxene, zircon, and baddeleyite) from Mbuji-Mayi and Kundelungu kimberlites (DRC) were investigated for their Hf and Nd isotope systematics. $\varepsilon Nd_{(t)}$ values of megacrysts are similar to those of their kimberlitic hosts, but their $\varepsilon Hf_{(t)}$ values appear to be slightly more radiogenic than worldwide kimberlites. These megacrysts do not simply result from fractional crystallization in the kimberlitic magma. A more complex scenario, implying metasomatism, is favored. In addition, garnet on the one hand and clinopyroxene, zircon, and baddeleyite on the other show different coupled Hf–Nd systematics, suggesting that they result from different processes. It is proposed that garnet recrystallized from an old garnet-bearing cratonic peridotite during interaction with the proto-kimberlite melt/fluid. On the contrary, clinopyroxene, zircon, and baddeleyite recently crystallized during metasomatic interaction of the proto-kimberlite melt/fluid percolating through these peridotites. The proto-kimberlite composition of the metasomatic agent has been confirmed through the modeling of rare earth element patterns of liquids in equilibrium with clinopyroxene and garnet megacrysts.

Acknowledgments We are grateful to the late Dr Carlos Fieremans (former executive officer of the MIBA) and to Dr Mark Fieremans, who generously provided the exceptional collection of megacrysts from Mbuji-Mayi. MP is supported by a FRIA-FNRS grant and Van Buuren funding. Ivan Petrov and Claude Maerschalk are thanked for helping on the Nu Plasma and in the clean laboratory, respectively. The detailed and constructive review of Dr S. Tappe is gratefully acknowledged.

References

Ballentine CJ, Lee D-C, Halliday AN (1997) Hafnium isotopic studies of the Cameroon line and new HIMU paradoxes. Chem Geol 139:111–124

Batumike JM, O'Reilly SY, Griffin WL, Belousova EA (2007) U-Pb and Hf-isotopes analyses of zircon from the Kundelungu kimberlites, DR Congo: implications for crustal evolution. Prec Res 156:195–225

Batumike JM, Griffin WL, Belousova EA, Pearson NJ, O'Reilly SY, Shee SR (2008) LAM-ICPMS U-Pb dating of kimberlitic perovskite: Eocene-Oligocene kimberlites from the Kundelungu Plateau, DR Congo. Earth Planet Sci Lett 267:609–619

Becker M, le Roex AP (2006) Geochemistry of South African on- and off-craton, Group I and Group II kimberlites: petrogenesis and source region evolution. J Petrol 47:673–703

Bedini R-M, Blichert-Toft J, Boyet M, Albarède F (2004) Isotopic constraints on the cooling of the continental lithosphere. Earth Planet Sci Lett 223:99–111

Begemann F, Ludwig KR, Lugmair GW, Min K, Nyquist LE, Patchett PJ, Renne PR, Shih C-Y, Villa IM, Walker RJ (2001) Call for an improved set of decay constants for geochronological use. Geochim Cosmochim Acta 65:111–121

Belousova EA, Griffin WL, Pearson NJ (1998) Trace element composition and cathodoluminescence properties of southern African kimberlitic zircons. Min Mag 62:355–366

Blichert-Toft J, Albarède F (1997) The Lu-Hf isotope geochemistry of chondrites and the evolution of the mantle-crust system. Earth Planet Sci Lett 148:243–258

Bouvier A, Vervoort JD, Patchett PJ (2008) The Lu-Hf and Sm-Nd isotopic composition of CHUR: constraints from unequilibrated chondrites and implications for the bulk composition of terrestrial planets. Earth Planet Sci Lett 273:48–57

Brey GP, Bulatov VK, Girnis AV, Lahaye Y (2008) Experimental melting of carbonated peridotite at 6–10 GPa. J Petrol 49:797–821

Chakhmouradian AR, Williams CT (2004) Mineralogy of high-field-strength elements (Ti, Nb, Zr, Ta, Hf) in phoscoritic and carbonatitic rocks of the Kola Peninsula, Russia. In: Wall F, Zaitsev AN (eds) Phoscorites and carbonatites from mantle to mine: the key example of the Kola Alkaline Province, vol 10. Mineralogical Society of Great Britain, London, pp 293–340

Chauvel C, Blichert-Toft J (2001) A hafnium isotope and trace element perspective on melting of the depleted mantle. Earth Planet Sci Lett 190:137–151

Coussaert N (2005) Evaluation du degré d'équilibre dans les péridotites mantéliques du Lesotho. PhD Thesis, ULB, Université de la Rochelle, p 321

Davies GR, Spriggs AJ, Nixon PH (2001) A non cognate origin for the Gibeon kimberlite megacryst suite, Namibia: implications for the origin of Namibian kimberlites. J Petrol 42:159–172

Dawson JB, Hill PG, Kinny PD (2001) Mineral chemistry of a zircon-bearing, composite, veined and metasomatised upper-mantle peridotite xenolith from kimberlite. Contrib Miner Petrol 140:720–733

Debaille V, Yin Q-Z, Brandon AD, Jacobsen B (2008) Martian mantle mineralogy investigated by the $^{176}Lu^{-176}Hf$ and $^{147}Sm^{-143}Nd$ systematics of shergottites. Earth Planet Sci Lett 269:186–199

Demaiffe D, Fieremans M (1981) Strontium isotopic geochemistry of the Mbuji-Mayi and Kundelungu kimberlites (Zaire, Central Africa). Chem Geol 31:311–323

Demaiffe D, Fieremans M, Fieremans C (1991) The kimberlites of Central Africa: a review. In: Kampunzu AB, Lubala RT (eds) Magmatism in extensional structural settings. The Phanerozoic African Plate. Springer, New York, pp 537–559

Eggler DH, Mccallum ME, Smith CB (1979) Megacryst assemblage in kimberlite from Northern Colorado and Southern Wyoming: petrology, geothermometry-barometry and a real distribution. In: Boyd FR, Meyer HOA (eds) Proceedings of the 2nd international kimberlite conference, AGU, pp 213–226

El Fadili S, Demaiffe D (1999) Petrology of eclogite and granulite nodules from the Mbuji-Mayi kimberlites (Kasai, Congo): the significance of kyanite-omphacite intergrowths. In: Gurney JJ, Gurney JL, Pascoe MD, Richardson SH (eds) Proceedings of the 7th international kimberlite conference, National Book Printer, Cape Town. Goodwood, South Africa, pp 205–213

Fieremans M, Ottenburgs R (1979) The occurrence of zircon and baddeleyite crystals in the kimberlites formations at Mbuji-Mayi (Bakwanga, Zaïre). Bull Soc Belge Geol:25–31

Fieremans M, Hertogen J, Demaiffe D (1984) Petrography, geochemistry and strontium isotopic composition of the Mbuji-Mayi and Kundelungu kimberlites (Zaire). In: Kornprobst J (ed) Kimberlites I: Kimberlites and related rocks. Elsevier, Amsterdam, pp 107–120

Foster RP, Leahy K, Hunns SR, Pelham DA, Lawrence SR, Harrison AE (2001) Pan-African terranes: realizing the metal potential. Trans Instn Min Metall (Sect B: Appl Earth Sci) 110:B15–B23

Garrison JR, Taylor LA (1980) Megacryst and xenoliths in kimberlite, Elliott County, Kentucky: a mantle sample from beneath the Permian Appalachian Plateau. Contrib Miner Petrol 75:27–42

Goolaerts A, Mattielli N, de Jong J, Weis D, Scoates JS (2004) Hf and Lu isotopic reference values for the zircon standard 91500 by MC-ICP-MS. Chem Geol 206:1–9

Green TH, Blundy JD, Adam J, Yaxley GM (2000) SIMS determination of trace element partition coefficients between garnet, clinopyroxene and hydrous basaltic liquids at 2–7.5 GPa and 1080–1200 °C. Lithos 53:165–187

Grégoire M, Bell DR, le Roex AP (2003) Garnet lherzolites from the Kaapvaal craton (South Africa): trace element evidence for a metasomatic history. J Petrol 44:629–657

Griffin WL, Pearson NJ, Belousova E, Jackson SE, van Achterbergh E, O'Reilly SY, Shee SR (2000) The Hf isotope composition of cratonic mantle: LAM-MC-ICPMS analyses of zircon megacrysts in kimberlites. Geochim Cosmochim Acta 64:133–147

Harte B, Hunter RH, Kinny PD (1993) Melt geometry, movement and crystallization, in relation to mantle dykes, veins and metasomatism. Phil Trans Roy Soc London 342:1–21

Hauri EH, Wagner TP, Grove TL (1994) Experimental and natural partitioning of Th, U, Pb and other trace elements between garnet, clinopyroxene and basaltic melts. Chem Geol 117:149–166

Heaman LM, Lecheminant AN (1993) Paragenesis and U-Pb systematics of baddeleyite (ZrO_2). Chem Geol 110:95–126

Hops JJ, Gurney JJ, Harte B (1992) The Jagersfontein Cr-poor megacryst suite: towards a model for megacryst petrogenesis. J Volc Geotherm Res 50:143–160

Ionov D, Blichert-Toft J, Weis D (2005) Hf isotope compositions and HREE variations in off-craton garnet and spinel peridotite xenoliths from central Asia. Geochim Cosmochim Acta 69:2399–2418

Janney PE, le Roex AP, Carlson RW, Viljoen KS (2002) A chemical and multi-isotope study of the Western Cape olivine Melilite Province, South Africa: implications for the sources of kimberlites and the origin of the HIMU signature in Africa. J Petrol 43:2339–2370

Jones RA (1987) Strontium and neodymium isotope and rare earth element evidence for the genesis of megacrysts in kimberlites of Southern Africa. In: Nixon PH (ed) Mantle xenoliths. Wiley, Chichester, pp 711–724

Kampata MD (1993) Minéralogie et géochimie des kimberlites du Haut plateau du Kundelungu (Shaba, Zaïre). PhD Thesis, Université Catholique de Louvain-La-Neuve, p 248

Kampata MD, Moreau J, Hertogen J, Demaiffe D, Condliffe E, Mvuemba NF (1995) Megacrysts and ultramafic xenoliths from Kundelungu kimberlites (Shaba, Zaire). Miner Mag 59:661–676

Kapustin YL (1980) Mineralogy of carbonatites. Amerind Publishing Co. Pvt. Ltd., New Delhi, p 259

Keshav S, Corgne A, Gudfinnsson GH, Bizimis M, McDonough WF, Fei Y (2005) Kimberlite petrogenesis: insights from clinopyroxene-melt partitioning experiments at 6 GPa in the CaO-MgO-Al_2O_3-SiO_2-CO_2 system. Geochim Cosmochim Acta 69:2829–2845

Konzett J, Armstrong RA, Gunther D (2000) Modal metasomatism in the Kaapvaal craton lithosphere: constraints on timing and genesis from U-Pb zircon dating of metasomatized peridotites and MARID-type xenoliths. Contrib Miner Petrol 139:704–719

Kopylova MG, Nowell GM, Pearson DG, Markovic G (2009) Crystallization of megacrysts from protokimberlitic fluids: geochemical evidence from high-Cr megacrysts in the Jericho kimberlite. Lithos 112S:284–295

Kostrovitsky SI, Mitchell RH, Ivanova RN, Suvorova LF (1997) Trends of variability of garnet megacryst composition from diamond-bearing and diamond-free kimberlite pipes (Yakutia, Russia). Russ Geol Geophys 38:444–453

Kostrovitsky SI, Morikiyo T, Serov IV, Yakovlev DA, Amirzhanov AA (2007) Isotope-geochemical systematics of kimberlites and related rocks from the Siberian Platform. Russ Geol Geophys 48:272–290

Kresten P, Fels P, Berggren G (1975) Kimberlitic zircons: a possible aid in prospecting for kimberlites. Miner Deposita 10:47–56

le Roex AP, Bell DR, Davis P (2003) Petrogenesis of group I kimberlites from Kimberley, South Africa: evidence from bulk-rock geochemistry. J Petrol 44:2261–2286

Lumpkin GR (1999) Physical and chemical characteristics of baddeleyite (monoclinic zirconia) in natural environments: an overview and case study. J Nuclear Materials 274:206–217

Malarkey J, Pearson DG, Kjarsgaard BA, Davidson JP, Nowell GM, Ottley CJ, Stammer J (2010) From source to crust: tracing magmatic evolution in a kimberlite and a melilitite using microsample geochemistry. Earth Planet Sci Lett 299:80–90

McDonough WF, Sun S (1995) The composition of the Earth. Chem Geol 120:223–254

Menzies MA, Hawkesworth CJ (1987) Mantle metasomatism. Academic Press, London, p 472

Moore A, Belousova E (2005) Crystallization of Cr poor and Cr rich megacryst suites from the host kimberlite magma: implications for mantle structure and the generation of kimberlite magmas. Contrib Miner Petrol 149:462–481

Moore RO, Griffin WL, Gurney JJ, Ryan CG, Cousens DR, Sie SH, Suter GF (1992) Trace element geochemistry of ilmenite megacrysts from the Monastery kimberlite, South Africa. Lithos 29:1–18

Nixon PH, Boyd FR (1973) The discrete nodule (megacryst) association in kimberlites from Northern Lesotho. In: Nixon PH (ed) Lesotho kimberlites. Lesotho National Development Corporation, Maseru, pp 67–76

Nowell GM, Pearson DG, Kempton PD, Noble SR, Smith CB (1999) Origins of kimberlites: a Hf isotope perspective. In: Gurney JJ, Gurney JL, Pascoe MD, Richardson SH (eds) Proceedings of the 7th international kimberlite conference, National Book Printer, Cape Town. Goodwood, South Africa, pp 616–624

Nowell GM, Pearson DG, Bell DR, Carlson RW, Smith CB, Kempton PD, Noble SR (2004) Hf isotope systematics of kimberlites and their megacrysts: new constraints on their source regions. J Petrol 45:1583–1612

Odling NWA (1995) An experimental replication of upper-mantle metasomatism. Nature 373:58–60

Page FZ, Fu B, Kita NT, Fournelle J, Spicuzza MJ, Schulze DJ, Viljoen F, Basei MAS, Valley JW (2007) Zircons from kimberlite: new insights from oxygen isotopes, trace elements, and Ti in zircon thermometry. Geochim Cosmochim Acta 71:3887–3903

Paton C, Hergt JM, Woodhead JD, Phillips D, Shee SR (2009) Identifying the asthenospheric component of kimberlite magmas from the Dharwar Craton, India. Lithos 112S:296–310

Pearson DG, Shirey SB, Carlson RW, Boyd FR, Pokhilenko NP, Shimizu N (1995) Re-Os, Sm-Nd, and Rb-Sr isotope evidence for thick Archaean lithospheric mantle beneath the Siberian craton modified by multistage metasomatism. Geochim Cosmochim Acta 59:959–977

Pearson DG, Nowell GM, Kjarsgaard BA, Dowall DP (2008) The genesis of kimberlite: geochemical constraints. In: Extended abstract from the 9th international kimberlite conference, 9IKC-A-00149

Pivin M, Féménias O, Demaiffe D (2009) Metasomatic mantle origin for Mbuji-Mayi and Kundelungu garnet and clinopyroxene megacrysts (Democratic Republic of Congo). Lithos 112S:951–960

Pivin M, Berger J, Demaiffe D (2011) Nature and origin of an exceptional Cr-rich kyanite-bearing clinopyroxenite xenolith from Mbuji-Mayi kimberlite (DRC). Eur J Miner 23:257–268

Pokhilenko NP (2009) Polymict breccia xenoliths: evidence for the complex character of kimberlite formation. Lithos 112S:934–941

Raber E, Haggerty SE (1979) Zircon-oxide reactions in diamond-bearing kimberlites. In: Boyd FR, Meyer HOA (eds) Proceedings of the 2nd international kimberlite conference, AGU, pp 229–240

Rehfeldt T, Foley SF, Jacob DE, Carlson RW, Lowry D (2008) Contrasting types of metasomatism in dunite, wehrlite and websterite xenoliths from Kimberley, South Africa. Geochim Cosmochim Acta 72:5722–5756

Reischmann T, Brügmann GE, Jochum KP, Todt WA (1995) Trace element and isotopic composition of baddeleyite. Miner Petrol 53:155–164

Richardson SH, Erlank AJ, Hart S (1985) Kimberlite-borne garnet peridotite xenoliths from old enriched subcontinental lithosphere. Earth Planet Sci Lett 75:116–128

Rubatto D, Scambelluri M (2003) U-Pb dating of magmatic zircon and metamorphic baddeleyite in the Ligurian eclogites (Voltri Massif, Western Alps). Contrib Miner Petrol 146:341–355

Salters VJM, White WM (1998) Hf isotope constraints on mantle evolution. Chem Geol 145:447–460

Scatena-Wachel DE, Jones AP (1984) Primary baddeleyite (ZrO_2) in kimberlite from Bentfontein, South Africa. Miner Mag 48:257–261

Schärer U, Corfu F, Demaiffe D (1997) U-Pb and Lu-Hf isotopes in baddeleyite and zircon megacrysts from the Mbuji-Mayi kimberlite: constraints on the subcontinental mantle. Chem Geol 143:1–16

Schärer U, Berndt J, Deutsch A (2011) The genesis of deep-mantle xenocrystic zircon and baddeleyite megacrysts (Mbuji-Mayi kimberlite): trace-element patterns. Eur J Miner 23:241–255

Scherer E, Münker C, Mezger K (2001) Calibration of the lutetium-hafnium clock. Science 293:683–687

Schmidberger SS, Simonetti A, Francis D (2001) Sr-Nd-Pb isotope systematics of mantle xenoliths from Somerset Island kimberlites: evidence for lithosphere stratification beneath Arctic Canada. Geochim Cosmochim Acta 65:4243–4255

Schmidberger SS, Simonetti A, Francis D, Gariépy C (2002) Probing Archean lithosphere using the Lu-Hf isotope systematics of peridotite xenoliths from Somerset Island kimberlites, Canada. Earth Planet Sci Lett 197:245–259

Simon NSC, Irvine GJ, Davies GR, Pearson DG, Carlson RW (2003) The origin of garnet and clinopyroxene in "depleted" Kaapvaal peridotites. Lithos 71:289–322

Simon NSC, Carlson RW, Pearson DG, Davies GR (2007) The origin and evolution of the Kaapvaal cratonic lithospheric mantle. J Petrol 48:589–625

Spetsius ZV, Belousova EA, Griffin WL, O'Reilly SY, Pearson NJ (2002) Archean sulfide inclusions in Paleozoic zircon megacrysts from the Mir kimberlite, Yakutia: implications for the dating of diamonds. Earth Planet Sci Lett 199:111–126

Tappe S, Steenfelt A, Heaman LM, Simonetti A (2009) The newly discovered Jurassic Tikiusaaq carbonatite-aillikite occurrence, West Greenland, and some remarks on carbonatite-kimberlite relationships. Lithos 112S:385–399

Tappe S, Pearson DG, Nowell G, Nielsen T, Milstead P, Muehlenbachs K (2011) A fresh isotopic look at Greenland kimberlites: cratonic mantle lithosphere imprint on deep source signal. Earth Planet Sci Lett 305:235–248

Tappe S, Steenfelt A, Nielsen T (2012) Asthenospheric source of Neoproterozoic and Mesozoic kimberlites from the North Atlantic craton, West Greenland: new high-precision U-Pb and Sr-Nd isotope data on perovskite. Chem Geol 320–321:113–127

van Achterbergh E, Griffin WL, Stiefenhofer J (2001) Metasomatism in mantle xenoliths from the Letlhakane kimberlites: estimation of element fluxes. Contrib Miner Petrol 141:397–414

Vervoort JD, Blichert-Toft J (1999) Evolution of the depleted mantle: Hf isotope evidence from juvenile rocks through time. Geochim Cosmochim Acta 63:533–556

Vervoort JD, Patchett PJ, Blichert-Toft J, Albarède F (1999) Relationships between Lu-Hf and Sm-Nd isotopic systems in the global sedimentary system. Earth Planet Sci Lett 168:79–99

Walker RJ, Carlson RW, Shirey SB, Boyd FR (1989) Os, Sr, Nd, and Pb isotope systematics of southern African peridotite xenoliths: implications for the chemical evolution of subcontinental mantle. Geochim Cosmochim Acta 53:1583–1595

Weis D, Demaiffe D (1985) A depleted mantle source for kimberlites from Zaire: Nd, Sr and Pb isotopic evidence. Earth Planet Sci Lett 73:269–277

Williams CT (1996) The occurrence of niobian zirconolite, pyrochlore and baddeleyite in the Kovdor carbonatites complex, Kola Peninsula, Russia. Miner Mag 60:639–646

Woodhead J, Hergt J, Phillips D, Paton C (2009) African kimberlites revisited: in situ Sr-isotope analysis of groundmass perovskite. Lithos 112S:311–317

Petrology of Lamproites from the Nuapada Lamproite Field, Bastar Craton, India

N. Sahu, T. Gupta, S. C. Patel, D. B. K. Khuntia, D. Behera, K. Pande, and S. K. Das

Abstract

This work presents the first mineralogical, geochemical and $^{40}Ar/^{39}Ar$ geochronological data on hypabyssal facies lamproites near Kalmidadar and Darlimunda in the Nuapada Lamproite Field of the Bastar Craton. The Kalmidadar lamproite is a diamondiferous intrusion with surface dimension of ~ 320 m $\times$ 160 m, whereas the Darlimunda lamproite is a dyke swarm comprising clusters of several narrow (<5 m wide) and elongated bodies. Indicator mineral suite around the Kalmidadar lamproite is marked by abundance of Cr-spinel, rarity of garnet and absence of Cr-diopside and picroilmenite. Mineralogically, the Kalmidadar lamproite comprises phenocrysts of olivine (pseudomorphed by calcite and talc) and microphenocrysts of phlogopite set in a groundmass of chlorite and calcite. The phlogopite is Ti rich (5.4–7.4 wt % TiO_2), and the relationship between its Ti content and octahedral site deficiency indicates two substitution mechanisms, viz. $Ti + \square \leftrightarrow 2\,Mg$ and $Ti + 2Al \leftrightarrow Mg + 2Si$. The Darlimunda lamproites have undergone pervasive hydrothermal and/or deuteric alteration, which has resulted in complete chloritisation of phlogopite and extensive silicification of the rocks. Tiny grains of rutile and apatite are commonly scattered in the groundmass of both Kalmidadar and Darlimunda lamproites. The Nuapada lamproites have high contents of compatible elements such as V, Cr and Ni and of incompatible elements such as Ba, Zr, Nb and Hf. They also show high abundance of REE and enrichment in LREE relative to HREE. The incompatible element distribution patterns of the lamproites are marked by Nb, Sr, P, Hf and Zr anomalies relative to REE. The observed petrological and geochemical characteristics of the Nuapada lamproites are consistent with the derivation of the magma from a metasomatised subcontinental lithospheric mantle source. Whole-rock $^{40}Ar/^{39}Ar$ isotopic data yields an age of 1055 $\pm$ 10 Ma for the Nuapada lamproites.

Keywords

Lamproite • Phlogopite • Diamond • Bastar Craton • Nuapada • Kalmidadar • India

Introduction

Kimberlites and lamproites in India are known to occur in the Dharwar Craton, Bastar Craton and Bundelkhand Craton (Fig. 1). In the Bastar Craton, kimberlites are distributed in two fields, viz. Mainpur Kimberlite Field (MKF, diamondiferous) and Tokapal Kimberlite Field (TKF, barren).

N. Sahu · S. K. Das
Directorate of Geology, Government of Odisha, Bhuvigyan Bhawan, Bhubaneshwar, 751001, India

T. Gupta · S. C. Patel (✉) · D. B. K. Khuntia · K. Pande
Department of Earth Sciences, Indian Institute of Technology, Powai, Mumbai, 400076, India
e-mail: scpatel@iitb.ac.in

D. Behera
Department of Earth Sciences, Sambalpur University, Burla, Odisha 768019, India

D. G. Pearson et al. (eds.), *Proceedings of 10th International Kimberlite Conference*, Volume 1, Special Issue of the Journal of the Geological Society of India, DOI: 10.1007/978-81-322-1170-9_9,

The MKF has six intrusions, while the TKF has two intrusions (Mainkar 2011). Lamproites are known from two areas in the Bastar Craton, viz. Nuapada in Odisha (Patnaik et al. 2004; Mukhopadhyay et al. 2004) and Khadka in Chhattisgarh (Yellappa et al. 2010). The Nuapada Lamproite Field (NLF) comprises hypabyssal facies lamproite dykes at Kalmidadar and Darlimunda, and lamproitoid dykes at Amlidadar, Parkom and Darlimunda. The term lamproitoid as used in this study refers to mineralogically lamproite-like rocks in which mica is biotite instead of phlogopite, and diopside, if present, is Al-rich instead of Al-poor. The objectives of the present study are to (1) examine the major element composition of mantle-derived spinel and garnet recovered from stream sediments and soil samples in the area around Kalmidadar lamproite; (2) present the mineralogical and geochemical data on the lamproites and lamproitoids of the NLF; and (3) establish the age of the Kalmidadar and Darlimunda lamproites by $^{40}Ar/^{39}Ar$ geochronology.

Geology of the Bastar Craton

The Bastar Craton, also known as the Bhandara Craton, covers an area of ~1,80,000 km^2 in east-central India and is separated from the Singhbhum Craton in the north-east and the Dharwar Craton in the south-west by the Mesoproterozoic Mahanadi rift and Godavari rift, respectively (Fig. 1). The craton is bordered by two Proterozoic mobile belts, viz. the Satpura Mobile Belt (also referred to as the Central Indian Tectonic Zone) in north-west and the Eastern Ghats Mobile Belt (EGMB) in the south-east. Kimberlites and lamproites of the craton are confined to the eastern part, towards its contact with the EGMB.

Tonalitic–trondhjemite–granodiorite (TTG) gneisses, granitoids, supracrustal sequences and mafic dyke swarms constitute the major lithologies of the Bastar Craton. Together, they constitute the basement for several Meso–Neoproterozoic cover sequences, which are virtually unmetamorphosed. The basement TTG gneisses and

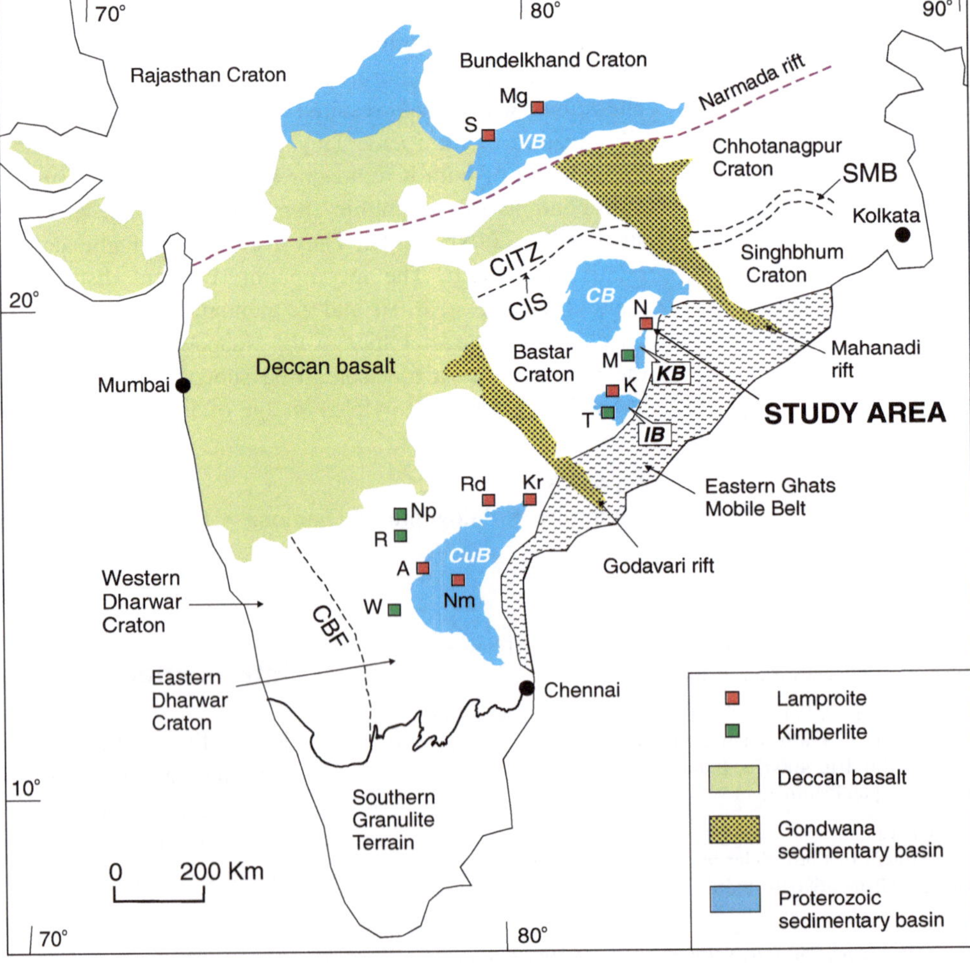

Fig. 1 Schematic geological map of India showing occurrences of kimberlites and lamproites. *A* Aliabad; *CB* Chhatisgarh basin; *CBF* Chitradurga Boundary Fault; *CIS* Central India Shear Zone; *CITZ* Central Indian Tectonic Zone; *CuB* Cuddapah basin; *IB* Indravati basin; *K* Khadka; *KB* Khariar basin; *Kr* Krishna; *M* Mainpur; *Mg* Majhgawan; *N* Nuapada; *Nm* Nallamalai; *Np* Narayanpet; *R* Raichur; *Rd* Ramadugu; *SMB* Singhbhum Mobile Belt; *S* Saptarshi; *T* Tokapal; *VB* Vindhyan basin; *W* Wajrakarur

granitoids have yielded U–Pb zircon ages of 3.5–3.6 Ga (Sarkar et al. 1993; Ghosh 2004; Rajesh et al. 2009). Granitic rocks dated to between 2,560 and 2,480 Ma (Sarkar et al. 1993; Stein et al. 2004) reflect a period of widespread granite plutonism in the craton. Supracrustal sequences range in age from Neoarchaean to Palaeoproterozoic and include rocks of varying metamorphic grades falling between greenschist and granulite facies. The craton contains two major sedimentary basins, the Chhattisgarh basin and the Indravati basin, and several minor basins including the Khariar basin which are filled with sandstone–shale–carbonate sequences. A comprehensive review of the geology of the craton can be found in Ramakrishna and Vaidyanadhan (2008).

Proterozoic mafic dyke swarms are ubiquitous within the Bastar Craton (Srivastava and Gautam (2009) and references therein). In the southern sector of the craton, the dykes have a dominant NW–SE to WNW–ESE trend parallel to the Godavari rift with subordinate N–S to NE–SW orientations. Some of these NW–SE-trending dykes are ~1.88 Gy old (French et al. 2008). In the north-eastern sector of the craton, a group of dykes, called the Lakhna dyke swarm, occurs close to the boundary of the craton with the EGMB and includes dykes of granophyre, syenite, alkali gabbro, dolerite and peridotite (Nanda et al. 2000). The NLF is located in this part of the craton. Most of the Lakhna dykes are oriented N–S, while a few trend E–W to NW–SE. The dykes have yielded ages in the range of 1450–1465 My (Ratre et al. 2010; Pisarevsky et al. 2012).

French et al. (2008) dated a mafic sill in the Cuddapah basin of the Dharwar Craton as ~1.9 Gy old and proposed that the Bastar–Cuddapah mafic igneous activity represented a large igneous province emplaced in the Dharwar and Bastar cratons at ~1.9 Ga. Palaeomagnetic results from mafic dykes lend credence to the notion of temporal and spatial link between the Dharwar and Bastar cratons by ~1.9 Ga (Meert et al. 2011) and between the Bastar and Bundelkhand cratons by ~2.0 Ga (Pradhan et al. 2012). The EGMB, which is a granulite facies, high-grade metamorphic belt, was juxtaposed against the Bastar Craton along a terrane boundary shear zone during the Pan-African orogeny at 550–500 Ma (Biswal et al. 2007; Das et al. 2008).

Nuapada Lamproite Field

The basement rocks of the NLF consist of two types of Precambrian granitoids, viz. an older grey granitoid (weakly foliated) and a younger pink granitoid (unfoliated) (Fig. 2a). The latter occupies the western part of the NLF and locally contains blocks of grey granitoid. Occasionally, gneissic xenoliths of up to several tens of metres in diameter occur within the granitoids. Two N–S-trending granophyre dykes occur in the eastern part of the study area. One of these dykes, which is located west of Saipala, is ~8 km long and has yielded a U–Pb zircon age of 1,450 ± 22 My (Ratre et al. 2010). The dyke contains a zircon xenocryst, which has yielded a Pb–Pb zircon age of 2,563 ± 22 My that could represent the intrusive age of the grey granitoid host. The other granophyre dyke occurs east of Saipala and extends over a strike length of nearly 50 km. Platformal cover rocks of the Khariar basin are exposed in the southern part of the area and include sandstone (Devdahra Sandstone) and shale (Kulharighat Formation), which constitute the basal lithounits of the Pairi Group (Das et al. 2001). Porcellanitic tuffaceous beds sandwiched between these lithounits have yielded a monazite U–Th–Pb age of 1455 ± 47 My, which is considered to mark the initial opening of the basin (Das et al. 2009).

All lamproites and lamproitoids of the NLF, except the Kalmidadar lamproite, occur as thin, elongated dykes. The Kalmidadar lamproite (KD) (82° 22′ 20″ : 20° 45′ 10″) is a broadly N–S-trending oval-shaped body with a surface dimension of ~320 m × 160 m. It is completely weathered at the surface, and weathering is persistent up to a depth of 12 m or more as revealed by drill cores. The intrusion is diamondiferous with a diamond grade of 9.95 cpht. The Darlimunda lamproite is a dyke swarm comprising several, mostly N–S-trending dykes, which are marked by isolated, patchy hardebank exposures (Fig. 2b). Mukhopadhyay et al. (2004) considered the isolated exposures as individual dykes of lamproite and reported eight such dykes (I–VIII) in the Darlimunda area. However, it appears that these small discontinuous exposures constitute parts of larger dyke systems. We have grouped the Darlimunda dykes III to VIII of Mukhopadhyay et al. (2004) into two lamproite dyke systems (DM1 and DM2). The DM1 lamproite (20° 50′ 10″ N : 82° 38′ 20″ E) is a N–S-trending dyke of ~5 m × 150 m size. The southern end of this dyke is marked by the occurrence of a small, NE–SW-trending dyke (20° 50′ 09″ N : 82° 38′ 13″ E) of ~2 m × 20 m size. This small dyke corresponds to the dyke VI, whereas the main DM1 dyke includes the dykes VII and VIII of Mukhopadhyay et al. (2004). Near the northern end of the dyke, a NNE–SSW-trending dyke of ~0.5 m × 100 m size has been found. This dyke, which was not reported by Mukhopadhyay et al. (2004), is considered to be a part of the DM1 dyke system. The DM2 lamproite (20° 50′ 00″ N : 82° 38′ 32″ E) is a broadly N–S-trending dyke with an average width of ~5 m and length of ~130 m. Hardebank outcrops are patchy and sparse in this dyke system, and the granitoid country rock is veined by thin, irregular apophyses of lamproite originating from the main intrusion. The DM2 intrusion includes the dykes III, IV and V of Mukhopadhyay et al. (2004).

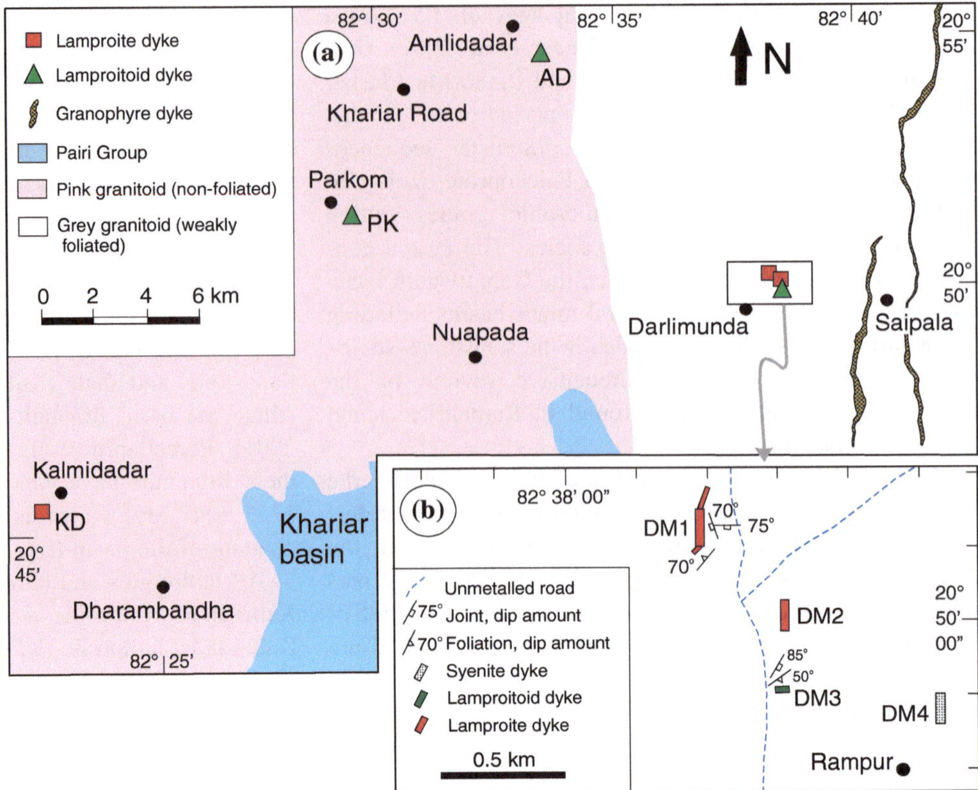

Fig. 2 **a** Generalised geological map of Nuapada Lamproite Field. *AD* Amlidadar; *KD* Kalmidadar; *PK* Parkom. **b** Map showing locations of Darlimunda dykes DM1–DM4. Dimensions of the dykes are exaggerated. Divisions of latitude and longitude are 10″ each

The Darlimunda dykes I and II of Mukhopadhyay et al. (2004) are not lamproites. While the dyke I is a lamproitoid and is named here DM3 (20° 49′ 52″ N : 82° 38′ 26″ E), the dyke II is a syenite, which is named here DM4 (20° 49′48″ N : 82°38′53″ E). The DM3 lamproitoid is an E–W-trending dyke with a dimension of ~3 m × 25 m. The DM4 syenite trends N–S and has a size of ~20 m × 120 m. The Amlidadar lamproitoid (AD) (20° 54′ 52″ N : 82° 33′ 11″ E) is a N–S-trending dyke of ~1 m × 40 m size. The Parkom lamproitoid (PK) (20° 51′ 32″ N : 82° 29′ 11″ E) is a thin (0.5 m wide), N–S-trending dyke of 15 m length. The dykes AD and PK correspond to the dykes IX and X, respectively, of Mukhopadhyay et al. (2004), who considered these bodies as lamproites. However, our mineralogical study shows that these are not lamproites, but lamproitoids.

Sampling Methods and Sample Descriptions

Indicator Minerals

A well-planned regional stream sediment sampling, follow-up sampling and large-scale geological mapping have led to the discovery of the Kalmidadar lamproite (Patnaik et al. 2004). The collected gravel and soil samples were processed for the recovery of heavy minerals. The concentrate of the size fraction −1.0 to +0.42 mm was subjected to magnetic separation. Cr-spinel was abundant in the intermediate magnetic fraction, whereas garnet was occasionally found in the non-magnetic fraction. Picroilmenite and Cr-diopside were not found in any of the samples. Cr-spinel grains are elongated to subrounded in shape with beveled edges and exhibit concave fracture and micropitted surface. They appear fresh worn to worn. Garnet grains are purple to wine red in colour with frosted to scaly frosted surface. They exhibit strong internal colour change (firing) and subconchoidal fracture. Patnaik et al. (2004) reported the compositions of 8 Cr-spinels, but did not attempt any interpretation of the data. The present study examines the compositions of 54 spinels and 9 garnets from the indicator mineral suite in and around the Kalmidadar lamproite.

Lamproites and Lamproitoids

Samples of the Kalmidadar lamproite are from bore holes (up to 30 m depth) drilled by the Directorate of Geology, Government of Odisha, whereas samples from all other lamproites and lamproitoids are from surface exposures. Petrography and whole-rock geochemistry were carried out on 8 samples including 2 from Kalmidadar lamproite, 1 from Darlimunda DM1 lamproite, 2 from Darlimunda DM2 lamproite and 1 each from Amlidadar, Parkom and

Table 1 Summary of salient petrographic features of studied samples of lamproites and lamproitoids. Kalmidadar lamproite samples are from bore holes, whereas all other samples are from surface outcrops

Sample No.	Dyke	Phenocrysts[a] (>0.5 mm)	Microphenocrysts (0.1–0.5 mm)	Groundmass	Alteration features
Kalmidadar lamproite					
KBH15/14	KD	Ol (25 %)	Phl, ol	Chl, cc, qtz, rt, tit	Ol pseudomorphed by cc and tlc. Partial chloritisation of phl
KBH42	KD	Ol (10 %)	Phl, ol	Cc, chl, qtz, rt, ap	Ol pseudomorphed by cc. Phl mostly fresh
Darlimunda lamproite					
DM1/2	DM1	Ol (5 %), phl (10 %)	Phl	Qtz, chl, rt, ap	Ol pseudomorphed by qtz and cc. Almost complete chloritisation of phl. Microscopic veins of quartz. Rutile altered to TiO_2–SiO_2 nodules
DM2/2	DM2	Ol (8 %), phl (5 %)	Phl	Qtz, chl, rt, ap	
DM2/2C	DM2	Ol (10 %)	Phl, ol	Qtz, chl, cc, rt, ap	
Darlimunda lamproitoid					
DM3/1	DM3	Ol (5 %), bt (15 %), mag (5 %)	Bt, mag	Qtz, chl, cc, ap	Ol pseudomorphed by qtz and cc. Bt completely chloritised
Amlidadar lamproitoid					
AD1/1	AD	Ol (20 %), Chl (10 %)	Ol, bt	Cc, chl, rt, ap	Ol pseudomorphed by cc and chl. Bt partly chloritised. Py present
Parkom lamproitoid					
PK1	PK	Ol (20 %), Di (5 %)	Ol, bt	Di, bt, ap	Ol pseudomorphed by tlc and mag. Bt mostly fresh

Mineral abbreviations: *Ap* apatite; *Bt* biotite; *Cc* calcite; *Chl* chlorite; *Di* diopside; *Mag* magnetite; *Ol* olivine; *Phl* phlogopite; *Py* pyrite; *Qtz* quartz; *Rt* rutile; *Tit* titanite; *Tlc* talc

[a] Modal proportions from visual estimation

Darlimunda DM3 lamproitoids. Whole-rock $^{40}Ar/^{39}Ar$ dating was carried out on two samples, one from the Kalmidadar lamproite (sample KBH15/14) and the other from the Darlimunda DM2 lamproite (sample DM2/2C). Sample details along with salient petrographic features are given in Table 1.

Mineralogically, the Kalmidadar lamproite comprises phenocrysts (0.5–1 mm) and microphenocrysts (0.1–0.5 mm) of olivine (pseudomorphed by calcite and talc) and microphenocrysts of phlogopite set in a groundmass of chlorite and calcite (Fig. 3a). The phlogopite grains show different degrees of chloritisation. The Darlimunda lamproites (DM1 and DM2) have undergone pervasive hydrothermal and/or deuteric alteration, which has resulted in almost complete chloritisation of phlogopite microphenocrysts and groundmass flakes and extensive silicification of the rocks (Fig. 3b). Phenocrysts of olivine in the Darlimunda lamproites are pseudomorphed mostly by quartz with minor amount of calcite. However, relicts of the primary textures are well-preserved, thereby providing important clues to the mineralogy and nature of the protolith. Tiny grains of rutile and apatite are commonly scattered in the groundmass of both Kalmidadar and Darlimunda lamproites. Rutile often occurs as relict within silicified micronodules of variable TiO_2–SiO_2 composition. A rare subhedral grain of zircon (0.25 mm long) was observed in a slide of the Kalmidadar lamproite sample KBH15/14. Titanite occurs locally in the groundmass of Kalmidadar lamproite. Based on the approximate modal abundances of minerals, the Kalmidadar and Darlimunda lamproites range from olivine phlogopite lamproite to phlogopite lamproite.

The Amlidadar lamproitoid is broadly similar in mineralogy to the Kalmidadar lamproite, except that (1) mica is mostly biotite instead of phlogopite; (2) secondary talc is absent; and (3) secondary pyrite is present. The Parkom lamproitoid comprises phenocrysts and microphenocrysts of olivine (pseudomorphed by talc) set in a groundmass of acicular diopside and flaky biotite (Fig. 3c). Diopside occasionally forms phenocrysts (Fig. 3d), whereas biotite locally forms microphenocrysts. Flow texture in the groundmass is observed around phenocrysts in this rock. The Darlimunda lamproitoid (DM3) contains phenocrysts of olivine, and both phenocrysts and microphenocrysts of biotite and magnetite set in a groundmass of chlorite, calcite and apatite. The rock is highly silicified, and biotite is almost completely chloritised.

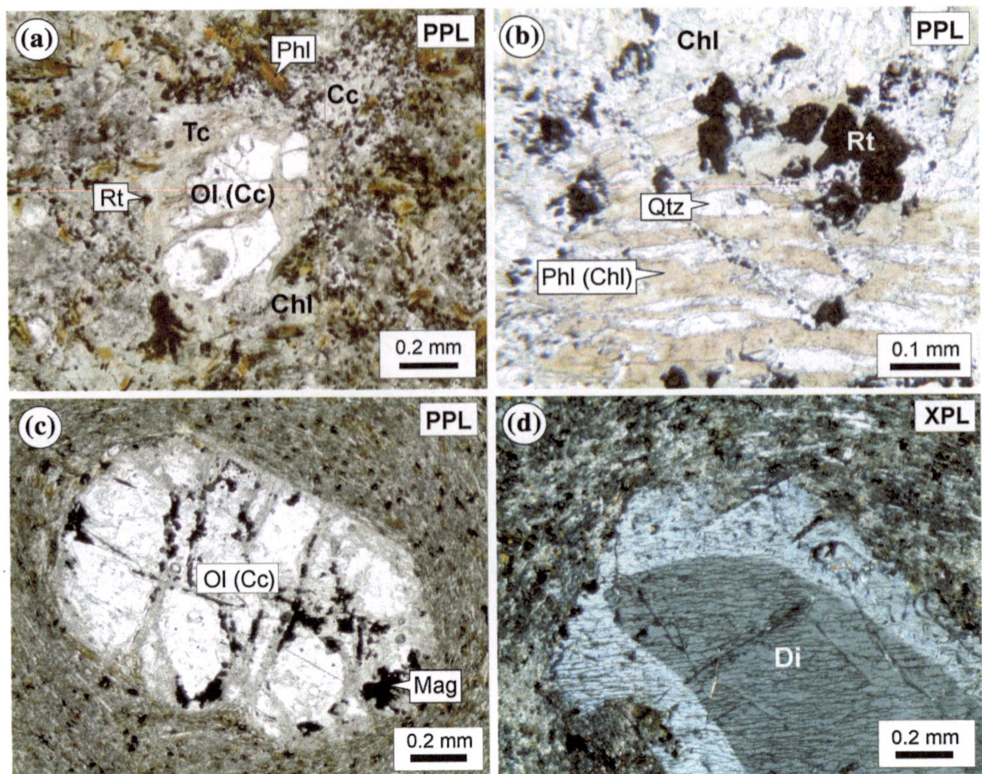

Fig. 3 Photomicrographs of under plane-polarised light (PPL) and crossed polars (XPL). **a** Kalmidadar lamproite showing olivine phenocryst pseudomorphed by calcite and talc. Phlogopite forms microphenocrysts. Groundmass comprises chlorite and calcite with scattered grains of rutile, which is altered to spongy micronodules of variable TiO_2–SiO_2 composition (sample KBH15/14). **b** Darlimunda lamproite showing alignment of phlogopite microphenocrysts, which are completely altered to chlorite. Tiny grains of rutile are scattered in the rock, but locally form clusters of up to 0.2 mm length. Secondary quartz is abundant (sample DM1/2). **c, d** Parkom lamproitoid showing olivine and diopside phenocrysts in a groundmass of acicular diopside and flaky biotite. Olivine is pseudomorphed by calcite and magnetite. Diopside phenocryst shows optical zoning under crossed polars. Circular flow layering is seen in the groundmass around phenocrysts (sample PK1). See Table 1 for mineral abbreviations

Analytical Techniques

Indicator minerals (Cr-spinel, garnet) were analysed by Oxford ISIS EDS attached to a JEOL6400 SEM at the Centre for Microscopy Characterisation and Analysis, University of Western Australia, Crawley, Western Australia. The operating conditions were 15 kV accelerating voltage and 3 nA beam current. The analyses were fully standardised using a range of natural minerals and synthetic standards. Minerals in the lamproites and lamproitoids were analysed by Cameca SX-100 electron microprobe at the Wadia Institute of Himalayan Geology, Dehradun. The operating parameters were acceleration voltage of 15 kV, probe currents of 20–50 nA and beam diameter of ~1 μm. Standards included both natural and synthetic minerals, and data reduction was done using the ZAF correction procedure.

Major elements and rare earth elements (preconcentrated via ion exchange resin) were analysed by Jobin Yvon Ultima-2 ICP-AES and other trace elements by Philips PW-2404 WD-XRF at the Indian Institute of Technology, Bombay. Visibly unaltered rock chips of up to ~5 mm size and ~100 g total weight were powdered to −230 ASTM mesh size. For analysis by ICP-AES, 0.25 g of sample powder was fused with a flux of 0.75 g lithium metaborate +0.50 g lithium tetraborate and dissolved in 1 M HCl. REE preconcentration was done in chromatographic glass column (2 cm internal diameter and 25 cm long) charged with 30 g of 200–400 mesh AG 50WX8 cation exchange resin. For analysis by WD-XRF, 4 g of sample powder and 1 g of cellulose were mixed with propanol. The mixture was evaporated under infrared lamp to dry powder, which was then pressed into pellet. The standards used for the analyses include SARM-39 (kimberlite) from MINTEK, South Africa; MY-4 (kimberlite) from IGEM, Russia; BIR-1 (basalt), BCR-2 (basalt) and BHVO-2 (basalt) from USGS, USA; and JG-1a (granodiorite) and JR-3 (rhyolite) from GSJ, Japan. Trace element concentrations in MY-4 and SARM-39 are taken from Roy et al. (2007).

Whole-rock $^{40}Ar/^{39}Ar$ dating was carried out with a Thermo Fisher ARGUS VI mass spectrometer at the Indian Institute of Technology, Bombay. Crushed rock of

120–180-μm size fraction, weighing 20–25 g, was leached with 1 % HCl solution and then cleaned with deionised water in an ultrasonic bath. About 0.2 g of sample was packaged in 0.5-mm aluminium capsule and irradiated in the light-water moderated CIRUS reactor at the Bhabha Atomic Research Centre, Mumbai, for 100 h. Irradiation flux was monitored using the Minnesota hornblende reference material (MMhb-1) with an age of 523.1 ± 2.6 Ma (Renne et al. 1998) and high-purity CaF_2 and K_2SO_4 salts. High-purity nickel wires were placed in both sample and monitor capsules to check the neutron fluence variation among the samples and the standard in the irradiation can, which was typically ~5 %. The ^{58}Co γ-activity that resulted from the Ni target during irradiation was used to monitor and correct for the J values of the samples. Argon was extracted from the irradiated samples by incremental heating up to 1,400 °C in an electrically heated ultrahigh vacuum furnace. The released gas was purified using Ti–Zr getters before isotopic measurements. The samples were analysed in multicollector mode using Faraday cups, and the checks for detector linearity were made using air shots. The average values for $^{40}Ar/^{36}Ar$ and $^{38}Ar/^{36}Ar$ for the period of analyses were 297.4 and 0.18985, respectively. Interference corrections for Ca- and K-produced Ar isotopes in the samples, based on analysis of CaF_2 and K_2SO_4 salts, were $(^{36}Ar/^{37}Ar)_{Ca} = 0.00040$, $(^{39}Ar/^{37}Ar)_{Ca} = 0.00898$ and $(^{40}Ar/^{39}Ar)_K = 0.00054$. ^{40}Ar blank contributions were 1–2 % or less for all temperature steps. The irradiation parameter J for each sample was corrected for neutron flux variation using the activity of nickel wires irradiated with the samples. The fluence-corrected J values for the samples are as follows: 0.002921 ± 0.000014 for KBH15/14 and 0.002869 ± 0.000015 for DM2/2C. The plateau ages reported are based on a minimum of 60 % of the total ^{39}Ar released and six or more successive degassing steps. Their mean ages overlap at the 2σ level including the error contribution from the J value. The Ar isotopic data were plotted using the program ISOPLOT version 2.49 (Ludwig 2001).

Results

Indicator Mineral Composition

Compositions of Cr-spinel and garnet in the Kalmidadar indicator suite are given in Tables 2 and 3, respectively. The Cr-spinels show wide variations in Cr_2O_3 (33–67 wt %, mostly above 50 wt %), Al_2O_3 (5–38 wt %.), MgO (11–18 wt %) and FeO^T (9–20 wt %). They have low TiO_2 content (<1 wt %) except for a few grains with up to 2.4 wt % TiO_2. Stoichiometric calculations following the method of Barnes and Roeder (2001) yield very low ferric iron in the Cr-spinels with $Fe^{3+}/(Cr + Al + Fe^{3+})$ ratios below 0.06. Their X_{Mg} (= $Mg/Mg + Fe^{2+}$) ratio falls in the range of 0.49–0.77, while the Cr# [= Cr/(Cr + Al)] varies from 0.52 to 0.91 except for a few grains with values between 0.38 and 0.48. Majority of the Cr-spinel population can be classified as Ti-poor, aluminous magnesiochromite. In the MgO–Cr_2O_3 plot, the Kalmidadar Cr-spinels exhibit a negatively sloping trend and extend into the diamond inclusion field consistent with their diamondiferous association (Fig. 4a). On a plot of Cr/(Cr + Al) versus $Fe^{2+}/(Mg + Fe^{2+})$, the Cr-spinels show an overall tendency of increasing Cr number with increasing $Fe^{2+}/(Mg + Fe^{2+})$ (Fig. 4b). This trend, known as the Cr–Al trend, could be a high- to low-pressure trend if spinel compositions in mantle peridotitic source were controlled by Al exchange with coexisting pyroxenes (Barnes and Roeder 2001).

The Kalmidadar garnets have Mg# [= $Mg/(Mg + Fe^T)$] in a narrow range of 0.83–0.88. Their Cr_2O_3 content varies significantly between 1.4 and 7.2 wt %, while the CaO content falls in the range of 4.6–6.1 wt %. In the MgO–CaO–Cr_2O_3 diagram of Dawson and Stephens (1975), the Kalmidadar pyropes fall in the G9 (lherzolitic garnet) field (Fig. 5a). On Cr_2O_3 versus CaO plot, the pyropes follow the typical lherzolitic trend (Fig. 5b).

Mineral Composition of Lamproites and Lamproitoids

Compositions of phlogopite in the Kalmidadar lamproite and of biotite in the Amlidadar and Parkom lamproitoids are given in Table 4, which can be downloaded from the journal website. All the micas have high TiO_2 content (5.4–8.6 wt %). Phlogopites in the Kalmidadar lamproite have Mg# in the range of 0.80–0.83, while their Al content is 8.6–11.2 wt % Al_2O_3 (Fig. 6a). The range of Mg# in the biotites of Amlidadar lamproitoid is 0.37–0.68, while that in the biotites of Parkom lamproitoid is 0.27–0.44. The Ba content of Kalmidadar phlogopites and Parkom biotites is low (<0.5 wt % BaO), whereas that of Amlidadar biotites is relatively high (2–3.6 wt % BaO). Diopside in the Parkom lamproitoid is rich in both Al (2.7−4.4 wt % Al_2O_3) and Ti (2.6–3.9 wt % TiO_2), while its Na_2O content is low (<0.5 wt %). Rutile in the lamproites and lamproitoids contains up to 2.6 wt % Nb_2O_5 and up to 0.5 wt % ZrO_2.

Mica grains in the lamproites and lamproitoids often show chemical zoning marked by a narrow Fe-rich rim relative to the core. All the micas have appreciable tetra-ferric component since their (Si + Al) content is <8 pfu (Fig. 6b). Relationship between Ti content and octahedral site deficiency indicates two important substitution mechanisms in accommodating Ti in the micas, viz. $Ti + \square \leftrightarrow 2\,Mg$ and $Ti + 2Al \leftrightarrow Mg + 2Si$ (Fig. 6c). In terms of compositional range of phlogopite, the Kalmidadar

Table 2 SEM–EDS analyses of spinels recovered during indicator mineral survey around Kalmidadar lamproite

Grain no.	1	2	3	4	5	6	7	8	9	10	11	12	13	14	15	16	17	18
SiO_2	0.10	0.00	0.12	0.09	0.00	0.19	0.03	0.00	0.02	0.11	0.01	0.09	0.05	0.00	0.05	0.00	0.06	0.08
TiO_2	0.00	0.13	0.48	0.02	0.19	0.19	0.07	0.46	0.06	1.28	0.14	0.15	0.62	0.19	0.30	0.27	0.03	0.03
Al_2O_3	11.59	23.18	11.89	11.22	23.24	9.05	14.70	12.15	19.88	9.78	15.75	9.38	9.60	19.97	11.19	25.32	17.00	10.76
Cr_2O_3	61.27	47.52	60.73	60.94	48.81	59.52	57.73	57.61	51.88	58.39	54.17	61.43	61.59	52.21	60.11	45.45	55.36	61.69
V_2O_3	0.28	0.19	0.17	0.19	0.22	0.19	0.25	0.29	0.25	0.21	0.27	0.22	0.23	0.26	0.24	0.21	0.24	0.21
Fe_2O_3	0.65	0.49	0.00	0.05	1.21	4.43	0.40	2.07	0.18	3.26	1.96	2.57	0.20	0.73	1.95	1.05	0.55	0.00
FeO	12.16	15.61	11.49	14.16	9.41	12.55	14.08	13.95	12.95	13.23	13.49	13.03	13.84	11.47	13.27	12.33	12.67	13.06
MnO	0.06	0.15	0.07	0.10	0.07	0.11	0.08	0.10	0.08	0.11	0.09	0.12	0.09	0.10	0.11	0.09	0.12	0.07
MgO	14.29	13.17	14.58	12.73	17.29	13.80	13.31	13.21	14.37	14.07	13.60	13.51	12.94	15.65	13.63	15.57	14.40	13.22
ZnO	0.01	0.24	0.12	0.00	0.17	0.00	0.16	0.02	0.01	0.09	0.04	0.04	0.11	0.00	0.05	0.08	0.20	0.08
NiO	0.09	0.05	0.12	0.10	0.10	0.17	0.11	0.09	0.11	0.13	0.08	0.11	0.14	0.11	0.13	0.11	0.09	0.10
K_2O	0.00	0.00	0.00	0.01	0.00	0.00	0.00	0.00	0.01	0.00	0.00	0.00	0.00	0.00	0.00	0.01	0.00	0.01
Na_2O	0.01	0.00	0.00	0.00	0.00	0.02	0.00	0.00	0.02	0.00	0.01	0.00	0.01	0.00	0.02	0.00	0.01	0.00
CaO	0.00	0.01	0.00	0.00	0.01	0.00	0.00	0.00	0.00	0.00	0.00	0.00	0.00	0.00	0.00	0.00	0.01	0.00
Nb_2O_5	0.02	0.10	0.00	0.00	0.06	0.09	0.00	0.02	0.02	0.00	0.00	0.00	0.00	0.02	0.00	0.01	0.06	0.01
Total	100.53	100.84	99.77	99.61	100.78	100.31	100.92	99.97	99.84	100.66	99.61	100.65	99.42	100.71	101.05	100.5	100.8	99.32
Cations normalised to 3																		
Si	0.003	0.000	0.004	0.003	0.000	0.006	0.001	0.000	0.001	0.004	0.000	0.003	0.002	0.000	0.002	0.000	0.002	0.003
Ti	0.000	0.003	0.011	0.000	0.004	0.005	0.002	0.011	0.001	0.031	0.003	0.004	0.015	0.004	0.007	0.006	0.001	0.001
Al	0.433	0.831	0.446	0.428	0.812	0.344	0.545	0.459	0.722	0.369	0.587	0.356	0.369	0.714	0.419	0.889	0.620	0.411
Cr	1.537	1.143	1.527	1.559	1.144	1.519	1.435	1.461	1.264	1.478	1.353	1.563	1.588	1.253	1.511	1.070	1.355	1.580
V	0.007	0.005	0.004	0.005	0.005	0.005	0.006	0.007	0.006	0.005	0.007	0.006	0.006	0.006	0.006	0.005	0.006	0.005
$Fe^{3+,*}$	0.016	0.011	0.000	0.001	0.027	0.108	0.009	0.050	0.004	0.079	0.047	0.062	0.005	0.017	0.047	0.024	0.013	0.000
Fe^{2+}	0.323	0.397	0.306	0.383	0.233	0.339	0.370	0.374	0.334	0.354	0.356	0.351	0.377	0.291	0.353	0.307	0.328	0.354
Mn	0.002	0.004	0.002	0.003	0.002	0.003	0.002	0.003	0.002	0.003	0.002	0.003	0.002	0.003	0.003	0.002	0.003	0.002
Mg	0.676	0.597	0.691	0.614	0.764	0.664	0.624	0.632	0.660	0.672	0.641	0.648	0.629	0.708	0.646	0.691	0.664	0.638
Zn	0.000	0.005	0.003	0.000	0.004	0.000	0.004	0.000	0.000	0.002	0.001	0.001	0.003	0.000	0.001	0.002	0.005	0.002
Ni	0.002	0.001	0.003	0.003	0.002	0.004	0.003	0.002	0.003	0.003	0.002	0.003	0.004	0.003	0.003	0.003	0.002	0.003
K	0.000	0.000	0.000	0.000	0.000	0.000	0.000	0.000	0.000	0.000	0.000	0.000	0.000	0.000	0.000	0.000	0.000	0.000
Na	0.001	0.000	0.000	0.000	0.000	0.001	0.000	0.000	0.001	0.000	0.001	0.000	0.001	0.000	0.001	0.000	0.001	0.000
Ca	0.000	0.000	0.000	0.000	0.000	0.000	0.000	0.000	0.000	0.000	0.000	0.000	0.000	0.000	0.000	0.000	0.000	0.000
Nb	0.000	0.001	0.000	0.000	0.001	0.001	0.000	0.000	0.000	0.000	0.000	0.000	0.000	0.000	0.000	0.000	0.001	0.000
Total	3.000	3.000	2.996	3.000	3.000	3.000	3.000	3.000	3.000	3.000	3.000	3.000	3.000	3.000	3.000	3.000	3.000	2.999

(continued)

Table 2 (continued)

Grain no.	1	2	3	4	5	6	7	8	9	10	11	12	13	14	15	16	17	18
$Fe^{2+}/Mg+Fe^{2+}$	0.32	0.40	0.31	0.38	0.23	0.34	0.37	0.37	0.34	0.35	0.36	0.35	0.37	0.29	0.35	0.31	0.33	0.36
Cr/Cr+Al	0.78	0.58	0.77	0.78	0.58	0.82	0.72	0.76	0.64	0.80	0.70	0.81	0.81	0.64	0.78	0.55	0.69	0.79
Grain no.	19	20	21	22	23	24	25	26	27	28	29	30	31	32	33	34	35	36
SiO_2	0.03	0.04	0.00	0.00	0.06	0.00	0.01	0.00	0.03	0.09	0.07	0.04	0.12	0.03	0.05	0.01	0.04	0.01
TiO_2	0.03	0.01	0.08	0.38	0.13	0.11	0.07	0.04	0.14	0.18	0.69	0.07	0.31	0.16	0.22	0.68	0.31	0.21
Al_2O_3	13.82	14.38	26.23	24.49	10.96	37.38	25.71	33.61	14.66	11.06	8.71	14.36	10.82	26.05	36.38	18.31	14.01	27.06
Cr_2O_3	57.33	58.08	44.22	45.29	60.94	33.60	44.76	36.91	55.79	60.03	60.65	56.97	59.86	44.26	33.16	50.48	57.48	43.57
V_2O_3	0.24	0.23	0.19	0.19	0.21	0.19	0.25	0.20	0.25	0.23	0.20	0.24	0.29	0.15	0.15	0.38	0.26	0.18
Fe_2O_3	1.83	0.00	1.13	0.99	0.52	0.35	1.17	0.41	1.76	1.38	1.57	1.48	0.98	0.97	1.35	2.12	0.71	0.81
FeO	12.75	13.69	13.57	14.02	13.75	9.61	13.94	11.89	14.62	13.09	14.54	12.69	14.15	11.91	10.41	15.37	13.77	12.62
MnO	0.12	0.10	0.11	0.11	0.09	0.06	0.12	0.12	0.11	0.09	0.10	0.10	0.07	0.09	0.07	0.13	0.10	0.12
MgO	13.95	13.07	14.72	14.22	12.95	18.36	14.39	16.34	12.80	13.51	12.41	14.08	12.67	15.71	17.77	13.15	13.47	15.35
ZnO	0.05	0.06	0.14	0.26	0.12	0.14	0.20	0.19	0.23	0.12	0.05	0.09	0.13	0.07	0.02	0.12	0.10	0.10
NiO	0.10	0.08	0.08	0.07	0.09	0.12	0.08	0.10	0.11	0.10	0.14	0.08	0.13	0.16	0.21	0.08	0.09	0.11
K_2O	0.00	0.00	0.00	0.00	0.00	0.00	0.00	0.00	0.00	0.00	0.00	0.01	0.00	0.00	0.01	0.00	0.00	0.01
Na_2O	0.00	0.01	0.00	0.02	0.02	0.00	0.02	0.02	0.01	0.00	0.01	0.00	0.04	0.00	0.02	0.00	0.01	0.03
CaO	0.00	0.00	0.00	0.02	0.01	0.02	0.00	0.00	0.00	0.00	0.01	0.00	0.01	0.00	0.00	0.00	0.01	0.00
Nb_2O_5	0.00	0.00	0.00	0.02	0.00	0.00	0.00	0.02	0.00	0.03	0.02	0.02	0.00	0.05	0.00	0.00	0.00	0.00
Total	100.25	99.75	100.47	100.08	99.85	99.94	100.72	99.85	100.51	99.91	99.17	100.23	99.58	99.61	99.82	100.83	100.36	100.18
Cations normalised to 3																		
Si	0.001	0.001	0.000	0.000	0.002	0.000	0.000	0.000	0.001	0.003	0.002	0.001	0.004	0.001	0.001	0.000	0.001	0.000
Ti	0.001	0.000	0.002	0.009	0.003	0.002	0.002	0.001	0.003	0.004	0.017	0.002	0.008	0.004	0.005	0.016	0.007	0.005
Al	0.515	0.539	0.923	0.873	0.417	1.238	0.906	1.143	0.547	0.419	0.338	0.533	0.414	0.918	1.214	0.670	0.522	0.947
Cr	1.433	1.460	1.044	1.083	1.556	0.746	1.059	0.842	1.397	1.526	1.579	1.419	1.535	1.046	0.742	1.239	1.438	1.023
V	0.006	0.006	0.005	0.005	0.005	0.004	0.006	0.005	0.006	0.006	0.005	0.006	0.008	0.004	0.003	0.009	0.007	0.004
$Fe^{3+,*}$	0.044	0.000	0.025	0.023	0.013	0.007	0.026	0.009	0.042	0.033	0.039	0.035	0.024	0.022	0.029	0.050	0.017	0.018
Fe^{2+}	0.337	0.364	0.339	0.355	0.371	0.226	0.349	0.287	0.387	0.352	0.400	0.334	0.384	0.298	0.247	0.399	0.364	0.313
Mn	0.003	0.003	0.003	0.003	0.002	0.001	0.003	0.003	0.003	0.002	0.003	0.003	0.002	0.002	0.002	0.003	0.003	0.003
Mg	0.657	0.619	0.655	0.641	0.623	0.769	0.642	0.703	0.604	0.648	0.609	0.661	0.613	0.700	0.750	0.609	0.635	0.679
Zn	0.001	0.001	0.003	0.006	0.003	0.003	0.004	0.004	0.005	0.003	0.001	0.002	0.003	0.002	0.000	0.003	0.002	0.002
Ni	0.003	0.002	0.002	0.002	0.002	0.003	0.002	0.002	0.003	0.003	0.004	0.002	0.003	0.004	0.005	0.002	0.002	0.003
K	0.000	0.000	0.000	0.000	0.000	0.000	0.000	0.000	0.000	0.000	0.000	0.000	0.000	0.000	0.000	0.000	0.000	0.000

(continued)

Table 2 (continued)

Grain no.	19	20	21	22	23	24	25	26	27	28	29	30	31	32	33	34	35	36
Na	0.000	0.001	0.000	0.001	0.001	0.000	0.001	0.001	0.001	0.000	0.001	0.000	0.003	0.000	0.001	0.000	0.001	0.002
Ca	0.000	0.000	0.000	0.001	0.000	0.001	0.000	0.000	0.000	0.000	0.000	0.000	0.000	0.000	0.000	0.000	0.000	0.000
Nb	0.000	0.000	0.000	0.000	0.000	0.000	0.000	0.000	0.000	0.000	0.000	0.000	0.000	0.001	0.000	0.000	0.000	0.000
Total	3.000	2.996	3.000	3.000	3.000	3.000	3.000	3.000	3.000	3.000	3.000	3.000	3.000	3.000	3.000	3.000	3.000	3.000
$Fe^{2+}/Mg+Fe^{2+}$	0.34	0.37	0.34	0.36	0.37	0.23	0.35	0.29	0.39	0.35	0.40	0.34	0.39	0.30	0.25	0.40	0.36	0.32
Cr/Cr+Al	0.74	0.73	0.53	0.55	0.79	0.38	0.54	0.42	0.72	0.78	0.82	0.73	0.79	0.53	0.38	0.65	0.73	0.52
Grain no.	37	38	39	40	41	42	43	44	45	46	47	48	49	50	51	52	53	54
SiO_2	0.13	0.07	0.12	0.07	0.03	0.03	0.01	0.08	0.06	0.03	0.15	0.18	0.03	0.12	0.10	0.12	0.39	0.03
TiO_2	0.13	0.03	0.03	0.26	0.13	0.01	0.24	0.03	0.05	2.37	1.20	0.09	0.37	0.28	0.14	0.37	0.30	0.01
Al_2O_3	10.24	12.56	10.34	10.92	36.18	12.35	29.07	13.78	13.91	11.59	9.66	10.79	12.83	10.48	9.82	9.44	4.73	25.71
Cr_2O_3	62.06	60.05	60.52	58.27	34.30	60.29	40.78	59.27	58.11	52.15	57.87	61.59	58.75	60.63	61.13	60.42	67.25	44.71
V_2O_3	0.24	0.24	0.22	0.26	0.20	0.18	0.21	0.18	0.23	0.42	0.15	0.24	0.28	0.16	0.19	0.14	0.13	0.18
Fe_2O_3	1.25	1.10	2.36	2.94	0.79	0.68	1.26	0.96	1.20	3.50	3.59	0.00	0.00	0.39	2.05	2.50	2.20	1.34
FeO	12.42	12.14	12.46	15.26	10.02	12.47	12.17	10.03	12.03	19.50	13.21	13.43	14.41	13.90	13.15	13.65	11.12	13.45
MnO	0.08	0.10	0.12	0.15	0.07	0.10	0.09	0.05	0.14	0.19	0.09	0.05	0.08	0.08	0.08	0.08	0.08	0.13
MgO	14.00	14.32	13.79	12.24	17.93	13.98	15.95	15.85	14.42	10.55	13.89	12.82	12.73	12.78	13.43	13.17	14.78	14.74
ZnO	0.10	0.13	0.10	0.02	0.14	0.00	0.14	0.05	0.11	0.22	0.07	0.07	0.15	0.03	0.08	0.00	0.00	0.11
NiO	0.11	0.15	0.14	0.09	0.18	0.11	0.11	0.13	0.12	0.17	0.17	0.13	0.10	0.13	0.12	0.12	0.16	0.08
K_2O	0.00	0.00	0.01	0.00	0.00	0.02	0.01	0.00	0.01	0.00	0.00	0.00	0.00	0.00	0.00	0.01	0.00	0.03
Na_2O	0.02	0.00	0.00	0.00	0.02	0.01	0.00	0.00	0.00	0.01	0.02	0.00	0.01	0.03	0.00	0.02	0.00	0.00
CaO	0.00	0.01	0.00	0.00	0.00	0.00	0.00	0.00	0.03	0.00	0.00	0.00	0.00	0.01	0.00	0.01	0.00	0.00
Nb_2O_5	0.00	0.00	0.00	0.00	0.00	0.00	0.04	0.03	0.00	0.00	0.00	0.00	0.00	0.00	0.03	0.08	0.05	0.07
Total	100.78	100.9	100.21	100.48	99.99	100.23	100.08	100.44	100.42	100.7	100.07	99.39	99.74	99.02	100.32	100.13	101.19	100.59
Cations normalised to 3																		
Si	0.004	0.002	0.004	0.002	0.001	0.001	0.000	0.002	0.002	0.001	0.005	0.006	0.001	0.004	0.003	0.004	0.013	0.001
Ti	0.003	0.001	0.001	0.006	0.003	0.000	0.005	0.001	0.001	0.058	0.029	0.002	0.009	0.007	0.003	0.009	0.007	0.000
Al	0.385	0.466	0.391	0.416	1.206	0.463	1.007	0.506	0.515	0.444	0.367	0.412	0.485	0.403	0.373	0.360	0.180	0.905
Cr	1.566	1.496	1.537	1.489	0.767	1.515	0.948	1.460	1.444	1.342	1.475	1.578	1.491	1.564	1.558	1.548	1.721	1.056
V	0.006	0.006	0.006	0.007	0.005	0.005	0.005	0.004	0.006	0.011	0.004	0.006	0.007	0.004	0.005	0.004	0.003	0.004
$Fe^{3+,*}$	0.030	0.026	0.057	0.071	0.017	0.016	0.028	0.023	0.028	0.086	0.087	0.000	0.000	0.010	0.050	0.061	0.054	0.030
Fe^{2+}	0.331	0.320	0.335	0.412	0.237	0.331	0.299	0.261	0.316	0.531	0.356	0.364	0.387	0.379	0.355	0.370	0.301	0.336
Mn	0.002	0.003	0.003	0.004	0.002	0.003	0.002	0.001	0.004	0.005	0.002	0.001	0.002	0.002	0.002	0.002	0.002	0.003

(continued)

Table 2 (continued)

Grain no.	37	38	39	40	41	42	43	44	45	46	47	48	49	50	51	52	53	54
Mg	0.666	0.673	0.660	0.590	0.756	0.662	0.699	0.736	0.676	0.512	0.668	0.619	0.609	0.621	0.645	0.636	0.713	0.657
Zn	0.002	0.003	0.002	0.000	0.003	0.000	0.003	0.001	0.003	0.005	0.002	0.002	0.004	0.001	0.002	0.000	0.000	0.002
Ni	0.003	0.004	0.004	0.002	0.004	0.003	0.003	0.003	0.003	0.004	0.004	0.003	0.003	0.003	0.003	0.003	0.004	0.002
K	0.000	0.000	0.000	0.000	0.000	0.001	0.000	0.000	0.000	0.000	0.000	0.000	0.000	0.000	0.000	0.000	0.000	0.001
Na	0.001	0.000	0.000	0.000	0.001	0.001	0.000	0.000	0.000	0.001	0.001	0.000	0.001	0.002	0.000	0.001	0.000	0.000
Ca	0.000	0.000	0.000	0.000	0.000	0.000	0.000	0.000	0.001	0.000	0.000	0.000	0.000	0.000	0.000	0.000	0.000	0.000
Nb	0.000	0.000	0.000	0.000	0.000	0.000	0.001	0.000	0.000	0.000	0.000	0.000	0.000	0.000	0.000	0.001	0.001	0.001
Total	3.000	3.000	3.000	3.000	3.000	3.000	3.000	3.000	3.000	3.000	3.000	2.994	2.999	3.000	3.000	3.000	3.000	3.000
$Fe^{2+}/Mg+Fe^{2+}$	0.33	0.32	0.34	0.41	0.24	0.33	0.30	0.26	0.32	0.51	0.35	0.37	0.39	0.38	0.35	0.37	0.30	0.34
Cr/Cr+Al	0.80	0.76	0.80	0.78	0.39	0.77	0.48	0.74	0.74	0.75	0.80	0.79	0.75	0.80	0.81	0.81	0.91	0.54

[a] Fe^{3+} calculated assuming stoichiometry and an ideal XY_2O_4 formula

lamproite is closer to the West Kimberly lamproites of Australia (Jaques et al. 1986; Mitchell and Bergman 1991) than to the Krishna lamproites in the Dharwar Craton of southern India (Reddy et al. 2003; Chalapathi Rao et al. 2010).

Whole-Rock Major and Trace Element Composition

The lamproites have variable SiO_2 content (44–60 wt %) (Table 5) with higher values observed in rocks that are more silicified as evidenced by the presence of quartz in olivine pseudomorphs and in microscopic veins. The K_2O content of the rocks is generally low (<0.4 wt %), which is attributed to chloritisation of phlogopite and consequent loss of K_2O to a hydrothermal fluid. The highest value of K_2O (1.34 wt %) is observed in a borehole sample from the Kalmidadar lamproite, which is rich in unaltered phlogopite. The Nuapada lamproites have highly variable contents of MgO (7.4–14.3 wt %), $Fe_2O_3^T$ (7.2–11.1 wt %), CaO (2.2–12.3 wt %), TiO_2 (2.6–7.4 wt %) and P_2O_5 (0.1–1.9 wt %), while their Al_2O_3 content falls within a narrow range of 5.3–7.2 wt %. The lamproitoids have similar ranges of SiO_2, Al_2O_3, CaO, TiO_2 and P_2O_5 as the lamproites, but are richer in $Fe_2O_3^T$ (14.6–17.2 wt %) while having relatively low but uniform value of MgO (~8.7 wt %) (Figs. 7a, b). Their K_2O content is low except for the Parkom lamproitoid (2.7 wt % K_2O), which is relatively rich in unaltered phlogopite.

The Nuapada lamproites have high contents of compatible elements such as V (215–347 ppm), Cr (257–478 ppm) and Ni (232–423 ppm) as well as of incompatible elements such as Ba (2,292–3,499 ppm), Zr (544–780 ppm), Nb (52–135 ppm) and Hf (13–19 ppm). The Nuapada lamproitoids have similar contents of V, Cr, Ba and Nb as the lamproites, but are relatively poor in Ni (166–221 ppm), Zr (218–349 ppm) and Hf (4–9 ppm). The Rb contents are highly variable with ranges of 1–38 ppm in lamproites and 1–79 ppm in lamproitoids. Low values of Rb correlate well with low K_2O contents in the rocks that resulted from petrographically observed chloritisation of phlogopite. When the fluid-mobile elements such as Rb and Ba are plotted against an immobile element such as La, the results vary from a weak negative correlation (Rb vs. La, Fig. 7c) to no correlation (Ba vs. La, Fig. 7d). The high field strength elements like Zr, Nb and Hf do not correlate with La, but show strong correlations among themselves on the plots of Nb versus Zr (Fig. 7e) and Hf versus Zr (Fig. 7f). The lamproites and lamproitoids can be clearly distinguished on these plots because of the higher values of Zr in the former. The Zr/Nb ratio in the lamproites is 7.9 ± 2.1, whereas that in the

Table 3 SEM–EDS analyses of garnets recovered during indicator mineral survey around Kalmidadar lamproite

Grain no.	1	2	3	4	5	6	7	8	9
SiO_2	42.79	42.20	42.51	42.02	41.92	41.52	41.49	41.69	41.58
TiO_2	0.14	0.12	0.16	0.20	0.16	0.16	0.06	0.25	0.20
Al_2O_3	23.35	21.85	23.19	20.00	22.05	18.36	19.87	19.19	19.53
Cr_2O_3	1.43	2.80	1.48	4.76	2.77	7.23	5.25	5.71	5.54
FeO	5.35	7.42	6.79	7.00	7.58	6.86	6.99	7.26	7.39
MnO	0.17	0.39	0.30	0.31	0.36	0.35	0.29	0.33	0.37
MgO	22.87	20.46	21.32	21.36	20.64	20.40	20.90	19.91	20.17
CaO	4.65	5.50	5.17	4.79	5.43	5.50	5.38	6.13	5.63
Na_2O	0.03	0.00	0.05	0.05	0.04	0.03	0.01	0.03	0.01
Total	100.78	100.74	100.97	100.49	100.95	100.41	100.24	100.50	100.42
Cations based on 12 oxygens									
Si	2.977	2.988	2.978	2.993	2.966	2.989	2.973	2.994	2.985
Ti	0.007	0.006	0.008	0.011	0.009	0.009	0.003	0.014	0.011
Al	1.915	1.824	1.914	1.679	1.839	1.558	1.678	1.624	1.652
Cr	0.079	0.157	0.082	0.268	0.155	0.412	0.297	0.324	0.314
Fe	0.311	0.439	0.398	0.417	0.448	0.413	0.419	0.436	0.444
Mn	0.010	0.023	0.018	0.019	0.022	0.021	0.018	0.020	0.022
Mg	2.372	2.160	2.226	2.268	2.177	2.189	2.233	2.132	2.159
Ca	0.347	0.417	0.388	0.366	0.412	0.424	0.413	0.472	0.433
Na	0.004	0.000	0.007	0.007	0.005	0.004	0.001	0.004	0.001
Total	8.021	8.015	8.019	8.026	8.032	8.020	8.036	8.020	8.022
Mg/Fe+Mg	0.88	0.83	0.85	0.84	0.83	0.84	0.84	0.83	0.83

lamproitoids is 3.3 ± 1.0. The lamproites display coherent Hf–Zr correlation with Zr/Hf = 42 ± 4, but the lamproitoids have a wide range of Zr/Hf ratios (38–71).

The lamproites show high abundance of REE ($\sum$REE = 790–1,335 ppm; average = 1,018 ppm) and enrichment in LREE relative to HREE [$(La/Yb)_N$ = 34–108] (Fig. 8a). The lamproitoids have lower average abundance of REE ($\sum$REE = 524–1,111 ppm; average = 723 ppm) than the lamproites. The REE patterns of both the rock types are broadly similar to those of Krishna lamproites (Vijesh 2010) and West Kimberly lamproites (Nixon et al. 1984; Fraser 1987; Lewis 1987). In the incompatible element distribution patterns normalised to primitive mantle (McDonough and Sun (1995), the Nuapada lamproites and lamproitoids are similar to Krishna lamproites (Vijesh 2010) and West Kimberly lamproites (Jaques et al. 1986) in terms of negative Nb, Sr and P anomalies relative to REE (Fig. 8b). However, the Nuapada lamproites and lamproitoids and Krishna lamproites are additionally marked by depletion in Rb, Hf and Zr, which is absent in the West Kimberly lamproites. While the Nuapada and Krishna lamproites have similar degrees of depletion in Hf and Zr relative to REE, the Nuapada lamproitoids exhibit somewhat higher degree of depletion in these elements. Despite the secondary silicification of rocks which has caused high silica values in some of the Nuapada lamproites and lamproitoids, their coherent REE and multielement patterns show that this contamination mostly has a dilutive effect on trace elements.

$^{40}Ar/^{39}Ar$ Dating

The complete set of analytical results for the two dated samples of lamproites is given in Table 6. The $^{40}Ar/^{39}Ar$ age spectra, isochrons and inverse isochrons of these samples are shown in Fig. 9. The $^{40}Ar/^{39}Ar$ age spectra display saddle-shaped patterns with a well-defined plateau at intermediate steps. The plateau ages for the samples KBH15/14 and DM2/2C for the intermediate temperature steps are 1,056 ± 7 My and 1,054 ± 14 My, respectively. The isochron plots of $^{40}Ar/^{36}Ar$ versus $^{39}Ar/^{36}Ar$ yield the ages of 1052 ± 17 My (KBH15/14) and 1,066 ± 300 My (DM2/2C). The ages obtained from the $^{36}Ar/^{40}Ar$ versus $^{39}Ar/^{40}Ar$ inverse isochron plots are 1,060 ± 6 My (KBH15/14) and 1,060 ± 20 My (DM2/2C). The trapped/initial argon isotopic compositions inferred from the isochron plots of both the samples are atmospheric. The concordance of the plateau and isochron ages and an atmospheric isotopic composition for the trapped

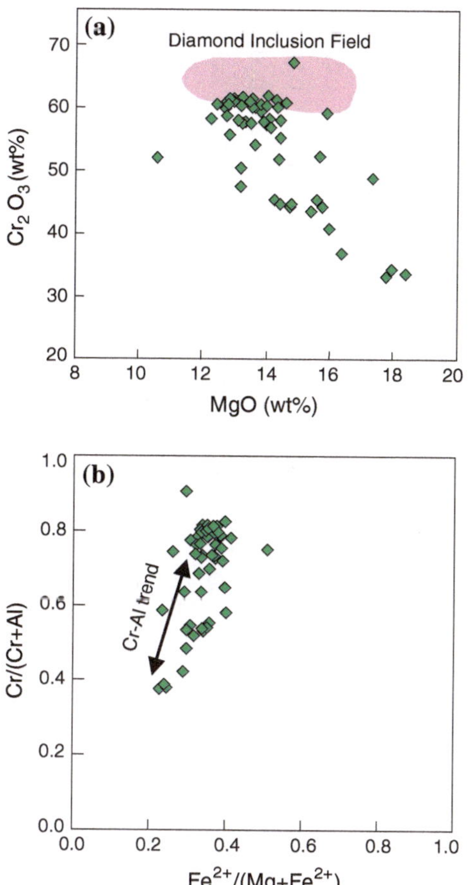

Fig. 4 Kalmidadar Cr-spinels on (**a**) MgO–Cr_2O_3 plot, with diamond inclusion field after Fipke et al. (1995); and (**b**) Cr/(Cr +Al) versus $Fe^{2+}/(Mg + Fe^{2+})$ diagram

component indicate that these could be the emplacement ages of the samples. The ages of the two samples fall in the range of 1,052–1,060 My (disregarding the isochron age of DM2/2C), with an average value of 1,055 ± 10 My, which is taken as the emplacement age of the lamproites.

Discussion

Age of Lamproite Magmatism

The $^{40}Ar/^{39}Ar$ whole-rock age of 1055 ± 10 My for the Nuapada lamproites overlaps, within error limits, with the well-constrained $^{40}Ar/^{39}Ar$ phlogopite age of 1073.5 ± 13.7 Ma (2σ) for the Majhgawan lamproite of the adjoining Bundelkhand Craton (Gregory et al. 2006). Two other ages of 1,044 ± 22 and 1,067 ± 31 Ma have been reported for the Majhgawan lamproite using Rb–Sr systematics (Smith 1992; Kumar et al. 1993, respectively), which match perfectly with the $^{40}Ar/^{39}Ar$ phlogopite age. These data lead us to believe that the $^{40}Ar/^{39}Ar$ age obtained in this study represents the time of emplacement of the Nuapada lamproites and that the lamproite magmatic events in the Bastar and Bundelkhand Cratons were contemporaneous. These events are some 50 My younger than the ca. 1,100 Ma that have been obtained for a number of kimberlites and lamproites of the Dharwar Craton (Chalapathi Rao et al. 1999; Kumar et al. 2007). The Krishna lamproites have yielded Rb–Sr phlogopite ages in the range of 1,100–1,400 My (Kumar et al. 2001) and K–Ar and $^{40}Ar/^{39}Ar$ phlogopite ages in the range of 1,400–1,500 My (Chalapathi Rao et al. 1996; Osborne et al. 2011). However, Phillips (2012) commented that the age of ~1500 Ma reported by Osborne et al. (2011) was an overestimate and that the studied lamproite was emplaced at <1,400 Ma. Hence, the likely age range of the Krishna lamproites is 1,100–1,400 My, the lower limit of which is still ~50 My older than the Nuapada and Majhgawan lamproites. The Chelima lamproite in the Nallamalai field of the Dharwar Craton, which has yielded a $^{40}Ar/^{39}Ar$ phlogopite age of 1,417 My (Chalapathi Rao et al. 1999), is significantly older than the Nuapada lamproites. The kimberlites of the Bastar Craton, which have been dated as 65 My old (Lehmann et al. 2010), are much younger than the Nuapada lamproites.

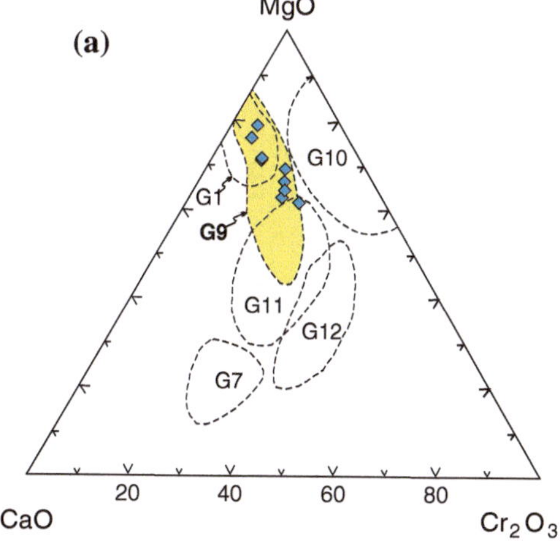

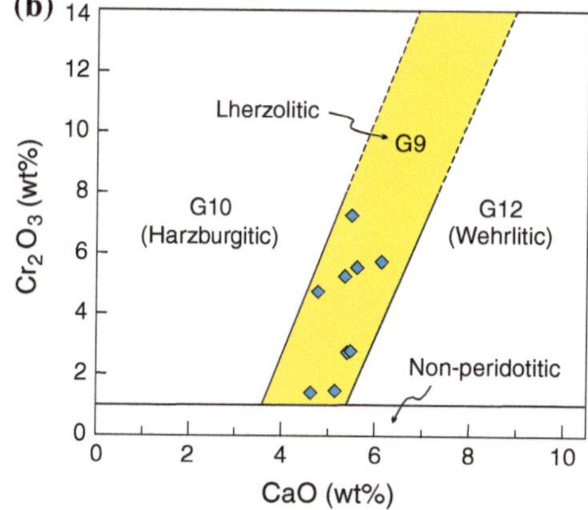

Fig. 5 Kalmidadar garnets on (**a**) MgO–CaO–Cr_2O_3 plot, with fields after Dawson and Stephens (1975); (**b**) CaO versus Cr_2O_3 plot, with fields after Grutter et al. (2004)

Table 4 Microprobe analyses of phlogopite in Kalmidadar lamproite (samples KBH15/14 and KBH42) and of biotite in Amlidadar lamproitoid (sample AM1) and Parkom lamproitoid (sample PK1). (r) and (c) indicate rim and core of grain, respectively

Sample No	KBH15/14 (Kalmidadar lamproite)														
	Grain 1					Grain 2	Grain 3		Grain 4		Grain 5		Grain 6		Grain 7
Analysis no.	1/1	1/2	1/3	1/4	1/5	2/1	3/1	3/2	4/1	4/2	5/1	5/2	6/1	6/2	7/1
SiO_2	41.07	40.59	40.66	40.11	40.07	39.82	39.31	39.56	40.26	39.80	38.33	38.64	39.05	38.74	39.01
TiO_2	6.13	6.43	5.45	6.20	5.70	6.95	5.73	5.89	6.68	6.87	7.39	7.39	6.74	6.33	7.02
Al_2O_3	9.71	8.61	10.60	10.16	9.90	10.48	11.17	10.42	10.67	10.48	10.25	9.83	10.01	10.05	10.67
Cr_2O_3	0.09	0.27	0.45	0.42	0.48	0.25	0.49	0.42	0.32	0.43	0.40	0.08	0.35	0.18	0.11
FeO	8.00	9.13	7.46	7.33	7.97	8.15	8.22	7.76	7.84	8.44	7.80	8.16	7.53	7.99	8.22
MnO	0.09	0.03	0.06	0.03	0.02	0.04	0.00	0.06	0.01	0.05	0.04	0.04	0.02	0.03	0.02
MgO	21.62	20.88	22.19	21.13	21.32	21.32	21.42	20.72	21.01	20.58	20.23	19.48	19.99	20.68	20.46
NiO	0.02	0.03	0.05	0.04	0.06	0.08	0.12	0.08	0.09	0.11	0.12	0.07	0.07	0.08	0.12
CaO	0.01	0.03	0.00	0.05	0.01	0.00	0.02	0.03	0.00	0.00	0.21	0.19	0.01	0.00	0.02
BaO	0.29	0.45	0.47	0.47	0.50	0.43	0.48	0.52	0.45	0.44	0.27	0.31	0.42	0.50	0.26
Na_2O	0.01	0.04	0.05	0.03	0.01	0.00	0.03	0.04	0.00	0.01	0.00	0.03	0.04	0.00	0.00
K_2O	10.13	10.01	9.98	10.22	10.16	9.92	9.06	9.59	9.99	9.99	9.64	9.43	9.36	9.48	9.13
Total	97.17	96.50	97.40	96.18	96.20	97.43	96.04	95.07	97.32	97.20	94.68	93.65	93.59	94.06	95.04
Cations based on 22 oxygen															
Si	5.858	5.880	5.779	5.788	5.800	5.685	5.671	5.773	5.740	5.709	5.634	5.736	5.777	5.727	5.683
Al^{total}	1.632	1.471	1.775	1.727	1.689	1.763	1.900	1.792	1.792	1.771	1.776	1.720	1.745	1.751	1.832
Fe^{3+}	0.510	0.649	0.446	0.485	0.511	0.552	0.429	0.435	0.468	0.520	0.590	0.544	0.477	0.522	0.485
Sum (T site)	8.000	8.000	8.000	8.000	8.000	8.000	8.000	8.000	8.000	8.000	8.000	8.000	8.000	8.000	8.000
Ti	0.658	0.701	0.583	0.672	0.620	0.746	0.622	0.646	0.716	0.741	0.817	0.825	0.750	0.704	0.769
Cr	0.010	0.031	0.050	0.048	0.055	0.029	0.056	0.048	0.037	0.049	0.046	0.009	0.041	0.021	0.013
Fe^{2+}	0.444	0.457	0.440	0.399	0.454	0.421	0.563	0.511	0.467	0.492	0.369	0.469	0.454	0.465	0.517
Mn	0.011	0.004	0.007	0.004	0.003	0.005	0.000	0.008	0.001	0.006	0.005	0.005	0.003	0.004	0.002
Mg	4.597	4.511	4.703	4.545	4.600	4.536	4.607	4.507	4.465	4.400	4.433	4.311	4.409	4.557	4.444
Ni	0.002	0.003	0.006	0.005	0.007	0.009	0.014	0.009	0.010	0.013	0.014	0.008	0.008	0.010	0.014
Sum (O site)	5.723	5.707	5.788	5.673	5.739	5.746	5.862	5.729	5.696	5.700	5.685	5.628	5.665	5.761	5.759
Ca	0.002	0.004	0.000	0.008	0.001	0.000	0.004	0.005	0.000	0.000	0.033	0.030	0.002	0.000	0.003
Ba	0.016	0.025	0.026	0.027	0.029	0.024	0.027	0.030	0.025	0.025	0.016	0.018	0.024	0.029	0.015
Na	0.003	0.011	0.014	0.009	0.004	0.000	0.007	0.011	0.000	0.004	0.000	0.009	0.011	0.000	0.000
K	1.843	1.851	1.809	1.881	1.875	1.806	1.667	1.785	1.816	1.828	1.808	1.786	1.766	1.788	1.697
Sum (X site)	1.864	1.891	1.849	1.924	1.908	1.830	1.705	1.830	1.841	1.856	1.856	1.843	1.804	1.817	1.715

(continued)

Table 4 (continued)

Sample No	KBH15/14 (Kalmidadar lamproite)														
	Grain 1					Grain 2	Grain 3		Grain 4		Grain 5		Grain 6		Grain 7
Analysis no.	1/1	1/2	1/3	1/4	1/5	2/1	3/1	3/2	4/1	4/2	5/1	5/2	6/1	6/2	7/1
All cations	15.586	15.598	15.638	15.597	15.648	15.576	15.566	15.559	15.538	15.556	15.541	15.471	15.469	15.577	15.474
$Mg/Mg+Fe^T$	82.81	80.31	84.14	83.72	82.66	82.35	82.29	82.65	82.68	81.30	82.22	80.97	82.55	82.19	81.61

Sample No	KBH42 (Kalmidadar lamproite)															
	Grain 1				Grain 2 (line scan from rim throgh core to rim)											
Analysis no.	1/1 (r)	1/2 (r)	1/3 (c)	1/4 (c)	2/1 (r)	2/2 (c)	2/3 (c)	2/4 (c)	2/5 (c)	2/6 (c)	2/7 (c)	2/8 (c)	2/9 (c)	2/10 (c)	2/11 (c)	2/12 (r)
SiO_2	40.74	40.58	39.48	39.74	40.20	39.75	39.79	40.72	40.10	40.79	38.71	39.65	40.80	40.25	40.38	40.15
TiO_2	6.04	5.98	4.84	5.60	5.61	6.27	5.89	6.35	5.79	5.82	6.41	5.98	5.41	5.40	6.14	5.82
Al_2O_3	9.26	9.03	10.96	9.99	9.75	9.57	10.05	9.00	10.06	9.80	9.24	8.87	9.50	9.81	9.69	9.63
Cr_2O_3	0.00	0.00	0.43	0.17	0.04	0.16	0.43	0.13	0.09	0.19	0.19	0.22	0.20	0.22	0.19	0.00
FeO	10.49	10.54	8.96	7.68	9.68	8.21	7.12	8.02	7.32	7.75	7.63	7.78	7.09	7.16	8.05	10.05
MnO	0.04	0.03	0.05	0.01	0.02	0.15	0.11	0.06	0.00	0.01	0.15	0.03	0.00	0.10	0.10	0.00
MgO	19.29	19.45	19.77	20.21	19.45	19.43	19.79	20.25	20.53	19.99	19.47	19.77	20.75	20.61	19.92	19.59
NiO	0.05	0.17	0.12	0.16	0.15	0.12	0.08	0.00	0.06	0.12	0.05	0.00	0.08	0.03	0.08	0.14
CaO	0.02	0.03	0.03	0.03	0.05	0.00	0.00	0.01	0.00	0.00	0.00	0.00	0.00	0.03	0.01	0.04
BaO	0.01	0.16	0.28	0.37	0.22	0.34	0.36	0.31	0.46	0.16	0.21	0.31	0.35	0.40	0.38	0.19
Na_2O	0.02	0.00	0.00	0.02	0.01	0.00	0.00	0.00	0.00	0.01	0.01	0.00	0.00	0.00	0.00	0.02
K_2O	9.13	8.98	8.79	9.98	8.96	10.08	10.23	10.24	10.28	10.14	10.10	10.17	10.21	10.24	10.12	9.00
Total	95.09	94.95	93.71	93.96	94.14	94.08	93.85	95.09	94.69	94.78	92.17	92.78	94.39	94.25	95.06	94.63
Cations based on 22 oxygen																
Si	5.959	5.955	5.830	5.867	5.927	5.883	5.875	5.946	5.869	5.950	5.844	5.940	5.970	5.912	5.901	5.901
Al^{total}	1.596	1.562	1.908	1.738	1.694	1.669	1.749	1.549	1.735	1.685	1.644	1.566	1.638	1.698	1.669	1.668
Fe^{3+}	0.445	0.483	0.262	0.395	0.379	0.448	0.376	0.505	0.396	0.365	0.512	0.494	0.391	0.389	0.430	0.430
Sum (T site)	8.000	8.000	8.000	8.000	8.000	8.000	8.000	8.000	8.000	8.000	8.000	8.000	8.000	8.000	8.000	8.000
Ti	0.664	0.660	0.537	0.622	0.622	0.698	0.654	0.697	0.637	0.638	0.728	0.674	0.595	0.597	0.675	0.643
Cr	0.000	0.000	0.050	0.020	0.005	0.019	0.050	0.015	0.010	0.022	0.023	0.026	0.023	0.026	0.022	0.000
Fe^{2+}	0.839	0.811	0.844	0.554	0.815	0.568	0.503	0.474	0.500	0.581	0.451	0.481	0.476	0.490	0.554	0.805
Mn	0.005	0.004	0.006	0.001	0.002	0.019	0.014	0.007	0.000	0.001	0.019	0.004	0.000	0.012	0.012	0.000
Mg	4.206	4.255	4.352	4.448	4.275	4.287	4.356	4.408	4.480	4.347	4.382	4.415	4.527	4.513	4.340	4.293
Ni	0.006	0.020	0.014	0.019	0.018	0.014	0.009	0.000	0.007	0.014	0.006	0.000	0.009	0.004	0.009	0.017
Sum (O site)	5.720	5.750	5.805	5.663	5.737	5.604	5.587	5.602	5.635	5.604	5.609	5.600	5.631	5.642	5.612	5.757
Ca	0.003	0.005	0.005	0.005	0.008	0.000	0.000	0.002	0.000	0.000	0.000	0.000	0.000	0.005	0.002	0.006

(continued)

Table 4 (continued)

Sample No	KBH42 (Kalmidadar lamproite)															
	Grain 1				Grain 2 (line scan from rim throgh core to rim)											
Analysis no.	1/1 (r)	1/2 (r)	1/3 (c)	1/4 (c)	2/1 (r)	2/2 (c)	2/3 (c)	2/4 (c)	2/5 (c)	2/6 (c)	2/7 (c)	2/8 (c)	2/9 (c)	2/10 (c)	2/11 (c)	2/12 (r)
Ba	0.001	0.009	0.016	0.021	0.013	0.020	0.021	0.018	0.026	0.009	0.012	0.018	0.020	0.023	0.022	0.011
Na	0.006	0.000	0.000	0.006	0.003	0.000	0.000	0.000	0.000	0.003	0.003	0.000	0.000	0.000	0.000	0.006
K	1.704	1.681	1.656	1.880	1.685	1.903	1.927	1.907	1.919	1.887	1.945	1.944	1.906	1.919	1.887	1.687
Sum (X site)	1.713	1.695	1.677	1.911	1.709	1.923	1.948	1.927	1.946	1.899	1.960	1.962	1.926	1.947	1.910	1.710
All cations	15.433	15.445	15.481	15.575	15.446	15.527	15.535	15.529	15.580	15.503	15.569	15.562	15.557	15.588	15.522	15.468
$Mg/Mg+Fe^T$	76.62	76.69	79.73	82.43	78.17	80.84	83.21	81.82	83.33	82.14	81.98	81.92	83.92	83.69	81.52	77.65

Sample no.	AD1/1 (Amlidadar lamproitoid)																	
	Grain 1				Grain 2			Grain 3 (line scan from rim throgh core to rim)										
Analysis no.	1/1 (r)	1/2 (r)	1/3 (r)	1/4 (c)	1/1 (r)	1/2 (c)	1/3 (c)	3/1 (r)	3/2 (r)	3/3 (c)	3/4 (c)	3/5 (c)	3/6 (c)	3/7 (c)	3/8 (c)	3/9 (c)	3/10 (c)	3/11 (r)
SiO_2	33.83	34.36	33.01	34.23	33.14	35.12	35.22	33.59	34.40	35.48	35.40	35.36	34.01	34.00	35.22	34.97	34.76	33.64
TiO_2	6.20	6.62	7.54	6.75	7.40	5.90	6.22	6.50	5.72	6.38	6.51	6.66	7.46	7.01	6.69	6.53	6.70	6.54
Al_2O_3	13.37	13.33	13.07	13.17	12.04	12.75	12.63	12.81	13.16	12.79	12.79	12.58	13.37	13.33	12.82	12.90	12.89	12.59
Cr_2O_3	0.02	0.00	0.00	0.00	0.01	0.02	0.06	0.00	0.00	0.00	0.00	0.01	0.01	0.07	0.01	0.02	0.01	0.00
FeO	15.80	15.33	15.10	13.37	22.52	13.61	13.36	17.59	15.54	12.71	12.80	11.91	12.28	12.52	12.17	13.76	14.29	18.64
MnO	0.05	0.08	0.00	0.11	0.14	0.14	0.11	0.12	0.13	0.17	0.11	0.11	0.10	0.09	0.06	0.10	0.14	0.07
MgO	12.53	12.46	12.13	13.43	7.33	13.91	14.10	11.14	13.25	14.59	14.53	14.33	14.03	14.01	14.49	13.77	13.02	9.93
NiO	0.11	0.00	0.00	0.00	0.04	0.00	0.07	0.00	0.01	0.00	0.03	0.01	0.00	0.00	0.00	0.00	0.00	0.06
CaO	0.11	0.07	0.41	0.02	0.11	0.00	0.01	0.04	0.02	0.00	0.01	0.00	0.04	0.02	0.00	0.04	0.01	0.04
BaO	2.85	3.22	2.78	3.04	2.03	2.68	2.78	2.42	2.33	2.63	2.54	2.60	3.56	3.47	2.85	2.46	2.54	2.56
Na_2O	0.30	0.36	0.37	0.37	0.15	0.33	0.34	0.26	0.33	0.36	0.37	0.37	0.45	0.40	0.36	0.32	0.36	0.25
K_2O	7.31	7.98	7.83	8.13	7.93	8.08	8.04	7.55	7.36	8.16	8.18	8.24	7.88	7.91	8.37	7.60	8.39	8.03
Total	92.48	93.81	92.24	92.62	92.84	92.54	92.94	92.02	92.25	93.27	93.27	92.18	93.19	92.83	93.04	92.47	93.11	92.35
Cations based on 22 oxygen																		
Si	5.360	5.381	5.260	5.381	5.411	5.502	5.492	5.391	5.426	5.486	5.474	5.515	5.304	5.326	5.464	5.461	5.440	5.430
Al^{total}	2.497	2.460	2.455	2.440	2.317	2.354	2.321	2.423	2.446	2.331	2.331	2.313	2.458	2.461	2.344	2.374	2.377	2.395
Fe^{3+}	0.144	0.159	0.285	0.179	0.272	0.144	0.187	0.186	0.128	0.183	0.195	0.172	0.238	0.213	0.192	0.165	0.183	0.174
Sum (T site)	8.000	8.000	8.000	8.000	8.000	8.000	8.000	8.000	8.000	8.000	8.000	8.000	8.000	8.000	8.000	8.000	8.000	8.000
Ti	0.739	0.780	0.904	0.798	0.909	0.695	0.729	0.784	0.678	0.742	0.757	0.781	0.875	0.826	0.780	0.767	0.788	0.794
Cr	0.003	0.000	0.000	0.000	0.001	0.002	0.007	0.000	0.000	0.000	0.000	0.001	0.001	0.009	0.001	0.002	0.001	0.000
Fe^{2+}	1.950	1.849	1.727	1.579	2.803	1.639	1.555	2.175	1.922	1.461	1.461	1.382	1.364	1.427	1.387	1.633	1.687	2.342
Mn	0.007	0.011	0.000	0.015	0.019	0.019	0.015	0.016	0.017	0.022	0.014	0.015	0.013	0.012	0.008	0.013	0.019	0.010
Mg	2.960	2.909	2.882	3.147	1.784	3.248	3.278	2.665	3.116	3.363	3.350	3.332	3.262	3.272	3.351	3.206	3.038	2.390

(continued)

Table 4 (continued)

Sample no.	AD1/1 (Amlidadar lamproitoid)																
	Grain 1				Grain 2			Grain 3 (line scan from rim throgh core to rim)									
Analysis no.	1/1 (r)	1/2 (r)	1/3 (r)	1/4 (c)	1/1 (r)	1/2 (c)	1/3 (c)	3/1 (r)	3/2 (r)	3/3 (c)	3/4 (c)	3/5 (c)	3/6 (c)	3/7 (c)	3/8 (c)	3/9 (c)	3/10 (c)	3/11 (r)
Ni	0.014	0.000	0.000	0.000	0.005	0.000	0.009	0.000	0.001	0.000	0.004	0.001	0.000	0.000	0.000	0.000	0.000	0.008
Sum (O site)	5.671	5.548	5.512	5.539	5.522	5.603	5.592	5.641	5.734	5.588	5.585	5.512	5.515	5.546	5.528	5.621	5.533	5.543
Ca	0.019	0.012	0.070	0.003	0.019	0.000	0.002	0.007	0.003	0.000	0.002	0.000	0.007	0.003	0.000	0.007	0.002	0.007
Ba	0.177	0.198	0.174	0.187	0.130	0.164	0.170	0.152	0.144	0.159	0.154	0.159	0.218	0.213	0.173	0.151	0.156	0.162
Na	0.092	0.109	0.114	0.113	0.047	0.100	0.103	0.081	0.101	0.108	0.111	0.112	0.136	0.121	0.108	0.097	0.109	0.078
K	1.477	1.594	1.592	1.630	1.652	1.615	1.599	1.546	1.481	1.610	1.614	1.640	1.568	1.581	1.656	1.514	1.675	1.654
Sum (X site)	1.765	1.913	1.950	1.934	1.848	1.879	1.873	1.786	1.729	1.877	1.880	1.910	1.928	1.918	1.938	1.768	1.942	1.901
All cations	15.437	15.461	15.462	15.473	15.371	15.483	15.466	15.426	15.464	15.465	15.466	15.422	0.000	15.464	15.465	15.389	15.475	15.444
Mg/Mg+Fe^T	58.57	59.16	58.88	64.17	36.72	64.56	65.29	53.03	60.32	67.17	66.93	68.20	67.07	66.61	67.97	64.08	61.89	48.71

Sample no.	PK1 (Parkom lamproitoid)				
	Grain 1		Grain 2	Grain 3	Grain 4
Analysis no.	1/1 (r)	1/2 (c)	2/1	3/1	4/1
SiO_2	35.15	37.50	36.77	35.06	35.16
TiO_2	6.28	8.61	5.65	7.36	6.52
$A_{12}O_3$	9.08	10.39	9.19	9.46	10.07
Cr_2O_3	0.05	0.00	0.00	0.00	0.00
FeO	29.30	22.70	27.72	23.38	23.07
MnO	0.10	0.06	0.17	0.17	0.13
MgO	6.12	8.44	7.22	7.57	10.16
NiO	0.00	0.03	0.01	0.00	0.09
CaO	0.09	0.19	0.14	0.26	0.35
BaO	0.48	0.27	0.46	0.25	0.25
Na_2O	0.03	0.07	0.04	0.08	0.09
K_2O	8.98	9.04	8.88	9.19	8.21
Total	95.67	97.29	96.24	92.77	94.11
Cations based on 22 oxygen					
Si	5.702	5.725	5.844	5.705	5.591
Al^{total}	1.736	1.870	1.722	1.814	1.887
Fe^{3+}	0.561	0.405	0.434	0.481	0.522
Sum (T site)	8.000	8.000	8.000	8.000	8.000
Ti	0.766	0.988	0.676	0.901	0.780

(continued)

Table 4 (continued)

Sample no.	PK1 (Parkom lamproitoid)				
	Grain 1		Grain 2	Grain 3	Grain 4
Analysis no.	1/1 (r)	1/2 (c)	2/1	3/1	4/1
Cr	0.006	0.000	0.000	0.000	0.000
Fe^{2+}	3.414	2.493	3.251	2.700	2.545
Mn	0.014	0.008	0.023	0.023	0.017
Mg	1.481	1.921	1.710	1.836	2.409
Ni	0.000	0.003	0.001	0.000	0.012
Sum (O site)	5.681	5.413	5.661	5.460	5.764
Ca	0.015	0.031	0.024	0.046	0.060
Ba	0.031	0.016	0.029	0.016	0.016
Na	0.010	0.021	0.012	0.025	0.028
K	1.858	1.760	1.800	1.907	1.665
Sum (X site)	1.913	1.829	1.864	1.993	1.769
All cations	15.594	15.242	15.525	15.453	15.532
$Mg/Mg+Fe^T$	27.14	39.86	31.69	36.59	43.99

Fe^{3+} calculated from the formula Si + Al^{IV} + Fe^{3+} = 8

Diamond Prospectivity from Indicator Minerals

The indicator mineral suite of the Kalmidadar lamproite is marked by abundance of Cr-spinel, rarity of pyrope garnet and absence of Cr-diopside and picroilmenite and is similar to the indicator suites from the diamondiferous lamproites at Majhgawan (Mathur and Singh 1971) and Saptarshi (Mason et al 2009) in the Bundelkhand Craton of central India. Mineral dispersion halos around diamondiferous lamproites in the Kimberly Province of Australia, including the Argyle, Aries, Ellendale and Pteropus Creek intrusions, are also dominated by Cr-spinel (Atkinson et al. 1984; Jaques et al. 1986; Lucas et al. 1989; Smith et al. 1991). One possible factor for the abundance of Cr-spinel is a high-degree weathering, which has destroyed other indicators such as Cr-diopside, pyrope garnet and ilmenite (e.g. Griffin et al. 1997). However, in the case of Kalmidadar lamproite, weathering does not seem to be responsible for the paucity of pyrope garnet and lack of Cr-diopside and picroilmenite. This observation is based on the fact that the stream sediments around kimberlites of the MKF are marked by abundance of Cr-diopside, pyrope garnet and picroilmenite along with Cr-spinel in the heavies (Sarkar et al. 2001; Mainkar 2011). Since the MKF is barely located 100 km south-west of the NLF in the same craton, a drastic increase in the intensity of weathering from the MKF to the NLF is unlikely. Therefore, the dominance of Cr-spinel over other indicators around the Kalmidadar lamproite is interpreted to reflect the original xenocryst population in the intrusion. This is an important aspect to be recognised while planning for future indicator mineral surveys in the region.

All the Kalmidadar pyropes are G9 type and are similar in composition to the pyropes that occur occasionally in the indicator mineral suites of the diamondiferous lamproites in the Kimberley Province (Jaques et al. 1986; Lucas et al. 1989). Subcalcic garnets (G10), which are traditionally considered as diamond indicators, have yet been reported from these areas. This implies that G9 garnets should not be excluded from being considered as diamond indicators.

Nature of Subcontinental Lithospheric Mantle

Indicator minerals such as pyrope garnet and Cr-spinel in lamproites and kimberlites give valuable information on the nature of subcontinental lithospheric mantle (SCLM) because they exhibit compositional variations, which reflect the physical, chemical and lithological environments in which they occur. The Ca-saturated composition of the Kalmidadar pyropes indicates that the pyropes have equilibrated with clinopyroxene in a lherzolitic mantle. Similar mantle lithology is indicated beneath the kimberlites of the Dharwar and Bastar cratons (Patel et al. 2010; Mainkar 2011,

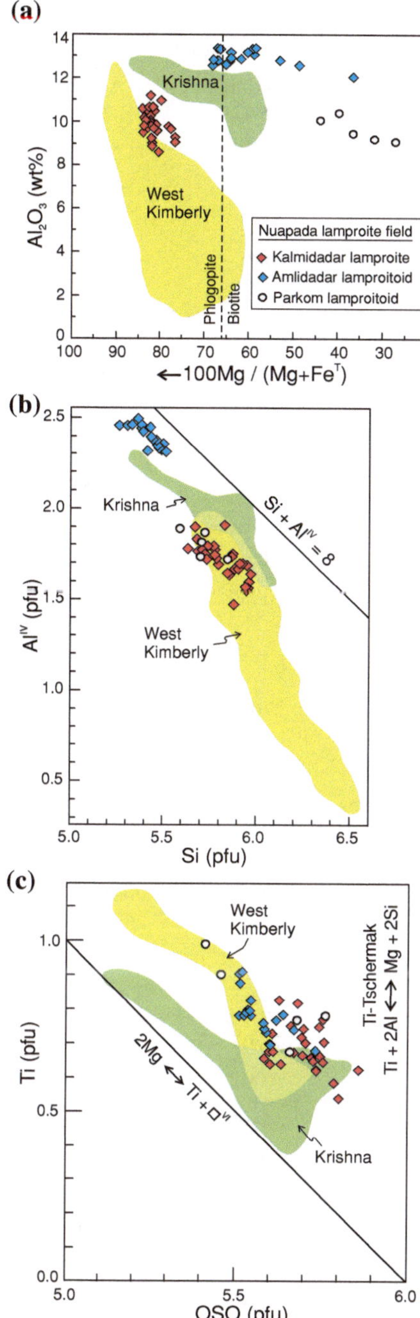

Fig. 6 Compositional variation of mica in Nuapada lamproites and lamproitoids compared with that in Krishna lamproites, southern India (Reddy et al. 2003; Chalapathi Rao et al. 2010) and West Kimberly lamproites (Jaques et al. 1986; Mitchell and Bergman 1991). **a** Cr_2O_3 versus Cr/(Cr +Al) versus 100 Mg/(Mg + Fe^T) plot; **b** Tetrahedral Al versus Si plot; **c** Ti versus octahedral site occupancy (OSO) plot

respectively) from the composition of pyropes in the respective indicator mineral suites.

Pressures for the Kalmidadar garnets have been calculated using the empirical Ca/Cr-in-garnet geobarometer of Grütter et al. (2006). Although the geobarometer is applicable for garnet coexisting with Cr-spinel, it provides an estimate of minimum pressure if the coexistence of garnet with Cr-spinel is unknown as is the case of discrete garnets in the indicator suites of lamproites and kimberlites. The knowledge of the mantle geotherm is required for the calculation of pressure from the geobarometer of Grütter et al. (2006). This information for the Bastar Craton is derived from the work of Babu et al. (2009), who reported pressure (P)–temperature (T) conditions of 43–47 kbar and 1,140–1,270 °C for xenoliths of garnet lherzolite in a kimberlite of the MKF. These P–T conditions are consistent with a mantle geotherm of 45 mW/m^2, following the model continental geotherms of Pollack and Chapman (1977). This is same as the mantle geotherm beneath the kimberlite fields of the Dharwar Craton reported by Patel et al. (2009), 2010). Using the geotherm of 45 mW/m^2, the minimum pressures obtained for the Kalmidadar garnets fall in the range of 16–32 kbar, which correspond to a minimum depth range of 50–100 km. The actual depths of the garnet source are likely to be higher in view of the pressures reported by Babu et al. (2009) for the garnet lherzolite xenoliths which correspond to a depth range of 130–140 km.

Mantle phase equilibria considerations show that the subsolidus transition from spinel peridotite to garnet peridotite occurs via the generalised reaction: spinel + pyroxene(s) = garnet + olivine. In the chemical system MgO–Al_2O_3–Cr_2O_3–SiO_2, a P–T field exists where garnet and spinel coexist. The width of this divariant field strongly depends on the Cr/(Cr + Al) of the system, and with increasing value of this ratio, the spinel–garnet transition progressively shifts to pressures as high as ~70 kbar (i.e. depth of ~200 km) (Klemme, 2004; Grutter et al. 2006). If we assume that the Kalmidadar spinels of high Cr# coexisted with garnet in the mantle source, then they could be derived from sources as deep as 200 km.

Geochemical Characteristics of the Mantle Source of Lamproites

Negative spikes in the primitive mantle-normalised element pattern (observed in the case of K, Rb, Sr, P, Nb, Zr and Hf in the present study) could result variously from (1) element mobility during secondary alteration, (2) residual mineralogy during partial melting at the source, (3) crystal fractionation during emplacement of magma or (4) could be inherited from the source region (le Roex et al. 2003). While the Darlimunda lamproites have low K values due to secondary alteration (chloritisation of phlogopite), marked depletion at K is present even in the Kalmidadar lamproite sample KBH42, which contains 'fresh' phlogopite. This indicates that the negative K anomaly is a primary feature of the Nuapada lamproites, although the absolute magnitudes have been affected by different degrees of

Table 5 Whole-rock geochemical analyses of Nuapada lamproites and lamproitoids. $Fe_2O_3^*$ denotes total iron

Oxide (wt %)	Lamproite					Lamproitoid		
	Kalmidadar (KD)		Darlimunda (DM1)	Darlimunda (DM2)		Darlimunda (DM3)	Amlidadar (AD)	Parkom (PK)
	KBH15/14	KBH42	DM1/2	DM2/2	DM2/2C	DM3/1	AD1/1	PK1
SiO_2	47.24	43.63	58.46	60.23	54.55	59.22	35.64	36.94
Al_2O_3	5.91	5.27	6.97	6.30	7.18	5.41	8.87	7.29
CaO	10.10	12.26	2.22	3.43	4.53	1.87	12.43	15.52
$Fe_2O_3^*$	7.21	7.86	10.29	10.36	11.11	14.56	15.87	17.22
K_2O	0.33	1.34	0.05	0.03	0.04	0.02	1.00	2.33
MgO	14.27	13.96	9.03	7.41	7.72	8.68	8.69	8.55
MnO	0.15	0.16	0.05	0.04	0.07	0.03	0.21	0.25
Na_2O	0.06	0.07	<0.01	<0.01	<0.01	0.02	0.02	0.44
P_2O_5	0.06	1.87	1.46	1.60	1.88	0.64	0.93	1.06
TiO_2	2.62	3.00	7.35	6.49	7.01	6.32	4.45	4.70
LOI	13.03	11.51	4.51	4.23	6.04	4.05	11.98	5.68
Total	100.98	100.93	100.39	100.12	100.13	100.82	100.09	99.98
Trace element (ppm)								
Sc	17.6	16.7	11.3	15.5	14.2	9.54	17.3	21.1
V	215	233	347	298	312	270	264	265
Cr	380	360	478	369	257	443	202	260
Ni	388	389	423	298	232	221	188	166
Rb	13.4	37.7	6.25	1.00	4.67	1.00	35.9	79.0
Sr	68.7	684	290	148	176	36.2	306	872
Y	20.7	31.8	16.3	21.3	28.1	22.4	29.8	38.1
Zr	719	780	544	581	683	219	335	349
Nb	108	135	55.0	65.0	73.7	51.7	132	116
Cs	3.82	4.60	5.25	5.13	4.56	5.23	2.82	6.15
Ba	2,292	2,689	3,499	2,983	3,061	2,705	3,574	3,564
Hf	16.5	18.5	12.9	15.2	15.1	4.00	8.73	4.92
Ta	2.6	2.6	2.2	3.3	2.7	<0.05	2.1	<0.05
La	143	190	205	292	284	236	115	103
Ce	313	402	422	601	530	521	241	227
Pr	48.7	26.9	32.4	48.5	41.0	41.4	18.9	19.3
Nd	254	166	175	253	214	220	104	109
Sm	15.3	22.6	37.6	54.0	47.3	41.5	21.8	24.2
Eu	2.94	4.13	7.34	11.2	9.27	9.54	5.35	6.44
Gd	7.10	8.01	21.7	32.2	34.2	23.3	13.5	16.5
Dy	2.44	2.58	12.5	23	28.5	11.1	8.31	10.2
Ho	0.35	0.49	2.72	4.68	6.44	1.86	1.47	1.70
Er	0.94	1.47	7.81	8.35	9.51	3.84	3.39	3.96
Yb	0.90	2.00	3.08	4.92	5.75	2.02	2.23	2.37
Lu	0.55	0.51	1.37	1.72	1.56	0.18	0.24	0.26

chloritisation of phlogopite. Negative K anomaly, in spite of high K_2O contents, is a feature common to a number of alkaline rocks (Foley et al. 1987) and is interpreted to be either inherited from the source region (e.g. Coe et al. 2008) or produced by equilibration with a residual potassic phase such as phlogopite (e.g. Rogers et al. 1992) or amphibole (e.g. Späth et al. 2001) during partial melting. We prefer the second alternative and choose phlogopite over amphibole as

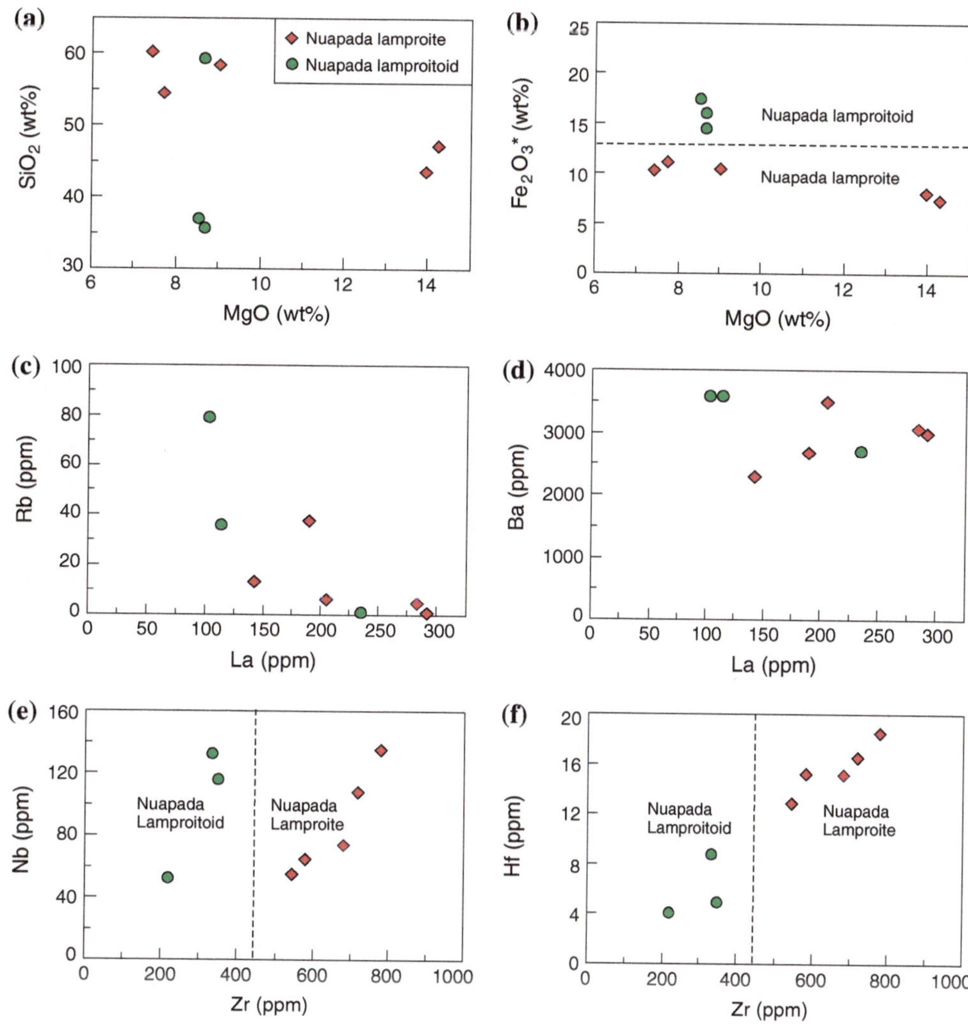

Fig. 7 Geochemical plots of Nuapada lamproites (Kalmidadar and Darlimunda) and lamproitoids (Amlidadar, Parkom and Darlimunda). **a** SiO_2 versus MgO; **b** Fe_2O_3 (total iron expressed as ferric iron) versus MgO; **c** Rb versus La; **d** Ba versus La; **e** Nb versus Zr; and **(f)** Hf versus Zr

the residual phase because the observed depletion of Rb relative to Ba and Cs in the Nuapada lamproites can be satisfactorily explained by the contrasting phlogopite–melt partition coefficients for these elements ($D_{Ba} = 1.03–1.61$; $D_{Cs} = 0.47–0.58$; $D_{Rb} = 1.23–1.44$; Schmidt et al. 1999).

Negative P anomaly in the Nuapada lamproites suggests a partial melting control in the presence of residual apatite. Negative Sr anomaly could be related to equilibration with residual clinopyroxene (Foley and Jenner 2004) or inherited from a mantle source that was already depleted in Sr before the extraction of lamproite melt. The cause of the depletion in Nb, Zr and Hf relative to REE is unclear, but can be considered to be intrinsic to the source region (e.g. Egorov et al. 2006).

Petrogenesis

Lamproites, along with kimberlites, are not only strongly enriched in incompatible elements but also have substantial concentrations of compatible elements (Foley et al. 1987; Mitchell and Bergman 1991). The high MgO and Ni contents, low Al_2O_3, and high concentrations of incompatible elements such as Ba, Nb, Zr, Hf and LREE in the Nuapada lamproites are consistent with the geochemistry of other lamproites worldwide and suggest derivation of these rocks from small degrees of partial melting of a mantle source that has undergone multiple events of depletion and enrichment (e.g. Fraser et al. 1985; Tainton and McKenzie 1994; Foley 1992a). The depletion event involves partial melting of lherzolite to produce a depleted lherzolitic or a harzburgitic substrate, while the enrichment process ranges from cryptic and/or modal metasomatism of the depleted substrate (references in Menzies and Hawkesworth 1987) to the formation of metasomatic veins cross-cutting the depleted substrate (Foley 1992b). Based on melting experiments of lamproites (Foley 1993; Mitchell 1995; Mitchell and Edgar 2002), the metasomatic assemblages present in the mantle source of the Nuapada lamproites are expected to be dominated by phlogopite with subordinate amounts of clinopyroxene, rutile and apatite.

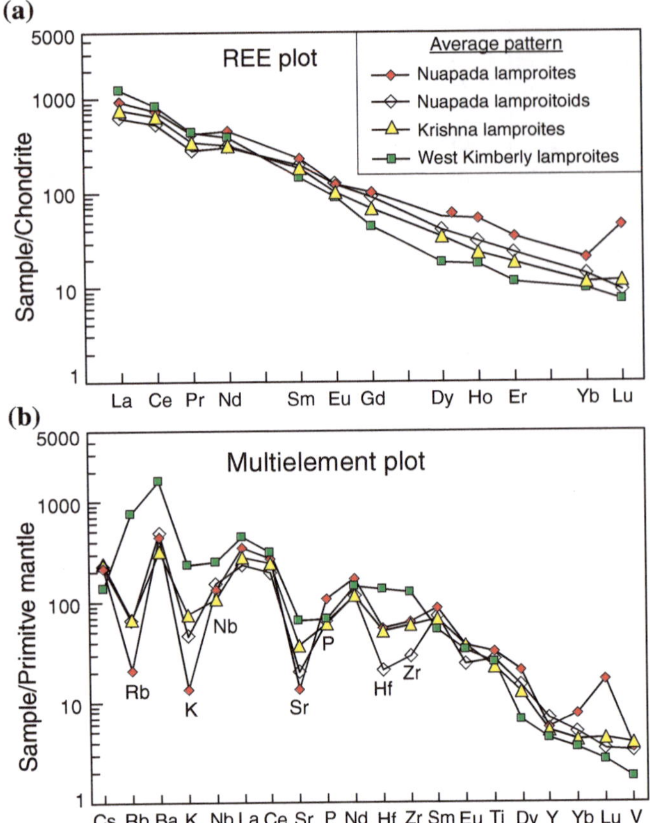

Fig. 8 Chondrite-normalised rare earth element patterns (**a**) and primitive mantle-normalised multielement patterns (**b**) of average Nuapada lamproites (Kalmidadar and Darlimunda) and lamproitoids (Amlidadar, Parkom and Darlimunda) compared with those of average Krishna lamproites (Vijesh 2010) and average West Kimberly lamproites (Mitchell and Bergman 1991 and references therein). Normalising values after McDonough and Sun (1995)

The diamondiferous nature of the Kalmidadar lamproite proves that the lamproite magma must have originated at depths of >155 km, considering a mantle geotherm of 45 mW/m^2. However, the actual depth of the source region remains unconstrained. A highly debated issue surrounding the petrogenesis of lamproites concerns whether their source region lies within the SCLM (Foley 1992b; Mitchell 2006) or sublithospheric mantle (Ringwood et al. 1992; Murphy et al. 2002). Following the experimental results of Foley (1993), Mitchell (1995) and Mitchell and Edgar (2002), which are permissive of derivation of lamproite magma at lithospheric pressures, we prefer to suggest that the Nuapada lamproites are derived from an SCLM source. The fluids or melts responsible for the development of metasomatic assemblages in the source can be derived from the dehydration or partial melting of subducted slabs (Davies et al. 2006; Prelevic et al. 2007; Akal 2008) or volatiles or melts that emanate from asthenospheric mantle upwellings (Tainton and McKenzie 1994; Choukroun et al. 2005; Mitchell 2006) or both (Mirnejad and Bell 2006). The Nuapada lamproites do not exhibit any large-scale subduction-related signatures such as negative Ti and Nb anomalies, suggesting that subducted materials may not have directly played a major role in the metasomatic input to the lamproite source. We, therefore, attribute the source region metasomatism to fluids or melts derived from mantle upwellings (e.g. Chalapathi Rao et al. 2011).

The genesis of lamproitoids in the NLF, which are relatively Fe rich compared to the lamproites, can be attributed to variations in the source mineralogy and the degree of melting of different mineral phases (e.g. Foley 1992b). The lamproitoid source must either have been enriched in Fe relative to the lamproite source or phases hosting this element melted completely in the lamproitoid source but were retained in the lamproite source.

Geodynamic Setting

A look at the spatial distribution of the lamproite and kimberlite fields of the Bastar Craton (Fig. 1) shows that the fields are located close to the margin of the Khariar and Indravati basins. It is quite likely that the basin margins were zones of lithospheric weakness which played a key role in the emplacement of the lamproites. Therefore, it is important to understand the temporal and tectonic evolution of these basins as well as of pertinent igneous activities in the region. The Khariar basin started opening at 1455 ± 47 Ma (Das et al. 2009), which is broadly synchronous with the intrusion of the 1450–1465 Ma Lakhna dykes to the east of the basin (Ratre et al. 2010; Pisarevsky et al. 2012). Age constraint on the upper limit of sedimentation comes from the Chhattisgarh basin, where felsic volcanism is recorded at ~1000 Ma in the form of rhyolitic tuffs in the upper part of the sedimentary succession (Patranabis-Deb et al. 2007; Bickford et al. 2011). Mukherjee and Ray (2010) contented that the sedimentation in the basin continued well beyond 1000 Ma and into the Neoproterozoic. Thus, the sedimentary basins of the Bastar Craton are Meso–Neoproterozoic in age and the emplacement of the Nuapada lamproites at 1055 ± 10 Ma occurred, while sedimentation was progressing in different parts of the craton. Therefore, a common geodynamic scenario can be envisaged for the evolution of the basins, and the intrusion of the Lakhna dykes and Nuapada lamproites.

In the absence of detailed geophysical data, the basement configuration and tectonic setting of the sedimentary basins of the Bastar Craton are highly debated topics. Nevertheless, different tectonic settings have been proposed for the basins which range from cratonic sag (Das et al. 1992) and intracratonic rift (Chaudhuri et al. 2002; Patranabis-Deb and Chaudhuri 2007) to foreland basin (Biswal et al. 2003) and foreland rift basin (Ratre et al. 2010). Pisarevsky et al. (2012)

Table 6 Argon isotopic composition (corrected for blank, mass discrimination and interference), age and percentage of nucleogenic and radiogenic argon. Errors on age are without and with error on J, respectively. Errors quoted are 2σ

Temperature (°C)	$^{36}Ar/^{39}Ar$ $\pm 2\sigma$	$^{40}Ar/^{39}Ar_k$ $\pm 2\sigma$	Apparent age (Ma) $\pm 2\sigma$	%Cum $^{39}Ar_k$	%$^{40}Ar^*$	$^{37}Ar/^{39}Ar$ $\pm 2\sigma$	$^{40}Ar/^{36}Ar$ $\pm 2\sigma$
Sample KBH15/14 (J = 0.002921 ± 0.000014)							
400	1.113	486.094	1,121.46	4.6	41.15	35.897	502.1
	0.010	2.453	18.01 (18.51)			0.232	63.7
450	1.022	506.954	1,240.25	1.5	47.87	31.550	566.8
	0.011	2.761	16.86 (17.48)			0.236	11.2
500	0.402	335.925	1,289.77	5.1	76.01	31.839	1,231.7
	0.001	1.709	7.29 (8.71)			0.231	10.9
550	0.422	356.390	1,342.77	4.4	75.75	31.916	1,218.7
	0.001	1.819	7.52 (8.96)			0.231	11.1
600	0.314	156.400	1,049.60	11.9	84.25	57.070	1,875.8
	0.001	1.011	7.89 (8.89)			0.230	38.1
650	0.296	157.040	1,052.75	8.7	86.54	55.586	2,195.4
	0.001	1.015	7.59 (8.63)			0.231	50.6
700	0.172	191.423	1,059.09	17.6	96.23	36.369	7,828.9
	0.001	1.043	5.42 (6.81)			0.233	468.7
750	0.222	254.722	1,055.84	17.3	84.94	22.914	1,962.7
	0.001	1.278	5.46 (6.84)			0.230	24.5
800	0.514	306.505	1,057.24	8.7	63.12	32.408	801.3
	0.001	1.543	7.70 (8.73)			0.231	5.7
850	0.500	298.361	1062.17	5.0	64.07	33.828	822.5
	0.001	1.525	7.74 (8.78)			0.231	6.6
900	0.522	429.261	1,501.58	3.4	73.16	32.726	1,101.1
	0.001	2.197	8.42 (9.92)			0.232	10.4
950	1.620	1,587.930	3,189.34	1.0	72.43	34.388	1,072.0
	0.011	8.091	13.15 (15.24)			0.240	12.5
1000	1.645	2,015.560	3,802.22	1.1	78.27	40.401	1,360.1
	0.011	10.155	12.77 (15.15)			0.239	25.3
1050	1.385	2,419.140	4,355.25	1.1	85.32	45.443	2,013.4
	0.011	12.151	12.25 (14.89)			0.240	37.5
1200	1.018	3,213.030	4,482.19	2.0	91.54	24.400	3,493.5
	0.002	16.089	10.30 (13.37)			0.236	44.4
1250	1.026	3,546.640	4,491.03	2.8	91.97	15.435	3,681.0
	0.002	17.746	9.02 (12.41)			0.031	37.9
1300	1.243	3,396.040	4,388.40	2.1	89.74	15.982	2,881.1
	0.010	17.004	9.35 (12.62)			0.032	44.0
1350	1.523	3,390.780	4,330.02	1.7	87.27	15.403	2,321.5
	0.011	16.994	9.59 (12.78)			0.031	39.8
Total	0.460	524.999	1,933.88	100.0	82.74	37.836	1,712.4
	0.001	0.742	2.58 (6.63)			0.075	5.3
Sample DM2/2C (J = 0.002869 ± 0.000015)							
400	1.403	1,385.870	2,974.40	0.4	73.04	34.379	1,096.3
	0.106	6.981	77.62 (77.99)			2.303	93.5
430	0.717	774.485	2,225.27	0.7	77.65	32.410	1,322.1
	0.103	3.913	81.92 (82.20)			2.280	234.4

(continued)

Table 6 (continued)

Temperature (°C)	$^{36}Ar/^{39}Ar$ ±2σ	$^{40}Ar/^{39}Ar_k$ ±2σ	Apparent age (Ma) ±2σ	%Cum $^{39}Ar_k$	%^{40}Ar*	$^{37}Ar/^{39}Ar$ ±2σ	$^{40}Ar/^{36}Ar$ ±2σ
460	0.715	813.849	2,297.48	0.6	78.7	31.856	1,387.3
	0.106	4.119	80.40 (80.69)			2.281	250.9
490	0.383	421.749	1,564.16	1.2	82.03	31.265	1,644.5
	0.010	2.133	38.84 (39.22)			2.255	89.0
520	0.359	385.646	1,448.98	1.4	81.68	29.723	1,613.3
	0.010	1.948	36.83 (37.20)			2.260	93.8
550	0.149	238.197	1,055.44	2.1	92.79	22.560	4,096.3
	0.010	1.207	25.13 (25.47)			2.249	967.3
580	0.233	238.698	1,052.55	2.6	85.67	28.903	2,061.5
	0.010	1.202	29.30 (29.59)			2.249	243.8
610	0.236	195.465	1,042.30	2.3	88.93	40.351	2,669.4
	0.010	1.015	33.34 (33.60)			2.254	500.1
700	0.080	256.879	1,065.82	12.5	96.74	12.801	9,060.6
	0.010	1.285	18.53 (19.00)			1.796	3,990.6
730	0.090	242.574	1,059.69	7.5	97.35	16.986	11,139.1
	0.010	1.214	19.23 (19.68)			1.796	6,408.7
760	0.210	228.756	1,067.15	3.4	88.99	30.747	2,685.1
	0.010	1.149	29.14 (29.44)			2.250	438.1
790	0.201	240.205	1,065.32	5.3	88.6	26.898	2,591.9
	0.010	1.203	27.81 (28.13)			2.253	385.0
820	0.104	295.818	1,046.05	5.5	90.61	2.394	3,146.2
	0.010	1.482	10.08 (10.90)			0.220	3,43.9
850	0.221	273.224	1,108.46	3.1	86.08	22.826	2,122.8
	0.010	1.373	25.45 (25.82)			2.023	216.6
880	0.299	394.222	1,396.02	2.2	83.93	20.874	1,838.7
	0.010	1.980	11.02 (12.14)			0.226	88.8
910	0.359	484.318	1,547.01	2.2	82.41	17.432	1,679.6
	0.010	2.430	10.58 (11.90)			0.226	60.9
940	0.315	543.051	1,651.20	2.7	85.56	12.242	2,046.9
	0.010	2.721	9.90 (11.41)			0.226	80.2
970	0.325	603.990	1,744.63	2.6	86	9.577	2,111.3
	0.010	3.025	9.64 (11.28)			0.225	76.7
1000	0.213	655.633	1,875.09	3.9	91.58	6.411	3,509.4
	0.010	3.281	9.05 (10.92)			0.225	192.5
1050	0.107	647.708	1,886.19	9.4	95.82	3.904	7,072.5
	0.001	3.281	6.71 (9.09)			0.225	121.2
1100	0.080	553.972	1,692.51	12.1	96.15	1.971	7,668.8
	0.001	2.801	5.81 (8.18)			0.023	123.0
1150	0.116	557.985	1,705.57	8.4	94.71	4.100	5,591.2
	0.001	2.791	6.35 (8.59)			0.227	91.7
1200	0.241	725.319	1,999.96	3.3	91.37	7.090	3,422.9
	0.010	3.629	9.05 (11.05)			0.225	166.1
1250	0.319	862.499	2,210.44	2.6	90.18	8.056	3,010.5

(continued)

Table 6 (continued)

Temperature (°C)	$^{36}Ar/^{39}Ar$ ±2σ	$^{40}Ar/^{39}Ar_k$ ±2σ	Apparent age (Ma) ±2σ	%Cum $^{39}Ar_k$	%$^{40}Ar^*$	$^{37}Ar/^{39}Ar$ ±2σ	$^{40}Ar/^{36}Ar$ ±2σ
	0.010	4.713	9.65 (11.73)			0.226	108.4
1300	0.878	1,023.370	2,321.95	0.9	76.52	15.974	1,258.4
	0.010	7.068	13.50 (15.14)			0.227	18.9
1350	0.751	976.209	2,258.17	1.1	78.85	12.866	1,397.4
	0.010	6.501	12.59 (14.28)			0.225	23.8
Total	0.195	442.788	1,519.72	100	90.8	14.212	3,210.3
	0.002	0.579	4.64 (7.11)			0.343	64.3

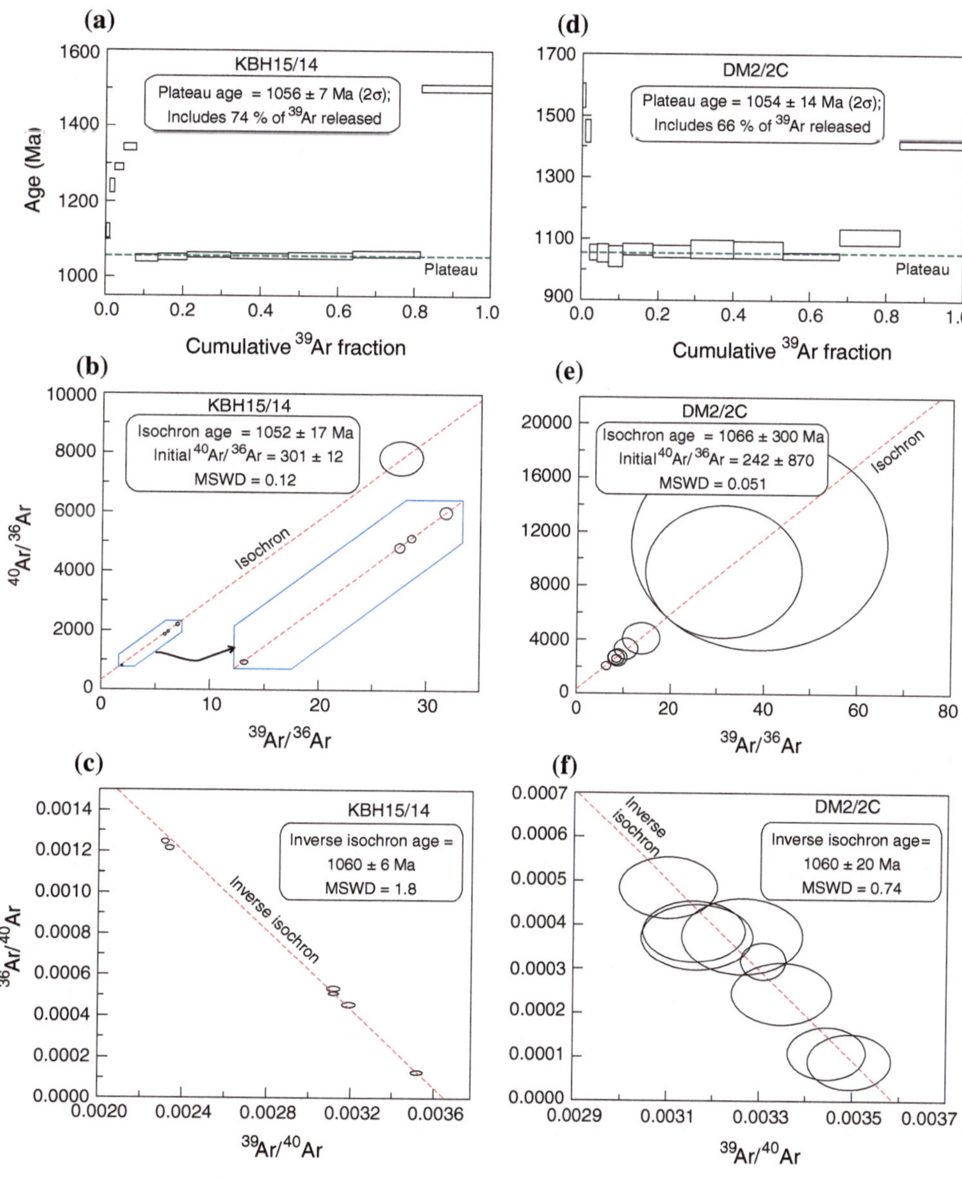

Fig. 9 $^{40}Ar/^{39}Ar$ age spectra (**a**, **d**), isochrons (**b**, **e**) and inverse isochrons (**c**, **f**) of Kalmidadar lamproite sample KBH15/14 and Darlimunda DM2 lamproite sample DM2/2C

ruled out a rift setting for the emplacement of the Lakhna dykes at 1450–1465 Ma based on the occurrence of orthogonal mafic dykes of the same ages and proposed a subduction-related continental margin setting based on the shoshonitic and high-K calc-alkaline affinities of some of the studied dykes. However, in our opinion, the contemporaneous intrusion of dykes in perpendicular directions does not negate the possibility of a rift setting because the orientations of some of the dykes could have been controlled by inherited structural fabrics in the region. In particular, the E–W- to NW–SE-trending Lakhna dykes are parallel or subparallel to the trend of the Mahandi and Godavari rifts, suggesting that they have exploited a long-lived weakness in the crust. Regarding the origin of the shoshonitic and high-K calc-alkaline dykes, Pisarevsky et al. (2012) rightly noted that a premetasomatised mantle source in an intraplate tectonic setting could be a viable alternative to direct contribution of melt from subducted sediments in an active continental margin setting. We favour the former setting because of two reasons. First, the existence of metasomatised mantle sources beneath the Bastar Craton, since the Palaeoproterozoic time or even earlier, has been proposed in several studies (e.g. Srivastava and Singh 2004; Hussain et al. 2008; Subba Rao et al. 2008; Srivastava and Gautam 2012), and hence, an active continental margin setting is not required for the genesis of the Lakhna dykes. Secondly, the sedimentary basins of the craton are satisfactorily explained as intracratonic rifts (Chaudhuri et al. 2002; Patranabis-Deb and Chaudhuri 2007). It is therefore reasonable to believe that the intrusion of the Nuapada lamproites took pace in a passive rift setting and that the lamproite magma was generated by small amount of lithospheric extension similar to that proposed for the origin of kimberlites and lamproites in the eastern Dharwar Craton (Chalapathi Rao et al. 2004, 2010).

Conclusions

The Nuapada Lamproite Field comprises several hypabyssal facies dykes of lamproites and lamproitoids in which Ti-rich mica is a major constituent phase. The mica is phlogopite in the lamproites and is biotite in the lamproitoids. One of the lamproites, namely the Kalmidadar lamproite, is diamondiferous. The indicator mineral suite around this intrusion comprises abundant Cr-spinel, rare garnet and no Cr-diopside and picroilmenite. It is argued that this suite reflects the original xenocryst population in the intrusion. The Kalmidadar garnets are calcium-saturated and indicate that the underlying mantle is lherzolitic. Geobarometric calculations indicate a minimum depth range of 50–100 km for these garnets.

The Nuapada lamproites have high MgO, V, Cr and Ni contents, low Al_2O_3 and high concentrations of incompatible elements such as Ba, Nb, Zr, Hf and LREE which indicate that the mantle source has undergone both depletion and enrichment events. Depletions in Nb, Hf and Zr relative to REE are observed in the trace element patterns of the lamproites and are considered to be inherited from the source region. The Nuapada lamproitoids, which are relatively Fe rich compared to the lamproites, are derived from a source that was either enriched in Fe relative to the lamproite source or phases hosting this element melted completely in the lamproitoid source but were retained in the lamproite source.

Whole-rock $^{40}Ar/^{39}Ar$ dating of the Nuapada lamproites gives an age of 1055 ± 10 Ma, which is interpreted as the emplacement age. The occurrence of the lamproites is attributed to extension of the lithospheric mantle.

Acknowledgments The paper is a contribution to the IGCP–557. Thanks are due to John Garlick of M/s Mackay and Schnellmann Pty. Ltd., Australia, who, as UNDP consultant, helped in diamond exploration in Odisha. B.C. Patnaik (Retd.) of the Directorate of Geology, Government of Odisha, is thanked for help in different stages of field work. Dr. L. S. Mombasawala and Mrs. Y. Y. Durve of the SAIF, IIT Bombay, are thanked for help in ED–XRF analysis. Dr. V. Balaram of NGRI, Hyderabad, is thanked for providing the MY-4 and SARM-39 standards. The FIST grant from the Department of Science and Technology (DST), Government of India, is acknowledged for funding the ICP-AES facility at IIT Bombay. Grant no. IR/S4/ESF-04/2003 from the DST for setting up of National Facility for $^{40}Ar/^{39}Ar$ Geothermochronology at IIT Bombay is gratefully acknowledged. Constructive comments from two anonymous reviewers significantly improved the quality of the paper.

References

Akal C (2008) K-richterite-olivine-phlogopite-diopside-sanidine lamproites from the Afyon volcanic province. Turkey Geol Mag 145:570–585

Atkinson WJ, Smith CB, Boxer GL (1984) A review of the kimberlitic rocks of Western Australia. In: Kornprobst J (ed) Kimberlites I: kimberlites and related rocks. Elsevier, Amsterdam, pp 195–224

Babu EVSSK, Bhaskar Rao YJ, Mainkar D, Pashine JK, Shrikant Rao R (2009) Mantle xenoliths from the Kodomali kimberlite pipe, Bastar Craton, central India: Evidence for decompression melting and crustal contamination in the mantle source. Geochim. Cosmochim. Acta Goldschmidt Abstracts, vol 73, p A66

Barnes SJ, Roedder PL (2001) The range of spinel compositions in terrestrial mafic and ultramafic rocks. J Petrol 42:2279–2302

Bickford ME, Basu A, Mukherjee A, Hietpas J, Schieber J, Patranabis-Deb S, Ray RK, Guhey R, Bhattacharya P, Dhang PC (2011) New U-Pb SHRIMP zircon ages of the Dhamda tuff in the Mesoproterozoic Chhattisgarh basin, peninsular India: stratigraphic implications and significance of a 1-Ga thermal-magmatic event. J Geol 119:535–548

Biswal TK, Sinha S, Mandal A, Ahuja H, Das MK (2003) Deformation pattern of Bastar Craton adjoining Eastern Ghat Mobile Belt, NW Orissa. Gondwana Geol Mag Spl Publ 7:101–108

Biswal TK, De Waele B, Ahuja H (2007) Timing and dynamics of the juxtaposition of the Eastern Ghats Mobile Belt against the Bhandara Craton, India: a structural and zircon U-Pb SHRIMP study of the fold-thrust belt and associated nepheline syenite plutons. Tectonics, vol 26, TC4006, doi:10.1029/2006TC002005

Chalapathi Rao NV, Gibson SA, Pyle DM, Dickin AP (2004) Petrogenesis of Proterozoic lamproites and kimberlites from the Cuddapah basin and Dharwar Craton, southern India. J Petrol 45:907–948

Chalapathi Rao NV, Kamde G, Kale HS, Dongre A (2010) Petrogenesis of the mesoproterozoic lamproites from the Krishna valley, Eastern Dharwar Craton, Southern India. Precambr Res 177:103–130

Chalapathi Rao NV, Lehmann B, Mainkar D, Belyatsky B (2011) Petrogenesis of the end-Cretaceous diamondiferous Behradih orangeite pipe: implication for mantle plume lithosphere interaction in the Bastar Craton, Central India. Contrib Miner Petrol 161:721–742

Chalapathi Rao NV, Miller JA, Gibson SA, Pyle DM, Madhavan V (1999) Precise 40Ar/39Ar age determinations of the Kotakonda kimberlite and Chelima lamproite, India: implication to the timing of mafic dyke swarm emplacement in the eastern Dharwar Craton. J Geol Soc India 53:425–432

Chalapathi Rao NV, Miller JA, Pyle DM, Madhavan V (1996) New proterozoic K–Ar ages for some kimberlites and lamproites from the Cuddapah Basin and Dharwar Craton, South India: evidence for non-contemporaneous emplacement. Precambr Res 79:363–369

Chaudhuri AK, Mukhopadhyay J, Patranabis-Deb S, Mukherjee MK, Ghosh G (2002) The Purana basins of southern cratonic province of India—a case for Mesoproterozoic fossil rifts. Gond Res 5:23–33

Choukroun M, O'reilly SY, Griffin WL, Pearson NJ, Dawson JB (2005) Hf isotopes of MARID (mica-amphibole-rutile-ilmenite-diopside) rutile trace metasomatic processes in the lithospheric mantle. Geology 33:45–48

Coe N, Le Roex A, Gurney JJ, Pearson GD, Nowell G (2008) Petrogenesis of Swartruggens and Star Group II kimberlite dyke swarms, South Africa: constraints from whole rock geochemistry. Contrib Miner Petrol 156:627–652

Das DP, Kundu A, Das N, Dutta DR, Kumaran K, Ramamurthy S, Thanavelu C, Rajaiya V (1992) Lithostratigraphy and sedimentation of Chhattisgarh basin. Indian Minerals 46:271–288

Das K, Yokoyama K, Chakraborty PP, Sarkar A (2009) Basal tuffs and contemporaneity of the Chattisgarh and Khariar basins based on new dates and geochemistry. J Geol 117:88–102

Das N, Dutta DR, Das DP (2001) Proterozoic cover sediments of southeastern Chhattisgarh state and adjoining parts of Orissa. Geol Surv India Spec Publ 55(2):237–262

Das S, Nasipuri P, Bhattacharya A, Swaminathan S (2008) The thrust-contact between the Eastern Ghats belt and the adjoining Bastar Craton (Eastern India): evidence from mafic granulites and tectonic implications. Precambr Res 162:70–85

Davies GR, Stolz AZ, Mahotkin IL, Mowell GM, Pearson DG (2006) Trace element and Sr-Pb-Nd-Hf evidence for ancient, fluid-dominated enrichment of the source of Aldan shield lamproites. J Petrol 47:1119–1146

Dawson JB, Stephens WE (1975) Statistical classification of garnets from kimberlite and associated xenoliths. J Geol 83:589–607

Egorov KN, Solov'eva LV, Kovach VP, Men'shagin YuV, Maslovskaya MN, Sekerin AP, Bankovskaya EV (2006) Petrological features of olivine–phlogopite lamproites of the Sayan region: evidence from Sr–Nd isotope and ICP-MS trace-element data. Geochem Int 44:729–735

Fipke CE, Gurney JJ, Moore RO (1995) Diamond exploration techniques emphasizing indicator mineral geochemistry and Canadian examples. Geol Surv Canada Bull 423

Foley SF (1992a) Petrological characterization of the source components of potassic magmas: geochemical and experimental constraints. Lithos 28:187–204

Foley SF (1992b) Vein-plus-wall-rock melting mechanisms in the lithosphere and the origin of potassic magmas. Lithos 28:435–453

Foley SF (1993) An experimental study of olivine lamproite: first results from the diamond stability field. Geochim Cosmochim Acta 57:483–489

Foley SF, Venturelli G, Green DH, Toscani L (1987) The ultrapotassic rocks: characteristics, classification and constraints for petrogenetic models. Earth Sci Rev 24:81–134

Foley SF, Jenner GA (2004) Trace element partitioning in lamproitic magmas—the Gaussberg olivine leucitite. Lithos 75:19–38

Fraser KJ (1987) Petrogenesis of kimberlites from South Africa and lamproites from Western Australia and North America. Ph.D. thesis, The Open University, Milton Keynes, UK

Fraser KJ, Hawkesworth CJ, Erlank AJ, Mitchell RH, Scott-Smith BH (1985) Sr, Nd and Pb isotope and minor element geochemistry of lamproites and kimberlites. Earth Planet Sci Lett 76:57–70

French JE, Heaman LM, Chacko T, Srivastava RK (2008) 1891–1883 Ma Southern Bastar–Cuddapah mafic igneous events, India: a newly recognized large igneous province. Precambr Res 160:308–322

Ghosh JG (2004) 3.56 Ga tonalite in the central part of the Bastar Craton, India: oldest Indian date. J Asian Earth Sci 23:359–364

Gregory LC, Meert JG, Pradhan V, Pandit M, Tamrat E, Malone SJ (2006) A paleomagnetic and geochronologic study of the Majhgawan kimberlite, India: implications for the age of the upper Vindhyan supergroup. Precambr Res 149:65–75

Griffin WL, Fisher NI, Friedman JH, Ryan CG (1997) Statistical techniques for the classification of chromites in diamond exploration samples. J Geochem Explor 59:233–249

Grütter HS, Gurney JJ, Menzies AH, Winter F (2004) An updated classification scheme for mantle-derived garnet, for use by diamond explorers. Lithos 77:841–857

Grütter HS, Latti D, Menzies A (2006) Cr-Saturation arrays in concentrate garnet compositions from kimberlite and their use in mantle barometry. J Petrol 47:801–820

Hussain MF, Ahmad T, Mondal MEA (2008) Geochemistry of the Precambrian mafic dyke swarms of the central and northeastern parts of Bastar Craton, central India: constraints of their enrichment processes. In: Srivastava RK, Sivaji Ch, Chalapathi Rao NV (eds) Indian dykes: geochemistry, geophysics and geochronology. Narosa Publishing House, New Delhi, pp 397–412

Jaques AL, Lewis JD, Smith CB (1986) The kimberlites and lamproites of Western Australia. Geol Surv Western Australia Bull 132:268p

Klemme S (2004) The influence of Cr on the garnet-spinel transition in the Earth's mantle: experiments in the system $MgO–Cr_2O_3–SiO_2$ and thermodynamic modeling. Lithos 77:639–646

Kumar A, Gopalan K, Rao KRP, Nayak SS (2001) Rb–Sr Age of kimberlites and lamproites from Eastern Dharwar Craton, South India. J Geol Soc India 58:135–141

Kumar A, Padmakumari VM, Dayal AM, Murthy DSN, Gopalan K (1993) Rb–Sr ages of Proterozoic kimberlites of India: evidence for contemporaneous emplacement. Precambr Res 62:227–232

Kumar A, Heaman LM, Manikyamba C (2007) Mesoproterozoic kimberlites in south India: A possible link to $\sim$1.1 Ga global magmatism. Precambr Res 154:192–204

le Roex AP, Bell DR, Davis P (2003) Petrogenesis of group I Kimberlites from Kimberley, South Africa: evidence from bulk-rock geochemistry. J Petrol 44:2261–2286

Lehmann B, Burgess R, Frei D, Belyatsky B, Mainkar D, Chalapathi Rao NV, Heaman LM (2010) Diamondiferous kimberlites in central India synchronous with Deccan flood basalts. Earth Planet Sci Lett 290:142–149

Lewis JD (1987) The geology and geochemistry of the West Kimberly lamproite province, Western Australia. M.Sc. thesis, University of Western Australia, Perth

Lucas H, Ramsay R, Hall AE, Smith CB, Sobolev NV (1989) Garnets from Western Australian kimberlites and related rocks. In: Ross J, Ferguson J, Green DH, O'Reilly SY, Danchin RV, Janse AJA (eds) Kimberlites and related rocks. Geol Soc Australia Spec Publ 14(2):809–819

Ludwig KR (2001) ISOPLOT 2.49: a geochronological toolkit for microsoft excel, Berkeley Geochronology Center, Berkeley, CA

Mainkar D (2011) Petrological and geochemical investigation of the Behradih kimberlite from the Bastar Craton, central India, with special reference to its diamond potential. Unpublished Ph.D. thesis, Pt. R.S. University, Raipur (Chhattisgarh), 175p

Masun K, Sthapak AV, Singh A, Vaidya A, Krishna C (2009) Exploration history and geology of the diamondiferous ultramafic Saptarshi intrusions, Madhya Pradesh, India. Lithos 112:142–154

Mathur SM, Singh HN (1971) Petrology of the Majhgawan pipe rock. Geol Surv India Misc Publ 19:78–85

Mc Donough WF, Sun SS (1995) The composition of the Earth. Chem Geol 120:223–253

Meert JG, Pandit MK, Pradhan VR, Kamenov G (2011) Preliminary report on the paleomagnetism of 1.8 Ga dykes from the Bastar and Dharwar cratons, Peninsular India. Gond Res 20:335–343

Menzies MA, Hawkesworth CJ (eds) (1987) Mantle metasomatism. Academic Press, London

Mirnejad H, Bell K (2006) Origin and source evolution of the Leucite Hills lamproites: evidence from Sr–Nd–Pb–O isotopic compositions. J Petrol 47:2463–2489

Mitchell RH (1995) Melting experiments on a sanidine phlogopite lamproite at 4–7 GPa and their bearing on the sources of lamproitic magmas. J Petrol 36:1455–1474

Mitchell RH (2006) Potassic magmas derived from metasomatised lithospheric mantle: nomenclature and relevance to exploration for diamond-bearing rocks. Geol J Soc India 67:317–327

Mitchell RH, Bergman SC (1991) Petrology of lamproites. Plenum Press, New York 447p

Mitchell RH, Edgar AD (2002) Melting experiments on SiO_2-rich lamproites to 6.4 GPa and their bearing on the sources of lamproite magmas. Miner Petrol 74:115–128

Mukherjee A, Ray RK (2010) An alternate view on the stratigraphic position of the ~1-Ga Sukhda tuff vis-a'-vis chronostratigraphy of the Precambrians of the Central Indian Craton. J Geol 118:325–332

Mukhopadhyay PK, Ghosh S, Rath SC, Swain RB, Shome S (2004) New finds of lamproite dykes in Nawapara district, Orissa. Indian Minerals 58:183–196

Murphy DT, Collerson KD, Kamber BS (2002) Lamproites from Gaussberg, Antarctica: possible transition zone melts of Archaean subducted sediments. J Petrol 43:981–1001

Nanda JK, Rath SC, Behera SN (2000) Alkaline and ultramafic magmatism in the contact zone between high and low grade terrains: Example from northwestern Orissa, India. Geol Surv India Spec Publ 57:122–130

Nixon PH, Thirwall MF, Buckley F, Davis CJ (1984) Spanish and Western Australian lamproites: Aspects of whole rock chemistry. In: Kornprobst J (ed) Kimberlites and related rocks. Proceedings of the Third International Kimberlite Conference, vol 1. pp. 285–296

Osborne I, Sherlock S, Anand M, Argles T (2011) New Ar–Ar ages of southern Indian kimberlites and a lamproite and their geochemical evolution. Precambr Res 189:91–103

Patel SC, Ravi S, Anilkumar Y, Naik A, Thakur SS, Pati JK, Nayak SS (2009) Mafic xenoliths in Proterozoic kimberlites from Eastern Dharwar Craton, India: mineralogy and P–T regime. J Asian Earth Sci 34:336–346

Patel SC, Ravi S, Anilkumar Y, Pati JK (2010) Major element composition of concentrate garnets in Proterozoic kimberlites from the Eastern Dharwar Craton, India: implications on sub-continental lithospheric mantle. J Asian Earth Sci 39:578–588

Patnaik BC, Sahu N, Mishra BP, Maharana RC (2004) Discovery of a diamondiferous olivine-lamproite pipe in Orissa. Bull Soc Geoscientists Allied Technol 5:34–38

Patranabis-Deb S, Bickford ME, Hill B, Chaudhury AK, Basu A (2007) SHRIMP ages of zircon in the uppermost tuff in Chattisgarh Basin in central India require ~500-Ma adjustment in Indian Proterozoic stratigraphy. J Geol 115:407–415

Patranabis-Deb S, Chaudhury AK (2007) A retreating fan-delta system in the Neoproterozoic Chhattisgarh rift basin; central India: major controls on its evolution. AAPG Bull 91:785–808

Phillips D (2012) Comment on "New Ar–Ar ages of southern Indian kimberlites and a lamproite and their geochemical evolution" by Osborne et al. [Precambrian Res. 189 (2011) 91–103]. Precambr Res 208–211:49–52

Pisarevsky SA, Biswal TK, Wang X, De Waele B, Ernst R, Söderlund U, Tait JA, Ratre K, Singh YK, Cleve M (2012) Palaeomagnetic, geochronological and geochemical study of Mesoproterozoic Lakhna Dykes in the Bastar Craton, India: Implications for the Mesoproterozoic supercontinent. Lithos, http://dx.doi.org/10.1016/j.lithos.2012.07.015

Pollack HN, Chapman DS (1977) On the regional variation of heat flow, geotherms, and lithospheric thickness. Tectonophysics 38:279–296

Pradhan VR, Meert JG, Pandit MK, Kamenov G, Mondal MEA (2012) Paleomagnetic and geochronological studies of the mafic dyke swarms of Bundelkhand craton, central India: implications for the tectonic evolution and paleogeographic reconstructions. Precambr Res 198–199:51–76

Prelevic D, Foley SF, Cvetkovic V (2007) A review of petrogenesis of Mediterranean Tertiary lamproites: a perspective from the Serbian ultrapotassic province In: Beccaluva L, Banchini G, Wilson M (eds) Cenozoic Volcanism in the Mediterranean area. Geol Soc Amer Spec Paper 418:113–129

Rajesh HM, Mukhopadhyay J, Beukes NJ, Gutzmer J, Belyanin GA, Armstrong RA (2009) Evidence for an early Archaean granite from Bastar Craton. India J Geol Soc 166:193–196

Ramakrishnan M, vaidyanadhan R (2008) Geology of India, vol 1. Geological Society of India, Bangalore, pp 556p

Ratre K, De Waele B, Biswal TK, Sinha S (2010) SHRIMP geochronology for the 1450 Ma Lakhna dyke swarm: Its implication for the presence of Eoarchaean crust in the Bastar Craton and 1450–517 Ma depositional age for Purana basin (Khariar), Eastern Indian Peninsula. J Asian Earth Sci 39:565–577

Reddy TAK, Sridhar M, Ravi S, Chakravarthi V, Neelakantam S (2003) Petrography and geochemistry of the Krishna Lamproite field, Andhra Pradesh. J Geol Soc India 61:131–146

Renne PR, Swisher CC, Deino AL, Karner DB, Owens TL, De Paolo DJ (1998) Intercalibration of standards, absolute ages and uncertainties in $^{40}Ar/^{39}Ar$ dating. Chem Geol 145:117–152

Ringwood AE, Kesson SE, Hibberson W, Ware N (1992) Origin of kimberlites and related magmas. Earth Planet Sci Lett 113:521–538

Rogers NW, Hawkesworth CJ, Palacz ZA (1992) Phlogopite in the generation of olivine-melilitites from Namaqualand, South Africa, and its implications for element fractionation processes in the upper mantle. Lithos 28:347–365

Roy P, Balaram V, Kumar A (2007) New REE and trace element data on two kimberlite reference materials by ICP-MS. Geostand Geoanal Res 31:261–273

Sarkar G, Corfu F, Paul DK, Mc Naughton NJ, Gupta SN, Bishui PK (1993) Early Archean crust in Bastar Craton, central India—a geochemical and isotopic study. Precambr Res 62:127–137

Sarkar SK, Shashidharan K, Mohanty AK, Mishra BK, Patel MK, Datta B, Ganvir DV (2001) Exploration for diamond and KCR in the Bastar Craton. Geol Surv India Spec Publ 64:425–434

Schmidt KH, Bottazzi PR, Vannucci R, Mengel K (1999) Trace element partitioning between phlogopite, clinopyroxene and leucite lamproite melt. Earth Planet Sci Lett 168:287–299

Smith CB (1992) The age of the Majhgawan Pipe, India. Scott Smith Petrology, p 9

Smith CB, Atkinson WJ, Tyler EWJ (1991) Diamond exploration in Western Australia, Northern Territory and South Australia. In: Glasson KR, Rattigan JH (eds) Geological aspects of the discovery of important minerals in Australia. Australas Inst Mineral Metall, Melbourne, pp 429–453

Späth A, Le Roex AP, Opiyo-Akech N (2001) Plume–lithosphere interaction and the origin of continental rift-related alkaline volcanism—the Chyulu Hills Volcanic Province, southern Kenya. J Petrol 42:765–787

Srivastava RK, Singh RK (2004) Trace element geochemistry and genesis of Precambrian sub-alkaline mafic dikes from the central Indian craton: evidence for mantle metasomatism. J Asian Earth Sci 23:373–389

Srivastava RK, Gautam GC (2009) Precambrian mafic magmatism in the Bastar Craton, central India. J Geol Soc India 73:52–72

Srivastava RK, Gautam GC (2012) Early Precambrian mafic dyke swarms from the central Archaean Bastar Craton, India: geochemistry, petrogenesis and tectonic implications. J Geol 47:144–160

Stein HJ, Hannah JL, Zimmerman A, Markey RJ, Sarkar SC, Pal AB (2004) A 2.5 Ga porphyry Cu–Mo–Au deposit at Malanjkhand, central India: implications for Late Archean continental assembly. Precambr Res 134:189–226

Subba Rao DV, Sridhar DN, Balara V, Nagaraju K, Rao TG, Keshavakrishna A, Singh UP (2008) Proterozoic mafic-ultramafic dyke swarms in the vicinity of Chhattisgarh–Khariar–Singhora basins in northern Bastar Craton, India. In: Srivastava RK, Sivaji CH, Chalapathi Rao NV (eds) Indian dykes: geochemistry, geophysics and geochronology. Narosa Publishing House, New Delhi, pp 377–396

Tainton KM, Mc Kenzie D (1994) The generation of kimberlites, lamproites and their source rocks. J Petrol 35:787–817

Vijesh VK (2010) Petrology of lamproites from the Krishna Lamproite field, Andhra Pradesh. Unpublished M.Tech (Geoexploration) thesis, IIT Bombay, 65p

Yellappa T, Chalapathi Rao NV, Chetty TRK (2010) Occurrence of lamproitic dykes at the northern margin of the Indravati Basin, Bastar Craton, central India. J Geol Soc India 75:632–643

The Geology and Geochemistry of the Wadagera Kimberlite and the Characteristics of the Underlying Subcontinental Lithospheric Mantle, Dharwar Craton, India

Michael Lynn, Sojen Joy, and Robin Preston

Abstract

The Wadagera kimberlite is situated near the northern bank of the Krishna River in northern Karnataka. It is the largest pipe in a cluster discovered by De Beers during exploration in 2002. The pipe has been dated at 1,083.8 ± 5.3 Ma, which corresponds to published ages of kimberlites in the adjacent Narayanpet cluster. The pipe was discovered during follow-up of an airborne multispectral scanner survey. Ground magnetic and drilling data indicate that the pipe has an area of approximately 6 ha. The kimberlite intrudes the Peninsular Gneiss Complex, and limited petrographic analyses suggest that both coherent and volcaniclastic kimberlites are preserved. Kimberlite surface samples show extreme calcretisation and yield few mantle garnets. Garnet xenocrysts are more abundant in drill samples below 20 m depth. Differential destruction of silicate and oxide minerals is observed, which impacts on the abundance and distribution of indicator minerals in the secondary environment. KIMs recovered from the Wadagera kimberlite reveal the SCLM architecture sampled by the kimberlite. The major element garnet mineral chemistry represents a dominant lherzolitic population. Diamond proxy composition garnets are rare or absent. The well-developed CCGE trend, together with the spinel data, suggests that the kimberlite has mainly sampled fertile mantle at relatively shallow depth. The nature of the SCLM at Wadagera contrasts with that reported further south and south-east on the EDC within the diamondiferous Wajrakarur kimberlite and Banganapalle lamproite clusters. The geochemical data, as well as limited petrographic data, are consistent with the apparent absence of diamond at Wadagera.

Keywords

Kimberlite • Wadagera • India • Indicator minerals • Geotherm • East Dharwar Craton • Garnet geochemistry

Introduction

The border area between northern Karnataka and Andhra Pradesh is host to a large number of kimberlite occurrences, which are intruded into the East Dharwar Craton (EDC). They were reported by the Geological Survey of India (GSI) during 1990s and are known as the Narayanpet cluster (Nayak et al. 1988; Babu 1998, pp. 137–140; Fareeduddin and Mitchell 2012, pp. 72–93). These kimberlites occur

M. Lynn (✉)
The MSA Group, Johannesburg, South Africa
e-mail: mikel@msagroupservices.com

S. Joy · R. Preston
De Beers Group Services, Group Exploration, Johannesburg, South Africa

D. G. Pearson et al. (eds.), *Proceedings of 10th International Kimberlite Conference*, Volume 1, Special Issue of the Journal of the Geological Society of India,
DOI: 10.1007/978-81-322-1170-9_10,

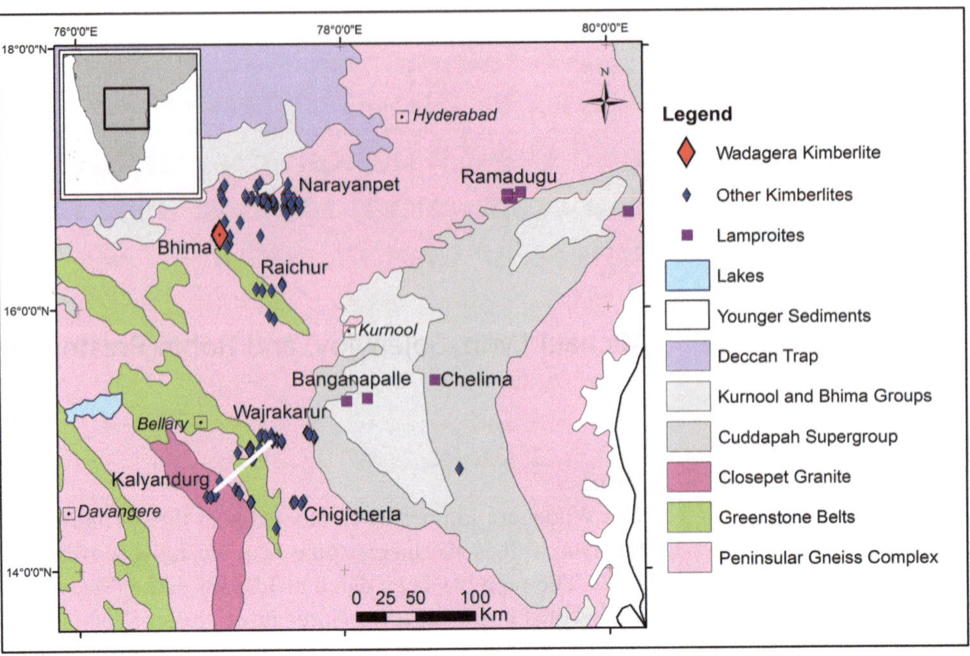

Fig. 1 Regional geological setting of the Bhima cluster and the Wadagera kimberlite pipe on the Dharwar Craton. The *blue triangles* are kimberlite occurrences, and the *magenta squares* are lamproite occurrences. The labels reflect the names of kimberlite/ lamproite clusters. The *white line* is the location of the Kalyandurg–Wajrakarur (K/W) transect described by Griffin et al. (2009)

within the catchment of the Krishna River basin where diamonds have historically been reported and were consequently targeted for exploration. The GSI reported a total of 30 kimberlite occurrences, but no diamonds were reported. Further exploration by De Beers during the period 2001–2002, using a variety of techniques, increased this number by 18 occurrences within the same cluster. De Beers also discovered a new cluster of kimberlite pipes to the south of the Narayanpet cluster (Fig. 1) near the confluence of the Krishna and Bhima rivers. This cluster comprises 13 occurrences and is named the Bhima cluster (Lynn 2005; Fareeduddin and Mitchell 2012, p. 93). The Wadagera occurrence is the largest of the Bhima cluster kimberlites and is an elongate pipe with a surface area of approximately 6 ha, estimated from geophysics and drilling. This paper summarises the discovery, geology and geochemistry of the kimberlite as elucidated from exploration data.

Geological Setting

The Dharwar Craton is divided into two tectonic blocks namely the West Dharwar Craton (WDC) and the EDC separated by the Chitradurga Shear Zone, on the western side of the Closepet Granite (Rogers 1986; Ramakrishnan and Vaidyanadhan 2008, p. 99). The Dharwar Craton is comprised of gneisses, greenstone belts and granitic intrusives of Archaean age. The oldest gneisses (~3,300 Ma) occur in the WDC and have been grouped as Peninsular Gneiss-I (Balasubrahmanyan 2006, p. 23) or as the Peninsular Gneiss sensu stricto (Chadwick et al. 2000). They are Na-rich tonalite–trondhjemite gneisses, which are less deformed and nearly massive when compared to younger gneisses (Balasubrahmanyan 2006, p. 23). Linear belts of ancient supracrustals (older greenstones) occur east of the Closepet Granite and as rafts in the EDC. These ancient supracrustals are named as the Sargur Group (Swami Nath and Ramakrishnan 1981; Radhakrishna and Naqvi 1986). They represent shelf facies sediments intercalated with mafic rocks and iron formations, which have been metamorphosed to a high grade. The Peninsular Gneiss–II (~3,000 Ma) is the basement gneiss to the Dharwar Supergroup (Balasubrahmanyan 2006, p. 26). This gneiss has also been called the Dharwar Batholith (Chadwick et al. 2000). The Dharwar Supergroup comprises younger greenstone belts including the Chitradurga, Bababudan, Shimoga, Gadag, Sandur, Kolar, Hutti and Ramagiri Belts. These younger greenstone belts occur on both the EDC and WDC. The Closepet Granite (~2,600 Ma) occurs as early silica-poor porphyritic granite and as later silica-rich anatectic grey to pink granites and marks a probable rift zone (Balasubrahmanyan 2006, pp. 33–34).

Proterozoic intracratonic basins cover parts of the Archaean rocks of the Dharwar Craton. The main basins are the Cuddapah Basin with its younger Kurnool–Palnad sub-basins to the east, the Kaladgi–Badami Basin to the north-west, the Bhima Basin to the north and the Pranhita–Godavari Basin to the north-east (Ramakrishnan and Vaidyanadhan 2008, p. 456).

The Dharwar Craton is covered in the north by basalts of the Deccan Traps (~65 Ma; Balasubrahmanyan 2006, p. 130). The basalts themselves, as well as their residual weathering products, which are preserved in some areas away from the mapped extent of the basalts, have implications for exploration. The basalts and residual clay-rich soils have the potential to cover kimberlite occurrences and to impact on the efficacy of kimberlite exploration techniques such as heavy mineral, spectral, magnetic and electromagnetic surveys.

The Wadagera kimberlite intrudes rocks of the Peninsular Gneiss Complex-II on the exposed shield of the EDC. Fifty-five kilometres to the north-west of the kimberlite, the craton is covered by the basalts of the Deccan Traps and the Bhima Group sedimentary sequence, which has been classified as Mesoproterozoic and correlated with the Kurnool Group (Chalapathi Rao et al. 2009). The kimberlite intrusion age has been dated at 1,083.8 ± 5.3 Ma by a combination of Rb–Sr and $^{40}Ar/^{39}Ar$ on phlogopite and U–Pb on in situ perovskite (Patton 2007; Patton et al. 2007), which broadly corresponds to the age of kimberlites in the Narayanpet cluster (Patton et al. 2007; Chalapathi Rao et al. 1999; Kumar et al. 2007).

Exploration Methodology

The kimberlites of the Bhima cluster were discovered by a variety of techniques including a spectral scanner survey, kimberlitic indicator mineral (KIM) stream and loam sampling and an airborne geophysical survey. The Wadagera kimberlite was discovered using the De Beers/Anglo-American airborne multispectral scanner (AMS). The kimberlite also gives a strong magnetic signature and occurs within a broad indicator mineral anomaly. However, no clear airborne frequency-domain electromagnetic (FDEM) anomaly was produced.

Airborne Multispectral Scanner Survey

The AMS system was developed in Australia and is a mineral mapping technology based on the spectral signatures of minerals occurring at the Earth's surface. The system collects data from up to 126 channels in the very near-infrared (VNIR) and short-wave infrared (SWIR) wavelengths across a swath of 512 pixels. When applied to diamond exploration, anomalies are selected by identifying the weathering products of kimberlite, including talc, saponite, serpentine, nontronite and carbonates. Matched filtering is used to distinguish anomalous minerals from the background. Matched filtering is a partial unmixing technique, which maps the proportion of a target signature in each pixel against a background calculated statistically from the survey scene. Pixels with a high proportion of target material will appear bright, while those with only background spectra will appear dark. An RGB colour composite can be constructed to produce a ternary mineral map showing relative distributions of targeted minerals. False positives are possible where pixels report material that is atypical of the statistical background and where minerals occur that are unrelated to kimberlite but have similar absorption wavelengths to the target minerals. In northern Karnataka, the contrast between Mg-rich minerals (saponite, serpentine and nontronite) and carbonate (calcretes) with the granitic country rocks proved particularly effective in identifying kimberlite. The limited presence of transported cover is also an important factor in the efficacy of the method.

The survey was flown on a fixed wing platform (P68C aircraft) at an altitude of approximately 3,000 m (providing a pixel resolution of ~10 m on the ground) along flight lines 2 km apart. The matched filter image was generated from the following combinations of absorption features (red, green and blue, respectively): talc and saponite (primary absorption near 2,314 nm with a secondary absorption near 2,390 nm), serpentine and carbonates (2,322 nm) and nontronite (2,290 nm). False positives were notable where Mg-rich tremolite and actinolite occur in the country rocks (such as amphibolite; red band) and where hydrous clays such as montmorillonite occur (green band). Anomalies were defined by drawing a polygon, outlining the anomalous pixels.

Follow-up of the anomalies generated by the AMS survey was conducted using false colour composite images for photo-location in the field, with the anomalous polygons superimposed on the image. Rock or soil samples were collected from the anomalous areas for spectral analysis at the field base to confirm that the material observed in the field was representative of the airborne anomaly. Field analysis was conducted with a portable infrared mineral analyser (PIMA), which measures absorption wavelengths between 1,300 and 2,500 nm. At Wadagera, an area of calcrete float was observed in the field during follow-up of a serpentine/carbonate spectral anomaly (Fig. 2). Pitting of the anomaly led to the initial kimberlite discovery and showed that the kimberlite is altered to a very mature calcrete and silcrete duricrust at surface. The calcrete is in the process of eroding to produce a deflation of calcrete nodules on surface.

Indicator Mineral Sampling

The Wadagera kimberlite was selected for orientation KIM sampling studies, because of its size and spatial association with drainages. This was achieved by collecting drill samples from percussion boreholes, soil samples along two north–south traverse lines across the pipe, and stream samples at intervals downstream of the pipe.

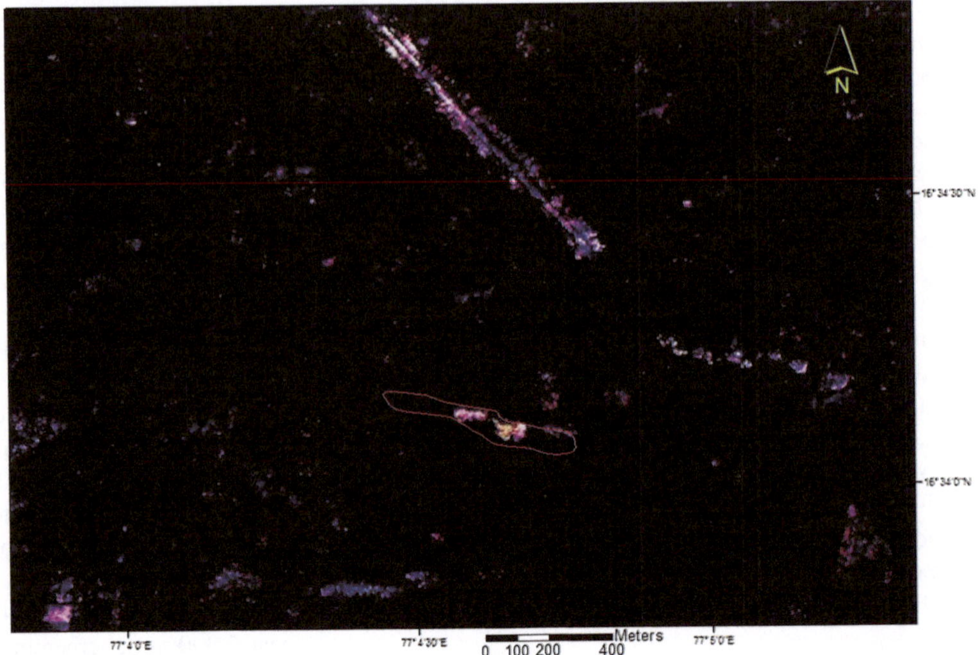

Fig. 2 The match-filtered AMS image for kimberlitic minerals. The outline of the Wadagera kimberlite, determined from geophysics, is shown by the *red* outline. The NW–SE-orientated linear feature north of the kimberlite is a canal. Note that the AMS will only map exposed and weathered portion of the kimberlite and therefore appears smaller than the geophysically mapped outline

Samples comprising 1 litre of drill chips were recovered at 1 m downhole intervals from two percussion boreholes on the pipe. The samples were processed at the De Beers processing plant in Bangalore to recover KIMs in the >0.3,<1.0 mm size interval. KIMs were visually extracted by analysts at the De Beers mineral laboratories in Bangalore and Melbourne. Both boreholes gave similar results, and only one (551/081/048/H3) is reported here (Fig. 3a). Results presented are visual counts, and all the grains were subsequently analysed and confirmed to be >95 % kimberlitic (refer to section on Chemistry of KIMs). It can be seen that garnet and chrome diopside counts are significantly reduced within approximately 20 m of surface, and it is deduced that the effect of weathering processes is to preferentially destroy the silicate KIMs (garnet + chrome diopside) relative to the oxides (chromite + ilmenite).

This differential destruction of KIMs also has an effect on indicator dispersion in soil and stream sampling results. In Fig. 3b, the results from the two north-south oriented soil sample traverses across the pipe are shown. Samples comprising 10 l of soil from below the deflation surface were collected at 20 m intervals along the two traverses. Apart from a few single grains, garnet and chrome diopside distribution is restricted to within a few tens of metres of the pipe surface, whereas the oxide phases are far more widely dispersed (hundreds of metres).

Four stream samples comprising 30 l of material, screened to <2.0 mm, were collected at 1 km intervals downstream of the kimberlite from the best available heavy mineral trap sites in the second-order seasonal stream that transects the kimberlite. An additional sample was collected approximately 8 km downstream of the kimberlite. The trap sites included transverse rock bars and the heads of gravel bars. Total indicator mineral recoveries in the >0.3, <2.0 mm size range are shown in Fig. 3c. Chrome diopside and garnet survive in the stream environment up to approximately 2 km from the pipe. In contrast, ilmenite and chromite may be found several kilometres from the source. It is notable that while chromite is recovered up to several kilometres from the source kimberlite, high counts are restricted to a short distance from the pipe. This may be due to the mineral's high density and brittleness. The five kimberlitic indicators (3 chromites and 2 ilmenites) recovered in the sample furthest south, close to Krishna River, could contain a component from a small kimberlite discovered later in the area (Fig. 3c). However, the absence of garnet and chrome diopside in that sample is still relevant for this discussion.

While these results to some extent reflect the primary abundance of KIMs in the source kimberlite, the preferential destruction of silicate minerals in the weathering profile is thought to have the most significant impact on their distribution. Ilmenite and spinel dispersion is such that these minerals indicate the presence of a kimberlite cluster, but they do not serve to determine the exact location of individual kimberlites within the cluster. Conversely, the presence of garnet and chrome diopside in sample results suggests a high probability of close proximity to a kimberlite occurrence.

Airborne Geophysical Survey

The area of the kimberlite cluster (as determined by kimberlite discoveries and the ilmenite/chromite KIM anomaly)

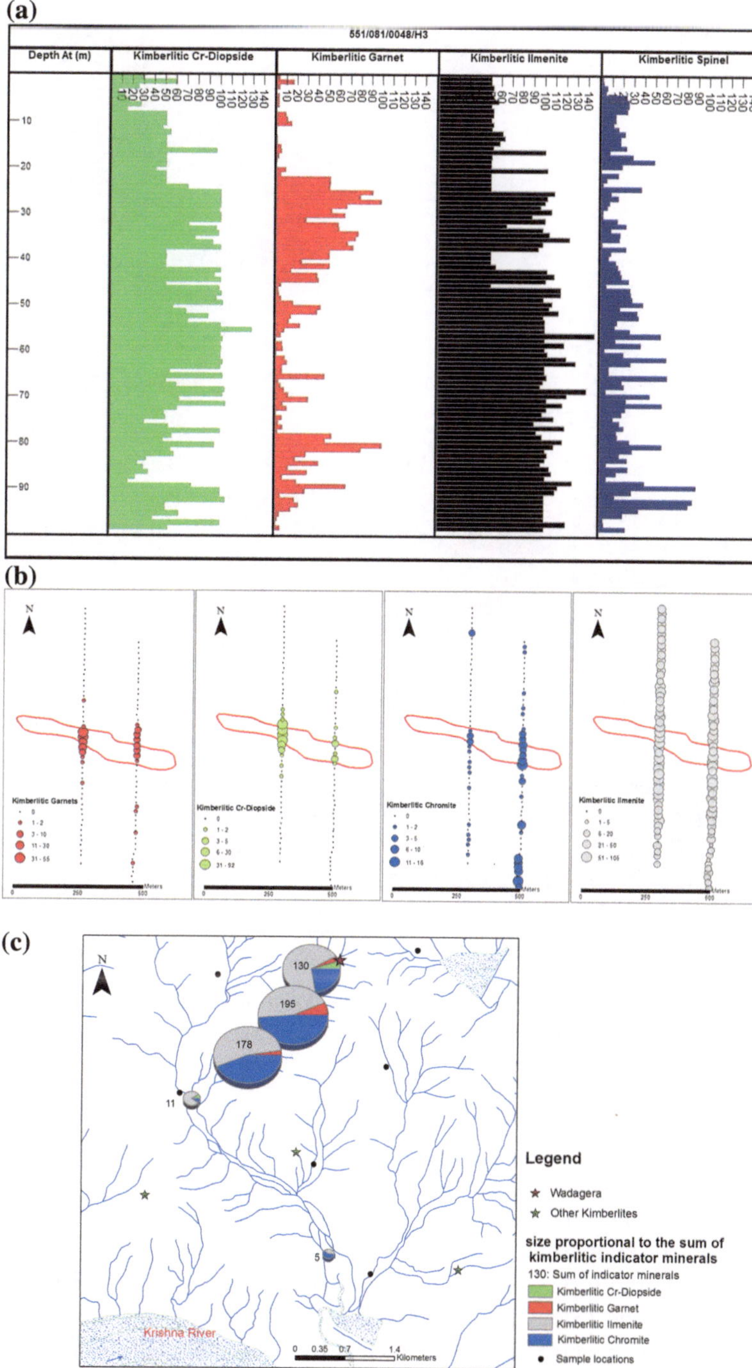

Fig. 3 **a** Downhole distribution of visually identified KIMs. The X-axis represents number of visually identifiedindicator minerals. Note destruction of silicate phases (garnet and Cr-Diopside) near surface. **b** Map view of thekimberlite with visually identified KIMs from two soil sampling traverses. Note that the silicate KIMs are rare beyonda few tens of metres from the kimberlite in soil samples. **c** Visually identified KIMs from stream samplingdownstream of the Wadagera kimberlite. Note the absence of silicate KIMs beyond a few km from the kimberlite

was subsequently flown using Fugro Airborne Surveys' DIGital helicopter-borne ElectroMagnetic (DIGHEM) and magnetics systems to determine the presence of further kimberlites (for technical details, visit www.fugroairborne.com). DIGHEM is a frequency-domain electromagnetic system and is configured in a cylindrical "bird", which is carried beneath a helicopter. DIGHEM employs five pairs of transmitting and receiving coils: three horizontal coplanar coil pairs and two vertical coaxial coil pairs. The coil separation is normally 8 m on all frequencies. The system uses multiple frequency electromagnetic fields to measure and map the electrical conductivity of the Earth in three dimensions at shallow depths to approximately 150 m. The coplanar coil pairs provide greatest sensitivity to the conductivity of the Earth, to layers and to subtle changes in conductivity. They are also least sensitive to the direction of the survey. The coaxial coil pairs produce the strongest anomalies from vertical conductors, such as kimberlite pipes. The magnetic system was a simple single sensor "stinger" towed above the DIGHEM system.

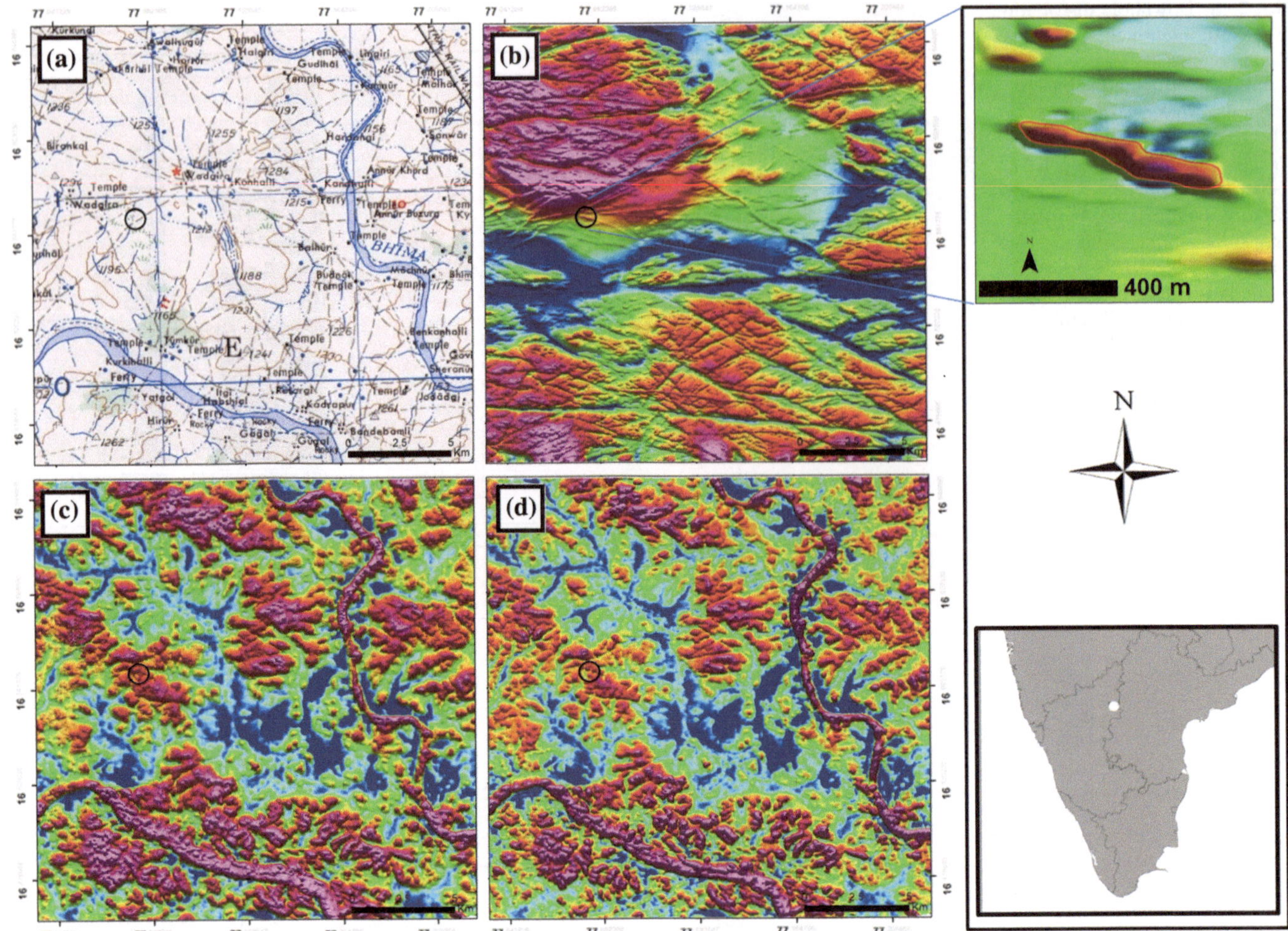

Fig. 4 Comparison of the Wadagera kimberlite's response to airborne magnetics and frequency-domain EM geophysical techniques. The *black circle* centre *left* of each image shows the location of the kimberlite. **a** Extract from the 1:250,000 map sheet NE 43-16 (Raichur). **b** Reduced-to-pole (*RTP*) magnetic image. **c** Resistivity (3,000 Hz). **d** Resistivity (15,000 Hz). The scale bar at the *bottom right* of each of the images is 5 km. The image in the *top right* corner is a close-up of the RTP magnetic image over the kimberlite

Figure 4 is an extract of the area flown and compares the EM and magnetic images, as well as an extract of the same area from the published 1:250,000 topographic map. The Wadagera kimberlite can be clearly seen on the magnetic image, and in fact, ground magnetics was later used to map the shape and extent of the kimberlite. However, the electromagnetic method was ineffective, partly due to the highly conductive clays present in the area, derived from the weathering of basalts to the north and concentrated in drainage valleys.

Drilling

The kimberlite is elongated in a WNW-ESE direction, with a maximum length of approximately 600 m. A total of five shallow (100 m) percussion holes were drilled into the kimberlite to test the areal extent and to obtain samples for mineral chemistry, petrographic and microdiamond analysis.

Petrography and Whole-Rock Geochemistry

Petrographic analyses of drill chips show that the kimberlite is very much altered and textures are generally difficult to interpret (Fig. 5). However, analysis of three separate samples suggests that the kimberlite may have both magmatic and fragmental (volcaniclastic) textures preserved. Coarse groundmass perovskite and groundmass spinel are notable and suggest relatively slow cooling of the kimberlite magma. The petrographical observations are summarised in Table 1.

Macroscopically the fragmental samples are described as light grey to orange brown and texturally inequigranular. Microscopically the rocks are pervasively altered to

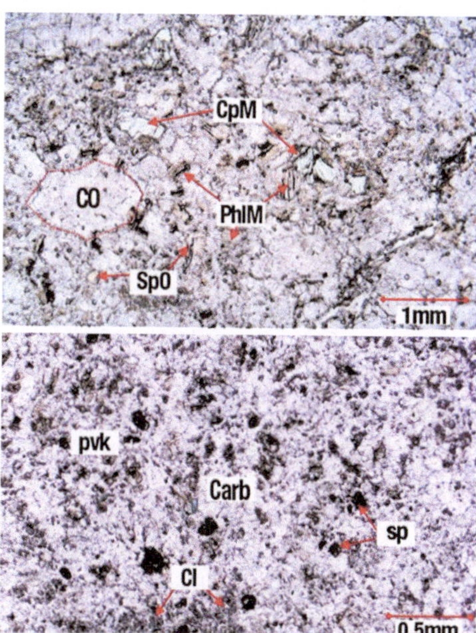

Fig. 5 Transmitted light thin sections from the Wadagera kimberlite. The rock is extensively calcretised. CO—carbonate pseudomorph after olivine; SpO—serpentine pseudomorph after olivine; CpM—clinopyroxene macrocryst; PhlM—phlogopite macrocryst; Carb—carbonate; Cl—clay. The rock is notable for coarse perovskite (pvk) and coarse groundmass spinel (sp). The *upper image* is interpreted as a possible volcaniclastic kimberlite, while the *lower image* is interpreted as a coherent kimberlite

carbonate, resulting in the overprinting of most of the original textural and mineralogical character. Resedimented volcaniclastic kimberlite (RVK) autoliths were identified in both fragmental samples. The magmatic component has uniform texture and comprises olivine macrocrysts and phenocrysts set in a groundmass of fresh to partly altered poikilitic phlogopite, perovskite, sphene, spinel, clay and secondary carbonate. Abundance and size estimates of the olivines were precluded by the extent of alteration in the first sample. Phlogopite macrocrysts occur as anhedral, fresh to partly altered chlorite and comprise approximately 5 % of the rock. Clinopyroxene macrocrysts are subhedral to anhedral and fresh to partly altered and comprise approximately 3 % of the rock. Spinel (average size ~0.05 mm) is uncommon and occurs both as anhedral interstitial aggregates and as euhedral grains which are partly altered to magnetite. In the second fragmental sample, serpentine and carbonate pseudomorphs after olivine macrocrysts have an average size of 1 mm (maximum 3.5 mm) and comprise approximately 15 % of the sample. The serpentine and carbonate pseudomorphs after olivine phenocrysts have an average size of 0.3 mm and comprise about 20 % of the sample. The perovskites are euhedral to subhedral and are fresh with an average size of 0.1 mm and comprise about 5 % of the sample. The groundmass spinels are euhedral to subhedral and partially altered to an average size of 0.04 mm and comprise about 4 % of the samples. Minor ilmenite is also present in the groundmass, and clay and carbonate occur interstitially between the olivine grains.

The whole-rock geochemistry, as determined by X-ray fluorescence (XRF), corroborates the highly altered nature of the rock, making geochemical interpretation difficult. The contamination index (Clement 1982) is over 2 for two samples (EBH393 and EBH390, Table 2). Sample EBZ584 has the lowest contamination index (Table 2) and is represented by the blue square in Fig. 6. The low analysis totals, especially for EBH393 and EBH 390, probably reflect the abundance of volatile-rich secondary minerals such as carbonates, clay and serpentine in the samples. Nevertheless, chondrite-normalised values of La, Ce, Nb, Sr and Ti show levels of enrichment (Table 2) along with elevated Zr/Nb ratios (Table 2), comparable to those in kimberlitic rocks. The Zr–Nb plot of Taylor et al. (1994) indicates that the Wadagera occurrence is kimberlite (Fig. 6). However, it should be noted that a detailed classification of the rock is very difficult at this stage and the authors would cautiously name this a Group I kimberlite based on the presently available data, including spatial and temporal association with other Group 1 kimberlites in the area and mineralogical and compositional data discussed here.

Chemistry of KIMs

Sampling and Methodology

KIM major element chemistry was obtained from 68 samples from a combination of drill chip, surface deflation, stream and rock grab samples (Fig. 3). Most samples were processed at De Beers India's heavy mineral separation laboratory in Bangalore. Sample preparation involved crushing (for rock samples) and screening into various size fractions. The recovery of heavy minerals, including the KIM suite, employed a combination of techniques including (1) dense-media separation (DMS) using ferrosilicon; (2) cleaning with diluted hydrochloric acid; (3) magnetic separation; and (4) heavy liquid density separation using lithium heteropolytungstate (LST). The drill chip and stream samples were reduced to two fractions (0.3–0.5 and 0.5–1.0 mm). The drill chip samples yielded the highest numbers of KIMs. The rock grab samples were reduced to one fraction in the size range 1.0–2.0 mm, while the surface deflation samples were reduced to 0.3–0.5 mm. KIMs were extracted manually from the resultant concentrates, under magnification (usually 10 times). A total of 723 garnet (including 249 crustal grains—almandine and spessartine), 2,765 chromite, 1,476 chrome diopside and 6,299 ilmenite grains were extracted for chemical analysis. All KIMs

Table 1 Summary of petrographical observations

Sample Number	Texture	Olivine macrocrysts		Groundmass spinel		Observations
		Size	Abundance	Size	Shape	
EBH390	Fragmental (RVK)	Sample is too altered	Sample is too altered	Average 0.04 mm	Euhedral to Subhedral	Magmaclasts and coarse perovskite
EBH393	Fragmental (RVK)	Sample is too altered	Sample is too altered	Average 0.05 mm	Euhedral to Anhedral	Phlogopite and clinopyroxene macrocrysts
EBZ584	Coherent	<3.5 mm	15 %	Average 0.04 mm	Euhedral to Subhedral	Coarse perovskite

recovered were analysed for their major element compositions, while a subset of 141 garnets were selected for trace element analysis.

Individual grains were mounted by hand in epoxy resin discs, which were ground and polished to expose the mineral surfaces for major and trace element analysis. Electron microprobe analysis (EPMA) for major elements (Al_2O_3, CaO, Cr_2O_3, K_2O, MgO, MnO, SiO_2, TiO_2, ZnO) was conducted at De Beers Group Exploration's central analytical facility in Johannesburg, South Africa, during 2003 and 2004. Major element data, reported in the form of the weight percent (wt%) of each oxide, were generated using Cameca SX50® electron microprobes with five vertical wavelength-dispersive spectrometers using a range of natural and synthetic standards. The electron beam parameters were 20 kV accelerating voltage and a beam current of 20 or 30 nA. Counting times were 10 s on peak and 5 s at one background position. Detection limits were better than 0.05 elemental wt% for all elements except zinc (which was 0.09 wt%). Matrix correction calculations were performed using the PAP procedure after Pouchou and Pichoir (1984). Trace elements in garnet (Ga, Ni, Y, Zr as well as the rare earth elements from La to Lu) for a subset of 141 grains (Fig. 7a, b) were acquired by laser ablation inductively coupled plasma mass spectrometer (ICP-MS) utilising a Thermo Fisher X-Series 2 quadrupole instrument coupled to a New Wave UP193-FX Fast Excimer laser ablation system with an excitation wavelength of 193 nm and analysis spot size of 100 μm. All analyses were generated at the same analytical facility as the major element data. All measurements were taken on the epoxy grain mounts that were prepared for EPMA. NIST SRM-610, a homogeneous, synthetic doped glass and a natural garnet megacryst reference grain, GHR-1, obtained from Monastery Mine in South Africa were used as external standards. ^{27}Al was used as an internal standard in order to compensate for ionisation and matrix effects (values for Al_2O_3 in the reference and sample grains were obtained from the electron microprobe analyses). GLITTER® version 4.4.2 (developed and supplied by GEMOC, Macquarie University) was used to reduce the time resolved analysis spectra to calculate the concentrations of the various analytes quoted in parts per million (ppm). Trace element analytical uncertainties are summarised in Table 3.

For comparison with the available published data (Griffin et al. 2009; Joy et al. 2012), the garnets collected from the study area were classified on the basis of their major element compositions, using the method developed by Griffin et al. (2002) to quantify the proportions of different paragenetic classes within the concentrate samples. These comprise depleted (garnets from low-calcium harzburgite and harzburgite) to fertile (lherzolite, wehrlite and low-chrome megacrysts) compositions. Garnet pressure–temperature (P–T) pairs were calculated for the subset of 141 grains using (1) the nickel-in-garnet thermometer of Ryan et al. (1996) and (2) chrome (Cr_2O_3)-in-garnet to estimate pressure using the Ryan et al. (1996) empirically derived barometer. The combination of garnet P–T pairs was used to attempt to estimate the palaeo-geotherm for the portion of lithosphere sampled by the kimberlite according to the methodology proposed by Ryan et al. (1996), Griffin et al. (2002) and Kobussen et al. (2008, 2009) using the set of model reference geotherms from Pollack and Chapman (1977).

Results

Garnet

The 723 garnets analysed show limited compositional variation when applying the conventional binary chrome–calcium (Fig. 7a and b) and chrome–titanium systematics (Fig. 8a and b). The data are dominated by garnets from low-titanium lherzolite (~36 %) with variable chrome content (2–10 wt% Cr_2O_3). Garnets from harzburgite (classified here as calcium harzburgite after Griffin et al. 2002) are rare, accounting for just 1 % of the concentrate sample. Sampling efforts to date have not recovered any low-Ca harzburgitic garnets that appear to be characteristic of the mantle samples further to the south and south-east on the EDC (Fig. 7a and b).

The garnet concentrate samples presented by Griffin et al. (2009) for selected kimberlite occurrences along the

Table 2 Whole-rock geochemistry results of three samples from the Wadagera kimberlite

	EBZ584	EBH393	EBH390
Major elements (wt %)			
SiO2	26.67	8.02	6.63
TiO2	3.2	0.42	0.69
Al2O3	3.66	1.29	1.09
Fe2O3	10.77	2.7	3.8
MnO	0.14	0.16	0.07
MgO	16	3.08	3.16
CaO	20.35	46.99	50.15
Na2O	0	0.25	0.23
K2O	0.24	0	0
P2O5	0.21	0.3	0.38
Cr2O3	0.12	0.3	0.38
LOI	18.3	NA	NA
Total	99.66	63.51	66.58
Trace elements (ppm)			
V	83	21	31
Cr	752	482	131
Co	65	56	47
Ni	684	179	151
Cu	62	bd	NA
Zn	63	bd	NA
As	NA	NA	20
Rb	19	8	10
Sr	410	228	298
Y	19	6	5
Zr	238	13	87
Nb	100	14	38
Mo	2	bd	NA
Ba	122	205	161
La	139	NA	NA
Ce	182	NA	NA
Th	12	29	53
U	NA	6	12
W	51	0	NA
Pb	9	16	NA
Contamination index	1.9	3.1	2.5
Zr/Nb	2.4	0.93	2.3
La_N	422	NA	NA
Ce_N	210	NA	NA
Nb_N	285	40	108
Sr_N	34.7	19.3	25.2
Ti_N	31	4.0	6.6

bd below detection; *NA* Not analysed

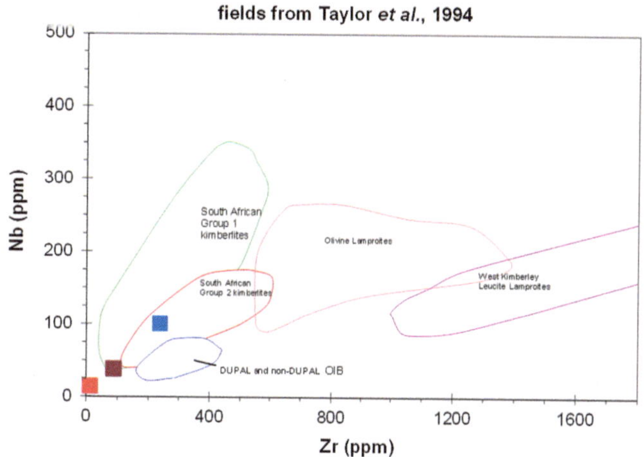

Fig. 6 Whole-rock Nb-Zr discrimination diagram for three samples from the Wadagera kimberlite. Fields after Taylor et al. (1994)

Kalyandurg/Wajrakarur (K/W) transect (Fig. 1) and by Joy et al. (2012) for the Banganapalle lamproite dykes (Fig. 1) represent the only representative published KIM data for the kimberlites/lamproites of the EDC in India. Comparison with the Griffin et al. (2009) data (in Cr–Ca compositional space Fig. 7a) indicates a significant change in the composition of the SCLM between the K/W transect in the south and the Bhima cluster in the north. The garnets from Wadagera are, on average, more calcium rich: the low-calcium edge of the lherzolite trend projects to between 4 and 5 wt% CaO compared with 3–3.5 for the K/W and Banganapalle samples. The low-chromium portion of the lherzolite trend forms a conspicuous "splay" oblique to the lherzolite trend extending to much higher calcium contents, exceeding 6.0 wt% CaO. This feature is indicative of sampling of fertile lherzolite/wehrlite at relatively shallow depth and is consistent with the so-called chromite–clinopyroxene–garnet equilibrium (CCGE) trend described by Kopylova et al. (2000). The extent of this feature (high-Ca, low-Cr garnets) in chrome–calcium compositional space clearly differentiates the Wadagera SCLM sample from those of other kimberlite clusters on the EDC. Comparison with the data from Joy et al. (2012) in Fig. 7b highlights the absence of low-Ca, high-Cr garnets of harzburgitic to dunitic composition in the Wadagera mantle sample, especially considering the number of grains analysed, and indicates the likely absence of depleted harzburgite in the section of SCLM sampled by the Wadagera kimberlite.

Ni-in-garnet thermometry, covering the entire compositional range of kimberlitic garnets (Fig. 7a and b) and including a number of crustal garnets, produces a bimodal temperature distribution (Fig. 9a and b). The low-chrome lherzolitic and wehrlitic components, constituting the CCGE trend, return consistently lower Ni temperatures (in

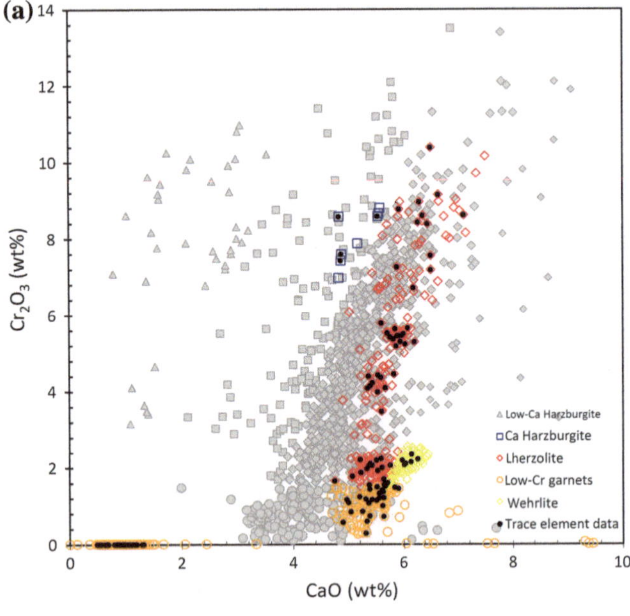

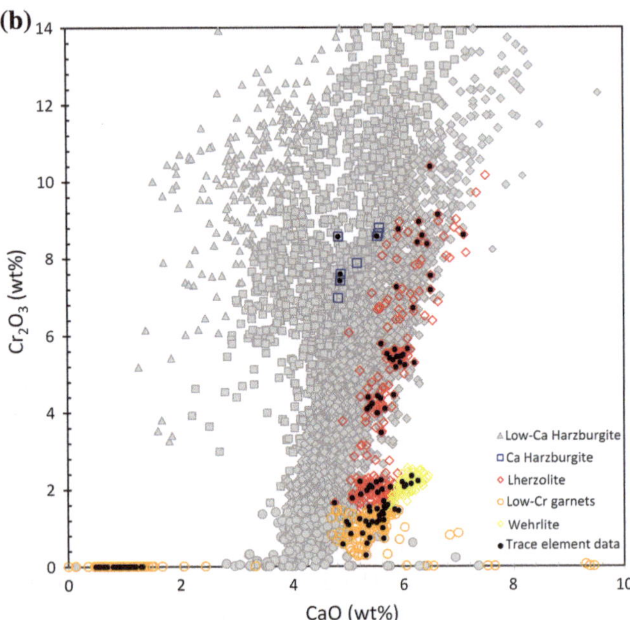

Fig. 7 a Chrome–calcium discrimination diagram for garnets (n = 723) recovered from the Wadagera kimberlite (coloured symbols). The data from Griffin et al. (2009) from the Kalyandurg–Wajrakarur kimberlite transect are included (n = 1,084; *grey* symbols) for comparison. The classification system used is that of Griffin et al. (2002)—the symbol shapes (see legend) apply to both datasets. Garnets analysed for trace element concentrations are indicated by the superimposed *black* symbols. **b** Chrome–calcium discrimination diagram for garnets (n = 723) recovered from the Wadagera kimberlite (coloured symbols). The data from Joy et al. (2012) for the Banganapalle lamproite area (n = 5,064; *grey* symbols) are included for comparison. The classification system used is that of Griffin et al. (2002), the symbol shapes (see legend) apply to both datasets. Garnets analysed for trace element concentrations are indicated by the superimposed *black* symbols

Table 3 Analytical uncertainties of LA-ICPMS

Element (n = 141)	Average 1σ error (ppm)	Average detection limit (ppm)
^{60}Ni	4.42	0.29
^{69}Ga	0.3	0.04
^{89}Y	4.92	0.17
^{90}Zr	0.73	0.09
^{93}Nb	0.04	0.01
^{139}La	0.14	0
^{140}Ce	0.26	0.01
^{141}Pr	0.04	0
^{146}Nd	0.27	0.03
^{147}Sm	0.18	0.03
^{153}Eu	0.04	0.01
^{157}Gd	0.29	0.05
^{159}Tb	0.08	0.01
^{163}Dy	0.74	0.1
^{165}Ho	0.22	0.03
^{166}Er	0.72	0.09
^{169}Tm	0.14	0.02
^{172}Yb	0.92	0.12
^{175}Lu	0.15	0.02

the range of 700–850 °C) as predicted (Kopylova et al. 2000) and are consistent with entrainment within the graphite stability field. Coupled with the yttrium data (Fig. 9a and b), the presence of this component in the Wadagera garnet concentrate sample indicates that the ascending magma sampled a more fertile, shallow component, which is inferred to have a negative impact on diamond potential. The higher-Cr lherzolitic and Ca harzburgitic components have significantly higher Ni temperatures (on the order of 1,200–1,400 °C) that are comparable to the data from the K/W transect and consistent with entrainment in the diamond stability field (if a similar palaeo-geotherm constraint is assumed for the SCLM beneath the Bhima cluster). Comparison of the Wadagera garnet yttrium data highlights the absence of a strongly developed yttrium depletion trend (Fig. 9a and b). Even the most calcium-poor (harzburgitic) garnets are enriched in yttrium relative to the reference datasets.

The absence of high-Cr, Ca-poor harzburgitic garnets precludes a robust estimation of the palaeo-geotherm for the Wadagera kimberlite. However, if a 40 mW/m^2 conductive geotherm for the region encompassing the Bhima and Narayanpet clusters is assumed, then the xenocryst data for the Wadagera kimberlite indicate that garnet-bearing

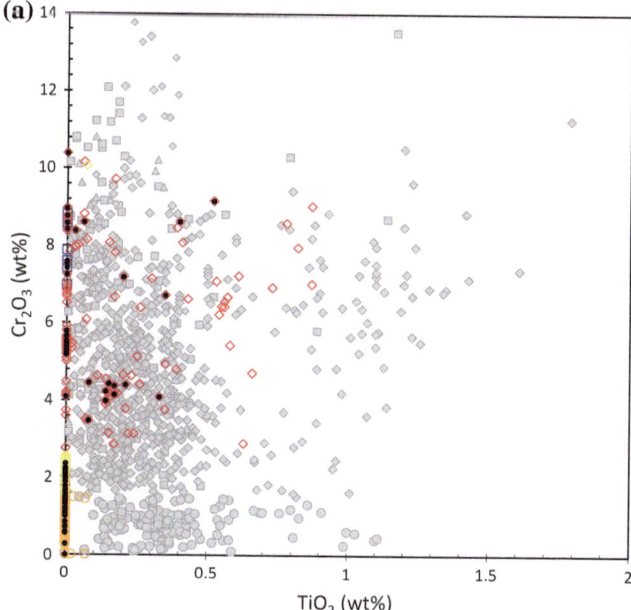

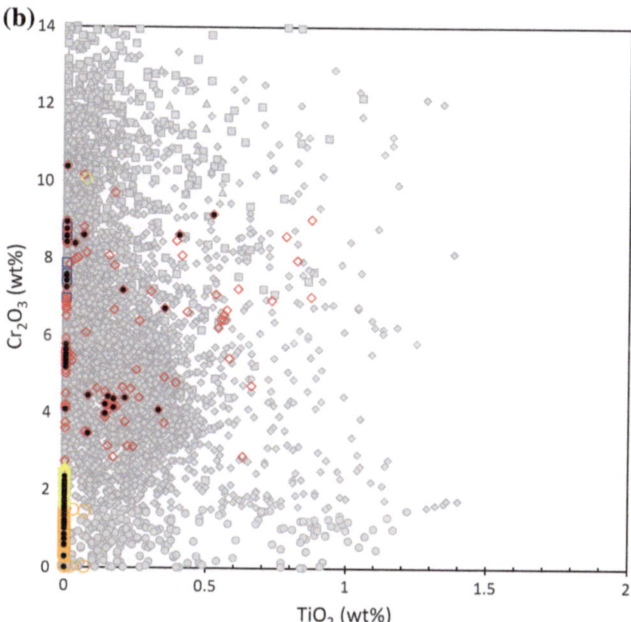

Fig. 8 **a** Chrome–titanium discrimination diagram for garnets (n = 723) recovered from the Wadagera kimberlite (coloured symbols). The data from Griffin et al. (2009) from the Kalyandurg–Wajrakarur kimberlites transect are included (n = 1,084; *grey* symbols) for comparison. Symbol legend is same as for Fig. 7a. **b** Chrome–titanium discrimination diagram for garnets (n = 723) recovered from the Wadagera kimberlite (coloured symbols). The data from Joy et al. (2012) for the Banganapalle lamproite area (n = 5,064; *grey* symbols) are included for comparison. Symbol legend is same as for Fig. 7b

lithologies (mainly fertile lherzolite) were entrained from within the diamond stability field close to the isentrope at temperatures in the range of 1,350–1,400 °C. Projection of the P–T pairs onto the Pollack and Chapman (1977) 40 mW/m^2 reference geotherm confirms the shallow origin of the CCGE component at geotherm-corrected pressures of less than 40 kbar (the graphite–diamond transition occurs at approximately 44 kbars). Conversely, P–T data from the Banganapalle cluster (Fig. 9b), where low-calcium harzburgitic garnets are abundant, confirm the presence of strongly depleted lithospheric mantle peridotites (Y-in-garnet $\ll$1.0 ppm) from mid levels ($\sim$40 kb) down to nearly the lithosphere–asthenosphere boundary ($\sim$60 kb) that equilibrated along a 39–40 mW/m^2 conductive geotherm. Data for the K/W transect indicate an even cooler palaeo-geotherm, well below 37 mW/m^2 (indicative of thickening of the SCLM to the south-west away from the Banganapalle lamproites) with Y-depleted garnets at both high (>1,200 °C) and low ($\leq$1,000 °C) temperatures.

Overall, comparison of the garnets from the Wadagera kimberlite with published data from various kimberlites and lamproites to the south and south-east highlights clear differences in the compositions of the garnet-bearing portions of the SCLM sampled. The absence of proxies for diamond (Ca-poor and incompatible element-depleted harzburgitic compositions) and strong evidence for entrainment of shallow, fertile lherzolite significantly downgrade the diamond potential of this occurrence. The kimberlite was tested with a 100-kg sample for microdiamonds with negative results. However, a larger sample could potentially indicate a very low-grade presence of diamonds.

Other Minerals (Chromite, Chrome Diopside and Ilmenite)

A total of 1,476 chrome diopside grains were recovered, the majority of which plot in the garnet peridotite field in Cr_2O_3–Al_2O_3 compositional space (Fig. 10) as defined by Ramsay (1992). Closer examination of this plot highlights a dominant low-Cr (<1.5 wt% Cr_2O_3), high-Al (up to 6 wt%) component straddling the boundary between the garnet–lherzolite and spinel–lherzolite fields. This cluster of data is consistent with entrainment of significant amounts of fertile lherzolite at shallow depth that is predicted by the low-T CCGE trend observed in the garnets and confirmed by Ni-in-garnet thermometry (see above).

A small proportion ($\sim$9 %) of the 2,765 chromites plot in the diamond inclusion and intergrowth field (58 wt% $\leq Cr_2O_3 \leq$69 wt%; $TiO_2 \leq$ 0.6 wt%) of Fipke et al. (1995). However, if the mantle component (TiO_2 $\leq$0.6 wt%; n = 974) is isolated, then the diamond inclusion component (Figs. 11 and 12) is significant ($\sim$28 %), indicating high diamond potential. Lower-Cr chromites are commonplace and concentrated in the range 40–58 wt% Cr_2O_3 and are indicative of entrainment of chromite-bearing lherzolite

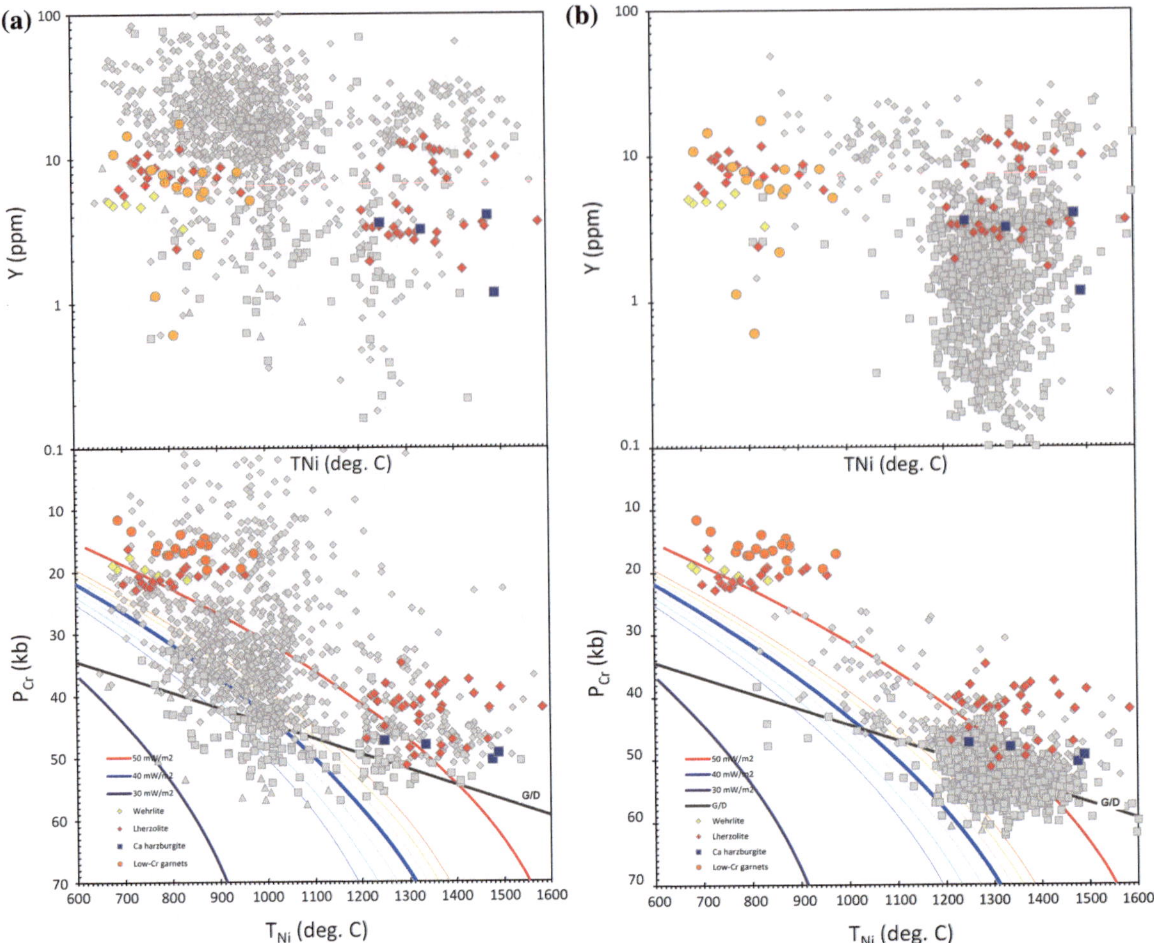

Fig. 9 **a** Garnet pressure–temperature array (*lower plot*) and the garnet yttrium content (*upper plot*) for a subset of 141 garnets selected from the Wadagera kimberlite concentrate sample (coloured symbols). Pressures and temperatures for single grains have been calculated following the methodology of Ryan et al. (1996). Data from Griffin et al. (2009) are shown in the background (*grey* symbols). G/D represents the graphite–diamond phase transition of Kennedy and Kennedy (1976). **b** Same as Fig. 9a with data from Joy et al. (2012) for Banganapalle shown in the background (*grey* symbols)

over a significant depth range, with the effect of diluting the potentially diamond-bearing component. The phenocryst (cognate) component shows significant compositional variation both in terms of Cr–Ti and in terms of Cr–Mg contents consistent with protracted evolution of the kimberlite magma. The abundant data points between 38 wt% $\leq Cr_2O_3 \leq 48$ and 2 wt%$\leq$ TiO_2 ≤ 6 wt% (Fig. 11) and 38 wt% $\leq Cr_2O_3$ ≤ 48 wt% and 4 wt% $\leq MgO \leq 12$ wt% (Fig. 12) suggest more evolved kimberlite (Mitchell 1986, p. 223) with a consequent probable negative effect on the preservation of any diamonds.

The 6,613 ilmenites are dominated by kimberlitic compositions as defined by the bounding reference line of Wyatt et al. (2004) in Ti–Mg compositional space (Fig. 13). The bulk of the data plot within the field for kimberlitic ilmenites. However, some degree of scatter towards the low-Mg, non-kimberlitic field is evident. Some of the ilmenite compositions that plot along the y-axis (n = 443) are likely to be of non-kimberlitic origin (these grains were recovered from soil samples collected over the kimberlite and are probably derived from the Deccan basalts exposed to the north-west; Fig. 1). The scatter of data points plotting between the K-NK reference line and the zero-Mg axis defines a parabolic trend in Mn–Mg space (Fig. 14) and represents convincing evidence for the replacement of magnesium by manganese in kimberlitic ilmenite that is postulated to result from the late-stage CO-CO_2 reaction in the kimberlite magma (Wyatt et al. 2004).

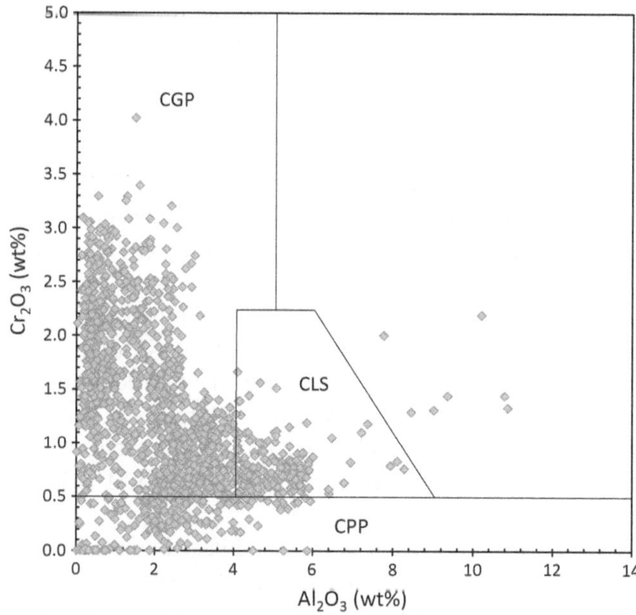

Fig. 10 Discrimination plot of chrome diopside grains recovered from Wadagera kimberlite after Ramsay (1992). CGP—chrome diopsides from garnet peridotites; CLS—chrome diopsides from spinel peridotites; and CPP—eclogitic, cognate and megacrystic chrome diopsides

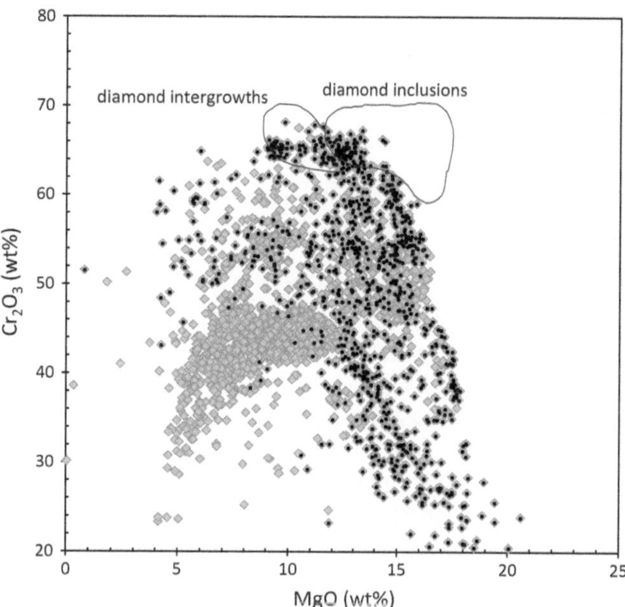

Fig. 12 Chromites from Wadagera in chrome–magnesium compositional space (*grey* symbols). The marked areas represent the fields of diamond intergrowth and diamond inclusion compositions after Fipke et al. (1995). The overlapping closed *black* symbols indicate low-titanium (TiO_2 <0.6 wt%) chromites of probable upper mantle origin (xenocrysts)

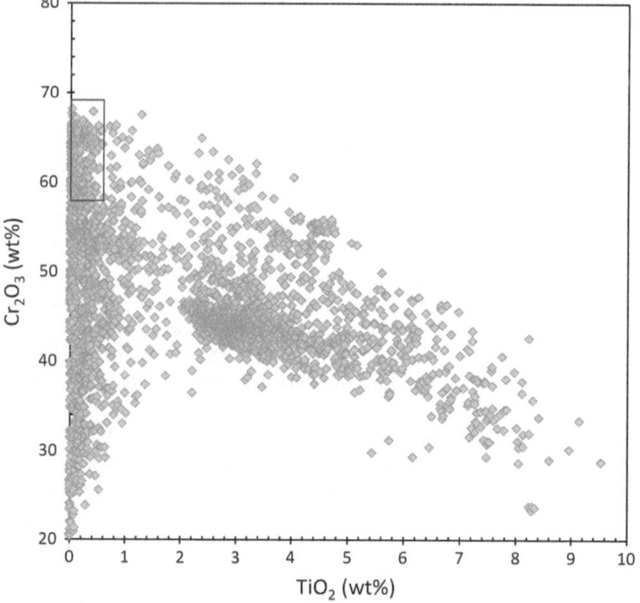

Fig. 11 Chromites from Wadagera in chrome–titanium compositional space. The *rectangle* represents the field of diamond inclusion compositions (Fipke et al. 1995)

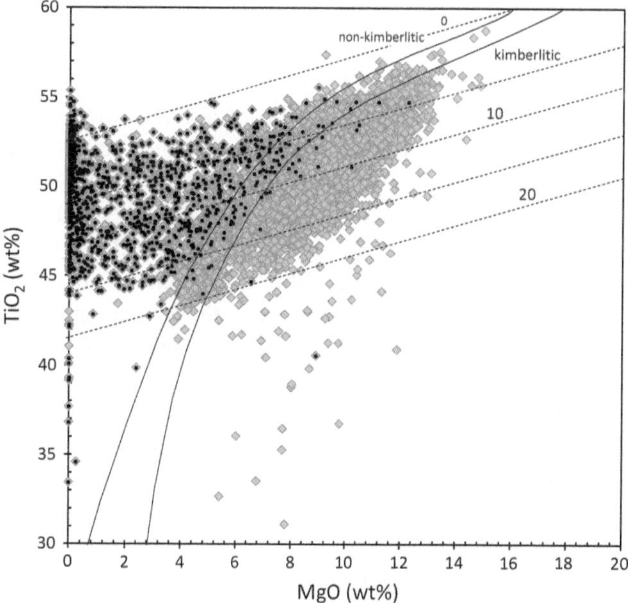

Fig. 13 Ilmenites from Wadagera in titanium–magnesium compositional space after Wyatt et al. (2004). The overlapping closed *black* symbols indicate high-manganese (MnO >1.0 wt%) ilmenites, which (see Fig. 14) represent altered ilmenites of original kimberlitic origin

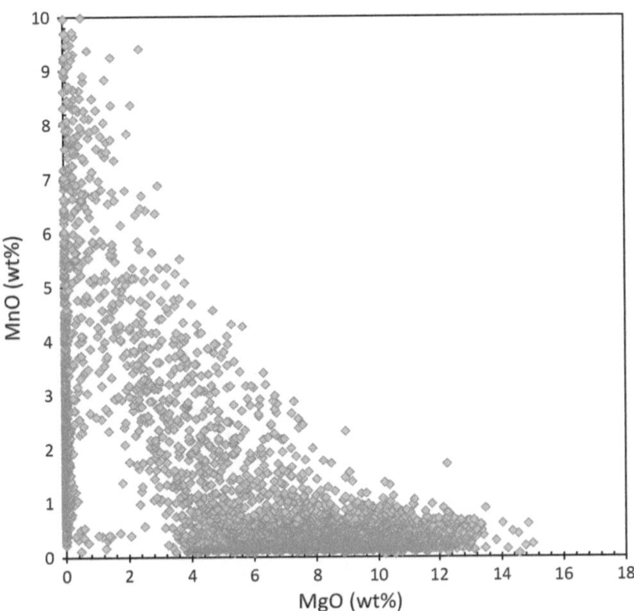

Fig. 14 MnO–MgO plot of the ilmenites from Wadagera kimberlite

Conclusions

The Wadagera kimberlite was discovered by an airborne multispectral scanner survey with high spectral resolution. The pipe has a surface area of 6 ha and is extensively calcretised at surface. It is clearly defined by an airborne magnetic survey, but frequency-domain airborne EM was unable to distinguish the kimberlite from the host gneissic basement. The pipe occurs within a regional ilmenite and chromite indicator mineral anomaly defined by stream sampling at a density of approximately 1 sample every 5 km^2. However, garnet and chrome diopside indicator minerals are only found within approximately 2 km of the pipe in stream samples and within a few tens of metres in soil samples. This is mainly because the silicate minerals (garnet and chrome diopside) are preferentially destroyed through the action of surface weathering processes, but it may also reflect the relative primary abundances of these minerals in the kimberlite. The weathered nature of the kimberlite samples recovered precludes detailed petrographic analysis. However, both coherent and volcaniclastic facies are apparent.

The chemistry of KIMs indicates minor sampling within the diamond stability field close to the lithosphere–asthenosphere boundary. However, the majority of data reflect sampling of a fertile, shallow component of the SCLM. This result indicates very different SCLM in this part of northern Karnataka when compared to the mantle samples at Banganapalle and along the K/W transect, which have sampled the diamond stability field far more extensively, and do not show evidence of having sampled the same shallow fertile shallow component evident at Wadagera. It is conceivable, although not likely, that the observed bimodality of the P–T array is a manifestation of preferential sampling of the SCLM by the ascending kimberlite magma (although a mechanism for this remains unconstrained and is beyond the scope of this paper). There is no record, to the authors' knowledge, of such selective sampling of a particular section of the SCLM by any kimberlite in the world. Thus, the authors strongly argue that the garnet sample obtained from the Wadagera kimberlite gives a representative vertical distribution of garnet-bearing lithologies in the continental lithospheric mantle section at the time of emplacement of the Wadagera kimberlite. Analysis of garnets from other kimberlites in the Bhima cluster may resolve the garnet-bearing portions of the SCLM at higher resolution in this part of the EDC.

Wadagera. This result is expected based on the inferred diamond potential from mantle-derived KIMs. The evolved nature of the kimberlite magma evidenced by the presence of relatively high-Ti chromites also has a negative effect on the diamond preservation potential.

Acknowledgments The authors gratefully acknowledge the permission of De Beers to publish this data. We also acknowledge Hugh O'Brien and Mark Hutchison for helpful editing and suggestions./ SimplePara>

References

Babu TM (1998) Diamonds in India. Geol Soc India Econ Geol Series Spl Publ 9:332

Balasubrahmanyan MN (2006) Geology and tectonics of India: an overview. Int Assoc Gondwana Res Mem 9:204

Chadwick B, Vasudev VN, Hegde GV (2000) The Chitradurga schist belt and its adjacent plutonic rocks, northwest of Tungabhandra, Karnataka: a duplex in late Archaean convergent setting of Dharwar craton. J Geol Soc India 61:645–663

Chalapathi Rao NV, Miller JA, Gibson SA, Pyle DM, Madhavan V (1999) Precise 40Ar/39Ar dating of Kotakonda kimberlite and Chelima lamproite, India: implication to the timing of mafic dyke swarm activity in the Eastern Dharwar craton. J Geol Soc India 53:425–433

Chalapathi Rao NV, Anad M, Dongre A, Osborne I (2009) Carbonate xenoliths hosted by the Mesoproterozoic Siddanpalli Kimberlite cluster (Eastern Dharwar Craton): implications for the geodynamic evolution of southern India and its diamond and uranium metallogenesis. Int J Earth Sci 99:1791–1804. doi:10.1007/s00531-009-0484-7

Clement CR (1982) A comparative geological study of some major kimberlite pipes in the northern Cape and Orange Free State. Unpublished Ph.D. Thesis, University of Cape Town, South Africa

Fareeduddin F, Mitchell, RH (2012) Diamonds and their source rocks in India. Geol Soc India, Bangalore, 434:72–93

Fipke CE, Gurney JJ, Moore RO (1995) Diamond exploration techniques emphasizing indicator mineral geochemistry and Canadian examples. Geol Surv Can Bull 423:86 p

Griffin WL, Fisher NI, Friedman JH, O'Reilly SY, Ryan CG (2002) Cr-pyrope garnets in the lithospheric mantle II. Compositional populations and their distribution in time and space. Geochem Geophys Geosyst 3(1):1–35. doi:10.1029/2002GC000298

Griffin WL, Kobussen AF, Babu EVSSK, O'Reilly SY, Norris R, Sengupta P (2009) A translithospheric suture in the vanished 1-Ga lithospheric root of South India: evidence from contrasting lithosphere sections in the Dharwar Craton. Lithos 112:1109–1119. doi:10.1016/j.lithos.2009.05.015

Joy S, Jelsma H, Preston R, Kota S (2012) Geology and diamond provenance of the Proterozoic Banganapalle conglomerates, Kurnool group, India. In: Majumdar R, D Saha D (eds) Palaeoproterozoic of India: geological society, vol 365. Special publication, London, pp 195–216

Kennedy CS, Kennedy GC (1976) The equilibrium boundary between graphite and diamond. J Geophys Res 81:2467–2470

Kobussen AF, Griffin WL, O'reilly SY, Shee S (2008) The ghosts of lithospheres past: imaging an evolving lithospheric mantle in southern Africa. Geology 36:515–518

Kobussen AF, Griffin WL, O'reilly SY (2009) Cretaceous thermo-chemical modification of the Kaapvaal cratonic lithosphere, South Africa. Lithos 112:886–895

Kopylova MG, Russell JK, Stanley C, Cookenboo H (2000) Garnet from Cr- and Ca-saturated mantle: implications for diamond exploration. J Geochem Explor 68:183–199

Kumar A, Creaser LM, Mankyanba C (2007) Mesoproterozoic kimberlite in South India: a possible link to ~1.1 Ga global magmatism. Precambr Res 154(3–4):192–204

Lynn MD (2005) The discovery of kimberlites in the Gulbarga and Raichur districts of Karnataka. Group discussion on Kimberlites and related rocks of India. Geol Soc India, Bangalore (November) 14:48–49

Mitchel RH (1986) Kimberlites: mineralogy, geochemistry and petrology. Plenum Press, New York, p 442

Nayak SS, Kasiviswanathan CV, Reddy TAK, Nagaraja Rao BK (1988) New find of kimberlite rocks in Andhra Pradesh, near Maddur, Mahabubnagar district, Andhra Pradesh. J Geol Soc India 31:343–346

Patton C (2007) The Petrogenesis of Indian Kimberlite Magmas. Unpublished PhD thesis, University of Melbourne

Patton C, Hergt J, Philips D, Woodhead J, Shee S (2007) New insights into the genesis of Indian Kimberlites from the Dharwar Craton via in situ Sr-isotope analysis of groundmass perovskite. Geology 35(11):1011–1014

Pollack HN, Chapman DS (1977) On the regional variation of heat flow, geotherm and lithospheric thickness. Tectonophysics 38:279–296

Pouchou JL, Pichoir F (1984) A new model for quantitative X-ray micro-analysis. I. Application to the analysis of homogeneous samples. La Recherche Aérospatiale 3:13–36

Radhakrishna BP, Naqvi SM (1986) Precambrian continental crust of India and its evolution. J Geol 94:145–166

Ramakrishnan M, Vaidyanadhan R (2008) Geology of India. Geol Soc India, Bangalore 1:556

Ramsay RR (1992) Geochemistry of diamond indicator minerals. Unpublished PhD thesis, University of Western Australia

Rogers JJM (1986) The Dharwar Craton and the assembly of Peninsular India. J Geol Soc India 94:129–143

Ryan CG, Griffin WL, Pearson NJ (1996) Garnet geotherms: pressure-temperature data from Cr-pyrope garnet xenocrysts in volcanic rocks. J Geophys Res 101:5611–5625

Swami Nath J, Ramakrishnan M (1981) Dharwar Supergroup; present classification and correlations. In early Precambrian Supracrustals of southern Karnataka. Geol Soc India 112:23–38

Taylor WR, Haggerty SE, Tompkins LA (1994) Comparative geochemistry of West African kimberlites: evidence for a micaceous kimberlite endmember of sublithospheric origin. Geochimica et Cosmochimica Acta 58(19):4017–4037

Wyatt BA, Baumgartner M, Anckar E, Grütter H (2004) Compositional classification of "kimberlitic" and "non-kimberlitic" ilmenite. Lithos 77:819–840

Petrology of P-5 and P-13 "Kimberlites" from Lattavaram Kimberlite Cluster, Wajrakarur Kimberlite Field, Andhra Pradesh, India: Reclassification as Lamproites

Gurmeet Kaur, M. M. Korakoppa, Fareeduddin, and K. L. Pruseth

Abstract

To date, all the potassic-ultramafic pipes from the Wajrakarur Kimberlite Field (WKF), Eastern Dharwar Craton (EDC), India, have been classified as kimberlites. Recent reclassification of a kimberlite pipe P-2 from the WKF as a lamproite necessitated a re-examination at the other kimberlite bodies in the region with a view to placing them in their appropriate slots in the mineralogic-genetic classification scheme of ultrapotassic rocks. The Pipes P-5 and P-13 of Lattavaram cluster in WKF were chosen for this present work. Both pipes exhibit macrocrystal textures and have two generations of olivine. The prominent groundmass minerals constitute spinel, perovskite, phlogopite, apatite, pyroxene, serpentine, chlorite and calcite/magnesite. The phlogopites from both the pipes are low in aluminium and barium, are high in total iron in comparison to archetypal kimberlites and follow lamproite trends. The pipes are characterized by the groundmass spinels of four varieties, for example, ulvöspinel, Mg-ulvöspinel, Mg-Cr-spinel and magnetites. Both the pipes have atoll and necklace-textured spinels besides the euhedral groundmass spinels. These spinels exhibit compositional trends identical to those found in kimberlite magmatic trend T2. Perovskites with bimodal size distribution occur as discrete grains throughout the rock. The two pipes represent polyphase intrusions that are in close proximity within Lattavaram region. The mineral chemistry data of phlogopites, spinels, perovskites and pyroxenes from these pipes are more akin to those in lamproite magmas rather than that in archetypal (Group-1) kimberlites. Thus, a careful reassessment of the petrological characteristics of all the 45 kimberlitic pipes so far discovered in the EDC is necessary keeping in view the economic and academic importance of these pipes due to their diamond and mantle xenolith potential.

Keywords

P-5 • P-13 • Lattavaram • Mineralogy • Lamproite • Archetypal (Group-1) kimberlite?

Introduction

Kimberlites and lamproites in India have been reported from the three Indian cratons, that is, Dharwar Craton, Bastar Craton and Bundelkhand Craton (Fareeduddin and Mitchell 2012). All the kimberlites and lamproites so far discovered in south India are on the Eastern Dharwar Craton (Fig. 1). Reddy (1987) has suggested previously P-2 and

G. Kaur (✉)
CAS in Geology, Panjab University, Chandigarh, 160014, India
e-mail: gurmeet28374@yahoo.co.in

M. M. Korakoppa · Fareeduddin
Petrology, Petrochemistry and Ore Dressing (PPOD) Division, Geological Survey of India, Bangalore, India

K. L. Pruseth
Department of Geology and Geophysics, Indian Institute of Technology, Kharagpur, 721302, India

D. G. Pearson et al. (eds.), *Proceedings of 10th International Kimberlite Conference*, Volume 1, Special Issue of the Journal of the Geological Society of India,
DOI: 10.1007/978-81-322-1170-9_11,

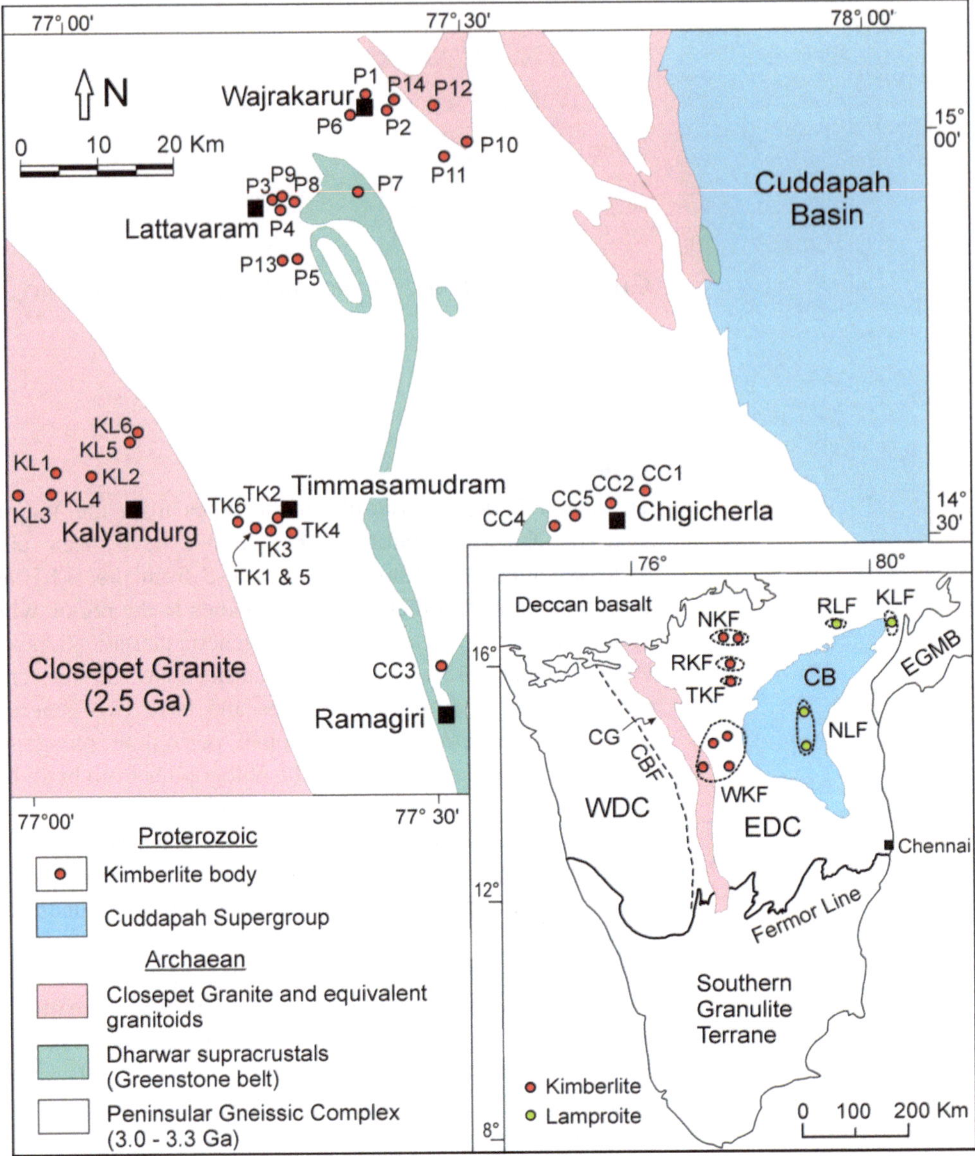

Fig. 1 Geological map of Wajrakarur Kimberlite Field, Anantpur district, Andhra Pradesh, India (after Ravi et al. 2009). Inset shows location of various kimberlite and lamproite fields in Eastern Dharwar Craton (*EDC*)

P-5 kimberlites to be lamproites and recent reports on detailed mineralogy of P-2 kimberlite from the Wajrakarur kimberlite field (Mitchell 2010; Fareeduddin and Mitchell 2012) indicates its affinity more towards a lamproitic magma in contrast to a archetypal kimberlite as suggested by earlier workers (Rao and Phadtre 1966; Murthy et al. 1980; Smith 1989; Chalapathi et al. 1996; Fareeduddin 2008; Ravi et al. 2009). These new results on P-2 kimberlite from Wajrakarur have prompted detailed petrographic studies of the other kimberlite bodies in the region with a view to classifying them in their appropriate slots in the mineralogic-genetic classification scheme of ultrapotassic rocks.

It has been established that kimberlites and lamproites are different rocks and originate from genetically diverse mantle–derived magmas (Mitchell 2006). Since kimberlites and lamproites are hybridized rocks with variable crustal and mantle xenoliths and xenocrysts, the best way to petrologically classify these rocks is on the basis of proper characterization of the primary minerals crystallizing in the groundmass. The groundmass minerals of kimberlites and lamproites reflect the vital characteristics of their parental magma (Mitchell 2006). Thus, the mineralogical studies on pipes P-5 and P-13 which form part of Lattavaram cluster in WKF were initiated through IGCP project 557: Diamonds, xenoliths and kimberlites. The samples were collected during one of the IGCP field trips to the WKF, Andhra Pradesh, India.

Geological Setting

The Dharwar Craton of India is divided into Eastern Dharwar Craton (EDC) and Western Dharwar Craton (WDC) by the Chitradurga Boundary Fault (CBF; Inset of

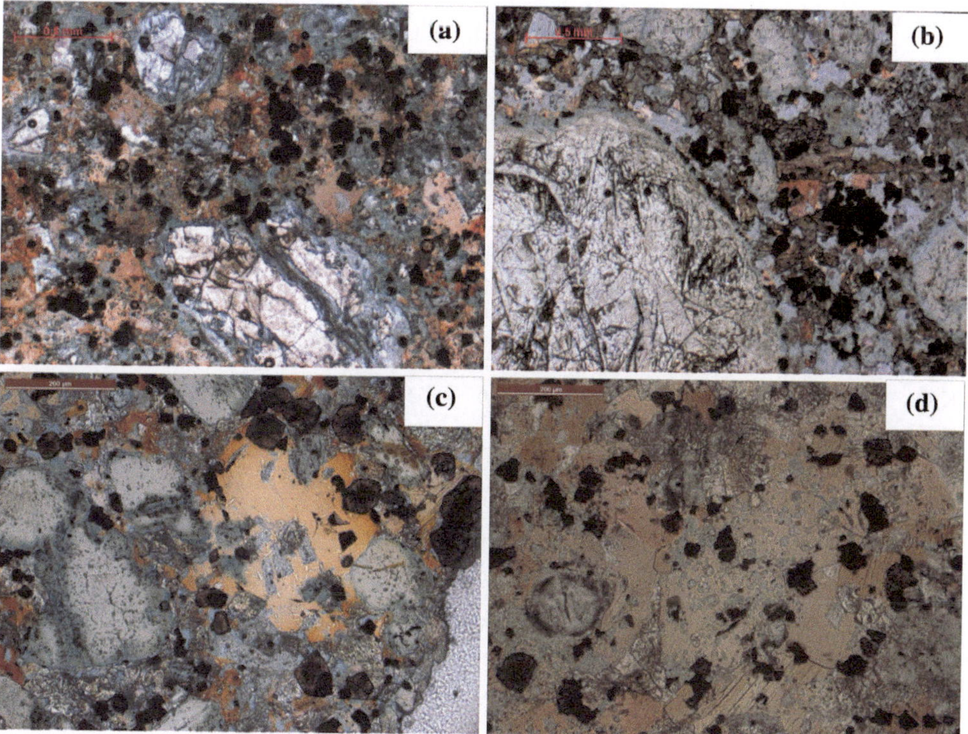

Fig. 2 **a** Macrocryst and groundmass olivines in P-5. Also seen are spinels, perovskites, phlogopites, serpentine and chlorite (PPL). **b** A macrocryst of olivine in P-13 (PPL). **c** Poikiolitic phlogopite in P-5 enclosing very fine needle-like inclusions and perovskites (PPL). **d** Poikiolitic phlogopite in P-13 enclosing spinels and perovskites (PPL)

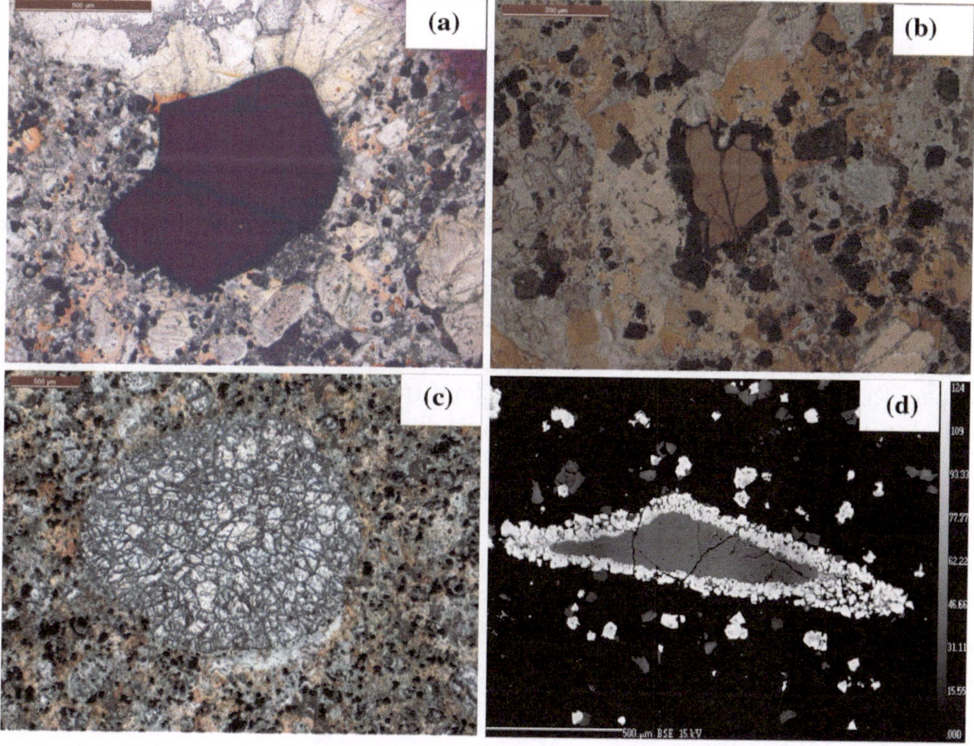

Fig. 3 **a** An ilmenite xenocryst in P-5. Also seen are the groundmass minerals such as olivines, phlogopites, spinels and perovskites (PPL). **b** An ilmenite xenocryst in P-13 surrounded by groundmass olivine, phlogopite, spinel and perovskites (PPL). **c** Polycrystalline olivine with resorbed outline in groundmass comprising spinels, perovskites, chlorite and serpentine (PPL). **d** An angular picro-ilmenite xenocryst in P-13. The ilmenite grain is rimmed by perovskites and spinels (BSE image)

Fig. 1) (Swami Nath et al. 1976; Drury et al. 1984; Chadwick et al. 2000; cf. Ravi et al. 2009). The kimberlites and lamproites in the EDC (Fig. 1) occur in clusters in four kimberlite fields, namely (1) Wajrakarur Kimberlite Field (WKF), (2) Narayanpet Kimberlite Field (NKF), (3) Raichur Kimberlite Field (RKF) and (4) Tungabhadra Kimberlite Field (TKF) and three lamproite fields, namely, (1) Krishna Lamproite Field (KLF), (2) Nallamalai Lamproite Field (NLF) and (3) Ramadugu Lamproite Field (RLF) (cf. Fareeduddin and Mitchell 2012).

Table 1 Representative compositions of phlogopites from P-5 (points 1–6) and P-13 (points 7–12)

	1	2	3	4	5	6	7	8	9	10	11	12
SiO_2	40.81	39.94	42.86	42.51	44.2	43.81	39.11	39.3	41.29	41.7	41.8	42.1
TiO_2	2.04	2.2	1.88	2.55	2.3	2.11	2.53	2.75	2.38	2.08	1.76	1.96
Al_2O_3	8.1	8.21	5.14	4.64	4.47	4.51	9.94	8.63	6.9	5.64	7.16	8.06
FeO (t)	8.74	10.73	8.98	9.84	8.88	9.24	8.51	10.57	11.51	11.79	7.62	7.4
MgO	23.13	22.09	23.58	22.47	22.63	23.17	23.14	22.51	21.83	22.25	24.69	23.93
MnO	0.14	0.08	0.03	0.16	0.02	0.06	0.21	0.14	0.14	0.11	0	0.09
BaO	0.66	0.8	0.33	0.43	0.5	0.45	1.14	1.06	0.44	0.26	0.35	0.49
Na_2O	0.43	0.21	0.85	0.94	1.44	1.34	0.05	0.07	0.18	0.07	0.52	0.6
K_2O	9.92	9.64	9.67	9.86	9.62	9.85	9.33	9.69	9.76	9.93	9.98	10.12
Total	93.97	93.9	93.32	93.4	94.06	94.54	93.96	94.72	94.43	93.83	93.88	94.75
Structural formula based on 22 atoms of oxygen												
Si	6.07	6.00	6.40	6.39	6.54	6.48	5.82	5.87	6.17	6.28	6.17	6.15
Ti	0.23	0.25	0.21	0.29	0.26	0.23	0.28	0.31	0.27	0.24	0.20	0.22
Al	1.42	1.45	0.90	0.82	0.78	0.79	1.74	1.52	1.22	1.00	1.25	1.39
Fe	1.09	1.35	1.12	1.24	1.10	1.14	1.06	1.32	1.44	1.48	0.94	0.90
Mg	5.13	4.95	5.25	5.04	4.99	5.11	5.13	5.02	4.86	4.99	5.43	5.21
Mn	0.02	0.01	0.00	0.02	0.00	0.01	0.03	0.02	0.02	0.01	0.00	0.01
Ba	0.04	0.05	0.02	0.03	0.03	0.03	0.07	0.06	0.03	0.02	0.02	0.03
Na	0.12	0.06	0.25	0.27	0.41	0.38	0.01	0.02	0.05	0.02	0.15	0.17
K	1.88	1.85	1.84	1.89	1.82	1.86	1.77	1.85	1.86	1.91	1.88	1.89

Wajrakarur Kimberlite Field

The WKF (120 km × 60 km) lies in Anantpur district, Andhra Pradesh, India. WKF contains 31 kimberlite intrusions distributed in five clusters (Fig. 1), viz., (1) Wajrakarur (7 pipes; P-1, P-2, P-6, P-10, P-11, P-12 and P-15), (2) Lattavaram (7 pipes; P-3, P-4, P-5, P-7, P-8, P-9, P-13), (3) Chigicherla (5 pipes; CC1–CC5), (4) Kalyandurg (6 pipes; KL-1 to KL-6), and (5) Timmasamudram (6 pipes; TK-1 to TK-6). Most of these 31 intrusions were discovered by the Geological Survey of India (Rao and Phadtre 1966; Reddy 1987; Nayak and Kudari 1999; Sravan Kumar et al. 2004; Mukherjee et al. 2007; Choudary et al. 2007; Ravi et al. 2009; Fareeduddin 2008; cf. Fareeduddin and Mitchell 2012). Most of the WKF kimberlites are poorly diamondiferous. The kimberlites of WKF are hosted by the granites and gneisses of the EDC (Fig. 1). A detailed account on the mode of occurrence and characteristics of the individual pipes of WKF is given by Neelkantam (2001). The kimberlites are capped by 0.5 m–1.5-m thick calcrete. Kimberlites of WKF are of middle to late Proterozoic age ranging from 840 to 1,153 Ma (Paul et al. 1975; Kumar et al. 1993, 2007; Chalapathi Rao et al. 1996, 2004; Paul et al. 2006; Ravi et al. 2009; Fareeduddin and Mitchell 2012).

Section Preparation and Analytical Techniques

Selected polished thin sections of both P-5 and P-13 kimberlites were prepared at the Petrology, Petrochemistry and Ore Dressing (PPOD) division, Geological survey of India (GSI), Southern Region, Bangalore and at GSI, Jaipur, and were analysed using a CAMECA SX-100 electron microprobe at PPOD Division, GSI, Bangalore. The analyses were done with a beam current of 20 nA, accelerating voltage of 15 kV and a static beam diameter of 0.5–1 μm. For the analysis of the REEs, an accelerating voltage of 20 kV and a beam current of 20 nA were used. The Kα emissions were used for all elements except Nb, Ba, REE and Sr, for which Lα emissions were used. Counting time was 10 s for both peak and background. Mineral and synthetic standards supplied by CAMECA were used for calibration, and the matrix correction was done by the PAP programme available in the operating software package of SX-100.

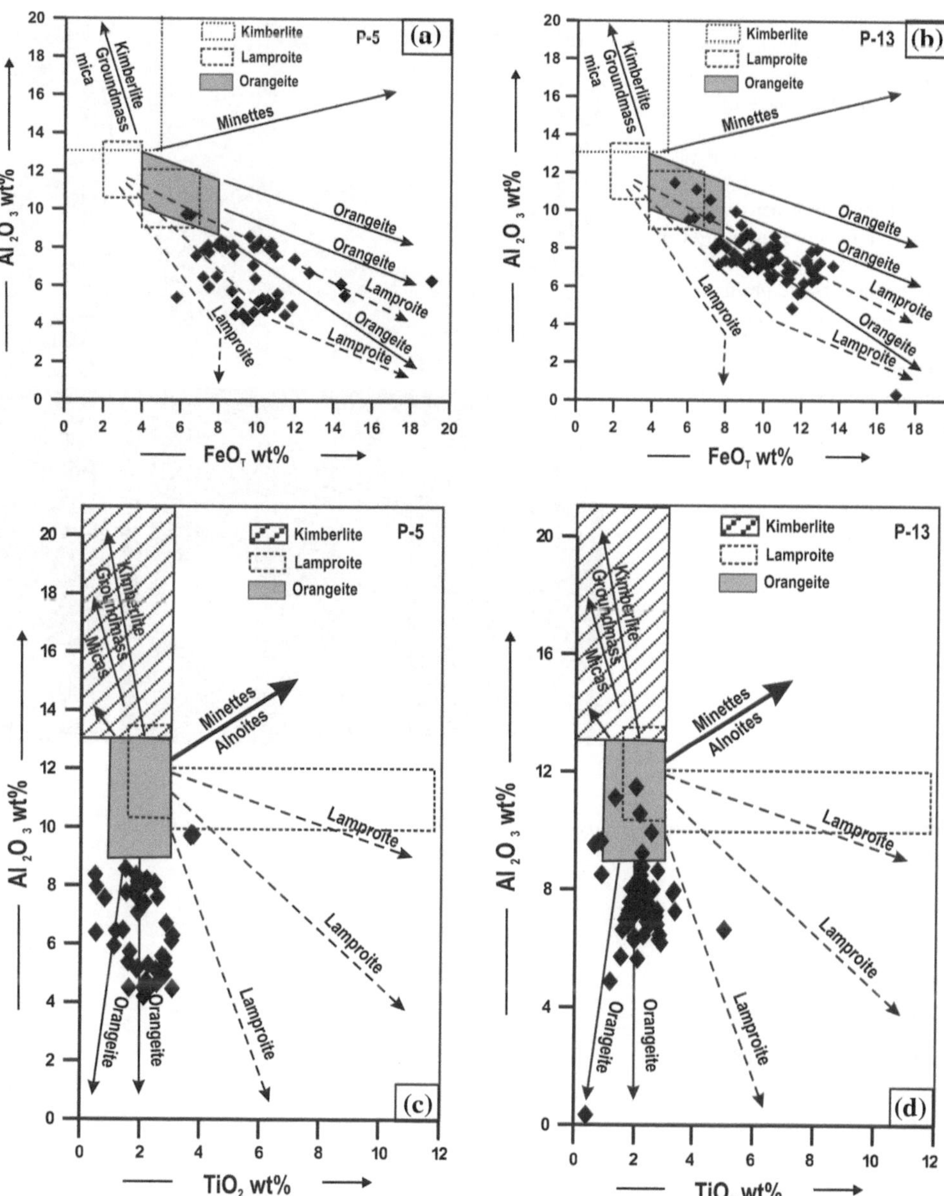

Fig. 4 **a**, **b**, Compositional trends of P-5 and P-13 phlogopites in Al_2O_3 versus FeO (t) plot (after Mitchell 1995). **c**, **d** Compositional trends of P-5 and P-13 phlogopites in Al_2O_3 versus TiO_2 plots (after Mitchell and Bergman 1991; Mitchell 1995)

Geology of Pipe-5 and Pipe-13

The pipe P-5 is situated approximately 2.5 km from P-13 in the Lattavaram cluster (Fig. 1). The kimberlite is medium to fine grained with autoliths and crustal xenoliths. The kimberlite has a less-than-a-metre calcrete cover. P-5 is an elongated body (240 m × 45 m). The crustal fragments/ xenoliths are of pink granite that surrounds the pipe rock. The rock is melanocratic, hard and compact (Fareeduddin 2008; Ravi et al. 2009).

Pipe P-13 is an oval-shaped body (125 m × 100 m) emplaced into the granitoids (Ravi et al. 1999; cf. Fareeduddin and Mitchell 2012). Most of the pipe P-13 exposures are weathered and carbonated except at a few places where fresh kimberlite has been identified. Crustal xenoliths of granites and basic rocks are common in P-13 kimberlite (Ravi et al. 1999, 2009; Fareeduddin 2008; Fareeduddin and Mitchell 2012). Hand specimens contain olivine macrocrysts and megacrysts up to 5 cm in diameter (Ravi et al. 1999, 2009).

Mineralogy of Pipe-5 and Pipe-13

The pipe P-5 and P-13 show inequigranular textures with macrocrysts and microphenocrysts of olivine and phlogopite in a fine grained segregated groundmass (Fig. 2a, b, c, d). Mantle xenoliths and xenocrysts are observed commonly in both the pipe rocks, although P-13 contains more mantle

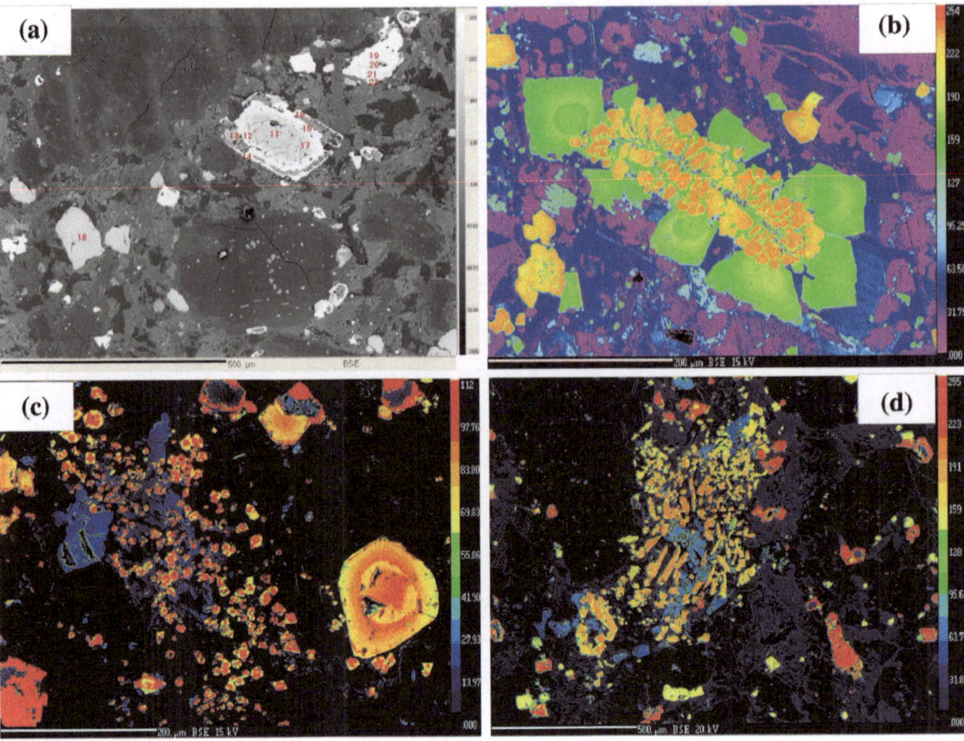

Fig. 5 **a** BSE image of atoll spinel in P-5 (analyses for the points on the figure are given in Table 2). **b** BSE image of intergrowth of spinel (*yellow*) and perovskite (*green*) in P-13. **c** BSE image showing bimodal size distribution of perovskites in P-5 (*red* in the core and *yellow* in the rim; Table 4 gives analyses for these perovskite grains). **d** BSE image of elongate perovskite grains in P-5 (Table 6 represents analyses for these elongate perovskites)

Table 2 Analyses for atoll spinel (analyses below correspond to points marked in Fig. 5a)

	11	12	13	14	15	16	17
MgO	13.92	0.55	0.09	0.31	2.03	0.22	12.88
CaO	0	0.17	38.98	0.48	2.15	0.3	0.03
MnO	0.24	2.01	0.03	1.29	2.17	1.4	0.24
FeO (t)	14.07	72.26	1.5	75.04	63.87	76.57	15.89
Al_2O_3	18.44	1.35	0.14	0.37	1.93	0.65	17.85
Cr_2O_3	50.85	4.75	0.03	0.27	5.17	0.39	51.55
SiO_2	0.09	0.19	0.06	0.83	3.53	0.19	0.13
TiO_2	0.1	12.62	56.13	12.03	14.56	12.99	0.15
Total	97.71	93.9	96.96	90.62	95.41	92.71	98.72

xenoliths and xenocrysts than P-5 (Fig. 3a, b, c, d). The predominant groundmass minerals in both the pipes are phlogopite, spinel, perovskite, apatite, monticellite, magnesian ilmenite and pyroxene along with serpentine, chlorite and carbonate (Fig. 2a, b, c, d). In the present study, monticellite was identified in one sample from P-13. No monticellite was found in P-5, although it has been reported previously (Smith 1989).

The P-5 and P-13 kimberlites have two generations of olivine with variable contents of macrocrystic, phenocrystic and microphenocrystic olivine (Fig. 2a, b). The macrocrystic and phenocrystic olivines are fresh compared to the groundmass olivines that are largely altered to serpentine. The volume percentage of macrocrystic species of olivine in P-5 kimberlite is 10–15 % whereas P-13 has slightly higher content of macrocrystic olivines, that is, 20–25 %.

Phlogopite, the second dominant mineral in both kimberlites, occurs as phenocrysts and as groundmass material. The phlogopites are reddish yellow to orange to honey yellow in colour (Fig. 2c, d). At places poikilitic phlogopites are observed (Fig. 2c). The phlogopites from P-5 are low in alumina ($Al_2O_3 < 8$ wt%; Table 1). Smith (1989) also reported P-5 phlogopites with low alumina contents. In P-13 phlogopites, the Al_2O_3 goes up to 12 wt%. Phlogopites in both the pipes are barium poor (BaO < 2 wt%). The titanium content varies between 1–4 wt% for P-5 and 1–6 wt% for P-13 (Table 1). Also observed are very fine small needle-like inclusions in phlogopites of both the pipes

Table 3 Representative compositions of spinels from P-5 and P-13

P-5 spinels														
	1	2	3	4	5	6	7	8	9	10	11	12	13	14
TiO_2	9.59	0.06	8.09	9.23	12.28	12.62	12.99	0.31	0.11	7.85	11.91	12.61	13.96	13.13
Al_2O_3	3.35	14.44	5.8	0.76	1.67	1.35	0.65	0.05	0.07	0.48	0.53	0.73	0.8	0.69
Cr_2O_3	15.14	55.68	31.36	7.23	6.53	4.75	0.39	0.04	0.25	0.01	0.05	3.27	0.4	1.03
FeO (t)	60.16	14.57	40.54	73.73	71.26	72.26	76.57	88.45	86.22	81.48	78.34	75.01	76.4	75.61
MnO	0.78	0.23	0.79	1.51	1.85	2.01	1.4	0.25	0.28	0.85	1.41	2	1.48	1.79
MgO	6.16	13.67	7.72	0.49	0.49	0.55	0.22	2.72	2.71	2.01	0.1	0.53	0.34	0.34
Total	95.18	98.65	94.3	92.95	94.08	93.54	92.22	91.82	89.64	92.68	92.34	94.15	93.38	92.59
Recalculated analyses														
FeO	30.05	12.86	26.12	36.68	39.49	39.34	40.33	29.31	28.55	33.84	39.59	39.52	41.31	39.59
Fe_2O_3	33.46	1.90	16.03	41.17	35.30	36.58	40.28	65.72	64.09	52.94	43.06	39.44	39.00	39.59
Total	98.53	98.84	95.91	97.08	97.62	97.21	96.26	98.40	96.06	97.98	96.65	98.10	97.29	96.56
P-13 spinels														
	1	2	3	4	5	6	7	8	9	10	11	12	13	14
TiO_2	0.07	7.8	8.36	11.63	15.73	19.26	10.18	12.16	20.68	11.82	14.08	11.24	0.05	0.1
Al_2O_3	17.7	5.05	3.23	1.31	1.39	1.4	0.99	1.55	0.48	0.84	2.08	0.79	0.04	0.07
Cr_2O_3	51.62	31.54	18.31	2.28	1.07	4.35	5.32	1.31	2.67	0.18	0.77	1	0.04	0.05
FeO (t)	13.88	43.57	58.47	76.45	71.1	62.64	72.56	74.07	66.22	78.28	71.9	77.66	87.68	86.51
MnO	0.23	0.7	0.76	1.55	2.15	2.44	1.26	1.43	2.07	1.41	1.77	1.7	0.24	0.32
MgO	14.36	8.32	6.2	0.51	2.88	5.28	3.51	3.11	1.8	0.38	2.89	0.38	2.76	3.63
Total	97.86	96.98	95.33	93.73	94.32	95.37	93.82	93.63	93.92	92.91	93.49	92.77	90.81	90.68
Recalculated analyses														
FeO	12.07	25.83	28.92	39.11	38.68	38.09	33.52	35.75	44.41	39.32	37.44	38.45	29.06	28.62
Fe_2O_3	2.01	19.71	32.84	41.49	36.03	27.29	43.39	42.58	24.24	43.30	38.29	43.57	65.14	64.34
Total	98.06	98.96	98.62	97.89	97.93	98.10	98.17	97.90	96.35	97.25	97.33	97.14	97.34	97.13

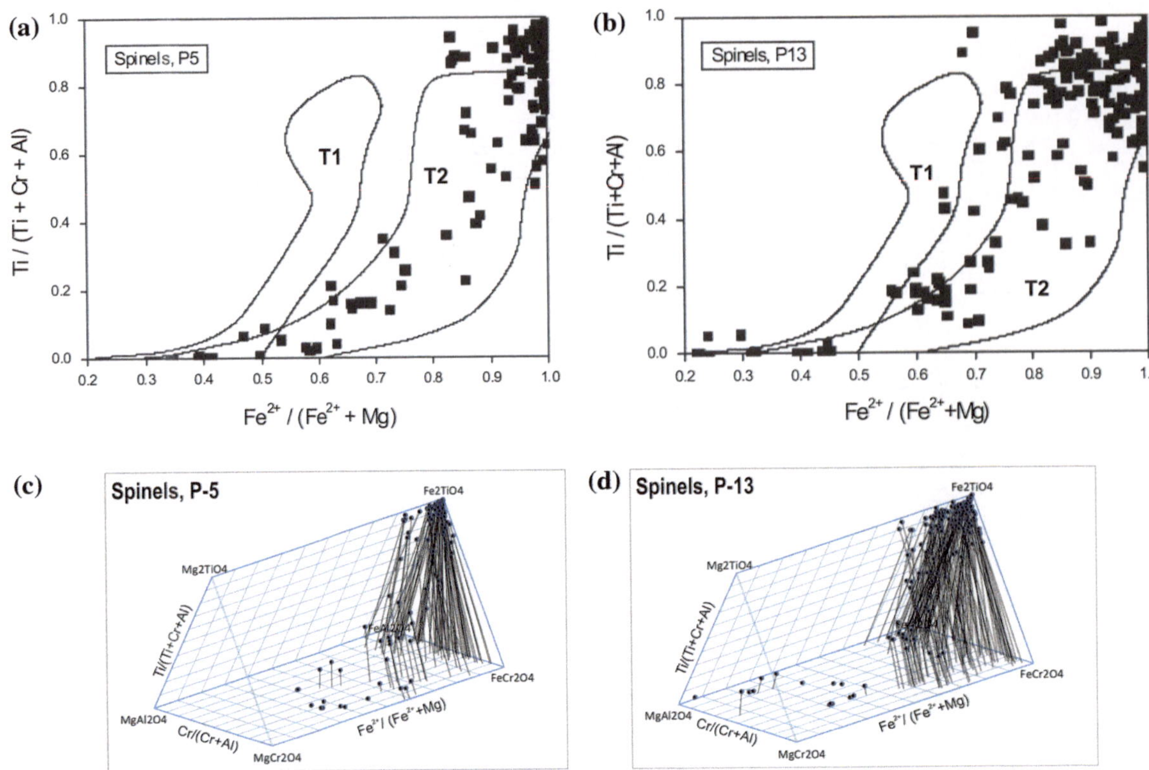

Fig. 6 **a**, **b** Compositional trends of P-5 and P-13 spinels in kimberlite and lamproite fields (Mitchell 1986, 1995; Mitchell and Bergman 1991). **c**, **d** Spinels from P-5 and P-13 plotted in the reduced spinel prism (Mitchell and Bergman 1991; Mitchell 1995)

Table 4 Compositions of perovskites showing bimodal size distribution in Fig. 5c

	1	2	3	4	5	6	7
TiO_2	55.94	55.36	55.56	44.73	54.04	42.35	51.51
CaO	38.98	38.96	37.86	24.48	36.88	27.35	33.87
Na_2O	0.29	0.3	0.46	2.11	0.56	1.2	0.83
K_2O	0.08	0.04	0.02	0.3	0.03	0.58	0.06
MgO	0.06	0.04	0.07	0.14	0.4	0.78	0.26
Fe_2O_3	1.96	1.28	1.06	4.17	1.15	4.25	2.38
SrO	0.32	0.3	0.23	0.47	0.26	0.27	0.38
La_2O_5	0.09	0.05	0.38	5.33	0.31	2.82	1.46
Ce_2O_3	0.67	1.05	1.7	10.49	1.18	6.13	3.42
Nb_2O_3	n.a.	n.a.	n.a.	n.a.	n.a.	n.a.	n.a.
Total	98.39	97.38	97.34	92.22	94.81	85.73	94.17
Structural formula based on 3 atoms of oxygen							
Ca	0.970	0.981	0.957	0.718	0.954	0.832	0.903
Na	0.013	0.014	0.021	0.112	0.026	0.066	0.040
K	0.002	0.001	0.001	0.010	0.001	0.021	0.002
Sr	0.004	0.004	0.003	0.007	0.004	0.004	0.005
La	0.001	0.000	0.003	0.054	0.003	0.030	0.013
Ce	0.006	0.009	0.015	0.105	0.010	0.064	0.031
Total A	0.996	1.010	0.999	1.006	0.998	1.017	0.995
Ti	0.977	0.979	0.986	0.921	0.982	0.905	0.964
Mg	0.002	0.001	0.002	0.006	0.014	0.033	0.010
Fe	0.034	0.023	0.019	0.086	0.021	0.091	0.045
Total B	1.014	1.003	1.007	1.012	1.017	1.029	1.018

n.a. Not analysed, *1–3* euhedral perovskite grain on the lower right-hand side
4–7 Analyses for the smaller grains scattered on the left side of euhedral perovskite grain

Table 5 Representative analyses of perovskites from P-5 (points 1–7) and P-13 (points 8–15)

	1	2	3	4	5	6	7	8	9	10	11	12	13	14	15
TiO_2	54.45	50.73	55.11	55.54	56.07	56.31	56.4	51.32	55.32	55.22	56.34	55.96	55.76	57.24	56.36
CaO	32.75	32.31	38.32	38.37	38.86	40.01	39.58	35.2	37.93	37.48	39.25	37.61	38.91	39.47	39.19
Na_2O	2.04	1.16	0.44	0.31	0.31	0.25	0.32	1.67	0.44	1	0.43	0.96	0.29	0.48	0.44
K_2O	0.09	0.07	0.04	0.02	0.06	0.03	0.04	0.13	0.01	0	0.03	0	0.14	0.05	0.03
MgO	0.01	0.08	0.07	0.07	0.06	0.05	0.03	2.19	0.09	0.02	0.03	0.03	0.04	0.03	0.05
Fe_2O_3	0.58	2.49	1.02	1.29	1.2	1.32	1.61	1.68	1.16	0.76	0.9	1.31	2.21	0.95	2.44
SrO	0.1	0.67	0.22	0.2	0.24	0.26	0.28	1.75	0.2	0.46	0.83	0.45	0.47	0.7	0.97
La_2O_3	n.d.	1.38	n.d.	n.d.	n.d.	n.d.	n.d.	n.d.	n.d.	n.d.	n.d.	n.d.	n.d.	n.d.	n.d.
Ce_2O_3	1.7	5.73	1.47	1.35	1.18	0.49	0.69	0.2	1.68	0.81	0.23	0.85	0.51	0.3	0.24
Nb_2O_5	0.42	0.5	0.25	0.23	0.27	0.29	0.22	0.25	0.3	1.97	0.24	1.48	0.33	0.25	0.37
Total	92.14	95.12	96.94	97.38	98.25	99.01	99.17	94.39	97.13	97.72	98.28	98.65	98.66	99.47	100.09
Structural formula based on 3 atoms of oxygen															
Ca	0.868	0.870	0.971	0.967	0.970	0.989	0.978	0.918	0.959	0.941	0.976	0.935	0.966	0.969	0.960
Na	0.098	0.057	0.020	0.014	0.014	0.011	0.014	0.079	0.020	0.045	0.019	0.043	0.013	0.021	0.020
K	0.003	0.002	0.001	0.001	0.002	0.001	0.001	0.004	0.000	0.000	0.001	0.000	0.004	0.001	0.001
Sr	0.001	0.010	0.003	0.003	0.003	0.003	0.004	0.025	0.003	0.006	0.011	0.006	0.006	0.009	0.013
La	0.000	0.013	0.000	0.000	0.000	0.000	0.000	0.000	0.000	0.000	0.000	0.000	0.000	0.000	0.000
Ce	0.015	0.053	0.013	0.012	0.010	0.004	0.006	0.002	0.015	0.007	0.002	0.007	0.004	0.003	0.002
Total A	0.986	1.004	1.008	0.996	0.999	1.009	1.003	1.028	0.997	1.000	1.010	0.991	0.994	1.004	0.996
Ti	1.014	0.959	0.980	0.982	0.983	0.978	0.979	0.940	0.982	0.973	0.984	0.976	0.972	0.987	0.970
Mg	0.000	0.003	0.002	0.002	0.002	0.002	0.001	0.079	0.003	0.001	0.001	0.001	0.001	0.001	0.002
Fe	0.012	0.052	0.020	0.025	0.023	0.025	0.031	0.031	0.021	0.013	0.016	0.023	0.039	0.016	0.042
Nb	0.005	0.006	0.003	0.003	0.003	0.003	0.003	0.003	0.003	0.021	0.003	0.016	0.003	0.003	0.004
Total B	1.031	1.021	1.006	1.013	1.012	1.008	1.013	1.053	1.009	1.008	1.003	1.016	1.016	1.007	1.017

n.d. Not detected

Table 6 Analyses for elongate perovskites from P-5 (points 1–3) representing core part and points 4–6 from rim part corresponding to Fig. 5d

	Core	Core	Core	Rim	Rim	Rim
	1	2	3	4	5	6
Na_2O	2.04	1.97	1.6	0.64	0.84	0.66
CaO	32.75	32.89	33.69	38.64	37.1	38.14
FeO (t)	0.58	0.53	0.62	0.69	0.67	0.85
SrO	0.1	0.09	0.06	0.34	0.27	0.3
BaO	0.91	1.02	0.95	0.87	0.92	0.96
Nb_2O_3	0.42	0.43	0.64	0.48	0.35	0.54
Ce_2O_3	1.7	1.63	1.63	0.47	0.68	0.78
TiO_2	54.45	54.83	54.41	55.92	55.44	55.1
P_2O_5	0.04	0.12	0.04	0.08	0.16	0.08
Total[a]	92.99	93.51	93.64	98.13	96.43	97.41

[a] Totals are low

(Fig. 2c). On the binary plot of wt% Al_2O_3 versus wt% total FeO, all the phlogopites from P-5 and P-13 clearly follow lamproitic trends (Fig. 4a, b). On the binary plot of wt% Al_2O_3 versus wt% TiO_2 the phlogopites from P-5 and P-13 straddle between the trends shown by orangeites and lamproites (Fig. 4c, d). In both the above-mentioned binary plots, the P-5 and P-13 phlogopites do not follow the Group-1 kimberlite trends but do follow the orangeites and

Table 7 Representative analyses of pyroxenes from P-5 (points 1–6) and P-13 (points 7–12)

	1	2	3	4	5	6	7	8	9	10	11	12
SiO_2	53.22	51.29	52.01	52.05	52.6	51.66	53.68	53.49	52.46	51.87	53.85	52.2
TiO_2	0.92	2.49	2.04	2.16	1.52	2.24	0.77	1.73	1.53	1.19	0.81	2.23
Al_2O_3	0.19	0.78	0.58	0.54	0.49	0.73	0.19	0.36	1.14	0.61	0.57	0.61
Cr_2O_3	0.04	0.02	0.01	0	0.01	0.03	0.31	0.22	0	0.02	0	0.02
FeO	5.07	6.26	5.57	6.03	5.98	5.97	5.16	4.57	4.56	5.08	4.13	6.83
MnO	0.15	0.26	0.13	0.17	0.24	0.09	0.12	0	0.17	0.09	0.16	0.17
MgO	15.22	14.13	14.52	14.35	14.43	14.38	15.12	15.29	15.2	15.5	15.54	13.54
CaO	23.29	22.29	22.84	22.41	22.69	22.94	22.88	22.81	23.86	23.62	24.66	21.42
Na_2O	1.09	1.68	1.33	1.54	1.28	1.38	1.39	1.42	0.75	0.99	0.57	2.08
Total	99.19	99.2	99.03	99.25	99.24	99.42	99.62	99.89	99.67	98.97	100.29	99.1
Structural formula based on 6 atoms of oxygen												
Si	1.980	1.926	1.947	1.965	1.965	1.932	1.988	1.970	1.941	1.941	1.974	1.959
Ti	0.026	0.070	0.057	0.043	0.043	0.063	0.021	0.048	0.043	0.034	0.022	0.063
Al	0.008	0.035	0.026	0.022	0.022	0.032	0.008	0.016	0.050	0.027	0.025	0.027
Cr	0.001	0.001	0.000	0.000	0.000	0.001	0.009	0.006	0.000	0.001	0.000	0.001
Fe	0.158	0.197	0.174	0.187	0.187	0.187	0.160	0.141	0.141	0.159	0.127	0.214
Mn	0.005	0.008	0.004	0.008	0.008	0.003	0.004	0.000	0.005	0.003	0.005	0.005
Mg	0.844	0.791	0.810	0.804	0.804	0.802	0.835	0.839	0.838	0.865	0.849	0.758
Ca	0.928	0.897	0.916	0.908	0.908	0.919	0.908	0.900	0.946	0.947	0.969	0.861
Na	0.079	0.122	0.097	0.093	0.093	0.100	0.100	0.101	0.054	0.072	0.041	0.151

lamproite trends (Mitchell 1986, 1995; Mitchell and Bergman 1991).

P-5 and P-13 kimberlites are characterized by the presence of euhedral to subhedral groundmass (0.001–0.1 mm) spinels throughout the rock (Figs. 2a, b, 3a, b). The atoll spinel in P-5 (Fig. 5a) has Ti-poor and chromium-, magnesium- and aluminium-enriched cores (analyses points 11 and 17 marked on Fig. 5a; Table 2) but titanium- and iron-rich and chromium-poor rim (analyses points 12, 14, 15 and 16 marked on Fig. 5a; Table 2). Four varieties of spinels, that is, ulvöspinel, Mg-ulvöspinel, Mg-Cr-Al-spinel and magnetites are recorded in both the pipes. The ulvöspinel, most dominant among spinels, are low in alumina (<3 wt%) and high in titanium (10–20 wt%) and iron (>60 wt%). The representative analyses of the spinels from both the pipes are given in Table 3. The spinels from P-5 and P-13 exhibit compositional trends identical to those found in kimberlite magmatic trend T2 in the binary plot Ti/(Ti + Cr + Al) versus $Fe^{2+}/(Fe^{2+} + Mg)$ (Fig. 6a, b; Mitchell 1986, 1995; Mitchell and Bergman 1991). Representative composition of spinels from P-5 and P-13 follows lamproitic trends in the reduced spinel prism (Fig. 6c, d; Mitchell and Bergman 1991; Mitchell 1995). Intergrowth of spinel and perovskite is observed at a number of places in both the pipe rocks (Fig. 5b).

Euhedral to subhedral grains of perovskites with bimodal size distribution occur throughout the P-5 and P-13 kimberlites (Figs. 2c, 5c). The perovskites in the Fig. 5c show bimodal size distribution and also exhibit difference in the total REE contents. The larger grain to the lower right side is impoverished in REEs compared to the smaller numerous scattered grains which are enriched in total REEs (Fig. 5c; Table 4). Perovskites from P-5 and P-13 in general have low LREE contents (<3 wt%), but in a few grains, the LREE content goes up to 16 wt% (Ce_2O_3-5-11 wt%; La_2O_3-2-6 wt%; Table 5). Perovskites at places form necklace around altered olivine grains. In one of the thin sections of P-5, unusual elongated shapes of perovskites have been observed. These perovskites are zoned with Na- enrichment in the rim compared to the core (Fig 5d; Table 6).

Both P-5 and P-13 show clinopyroxene laths which form clusters and have a patchy appearance. Volumetrically P-5 contains more clinopyroxene compared to P-13. Representative pyroxene analyses from P-5 and P-13 are given in Table 7. Most of the clinopyroxenes, both from P-5 and P-13, occupy lamproite field in Fig. 7a in the cationic Ti versus Al binary plot (Mitchell 1995). The low alumina clinopyroxenes mostly occupy the lamproite field in Fig. 7b (after Mitchell 1995).

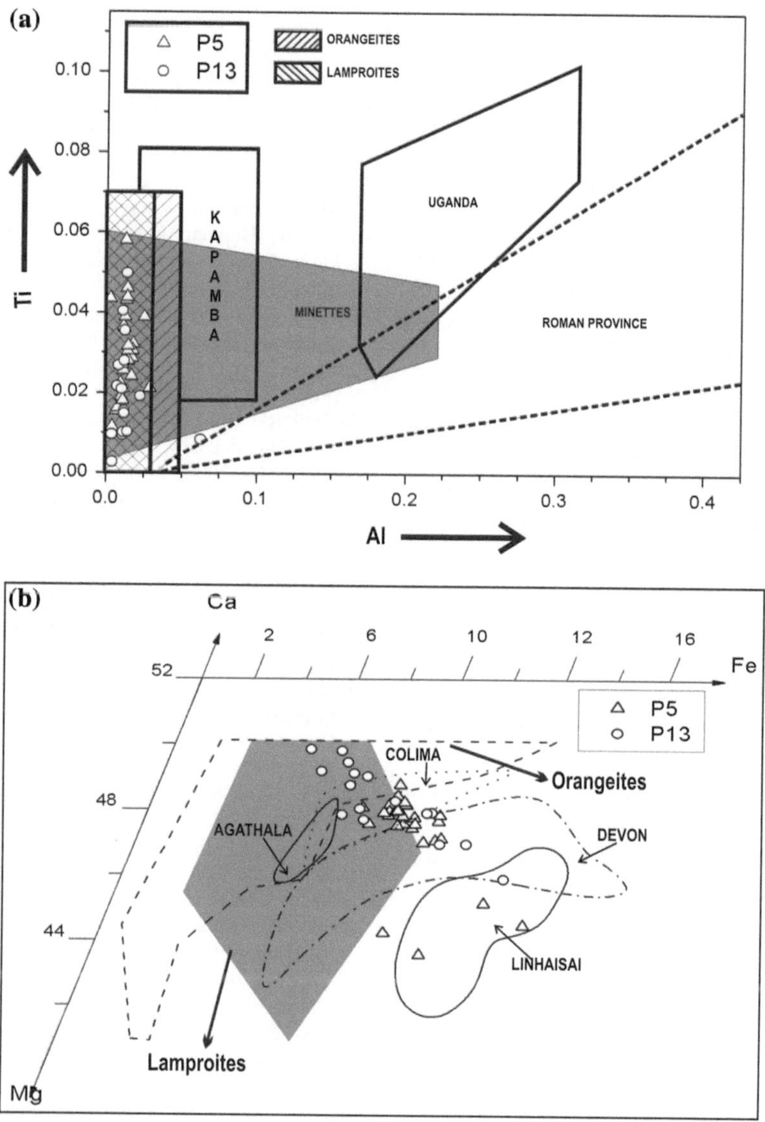

Fig. 7 **a** P-5 and P-13 clinopyroxenes plotted in Al versus Ti (atomic) plot. The clinopyroxenes occupy the lamproite field in this plot after Mitchell 1995. **b** Composition of clinopyroxenes from P-5 and P-13 compared with those of clinopyroxenes in orangeites, lamproites (*grey colour*) and minettes (*Agathla*, *Devon*, *Colima* and *Linhaisai*, all these minettes are marked in the plot). Most of P-5 and P-13 clinopyroxenes occupy lamproite field (Mitchell 1995)

Discussion and Conclusions

"As the two main sources of diamonds, kimberlites and lamproites, can occur in the same craton, it is evident that rock nomenclature is not strictly an academic concern, as the correct classification of rocks is important with respect to the exploration for, and evaluation of, potential diamond deposits" (Mitchell 2006). This statement justifies, apart from the style of emplacement, identification of indicator minerals, nature of mantle xenoliths and xenocrysts, source rocks, that is, metasomatized lithospheric mantle or depleted asthenospheric mantle, ideas about the depths of magma generation (using isotopic systematics), diamond incidence and tectonic environment, the need for proper petrographic classification of these diamond-bearing rocks. Petrographic study of textural relations and chemistry of groundmass minerals are valuable tools in this context.

A reassessment of the petrological characteristics and groundmass mineral chemistries of the P-5 and P-13 kimberlitic pipes has put a question mark on their being classified as archetypal (Group-1) kimberlites. Phlogopites in the two pipes are impoverished in alumina and enriched in iron compared to phlogopites of archetypal/Group-1 kimberlites. The clinopyroxenes are typically low in alumina. Clinopyroxenes are characteristic of lamproites and are generally non-existent in Group-1 kimberlites. Spinels from both P-5 and P-13 follow trends defined by lamproitic spinels. The spinels are impoverished in aluminium contrary to Group-1 kimberlite spinels. The spinels follow magnesiochromite-ulvöspinel-magnetite trend. Some perovskites from P-5 and P-13 are enriched in LREEs (up to 16 wt%; Table 4) contrary to true kimberlitic perovskites (Mitchell 2002). The compositional trends of phlogopite, spinel and clinopyroxene as seen in standard binary plots are not indicative of kimberlitic affinities; rather, they are more suggestive of affinity towards lamproitic magmas. Smith (1989) noted about features atypical of kimberlites in P-5 and suggested that P-5 could be a more evolved or extreme variety of kimberlite.

What can be concluded at the moment is as follows:

1. The two pipes represent polyphase intrusions in close proximity within Lattavaram region and they represent two different phases of kimberlite intrusion. Macrocrysts of olivine, mantle xenoliths and xenocrysts are present in a higher proportion in P-13 compared to P-5.
2. Phlogopite, spinels and perovskite are volumetrically predominant in P-13, while clinopyroxenes predominate in P-5. Perovskites in P-5 have marginally higher content of LREEs compared to P-13.
3. P-13 has more volume percentage of macrocrysts of olivine, mantle xenoliths and xenocrysts in comparison to P-5.
4. Despite the megascopic differences, groundmass mineral chemical characteristics show that neither of the two is an archetypal (Group-1) kimberlite. Both of them rather show trends akin to lamproites.

Acknowledgments Dr. Ayyasami and Dr. J. N. Das of GSI are thanked for their kind help in getting the samples analysed by EPMA at PPOD, GSI, Bangalore. Prof. S. C. Patel, Prof. Chalapathi Rao and Prof. N. C. Pant gave valuable inputs during the initial stage of data generation for this work. The work has immensely benefitted by suggestions from Prof. Roger H. Mitchell. The manuscript has been greatly improved by constructive comments from Prof. Barry Dawson, Prof. Wall and Prof. Stephen Sparks. IGCP-UNESCO project 557 grants and PURSE grant from DST, New Delhi, to GK are gratefully acknowledged.

References

Chadwick B, Vasudev VN, Hegde GV (2000) The Dharwar craton, southern India, interpreted as the result of Late Archaean oblique convergence. Precambr Res 99:91–111

Chalapathi Rao NV, Miller JA, Pyle DM, Madhavan V (1996) New Proterozoic K-Ar ages for some kimberlites and lamproites from the Cuddapah basin and Dharwar craton, south India: evidence for non-contemporaneous emplacement. Precambr Res 79:363–369

Choudary VS, Rau TK, Rao KSB, Sridhar M, Sinha KK (2007) Timmasamudram kimberlite cluster, Wajrakarur kimberlite Field, Anantapur district, Andhra Pradesh. J Geol Soc India 69:597–609

Drury SA, Harris NBW, Holtz RW, Reeves-Smith GJ, Wightman RT (1984) Precambrian tectonics and crustal evolution in south India. J Geol 92:3–20

Fareeduddin (2008) Training course notes on "An introduction to the petrology of diamond bearing rocks and modern methods in exploration for, and evaluation of, primary diamond deposits" conducted by Geological society of India Bangalore between 14th and 22 Jan 2008, p 42

Fareeduddin, Mitchell RH (2012) Diamonds and their source rocks in India. Geological Society of India, Bangalore, p 434

Kumar A, Padma Kumari VM, Dayal AM, Murthy DSN, Gopalan (1993) Rb-Sr ages for Proterozoic kimberlites of India: evidence for contemporaneous emplacement. Precambr Res 62:227–237

Kumar A, Heaman IM, Manikyamba C (2007) Mesoproterozoic kimberlites in south India: A possible link to ~1.1 Ga global magmatism. Precambr Res 154:192–204

Mitchell RH (1986) Kimberlites: mineralogy, geochemistry and petrology. Plenum press, New York, London, p 442

Mitchell RH (1995) Kimberlites, orangeites and related rocks. Plenum press, New York, p 410

Mitchell RH (2002) Perovskites: modern and ancient. Almaz Press Inc., Thunder Bay, p 319

Mitchell RH (2006) Potassic magmas derived from metasomatized lithospheric mantle: Nomenclature and relevance to exploration for diamond-bearing rocks. J Geol Soc India 67:317–327

Mitchell RH (2010) Mineralogy of the P2-west "kimberlite", Wajrakarur, AP, India. (Abst.) 6th International dyke conference, Varanasi, India, 4–7 Feb 2010, p 91

Mitchell RH, Bergman SC (1991) Petrology of lamproites. Plenum Press, New York, p 408

Mukherjee A, Sravan Kumar C, Reddy KK (2007) Two new kimberlites in Bommaganapalli area, Anantpur district, Andhra Pradesh, based on systematic stream sediment sampling and ground magnetic survey. J Geol Soc India 69:625–640

Murthy YGK, Rao MG, Misra RC, Reddy TAK (1980) Kimberlite diatremes of Andhra Pradesh-their assessment and search for concealed bodies. In: Papers and proceedings of the seminar: the many facets of the Indian Diamond industry, Bombay. Hindustan Diamond Co. Ltd. (unpaged-12 pages)

Nayak SS, Kudari SAD (1999) Discovery of diamond-bearing kimberlite in Kalyandurg area, Anantpur district, Andhra Pradesh. Curr Sci 76:1077–1079

Neelkantam S (2001) Exploration for diamonds in southern India. Geol Surv India Spec Publ 58:521–555

Paul DK, Rex DC, Harris PG (1975) Chemical characteristics and K-Ar ages of Indian Kimberlites. Bull Geol Soc Am 86:364–366

Paul DK, Nayak SS, Pant NC (2006) Indian kimberlites and related rocks: Petrology and geochemistry. J Geol Soc India 67(3):328–355

Rao SR, Phadtre PN (1966) Kimberlite pipe rocks of Wajrakarur area, Anantpur district, Andhra Pradesh. J Geol Soc India 7:118–123

Rao NVC, Gibson SA, Pyle DM, Dickin AP (2004) Petrogenesis of Proterozoic lamproites and kimberlites from the Cuddapah Basin and Dharwar Craton, Southern India. J Petrol 45:907–948

Ravi S, Bhaskara Rao KS, Rao KRP (1999) Search for kimberlites in the granite green stone terrain in the central segment of the Wajrakarur Kimberlite Field, Anantpur district, Andhra Pradesh. Rec Geol Surv India 132:40–43

Ravi S, Vaideswaran T, Rao KSB (2009) Field guide to Wajrakarur kimberlite field, Anantpur district, Andhra Pradesh. Geological Survey of India, pp 1–43

Reddy TAK (1987) Kimberlite and lamproitic rocks of Wajrakarur area, Andhra Pradesh. J Geol Soc India 61:131–146

Smith BHS (1989) Lamproites and kimberlites in India. Neues Jahrbuch Miner Abh 161:193–225

Sravan Kumar C, Mukherjee A, Vishwakarma RK (2004) Discovery of a new kimberlite pipe using multidisciplinary approach at Kalyandurg, Anantpur district, Andhra Pradesh. J Geol Soc India 64:813–817

Swami Nath J, Ramakrishnan M, Viswanatha MN (1976) Dharwar stratigraphic model and Karnataka cratonic evolution. Rec Geol Surv India 107:149–175

Kimberlite: Rapid Ascent of Lithospherically Modified Carbonatitic Melts

J. K. Russell, L. A. Porritt, and L. Hilchie

Abstract

Kimberlite deposits commonly contain in excess of 25 vol. % solid material in the form of mantle-derived macrocrysts and xenoliths, implying that the original magma was enriched in solids and, thus, dense. Numerous studies on this mantle-derived cargo, including diamond, suggest that the parental magmas derive from depths ≥200 km and that they are transported rapidly to the Earth's surface. Virtually, all models for kimberlite ascent invoke the presence of an exsolved fluid (i.e. CO_2 and H_2O) phase for buoyancy, yet the cause and nature of fluid exsolution remain elusive. Here, we present high-temperature analogue melting experiments to demonstrate a new mechanism for the efficient, continuous, and spontaneous production of a volatile phase within ascending kimberlite. We suggest that the parent melts to kimberlite range in composition from carbonatitic to carbonate-rich silica-undersaturated melts. In transit through the cratonic mantle lithosphere (CML), these silica-undersaturated melts assimilate mantle minerals, especially orthopyroxene, driving the melt to more silicic compositions and causing a marked drop in CO_2 solubility. The solubility drop manifests itself immediately by continuous and vigorous exsolution of a fluid phase, thereby reducing magma density, increasing buoyancy, and driving the rapid and accelerating ascent of the resultant kimberlite magma. Ultimately, continued orthopyroxene assimilation drives the melt towards silicate saturation allowing for late-stage magmatic crystallization of olivine as overgrowths on macrocrysts and as (micro-)phenocrysts.

Keywords

Kimberlite • Magma ascent • Carbonatite • Assimilation • Orthopyroxene • Buoyancy • Volatiles • Fluid

Introduction

Kimberlite magmas derive from ultra-deep (≥200 km) mantle sources (200–600 km; Mitchell 2008; Stachel 2001) and transport high loads (>25 vol. %) of mantle-derived xenoliths and xenocrysts from the base of the mantle lithosphere to the Earth's surface (e.g. Clement 1982; Mitchell 1986; Mitchell 2008; Shee 1985). Their ascent through ~150–180 km of cool cratonic mantle lithosphere (CML) is reputedly faster (Anderson 1979; Canil and

J. K. Russell (✉) · L. A. Porritt · L. Hilchie
Volcanology and Petrology Laboratory, Department of Earth, Ocean and Atmospheric Sciences, University of British Columbia, Vancouver, V6T 1Z4, Canada
e-mail: krussell@eos.ubc.ca

D. G. Pearson et al. (eds.), *Proceedings of 10th International Kimberlite Conference*, Volume 1, Special Issue of the Journal of the Geological Society of India, DOI: 10.1007/978-81-322-1170-9_12,

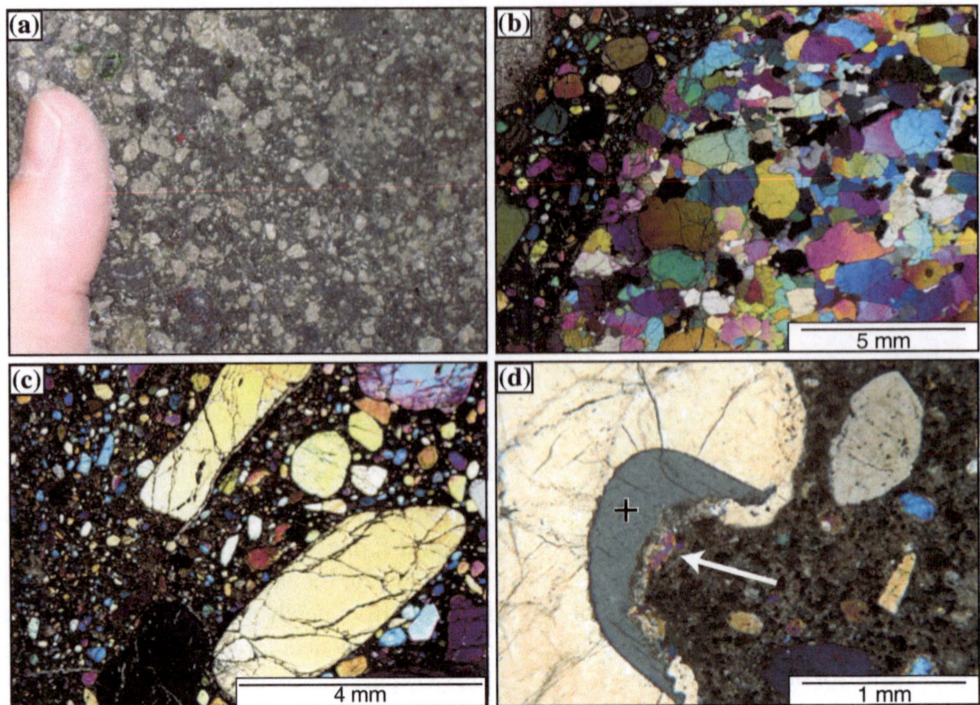

Fig. 1 Petrographic images of kimberlite samples, including **a** polished slab of kimberlite showing abundance of mantle-derived xenocrysts of olivine, clinopyroxene, garnet, and ilmenite, **b** photomicrograph of kimberlite containing fragment of peridotitic xenolith and illustrating grain shape variations, **c** photomicrograph of kimberlite showing high abundance and rounded shapes of mantle-derived olivine grains (courtesy of Stephen Moss; Russell et al. 2012), and **d** photomicrograph of kimberlite containing fragment of disaggregated peridotite comprising a grain of partially dissolved orthopyroxene (+) within larger olivine grain (courtesy of C. Brett; Russell et al. 2012). *Arrow* shows the reaction interface between orthopyroxene and melt. Note olivine overgrowth on the orthopyroxene grain. All thin-section images are taken under crossed Nicols

Fedortchouk 1999; Fedortchouk et al. 2007; Sparks et al. 2006; Wilson and Head 2007) than ascent rates estimated for other xenolith-bearing magmas (Rutherford and Gardner 2000; Sparks et al. 1977). Any holistic explanation for the rapid ascent of kimberlite should also explain the worldwide occurrence of kimberlite within, and on the margins of, cratons (Gurney et al. 2005; Janse and Sheahan 1995; Kjarsgaard 2007; Mitchell 1986; Nixon 1995).

The challenge has always been to provide the requisite buoyancy for these deep-seated magmas to ascend continuously and rapidly through the cool mantle lithosphere whilst sampling and entraining up to 25–50 vol. % dense mantle cargo. The exsolution of dissolved volatiles (CO_2 and H_2O) is, without exception, considered to play a critical role in contributing to this buoyancy; however, the *source* of the volatiles and the physical–chemical process behind their liberation is rarely specified (Anderson 1979; Wilson and Head 2007).

Russell et al. (2012) provided a simple but unique explanation for the rapid ascent of these enigmatic magmas. Specifically, their model provides a mechanism for efficient, continuous, and spontaneous production of a volatile phase within the ascending magma. One of the main attributes of this mechanism for volatile exsolution is its potential for operating at the very base of the lithosphere. Here, we recapitulate the main elements of this newly proposed model for kimberlite ascent (Russell et al. 2012). We then expand upon the original model by exploring its broader implications for the ascent and origin of kimberlite.

Geological Observations

The greatest impediment to understanding the ascent of kimberlite magma is that the primary compositions of kimberlite melts are unknown. The composition of the melt remains elusive because: (1) kimberlite rocks and deposits contain abundant (>30 %) mantle and crustal material (Fig. 1), (2) quenched kimberlite melt preserved as volcanic glass is unobserved, and (3) kimberlites are highly susceptible to alteration (Davidson 1943; Dawson 1971; Dawson 1980; Heaman et al. 2004; Kjarsgaard 2007; Nixon 1995; Verhoogen 1938) due to their ultrabasic composition (Hayman et al. 2009; Mitchell 1986; Stripp et al. 2006). Alteration is generally pervasive because of their age (Heaman et al. 2004; Kjarsgaard 2007; Nixon 1995) and,

Table 1 Measured and recalculated compositions of kimberlite rocks and melts based on bulk analyses of Igwisi Hills kimberlite (Dawson 1994; Wilcox et al. 2012) and experimental melts produced by high-pressure partial melting of mantle in the presence of CO_2 (Dalton and Presnall 1998a, b) and $CO_2 + H_2O$ (Foley et al. 2009)

	Igwisi Hills Lavas						Experimental melts		
No.	2	3	6	7	IHV[d]		KM14[e]	MP50[f]	HP50[f]
Label	BD864[a]	BD853[a]	BD864[b]	BD864[c]	As reported	Normalized	Solidus melt	~8 % Melt	~10 % Melt
							1,380 °C	1,180 °C	1,180 °C
SiO_2	19.88	17.38	15.27	17.65	21.46	18.43	5.79	5.19	7.46
TiO2	1.86	1.78	2.53	2.31	3.87	3.32		0.08	0.91
Al_2O_3	3.20	2.87	4.35	3.93	5.84	5.02	0.66	1.43	1.41
FeO(T)	9.81	9.73	10.91	10.80	18.75	16.11		5.92	7.28
MnO	0.25	0.23	0.31	0.30	0.41	0.35			
MgO	23.33	22.76	17.53	20.59	22.74	19.53	20.91	16.07	15.25
CaO	18.93	21.27	25.71	23.26	22.39	19.23	28.03	23.85	23.97
Na_2O	0.04	0.00	0.05	0.05	0.12	0.10		0.36	1.69
K_2O	0.03	0.01	0.04	0.04	0.06	0.05		0.37	0.54
P_2O_5	2.20	2.19	2.99	2.70	4.17	3.58			
H_2O	4.62	5.92	0.00	0.00				46.7	41.5
CO_2	14.24	14.51	19.37	17.52	14.10	14.10	44.61		

[a] Composiitons of matrix-rich tops of Igwisi Hills kimberlite lavas (Dawson 1994); [b] calculated composition based on removing ~21 % mantle olivine from BD864, [c] calculated composition based on sample BD864 and removing mantle olivine and adding ~ 10 % phenocrystic olivine. [d] Estimate of kimberlite melt composition calculated by removing 25 % mantle xenocrysts from Igwisi Hills lava composition (Wilcox et al. 2012) and then normalized to include CO_2 content. [e] Solidus melt composition in equilibrium with carbonated peridotite (Dalton and Presnall 1998a, b) and compositions of melts[f] from melting of pyrolite sources at 5 GPa (combined $H_2O + CO_2$ is by difference; Foley et al. 2009)

perhaps more importantly, because kimberlite bodies represent subsurface volcanic deposits that cooled relatively slowly in volatile-rich environments (i.e. rather than quenching as glasses). An abbreviated suite of compositions estimated for kimberlite melt is reported in Table 1; for more complete reviews, we refer readers to Price et al. (2000; Table 5), Le Roex et al. (2003; Table 1), and Sparks et al. (2006; Table 1, 2009; Table 1). Many different strategies have been used to estimate kimberlite melt compositions (cf. Canil and Bellis 2008; Patterson et al. 2009; Sparks et al. 2009), and there is general agreement that they are SiO_2-poor (~15–30 %) and MgO-rich (15–26 %) and that the volatile contents are high and dominated by CO_2 and H_2O (e.g. Kjarsgaard et al. 2009; Mitchell 1986, 2008; Patterson et al. 2009; Price et al. 2000; Sparks et al. 2009). However, the primary composition of kimberlite melts, especially their volatile contents, remains unmeasured. Physical properties estimated for the range of putative melt compositions include densities similar to basaltic to komatiitic melts (2,800–2,900 kg m^{-3}) and low viscosities (10^{-2}–10^2 Pa s) (e.g. Sparks et al. 2006, 2009).

Geothermobarometric studies on mantle-derived xenoliths of peridotite and eclogite transported by kimberlite magma constrain their depths of origin to the base of the CML or deeper (Boyd and Nixon 1973; Kopylova et al. 1999). Mechanical disaggregation of the xenoliths (Fig. 1b) produces the large volume of xenocrysts characteristic of most kimberlites (Figs. 1a, c; Brett 2009; Roedder 1965). The xenocryst suite is dominated by olivine (Ol > 25 vol. %) and accessory amounts of clinopyroxene (Cpx), garnet, ilmenite, phlogopite, Cr-spinel (Figs. 1a, c), and, in rare cases, diamond. Notably, orthopyroxene (Opx), which comprises ~15–27 % of the CML (1989; Kopylova and Russell 2000; McDonough 1990) and is the most silicic mineral present, is rare to absent in kimberlite rocks (e.g. Luth 2009; Mitchell 1973, 1986, 2008). Where observed, Opx is clearly in severe disequilibrium with the host melt (Fig. 1d).

Olivine xenocrysts record an early stage of grain size reduction and rounding (Fig. 1b, c); however, Ol overgrowths and micro-phenocrysts (Figs. 1d, 2c, d) indicate late-stage equilibrium with the melt and crystallization (e.g. Boyd and Clement 1977; Brett et al. 2009; Dawson 1994; Fedortchouk and Canil 2004; Kamenetsky et al. 2008; Mitchell 1973, 1986). Most Ol is xenocrystic and only a minor amount of Ol is cognate, expressed as subordinate overgrowths on xenocrysts and as micro-phenocrysts. Mitchell (2008) and Moore (2011) have dissenting respective views that half or more of the Ol is cognate and early-crystallized. However, regardless of the ultimate origin of Ol (xenocryst or phenocryst), the main effect of Ol content is to increase magma density during ascent. The observed range of solid volume fractions for kimberlite suggests bulk

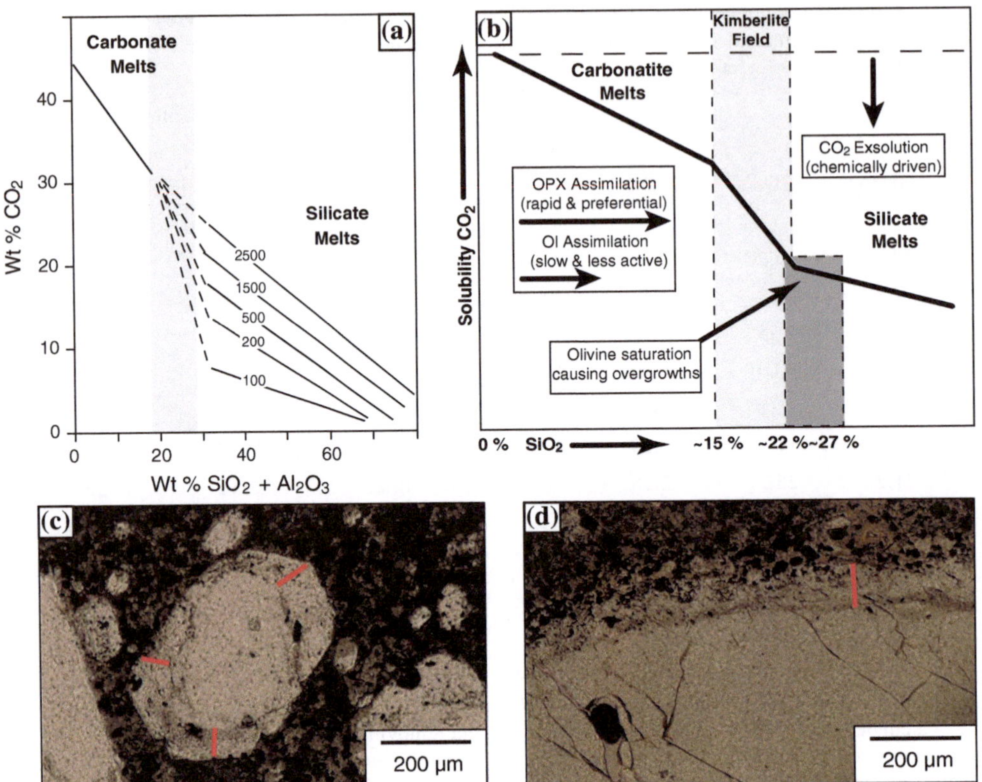

Fig. 2 Summary of CO_2 solubilities in silicic to carbonate melts and potential implications for late-stage crystallization of olivine in kimberlitic melts (after Russell et al. 2012). **a** Pressure and compositional dependence of CO_2 solubility across the carbonate–silicate transition (*shaded field*). *Solid* and *dashed lines* illustrate limited effects of pressure (MPa) on CO_2 solubility versus the strong effects of composition (i.e. $SiO_2 + Al_2O_3$ wt. %) on solubility and exsolution. Modified from Brooker et al. (2011) and reproduced from Russell et al. (2012; Fig. 2b). **b** Schematic model for evolution of carbonatitic melt due to assimilation of silicate mantle minerals and concomitant exsolution of CO_2. Assimilation produces an ever more silica-rich melt, causing a drop in CO_2 solubility expressed by decarbonation; ultimately, the process produces a kimberlitic melt that reaches silicate mineral saturation and allows for crystallization of olivine (overgrowths and phenocrysts). Reproduced from Russell et al. (2012; Fig. 2c). **c** Photomicrograph of small (300–400 μm) "phenocrystic" olivine (Brett et al. 2009) in kimberlite from Diavik Diamond Mines showing euhedral overgrowth of olivine on rounded xenocrystic core and partly defined by increase in concentration of inclusions. Modified from Brett et al. (2009). **d** Photomicrograph of thin olivine overgrowth on large (>1,000 μm) xenocrystic olivine. Modified from Brett et al. (2009)

densities of the magma (melt and crystal "cargo") in the range of 3,000–3,100 kg m^{-3}.

Kimberlite magmas reputedly ascend from the base of the mantle lithosphere (170–200 km) to the Earth's surface substantially faster (e.g. 1–30 m/s; Anderson 1979; Canil and Fedortchouk 1999; Downes et al. 2006; Kelley and Wartho 2000; Sparks et al. 2006; Wartho and Kelley 2003; Wilson and Head 2007) than other xenolith-bearing magmas (Rutherford 2008; Rutherford and Gardner 2000; 0.008–5 m/s; Sparks et al. 1977). As discussed immediately above, the componentry of most kimberlite suggests high aggregate magma densities that are 3–5 % less dense than the surrounding mantle lithosphere. However, Sparks et al. (2006) show that kimberlite ascent rates, driven by mantle–magma density contrasts as low as $\sim$100 kg m^{-3} and limited by dike widths of 1 m, could still be as high as $\sim$4 m s^{-1} (14 km h^{-1}). This ascent rate adequately compensates for the average sinking rates of $\sim$0.5 m mantle xenoliths within the magma (1 m s^{-1} or 4 km h^{-1}) (Sparks et al. 2006) and suggests transit times from the base of the mantle lithosphere of $\sim$15 h. One possible implication of the velocity differential between magma rise and xenolith sinking is that the first 5 h of eruption will be devoid of larger deep-seated mantle xenoliths. These calculations are based on kimberlite magmas having a Newtonian viscous rheology. However, as solid contents, gas fractions, and rise rates increase, these magmas may well begin to develop non-Newtonian behaviours (e.g. yield strength or strain hardening) with the attendant consequence of allowing for slower rates of ascent (Caricchi et al. 2007; Lavallée et al. 2007; Sparks et al. 1977).

The buoyant and rapid ascent of kimberlite from depths >200 km can be greatly facilitated by the presence of a (relatively) low-density ($\rho_{CO_2} \sim 1{,}200\,\text{kg m}^{-3}$ at P $\sim$ 2 GPa) supercritical fluid phase (Dobson et al. 1996). Deep-seated exsolution of volatiles would ensure sufficient magma–mantle density contrast to provide the excess buoyancy needed for sampling and entrainment of the deepest (~200 km) mantle lithosphere and for rapid, continuous ascent to the Earth's surface. Indeed, volatile (CO_2 and H_2O) exsolution is commonly invoked as an aid to kimberlite ascent, yet the source of volatiles and the physical–chemical process behind their liberation are rarely specified (e.g. Sparks et al. 2007; Wilson and Head 2007).

A Model for Kimberlite Ascent

Our original idea for kimberlite ascent (Russell et al. 2012) builds on the work of Brooker et al. (2011) and Brooker and Kjarsgaard (2011) who provided new experimental data on the solubility of volatiles in melts across the silicate–carbonate transition (Fig. 2a; after Russell et al. 2012). These experiments elucidate the pressure and compositional controls on CO_2 solubility in silicic (e.g. rhyolite to basalt to melilitite) to carbonate melts (Fig. 2a).

Our model is inspired by three main observations from these and other studies. Firstly, the solubility limits of CO_2 (and H_2O) in silicic melts are mainly dictated by pressure. Importantly, in magmas with $SiO_2 + Al_2O_3 > 35$ wt. %, CO_2 solubility is limited to ~10–15 % (cf. Brey 1976; Brey and Green 1976a, b; Brey and Ryabchikov 1994; Eggler and Mysen 1976; Mysen et al. 1975; Wyllie and Huang 1975). This limit precludes extraordinary amounts of volatile from being sequestered in kimberlitic melts even at mantle pressures (Sparks et al. 2007; Wilson and Head 2007). However, there are a plethora of papers that have shown that CO_2 solubility in melts increases with alkalinity and silica undersaturation (e.g. Brey and Ryabchikov 1994; Eggler and Mysen 1976; Girnis et al. 2011; Mysen et al. 1975). Thus, the total amount of CO_2 dissolved in "kimberlite" magma could be substantially higher if it was poorer in SiO_2 than commonly suggested (i.e. more carbonatitic or very silica-undersaturated) (e.g. <30–38 wt.% SiO_2; Girnis et al. 2011).

Secondly, carbonate melts (i.e. $SiO_2 + Al_2O_3 < 20$ wt. %) feature substantially higher CO_2 solubilities that are essentially independent of pressure (Fig. 2a). For example, CO_2 solubility in stoichiometric carbonate melts can be >40 % (Dalton and Presnall 1998a, b; Dobson et al. 1996; Foley et al. 2009; Hrma 1985), and there are no data indicating that CO_2 solubility in these melts is limited by carbonate stoichiometry (Keppler 2003). Carbonatite melts also have higher H_2O solubility than anhydrous silicate melts (e.g. Keppler 2003; Mysen et al. 1975); at just 0.25 GPa and 900 °C, carbonatite can dissolve ~15 wt. % H_2O without compromising the solubility of CO_2 (Keppler 2003). Thirdly, and lastly, the transition from carbonatitic to silicic (i.e. kimberlitic ~18–24 wt % SiO_2) melt compositions appears to be accompanied by a sharp, potentially nonlinear, decrease in CO_2 solubility (Fig. 2).

In our model, we argue that kimberlites begin as carbonatitic (≤5 wt. % SiO_2) or carbonate-rich, silica-undersaturated melts (≤15 wt. % SiO_2) (e.g. Foley et al. 2009). There are two critical elements inherent to these putative parental melts: (1) they contain substantial dissolved CO_2, and (2) they are extremely silica-undersaturated, having very low activities of SiO_2 (a_{SiO_2}); (Luth 2009; Mitchell 1973; Nicholls et al. 1971). Many previous workers have marshalled a diverse array of evidence suggesting that the primary melts to kimberlitic magmas are carbonatitic in composition (e.g. Canil and Bellis 2008; Francis and Patterson 2009; Kamenetsky et al. 2008; Nielsen and Sand 2008; Patterson et al. 2009). For example, Francis and Patterson (2009) and Patterson et al. (2009) showed that the bulk chemical compositions of kimberlites can be treated as simple mixtures of Ol and Opx and a carbonate-rich liquid end member having ~24 wt. % SiO_2 and ~15 wt. % MgO on a water-free basis. Their postulate is strongly supported by observations on carbonate-bearing melt (trapped early) and fluid inclusions (trapped late) hosted by Ol (Brett 2009; Brett et al. 2009; Kamenetsky et al. 2008, 2009). Mitchell and Tappe (2010) disagree with the Francis and Patterson (2009) model and make a genetic connection between carbonated ultramafic lamprophyres and carbonatites in rifted cratonic environments, rather than kimberlite.

The main attribute of adopting carbonatitic to carbonate-rich silica-undersaturated melts as parental to kimberlites is that such melts originate with dissolved CO_2 contents in excess of 40 wt. % and are initially buoyant relative to the mantle (~2,800 versus ~ 3,300 kg/m^3 at 1–3 GPa) (Dobson et al. 1996; Genge et al. 1995). Carbonatitic magma entering the CML initiates the mechanical incorporation and disaggregation of xenoliths (e.g. Lensky et al. 2006), producing the suite of mantle-derived xenocrysts and ensuring that newly liberated minerals are continuously fed into the melt.

In an elegant paper, Luth (2009) provides a clear rationale for the mineral–melt reactions expected in rising mantle melts (Fig. 3). His arguments are based on the limited effect of pressure on the a_{SiO_2} of the melt (fixed composition) versus the strong effects of pressure on a_{SiO_2}-buffering reactions. Initially in the source, the melt is in equilibrium with a peridotitic mineral assemblage, and the a_{SiO_2} of the melt is defined by the Ol–Opx buffer. However, the pressure dependence of that a_{SiO_2}-buffer ensures that as the magma rises, the melt resides in the Ol saturation field (on the basis of a_{SiO_2}); the melt is never in equilibrium with Opx (Fig. 3). In fact, at these pressures (~6–7 GPa) and temperatures (>1,300 °C), all mantle minerals in the original source will tend to be out of equilibrium and reacting

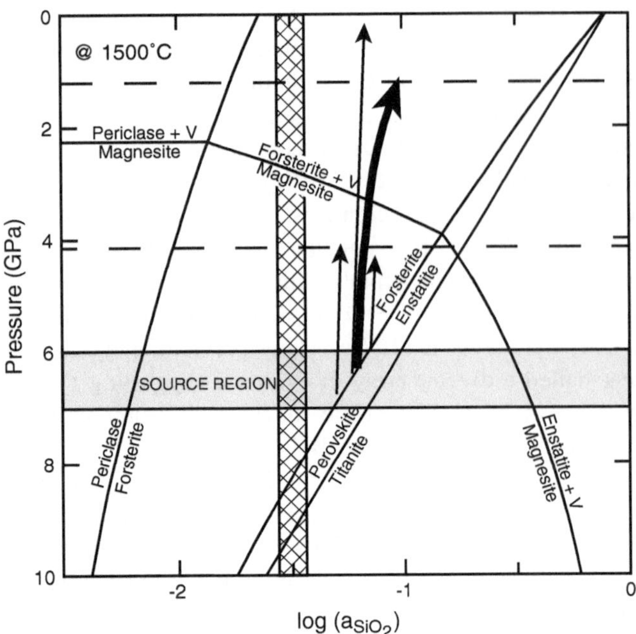

Fig. 3 Isothermal (1,500 °C) phase diagram showing relative stabilities and buffering capacities of mantle assemblages as a function of pressure (P; GPa) and silica activity (log a_{SiO_2}). Diagram is modified from Luth (2009) to explore the relative stabilities of forsterite and enstatite in melts derived from melting of mantle peridotite. Shaded area highlights the predicted source region for the primary carbonatitic-kimberlitic melts (e.g. Dalton and Presnall 1998a, b; Foley et al. 2009). Hashed region demarcates the range of a_{SiO_2} calculated or estimated for kimberlite melts (cf. Mitchell 1973; Barker 2001). *Arrows* indicate ascent paths of primary melts leaving the source region; the *thick arrow* represents a carbonatitic melt becoming more enriched with SiO_2 by assimilation of mantle-derived orthopyroxene. All ascent paths cause the melt to leave the forsterite–enstatite silica buffer curve; the consequence is that, in terms of a_{SiO_2}, orthopyroxene is unstable and silicate saturation would be expressed by crystallization of forsterite (i.e. olivine)

with this ascending carbonate-rich, silica-undersaturated melt; the lower the a_{SiO_2} of the melt (more carbonatitic), the greater the reactivity (Barker 2001; Luth 2009; Mitchell 1973; Nicholls et al. 1971). Orthopyroxene, as the most silica-saturated phase, has the highest affinity for dissolution and is assimilated faster and preferentially over other phases (Edwards and Russell 1996, 1998; Luth 2009; Mitchell 1973, 2008; Shaw 1999). The assimilation of Opx xenocrysts drives the melt to more silicic compositions, turning the system "kimberlitic" (Kopylova et al. 2007; Mitchell 2008; Patterson et al. 2009). Shaw (1999) performed a series of dissolution experiments involving basanite melt and Opx. He found that, even in these more silica-rich (~40 wt %) undersaturated melts, Opx dissolved in these low-viscosity, low a_{SiO_2} melts on a timescale of hours. In melts that have even lower a_{SiO_2}, we can expect dissolution rates to be substantially faster (Edwards and Russell 1996, 1998).

The key to successful ascent of kimberlite melt is the effect of this assimilation on volatile solubility (Fig. 2). The increased SiO_2 content causes a decrease in CO_2 solubility and immediate exsolution of a fluid phase deep within the CML. The process appears to be gradual until Opx assimilation produces a "kimberlitic" melt composition in excess of ~18 wt. % SiO_2 and causes a marked drop in CO_2 solubility (Fig. 2). This results in massive exsolution of a volatile phase that reduces magma density, increases buoyancy, and supports rapid and accelerating ascent. Our model differs from that of Ryabchikov and Girnis (2005) in that they drive CO_2-undersaturated kimberlite-like silicic magmas to CO_2 saturation by dissolving carbonate from garnet lherzolite in the overlying mantle lithosphere.

High-Temperature Analogue Experiments

A set of high-temperature (high-T) melting experiments were conducted as a test of the inductive ideas advanced above (Russell et al. 2012). The experiments (e.g. conditions, melt compositions, and scales) simulate the high-temperature (T ~ 1,000–1,100 °C) assimilation of natural finely crushed Opx ($MgSiO_3$) by carbonate melt (Na_2CO_3). They serve as an analogue to the natural process, in that they are designed to demonstrate the viability of the process and to illustrate how quickly and efficiently the process is likely to operate during magma transport (cf. Hrma 1985; Table 2; Fig. 4). Similar melts have also been used as analogue melts to explore melt infiltration processes in mantle peridotite (Hammouda and Laporte 2000).

In contrast to these analogue experiments, mantle-derived carbonate melts are Ca–Mg-rich, rather than sodic. Na_2CO_3 melt was chosen for these demonstration experiments because it offers a low melting temperature (T_m = 849 °C) and is stable at atmospheric pressure (i.e. does not decompose). The high temperatures (i.e. superliquidus) of the analogue experiments ensured fusion and reaction whilst suppressing crystallization of new carbonate, oxide, or silicate minerals that are not relevant to the natural process (i.e. kimberlite ascent). The preparation of samples for these experiments (Fig. 4a) is fully described in Russell et al. (2012).

Russell et al. (2012) described two types of experiments: transient and static (Table 2). Transient experiments demonstrated the rates of decarbonation of Na_2CO_3 due to dissolution of Opx. Each transient experiment involved a sample of known mass and composition (e.g. Na_2CO_3 + 0, 25, and 40 wt. % Opx) being placed in a platinum crucible and lowered into a preheated furnace from a balance (Russell et al. 2012). In these experiments, the Na_2CO_3 powder melts and reacts with (assimilates) the admixed crushed Opx (<25 min) creating a more silicic melt that is

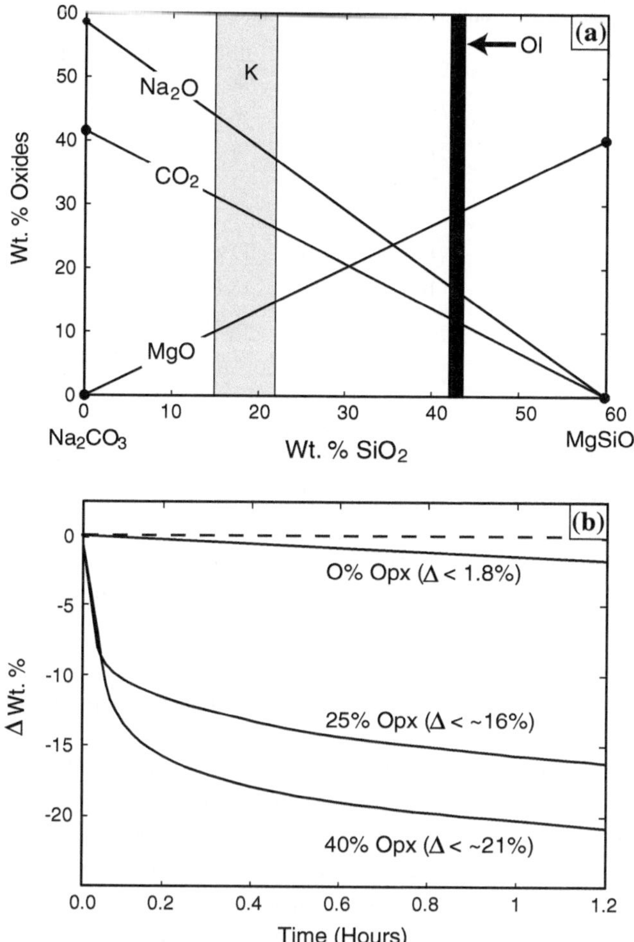

Fig. 4 Compositions of melts used in high-T analogue melting experiments and results of transient experiments (modified after Russell et al. 2012; Fig. 3a, b). **a** Solid lines indicate the range of bulk compositions (wt. % SiO_2) created by mechanical mixing of sodium carbonate (Na_2CO_3) and orthopyroxene ($MgSiO_3$). Shaded field (K) marks range of SiO_2 contents for putative kimberlite; thick vertical black line marks the SiO_2 content of forsteritic olivine. **b** Results of transient melting experiments expressed as mass loss (wt. %) versus time (hours). Mass loss is due to decarbonation during melting of Na_2CO_3 mixed with 0, 25 and 40 wt % Opx. Percentage weight loss (Δ) is reported in *brackets*

now oversaturated in dissolved CO_2. This promotes strong and immediate effervescence as the hybrid melt exsolves CO_2 (i.e. decarbonation), causing measurable weight loss (Fig. 4b). The gas exsolution reaction continues until the solubility limit of the new melt composition is reached and the loss of weight with time ceases. Each experiment is a record of total change in mass of the crucible plus sample as a function of time. The same transient experiments (>1 h at 1,100 °C) performed on pure Na_2CO_3 melt (i.e. 0 % Opx) resulted in only slight (<1.8 %) loss of mass (Fig. 4b), demonstrating the long-term stability of stoichiometric amounts of dissolved CO_2 (~40 wt. %) in carbonate melts. Even though performed in air, there is little to no dissociation of the melt or degassing. This stability of the Na_2CO_3 melt is also shown in static experiments (Table 2).

The transient experiments provide two insights. Firstly, they measure the total CO_2 lost and define the residual CO_2 in the melt; that value represents the new equilibrium CO_2 solubility of the hybrid melt. Secondly, the tangent to the mass loss versus time curve is a record of the relative rates of assimilation and degassing. These experiments illustrate the spontaneous nature and rapid rate of the assimilation and decarbonation reactions; virtually, all of the loss in mass occurs in the first 20 min after which there is little change, indicating that the assimilation-induced decarbonation process is over. In essence, the experiments rapidly converged to an equilibrium state wherein all Opx was dissolved and the melt reached its maximum SiO_2 content. At this point the melt retains the amount of CO_2 dictated by its new solubility limit; the addition of more Opx would foster more reaction, more weight loss, and a new more silicic melt further depleted in CO_2.

Russell et al. (2012) also performed a suite of static assimilation experiments wherein known masses and proportions of Na_2CO_3 + Opx were placed in a preheated oven (1,000–1,100 °C) for fixed times (>35–75 min; Table 2; Fig. 5). The observed weight losses were ascribed to decarbonation (i.e. Na_2O volatilization is negligible; Hrma 1985) driven by assimilation, and the residual CO_2 established the solubility limits of the hybrid melts. CO_2 content decreases dramatically with increasing $MgSiO_3$, and decarbonation is complete when the melt reaches ~37 wt. % SiO_2 (~42 wt. % Opx; Fig. 5b). The computed hybrid melt compositions assume that all available Opx dissolved into the melt, creating a more siliceous daughter melt (Table 2). Together, these experiments (Fig. 5; Table 2) map the progress of assimilation-induced decarbonation for Na_2CO_3-Opx melts (Fig. 5b) and quantify both the extent of decarbonation (i.e. Δwt. %) and the reduction in CO_2 solubility due to SiO_2 enrichment of the melt (Fig. 5c).

The decarbonation diagram is only for simple analogue Na_2CO_3-Opx melts, rather than multicomponent natural carbonatitic melts. Nevertheless, we have projected two compositional fields, relevant to our kimberlite ascent model, onto the decarbonation diagram. Firstly, we have plotted a field for putative kimberlite (Table 1) based on SiO_2 contents of 15–22 wt. % (light grey box; Fig. 5b); these melts are produced by assimilation of ~20–29 wt. % Opx. To reach this SiO_2 content, the assimilation process produces 15–20 wt. % exsolved CO_2 to fuel the ascent of kimberlite; the hybrid melt retains 10–18 wt. % CO_2. In the ascending magma, additional assimilation will promote continued exsolution, produce a more silica-rich (i.e. kimberlitic) melt, and will potentially drive the original (carbonatitic) melt to silicate saturation (Fig. 2). Olivine saturation allows for and explains late-stage magmatic

Table 2 Summary of static experiments involving Na_2CO_3 melts and assimilation of orthopyroxene (Russell et al. 2012)[a]

Label	JKR-1b	YL-250	JKR-2a	JKR-3a	JKR-4b	JKR-9a	JKR-5a	JKR-5b	YL-252	JKR-8a	JKR-6a	JKR-7a	YL-253
Wt. % Opx	0.00	0.00	4.66	9.03	16.45	20.46	23.08	23.08	25.00	25.59	33.38	37.42	40.01
T_{max} (°C)	1000	1100	1000	1025	1025	1075	1050	1075	1100	1100	1075	1100	1100
Time (min)	60	85	37	40	45	40	40	60	75	60	50	50	75
Original bulk compositions (Wt. % oxide)													
SiO_2	0.00	0.00	2.79	5.41	9.85	12.25	13.81	13.81	14.96	15.31	19.98	22.39	23.94
MgO	0.00	0.00	1.87	3.63	6.60	8.21	9.27	9.27	10.04	10.27	13.40	15.02	16.06
Na_2O	58.48	58.48	55.75	53.19	48.86	46.51	44.98	44.98	43.86	43.52	38.96	36.60	35.08
CO_2	41.52	41.52	39.59	37.77	34.69	33.03	31.94	31.94	31.14	30.90	27.66	25.99	24.91
Experimental results and evolved melt compositions													
Wt. Loss (%)	0.90	1.84	5.20	20.43	20.56	18.29	16.17	16.21	16.45	23.21	22.30	23.07	20.99
SiO_2	0.00	0.00	2.94	6.80	12.39	14.99	16.48	16.49	17.91	19.94	25.71	29.11	30.30
MgO	0.00	0.00	1.97	4.56	8.31	10.05	11.05	11.06	12.01	13.38	17.25	19.53	20.33
Na_2O	59.01	59.57	58.81	66.85	61.50	56.93	53.66	53.68	52.49	56.67	50.14	47.57	44.40
CO_2	40.99	40.43	36.28	21.80	17.79	18.03	18.81	18.77	17.59	10.02	6.90	3.79	4.97
ΔCO_2 (%)	−0.53	−1.09	−3.31	−15.97	−16.90	−14.99	−13.16	−13.13	−13.55	−20.88	−20.76	−22.20	−19.95

[a] The residual CO_2 contents in the melt are related to the amount of Opx assimilated by a quadratic expression (see Eq. 1)

crystallization of Ol as overgrowths on macrocrysts and as (micro-) phenocrysts (Figs. 1d and 2; Brett et al. 2009; Kamenetsky et al. 2008; Luth 2009; Mitchell 2008; Shaw 1999). Thus, we have added a compositional field where a natural carbonatitic melt might reach silicate saturation (dark grey box at $SiO_2 \sim$ 22–27 wt. %; Fig. 5b) and crystallize Ol depending on pressure and magma temperature (Brey and Ryabchikov 1994).

Consequences and Implications

Exsolution-Driven Ascent of Kimberlite

We suggest that kimberlite magmas originate as primary mantle-derived carbonatitic to carbonate-rich silica-undersaturated melts. Carbonatite melts are accepted as a common product of partial melting of carbonated ($\pm H_2O$) mantle sources at pressures >2.5 GPa (Eggler 1978; Sweeney 1994; Wallace and Green 1988; Wyllie 1977, 1980; Wyllie and Huang 1975; Yaxley et al. 1991). Indeed, experimental work shows that the solidus melts of natural and synthetic carbonated (~0.15 %) peridotite at >2.5 GPa will always be CO_2 rich and SiO_2 poor as long as carbonate is stable in the mantle assemblage (Dalton and Wood 1993; Falloon and Green 1989; Sweeney 1994). The rarity of mantle-derived extrusive carbonatite at the Earth's surface probably reflects the potential for forced crystallization of carbonatite within the lithospheric mantle at <2.5 GPa where the solidus to the melt rises substantially (e.g. Barker 1996; Dalton and Presnall 1998a). Carbonatitic melts can contain in excess of 40 % dissolved CO_2, and we suggest that the asthenospheric production of such melts marks the onset of kimberlite ascent.

Initially, the primary carbonate-rich, silica-undersaturated melts infiltrate the overlying mantle along grain boundaries (e.g. Hammouda and Laporte 2000) until sufficient melt volume accumulates. At this point, the melt transport mechanisms change from more ductile processes involving intergranular flow to more rapid, brittle processes leading to upward transport by dyke propagation. The properties of entrained mantle xenoliths and mantle xenocrysts indicate that this transition must occur at the base of the mantle lithosphere or deeper. At this stage, the rate of ascent is too rapid to sustain equilibrium between melt and host rock. Whereas Ol-Opx equilibrium governs silica activity in the initial stage, the silica activity of the liquid remains unbuffered during this disequilibrium stage (Luth 2009) and Opx xenocrysts invariably dissolve.

The high dissolved CO_2 content is stable in the melt until assimilation of Opx pushes the melt to >15 wt. % SiO_2. At this melt composition, CO_2 solubility decreases causing exsolution of a deep-seated CO_2-rich fluid phase. The assimilation of Opx is a key to this process because it dissolves on timescales that compete with kimberlite ascent rates (Edwards and Russell 1998; Shaw 1999). The rapid and efficient dissolution of Opx, relative to the other mantle macrocrysts, explains the lack of xenocrystic Opx in kimberlite (Brett et al. 2009; Kopylova et al. 2007; 2003; Mitchell 2008; Patterson et al. 2009).

Our model for kimberlite ascent is unique in that it begins with a melt having near-stoichiometric amounts of dissolved CO_2. The main consequence of Opx assimilation is to decrease the CO_2 solubility of the melt, thereby driving

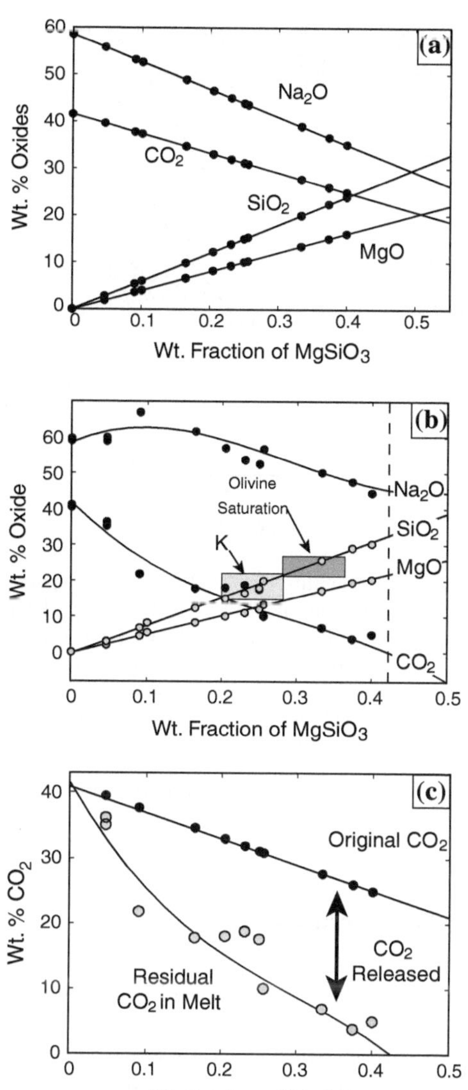

Fig. 5 Results of static melting experiments on mixtures of Na_2CO_3 and orthopyroxene ($MgSiO_3$). Revised from Russell et al. (2012; Fig. 3). **a** Specific compositions (*filled circles*) of mechanical mixtures used in static melting experiments versus the weight fraction of Opx added. **b** Melt compositions produced by melting of Na_2CO_3-Opx mixtures. Shaded boxes denote approximate fields for kimberlite melts (K) and the melt compositions for potential olivine saturation and crystallization. **c** Distribution of CO_2 in hybrid system (Na_2CO_3–$MgSiO_3$) showing CO_2 content of the partially decarbonated melt as a function of dissolved Opx. Distance between two lines is total amount of exsolved CO_2

deep-seated exsolution and effervescence of a CO_2-rich volatile phase and providing buoyancy. In the absence of Opx assimilation, the dissolved CO_2 is stable in the melt across a wide range of P–T conditions (Figs. 2, 5). Thus, the primary, deep-seated exsolution of the fluid phase is not driven by depressurization as it is in most magma ascent models (e.g. Brey and Ryabchikov 1994; Spera 1984). Rather, the CO_2 exsolution is driven only by an inevitable chemical reaction (i.e. assimilation) that creates a new more silica-rich melt. This production of the fluid phase does not depend on depressurization as the magma transits the lithosphere; rather it depends on access to, and assimilation of, Opx.

Many of these concepts are implicit in diagrams published by Brey and Ryabchikov (1994). We have reproduced their model CO_2 solubility curves for silicic melts containing 16–37 wt. % SiO_2 (Fig. 6a); these curves define a CO_2 solubility ledge above 5–3 GPa where CO_2 solubility drops. The CO_2 solubility ledge is dependent on SiO_2 content and is situated at the top of (for SiO_2 contents greater than putative kimberlite melts) or above the diamond stability field (for SiO_2 contents less than putative kimberlite melts; Fig. 5a). Thus, the CO_2 solubility ledge alone does not provide a convenient means of sampling the deepest and diamondiferous mantle lithosphere. The magnitude of the solubility drop, as well as the pressure interval it operates over, depends on the melt composition. The drop in CO_2 solubility is large (~ 20 %) for the most silicic melts and smaller (<10 %) for silica-poor melts. The solubility drop occurs at higher pressures (i.e. deeper mantle depths) in silicic melts than in more silica-undersaturated melts. Carbonatitic melts contain stoichiometric (or higher) amounts of CO_2 and are more or less independent of pressure (Fig. 2).

The relative effects of composition and pressure on CO_2 solubility in silicic melts are compared in Fig. 6b (modified from Brey and Ryabchikov 1994). The drop in CO_2 solubility as a function of composition (from ~ 5 to ~ 37 wt. % SiO_2) is 37 and 42 wt. % at 3 and 1 GPa, respectively. In comparison, a melt of fixed composition (i.e. ~ 20 wt. % SiO_2) shows a pressure-driven drop in CO_2 solubility of ~ 10 wt. % from 3 to 1 GPa. Thus, shifting melt composition from carbonatitic to more silicic offers greater potential for CO_2 exsolution than does depressurization.

In our model, assimilation of Opx increases the silica content of melt, driving it towards a "kimberlitic" composition, causing a decrease in CO_2 solubility, and promoting continued volatile exsolution. Although depressurization does not drive the initial fluid production in our carbonatitic magma model, the buoyancy of the fluid-saturated magma is greatly enhanced by it. Depressurization will cause volumetric expansion of any exsolved fluids. This secondary process would allow for continuous acceleration of the magma, entrainment of greater quantities of mantle material, and decoupling of a CO_2-rich fluid phase from the ascending magma. In addition, the CO_2 solubility of the more silicic melts created by Opx assimilation has greater pressure dependence (Fig. 6). This means that pressure-driven exsolution can begin to dominate, which when combined with decompression-driven volatile expansion drives the upward acceleration of the magma. Decoupling of the CO_2-rich fluid phase from the ascending

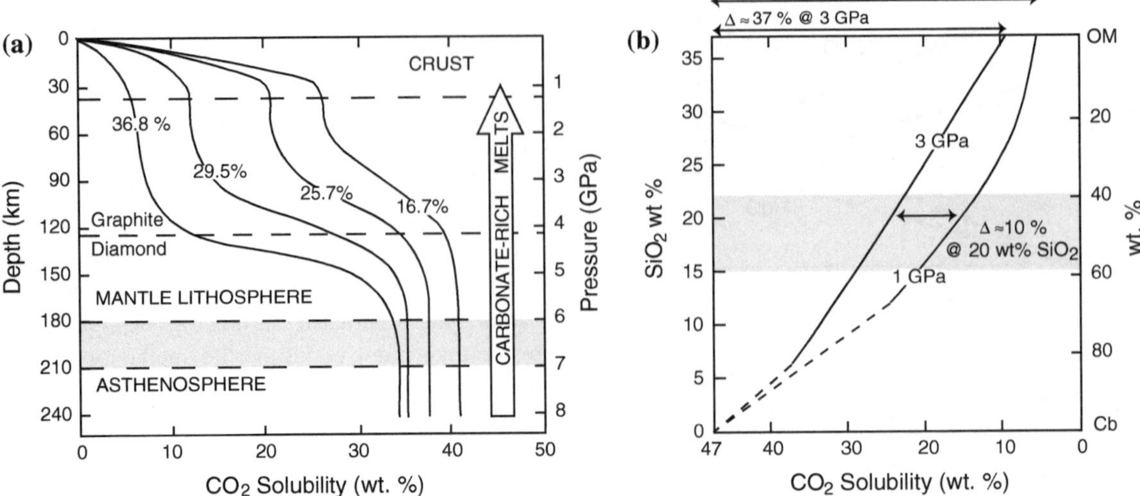

Fig. 6 Compositional and pressure dependencies of CO_2 solubility in silica-undersaturated melts reconstructed from Brey and Ryabchikov (1994). **a** Model pressure-dependent CO_2 solubilities for melts with increasing SiO_2 contents (wt. %) as computed by Brey and Ryabchikov (1994) defining a pressure-dependent CO_2 "solubility-shelf". CO_2 solubility in carbonatitic melts is presumably limited only by melt stoichiometry and shows little dependence on pressure. **b** Measured CO_2 solubility for melts spanning pure $Ca_{58}Mg_{42}$ carbonate (*Cb*) and olivine melilitite (*OM*) at 3 and 1 GPa. Extrapolation of the data from Brey and Ryabchikov (1994) shows profound decrease in CO_2 solubility (37–42 wt. %) attending an increase (0–37 wt. %) in SiO_2 content of the melt versus a more moderate drop in CO_2 solubility due to pressure at constant SiO_2 content ($\Delta P \sim$ 3–1 GPa yields a $\Delta CO_2 \sim$ 10 wt. % at 20 wt. % SiO_2)

magma may explain precursory fenitization (Stachel et al. 1995) events recorded at some kimberlite pipes (Ferguson et al. 1973; Masun et al. 2004; Smith et al. 2004).

Ultimately, the Opx assimilation process can drive these carbonatitic melts to silicate mineral saturation (Figs. 2 and 4b). The identity of the liquidus silicate phase(s) (i.e. Ol + Cpx versus Ol) is controlled by the pressure and SiO_2 content of the melt at the point at which silicate saturation first occurs (Fig. 7). Brey and Ryabchikov (1994) mapped the liquidus surface of carbonate to silicic (37 wt. % SiO_2) melts at 3, 1, and 0.5 GPa (Fig. 7). The silicate saturation surface is plotted as a vertical dashed line in each panel. At higher pressure (3 GPa) silicate saturation occurs at >12 wt. % SiO_2 and ~1,300 °C and would be expressed by crystallization of Ol and Cpx. Kimberlite magmas crystallize Ol but never crystallize augite. The implication is that at 3 GPa pressure the ascending carbonatite either has a lower SiO_2 content than 12 wt. % or, more likely, is superliquidus to these phase relationships (Fig. 7c).

At pressures below 3 GPa, and shallower mantle depths, the only silicate phase to saturate will be Ol (Fig. 7a, b). Once the melt exceeds 12 wt. % SiO_2, the continued rise of the magma above ~90 km depth allows for late-stage crystallization of Ol. This explains the pervasive Ol overgrowths observed on rounded Ol xenocrysts and the pervasive micro-phenocrysts of Ol in the groundmass (Brett et al. 2009; Fedortchouk and Canil 2004; Kamenetsky et al. 2008; Luth 2009; Mitchell 1986, 2008). This crystallization event would logically coincide with the later stages of kimberlite ascent and emplacement and clearly follows on from an earlier and more extensive period of Ol disequilibrium (Figs. 1, 2).

Cratonic Connections

Our model provides an explanation for the linkage between kimberlites and cratons. It is clear that if the parental magmas are carbonatitic, a ready source of highly reactive Opx is needed to initiate the assimilation-induced vesiculation in the deep mantle lithosphere and provide buoyancy to the evolving magma (melt + mantle cargo). The CML has two attributes that support this transformation (carbonatite to kimberlite): (1) firstly, it is enriched with Opx (15–30 %) relative to other mantle lithosphere (Boyd 1989; Kopylova and Russell 2000; McDonough 1990), and (2) secondly, it is thick (>160–180 km) and, thus, provides ample opportunity for continuous entrainment of Opx-enriched mantle material. As long as the evolving carbonatitic melt is exposed to new Opx, the assimilation-induced exsolution of CO_2 will continue maintaining or increasing the magma buoyancy. Therefore, we might expect the diversity of kimberlite at the Earth's surface to actually be reflective of the magma pathway (i.e. the mineralogical composition), thermal state, and the thickness of the underlying mantle lithosphere rather than the source region to the melt (Francis and Patterson 2009; Tappe et al. 2011).

The corollary is that if Opx were absent, this process would not be viable because other mantle silicate minerals (e.g. Cpx, Ol) may remain stable throughout much of the

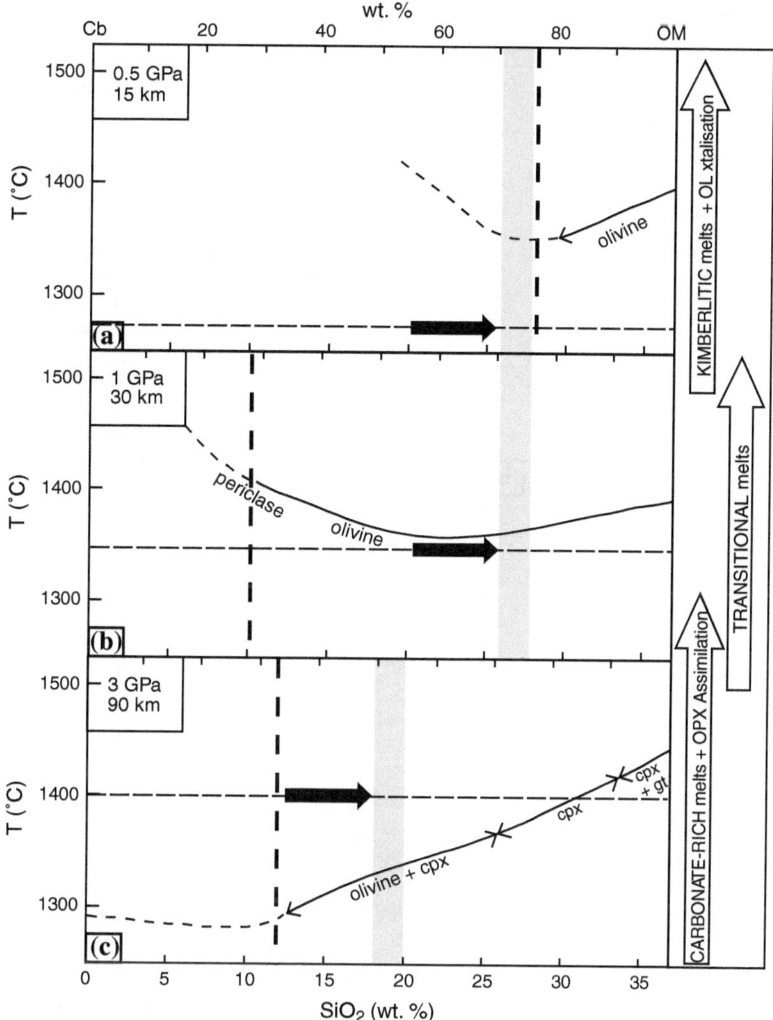

Fig. 7 Phase equilibria constraints on liquidus relationships for melts spanning $Ca_{58}Mg_{42}$ carbonate (*Cb*) to olivine melilitite (*OM*) from Brey and Ryabchikov (1994) at pressures of: **a** 0.5 GPa, **b** 1 GPa, and **c** 3 GPa. The horizontal dashed line indicates the approximate temperature for ascending kimberlite magma at each depth (calculated from Sparks et al. 2009). At 3 GPa the liquidus surface rises from 1,290 to 1,450 °C as melt composition changes from carbonatitic to silicic (37 % SiO_2). The manifestation of silicate mineral saturation (*dashed vertical line*) changes with increasing SiO_2 content from olivine and clinopyroxene (cpx; ~12 wt. %) to cpx (~25 wt. %) to cpx and garnet (gt; ~34 wt. %). At 1 GPa, the same range of melt compositions has higher silicate saturation temperatures (1,400–1,350 °C) beginning at ~12 wt. % SiO_2 and saturates with olivine. At 0.5 GPa, the silicate saturation surface is olivine and is reduced to melts with >25 wt. % SiO_2. Our model suggests that an initial carbonatitic melt rises from the asthenosphere containing <5 wt. % SiO_2 but continuously assimilates orthopyroxene during ascent through the mantle lithosphere. Assimilation (*black arrows*) increases the SiO_2 content of the melt (indicated by *shaded grey* fields) and causes silicate mineral saturation. The identity of the liquidus silicate phase(s) (Ol + Cpx versus Ol) is controlled by the pressure at which saturation occurs. The P (Depth)-T-X_{SiO_2} paths of these magmas suggest that the silicate saturation surface is met at depths less than 90 km and promotes olivine crystallization

ascent (Luth 2009) or dissolve too slowly to keep pace with ascent. Although dissolution of any silicate mineral into the carbonatitic melt will cause a reduction in CO_2 solubility and, hence, fluid exsolution, the affinity for Opx dissolution is substantially higher than for Ol and Cpx (Edwards and Russell 1998; Luth 2009; Shaw 1999). The abundance of Opx in the CML may therefore influence the presence or absence of kimberlites at the surface, as well as the transit time and hence diamond preservation on route. Pearson and Wittig (2008) show differences in the abundance of Opx within each craton, the Kaapvaal being the most Opx-rich (or Ol-poor with ~64 wt. % Ol), followed by the Slave, with Tanzania and then the North Atlantic craton being the most Opx-poor. Interestingly, when comparing the numbers of diamond mines on these cratons, the same order is observed (Janse 2007), with the Kaapvaal hosting 14 mines, the Slave 2, Tanzania with 1 and none in the North Atlantic craton. Other factors also influence the presence of diamond

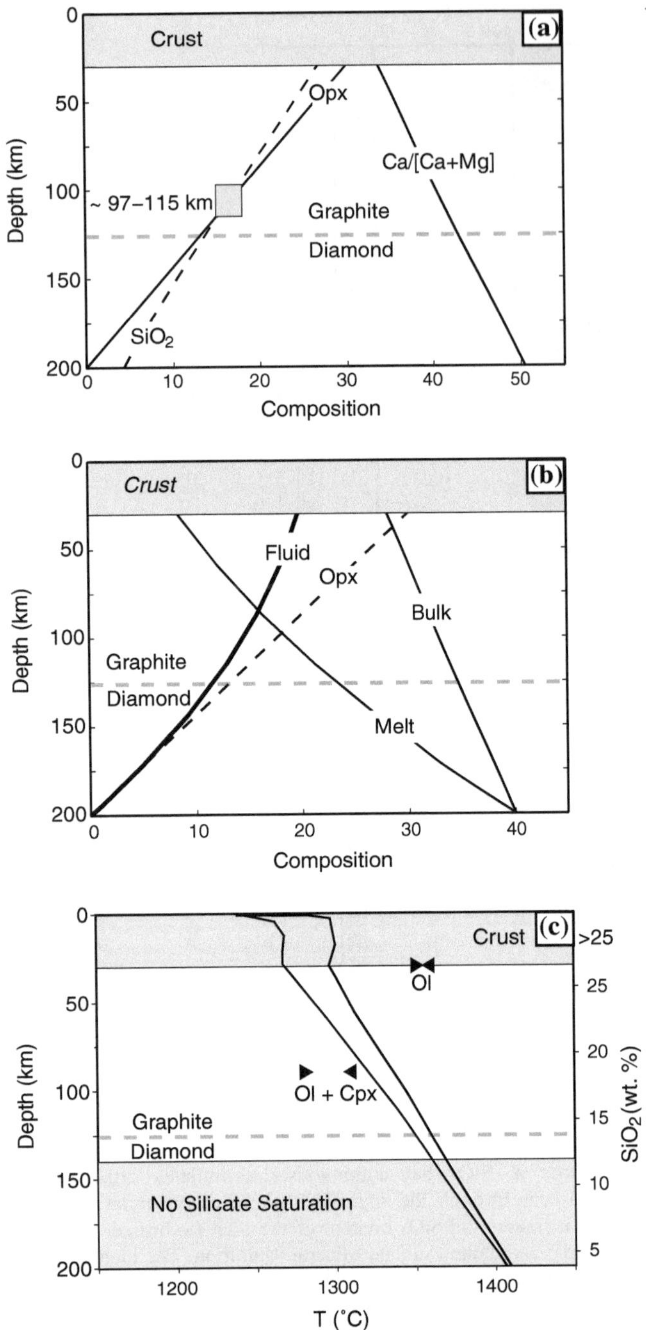

Fig. 8 Chemical evolution of carbonatitic melt as a function of depth (km) during its ascent (Table 3). **a** The model assumes orthopyroxene assimilation (Opx; wt. %) increases linearly with transport distance. Melt composition rises to a maximum of 30 wt. %. SiO_2 content (*dashed line*), whilst the 100Ca/[Ca + Mg] value of the melt decreases. First production of a silicic kimberlite melt (*grey box*; SiO_2 ~ 15–18 wt. %) occurs at depths of 115–97 km, above the diamond window. **b** The CO_2 budget for the evolving magma, including bulk CO_2 content of the system, the residual CO_2 content of the melt phase, and the total CO_2 released to the fluid phase. Dashed line denotes Opx assimilated. **c** Expected mineral saturation in evolving melt composition based on model temperature ascent paths of Kavanagh and Sparks (2009) and phase equilibria of Brey and Ryabchikov (1994; triangles). At depth the melt is not silicic enough to sustain silicate saturation. Above the diamond–graphite transition, the melt has 15–20 wt. % SiO_2 and could saturate with silicates; however, this would require kimberlite magma temperatures below the model ascent temperatures of Sparks et al. (2009; e.g. *solid lines*). Kimberlite melts never crystallize clinopyroxene (Cpx) and nearly always crystallize olivine (Ol), implying that silicate saturation and olivine crystallization begin at pressures <3 GPa and depths <90 km. Olivine overgrowths and phenocrysts, therefore, form above the diamond window

mines, that is, economics of mining and ease of exploration; however, this is a noteworthy observation.

In addition, the enthalpy of Opx dissolution is less than the heats of crystallization for Ol, which means that Opx assimilation is energetically supported by smaller quantities of Ol crystallization and little cooling (Brett et al. 2009). In essence, if the assimilation-induced vesiculation process were to rely on dissolution of OL (e.g. not OPX), longer transit times would be required, which would also result in substantial cooling of the magma. In parts of the planet where carbonatitic magmas do not transit CML (e.g. off-craton or rift environments), there are three possible outcomes. The carbonatitic melt successfully travels through a relatively thin hot mantle lithosphere to be emplaced or erupted as carbonatite (e.g. Bailey 1989; Stoppa and Woolley 1997; Woolley and Church 2005), or the carbonatite melt incorporates Opx-poor mantle and crystallizes at depth because it fails to assimilate mantle silicate phases efficiently enough to promote a deep-seated fluid phase and the requisite buoyancy. A third possibility is that the melt transits a relatively thin, Opx-rich mantle lithosphere. In this scenario, Opx assimilation would be efficient and substantial (i.e. producing higher SiO_2 content melts), but the assimilation-driven exsolution of volatiles would initiate at shallower depths (above the diamond window) leading to emplacement and eruption of non-diamondiferous basic melts (e.g. ultramafic lamprophyres, etc.). Such magmas are typical of cratonic margins during early stages of rifting (Foley 2008). Carbonatites that occur on cratons likely result either from slow initial ascent (rendering them unable to break up and incorporate mantle lithosphere) or from passing through regions of the mantle lithosphere that contain insufficient Opx for the transition to kimberlite to occur.

Reproducing Igwisi Hills Kimberlite

We conclude by using the results of the analogue experiments to predict the chemical evolution of an idealized carbonatitic melt ascending through the mantle lithosphere and preferentially assimilating Opx (Fig. 8; Table 3). Using the experimental data set (Fig. 4; Table 2), we have modelled the residual CO_2 content (wt. %) of the evolving melt (W^f) as a function of the amount (wt. %) of Opx assimilated

Table 3 Compositions of melts predicted to result from carbonatite assimilation of orthopyroxene

	Model Carbonatite		Opx	Model compositions							
No.	Sweeney (1994)			Opx assimilated (wt. %)							
Label	Original	Normalized	Ideal	0[a]	5	10	15	20	25	30	40
SiO_2	3.17	4.3	59.85	4.3	7.65	11.29	15.12	19.02	22.92	26.74	33.87
TiO_2	0.50	0.5	–	0.5	0.51	0.51	0.51	0.49	0.47	0.45	0.38
Al_2O_3	2.10	2.3	–	2.25	2.31	2.32	2.28	2.22	2.12	2.00	1.72
FeO(T)	4.96	5.5	–	5.5	5.64	5.66	5.58	5.42	5.19	4.90	4.20
MnO	0.00	0.0	–	0	0.00	0.00	0.00	0.00	0.00	0.00	0.00
MgO	15.70	17.0	40.15	17	19.60	22.08	24.43	26.61	28.61	30.42	33.38
CaO	22.50	24.0	–	24	24.62	24.71	24.37	23.65	22.64	21.38	18.35
Na_2O	5.40	5.5	–	5.5	5.64	5.66	5.58	5.42	5.19	4.90	4.20
K_2O	0.38	0.4	–	0.4	0.41	0.41	0.41	0.39	0.38	0.36	0.31
P_2O_5	0.52	0.6	–	0.55	0.56	0.57	0.56	0.54	0.52	0.49	0.42
CO_2	–	40.0	–	40	33.05	26.77	21.16	16.22	11.96	8.36	3.17
100 Ca/(Ca + Mg)				50.4	47.5	44.6	41.8	39.0	36.3	33.6	28.4
Mg#				0.76	0.78	0.80	0.81	0.83	0.85	0.86	0.89
Depth (km)[b]				200	172	143	115	87	58	30	–

[a] Starting carbonatitic melt and evolving melt compositions as a function of Opx assimilation and decarbonation. [b] Model depths are ascribed to sequential melt compositions assuming: i) a maximum potential assimilation of 30 % Opx and ii) all assimilation occurs within the mantle lithosphere (200–30 km depth)

(W_{Opx}) and the CO_2 content (wt. %) of the idealized parental carbonatitic melt (W^i):

$$W^f_{CO_2} = W^i_{CO_2} - 1.457\,W_{Opx} + 0.0134\,W^2_{Opx} \quad (1)$$

To model the chemical consequences of the Opx assimilation attending ascent, we assume an idealized parental carbonatitic melt composition from Sweeney (1994) (Table 3). We adopt a linear model for the amount of Opx that is assimilated during transit through the mantle lithosphere (200–30 km; Fig. 8a) to a maximum of 30 wt. % at the point where the magma reaches the MOHO. The range of melt compositions generated in this fashion is reported in Table 3.

We have elected to compare our model magma compositions produced by Opx assimilation (Table 3; Eq. 1) to the melt compositions: (1) recovered from phase equilibria melting experiments (Dalton and Presnall 1998a, Table 1; Foley et al. 2009, Table 3; Sweeney 1994) and (2) reported for the Igwisi Hills kimberlite (Table 1; Dawson 1994; Wilcox et al. 2012). Our rationale for choosing the compositions of the Igwisi Hills kimberlite as our target, rather than any of the many other published kimberlite rock compositions, is as follows. Firstly, the Igwisi Hills kimberlite volcano is the youngest kimberlite body on the planet (Brown et al. 2012; Dawson 1994). Secondly, the Igwisi Hills kimberlite deposits are volcanic in origin and include both extra-crater pyroclastic fallout deposits and lavas. Thus, the resulting rocks have formed via moderate to rapid quenching at the Earth's surface during eruption, rather than by slower cooling in the fluid-saturated, subsurface conduit/feeder system. Thirdly, the Tanzanian environment has been dry to arid over the late-Pleistocene to Holocene era. The integrated effect of these facts is that although the Igwisi Hills show slight signs of posteruption modification (Wilcox et al. 2012), these kimberlites offer us the truest estimates of the original composition of kimberlite rocks.

Our compiled compositions (Table 1) include two measurements on matrix-rich portions of kimberlite lava and two model compositions produced by correcting the lava compositions for the effects of xenocrystic and phenocrystic Ol (Dawson 1994). Also reported are new data on lava compositions from Wilcox et al. (2012). The best agreement between our model melt compositions resulting from progressive assimilation of Opx (Table 3) and the Igwisi Hills kimberlite melt estimates occurs at between 15 and 20 wt. % assimilation (cf. Tables 1 and 3). These model melts are very close in SiO_2, MgO, CaO, and CO_2 contents.

Figure 8 summarizes the range of compositions resulting from carbonatitic assimilation of Opx. The figure illustrates the rise in SiO_2 content and decrease in 100Ca/(Ca + Mg) ratio as a function of ascent (Fig. 8a). The earliest silicic kimberlite melt (SiO_2 ~ 15–18 wt. %) would be formed on this ascent path at depths of 115–97 km, which is situated above the diamond window. Mg #'s for the optimal melts are between 81 and 83 (Table 3). The implication is that the diamond-bearing mantle xenoliths are sampled by carbonatitic melts, rather than idealized kimberlite melts.

The CO_2 budget during ascent (Fig. 8b) illustrates the partitioning of CO_2 between the melt and an exsolved fluid phase driven by assimilation only (i.e. ignoring ΔP effects). By the time the magma reaches depths of 115–97 km, it would have released >20 wt. % of the original CO_2. The hybrid residual melt at this point will have retained ~15 wt. % CO_2, which reflects the combined effects of exsolution and dilution (Fig. 8b). The residual CO_2 in the melt could be exsolved by further depressurization as ascent of the now "silicic-kimberlitic" melt continued to assimilate Opx. A minimum of 20 wt. % CO_2 has been released by the time the magma reaches the MOHO.

We have explored the potential for silicate mineral saturation in the model melt compositions as a function of transport distance (Fig. 8c). The critical condition for silicate mineral saturation in the carbonatite-silicic melt continuum is SiO_2 content >12 wt. % (Figs. 2, 5, 7). That condition is met in the ascending magma at depths <125 km (Fig. 8a), and at those depths (~ 3 GPa) would crystallize both Ol and Cpx (Fig. 7). We have compared these phase equilibria constraints of Brey and Ryabchikov (1994) to the model temperature ascent paths of Kavanagh and Sparks (2009). Below the diamond–graphite transition, the melt is not silicic enough to sustain silicate saturation. Above the diamond–graphite transition, the melt has evolved to >12 wt. % SiO_2 and could saturate with silicates, except the ascending magma is likely superliquidus to Ol and Cpx (Fig. 7c). Kimberlite melts never crystallize Cpx and nearly always crystallize Ol, implying that the ascending magma reaches silicate saturation (i.e. Ol crystallization) in terms of both composition and temperature at pressures <3 GPa and depths <90 km. One other implication of this is that all Ol crystallization of overgrowths (e.g. Boyd and Clement 1977; Brett et al. 2009; Kamenetsky et al. 2008) and phenocrysts (e.g. Fedortchouk and Canil 2004; Mitchell 1986, 2008) takes place outside (i.e. above) of the diamond window.

Conclusions

Our research supports the premise that all kimberlites begin as carbonatitic melts produced through partial melting of carbonated peridotite in the subcratonic mantle. These primary carbonate-rich melts may contain nearly stoichiometric CO_2 contents (~40 wt. %) and have very low a_{SiO_2}. As the melts coalesce and ascend, they begin to mechanically sample the mantle lithosphere incorporating peridotite xenoliths. These xenoliths disaggregate, and the low a_{SiO2} of the carbonatitic melt ensures that Opx dissolves rapidly and preferentially relative to the other silicate mantle minerals. This increases the silica content of the melt and triggers immediate exsolution of CO_2, reducing the density of the magma and facilitating rapid ascent. The faster the magma rises, the more mantle material will be entrained and the more Opx will be dissolved. Providing Opx is available, this mechanism enables the continuous and accelerating ascent of the magma and the evolution of the melt from carbonatitic to kimberlitic compositions.

Acknowledgments We thank both Diavik Diamond Mines and Peregrine Diamonds for their continued support of kimberlite research in the Volcanology and Petrology Laboratory at UBC. JKR acknowledges financial support via the NSERC Discovery grant and NSERC Collaborative Research and Development grant programs. LAP acknowledges financial support from the European Union in the form of a Marie Curie Postdoctoral Fellowship. LH acknowledges financial support from a doctoral postgraduate scholarship awarded by NSERC. The clarity and conciseness of the manuscript benefitted from D.G. Pearson's editorial advice and reviews from S.F. Foley and one anonymous reviewer.

References

Anderson OL (1979) The role of fracture dynamics in kimberlite pipe formation. In: Meyer HOA, Boyd FR (eds) Proceedings of the 2nd international kimberlite conference. American Geophysical Union, Washington, pp 344–353

Bailey DK (1989) Carbonate melt from the mantle in the volcanoes of south-east Zambia. Nature 338:415–418

Barker D (2001) Calculated silica activities in carbonatite liquids. Contrib Miner Petrol 141:704–709

Barker DS (1996) Consequences of recycled carbon in carbonatites. Can Miner 34:373–387

Boyd FR (1989) Compositional distinction between oceanic and cratonic lithosphere. Earth Planet Sci Lett 96:15–26

Boyd FR, Clement CR (1977) Compositional zoning of olivine in kimberlite from the De Beers Mine, Kimberly, South Africa. Carnegie Institute of Washington Yearbook, Washington, pp 485-493

Boyd FR, Nixon PH (1973) Origin of the ilmenite-silicate nodules in kimberlites from Lesotho and South Africa. In: Nixon PH (ed) Lesotho Kimberlites. Lesotho National Development Corp., Maseru, pp 254–268

Brett RC (2009) Kimberlitic olivine. University of British Columbia

Brett RC, Russell JK, Moss S (2009) Origin of olivine in kimberlite: phenocryst or imposter? Lithos 112:201–212

Brey G (1976) CO_2 solubility and solubility mechanisms in silicate melts at high pressures. Contrib Miner Petrol 57:215–221

Brey GP, Green DH (1976a) Role of CO_2 in genesis of olivine melilitite—discussion—reply. Contrib Miner Petrol 55:237–239

Brey GP, Green DH (1976b) Solubility of CO_2 in olivine melilitite at high-pressures and role of CO_2 in Earth's upper mantle. Contrib Miner Petrol 55:217–230

Brey GP, Ryabchikov ID (1994) Carbon dioxide in strongly silica undersaturated melts and origin of kimberlite magmas. Neues Jahrbuch Fur Mineralogie-Monatshefte, pp 449–463

Brooker R, Sparks R, Kavanagh J, Field M (2011) The volatile content of hypabyssal kimberlite magmas: some constraints from experiments on natural rock compositions. Bull Volcanol 73:959–981

Brooker RA, Kjarsgaard BA (2011) Silicate-carbonate liquid immiscibility and phase relations in the system SiO_2-Na_2O-Al_2O_3-CaO-

CO_2 at 0.1–2.5 GPa with applications to carbonatite genesis. J Petrol 52:1281–1305

Brown RJ, Manya S, Buisman I, Fontana G, Field M, Mac Niocaill C, Sparks RSJ, Stuart FM (2012) Eruption of kimberlite magmas: physical volcanology, geomorphology and age of the youngest kimberlitic volcanoes known on earth (the Upper Pleistocene/Holocene Igwisi Hills volcanoes, Tanzania). Bull Volcanol 74: 1621–1643

Canil D, Bellis AJ (2008) Phase equilibria in a volatile-free kimberlite at 0.1 MPa and the search for primary kimberlite magma. Lithos 105:111–117

Canil D, Fedortchouk Y (1999) Garnet dissolution and the emplacement of kimberlites. Earth Planet Sci Lett 167:227–237

Caricchi L, Burlini L, Ulmer P, Gerya T, Vassalli M, Papale P (2007) Non-Newtonian rheology of crystal-bearing magmas and implications for magma ascent dynamics. Earth Planet Sci Lett 264:402

Clement CR (1982) A comparative geologic study of some major kimberlite pipes in the Northern Cape and Orange Free State. University of Cape Town

Dalton JA, Presnall DC (1998a) Carbonatitic melts along solidus of model lherzolite in the system $CaO–MgO–Al_2O_3–SiO_2–CO_2$ from 3 to 7 GPa. Contrib Miner Petrol 131:123–135

Dalton JA, Presnall DC (1998b) The continuum of primary carbonatitic-kimberlitic melt compositions in equilibrium with lherzolite: data from the system $CaO-MgO-Al_2O_3-SiO_2-CO_2$ at 6 GPa. J Petrol 39:1953–1964

Dalton JA, Wood BJ (1993) The compositions of primary carbonate melts and their evolution through wallrock reaction in the mantle. Earth Planet Sci Lett 119:511–525

Davidson CF (1943) The Archean rocks of the Rodil district, South Harris, Outer Hebrides. Trans R Soc Edinb 61:71–112

Dawson JB (1971) Advances in kimberlite geology. Earth-Sci Rev 7:187–214

Dawson JB (1980) Kimberlites and their xenoliths. Springer, Berlin

Dawson JB (1994) Quaternary kimberlitic volcanism on the Tanzania craton. Contrib Miner Petrol 116:473–485

Dobson DP, Jones AP, Rabe R, Sekine T, Kurita K, Taniguchi T, Kondo T, Kato T, Shimomura O, Urakawa S (1996) In-situ measurement of viscosity and density of carbonate melts at high pressure. Earth Planet Sci Lett 143:207–215

Downes PJ, Wartho JA, Griffin BJ (2006) Magmatic evolution and ascent history of the aries micaceous kimberlite, Central Kimberley Basin, Western Australia: Evidence from zoned phlogopite phenocrysts, and UV laser Ar-40/Ar-39 analysis of phlogopite-biotite. J Petrol 47:1751–1783

Edwards BR, Russell JK (1996) Influence of magmatic assimilation on minerals growth and zoning. Can Mineral 34:1149–1162

Edwards BR, Russell JK (1998) Time scales of magmatic processes: new insights from dynamic models for magmatic assimilation. Geology 26:1103–1106

Eggler DH (1978) The effect of CO_2 on partial melting of peridotite in the system Na_2O- $CaO-Al_2O_3-MgO-SiO_2-CO_2$ to 35 kb, with an analysis of melting in a peridotite-H_2O-CO_2 system. Am J Sci 278:305–343

Eggler DH, Mysen BO (1976) Role of CO_2 in genesis of olivine melilitite—discussion. Contrib Miner Petrol 55:231–236

Falloon TJ, Green DH (1989) The solidus of carbonated, fertile peridotite. Earth Planet Sci Lett 94:364–370

Fedortchouk Y, Canil D (2004) Intensive variables in kimberlite magmas, Lac de Gras, Canada and implications for diamond survival. J Petrol 45:1725–1745

Fedortchouk Y, Canil D, Semenets E (2007) Mechanisms of diamond oxidation and their bearing on the fluid composition in kimberlite magmas. Am Miner 92:1200–1212

Ferguson J, Danchin RV, Nixon PH (1973) Fenitization associated with kimberlite magmas. In: Nixon PH (ed) Lesotho Kimberlites. Lesotho National Development Corporation, pp 207–213

Foley SF (2008) Rejuvenation and erosion of the cratonic lithosphere. Nat Geosci 1:503–510

Foley SF, Yaxley GM, Rosenthal A, Buhre S, Kiseeva ES, Rapp RP, Jacob DE (2009) The composition of near-solidus melts of peridotite in the presence of CO_2 and H_2O between 40 and 60 kbar. Lithos 112:274–283

Francis D, Patterson M (2009) Kimberlites and aillikites as probes of the continental lithospheric mantle. Lithos 109:72–80

Genge MJ, Price GD, Jones AP (1995) Molecular dynamics simulations of $CaCO_3$ melts to mantle pressures and temperatures: implications for carbonatite magmas. Earth Planet Sci Lett 131:225–238

Girnis AV, Bulatov VK, Brey GP (2011) Formation of primary kimberlite melts—constraints from experiments at 6–12 GPa and variable CO_2/H_2O. Lithos 127:401–413

Gurney JJ, Helmstaedt HH, Le Roex AP, Nowicki TE, Richardson SH, Westerlund K (2005) Diamonds: crustal distribution and formation processes in time and space and an integrated deposit model. Econ Geol, 100th Anniversary Volume, pp 143–177

Hammouda T, Laporte D (2000) Ultrafast mantle impregnation by carbonatite melts. Geology 28:283–285

Hayman P, Cas RAF, Johnson M (2009) Characteristics and alteration origins of matrix minerals in volcaniclastic kimberlite of the Muskox pipe (Nunavut, Canada). Lithos 112:473–487

Heaman LM, Kjarsgaard BA, R.A C (2004) The temporal evolution of North American kimberlites. Lithos 76:377–397

Hrma P (1985) Reaction between sodium carbonate and silica sand at $874°C < T < 1022°C$. J Am Ceram Soc 68:337–341

Janse AJA (2007) Global rough diamond production since 1870. Gems Gemology 43:98–119

Janse AJA, Sheahan PA (1995) Catalogue of world wide diamond and kimberlite occurrences: a selective and annotative approach. J Geochem Explor 53:73–111

Kamenetsky VS, Kamenetsky MB, Sobolev AV, Golovin AV, Demouchy S, Faure K, Sharygin VV, Kuzmin DV (2008) Olivine in the Udachnaya-East kimberlite (Yakutia, Russia): types, compositions and origins. J Petrol 49:823–839

Kamenetsky VS, Kamenetsky MB, Weiss Y, Navon O, Nielsen TFD, Mernagh TP (2009) How unique is the Udachnaya-East kimberlite? Comparison with kimberlites from the Slave Craton (Canada) and SW Greenland. Lithos 112:334–346

Kavanagh JL, Sparks RSJ (2009) Temperature changes in ascending kimberlite magma. Earth Planet Sci Lett 286:404–413

Kelley SP, Wartho JA (2000) Rapid kimberlite ascent and the significance of Ar–Ar ages in xenolith phlogopites. Science 289:609–611

Keppler H (2003) Water solubility in carbonatite melts. Am Miner 88:1822–1824

Kjarsgaard BA (2007) Kimberlite diamond deposits. In: Goodfellow WD (ed) Mineral deposits of Canada: a synthesis of major deposit-types, district metallogeny, the evolution of geological provinces, and exploration methods. Geological Association of Canada, Mineral Deposits Division, pp 245–271

Kjarsgaard BA, Pearson DG, Tappe S, Nowell GM, Dowall DP (2009) Geochemistry of hypabyssal kimberlites from Lac de Gras, Canada: Comparisons to a global database and applications to the parent magma problem. Lithos 112:236–248

Kopylova MG, Matveev S, Raudsepp M (2007) Searching for parental kimberlite melt. Geochim Cosmochim Acta 71:3616–3629

Kopylova MG, Russell JK (2000) Chemical stratification of cratonic lithosphere: constraints from the Northern Slave craton, Canada. Earth Planet Sci Lett 181:71–87

Kopylova MG, Russell JK, Cookenboo H (1999) Petrology of peridotite and pyroxenite xenoliths from the Jericho kimberlite: implications for the thermal state of the mantle beneath the Slave Craton, Northern Canada. J Petrol 40:79–104

Lavallée Y, Hess K-U, Cordonnier B, Dingwell DB (2007) Non-Newtonian rheological law for highly crystalline dome lavas. Geology 35:843–846

Le Roex AP, Bell DR, Davis P (2003) Petrogenesis of group I kimberlites from Kimberley, South Africa: evidence from bulk-rock geochemistry. J Petrol 44:2261–2286

Lensky NG, Niebo RW, Holloway JR, Lyakhovsky V, Navon O (2006) Bubble nucleation as a trigger for xenolith entrapment in mantle melts. Earth Planet Sci Lett 245:278–288

Luth RW (2009) The activity of silica in kimberlites, revisited. Contrib Miner Petrol 158:283–294

Masun KM, Doyle BJ, Ball S, Walker S (2004) The geology and mineralogy of the Anuri kimberlite, Nunavut, Canada. Lithos 76:75–97

McDonough WF (1990) Constraints on the composition of the continental lithospheric mantle. Earth Planet Sci Lett 101:1–18

Mitchell RH (1973) Composition of olivine, silica activity and oxygen fugacity in kimberlite. Lithos 6:65–81

Mitchell RH (1986) Kimberlites: mineralogy, geochemistry, and petrology. Plenum Press, New York

Mitchell RH (2008) Petrology of hypabyssal kimberlites: relevance to primary magma compositions. J Volcanol Geoth Res 174:1–8

Mitchell RH, Tappe S (2010) Discussion of "Kimberlites and aillikites as probes of the continental lithospheric mantle", by Francis D, Patterson M (Lithos 109:72–80). Lithos, 115:288–292

Moore AE (2011) The case for a cognate, polybaric origin for kimberlitic olivines. Lithos 128–131:1–10

Mysen BO, Arculus RJ, Eggler DH (1975) Solubility of carbon dioxide in melts of andesite, tholeiite, and olivine nephelinite composition to 30 Kbar pressure. Contrib Miner Petrol 53:227–239

Nicholls J, Carmichael ISE, Stomer JC (1971) Silica activity and P_{total} in igneous rocks. Contrib Miner Petrol 33:1–20

Nielsen TFD, Sand KK (2008) The Majuagaa kimberlite dike, Maniitsoq region, West Greenland: Constraints on an Mg-rich silicocarbonatitic melt composition from groundmass mineralogy and bulk compositions. Can Miner 46:1043–1061

Nixon PH (1995) The morphology and nature of primary diamondiferous occurrences. J Geochem Explor 53:41–71

Patterson M, Francis D, McCandless T (2009) Kimberlites: magmas of mixtures? Lithos 112:191–200

Pearson DG, Wittig N (2008) Formation of Archean continental lithosphere and its diamonds: the root of the problem. J Geol Soc 165:895–914

Price SE, Russell JK, Kopylova MG (2000) Primitive magma from the Jericho Pipe, NWT, Canada: constraints on primary kimberlite melt chemistry. J Petrol 41:789–808

Roedder E (1965) Liquid CO_2 inclusions in olivine-bearing nodules and phenocrysts form basalts. Am Miner 50:1746–1782

Russell JK, Porritt LA, Lavallee Y, Dingwell DB (2012) Kimberlite ascent by assimilation-fuelled buoyancy. Nature 481:352–356

Rutherford MJ (2008) Magma ascent rates. Rev Miner Geochem 69:241–271

Rutherford MJ, Gardner JE (2000) Rates of magma ascent. In: Sigurdsson H (ed) Encyclopedia of volcanoes. Elsevier, New York, pp 207–217

Ryabchikov ID, Girnis AV (2005) Genesis of low-calcium kimberlite magmas. Russ Geol Geophys 46:1202–1212

Shaw CSJ (1999) Dissolution of orthopyroxene in basanitic magma between 0.4 and 2 GPa: further implications for the origin of Si-rich alkaline glass inclusions in mantle xenoliths. Contrib Miner Petrol 135:114–132

Shee SR (1985) The petrogenesis of the Wesselton Mine Kimberlite, Kimberley, Cape Province, RSA. Ph.D., University of Cape Town

Smith CB, Sims K, Chimuka L, Duffin A, Beard AD, Townend R (2004) Kimberlite metasomatism at Murowa and Sese pipes, Zimbabwe. Lithos 76:219–232

Sparks RSJ, Baker L, Brown RJ, Field M, Schumacher J, Stripp G, Walters A (2006) Dynamical constraints on kimberlite volcanism. J Volcanol Geoth Res 155:18–48

Sparks RSJ, Brooker RA, Field M, Kavanagh J, Schumacher JC, Walter MJ, White J (2009) The nature of erupting kimberlite melts. Lithos 112:429–438

Sparks RSJ, Brown RJ, Field M, Gilbertson M (2007) Kimberlite ascent and eruption. Nature 450:E21–E21

Sparks RSJ, Pinkerton H, Macdonald R (1977) The transport of xenoliths in magmas. Earth Planet Sci Lett 35:234–238

Spera FJ (1984) Carbon dioxide in petrogenesis III: role of volatiles in the ascent of alkaline magma with special reference to xenolith-bearing mafic lavas. Contrib Miner Petrol 88:217–232

Stachel T (2001) Diamonds from the asthenosphere and the transition zone. Eur J Miner 13:883–892

Stachel T, Brey G, Lorenz V (1995) Carbonatite magmatism and fenitization of the epiclastic caldera-fill at Gross-Brukkaros (Namibia). Bull Volcanol 57:185–196

Stoppa F, Woolley AR (1997) The Italian carbonatites: field occurrence, petrology and regional significance. Miner Petrol 59:43–67

Stripp G, Field M, Schumacher J, Sparks RSJ, Cressey G (2006) Post-emplacement serpentinization and related hydrothermal metamorphism in a kimberlite from Venetia, South Africa. J Metamorph Geol 24:515–534

Sweeney RJ (1994) Carbonatite melt compositions in the Earth's mantle. Earth Planet Sci Lett 128:259–270

Tappe S, Pearson DG, Nowell G, Nielsen T, Milstead P, Muehlenbachs K (2011) A fresh isotopic look at Greenland kimberlites: cratonic mantle lithosphere imprint on deep source signal. Earth Planet Sci Lett 305:235–248

Verhoogen J (1938) Les pipes de kimberlite du Katanga. Commun Spéc Katunga Ann Services Mines 9:1–15

Wallace ME, Green DH (1988) An experimental determination of primary carbonatite magma composition. Nature 335:343–346

Wartho JA, Kelley SP (2003) Ar-40/Ar-39 ages in mantle xenolith phlogopites: determining the ages of multiple lithospheric mantle events and diatreme ascent rates in southern Africa and Malaita, Solomon Islands. In: Geochronology: linking the isotopic record with petrology and textures, London, pp 231–248

Wilcox A, Buisman I, Sparks S, Brown R, Manya S, Schumacher, JC, Tuffen H (2012) Petrology, geochemistry and low-temperature alteration of extrusive lavas and pyroclastic rocks of the Igwisi Hills kimberlites, Tanzania. Chem Geol. doi:10.1016/j.pce.2011.03.002

Wilson L, Head JW (2007) An integrated model of kimberlite ascent and eruption. Nature 447:53–57

Woolley AR, Church AA (2005) Extrusive carbonatites: a brief review. Lithos 85:1–14

Wyllie PJ (1977) Peridotite-CO_2-H_2O, and carbonatitic liquids in the asthenosphere. Nature 266:45–47

Wyllie PJ (1980) The origin of kimberlites. J Geophys Res 85:6902–6910

Wyllie PJ, Huang WL (1975) Peridotite, kimberlite, and carbonatite explained in the system CaO-MgO-SiO_2-CO_2. Geology 3:621–624

Yaxley GM, Crawford AJ, Green DH (1991) Evidence for carbonatite metasomatism in spinel peridotite xenoliths from western Victoria, Australia. Earth Planet Sci Lett 107:305–317

Detailed Protracted Crystallization History of Perovskite in Orapa Kimberlite

Chiranjeeb Sarkar, Craig D. Storey, and Chris J. Hawkesworth

Abstract

Detailed petrographic and geochemical studies of perovskites from different lithofacies of Orapa kimberlite, a typical crater facies kimberlite from Botswana, have been conducted to investigate the crystallization and paragenesis of this very important groundmass phase. We suggest that there is no universal paragenetic sequence of mineral crystallization in the groundmass of kimberlite as it depends on the magma composition, which is highly variable. Our study reveals that most of the perovskite grains in Orapa grew after the macrocrystal phases such as olivine and Cr-bearing spinel, and simultaneously with "reaction" Fe-rich and groundmass spinel from ulvöspinel-magnetite group, as suggested earlier in the literature. However, certain perovskite grains contain inclusions of phlogopite and apatite, which are generally very late-crystallizing phases in kimberlite. Some perovskite grains are also found to appear as late-crystallizing rims around partially resorbed spinel. These textural features suggest protracted perovskite crystallization over a range of P and T from an evolving kimberlite magma. Previous O isotope data have also been used to suggest that perovskite crystallization succeeded late-stage magmatic degassing. Minor and trace element concentrations of Orapa perovskites also support this longer crystallization history as the post-degassed phase perovskite contain less Nb and Zr, which have preferentially partitioned into rutile, an alteration product of early-crystallizing perovskite. Calculated oxygen fugacities from Orapa perovskites range from −5.5 NNO to −0.2 NNO, emphasizing perovskite crystallization in an evolving magmatic system. Sudden degassing, mainly CO_2 release, prompted a change in the oxidation state of the magma, which was recorded by the late-crystallizing perovskites as this group shows an $f(O)_2$ value much higher (−2.3 NNO to −0.2 NNO) than the rest of the perovskite grains. All different lithofacies contain perovskites of different paragenesis with varying quantities while the pyroclastic kimberlite has the maximum abundance of late-stage post-degassing phase perovskites.

C. Sarkar (✉)
Department of Earth and Atmospheric Sciences, University of Alberta, Edmonton, T6G 2E3, Canada
e-mail: chiranjeeb.sarkar@ualberta.ca

C. Sarkar
Department of Earth Sciences, University of Bristol, Bristol, BS8 1RJ, UK

C. D. Storey
School of Earth and Environmental Sciences, Burnaby Building, University of Portsmouth, Portsmouth, PO1 3QL, UK

C. J. Hawkesworth
Department of Earth Sciences, University of St. Andrews, North Street, St. Andrews, KY16 9AL, Scotland

D. G. Pearson et al. (eds.), *Proceedings of 10th International Kimberlite Conference*, Volume 1, Special Issue of the Journal of the Geological Society of India,
DOI: 10.1007/978-81-322-1170-9_13, © Geological Society of India 2013

Keywords
Kimberlite • Perovskite • Paragenesis • Oxygen fugacity • Degassing

Introduction

Perovskite ($CaTiO_3$) is a common primary accessory mineral phase in kimberlites. It typically constitutes up to 10 vol. % of the kimberlitic groundmass. It is an important geochemical repository of incompatible trace elements such as Nb, Ba, Sr and rare earth elements (REE) in kimberlitic magma (Mitchell 1972; Boctor and Boyd 1980; Jones and Wyllie 1984; Chakhmouradian and Mitchell 2000; Armstrong et al. 2004). Although most kimberlitic perovskite is stoichiometrically close to pure $CaTiO_3$, Nb and REE oxides can contain between 5 and 9 wt % (Chakhmouradian and Mitchell 2000, 2001; Ogilvie-Harris et al. 2010). Perovskite from kimberlite is generally enriched in LREE, particularly Ce, Nd and La, relative to HREE (Boctor and Boyd 1980; Jones and Wyllie 1984). There is minimal zoning in many grains. However, in "normally" zoned grains, the LREE, Th and occasionally Nb and Na decrease from core to rim, sometimes with increasing Fe (Chakhmouradian and Mitchell 2000). There is negligible variation in perovskite chemical compositions between different kimberlite bodies (Jones and Wyllie 1984; Mitchell 1986) or within same kimberlite (Boctor and Boyd 1981).

Perovskite is one of the main carriers of U and Th in kimberlitic magma. This and the fact that it crystallizes directly from the magma make perovskite a good geochronometer in kimberlite and other alkaline and undersaturated rocks (Heaman 1989; Heaman et al. 2003, 2004; Batumike et al. 2008). Perovskite has an extremely low Rb/Sr value (<0.001) and high Sr concentration (typically up to 0.3 wt %), which make it a potentially useful phase for Sr isotope studies and recording the initial Sr isotope composition of kimberlites (Heaman 1989). Such low Rb/Sr value means that the measured isotope ratio needs very small age correction (due to in situ decay of Rb) in order to determine the initial Sr isotope composition. Perovskite also contains a considerable amount of Sm (up to 2,000 ppm) and Nd (up to 1 wt %) with variable Sm/Nd ratios (Heaman 1989). Thus, perovskite is an excellent geochemical indicator of the kimberlite magma. Recently, perovskite has been used to determine more reliable isotopic signatures (Sr, Nd and Hf) of the kimberlite magma than those from the bulk-rock analyses (Paton et al. 2007; Yang et al. 2009; Wu et al. 2010). Moreover, Canil and Bellis (2007) have shown that Fe and Nb contents of perovskite can be used to calculate the oxygen fugacity of the magma during perovskite crystallization, and it can give important insight into diamond resorption and late-stage processes in kimberlite magma evolution.

In spite of the wealth of information that can be obtained from kimberlitic perovskite, it has not been studied as extensively as groundmass spinel. The temperature and pressure range over which perovskite crystallizes is poorly constrained. It has been suggested from initial textural studies that perovskite crystallizes at 800−600 °C after macrocrystal spinel (aluminous magnesian chromite), and nearly simultaneously with Fe-rich spinel (Mitchell 1986; Chakhmouradian and Mitchell 2000). It has also been argued that in most cases, perovskite crystallization ceases prior to the resorption of groundmass spinel and phlogopite crystallization (Chakhmouradian and Mitchell 2001). However, the situation can be more complex as crustal assimilation, and multiple batches of magma mixing are common in kimberlites. In addition, it has been shown that kimberlitic perovskite records a wide range of $f(O)_2$ (Canil and Bellis 2007; Ogilvie-Harris et al. 2010) and $^{87}Sr/^{86}Sr$ compositions (Malarkey et al. 2010) suggesting a complex paragenesis. In this contribution, we comment on perovskite crystallization and evolution following our detail petrographic and microanalytical studies of perovskite from Orapa A/K1 kimberlite pipe from Botswana.

Geological Setting

The Orapa kimberlite is located in north-eastern Botswana, east of the Central Kalahari Basin. It is a cluster of ~60 pipes and dykes, among which the Orapa A/K1 diamond mine is the largest. It provides a 100–150-m thick-section of volcaniclastic deposits of two kimberlite pipes (North and South Pipe) that have coalesced near surface (Field et al. 1997). Orapa A/K1 has been dated as ca. 93 Ma old, and it has erupted through deformed Archaean basement overlain by volcanic and sedimentary rocks of the Karoo Supergroup (Permian–Triassic) (Davis 1977; Allsopp et al. 1989; Field et al. 1997, 2008). Orapa A/K1 is a typical example containing both crater and diatreme facies kimberlites. The Orapa North Pipe is filled with typical massive volcaniclastic kimberlite (MVK), and it is at a deeper level of erosion than the Orapa South pipe, which contains a stratiform sequence of volcaniclastic kimberlite (VK) overlain by a pyroclastic kimberlite (PK) unit (Field et al. 1997; Gernon et al. 2009). The Orapa North Pipe MVK is very well mixed due to extensive fluidization. Orapa kimberlite contains abundant clasts of basement, basalts and sedimentary rocks, and it has distinct gas escape structures (Gernon et al. 2008, 2009). Samples used in this study were collected from MVK from Orapa North Pipe and from VK and PK from Orapa South pipes. Samples were selected from both mine surfaces and drill cores.

Analytical Techniques

Petrography and Geochemistry

Polished thin sections were prepared from selected hand specimens and were examined to identify perovskite crystals for further analysis. Back-scattered electron (BSE) images were taken using a Hitachi S-3500 N scanning electron microscope (SEM) with a 15–20 kV accelerating voltage at the University of Bristol. Concentrations of 19 major, minor and trace elements were measured in a Cameca SX100 EPMA with an accelerating voltage of 15–20 kV and emission current of 60 nA at Bristol. The counting times were 20 s for major elements, 30 s for LREEs, 60 s for Fe, Zr, Nb, Ta and 90 s for Th and U. Conventional ZAF correction was used to reduce counts into elemental abundances. The following standard reference materials were used to determine the chemical compositions: albite (Na), olivine (Mg, Si), sanidine (Al), wollastonite (Ca), ilmenite (Ti, Fe), $SrTiO_3$ (Sr), Y-glass (Y), zircon (Zr), $LiNbO_3$ (Nb), La-glass (La), Ce-glass (Ce), Pr-glass (Pr), Nd-glass (Nd), Sm-glass (Sm), galena (Pb), ThO_2 (Th) and UO_2 (U).

Oxygen Fugacity

Oxygen fugacity was calculated using the empirical oxygen barometer calibration developed by (Bellis and Canil 2007). Based on the arguments put forward by Mitchell (2002), it can be assumed that all Fe in kimberlitic perovskite exists as Fe^{+3}. Thus, it is possible to determine the oxygen fugacity of the magma, in equilibrium with perovskite, by calculating the total Fe content of perovskite as Fe increasingly occurs as Fe^{3+} in kimberlite melts with increasing $f(O)_2$. The Nb content at a given $f(O)_2$ also affects the Fe content of perovskite, so the empirical oxygen barometer is defined as follows:

$$\Delta NNO = -[0.50(\pm 0.021) \times Nb - Fe(\pm 0.031) + 0.030(\pm 0.001)]/0.004(\pm 0.0002) \quad (1)$$

where Fe and Nb are in cation units per three oxygens, and uncertainties are given at the 2σ level (Bellis and Canil 2007). It is the $f(O)_2$ of the magma that determines the total Fe content of perovskite rather than the bulk Fe content of the melt (Bellis and Canil 2007).

Results

Petrography

Perovskite grains from Orapa kimberlite are in general similar to other perovskites reported from group I kimberlites worldwide (Chakhmouradian and Mitchell 2000). There is no obvious petrographic difference between perovskites from the Orapa North and South Pipe. It appears as major groundmass phase and comprises up to 10 vol % of the groundmass. The grains are euhedral to subhedral in shape, and grain sizes range between 20 and 60 μm (Fig. 1a–d) while occasionally bigger than 100 μm. Many grains are homogeneous although some show normal zonation with brighter cores and relatively darker rims in BSE images. Although rare, some grains in the studied perovskite suite show weak oscillatory zoning (Fig. 1c, d). Some complex zonation patterns were also observed where it appeared that more than one nucleus coexisted within the same crystal. Overall, perovskite crystals were classified into the following four parageneses; (1) discrete grains within the groundmass or matrix (Fig. 1c), (2) contained within ghost-lapilli or along grain boundaries of olivine macrocrysts and phenocrysts forming a "garland" (Fig. 1b), (3) intimate and complex intergrowths with groundmass spinel (Fig. 1f, h) (which often form an atoll rim around) and (4) as a reaction mantle around Ti bearing phase, for example, ilmenite macrocrysts (Fig. 1e). All four parageneses were observed in all three lithofacies; however, type (4) is rare in PK (Table 1). In places, perovskite also appears as a late-crystallizing rim around the Fe–Ti-rich spinel that has reacted to form titanite (sphene) around its grain boundary (Fig. 2e, f). Euhedral inclusions of serpentinized olivine and Cr spinel are common (Fig. 1a, g). As observed by others, perovskite and mica sometimes form a poikilitic texture where tiny perovskite crystals are included within a larger phlogopite grain (Chakhmouradian and Mitchell 2000). However, some Orapa perovskite crystals contain inclusions of phlogopite mica and apatite as well (Fig. 2b, h). Although rare, older relict perovskite grains can be found in the core of late growing perovskite rims (Fig. 2a). Based on their general appearance, perovskite grains from Orapa can be divided into two types. Some show grain boundary resorption, as there are rims of rutile around their grain boundary (Fig. 2c, d). These grains are often fractured and replaced by calcite and rutile that have filled the cracks. Sometimes, they are completely replaced

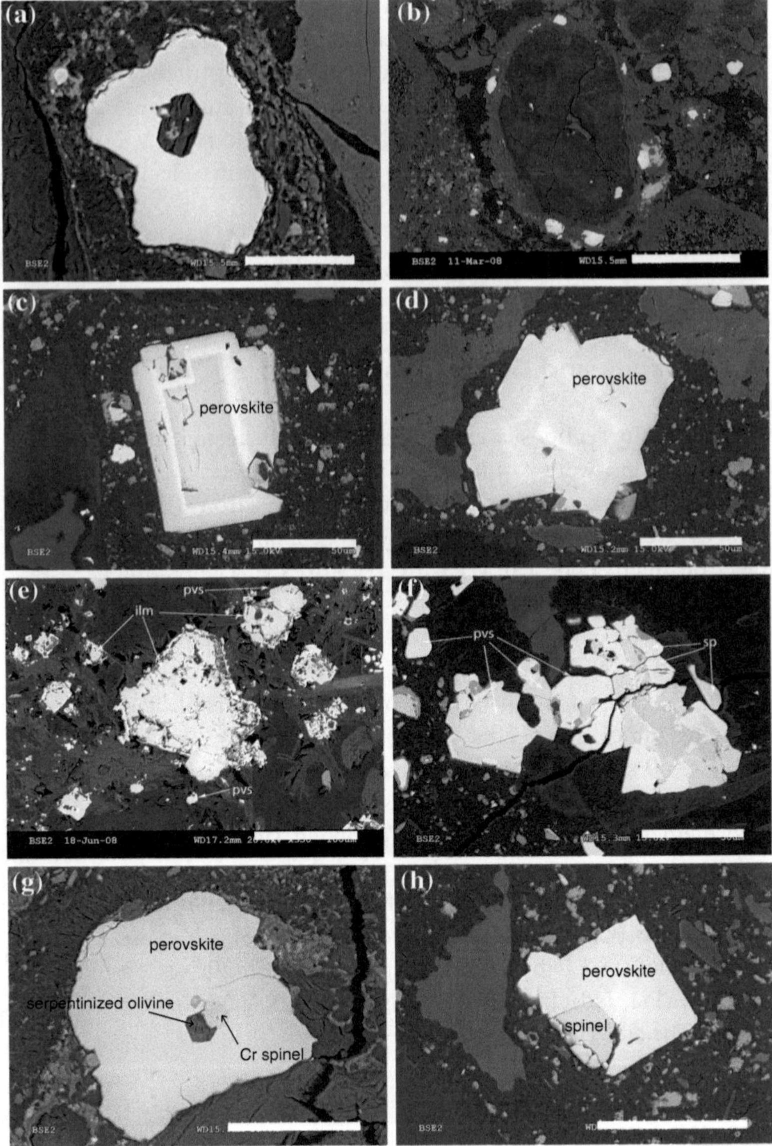

Fig. 1 Back-scattered electron images of perovskites from Orapa. **a** Anhedral perovskite with an inclusion of euhedral serpentinized olivine. **b** Perovskite crystallizing around the grain boundary of serpentinized olivine. **c** Euhedral perovskite showing oscillatory zonation. **d** Euhedral perovskite (probably twinned) showing faint oscillatory zonation. **e** Ilmenite macrocrysts with a reaction mantle of perovskite. **f** Complex association of spinel and perovskite suggesting simultaneous growth. **g** Euhedral perovskite with inclusions of euhedral serpentinized olivine and Cr spinel. **h** Intergrowth of euhedral perovskite and spinel with straight inter-grain boundaries. (Scale bar is 50 μm in **c**, **d**, **f**, **g**, **h**; 30 μm in **a**; 300 μm in **b** and 100 μm in **e**)

Table 1 Relative abundance of perovskites with different parageneses found in various lithofacies

Perovskite paragenesis		Pyroclastic kimberlite (PK)	Volcaniclastic kimberlite (VK)	Massive volcaniclastic kimberlite (MVK)
Major parageneses	Discrete groundmass	Abundant	Abundant	Abundant
	Along olivine grain boundary	Intermediate	Abundant	Abundant
	Intergrowth with spinel	Intermediate	Intermediate	Abundant
	Reaction mantle around Ilmenite	Very rare	Intermediate	Abundant
Minor paragenesis	Pvs with Olv and Cr spinel inclusion	Rare	Intermediate	Intermediate
	Pvs inclusion within phlogopite	Very rare	Rare	Rare
	Pvs containing inclusions of mica and apatite	Intermediate	Very rare	Rare
	Pvs rim around resorbed Fe–Ti spinel	Intermediate	Rare	Rare
Pvs with Resorbed grain boundary[a]		Abundant	Intermediate	Intermediate
Fresh lozenge-shaped perovskite		Intermediate	Intermediate	Intermediate

[a] Denotes a secondary alteration feature

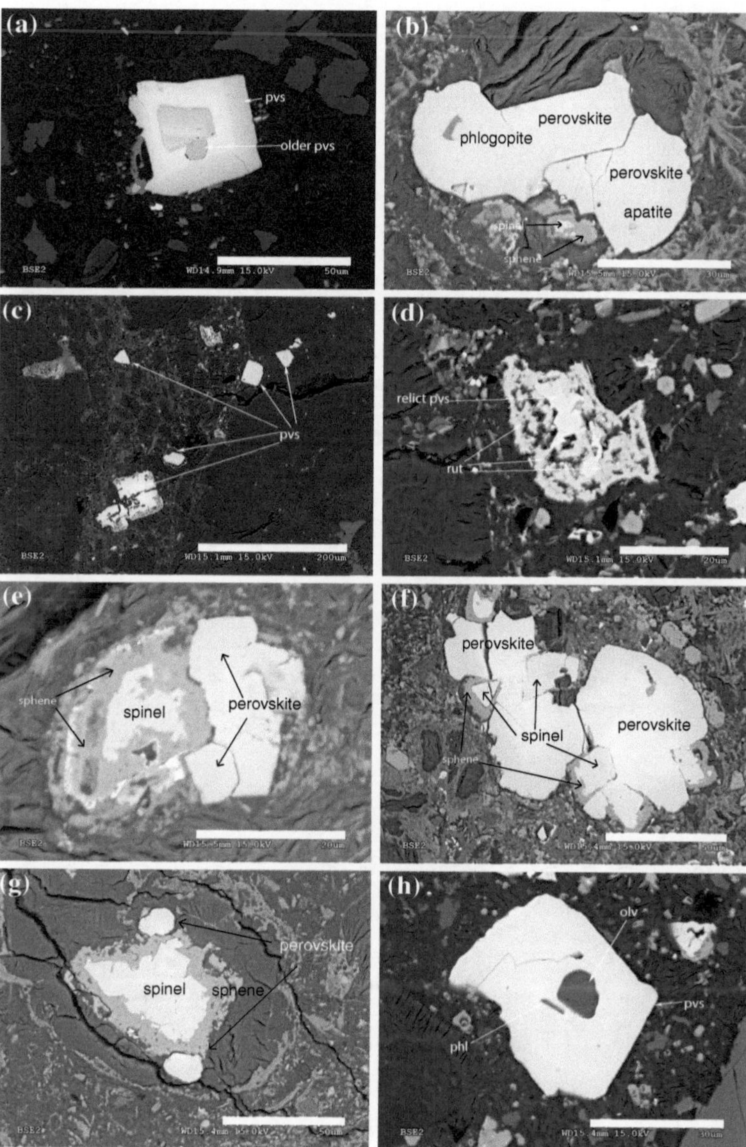

Fig. 2 Back-scattered electron images of perovskites from Orapa. **a** Euhedral zoned perovskite grain with a core of relict older perovskite. **b** Anhedral perovskite grain with inclusions of phlogopite and apatite. **c** Distribution of fresh unaltered and resorbed perovskite at the hand specimen scale. **d** A perovskite grain with a rim of rutile around it indicating a reaction between perovskite and CO_2 to produce the rim of TiO_2 around the grain boundary. A small part of the perovskite remains as a relict phase within rutile. **e** Perovskite growing after Fe–Ti spinel, which has reacted to form titanite around its grain boundary. **f** Euhedral spinel with rounded, anhedral perovskite intergrowth. A band of titanite lies between spinel and perovskite. **g** perovskite crystallizing after spinel resorption into titanite. **h** A subhedral perovskite grain with inclusions of serpentinized olivine and phlogopite. (Scale bar is 50 μm in **a**, **f**, **g**; 30 μm in **b**, **h**; 200 μm in **c** and 20 μm in **d**, **e**)

by rutile, although the original shapes are preserved as pseudomorphs. However, the core of the altered grains remains unaffected in case of partial replacement. On the other hand, some perovskite grains appear to be quite resistant and fresh. They often form a lozenge-shaped grains around spinel or olivine or as discrete grains (Fig. 2c, g). All textural varieties were found at variable abundance in each lithofacies of the Orapa kimberlite. However, the pyroclastic kimberlite is dominated by perovskites appearing as discrete groundmass phases. In contrast, volcaniclastic kimberlite contains perovskites appearing mostly as intergrowths with spinel and ilmenite. Both pipes and all lithofacies contain altered and fresh perovskite grains. In fact, the fresh and altered perovskite are so well mixed that almost every thin section studied contains both types of perovskite grains (Fig. 2c).

Chemical Compositions

Major, minor and trace element compositions of Orapa perovskites are reported in Table 2. Major element compositions of these perovskite grains remain close to ideal $CaTiO_3$ with CaO (36–37.2 wt %), TiO_2 (52–54 wt %) and minor amounts of FeO (0.9–1.5 wt %), Nb_2O_5 (0.9–1.5 wt %), Na_2O (0.5–0.9 wt %) and REE oxides (6–9 wt %). This range of compositions is very similar to the perovskites reported in kimberlites worldwide (Chakhmouradian and Mitchell 2000, 2001). Niobium and REE are the primary substitutions for Ti while Na replaces Ca for charge balance (Boctor and Boyd 1980, 1981). This is demonstrated by a strong negative correlation of Ca with Na and trace elements (REE, SrO, Nb_2O_5 and ThO_2; Fig. 3b). Orapa perovskite has low abundance of Na (0.3–1 wt %),

Table 2 Representative EPMA compositions of perovskite of different textural varieties in Orapa kimberlites

Sample texture	NP19	NP19	NP19	NP19	SP1	SP1	SP1	NP1	NP1	NP1
Discrete grain										
Na	0.55	0.52	0.71	0.55	0.48	0.68	0.59	0.49	0.58	0.52
Mg	–	–	–	–	0.19	0.01	–	–	0.01	–
Al	–	–	–	0.04	0.04	0.03	–	–	0.02	–
Si	–	–	–	–	0.43	0.04	–	–	0.03	–
Ca	25.98	25.76	24.24	25.75	25.00	24.66	25.80	26.26	25.29	26.17
Nd	0.95	0.92	1.09	1.00	0.92	1.30	0.94	0.79	0.96	0.81
Fe	0.81	0.76	0.84	0.89	1.05	0.69	0.69	0.74	0.83	0.74
Ti	32.50	31.99	31.36	32.27	33.11	31.71	32.79	31.96	32.10	32.05
Th	0.59	0.59	2.54	0.74	0.97	0.37	0.61	0.30	0.90	0.34
Sr	0.21	0.15	0.22	0.18	0.15	0.16	0.19	0.22	0.24	0.15
Nb	1.06	0.94	1.62	0.90	1.17	1.94	0.93	0.91	1.22	0.92
Zr	0.09	0.08	0.22	0.04	0.12	0.38	0.09	0.08	0.11	0.11
Y	–	–	–	–	–	0.02	–	0.01	0.03	0.02
Pr	0.17	0.11	0.23	0.17	0.18	0.28	0.14	0.19	0.20	0.21
La	1.01	0.88	0.88	0.84	0.98	1.11	0.87	0.91	0.88	0.83
Ce	2.09	2.11	2.55	2.16	2.22	3.03	2.11	1.96	2.31	1.95
Sm	0.12	0.10	0.10	0.10	0.10	0.18	0.11	0.12	0.13	0.12
O	33.82	33.25	33.04	33.54	33.65	33.56	33.80	33.34	33.53	33.40
Total	99.94	98.18	99.64	99.18	100.76	100.13	99.68	98.29	99.37	98.34
Fe/Nb	0.99	1.04	0.66	1.27	1.16	0.46	0.96	1.04	0.88	1.04
ΔNNO	−1.57	−1.65	−3.26	−0.01	−2.65	−6.19	−2.51	−1.80	−1.90	−1.79
	NP1	SP7	SP7	SP7	SP7	NP28	NP28	NP28	NP28	NP28
Mantled with spinel										
Na	0.46	0.50	0.66	0.64	0.64	0.65	0.70	0.71	0.70	0.67
Mg	–	–	–	–	–	–	–	–	–	–
Al	0.01	–	0.01	–	–	–	0.01	0.02	0.03	0.02
Si	0.01	–	–	–	–	–	–	–	–	–
Ca	26.49	26.41	25.26	25.40	25.63	25.37	24.93	25.05	24.71	24.88
Nd	0.62	0.68	1.08	0.99	0.95	0.97	1.15	1.07	1.17	1.08
Fe	0.70	0.63	0.70	0.70	0.68	0.68	0.71	0.73	0.71	0.73

(continued)

Table 2 (continued)

	NP1	SP7	SP7	SP7	SP7	NP28	NP28	NP28	NP28	NP28
Ti	32.71	31.98	32.06	32.55	32.30	32.26	31.54	31.81	31.43	31.85
Th	0.01	0.06	0.87	0.75	0.64	0.77	0.49	0.85	1.43	1.35
Sr	0.18	0.19	0.17	0.15	0.24	0.20	0.21	0.19	0.20	0.17
Nb	1.29	1.33	1.30	1.25	1.48	1.43	1.83	1.53	1.59	1.46
Zr	0.31	0.23	0.17	0.22	0.20	0.21	0.27	0.26	0.21	0.20
Y	0.00	0.02	0.04	0.02	0.07	0.03	0.02	–	0.02	0.02
Pr	0.05	0.17	0.23	0.12	0.21	0.21	0.24	0.27	0.24	0.19
La	0.81	0.84	1.00	–	0.86	1.09	1.23	1.07	1.05	0.88
Ce	1.75	1.87	2.36	2.12	2.21	2.31	2.67	2.74	2.65	2.48
Sm	0.11	0.12	0.09	0.16	0.13	0.13	0.12	0.09	0.11	0.12
O	34.03	33.55	33.49	33.58	33.77	33.68	33.33	33.46	33.16	33.35
Total	99.54	98.57	99.50	98.67	100.00	100.01	99.45	99.85	99.42	99.48
Fe/Nb	0.69	0.61	0.70	0.72	0.59	0.61	0.50	0.61	0.58	0.65
ΔNNO	−3.78	−4.70	−3.72	−3.52	−4.69	−4.46	−5.51	−4.32	−4.66	−3.91

	NP1	NP1	NP1	NP1	SP1	SP1	SP1	SP1	NP35	NP35
Around olivine										
Na	0.52	0.70	0.55	0.51	0.63	0.61	0.27	0.55	0.52	0.61
Mg	–	0.03	–	–	–	–	–	0.06	–	–
Al	0.01	0.03	0.03	0.01	0.05	0.03	0.08	0.03	0.01	0.03
Si	–	0.08	–	–	–	–	–	0.13	0.01	–
Ca	25.68	24.51	25.89	26.21	25.58	25.53	27.41	25.56	25.98	25.77
Nd	0.98	0.95	0.91	0.81	0.91	0.99	0.45	0.97	0.92	0.96
Fe	0.76	0.84	0.79	0.71	0.89	0.81	0.92	0.79	0.76	0.84
Ti	32.71	32.67	32.69	32.52	30.93	32.05	32.61	32.18	32.56	32.09
Th	0.77	1.36	0.58	0.24	0.94	0.73	0.03	0.70	0.48	0.81
Sr	0.24	0.19	0.18	0.20	0.20	0.19	0.24	0.24	0.18	0.19
Nb	0.81	1.31	0.90	0.84	1.43	1.16	1.25	0.98	0.99	1.34
Zr	0.09	0.14	0.13	0.09	0.20	0.13	0.23	0.09	0.10	0.11
Y	0.02	–	–	0.04	–	–	0.01	–	–	0.01
Pr	0.13	0.23	0.18	0.16	0.16	0.15	0.05	0.16	0.17	0.23
La	0.99	0.82	0.85	0.82	1.03	0.84	0.79	0.97	–	0.88
Ce	2.28	2.21	2.13	1.88	2.03	2.34	1.23	2.17	2.10	2.01

(continued)

Table 2 (continued)

	NP1	NP1	NP1	NP1	SP1	SP1	SP1	SP1	NP35	NP35
Sm	0.12	0.11	0.12	0.13	0.20	0.11	0.03	0.13	0.17	0.16
O	33.80	33.78	33.83	33.65	32.92	33.46	34.22	33.65	33.48	33.62
Total	99.90	99.95	99.75	98.82	98.09	99.14	99.81	99.35	98.43	99.64
Fe/Nb	1.21	0.82	1.14	1.08	0.80	0.89	0.94	1.04	0.99	0.81
ΔNNO	−1.21	−2.18	−1.20	−1.97	−1.83	−1.96	−1.02	−2.44	−2.89	−3.26
	NP35	NP35	NP19	NP19	NP19	NP19	SP7	SP7	SP7	SP7
Inclusion										
Na	0.52	0.41	0.71	0.54	0.44	0.53	0.53	0.37	0.52	0.41
Mg	–	–	–	–	–	–	–	–	–	–
Al	–	0.01	0.03	0.01	0.06	–	–	0.03	0.01	0.01
Si	0.01	0.01	–	–	–	–	–	0.02	–	–
Ca	26.51	26.51	25.16	25.94	26.19	26.37	26.13	26.97	26.44	26.72
Nd	0.75	0.72	1.06	0.93	0.88	0.81	0.82	0.61	0.82	0.69
Fe	0.71	0.74	0.82	0.76	0.84	0.69	0.69	0.71	0.72	0.76
Ti	33.07	33.16	31.88	32.64	31.44	32.62	32.87	32.71	32.98	32.33
Th	0.23	0.12	1.54	0.51	0.50	0.14	0.26	–	0.31	0.09
Sr	0.18	0.18	0.17	0.15	0.21	0.21	0.20	0.22	0.20	0.25
Nb	0.89	0.84	0.87	0.76	0.79	0.85	0.89	0.79	0.87	0.85
Zr	0.06	0.13	0.06	0.11	0.07	0.09	0.06	0.05	0.11	0.13
Y	0.01	0.04	0.02	0.01	0.05	–	–	0.01	0.05	0.04
Pr	0.13	0.10	0.23	0.14	0.14	0.14	0.16	0.06	0.11	0.03
La	0.73	0.80	0.97	0.98	0.86	0.82	0.91	0.61	0.75	0.74
Ce	1.96	1.90	2.58	2.29	2.04	2.00	1.97	1.55	1.97	1.59
Sm	0.11	0.09	0.11	0.17	0.13	0.13	0.13	0.11	0.10	0.10
O	34.11	34.16	33.40	33.76	33.04	33.77	33.88	33.86	34.12	33.62
Total	99.99	99.92	99.61	99.69	97.66	99.17	99.51	98.68	100.06	98.37
Fe/Nb	1.03	1.13	1.23	1.29	1.38	1.04	1.00	1.16	1.06	1.15
ΔNNO	−2.13	−1.66	−0.63	−1.10	−0.02	−2.23	−2.37	−1.75	−2.01	−1.30

Elemental abundances are reported in wt%. Analytical uncertainties are 10–15 % (2SD) depending on the concentration

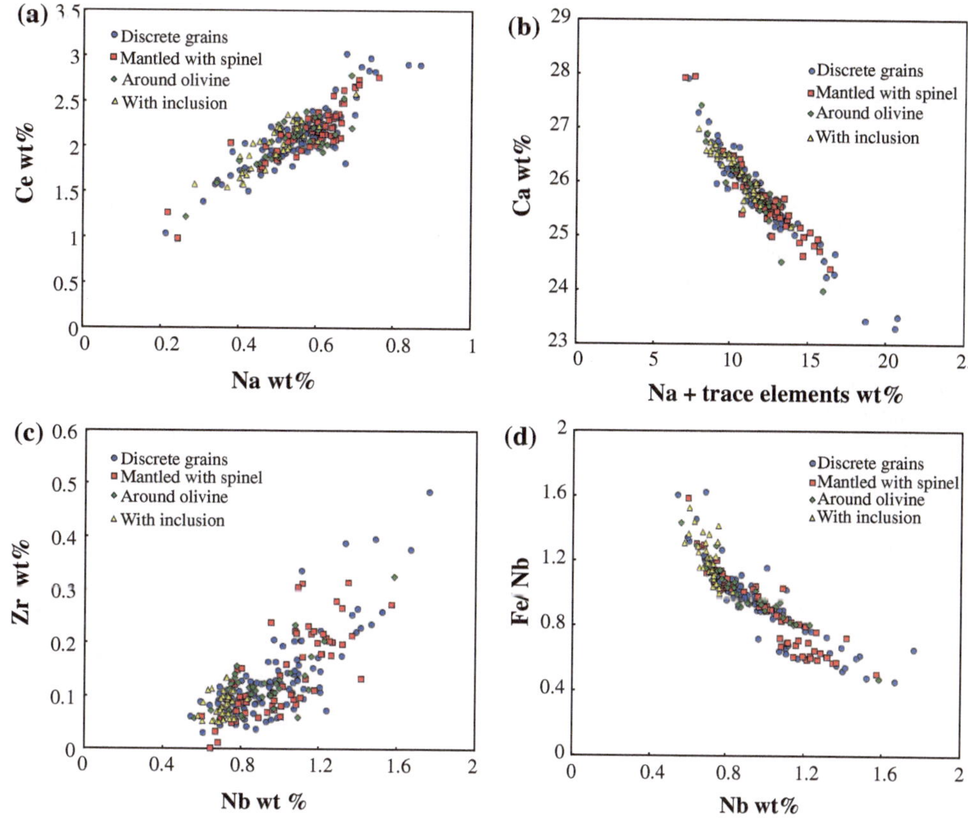

Fig. 3 Bivariate plot of different varieties of perovskite from Orapa. **a** Ce versus Na showing a positive correlation. **b** Strong negative correlation between Ca and Na + other trace elements. **c** Distribution of Zr and Nb. Note that the perovskite population with inclusions of late-stage minerals (phlogopite ± apatite) has relatively low Zr and Nb contents. **d** Variation of Nb content of perovskite with varying Fe/Nb values

but it has a positive correlation with LREE (especially Ce) content (1–3.5 wt %; Fig. 3a). Perovskites from all lithofacies contain very little Al_2O_3 (below detection to 0.3 wt %). In regular zoned perovskite, the core of the grain contains higher LREE, Na and Th while the rim has slightly higher Fe. However, there are no apparent compositional differences between perovskites from different parageneses or from different lithofacies of the Orapa kimberlite. Average major and trace element compositions of perovskites from the North and South Pipes are also similar. Strontium contents of all perovskite grains are remarkably consistent (2,000–2,300 ppm) while Ba appears to be absent (below detection limit). Although certain elements (e.g. Nb, Ta, Zr, Nd, Sm, Th) show some variation, no perovskite was found to be unusually enriched in any element, such as type III perovskite from Lac de Gras area, which have high in Na, Sr, Nb and LREE abundances (Chakhmouradian and Mitchell 2001). Although Fe does not show much variation (0.6–1 wt %), Nb and Zr contents of Orapa perovskites vary considerably (0.5–1.8 and 0.05–0.5 wt %, respectively). Variation of these elements does not correlate with any particular textural type of perovskite. All perovskite grains with inclusions of phlogopite, apatite or those appearing along the grain boundary of resorbed spinel have lower concentrations of Nb, Ta and Zr than other paragenetic varieties (Fig. 3c, d). This group of perovskites stands alone from the rest of the grains as they have low and less variable Nb and Zr contents (0.5–0.7 and 0.05–1.13 wt %, respectively).

Based on the oxybarometer equation given by Bellis and Canil (2007), the oxygen fugacity of the Orapa perovskite was calculated. Overall, the $f(O)_2$ of perovskite ranges between −5.5 and −0.2 with respect to NNO buffer (Fig. 4). The range displayed by discrete groundmass grains, and those that occur as intergrowths with spinel and ilmenite, overlap one another. Perovskites, contained within peletal lapilli or along olivine grain boundaries, record relatively restricted range of $f(O)_2$ (−3.4 to −1.1 NNO). Interestingly, all perovskite grains containing inclusions of phlogopite or apatite and those appearing along the grain boundary of resorbed spinel have relatively high NNO values (−2.3 to −0.2; Fig. 4).

Discussion

Crystallization of Perovskite

Petrographic and compositional analyses during this study emphasize the fact that discrete groundmass phase is the major mode of occurrence of perovskite in Orapa kimberlite. Perovskite is a relatively late-crystallizing mineral forming the groundmass. Large nucleation densities produce many small crystals which require elements for growth

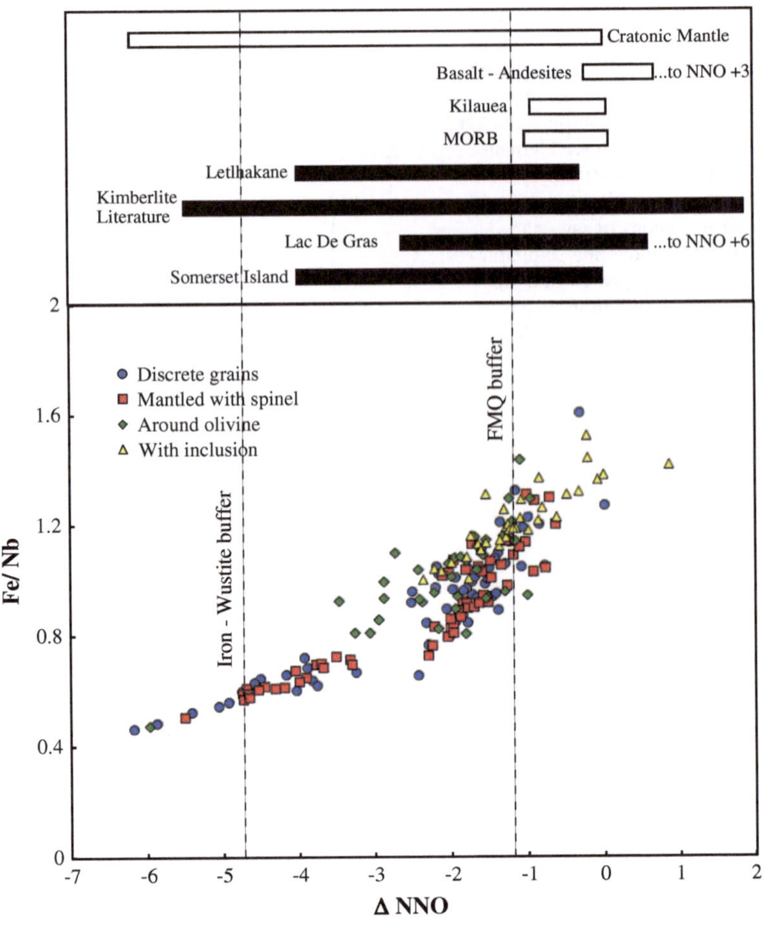

Fig. 4 Calculated oxygen fugacities (ΔNNO) of different perovskite grains with varying Fe/Nb values. Note that the perovskite population with inclusions of phlogopite ± apatite crystallized from a relatively oxidized magma. Oxygen fugacities (ΔNNO) of global kimberlite bodies, and other mantle-derived magmas are also plotted for reference. $f(O)_2$ ranges of different kimberlites and other magmatic rocks have been taken from Canil and Bellis (2007) while that of Letlhakane has been taken from Trickett (2007). *IW* iron-wustite, *FMQ* fayalite-quartz-magnetite

to be quickly depleted from the melt. It crystallized after most of the macrocrystal phases, such as olivine and Cr spinel, as previously observed by Clement (1982), Mitchell (1986, 2008), Chakhmouradian and Mitchell (2000). Intergrowth of perovskite with Fe–Ti spinel and magnetite suggests simultaneous crystallization (Fig. 1f). Sometimes, perovskite and spinel grew together for a relatively long time period allowing the contact faces to be well preserved (Fig. 1h). The formation of the atoll rim on the spinel occurred at a later stage when the unprotected surface of the grain came into contact with kimberlitic liquid. Sometimes, perovskite preferentially grew by heterogeneous crystallization at pre-existing grain boundaries of olivine, spinel and ilmenite where the activation enthalpy of crystallization was lower. Although it has been reported that perovskite crystallization ceases prior to resorption and development of an atoll rim on these spinels (Mitchell 1986, 2008; Chakhmouradian and Mitchell 2000), petrographic observations of a subset of Orapa perovskite indicate perovskite continues to crystallize during and possibly after the development of atoll rims on spinel. In some places, perovskite grew as a late-crystallizing rim around spinel resorbed into titanite (Fig. 2e, f, g). Titanite is not a primary mineral of kimberlites, and its precipitation indicates some increase in $a_{(SiO_2)}$ during a late evolutionary stage, probably due to contamination of upper crustal material. Although titanite has been reported as an altered phase of perovskite (Mitchell and Chakhmouradian 1998), it only appears to be replacing the spinel phases in Orapa kimberlite (Fig. 2e, f, g). The presence of perovskite rims around resorbed spinel indicates that perovskite crystallization continued at least up to partial resorption of spinel. Inclusions of phlogopite and apatite, which are generally very late-stage minerals in kimberlite (although Malarkey et al. (2010) suggested earlier crystallization of apatite), also support the protracted crystallization history of Orapa perovskite (Fig. 2b, h). Perovskite becomes very unstable during the final evolutionary stages when there is excess CO_2 in the system (Mitchell and Chakhmouradian 1998). It then reacts with the CO_2-rich fluid to form a TiO_2 polymorph (rutile) and calcite that precipitate in the vicinity of perovskite giving a spongy appearance to the assemblage (Fig. 2d). This alteration reaction of perovskite into a TiO_2 phase is dependant on temperature and $P_{(CO_2)}$. Nesbitt et al. (1981), Chakhmouradian and Mitchell (2000) have shown that, in the range of reasonable $P_{(CO_2)}$ values, this replacement reaction is confined to relatively low temperatures (<350 °C) and pressures (<2 kbar). However, certain Orapa

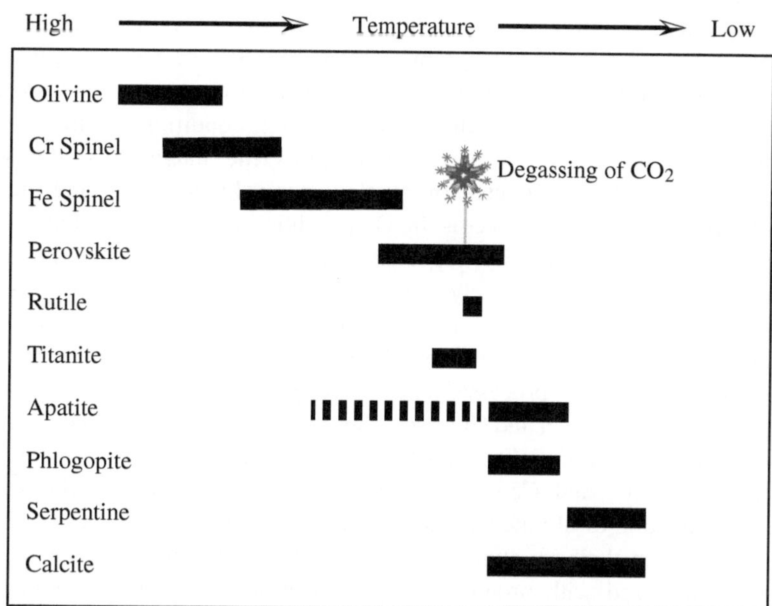

Fig. 5 Relative crystallization sequence of phenocryst and groundmass minerals (excluding megacrysts/macrocrysts suite) in Orapa kimberlite. The *dashed line* indicates earlier crystallization of apatite as suggested by Malarkey et al. (2010)

perovskite crystals appear to be very fresh and are not converted into rutile, suggesting that they probably did not react with a CO_2 fluid (Fig. 2c).

Oxygen isotope data of perovskite from the Orapa kimberlite also suggest a two-stage crystallization (Sarkar et al. 2011). Early-crystallizing perovskites have a mantle-like $\delta^{18}O$ compositions while the second phase of perovskite, which crystallized after degassing, shows very low $\delta^{18}O$ values due to preferential partitioning of ^{18}O into the gaseous phase. This two-stage crystallization explains the perovskite alteration pattern nicely. Early-crystallizing perovskites reacted with the CO_2 fluid during degassing and precipitated rutile and calcite along their grain boundaries (Fig. 2d), while those crystallizing in post-degassing phase stayed relatively fresh and unaltered as the gaseous phase might have escaped from the system. Minor and trace element concentrations of Orapa perovskite also support this two-stage crystallization model. Abundances of the minor and trace elements in perovskites of different textural types overlap. Nb and Zr are exceptions to this rule; these elements are low in perovskite grains containing inclusions of phlogopite ± apatite or those appearing as a rim around resorbed spinel (Fig. 3c). These textures are indicative of their relatively late crystallization with respect to the other perovskite paragenetic varieties. Low Nb and Zr contents of this textural variety may be attributed to their crystallization after rutile (alteration product of early perovskite), which is a major sink of Nb and Zr. However, it is unclear whether perovskite crystallization paused during degassing as the $\delta^{18}O$ values of the pre- and post-degassing phase perovskite have two distinct clusters (Sarkar et al. 2011). It is unlikely that no perovskite crystallized during rapid degassing accompanying the final stages of kimberlite ascent. Degassing might have a greater impact on oxygen isotope fractionations resulting in two distinctly different clusters, whereas other minor and trace elements generally show a continuum with the late-crystallizing perovskite phase that show limited variability in certain elements (Nb, Zr) due to their preferential partitioning into rutile. The crystallization sequence of Orapa perovskite with respect to other kimberlitic phases is shown in Fig. 5.

Oxygen Fugacity and Effect of Degassing

The range of oxygen fugacity calculated from Orapa perovskite is NNO −5.5 to NNO −0.2 (Fig. 4). This is consistent with $f(O)_2$ of other kimberlites worldwide (Canil and Bellis 2007; Trickett 2007). It is also evident that kimberlites have the largest range in $f(O)_2$ of all igneous rocks (Carmichael 1991). It has been reported that the oxygen fugacity, shown by perovskite, overlaps with the $f(O)_2$ calculated from groundmass spinel from nearby Letlhakane kimberlite (Trickett 2007). It has also been observed that spinel inclusions in olivine lie towards the more reduced end of the scale (near the iron-wustite buffer) as they grow at a greater depth in the upper mantle and earlier than perovskite. Thus, Orapa perovskite probably started crystallizing under slightly more oxidizing conditions than the IW buffer (~ −5 NNO) and continued till FMQ (fayalite-magnetite-quartz) buffer (~ −1 NNO) and beyond (Fig. 4). This interpretation is supported by the calculated $f(O)_2$ values of late-stage Fe–Ti spinel and magnetite, which grow simultaneously with most perovskite

(Trickett 2007; Roeder and Schulze 2008). Some of the highly reduced perovskite grains (lower than IW buffer) probably limit the applicability of this oxygen barometer as it will form metallic iron under such reducing conditions, which is unlikely to be achieved in kimberlite magma. Although there is a degree of overlap, perovskites from different textural parageneses in Orapa show some variability in oxygen fugacity; especially the late-crystallizing ones, which have definitely crystallized from relatively oxidized conditions (−2.3 to −0.2 NNO; Fig. 4). Variability in $f(O)_2$ values shown by Orapa perovskites can be caused by several processes such as degassing, decompression (Carmichael and Ghiorso 1986; Sparks et al. 2006), crystallization (Carmichael and Nicholls 1967) and magma mixing. Bellis and Canil (2007) have argued that early crystallization of olivine and monticellite, which favor Fe^{+2} over Fe^{+3} in their structure, would make the melt increasingly oxidized with progressive crystallization. Thus, the larger range of $f(O)_2$ calculated from Orapa perovskites points towards perovskite crystallization over a wide range of P–T conditions in an evolving kimberlite melt.

Volatile degassing during magma ascent also has a significant impact on the oxidation state of the magma. Sudden degassing of volatiles from a kimberlite magma can trigger large amounts of crystallization (Wyllie and Tuttle 1960; Dalton and Presnall 1998; Dasgupta and Hirschmann 2006; Sparks et al. 2009), which can deplete the melt in ferrous iron, thereby increasing Fe^{+3}/Fe^{+2} ratio in the melt phase (Canil and Bellis 2007). In a volatile-rich rock like kimberlite, various degassing mechanisms involving several gaseous phases would control the oxygen fugacity of the magma. One such mechanism could be the continuous outward degassing of H from the magma, which would dissociate H_2O and increase the O concentration of the magma, thus by increasing the $f(O)_2$ (Carmichael and Ghiorso 1986; Cortés et al. 2006).

$$H_2O_{(m)} \rightarrow H_{2\,(m)} + 1/2O_{2\,(m)} \quad (2)$$

Mathez (1984) has shown that C-rich volatile species would dominate the degassing process. CO_2 would be the first vapor species to separate from the melt at shallow pressure. Carbon is dissolved in the magma as CO_3^{-2}. On the other hand, solubility of CO and CO_2 is independent of the partial pressure of CO and CO_2, and the oxidation state of the system (Mathez 1984). Thus, the following two degassing reactions would involve the C species.

$$CO^{-2}_{3\,(m)} \rightarrow CO_{(v)} + O^{-2}_{(m)} + 1/2O_{2\,(m)} \quad (3)$$

$$CO^{-2}_{3(m)} \rightarrow CO_{2(v)} + O^{-2}_{(m)} \quad (4)$$

Reaction (3) produces CO vapor and also increases the O content of the magma, thus oxidizes it, whereas reaction (4) does not involve any change in the oxidation state of the magma. The initial vapor exsolving through reactions (3) and (4) from the magma would be CO rich. Consequently, this will oxidize the melt, and succeeding vapor fractions will be CO_2 rich. Gradually, equation (4) would replace (3), and the $f(O)_2$ of the magma will increase slowly. Similar changes to the oxidation state of an ascending magma containing gaseous species including H_2S, S_2, SO_2, O_2 and H_2O have been observed by Burgisser and Scaillet (2007). Crystallization of some highly oxidized minerals such as mica and garnet in mela-aillikite has been attributed to exsolution of CO_2 gases which has resulted in an increase in Fe^{+3} in the melt (Upton et al. 2006). Another possibility suggested by Ogilvie-Harris et al. (2010) involves monticellite formation from a Ca-rich kimberlite, which releases CO_2 while oxidizing the magma. Thus, it can be concluded that several degassing reactions could make the residual magma more oxidized that has been recorded by post-degassing phase perovskite in the form of higher $f(O)_2$ than the early-crystallizing ones (Fig. 4).

Conclusions

Paragenetic sequence of groundmass minerals varies from one kimberlite to another due to variable magma composition. In Orapa, the chemical composition and textures of perovskites are similar to those reported from other worldwide kimberlites suggesting their crystallization after macrocrystal olivine and Cr spinel but simultaneously with Fe–Ti spinel and magnetite. However, certain petrographic features from Orapa indicate that the perovskite from Orapa kimberlite continued to crystallize until quite late, as they have inclusions of groundmass phlogopite ± apatite and also appear as rims around the Fe–Ti-rich spinel that has reacted to form titanite around its grain boundary. This late-crystallizing perovskite group has lower Nb and Zr abundances indicating their growth after degassing that triggered crystallization of rutile, a sink of Nb and Zr. These perovskite grains also have very low $\delta^{18}O$ values, and they have crystallized from a magma significantly more oxidized than that parental to early formed perovskite.

Acknowledgements We are grateful to the anonymous reviewer whose comments have made this manuscript stronger. We thank Steve Sparks, Horst Marschall, Mike Walter and A. P. Jones for their encouragement and important discussions during this project. Stuart Kearns is thanked for his help with the analytical work. We acknowledge De Beers Consolidated Mines and Debswana for giving access to the Orapa mines and samples. This work is part of the PhD research of Chiranjeeb Sarkar, who was funded by the Overseas Research Scholarship and University of Bristol Postgraduate Scholarship along with financial support from De Beers. Chiranjeeb Sarkar is also grateful to the financial assistance provided by the 10th International Kimberlite Conference. CD Storey acknowledges NERC fellowship NE/D008891/2.

References

Allsopp HA, Bristow JW, Smith CB, Brown R, Gleadow AJW, Kramers JD, Garvie OG (1989) A summary of radiometric dating methods applicable to kimberlites and related rocks. In: Ross J, Jacques AL, Ferguson J, Green DH, O'Reilly SY, Danchin RV, Janse AJA (eds) Kimberlites, Related Rocks., v.1, Proceedings of the fourth international kimberlite conference, Geological Society of Australia Special Publication 14, Perth, Australia, pp 343–357

Armstrong JP, Wilson M, Barnett RL, Nowicki T, Kjarsgaard BA (2004) Mineralogy of primary carbonate-bearing hypabyssal kimberlite, Lac de Gras, Slave Province, Northwest Territories, Canada. Lithos 76:415–433

Batumike JM, Griffin WL, Belousova EA, Pearson NJ, O'reilly SY, Shee SR (2008) LAM-ICPMS U-Pb dating of kimberlitic perovskite: Eocene-Oligocene kimberlites from the Kundelungu Plateau, DR Congo. Earth Planet Sci Lett 267:609–619

Bellis A, Canil D (2007) Ferric iron in $CaTiO_3$ perovskite as an oxygen barometer for kimberlitic magmas I: experimental calibration. J Petrol 48:219–230

Boctor NZ, Boyd FR (1980) Oxide minerals in the Liqhobong Kimberlite, Lesotho. Am Mineral 65:631–638

Boctor NZ, Boyd FR (1981) Oxide minerals in a layered kimberlite-carbonate Sill from Benfontein, South-Africa. Contrib Miner Petrol 76:253–259

Burgisser A, Scaillet B (2007) Redox evolution of a degassing magma rising to the surface. Nature 445:194–197

Canil D, Bellis AJ (2007) Ferric iron in $CaTiO_3$ perovskite as an oxygen barometer for kimberlite magmas II: applications. J Petrol 48:231–252

Carmichael ISE (1991) The redox states of basic and silicic magmas: a reflection of their source regions? Contrib Miner Petrol 106:129–141

Carmichael ISE, Ghiorso MS (1986) Oxidation-reduction relations in basic magma: a case for homogeneous equilibria. Earth Planet Sci Lett 78:200–210

Carmichael ISE, Nicholls J (1967) Iron-titanium oxides and oxygen fugacities in volcanic rocks. J Geophys Res 72:4665–4687

Chakhmouradian AR, Mitchell RH (2000) Occurrence, alteration patterns and compositional variation of perovskite in kimberlites. Can Mineral 38:975–994

Chakhmouradian AR, Mitchell RH (2001) Three compositional varieties of perovskite from kimberlites of the Lac de Gras field (Northwest Territories, Canada). Mineral Mag 65:133–148

Clement C (1982) A comparative geological study of some major kimberlite pipes in the Northern Cape and orange free state. Unpublished Ph.D. thesis, vol 431. University of Cape Town, South Africa

Cortés JA, Wilson M, Condliffe E, Francalanci L (2006) The occurrence of forsterite and highly oxidizing conditions in basaltic lavas from Stromboli volcano, Italy. J Petrol 47:1345–1373

Dalton JA, Presnall DC (1998) The continuum of primary carbonatitic, kimberlitic melt compositions in equilibrium with lherzolite: data from the system CaO, MgO, $Al2O_3$, SiO_2, CO_2 at 6 GPa. J Petrol 39:1953–1964

Dasgupta R, Hirschmann MM (2006) Melting in the Earth's deep upper mantle caused by carbon dioxide. Nature 440:659–662

Davis GL (1977) The ages and uranium contents of zircons from kimberlites and associated rocks. In: Second international kimberlite conference, Santa Fe, New Mexico (extended abstracts)

Field M, Gibson JG, Wilkes TA, Gababotse J, Khutjwe P (1997) The geology of the Orapa A/K1 kimberlite Botswana: further insight into the emplacement of kimberlite pipes. Geol Geofiz 38:25–41

Field M, Stiefenhofer J, Robey J, Kurszlaukis S (2008) Kimberlite-hosted diamond deposits of southern Africa: a review. Ore Geology Reviews 34:33–75

Gernon TM, Field M, Sparks RSJ (2009) Depositional processes in a kimberlite crater: the upper cretaceous Orapa South Pipe (Botswana). Sedimentology 56:623–643

Gernon TM, Sparks RSJ, Field M (2008) Degassing structures in volcaniclastic kimberlite: examples from southern African kimberlite pipes. J Volcanol Geoth Res 174:186–194

Heaman LM (1989) The nature of the subcontinental mantle from Sr-Nd-Pb isotopic studies on kimberlitic perovskite. Earth Planet Sci Lett 92:323–334

Heaman LM, Kjarsgaard BA, Creaser RA (2003) The timing of kimberlite magmatism in North America: implications for global kimberlite genesis and diamond exploration. Lithos 71:153–184

Heaman LM, Kjarsgaard BA, Creaser RA (2004) The temporal evolution of North American kimberlites. Lithos 76:377–397

Jones AP, Wyllie PJ (1984) Minor elements in perovskite from kimberlites and distribution of the rare-earth elements: an electron-probe study. Earth Planet Sci Lett 69:128–140

Malarkey J, Pearson DG, Kjarsgaard BA, Davidson JP, Nowell GM, Ottley CJ, Stammer J (2010) From source to crust: tracing magmatic evolution in a kimberlite and a melilitite using microsample geochemistry. Earth Planet Sci Lett 299:80–90

Mathez E (1984) Influence of degassing on oxidation states of basaltic magmas. Nature 310:371–375

Mitchell RH (1972) Composition of perovskite in kimberlite. Am Mineral 57:1748–1753

Mitchell RH (1986) Kimberlites: mineralogy, geochemistry and petrology. Plenum Press, New York

Mitchell RH (2002) Perovskites: modern and ancient. Almaz Press Ontario, Thunder Bay

Mitchell RH (2008) Petrology of hypabyssal kimberlites: relevance to primary magma compositions. J Volcanol Geoth Res 174:1–8

Mitchell RH, Chakhmouradian AR (1998) Instability of perovskite in a CO_2-rich environment: examples from carbonatite and kimberlite. Can Mineral 36:939–951

Nesbitt HW, Bancroft GM, Fyfe WS, Karkhanis SN, Nishijima A (1981) Thermodynamic stability and kinetics of perovskite dissolution. Nature 289:358–362

Ogilvie-Harris R, Field M, Sparks R, Walter M (2010) Perovskite from the Dutoitspan kimberlite, Kimberley, South Africa: implications for magmatic processes. Mineral Mag 73:915

Paton C, Hergt JM, Phillips D, Woodhead JD, Shee SR (2007) New insights into the genesis of Indian kimberlites from the Dharwar Craton via in situ Sr isotope analysis of groundmass perovskite. Geology 35:1011–1014

Roeder PL, Schulze DJ (2008) Crystallization of groundmass spinel in kimberlite. J Petrol 49:1473–1495

Sarkar C, Storey CD, Hawkesworth CJ, Sparks RSJ (2011) Degassing in kimberlite: oxygen isotope ratios in perovskites from explosive and hypabyssal kimberlites. Earth Planet Sci Lett 312:291–299

Sparks R, Brooker R, Field M, Kavanagh J, Schumacher J, Walter M, White J (2009) The nature of erupting kimberlite melts. Lithos 112:429–438

Sparks RSJ, Baker L, Brown RJ, Field M, Schumacher J, Stripp G, Walters A (2006) Dynamical constraints on kimberlite volcanism. J Volcanol Geoth Res 155:18–48

Trickett SK (2007) Mapping lithofacies within the D/K1 kimberlite pipe at Letlhakane, Botswana: an assessment of petrographic, geochemical and mineralogical indicators. PhD Thesis, University College London

Upton B, Craven J, Kirstein L (2006) Crystallisation of mela-aillikites of the Narsaq region, Gardar alkaline province, south Greenland

and relationships to other aillikitic-carbonatitic associations in the province. Lithos 92:300–319

Wu FY, Yang YH, Mitchell RH, Li QL, Yang JH, Zhang YB (2010) In situ U-Pb age determination and Nd isotopic analysis of perovskites from kimberlites in southern Africa and Somerset Island, Canada. Lithos 115:205–222

Wyllie P, Tuttle O (1960) The system CaO, CO_2, H_2O and the origin of carbonatites. J Petrol 1:1–46

Yang YH, Wu FY, Wilde SA, Liu XM, Zhang YB, Xie LW, Yang JH (2009) In situ perovskite Sr-Nd isotopic constraints on the petrogenesis of the Ordovician Mengyin kimberlites in the North China Craton. Chem Geol 264:24–42

The Age and Localization of Kimberlite Magmatism in the Yakutian Kimberlite Province: Constraints from Isotope Geochronology—An Overview

A. P. Smelov and A. I. Zaitsev

Abstract

Over the last 50 years of geologic prospecting on the territory of Yakutia, some 1,070 kimberlite bodies have been discovered, which are united into the Yakutian kimberlite province (YKP). Hundreds of scientific papers on the mineral content of diamondiferous kimberlites, their tectonic setting, composition of the mantle substrate, physical properties of diamonds, and chemical composition of their mineral inclusions have been published over this period. However, little attention was given to the analysis of isotope data available for kimberlites. The authors made a comparative analysis of over 900 isotope age determinations of kimberlites from 24 fields. During this period, the age of 350 kimberlite bodies was determined, and it makes about 33 % of all known kimberlite bodies in YKP. Statistical analysis of maximum correlations of the age values, obtained by different isotopic methods, allowed the recognition of 7 epochs of kimberlite magmatism in YKP. The article examines the relationship between the time of kimberlite magmatism manifestation (including diamond-bearing kimberlites) and age and composition of Precambrian basement terranes, as well as its relationship with formation of the Late Precambrian and Phanerozoic tectonic structures of the Siberian platform.

Keywords

Kimberlite • Isotopic data • Diamond • Yakutian kimberlite province • Siberian platform

Introduction

The Yakutian kimberlite province (YKP) occupies the northeastern part of the Siberian platform within the North Asian craton (Fig. 1). The YKP extends northwards for about 1,500 km from the Vilyui R. to the Laptev Sea and easterly for 1,000 km from the Olenek R. head in Krasnoyarsk Territory to the Lena R. in Yakutia (Russia) (Khar'kiv et al. 1998). The province is bounded on the north and east by the Lena-Anabar and Priverkhoyansk basins and on the southeast by the Vilyui syneclise and the Angara-Vilyui basin. The western boundary is less well defined and runs along the eastern slope of the Tunguska syneclise. Over the last 50 years, some 1,070 kimberlite bodies have been discovered within the YKP territory, which are united into 24 kimberlite fields (Brakhfogel' 1984; Ilupin et al. 1990; Khar'kiv et al. 1998). Hundreds of scientific papers on the mineral content of diamondiferous kimberlites, their tectonic setting, composition of the mantle substrate, physical properties of diamonds, and composition of their mineral inclusions have been

Electronic supplementary material The online version of this article (doi:10.1007/978-81-322-1170-9_14) contains supplementary material, which is available to authorized users.

A. P. Smelov (✉) · A. I. Zaitsev
Diamond and Precious Metal Geology Institute, Siberian Branch, Russian Academy of Sciences, 39 Lenin Avenue, Yakutsk, 677980, Russian Federation
e-mail: a.p.smelov@diamond.ysn.ru

D. G. Pearson et al. (eds.), *Proceedings of 10th International Kimberlite Conference*, Volume 1, Special Issue of the Journal of the Geological Society of India, DOI: 10.1007/978-81-322-1170-9_14,

published. However, little attention was given to the analysis of isotope data available for kimberlites, though they yield important information not only about the timing of kimberlite magmatism but about the isotopic composition of mantle protoliths and the processes responsible for the diamond tenor and occurrences of kimberlites too.

Geology

The YKP is divided into two subprovinces: the southern Vilyui subprovince in the center of the Siberian platform and the northern Anabar-Olenek subprovince in the northeastern margin of the platform (Fig. 1). Their kimberlites differ in a number of features. Kimberlites from the Vilyui subprovince are mainly of explosive type. Northward, there is an increase in the number of intrusive bodies unevenly distributed amidst the explosive ones. The overall amount of kimberlite bodies in the Anabar-Olenek subprovince is 4.5 times higher than in the Vilyui one, but their diamond tenor is 90 % lower. The diamond morphology and size also differ in the two subprovinces. For both subprovinces, excluding Anabar shield, it is typical that Upper and Middle Paleozoic sediments are destroyed by erosion almost completely. Sediments of Carboniferous and Permian age preserved only in the areas of trappean magmatism development (Khar'kiv et al. 1998). The main fields of diamondiferous and non-diamondiferous kimberlites are located within the limits of several kimberlite-controlling zones, two of which—Vilyusk-Markha and Daldyn-Olenek have northeast strike, and the Anabar zone—north–northwest strike. It is proposed that the existence of the Arga-Sala zone of sublatitudinal strike controls the locations of kimberlite fields of the western slope of Anabar shield (Khar'kiv et al. 1998; Brakhfogel' 1984; Ilupin et al. 1990). Such location of kimberlite-controlling zones was explained

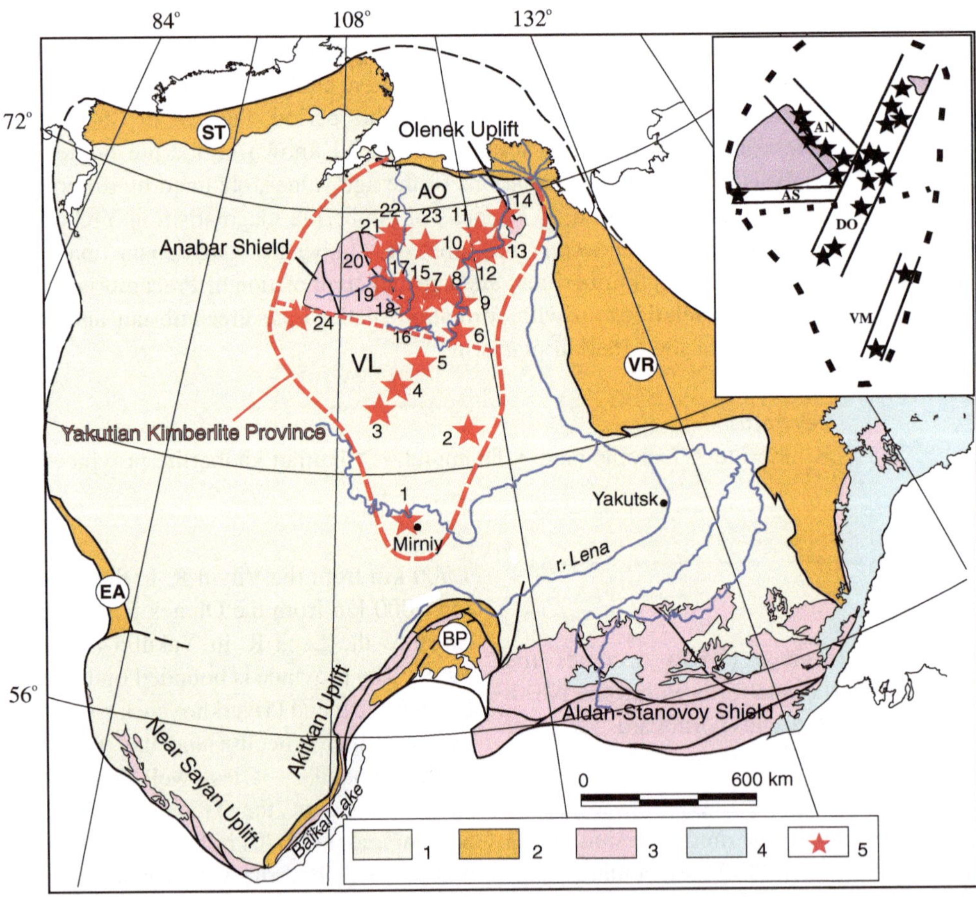

Fig. 1 Tectonic sketch map of the North Asian craton (Smelov and Timofeev 2007) showing position of the Yakutian kimberlite province: *1*—Siberian platform; *2*—subsided craton margin, fold-and-thrust belts (*ST* South Taimyr, *EA* East Angara, *BP* Baikal-Patom, *VR* Verkhoyansk); *3*—Precambrian basement (shields and uplifts); *4*—Mesozoic volcanogenic belt; *5*—kimberlite fields: *1*—Malo-Botuobiya, *2*—Nakyn, *3*—Alakit-Markha, *4*—Daldyn, *5*—Upper Muna, *6*—Chomurdakh, *7*—West Ukukit, *8*—East Ukukit, *9*—Ogoner-Yuryakh, *10*—Merchemde, *11*—Kuoyka, *12*—Molodo, *13*—Tolupka, *14*—Khorbosuonka, *15*—Luchakan, *16*—Kuranakh, *17*—Dyuken, *18*—Biriginde, *19*—Ary-Mastakh, *20*—Starorechenskoye, *21*—Orto-Yarga, *22*—Ebelyakh, *23*—Tomtor, *24*—Kharamay. *VL*—Vilyui and *AO*—Anabar-Olenek subprovinces. Inset shows the kimberlite-controlling zones: *VM*—Vilyui-Markha, *DO*—Daldyn-Olenek, *AN*—Anabar zone, *AG*—Arga-Sala

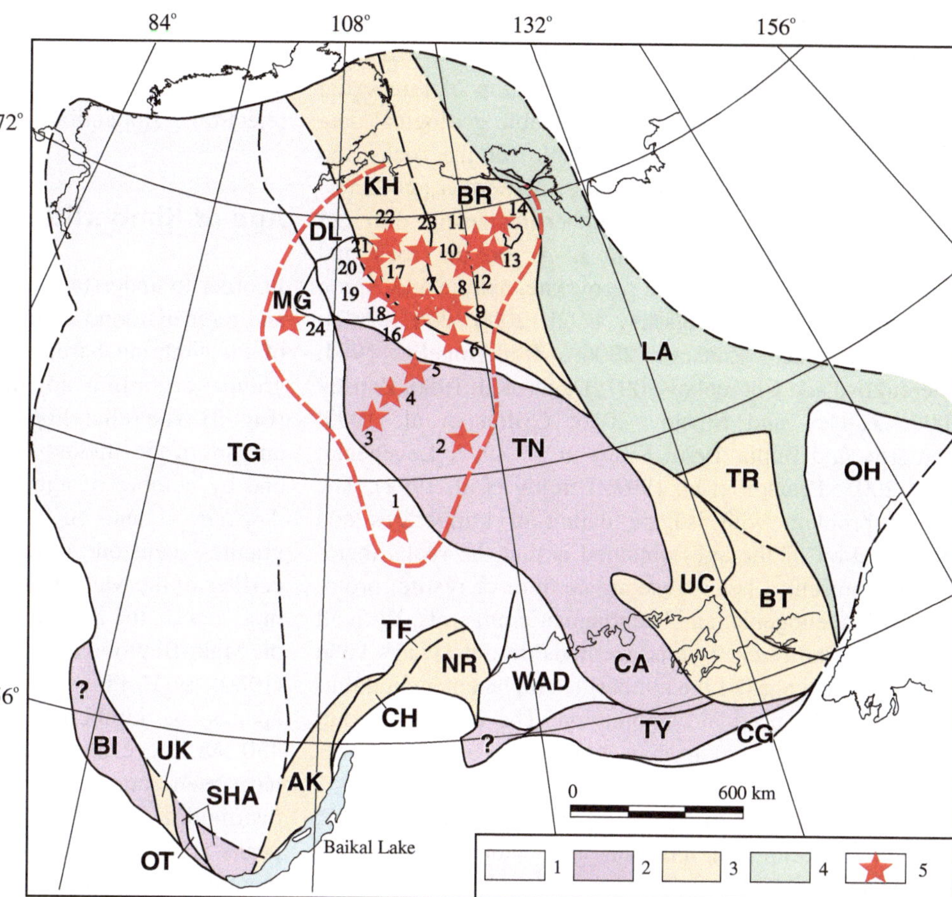

Fig. 2 Geologic sketch map of the North Asian craton basement (Smelov and Timofeev 2007) and position of kimberlite fields: *1*—Archean, *2*—Archean and Paleoproterozoic, undifferentiated, *3*—Paleoproterozoic, *4*—Mesoproterozoic; *5*—kimberlite fields (see Fig. 1 for numeration). Precambrian terranes: West Aldan (WA), Central Aldan (CA), Uchur (UC), Batomga (BT), Chogar (CG), Tynda (TY), Daldyn (DL), Khapchan (KH), Magan (MG), Akitkan (AK), Chuya (CH), Nechera (NR), Tondo (TF), Sharyzhalgay (SHA), Onot (OT), Urik-Liya (UL), Biryusa (BI), Tunguska (TG), Tyung (TN), Birekta (BR), Tyryn (TR), Lena-Aldan (LA). Names of kimberlite fields are given in Fig. 1

by tectonic and geodynamic concepts of the regulation of kimberlite localization, existed in different years (Koval'skiy 1963; Mokshantsev et al. 1974; Nikishov 1984; Milashev 1974; Williams and Williams 1977; Crockett et al. 1968; England and Houseman 1984; Crockett et al. 1968). However, these hypotheses do not explain regularities of kimberlite magmatism manifestation on the Siberian platform (Nikulin et al. 2001; Safronov et al. 2001).

Clifford (1966) showed that areas of kimberlite magmatism are located within the limits of ancient cratons. Cratons consist of nuclei and made up of rocks of consolidation age about 1.5–2.0 Ga and older, which are surrounded by mobile belts of 1.0–0.8 Ga. Kimberlites located in young mobile belts are not diamondiferous, and kimberlites in ancient nuclei are diamondiferous. According to this rule, location of kimberlites can be regulated by any folded or fault structures within the craton, but the diamond content of the rocks is determined by their location relative to structures of the ancient platform basement.

The Siberian platform basement is a nucleus of the North Asian craton, which was formed during considerable period of time, from 3.5 to 1.9 Ga (Rosen et al. 1994; Smelov and Timofeev 2007). Granite-greenstone, tonalite-trondhjemite gneiss, and granulite-gneiss terranes, superterranes, composite terranes (further "terrane") with Archean, Archean-Paleoproterozoic and Paleoproterozoic crust make up Siberian platform basement (Fig. 2). Granite-greenstone and tonalite-trondhjemite gneiss terranes are interpreted as stable areas of the earth's crust, and granulite-gneiss terranes as fragments of the root parts of collision belts. The main tectonic structures of the North Asian craton basement are granite-greenstone and tonalite-trondhjemite Archean (West Aldan, Tunguska, Sharyzhalgay, and probably Magan terranes) and Paleoproterozoic (Brekta, Tyryn, and Batomga terranes) terranes of consolidation age over 2.6–2.5 and 2.4–2.1 Ga, respectively. They are separated by terranes of granulite-gneiss belts of different age and composition (Central Aldan, Tyung and Daldyn, Khapchan, and Uchur terranes). The age of granulite metamorphism is dated at 2.1–1.9 Ga. In the east of the basement of Verkhoyansk, fold-and-thrust belt records Lena-Aldan orogenic belt of the age 1.4–1.0 Ga (Fig. 2).

Methods

The number of kimberlite magmatism epochs within the YKP has been discussed repeatedly. Different authors define them differently. Brakhfogel' (1984) was the first to review the existing data on the age of kimberlite magmatism within the YKP. He found that according to different researchers, the number of epochs varied from 1 to 5 (from Late Ordovician through Cretaceous). At that time, it was difficult to determine more precisely the timing of kimberlite magmatism because of the lack of reliable geological data and insufficient isotope dates available for the rocks and minerals. Over the last two decades, numerous age determinations were obtained for kimberlites using different radiologic methods: Rb–Sr and K–Ar on the rock and minerals, U–Pb on zircon and perovskite, apatite and zircon fission-track analysis (Agashev et al. 2000, 2001, 2004; Altukhova and Zaitsev 2006; Brakhfogel' 1984; Cherenkova and Cherenkov 2007; Davis et al. 1980; Zaitsev 2007; Zaitsev and Smelov 2010; Griffin et al. 1999; Komarov and Ilupin 1990; Kinny et al. 1997; Levchenko et al. 2005; Pearson et al. 1997; Fefelov et al. 1992). The central problem with isotope dating of kimberlites and interpretation of the ages obtained is that the rocks themselves represent a heterogeneous mixture of restite, protomagmatic, xenogenic, and late magmatic minerals. With all their merits, different isotope methods do not always yield real geological ages of the kimberlite emplacement and thus they should be used in combination. The results of comparative analysis of more than 900 isotope age determinations of kimberlites from 24 fields are shown on Fig. 3 (Zaitsev and Smelov 2010). During the entire study of YKP, the age of rocks and minerals was determined for 350 kimberlite bodies (Table 1) and it makes about 33 % of all known kimberlite bodies in YKP. Isotopic data for kimberlites from 24 kimberlite fields obtained by different methods are submitted in the electronic Appendix-1 to this paper. Maximum correlations of the age values, obtained by different methods for one kimberlite body or field, allowed us to recognize possible epochs of kimberlite magmatism (electronic Appendix-2).

Age of Kimberlite Magmatism

In order to understand whether age, location of kimberlites, and their diamond content depend on composition and age of Precambrian basement, we divided all kimberlites into groups according to the age of Precambrian terranes (Fig. 2). The reliability of our conclusions can be judged by amount of the studied kimberlite bodies in individual fields and by number of age determinations (Table 1, electronic Appendix-1 and 2). I. Within the limits of the Tunguska granite-greenstone terrane of Archean age, there are kimberlites of the Malo-Botuobiya and Kharamay field: (1) The first data on the age of minable diamondiferous kimberlites of Malo-Botuobiya field were obtained by Davis et al. (1980) by U–Pb method on zircons. They revealed three epochs of kimberlite magmatism: 352–367, 403, and 450 Ma. Since then, isotopic data using different methods have been obtained for several pipes in the field. Statistical distribution of isotopic data of kimberlites shows two major peaks at 340–400 and 420–460 Ma, reflecting the time of manifestation of kimberlite magmatism. (2) In the Kharamay field, the age was determined only for low diamondiferous kimberlites of the Evenkaya pipe by different isotopic methods. According to the Rb–Sr method, isotope

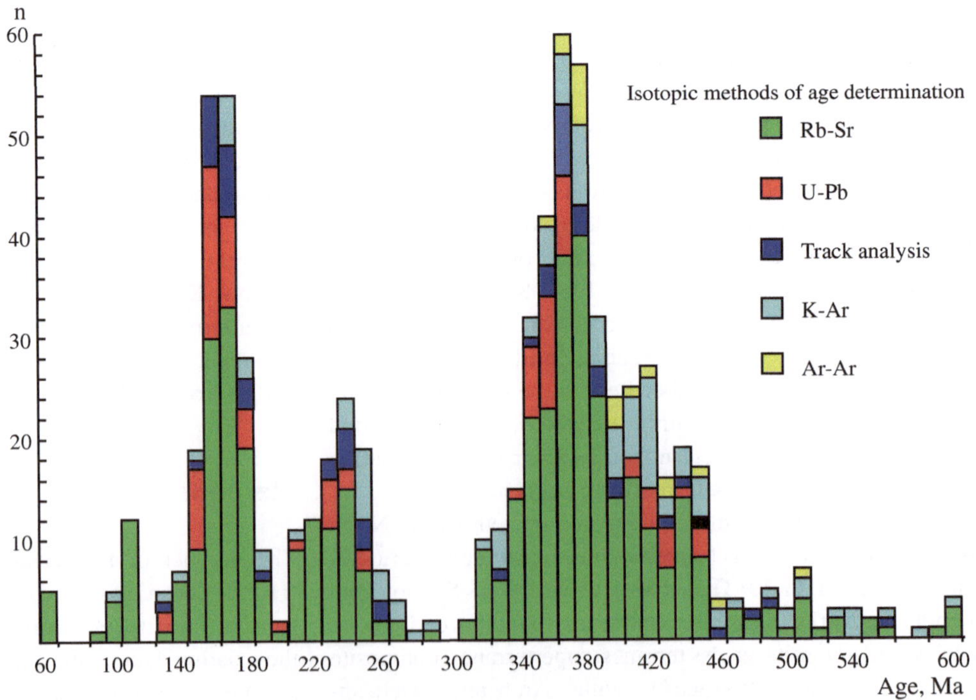

Fig. 3 Distribution of kimberlite isotope ages determined by different methods in the whole of the Yakutian province (electronic Appendix-1 and 2)

Table 1 Location of kimberlite fields (KF) relative to different terranes of Early Precambrian Siberian platform basement

Composition of the terranes	Age of the terranes		
	Archean	Archean and Paleoproterozoic, undifferentiated	Paleoproterozoic
Granite-greenstone terrane	Malo-Botuobiya KF (1)[c]—28/6[a],		Merchemde KF (10)—40/20,
	Kharamay KF (24)[b]—28/18		Molodo KF (12)[b]—23/13,
	(Tunguska terrane—TG)		Toluopka KF(13)—16/5,
			Khorbosuonka KF (14)—7/2,
			Kuoyka KF (11)[b]—73/31,
			(Birekta terrane-BR)
Granulite-orthogneiss terrane	Chomurdakh KF (6)[b]—60/22,	Nakyn KF (2)[c]—41/2,	
	Biriginde KF (18)—49/21,	Alakit-Markha KF (3)[c]—59/26,	
	Kuranakh KF (16)[b]—20/8	Daldyn KF (4)[c]—110/17,	
	(Daldyn terrane—DL)	Upper Muna KF (5)[c]—24/10	
		(Tyung terrane-TN)	
Granulite–paragneiss terrane			West Ukukit KF (7)[b]—37/11,
			East Ukukit KF (8)—60/19,
			Ogoner-Yuryakh KF(9)[b]—4/2,
			Luchakan KF (15)[b]—14/9,
			Dyuken KF (17)—32/25,
			Starorechenskoye KF (20)[b]—38/17,
			Orto-Yarga KF (21)—23/10,
			Ary-Mastakh KF (19)[b]—50/25,
			Ebelyakh KF (22)[b]—7/5,
			Tomtor KF (23)[b]—55/25
			(Khapchan terrane—KH)

Note [a] Number of age datings/number of studied kimberlite bodies, [b] mineralogically diamondiferous kimberlite field, [c] *commercially* diamondiferous kimberlite field (after Argunov 2005). Names of kimberlite fields are given in Fig. 1

ages of kimberlites vary from 156 to 245 Ma, and according to the fission-track method, the age of kimberlites correlates with the time interval 224–235 Ma. As a rule, age data on kimberlites all over the field obtained by the fission-track method yield more ancient values (224–290 Ma), than Rb–Sr ages (136–245 Ma). Statistical distribution of isotope data of kimberlites from Kharamay field forms two peaks 140–160 and 220–240 Ma. In general, within the limits of Daldyn terrane, 5 epochs of kimberlite magmatism can be recognized: 460–420, 400–340, 240–220, and 160–140 Ma.

II. Three kimberlite fields (Chomurdakh, Kuranakh, and Biriginde) are located within the limits of Archean Daldyn granulite-orthogneiss terrane (Table 1) and are characterized by low diamond content. (1) The distribution of isotope ages of kimberlites of Chomurdakh filed, performed by different methods, show that two maximums, 360–400 Ma and 420–460 Ma, are revealed among them. (2) Kimberlites of the Kuranakh field were mainly formed in Triassic, only the Malo-Kuonamskaya pipe is an exception. This pipe is represented by two stages of kimberlite emplacement. The early stage has northeast strike and an age of 223–269 Ma, obtained by different isotopic methods. The later-stage kimberlite bodies intersect earlier kimberlites and have a sublatitudinal strike. The age of these kimberlites is 170–188 Ma, determined by the Rb–Sr method. Undoubtedly, these data should be verified by other isotopic methods. (3) Three peaks of age values (80–100, 160–180, and 220–240 Ma) constitute data on kimberlites of Biriginde field. Thus, within the limits of Daldyn granulite-orthogneiss terrane, rejuvenation of kimberlite magmatism begins to appear from the southeast (Chomurdakh field) to the northwest (Biriginde field) from Middle Paleozoic to Cretaceous age. Moreover, if in Chomurdakh and Kuranakh fields two epochs of kimberlite magmatism each are supposed, then within the limits of the Biriginde field, this magmatism occurred in three epochs. In general, within the limits of Daldyn terrane, 5 epochs of kimberlite magmatism can be recognized: 460–420, 400–360, 240–220, 180–160, and 100–80 Ma.

III. Kimberlites of Nakyn, Alakit-Markha, Daldyn, and Upper Muna fields are located within the limits of Tyung granulite-orthogneiss terrane of Archean-Paleoproterozoic age. Almost all fields contain kimberlites with commercial minable diamond content. (1) According to isotope data, kimberlites of the Nakyn field clearly yield presence of two epochs of magmatism (460–440 and 360–400 Ma). According to paleomagnetic studies, remnant magnetization of kimberlites from Botuobiya pipe determines their age as 420 Ma. It seems clear that kimberlites of the Nakyn field were formed in the Late Ordovician—Early Silurian time (450–440 Ma). (2) Two maxima in the ages of kimberlites from the Alakit-Markha field can be recognized (380–340 Ma and 440–420 Ma). Ages of 431 Ma (Rb–Sr method) and 358 Ma (U–Pb method on perovskite) have been determined for the different stages of kimberlites in the Druzhba pipe. It is possible that manifestations of kimberlites of the Permian–Triassic epoch occurred in the field; this epoch is recorded in kimberlites of the Radiovolnovaya pipe by U–Pb (229 Ma) and K–Ar (272–242 Ma) methods. (3) Considerable variation of isotope ages is typical for kimberlites of the Daldyn field. The most distinct maximum of their values falls in the time span 340–360 Ma whereas minor peaks are recorded in the interval 420–400 and 600–500 Ma. Probably, just two of the first maxima (360–340 and 420–400 Ma) are likely to correlate with kimberlite activity. More ancient ages probably result from ancient mantle protolith material within the bulk kimberlite. (4) Two epochs of kimberlite magmatism are recognized in the Upper Muna field. One of them correlates with time span 360–340 Ma and the second one with 460–420 Ma, thus isochronous ages of kimberlites of the Tyun terrane field. The most frequent period of eruption was in the time span 380–340 and 460–420 Ma. It is possible that a Permian–Triassic epoch of kimberlite magmatism occurred here.

IV. Ten kimberlite fields are located within the limits of Khapchan granulite-paragneiss terrane with Paleoproterozoic age of the crust. In terms of their age, they are divided into two groups: Paleozoic and Mesozoic. Kimberlites of West Ukukit, East Ukukit, Ogoner-Yuryakh, and Tomtor fields belong to the Paleozoic group. Usually, it is accepted that kimberlites of Luchakan, Dyuken, Starorechenskoye, Orto-Yarga, and Ebelyakh fields belong to the Mesozoic group. However, recently obtained isotope ages of kimberlites of these fields define a more complicated geochronologic history of kimberlite magmatism. (1) In the West Ukukit field, the main peak of kimberlite formation correlates with interval the 380–360 Ma, with a minor peak recorded in the interval 430–410 Ma. (2) In the Ogoner-Yuryakh field, there are a few isotope ages showing an interval of kimberlite formation from 409 to 453 Ma with a probable peak at 430–410 Ma. (3) The most complex type of distribution of isotope data is observed for kimberlites of West Ukukit and Tomtor fields. A wide range of available isotope ages (from Triassic to Precambrian) suggests a long and multi-stage activity of mantle magmatism. Regarding kimberlites of West Ukukit field, age distribution maxima occur at 400–380, 440–420, and 520–460 Ma. Also, Permian–Triassic ages (257–235 Ma) are determined for kimberlites of the Khayargastakh pipe by the Rb–Sr method. The Tomtor field is characterized by a wide range magmatic composition, from carbonatites and kimberlites to picrites. Statistical analysis of the distribution of isotope ages for the Tomtor field magmatic rocks clearly yields several time intervals of eruption. The earliest formation of kimberlite-like rocks occurs in time the span 620–480 Ma, with a poorly expressed peak of activity at 540–520 Ma. There is a prominent mode of ages in interval 420–360 Ma, typical for Middle Paleozoic kimberlites of the YKP. A minor age mode is recorded in the interval 320–300 Ma. Some ages that occur within the Permian–Triassic period (260–200 Ma) are typical for picrite bodies. Thus, on the basis of all isotope ages of the Paleozoic group of Khapchan terrane kimberlites, two epochs of kimberlite magmatism are most clearly recorded: 360–400 and 420–460 Ma.

The Mesozoic kimberlite fields contain bodies of highly variable age, without a major mode in age. At the same time, radiometric ages show that kimberlites of different age can exist in different fields. (1) Isotope dating of the Luchakan field kimberlites is grouped into two age intervals 253–217 and 177–158 Ma and is characteristic of Permian–Triassic and Jurassic epochs. (2) Kimberlites of the Dyuken field are characterized only by Rb–Sr ages, which show three stages of formation. The main age mode correlates with the interval 180–160 Ma, with a minor peak at—240–200 Ma. A clear peak at 340–320 Ma indicates formation of some kimberlite bodies in the Middle Paleozoic. (3) Kimberlites and related rocks of the Ary-Mastakh fields were intruded during five epochs (Altukhova and Zaitsev 2006; Zaitsev 2007). The earliest manifestation of kimberlite magmatism, probably, occurred in Middle Paleozoic epoch. This is defined by a few Rb–Sr and K–Ar determinations of whole rock and phlogopites. In particular, an age of 342 Ma is determined for phlogopite phenocrysts from the Alfa pipe by the K–Ar method, whereas the phlogopite microlites from the groundmass give an age of 242 Ma. Kimberlites of the Perm pipe yield an Rb–Sr age on the groundmass of 99 Ma, contrasting with an age of 370 Ma for the phlogopite phenocrysts. Additionally, detailed isotopic–geochronologic studies are necessary for reliable determination of the presence of Middle Paleozoic magmatism here. Kimberlite magmatism of Mid—Late Triassic (235–205 Ma) and Jurassic (168–149 Ma) ages are most commonly present. The Cretaceous epoch is represented by kimberlites of two pipes: Perm (99 and 104 Ma) and Polyarnaya (100–109 Ma). Sufficient data are available to propose the inception of alkali-ultrabasic magmatism in the Paleogene. This is recorded by Rb–Sr dating of alnoites from Beta-2 (51–55 Ma), the Bargydamalakh-1 (53 Ma) pipes and from autoliths of kimberlite breccia derived from the Bargydamalakh-2 pipe (52–55 Ma). (4) Alkaline-ultrabasic rocks of Starorechenskoye field were formed in the Triassic (246–209 Ma) and Jurassic (170–151 Ma). (5) According to zircon U–Pb isotope dating, alkali-ultrabasic rocks of the Ogoner-Yuryakh field were formed mainly in the Jurassic epoch (202–147 Ma). A single Rb–Sr determination for the An-20 kimberlite bodies indicates their age as Middle Paleozoic (389 Ma) but needs to be verified by other isotopic methods. (6) The few age determinations of alkaline-ultrabasic rocks of the Ebelyakh field define it as polychronal one, formed in three epochs: 374–342, 235–216 and 159 Ma. Thus, 7 epochs of kimberlite formation can be conditionally recognized within the limits of Khapchan terrane: 460–440, 430–410, 380–350, 240–210, 170–140, 110–95, and 60–50 Ma. In this terrane area, the age of kimberlite eruption rejuvenates from the south to the north, and the proportion of anomalously ancient ages decreases.

V. Two age groups of kimberlite fields are known in the Berekta granite-greenstone terrane of Paleoproterozoic age. Paleozoic kimberlites are typical for the Merchemde and Toluopka fields, whereas Mesozoic kimberlites predominate in the Molodo, Kuoyka, and Khorbosuonka fields. Indeed, analysis of the type of distribution of isotope ages for kimberlites from the Paleozoic fields yields a Middle Paleozoic mode of isotope ages. This mode can be split into a 340–360 Ma period for kimberlites of the Toluopka field, and a 400–360 Ma period for the kimberlites of Merhcemden. At the same time, U–Pb determinations on zircons give ages of 428–419 Ma, and some Rb–Sr age determinations fall within the interval 438–415 Ma for other kimberlites in this field suggesting that eruption kimberlite magmatism in the Silurian epoch occurred here also.

Kimberlite activity of the Jurassic epoch is most clearly manifested for in the Mesozoic group of kimberlite fields. However, the real range of kimberlite ages is rather wide, meaning that a polychronal history of kimberlite magmatism is possible. So if, within the limits of the Khorbosuonka field, kimberlites were formed in the time span 177–142 Ma, then separate ages are recorded for kimberlites of other fields, indicating possible kimberlite eruption in other epochs. In particular, there are ages of 246 Ma for kimberlites of the Molodo field, obtained by the K–Ar method for Molodo-2 pipe rocks and of 332 Ma for An-60/90 (Rb–Sr method). The biggest variation of ages is recorded for kimberlites of Kuoyka field. Pridorozhnaya pipe kimberlites yield an Rb–Sr age of 338 Ma. Middle Paleozoic K–Ar ages of kimberlites were obtained also for the Obnazhennaya, Ruslovaya, and Snezhnaya-1 pipes. Ages obtained by the Rb–Sr method for kimberlites of the Obnazhennaya (161–135 Ma) and Ruslovaya (179–159 Ma) pipes establish their formation in Jurassic time. Generally, K–Ar ages for these and other pipes of the field were probably affected by xenogenic mantle material, which to some extent, preserved its isotope systematics, and inherited from previous mantle processes.

According to modern isotope dating, 7 epochs of kimberlite formation can confidently be recognized within the YKP: 450–430, 420–400, 380–350, 250–230, 170–150, 110–100, and 60–50 Ma (Fig. 4). Older isotope dates of kimberlites are also available, which together with the U–Pb zircon ages obtained from diamondiferous intermediate sedimentary collectors are indicative of the Vendian and Early Ordovician epochs of kimberlite magmatism in some regions of the province.

Localization of Kimberlites

Figure 4 shows that in different parts of the Siberian platform, irrespective of the age of the basement, the number of kimberlite formation epochs (including diamondiferous kimberlites) varies from 2 to 7. The least number of epochs (Ordovician–Silurian and Devonian–Carboniferous) is observed in the Vilyui subprovince, while in the Anabar-Olenek subprovince from 2 to 7 stages are established in different fields. Distribution areas of kimberlite magmatism (including diamondiferous kimberlites) do not correlate with the age and composition of the Early Precambrian basement (Figs. 2 and 4). However, it should be noted that a significant part of the kimberlite fields, regardless of their age, is located close to collisional sutures of 2.1–1.9 Ga, separating Early Precambrian terranes of different age and composition (Fig. 2).

Figure 5 shows the distribution areas of kimberlite magmatism and diamond tenor of kimberlites in different periods of the Siberian platform development. Analysis of the kimberlite magmatism distribution in different time spans shows that Paleozoic epochs of kimberlite magmatism are mainly characteristic of the southern and northern parts of the YKP (Fig. 5a–b). It should be noted that the eastern boundary of the kimberlite magmatism area coincides with or is close to the western boundary of the Devonian rift-related structures and dike swarms developed in the eastern Siberian platform. It is established that Paleozoic high-grade diamondiferous kimberlites occur to the south of the Meso- and Neoproterozoic paleorifts boundary (Fig. 5a). It is likely that endogenic activities related to the breakup of the Rodinia supercontinent in the Precambrian led to degradation of the diamondiferous mantle keel.

Mesozoic kimberlite magmatism manifested itself in the northern Siberian platform northeastward of the areas of intense trap magmatism (Fig. 5c–e). Different configuration,

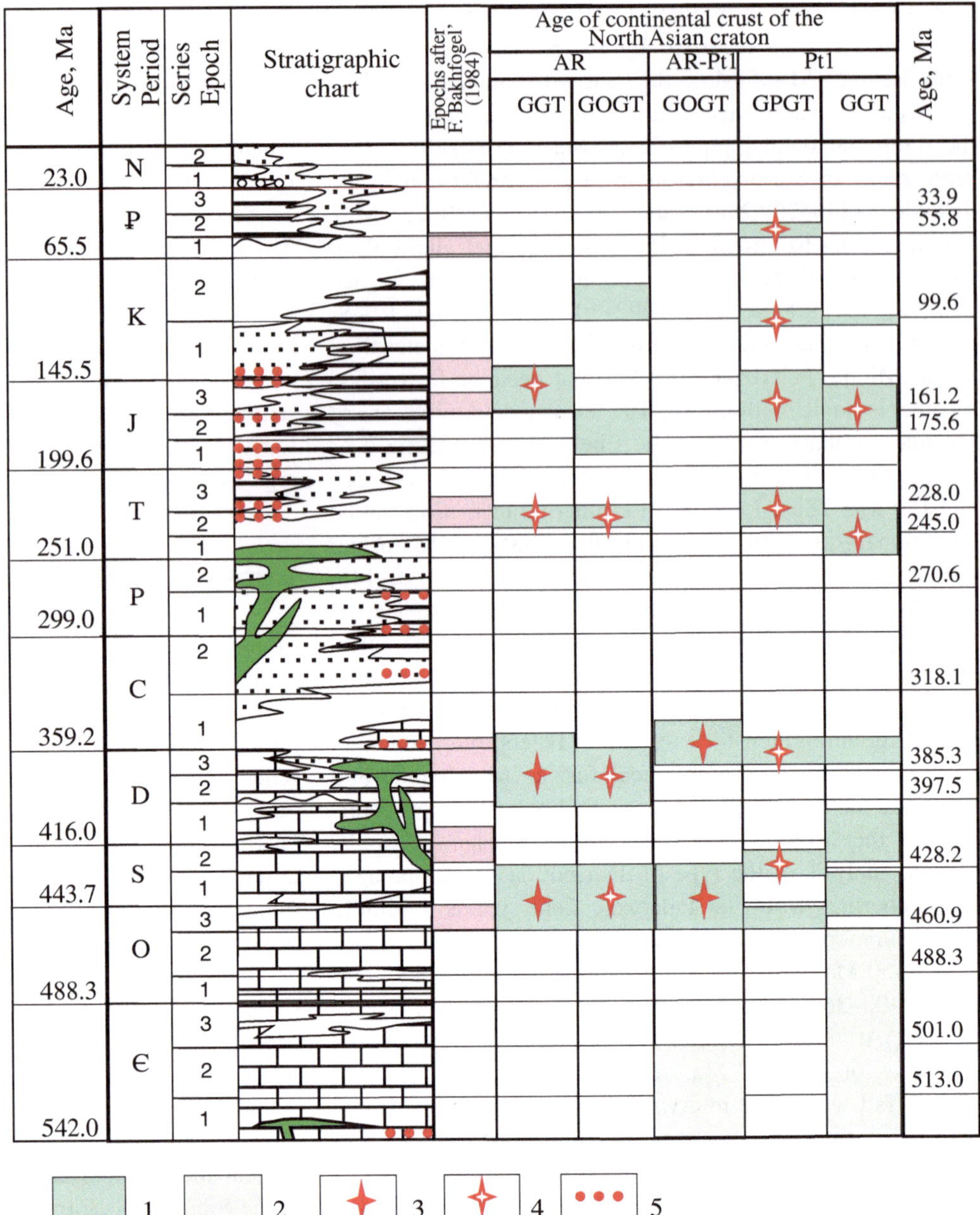

Fig. 4 Age epochs of kimberlite formation, diamond tenor, age, and composition of terranes of the basement of the Siberian platform: epochs of kimberlite formation (1–2): *1*—defined, *2*—assumed; epochs of diamondiferous kimberlite magmatism (3–4): *3*—high-grade diamond (commercial) kimberlites, *4*—low-grade diamond kimberlites, *5*—occurrences of kimberlite indicator minerals. Terranes of the Siberian platform basement: *GGT* granite-greenstone, *GOGT* granulite-orthogneiss, *GPGT* granulite-paragneiss

size, and location of the areas of different-aged kimberlite magmatism (Fig. 5a–e) likely indicate their different geodynamic nature. Discussion of the problems of geodynamic nature of kimberlite magmatism is beyond the scope of this paper. The presence of different-aged kimberlites within the same kimberlite-controlling zones may be related to the formation of the basement relief because the zones themselves are confined to peripheral parts of the basement uplifts (Fig. 5a–e). The Anabar and northern parts of the Daldyn-Olenek kimberlite-controlling zones are located along the boundaries of Anabar shield and Olenek uplift with Meso- and Neoproterozoic rifts, Vilyuisk-Markha kimberlite-controlling zone—along the Muna uplift boundary and Paleozoic rift. Thus, the location of kimberlite-controlling zones is determined mainly by combination of two tectonic factors: (1) Location of collision sutures between Early Precambrian terranes of the basement and (2) Location of boundaries of the basement uplifts. The influence of the third factor, the manifestation of synchronous basic magmatism (formation of dike belts and trappean fields), affecting the location of kimberlite fields is not defined yet.

Conclusion

We have made a comparative analysis of over 900 isotope age determinations of kimberlites from 24 fields and came to the following conclusions:

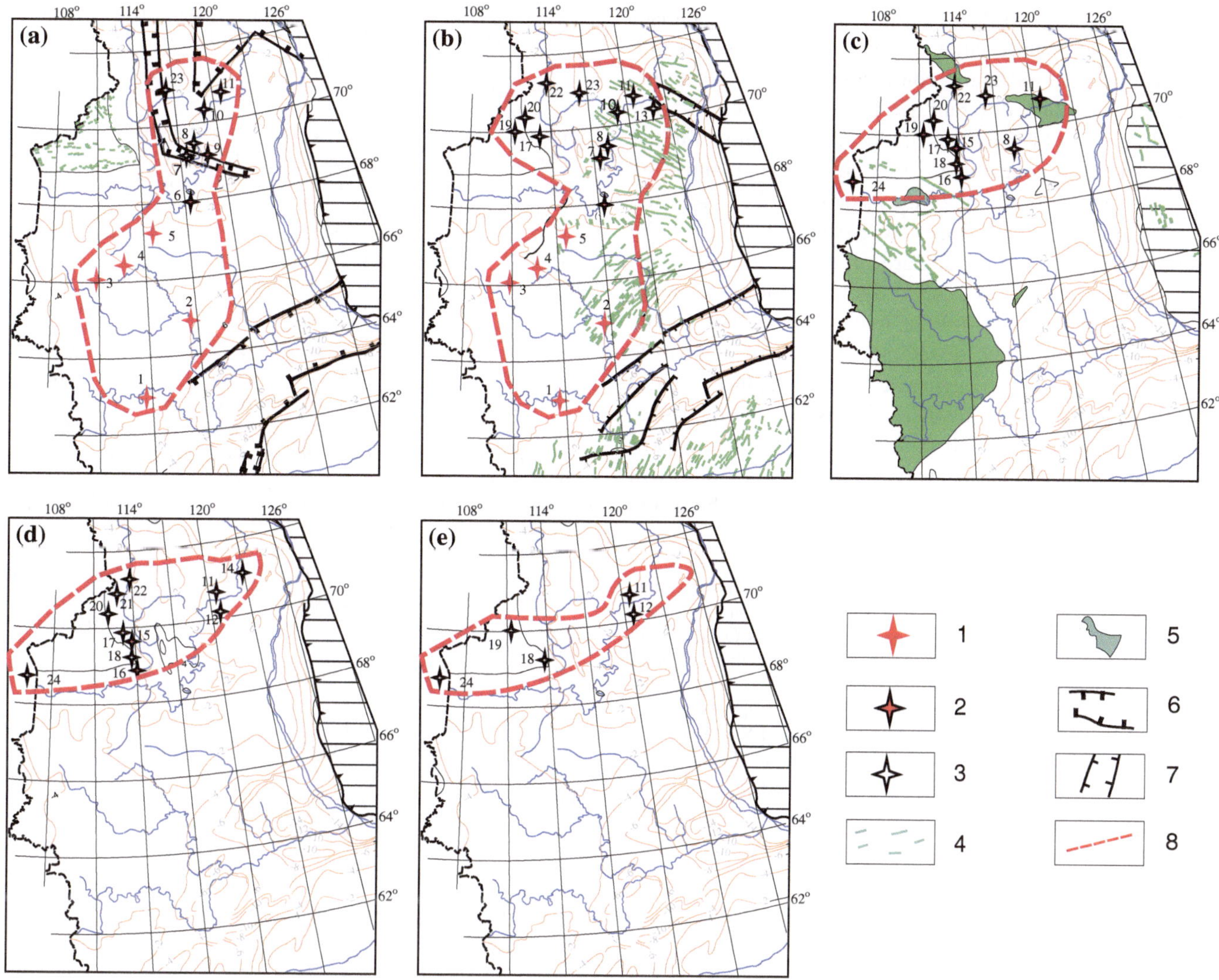

Fig. 5 Extent of kimberlite magmatism in different time spans. (**a**) Ordovician–Silurian. (**b**) Devonian–Carboniferous. (**c**) Triassic. (**d**) Jurassic. (**e**) Cretaceous. Kimberlite fields (1–3): *1*—high-grade, *2*—low-grade, *3*—barren; *4*—dikes of Riphean, Paleozoic, and Mesozoic diabase and dolerite; *5*—Permo–Triassic traps, *6*—Meso- and Neoproterozoic rift boundaries, *7*—Devonian rift boundaries, *8*—boundaries of kimberlite magmatism areas. Shown are depth contours of the crystalline basement surface (in km). Names of kimberlite fields are given in Fig. 1

Modern isotope dating has revealed seven epochs of kimberlite magmatism within YKP: 450–430, 420–400, 380–350, 250–230, 170–150, 110–100, and 60–50 Ma. Older isotope dates are also available, which together with the zircon U–Pb ages obtained from intermediate diamond reservoirs in some regions of the province are indicative of the Vendian and Early Ordovician epochs of kimberlite magmatism. It is established that in different parts of the Siberian platform, irrespective of the age of the basement, the number of the epochs of kimberlite formation (including diamond-bearing kimberlites) varies from 2 to 7. The least number of the epochs (Ordovician–Silurian, Devonian–Carboniferous) is observed in the Vilyui subprovince, while the Anabar-Olenek subprovince exhibits 2–7 epochs. Manifestation of the kimberlite magmatism (including diamond-bearing kimberlites) does not correlate with the age and composition of the terranes of Early Precambrian basement.

Maps have been constructed showing the extent of kimberlite magmatism and its diamond tenor in different periods of the development of the Siberian platform. Analysis of the maps available for different time spans showed that Paleozoic epochs of kimberlite formation are mainly characteristic of the southern and northern parts of the YKP. The eastern boundary of kimberlite magmatism coincides with or is close to the western termination of the Devonian rift-related structures and dike swarms developed in the eastern Siberian platform. The southern boundary of the Meso- and Neoproterozoic rifts is characterized by a high diamond tenor of Paleozoic kimberlites. Mesozoic kimberlite magmatism manifested itself in the northern

Siberian platform, northeastward of the trap magmatism area. Areas of manifestation of different-aged kimberlite magmatism have different configurations and locations, which suggests their different geodynamic nature. The fact that different-aged kimberlites are localized within a limited number of their controlling zones may be explained by different times of formation of the basement relief, since the zones themselves are confined to peripheral parts of the basement uplifts. Another feature of the location of kimberlite-controlling zones is their spatial relationship with collisional sutures of 2.1–1.9 Ga. In conclusion, it should be noted that the problems of tectonic control of kimberlite magmatism and its geodynamic nature require further studies and analysis of diamondiferous provinces worldwide. We think that Lu–Hf dating techniques of the kimberlites and their low-Cr megacrysts (Novel et al. 2004) will allow to solve many of the existing problems of the origin for different group of the kimberlites.

Acknowledgments The work has been completed under the framework of the Project № 27.1 Presidium RAN program. We are grateful to the Siberian Branch, RAS for financial support. We thank O. B. Oleynikov, A.Ya. Biller, I. Krivoshapkin, and V.F. Timofeev for their technical assistance in compilation of the complete catalog of all kimberlite bodies of YKP, and YKP geographic, geologic, and tectonic maps. Thanks are also to G.P. Bulanova, C.B. Smith, O.V. Koroleva, and N.V. Popov for their critical comments. We are grateful to David Graham Pearson, who spent considerable editorial effort to improve our paper.

References

Agashev AM, Orikashi Yu, Watanabe T, Pokhilenko NP, Serenko VP (2000) Isotope-geochemical characteristics of kimberlites of the Siberian platform in connection with the problem of their origin. Geol Geophys 41(1):90–99 in Russian

Agashev AM, Pokhilenko NP, Tolstov AV, Polyanichko NP, Mal'kovets VG, Sobolev NV (2004) New data on age of kimberlites from Yakutian kimberlite province. Dokl RAN 399(1):95–99 in Russian

Agashev AM, Watanabe T, Budaev DA, Pokhilenko NP, Fomin AS, Maehara K, Maeda J (2001) Geochemistry of kimberlites from the Nakyn field, Siberia: evidence for unique source composition. Geology 29(3):267–270

Altukhova ZA, Zaitsev AI (2006) Features of material composition and age of kimberlite rocks from Dyuken, Luchakan and Ary-Mastakh fields of Yakutian province. Lithosphere 2:34–64 in Russian

Argunov KP (2005) Diamonds of Yakutia: physical, morphologic, gemologic features. Geo, Novosibirsk, p 402 (in Russian)

Brakhfogel' FF (1984) Geological aspects of kimberlite magmatism in the northeastern Siberian platform. YaF SO AN SSSR, Yakutsk, p 128 (in Russian)

Cherenkova AF, Cherenkov VG (2007) Kimberlites of Kharamay field and depth xenoliths in them. Geoinformark, Moscow, p. 236 (in Russian)

Clifford TN (1966) Tectono-metallogenic units and metallogenic provinces of Africa. Earth and planet. Sci Lett 1:421–434

Crockett RB, Mason R (1968) Foci of mantle disturbance in Southern Africa and their economic significance. Econ Geol 63:522–540

Davis GL, Sobolev NV, Khar'kiv AD (1980) New data on age of Yakutian kimberlites obtained by uranium-lead age method on zircons. Dokl Akad Nauk SSSR 254(1):175–179 (in Russian)

England PH, Houseman G (1984) On the geodynamic setting of kimberlite genesis. Earth and planet. Sci Lett 67:109–122

Fefelov NN, Kostrovitskiy SI, Zarudneva NV (1992) Isotopic composition Pb in kimberlites of Russia. Geol Geophys 33(11):102–107 (in Russian)

Griffin WL, Ryan CG, Kaminsky FV, O'Reilly SY, Natapov LM, Win TT, Kinny PD, Jenpin IP (1999) The Siberian lithosphere traverse: mantle terranes and the assembly of the Siberian Craton. Tectonophysics 310:1–35

Ilupin IP, Vaganov NN, Prokopchuk BI (1990) Kimberlites: reference book. Nedra, Moscow, p 238 (in Russian)

Khar'kiv AD, Zinchuk NN, Kryuchkov AI (1998. Primary diamond deposits of the world. Nedra, Moscow, p 555 (in Russian)

Kinny PD, Griffin BJ, Heaman LM, Brakhfogel' FF, Spetsius ZV (1997) Determination of U-Pb age of perovskites from Yakutian kimberlites by Super High Resolution Ion Microprobe (SHRIMP) method. Geol Geophys 38(1):91–99 in Russian

Komarov AN, Ilupin IP (1990) Geochronology of from the Siberian Platform according to track method data. Geochemiya 3:365–372 in Russian

Koval'skiy VV (1963) Kimberlite rocks of Yakutia and basic principles of their petrogenetic classification. AN SSSR, Moscow, p 183 (in Russian)

Levchenkov OA, Gaidamako IM, Levskiy LK, Komarov AK, Yakovliva SZ, Rezvanova NG, Makeev AF (2005) U-Pb-age of zircon from Mir and 325 let Yakutii kimberlite pipes. Dokl RAN 400(2):233–235 in Russian

Milashev VA (1974) Kimberlite provinces. Nedra, Leningrad, p 238 (in Russian)

Mokshantsev KB, Elovskikh VV, Koval'skiy BB (1974) Structural control of kimberlite magmatism manifestations in the northeastern Siberian Platform. Science, Novosibirsk, p 98 (in Russian)

Nikishov KN (1984) Petrologic-mineralogic model of kimberlite process. Science, Москва, p 213 (in Russian)

Nikulin VI, Lelyukh MI, Fon-der-Flaass GS (2001) Diamond forecast (concept and methodology). Glazkovskaya tipografia, Irkutsk, p 320 (in Russian)

Nowell GM, Pearson DG, Bell DR, Carlson RW, Smith CB, Kempton PD, Noble SR (2004) Hf isotope systematics of kimberlites and their megacrysts: New constraints on their source region. J Petrol 45:1583–1612

Pearson DG, Kelley SP, Pokhilenko NP, Boyd FR (1997) Age determination of phlogopites from South African and Siberian kimberlites and their xenoliths by laser $^{40}Ar/^{39}Ar$ method: modeling the age of eruption, degassing of melt and composition of mantle fluid. Geol Geophys 38(1):100–111 in Russian

Rosen OM, Condie KC, Natapov LM, Nozhkin AD (1994) Archean and Early Proterozoic evolution of the Siberian craton: a preliminary assessment. In: Condie KC (ed) Archean crustal evolution. Elsevier, Amsterdam, pp 411–459

Safronov AF, Smelov AP, Zaitsev AI (2001) Issues of tectonic control of diamondiferous kimberlites of the Siberian Platform. Otechestvennaya Geologiya 5:3–6 in Russian

Smelov AP, Timofeev VF (2007) The age of the North Asian cratonic basement: an overview. Gondwana Res 12:279–288

Williams HR, Williams RA (1977) Kimberlites and plate tectonics in West Africa. Nature 270:507–508

Zaitsev AI (2007) Rubidium-strontium isotopic geochemistry and age of kimberlite rocks of Ary-Mastakh field. Otechestvennaya Geologiya 11:94–105 in Russian

Zaitsev AI, Smelov AP (2010) Isotope geochronology of rocks of kimberlite formation in the Yakutian province. Offset, Yakutsk, p 108 (in Russian)

Mineral Associations in Diamonds from the Lowermost Upper Mantle and Uppermost Lower Mantle

Ben Harte and Neil F. C. Hudson

Abstract

The chapter reviews the worldwide occurrence of inclusions in diamonds which involve Si-rich minerals and which appear to come mainly from depths in the range ca 550 to 800 km. They are referred to collectively as the Perovskite and Periclase Suite. Attention is focussed upon distinguishing retrograde and primary minerals, and upon identifying associations of different primary minerals within single diamonds; this provides potential equilibrium inclusion assemblages whose depths of formation may be estimated with reference to experimental studies. Associations in the same diamond of separate inclusions of $(Mg,Fe)SiO_3$, $(Mg,Fe)O$ and $(Mg,Fe)_2SiO_4$—potentially representing the original phases MgSi-perovskite(mPv), ferropericlase (fPer) and ringwoodite (rw)—indicate formation at the Upper/Lower Mantle boundary. Associations involving MgSi-perovskite and ferropericlase, without ringwoodite, are taken to indicate Lower Mantle assemblages of ultrabasic bulk composition, and these are divided into two: those with low-Al MgSi-perovskite, mPv, from the shallowest Lower Mantle, and those with high-Al MgSi-perovskite, mPv(Al), from greater depths. In assemblages of basic bulk composition, the primary phases of mPv(Al), sodic majoritic garnet (maj-grt), new Al–silicate phase (NAL) and the calcium ferrite structured phase (CF) are all represented by composite inclusions, which include a variety of retrograde products such as olivine, spinel, tetragonal almandine-pyrope phase (TAPP), NaAl-rich pyroxene phase (NaAl-pyrox) and nepheline. In both ultrabasic and basic bulk compositions, the principal Ca-bearing phase appears to be CaSi-perovskite (cPv). The chemical compositions of the primary phases show considerable coherency, including relatively constant Fe–Mg partition coefficients. The postulated mineral assemblages conform well with experimental investigations, and a series of different depths of formation

Electronic Supplementary Material The online version of this article (doi:10.1007/978-81-322-1170-9_15) contains supplementary material (with mineral analytical data), which is available to authorized users.

B. Harte (✉)
Centre for Science at Extreme Conditions (CSEC), School of GeoSciences, University of Edinburgh, Edinburgh, Scotland EH9 3JW, UK
e-mail: ben.harte@ed.ac.uk

N. F. C. Hudson
Geological Sciences, Kedleston Road, University of Derby, Derby, DE22 1GB, UK

D. G. Pearson et al. (eds.), *Proceedings of 10th International Kimberlite Conference*, Volume 1, Special Issue of the Journal of the Geological Society of India, DOI: 10.1007/978-81-322-1170-9_15, © Geological Society of India 2013

are indicated. In the inclusion assemblages of basic bulk composition, a transition from assemblages with sodic majoritic garnet (maj-grt) to ones with mPv(Al), NAL and CF is believed to approximately coincide with the change from mPv to mPv(Al) in ultrabasic assemblages. The eight minerals mPv–mPv(Al), fPer-mW, maj-grt, NAL, CF, cPv, corundum (crn) and stishovite (stv) [where mPv–mPv(Al) and fPer-mW indicate solid solutions] potentially define an invariant point in the 6-component composition space MgO–FeO–Al_2O_3–SiO_2–CaO–Na_2O, giving rise to a series of univariant and divariant mineral assemblages in pressure–temperature space. Consideration of the available experimental data indicates that this invariant point is probably in the pressure range 24–28 GPa under the range of temperatures expected in the uppermost Lower Mantle.

Keywords
MgSi-perovskite • Ferropericlase • Majoritic garnet • CaSi-perovskite

Introduction

The purpose of this chapter is to review the worldwide occurrence of sets of inclusions in diamonds which involve (Mg, Fe)Si-perovskite, ferropericlase and other phases suggestive of depths of origin near the base of the Upper Mantle and in the Lower Mantle (Fig. 1; note that in this chapter the Transition Zone is considered to be part of the Upper Mantle). Emphasis will be placed upon cases with Si-bearing (silicate) minerals and where several minerals are associated together in the same diamond, thereby giving mineral assemblages that may be assigned to depth ranges on the basis of high-P–T experimental studies. The chemical compositions of the silicate mineral inclusions will be compared particularly with those found in experimental studies for pressures of 19–37 GPa.

The suite of diamonds and inclusions to be described will be referred to as the Perovskite and Periclase Suite. This suite essentially corresponds with 'MgSi-perovskite and ferropericlase' (mPv + fPer) suite of Harte and Richardson (2012), and for the most part includes the 'Lower Mantle' suite of Stachel et al. (2005) and the 'Juvenile ultramafic mineral association' of Kaminsky (2012). The Perovskite and Periclase Suite is clearly distinct from the common suites of inclusions with peridotitic and eclogitic minerals found in diamonds from the continental lithosphere (Stachel and Harris 2008). It is also distinct from other suites of unusually 'deep' diamonds, such as the widely found maj-grt inclusions, largely from the depth range 250–550 km (e.g. Moore and Gurney 1989; Tappert et al. 2005a; Harte and Cayzer 2007; Harte 2010). In the case of the exceptional Juina province in Brazil, diamonds with inclusions of the Perovskite and Periclase Suite are found alongside other 'deep' diamond suites, in particular a suite of Ca-rich and Al-rich inclusions with 'deep' Upper Mantle mineral compositions (Wirth et al. 2007; Walter et al. 2008, 2011; Bulanova et al. 2010), and also a diverse volatile-rich (natro)-carbonatitic suite of micro- and nano-metric inclusions of potential Lower Mantle origin (Kaminsky et al. 2009a, b; Wirth et al. 2010; Kaminsky 2012).

Several papers on all the suites of 'deep' diamonds have suggested the connection of the diamonds and their inclusions to subduction processes (Harte et al. 1999; Stachel et al. 2000, 2005; Tappert et al. 2005a, b, 2009; Harte 2010; Bulanova et al. 2010; Walter et al. 2011). In the case of the Juina diamonds and inclusions, Harte (2011) drew attention to the possible connection of diamond formation with development of a subducted stagnant slab lying near the Upper/Lower Mantle boundary. Harte and Richardson (2012) linked all the Juina 'deep' diamond suites to a series of events in the evolution of a slab of oceanic lithosphere subducted beneath South America during the Mesozoic era.

Methods

Selection of Mineral Composition Data

Many of the diamonds carrying the 'deep' mantle suites of silicate inclusions show several separate inclusions, which may involve different mineral phases (Fig. 2). In the case of *multiple single-phase inclusions* within one diamond, it is usual to initially assume that the different phases represented were in equilibrium with one another at the time of crystallisation of the diamonds. Harte et al. (1999) referred to such a set of inclusions within one diamond as a *mineral association,* and found that treating them as an equilibrated mineral assemblage usually showed consistent results, and that careful comparison of the mineral chemical compositions could provide evidence of disequilibrium where it occurred.

The use of the term 'single-phase' in the above paragraph needs qualification. In effect, it is used for inclusions

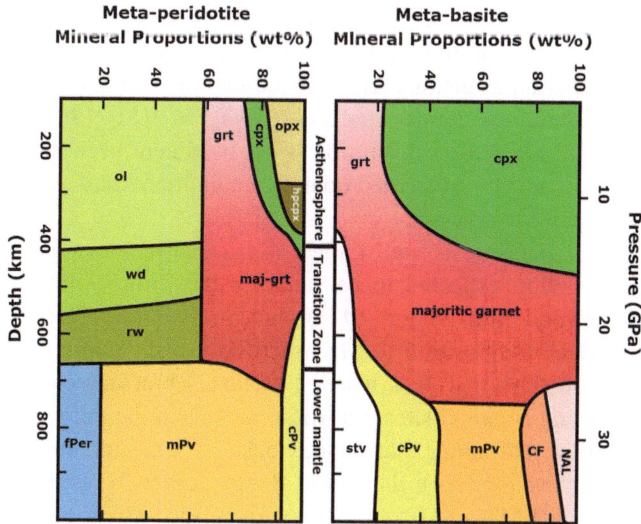

Fig. 1 Mineral proportions present in average ultrabasic or meta-peridotite rocks and basic or meta-basite rocks as a function of depth from 100 to 1,000 km depth; both asthenosphere and Transition Zone are considered to belong to the Upper Mantle. Modified from Harte (2010) and based upon Stixrude and Lithgow-Bertelloni (2007) for ultrabasic compositions and Ricolleau et al. (2010) for basic compositions; correlation of pressure with depth based on Dziewonski and Anderson (1981)

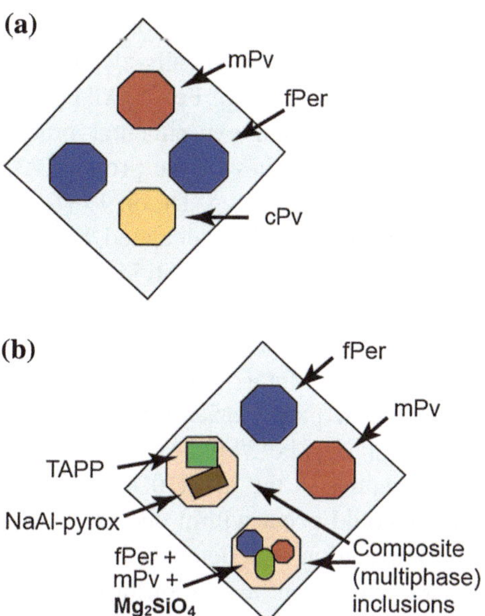

Fig. 2 Schematic illustration and interpretation of different sets of inclusions within single diamonds. In case **a**, individual inclusions are homogenous in composition, and each inclusion appears to represent a single phase. If several (*multiple*) inclusions of the same phase occur, then they are expected to show constancy of composition. In this case, the set of minerals is described as a *mineral association*, and the isolated inclusion compositions are assumed to represent those of an initial equilibrium mineral assemblage at the time of diamond formation. In case **b**, some inclusions are composite, that is, they comprise several distinct phases. It is assumed most likely that the composite inclusions have been affected by down-pressure retrogression; thus, the inclusion with TAPP + NaAl-pyroxene probably represents an original majoritic garnet, while the inclusion with mPv + fPer + Mg_2SiO_4 inclusion may have initially been mPv + fPer (Brey et al. 2004; Hayman et al. 2005). As a consequence, the original nature of the composite inclusions must be reconstructed; but the isolated individual inclusions of mPv and fPer composition may represent initial mineral compositions

showing constant composition when analysed by the electron microprobe. This criterion commonly applies to the (Mg, Fe)Si-perovskite (mPv) and CaSi-perovskite (cPv) inclusions, even though X-ray diffraction investigations commonly show that these phases have inverted to phases stable at lower pressures (Harte et al. 1999; Stachel et al. 2000; Hutchison et al. 2001).

In contrast to the above inclusions are ones referred to as '*composite*' because several distinct phases may be detected within one inclusion. These constitute an assemblage of minerals in contact with one another, but it seems unlikely that they represent an original equilibrium mineral assemblage. If the inclusion was composite at the time of its origin, then it seems likely that the close proximity of the original minerals would lead to them reacting during decompression (e.g. Brey et al. 2004; Tappert et al. 2005b; Hayman et al. 2005). Alternatively, a composite inclusion may be the multi-phase decompression product of an original single-phase inclusion (Walter et al. 2011). In either case, the phase(s) of the original inclusion has to be reconstructed using evidence from chemical compositional and textural features (Hayman et al. 2005; Harte and Cayzer 2007; Walter et al. 2011).

Therefore, in seeking original mineral compositions, the 'single-phase' inclusions showing homogenous chemical composition are given a high priority, because such inclusions have been protected from chemical alteration by the encapsulating diamond though this does not imply that they have not undergone inversion to lower-pressure phases. Similarly, associations of such single-phase inclusions within the same diamond are considered a good guide to original mineral assemblages at the time of diamond formation, though one has to consider the possibility of there being more than one period of diamond growth. In general, we shall avoid using the mineral compositions found in composite inclusions as evidence of original mineral chemistry though composite inclusions may be used as evidence of original phases if the original composition can be reconstructed.

Re-calculation of Analytical Data

All the analytical data in this chapter have been previously published elsewhere, but all such information (whether natural mineral inclusions or experimental phases) has been re-calculated to cations per formula unit (pfu). Considerable attention was given to the question of Fe^{2+}–Fe^{3+} proportions, given that analyses are typically electron microprobe data and only have the amount of Fe^{total}. All cation plots, for both natural inclusions and experimental products, were initially made using estimates of the proportions of Fe^{2+}–Fe^{3+} based on stoichiometry (e.g. Droop 1987). For several phases (e.g. fPer and rw), the calculated Fe^{3+}/Fe^{total} were consistently small, and other calculations based on total Fe (calculated as Fe^{2+}) differed little from those based on calculated Fe^{2+} and Fe^{3+}. However, the mPv and CF compositions (from both natural inclusions and experimental charges) show much more variability in calculated Fe^{3+}/Fe^{total}. For these phases, Fe^{3+}/Fe^{total} was found to vary from 0.0 to 1.0, and values in the range 0.4–0.8 were common. Such variations could occur within one set of experimental data, and no correlation was detected between Fe^{3+}/Fe^{total} and other parameters (e.g. a relationship to Al content in mPv as suggested by McCammon et al. 2004). It is also notable that Fe/(Mg + Fe) ratios calculated using Fe^{total} show more regular variations and yield more uniform mPv/fPer partition coefficients than those calculated using Fe^{2+}. Overall, we concluded that the use of Fe^{3+} and Fe^{2+} calculated by stoichiometry produced irregular scattering in the data, and that they may well be affected by the fact that most phases have undergone retrograde alteration. We have therefore calculated all Fe as being Fe^{2+} and use this as Fe^{total} in all parameterisation and plots.

Spreadsheets (entered in Microsoft Excel) with all the data (both natural inclusion and experimental compositions) used in this chapter are given in the Supplementary Information available online (see page 1).

Garnet–Perovskite End-member Composition Plot

In order to compare a variety of mPv and garnet compositions, and also put in perspective the compositions of retrograde phases (TAPP, NaAl-pyroxene), a plot is used based on mPv and garnet ideal end-member compositions in MgO–FeO–Al_2O_3–Na_2O–SiO_2 space. Three garnet end-member compositions are used, two of which coincide with mPv compositions. In presenting this, Mg and Fe^{2+} are assumed to readily substitute for one another in the natural phases concerned, and the two cations are represented together in the following formulae with the symbol 'FM'.

The garnet molecular compositions used are the following: $(FM_3)\{Al_2\}[Si_3]O_{12}$ (pyrope-almandine), $(FM_3)\{FM_1Si_1\}[Si_3]O_{12}$ (pure majorite) and $(Na_1FM_2)\{Al_1Si_1\}[Si_3]O_{12}$ (Na-garnet molecule), where garnet X site cations are in ( ) brackets, garnet Y site cations are in { } brackets and garnet Z site cations are in [ ] brackets. The second and third of these compositions are 'majoritic' in that Si is substituting on the Y site where it is in octahedral coordination as distinct from the tetrahedral Si in the Z site. A potential Na-garnet molecule, $(Na_2FM_1)\{Si_2\}[Si_3]O_{12}$, with full majoritic substitution is not used here because the natural maj-grts concerned do not contain more than one Na atom per 12 oxygens.

Pure majorite is well known as an end-member garnet composition expected to be stable in the Transition Zone (Ringwood and Major 1971), but it also has identical chemical composition to MgSi-peroskite (mPv) with composition $(FM_1)\{Si_1\}O_3$, where ( ) and { } brackets enclose the A and B cation sites. Substituting Al into the perovskite structure, assuming that charge balance is maintained by equal substitution on the A and B sites, gives the potential formula $(FM_{0.75}, Al_{0.25})\ \{Si_{0.75}, Al_{0.25}\}O_3$, which has the same chemical composition as pyrope-almandine garnet. Thus, the pyrope-almandine to majorite chemical composition range is the same as the Al-rich (Mg,Fe)Si-perovskite to Al-free (Mg, Fe)Si-perovskite range. This convergence is used in comparing mPv, garnet, TAPP and NaAl-pyroxene compositions in several sections of the chapter.

In using the garnet-perovskite end-member plot, mineral cation proportions are allocated to the three end members, but consideration must be given to the chemical components in natural minerals falling outside MgO–FeO–Al_2O_3–Na_2O–SiO_2 composition space. For the most part, the amounts of these other components (Ti, Cr, Mn, Ni, Ca, Na, K) are very small and will substitute in the sites of major cations: Cr may be added to Al, Mn and Ni to Mg–Fe (FM), and K to Na. It should be noted that Na + K are typically very minor in mPv. Although quadrivalent Ti substitutes in garnet octahedral sites (with Al, Cr) at low pressure, it has been included with quadrivalent Si on the basis that Si enters the octahedral as well as the tetrahedral sites in maj-grt. The principal additional cation that may be relatively significant in garnets is Ca, and the amount of Ca and other cations associated with Ca end-member molecules should be determined and not included in the plotting parameters. Thus, the proportion of the garnet molecule $Ca_3Al_2Si_3O_{12}$ is calculated on the basis of Ca cation content and set aside. Other cations are then allocated to $(Na_1FM_2)\{Al_1Si_1\}[Si_3]O_{12}$ (Na-garnet mol), $(FM_3)\{Al_2\}[Si_3]O_{12}$ (pyrope-almandine mol) and $(FM_3)\{FM_1Si_1\}[Si_3]O_{12}$ (pure majorite mol) on the basis, respectively, of: the Na cation content, the remaining Al content (after allocation to Ca and Na molecules), and the remaining Mg–Fe(FM) cation proportions (after previous allocation to Na-garnet and almandine-pyrope molecules). The relative proportions of the three ideal garnet molecules are determined and plotted using the proportion of majorite (equivalent to mPv) on the x-axis and that of Na-grt mol on the y-axis, with pyrope-almandine (or Al-rich mPv) at the origin.

Mineral Associations of the Perovskite and Periclase Suite

The name Perovskite and Periclase Suite is used in this chapter as a general name for the sets of mineral inclusions in diamonds, which involve combinations of the phases MgSi-perovskite, cPv and ferropericlase and appear to derive from broadly ultrabasic (meta-peridotite) and basic (meta-basite) bulk rock compositions. MgSi-perovskite plays a key role in the inclusion associations of the suite. Under P–T conditions expected along typical mantle geotherms, in basic and ultrabasic bulk compositions, MgSi-perovskite is not expected to be stable until the depth of the Upper Mantle/Lower Mantle boundary (UM/LM boundary at 660 km) is reached, where ringwoodite ($MgSi_2O_4$) is replaced by MgSi-perovskite and ferropericlase (Fig. 1). cPv is expected to become stable in ultrabasic and basic bulk compositions at somewhat shallower depths (ca 550 km) approximately in the middle of the Transition Zone (Fig. 1). Ferropericlase alone may occur at much shallower depths (e.g. Brey 2004), and so we do not automatically attribute inclusions of ferropericlase alone to the Perovskite and Periclase Suite. Neither ferropericlase nor cPv is recorded as members of the common suites of maj-grts (e.g. Moore and Gurney 1989; Tappert et al. 2005a; Harte and Cayzer 2007; Harte 2010) which appear to derive from depths of the ca 250–550 km. However, cPv does occur as a member of the Upper Mantle Ca-rich suite of inclusions described by Brenker et al. (2007), Walter et al. (2008) and Bulanova et al. (2010), and which indicate connections to carbonatitic melts (see also, Harte and Richardson 2012). In this chapter, our attention is on cPv inclusions associated with inclusions of MgSi-perovskite and/or ferropericlase.

The full list of primary phases attributed to the Perovskite and Periclase Suite in this chapter is given in Table 1, together with abbreviations adopted. Interpretation of the inclusions is complicated by the fact that many of the primary phases have undergone retrogression, although the original bulk chemical composition of each inclusion is believed to have been preserved by virtue of being encapsulated within diamond. Therefore, Table 1 also lists the secondary or retrograde phases typically found in the inclusions. Of the primary phases listed in Table 1, ferropericlase (fPer) and corundum (crn) are found as the original inclusions, but all others have undergone retrogression. In the case of many MgSi-perovskites with low-Al content (mPv), cPv and ringwoodite (rw) inclusions, electron microprobe analysis often shows homogeneity of chemical composition and apparently gives the composition of the original phase despite inversion to low-pressure phases (e.g. Harte et al. 1999; Davies et al. 1999; Stachel et al. 2000; Hayman et al. 2005). In the case of mPv, a useful guide to its original nature is usually provided by its low Ni contents by comparison with otherwise similar $(Mg,Fe)SiO_3$ pyroxene phases found in lower-pressure inclusions (Harte et al. 1999; Stachel et al. 2000). Where the original MgSi-perovskite inclusion was Al-rich [designated mPv(Al) in Table 1], it appears to be more usual for it to be retrogressed to complex intergrowths involving orthopyroxene, spinel, olivine and the TAPP phase (see below). The new Al–silicate phase (NAL) and the calcium ferrite–structured phase (CF) have only recently been identified, and they occur as retrogressed composite inclusions of spinel with Na(K)Al-silicates (Walter et al. 2011).

Although maj-grts, typically with moderate Ca content, have been widely found as inclusions attributed to the Upper Mantle, garnets with a large majorite molecule component have not previously been clearly identified in association with mPv or mPv(Al) inclusions. However, two unusual minerals have been found which need to be taken into account, namely a tetragonal almandine-pyrope phase (TAPP) and a NaAl-pyroxene phase (NaAl-pyrox). The occurrence of TAPP in apparently homogenous single-phase inclusions (Harris et al. 1997; Harte et al. 1999; Hutchison et al. 2001) suggested the possibility that it might be a primary phase occurring at high pressure as an alternative to garnet. Recent experimental evidence now suggests that it is always the product of retrogression and the original phase might have been an aluminous ferromagnesian perovskite, mPv(Al), or a maj-grt (Armstrong and Walter 2012). NaAl-pyroxene is rare, but experimental evidence suggests that it is product of retrogression of a sodic maj-grt (Gasparik and Hutchison 2000). Therefore, in Table 1, we list maj-grt as a primary phase belonging to the Perovskite and Periclase Suite, while TAPP and NaAl-pyroxene are treated as retrograde phases. These relationships and the detailed mineral compositions will be examined more closely in later sections on mineral compositions.

Table 2 lists the postulated critical mineral assemblages belonging to the Perovskite and Periclase suite; as discussed in the 'Methods' section of the chapter, these are typically 'associations' of separate (single-phase) inclusions occurring together in the same diamond. In general in listing these mineral associations, we give the original UM/LM boundary or Lower Mantle minerals believed to be represented by the inclusions and not the products of inversion or retrogression. An exception to this is made in the case of some TAPP and NaAl-pyrox inclusions where it is not clear whether original mPv or maj-grt is represented; in such cases, TAPP and NaAl-pyrox are listed in square brackets in Table 2, and their significance will be examined later.

Table 1 List of minerals occurring in the Perovskite and Periclase suite of deep diamonds inclusions

Original phase	Original composition	Found as	Abbreviation
MgSi-perovskite (low-Al)	$(Mg, Fe)SiO_3$	Inverted to pyroxene	mPv
MgSi-perovskite (high-Al)	$(Mg, Fe, Al)(Al, Si)O_3$	Inverted to pyroxene, and may be retrogressed to olivine, TAPP and spinel	mPv(Al)
Ferropericlase—(magnesiowustite)	$(Mg, Fe)O$	Ferropericlase—(magnesiowustite)	fPer
CaSi-perovskite	$CaSiO_3$	Inverted to low-P CaSi phases	cPv
Ringwoodite	$(Mg, Fe)_2SiO_4$	Inverted to olivine	rw
Corundum	Al_2O_3	Corundum	crn
Stishovite	SiO_2	Quartz or coesite	stv
Majoritic garnet	$(Na, Mg, Fe, Ca)_3$ $(Mg, Fe, Al, Si)_2$ Si_3O_{12}	Retrogressed to tetragonal almandine-pyrope phase (TAPP) and/or Na–Ca–Mg–Fe–Al-pyroxene (NaAl-pyrox)	maj-grt
New Al-silicate phase	(Na, K)–Mg–Fe–Al–Si–O	Retrogressed to spinel and Na- and K-silicates	NAL
Calcium ferrite structured phase	Mg–Fe–Na–Al–Si–O	Retrogressed to spinel and nepheline	CF

Note that fPer is used as an abbreviation for (Mg, Fe)O inclusions irrespective of Mg/Fe ratio

Diamonds with inclusions of fPer alone, or fPer with $(Mg,Fe)_2SiO_4$ or fPer with other oxide phases (e.g. Kaminsky et al. 2001) are omitted, because of uncertainties over their depth of origin. Likewise, diamonds containing a only one inclusion are usually omitted because of the lack of further evidence of the mineral assemblage. An exception to this is made in the case of diamonds with only inclusions of either of NAL, or CF or mPv(Al), because experimental data shows that these phases are only stable under Lower Mantle conditions.

We shall consider the detailed evidence concerning the depths of formation of the mineral associations in a later section of this chapter, but at this stage, we follow previous studies in placing the associations in Table 2 into three depth-related groups and one uncertain group. The three depth-related groups were previously recognised principally on the basis of ultrabasic associations containing ferropericlase (and sometimes olivine) in addition to MgSi-perovskite (Harte et al. 1999; Harte 2010). The place of inclusion associations of basic bulk composition in these groups has become clearer as a result of recent discoveries, and better recognition of the role of TAPP as a retrograde phase (Walter et al. 2011; Armstrong and Walter 2012).

1. *UM/LM boundary associations marked by the occurrence of fPer, rw, and mPv*, that is, an assemblage of minerals which defines the UM/LM boundary with the breakdown of rw to mPv + fPer. The three phases may occur as separate inclusions within the same diamond, or sometimes as composite inclusions, or sometimes as both single-phase and composite inclusions (Fig. 2). Some uncertainties exist over the effects of retrograde reaction (see Methods section), and we have chosen to ignore the composite inclusions, believing that an association of single-phase inclusions in the same diamond is a more reliable guide to mineral compositions at the time of diamond formation. Inclusion associations of cPv alongside fPer + rw + mPv have been found (Table 2); and the occasional occurrence of TAPP may indicate that maj-grt can also form part of this assemblage.
2. *Lower Mantle associations involving low-Al MgSi-perovskite, mPv*. These associations lack Mg_2SiO_4. Most have low-Al MgSi-perovskite with fPer and on this basis are allocated to the uppermost Lower Mantle. Three associations with low-Al MgSi-perovskite, but without fPer, are also placed here.
3. *Lower Mantle associations involving high-Al MgSi-perovskite, mPv(Al)*. Following Harte et al. (1999), the high-Al content of the MgSi-perovskite in these mineral associations indicates greater depths of formation than group 2 (see also later discussion on mineral compositions). Some inclusions are listed here where only one phase appears to have been included during diamond formation. They are placed in group 3 because experimental evidence has shown that the NAL and CF phases only occur with mPv(Al) and not with (mPv). Also, the mPv(Al), CF, NAL and cPv(basic) inclusions noted have been found together at a single locality (Juina 5 kimberlite) and provide the best evidence of a natural basic, rather than ultrabasic, assemblage which definitely comes from the Lower Mantle (Walter et al. 2011).
4. *Uncertain UM/LM boundary region associations*. These associations may contain fPer and cPv, but not mPv or mPv(Al), and so formation at the UM/LM boundary or within the Lower Mantle is uncertain. However, the TAPP and/or NaAl-pyrox phases listed, potentially do represent retrograde products either from original ultra-high-pressure maj-grt, or original mPv(Al) (see above and Table 1). Therefore, these associations could have

Table 2 Perovskite and periclase suite inclusion associations

Mineral associations	No[a]	Localities[b]
(1) UM/LM boundary		
fPer + mPv + rw	2	J-RS, DO27
fPer +mPv + rw + cPv	1	Ka
fPer + mPv + rw + [TAPP]	1	J-RS
fPer + mPv + rw + cPv + [TAPP]	1	J-RS
(2) LM (low-Al MgSi-perovskite)		
fPer + mPv	4	Or, J-SL, Ko, Ka
fPer + mPv + [TAPP]	1	J-SL, Ka
fPer + mPv + cPv	2	Ka, DO27
fPer + mPv + cPv + [Si–Mg phase]	1	DO27
fPer + mPv + Ni	1	DO27
fPer + mPv + $FeCO_3$	1	Ka
mPv + [TAPP]	2	J, J-SL
mPv + $KAlSi_3O_8$	1	Ka
(3) LM (high-Al MgSi-perovskite)		
fPer + mPv(Al)	1	J-SL
fPer + mPv(Al)[c]	2	J-SL, J-RS
fPer + mPv(Al) + crn	1	J-SL
mPv(Al)[c]	2	J-5
CF[c]	1	J-5
NAL[c]	2	J-5
cPv (basic)[c]	1	J-5
(4) Uncertain UM/LM boundary region		
maj-grt[c] + [TAPP]	1	J-SL
maj-grt[c] + [NaAl-pyrox]	1	J-SL
fPer + [NaAl-pyrox]	1	J-SL
fPer + rw + [TAPP]	1	J-SL
fPer + [TAPP]	3	J-SL
fPer + cPv	9	J-SL, J-RS, Ka, Pa
fPer + stv	10	J-SL, J-RS, J, Ka, DO27, Pa

A phase enclosed in square brackets [ ] is believed to be retrograde and suggests either mPv(Al) or maj-grt

[a] No is the number of diamonds containing the given association of inclusions

[b] Locality abbreviations as follows: J-Juina with RS (Rio Soriso), SL (São Luiz), 5 (Juina 5); Or-Orrorroo; DO27 in Slave craton; Ka-Kankan; Ko-Koffiefontein; Pa-Panda. Principal references for the localities are as follows: *São Luiz*—Harte et al. 1999; Hutchison et al. 2001. *Kankan*—Stachel et al. 2000. *Juina*—Kaminsky et al. 2001; Bulanova et al. 2010. Juina 5—Walter et al. 2011; *Rio Soriso*—Hayman et al. 2005. *Koffiefontein*—Moore et al. 1986. *DO27*—Davies et al. 1999, 2004. *Panda*—Tappert et al. 2005b

[c] Denotes primary phase identified from composite inclusion

Mineral abbreviations are listed in Table 1

come from near the base of the Upper Mantle, or the UM/LM boundary, or the Lower Mantle. In ultrabasic and basic bulk compositions, cPv is expected to have a minimum depth of formation around 550–600 km, in the lower half of the Upper Mantle Transition Zone. The association fPer + cPv is curiously abundant (e.g. Hayman et al. 2005), and it is possible that this cPv association, and other occurrences of cPv alone or with Mg_2SiO_4 and SiO_2 (listed in Harte, 2010) belong to the Ca-rich suite inclusions believed to have crystallised from carbonatitic melts in the Upper Mantle (Walter et al. 2008; Bulanova et al. 2010; Harte and Richardson 2012). The association fPer + stv is also curiously abundant and is listed in Table 2 with some uncertainties. SiO_2 may occur in equilibrium with Fe-rich ferropericlase (Harte et al. 1999); but doubts on some of the fPer + stv associations have been cast because the fPer inclusions are Mg-rich (Stachel et al. 2000). Also,

Hayman et al. (2005) report the occurrence of SiO_2 on fracture surfaces, and perhaps such occurrences reflect the presence of a fluid/melt phase (Harte 2010). However, the depths of occurrence of these associations are widely uncertain.

The Lower Mantle associations of group 2 (Table 2) suggest assemblages from very shallow depths in the Lower Mantle because the mPv compositions are very poor in Al (e.g. Irifune et al. 1996; Harte et al. 1999). The depths of formation of assemblages of deeper origin, with mPv(Al), CF and NAL (group 3 minerals, Table 2), potentially could extend deep into the Lower Mantle; however, their minimum depths of formation are close to the UM/LM boundary (Ono et al. 2001; Hirose et al. 2001; Guignot and Andrault 2004; Perillat et al. 2006; Ricolleau et al. 2010). Thus, as suggested by Harte (2010), the overall set of mineral assemblages shown by Table 2 might form in a narrow depth range of ca 550–800 km near the UM/LM boundary, and we shall discuss this further after presenting data on the mineral compositions.

Mineral Compositions in the Perovskite and Periclase Suite

Ferropericlase—Magnesiowustite

Of all the phases of potential origin at the UM/LM boundary and in the Lower Mantle, ferropericlase (fPer) is the one found most commonly as inclusions in diamonds. The wide involvement of ferropericlase in associations with MgSi-perovskite is evident in Table 2, but there are also many diamonds in which ferropericlase is the only included phase (ca 100 diamonds from the locations itemised in Table 2 are reported as containing ferropericlase alone). Since fPer is stable in both crust and mantle, the fPer inclusions alone *do not necessarily* indicate deep provenances (Stachel et al. 2000; Brey et al. 2004) and are not listed in Table 2.

The chemical compositions of fPer inclusions are typically close to stoichiometric (Mg, Fe)O and the principal chemical variation seen is in Fe/(Mg + Fe) ratio. Overall, considering inclusions of fPer alone as well as those associated with silicate phases, the Fe/(Mg + Fe) ratios are usually highly magnesian. In the Slave province (Canada), southern Africa and Australia, fPer compositions are in the range of 11.4–19.9 atomic % Fe/(Mg + Fe), with the exception of a very unusual value of 87.7 for one-magnesiowustite inclusion from Monastery, South Africa (Moore et al. 1986). At Kankan, (West Africa (Stachel et al. 2000) the range is 6.2–25.3 % (Fe/(Mg + Fe), but at Juina, the range is from 11.1 to 74.1 and therefore extends well into the magnesiowustite (mW) field. The maximum Fe/(Mg + Fe) values found for ferropericlase-magnesiowustite associated with mPv or cPv in the same diamond are, respectively, 35.3 and 54.9 (Hayman et al. 2005). Note that the abbreviation 'fPer' is used for all ferropericlase-magnesiowustite compositions.

Despite the wide range in Fe/(Mg + Fe), the (Mg,Fe)O inclusions show very limited variation in other chemical elements. Ti is usually less than 0.001 cations pfu and Al <0.002. Mn abundances are often <0.004 cations pfu, but occasionally approaches 0.010. The most abundant additional elements are Cr, Ni and Na, but even these are typically <0.01 and only rarely reach 0.02 cations pfu. The compositions are illustrated in Fig. 3 for the São Luiz (Juina province) inclusions, which are the ones showing the widest range in Fe/(Mg + Fe), but similar ranges of composition of minor elements are seen at the other localities. The Na is probably charge balanced by Cr and Fe^{3+} (Stachel et al. 2000). There is a suggestion of decreasing Ni and Cr with increasing Fe/(Mg + Fe) in the main range of inclusions shown in Fig. 3. Kaminsky et al. (2001) and Davies et al. (2004) also found Ni and Fe/(Mg + Fe) to be negatively correlated for the Juina and Lac de Gras (Slave) localities. Davies et al. (2004) note that the compositions of all fPer inclusions show a similar range of minor element compositions irrespective of whether they occur as the only inclusion in one diamond or are associated with other phases in the same diamond. This observation appears to be true for all localities, and in Fig. 3, it can be seen that fPer inclusions associated with MgSi-perovskite in the same diamond do not have distinct compositions from those without MgSi-perovskite.

There has been some debate about the origin of the more Fe-rich (Mg,Fe)O inclusions extending into the compositional range of magnesiowustite (Fig. 3). Some of these inclusions are also rich in Fe^{3+} as magnesioferrite (Harte et al. 1999). Suggested origins for these Fe-rich inclusions have included Fe-rich differentiation near the Upper/Lower Mantle boundary, and core contamination near the Core/Mantle boundary (Harte et al. 1999; Hayman et al. 2005). Recent data may support both places of derivation (Holland et al. in preparation) show Fe-rich (Fe, Mg)O occurring in calculated phase equilibria for the Transition Zone; while Wirth et al. (in preparation) suggest Fe^{3+}-rich varieties may be attributed to stabilisation of low-spin Fe^{3+} at the base of the mantle. Rarely, Fe-rich and Ni-rich bodies are also found within ferropericlase inclusions; again, these might indicate a core-mantle boundary derivation, but Davies et al. (1999) suggest the possibility of a disproportionation mechanism within ferropericlase resulting in the coupled formation of small bodies of Ni or Fe–Ni metal (see also Frost et al. 2003).

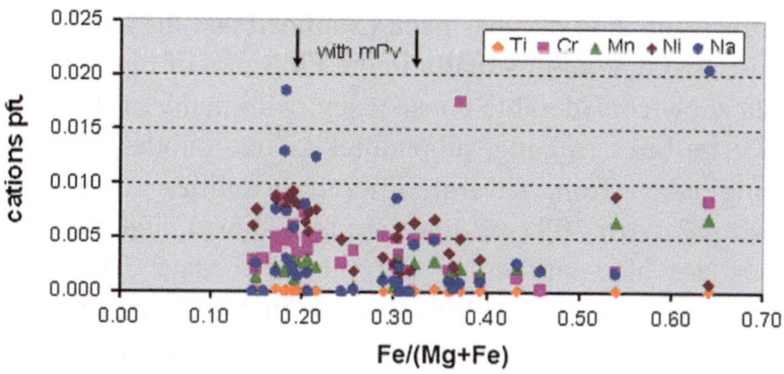

Fig. 3 Minor element compositions (cations per formula unit, pfu) in ferropericlase-magnesiowustite inclusions from São Luiz diamonds, plotted against Fe/(Mg + Fe), with Fe = Fe^{total}. Single-ferropericlase inclusions are plotted as well as ones coexisting with MgSi-perovskite (mPv) and other phases. The *black arrows* show the range of Fe/Mg + Fe) from 0.178 to 0.306 of fPer inclusions coexisting with mPv. Data from Harte et al. (1999) and Hutchison et al. (2001)

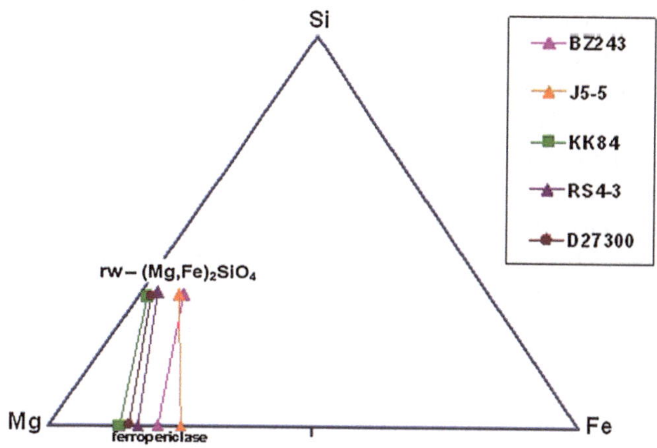

Fig. 4 Atomic FMS plot of assumed ringwoodite [rw – $(Fe,Mg)_2SiO_4$] and ferropericlase [fPer – (Fe,Mg)O] inclusion compositions found as single individual inclusions in the same diamond (composite inclusions not shown). Data from: BZ prefix, São Luiz (Hutchison et al. 2001); J prefix, Juina (Kaminsky et al. 2001), KK prefix, Kankan (Stachel et al. 2000); RS prefix Rio Soriso (Hayman et al. 2005); D prefix Lac de Gras (Davies et al. 2004)

Mg_2SiO_4

In its occurrence, Mg_2SiO_4 is unique among the inclusion minerals in occurring more frequently in composite (polymineralic) inclusions than in single-phase inclusions, 12 out of the 20 we tabulated being composite. Some of the Mg_2SiO_4 in these composite inclusions may be a product of retrograde reaction, and we have only listed in Table 2 associations of Mg_2SiO_4 and other phases where the Mg_2SiO_4 appears to be in a separate single-phase inclusion (see Methods) and therefore possibly an original ringwoodite (rw) inclusion. The overall range of Mg/(Mg + Fe) composition is similar for single-phase and multi-phase inclusions, being 96.5–87.4 for single-phase inclusions and 96.5–89.1 for composite inclusions. These ranges are a little more Fe-rich than those of Mg_2SiO_4 inclusions found in lithospheric diamonds with a dominant range of 91–95 (Stachel and Harris 2008).

Where the crystal structure of the Mg_2SiO_4 inclusions has been determined, it has been found to be that of olivine (Stachel et al. 2000), which is presumed to be a result of inversion from ringwoodite. The possibility of distinguishing the Mg_2SiO_4 phases (olivine, wadsleyite, ringwoodite) using geochemical criteria has been extensively considered, and the possibility of using minor concentrations of Al, Cr, Ni and Mn explored. However, no consistent relationships have been found (e.g. Davies et al. 2004; Tappert et al. 2005b; Hayman et al. 2005). Where the Mg_2SiO_4 is a product of the retrograde reaction of other phases, then its composition may well reflect that of the original mineral (e.g. mPv) rather than the original nature of the Mg_2SiO_4 polymorph crystallised (Brey et al. 2004).

Figure 4 shows Mg_2SiO_4 compositions plotted in FeO–MgO–SiO_2 with tie-line connections to ferropericlase found in association in the same diamonds. Composite inclusions have been omitted from this plot, because of the possibility of retrograde olivine formation and disequilibrium mineral compositions (see above). Many of the tie-lines in Fig. 4 show close consistency of orientation, indicating similar partition coefficients and equilibration conditions for sets of inclusions from different localities (Brazil, West Africa and Canada). However, two tie-lines from Brazil markedly cross one another and either indicate different conditions of equilibration or a case where single-phase inclusions in one diamond are not in equilibrium.

MgSi-Perovskite

MgSi-perovskite inclusions are dominantly of $(Mg, Fe)SiO_3$ composition, but show a clear distinction to Upper Mantle $(MgFe)SiO_3$ (orthopyroxene) phases in that they show very low Ni contents (typically < 0.03 wt%). This is believed to reflect the partitioning of Ni into ferropericlase in preference to MgSi-perovskite in mPv + fPer assemblages (Harte et al. 1999; Stachel et al. 2000). Al abundances in MgSi-perovskite inclusions show a bimodal distribution, with most inclusions (designated mPv in Tables 1 and 2) showing a range of Al_2O_3 of 0.55 to 3.37 wt%, while other inclusions [designated mPv(Al) in Tables 1 and 2] show >7.5 wt%. Within the mPv(Al) group, there is further distinct separation between the inclusions associated with fPer and those without fPer. High-Al MgSi-perovskites, mPv(Al), associated with fPer have relatively low TiO_2 and FeO^{total} (below 0.22 and 9.0 wt%, respectively), and are thought to come from ultrabasic bulk compositions. The mPv(Al) inclusions not associated with fPer have $TiO_2 > 4.0$ and $FeO^{total} > 20.3$ (wt%) and are thought to be of basic rather than ultrabasic origin (Walter et al. 2011).

Sufficient low-Al MgSi-perovskite (mPv) inclusions have been found worldwide to allow reasonable comparison of their ranges in Fe/(Mg + Fe) composition. As with the fPer compositions, the mPv Fe/(Mg + Fe) ratios are limited to highly magnesian values at all localities except for the Juina area. At the various Juina localities (listed in Table 2), the range of compositions is similar with an overall range of 4.9–13.6 Fe/(Mg + Fe). To further document phase relations and test consistency in their mutual compositions, Mg/Fe partitioning between fPer and mPv inclusions associated in the same diamond has been examined and the results are illustrated diagrammatically in the FMS plot of Fig. 5a. The data show considerable consistency, with many of the fPer-mPv tie-lines running subparallel to one another, despite being taken from several different localities worldwide. Two mPv(Al)—fPer tie-lines are also shown, and their tie-lines are also similar to the mPv-fPer data. Thus, the inclusion data suggest little change in Mg/Fe partitioning over the total depth range concerned. These results are largely in accord with the available experimental data

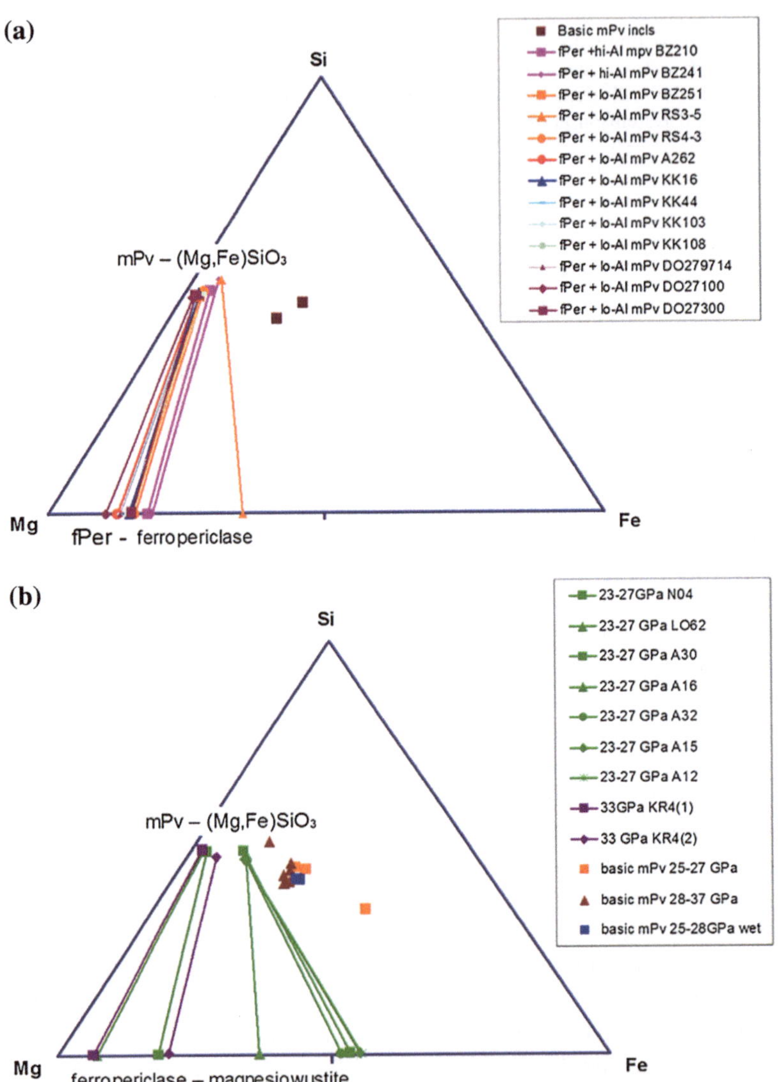

Fig. 5 Atomic FMS plots of mPv and fPer mineral compositions (Fe = Fe^{total}). **a** shows pairs of mPv and fPer occurring as separate inclusions in one diamond (associated inclusions—see Fig. 2). In these pairs, all the mPv inclusions have low-Al (see text) except for BZ210 and BZ241 (with *magenta colour*). Data from a wide variety of locations: prefix BZ are from São Luiz, Brazil (Harte et al. 1999); prefix A is from Koffiefontein, South Africa (Moore et al. 1986; Harte et al. 1999); prefix KK are from Kankan, West Africa (Stachel et al. 2000); prefix RS are from Rio Soriso, Brazil (Hayman et al. 2005); prefix DO are from Lac de Gras, Canada (Davies et al. 1999, 2004). The single *brown squares* are mPv inclusions with high-Al and indicating a basic bulk composition (Walter et al. 2011). **b** shows coexisting pairs of mPv and ferropericlase-magnesiowustite in experiments on ultrabasic and chondritic bulk compositions at the pressures indicated (Hirose and Fei 2002; Litasov et al. 2004; Nishiyama et al. 2004; Asahara et al. 2004). The *separate points* give mPv compositions from experiments on basic (MORB) bulk compositions at the pressures indicated (Ono et al. 2001; Hirose and Fei 2002; Litasov et al. 2004; Litasov and Ohtani, 2005; Ricolleau et al. 2010)

(Fig. 5b), where magnesian compositions show similar tie-line orientations to the majority of the inclusion data, but extend to more Fe-rich compositions.

The experimental fPer-mPv data plotted in Fig. 5b are for a range of 23–33 GPa, corresponding to a depth range of approaching 300 km, and the broad consistency of the compositional relationships seen in the experimental data for this significant depth range, is in accord with the fact that the natural data also show considerable consistency despite coming from widely separated worldwide localities. There is also a good correspondence of the maximum Fe/(Mg + Fe) in MgSi-perovskite with the experimental data of Fei et al. (1996) for the position of the fPer + mPv + stv 3-phase field.

In Fig. 5, the mPv basic inclusions (not co-existing with fPer), and the mPv from basic (MORB) experiments all plot close to the centre of FMS space, with obviously higher Fe/(Mg + Fe) than the mPv associated with fPer. They also tend to show lower Si contents because of their greater Al contents. The compositional relationships between mPv in various natural and experimental bulk compositions are further compared in the garnet-perovskite end-member plot (see Methods section for explanation) of Fig. 6a. All compositions plot close to the x-axis because of the generally limited Na contents of MgSi-perovskite, but there is a considerable spread from near pure $(FM_1)\{Si_1\}O_3$ to $(FM_{0.75},Al_{0.25})\{Si_{0.75},Al_{0.25}\}O_3$, where A- and B-site cations are in ( ) and { } brackets, respectively, and FM represents $(Mg + Fe^{total})$. As noted in the Methods section, this composition range corresponds to that between pure majorite garnet and pyrope-almandine garnet, and it is notable that the more Al-rich MgSi-perovskite compositions overlap in Fig. 6 with those of natural garnet inclusions with a majoritic component believed to have formed in the depth range 250–550 km (e.g. Harte 2010). This is in agreement with the extensive experimental evidence (e.g. Irifune et al. 1996; Hirose and Fei 2002; Stixrude and Lithgow-Bertelloni 2007; Perillat et al. 2006; Ricolleau et al. 2010) of replacement of a maj-grt stability field in the Upper Mantle with one of Al-bearing MgSi-perovskite in the Lower Mantle.

A broad consistency of natural inclusion and experimental MgSi-perovskite compositions is also shown in Fig. 6a. Natural MgSi-perovskite compositions associated with fPer in the same diamond are usually believed to be from ultrabasic bulk compositions; and in Fig. 6a, they are seen to typically plot near the end of the x-axis (close to pure MgSi-perovskite) and adjacent to MgSi-perovskite compositions from experiments on ultrabasic bulk (e.g. pyrolite) compositions. Two exceptions to this are given by the two known high-Al MgSi-perovskite inclusions associated with fPer, and which are believed to have formed at higher pressure (see above). For basic bulk compositions, the natural inclusion and experimental bulk compositions overlap. Interestingly, within the range of experimental pressure conditions (22–37 GPa) illustrated, no systematic changes in composition with pressure are evident. Nor do the mineral data from 'wet' experiments differ significantly from those done without H_2O (Litasov and Ohtani 2005); and this is consistent with Harte's (2010) suggestion that the formation of the diamonds and their inclusions may be associated with the dehydration reactions of dense hydrous Mg-silicates in subducted slabs.

TAPP, NaAl-pyroxene and Majoritic Garnet

The phase listed as TAPP in Table 2 is a Fe–Mg–Al-silicate phase that until very recently had only been found as inclusions in diamonds and never in high-P–T experimental studies. It has a tetragonal crystal structure, distinct to that of garnet, but its compositions may be recast to conform with cation-site occupancies found in garnets of pyrope-almandine composition; and it was for this reason that it was given the provisional name of tetragonal almandine-pyrope phase and the acronym TAPP (Harris et al. 1997; Harte et al. 1999). The mineral was considered to be part of the Upper Mantle/Lower Mantle boundary and Lower Mantle assemblages because it was found associated with fPer and mPv inclusions in the same diamond, and sometimes appeared to occur as a single-phase inclusion. Harte (2010) noted that by comparison with experimental data for ca UM/LM boundary P–T conditions, TAPP seemed to be occurring in place of garnet. However, the placing of TAPP in the UM/LM boundary assemblages has been enigmatic because its molar volume differs little from garnet and its Si is in [4]-fold coordination (Harris et al. 1997; Finger and Conrad 2000; Armstrong and Walter 2012). This enigma has been largely resolved by Armstrong and Walter (2012) who recently synthesised TAPP for the first time and found that it only had a stability field at relatively low pressures (6–10 GPa); therefore, leading them to the conclusion that TAPP must be a retrograde product, probably forming mainly from mPv(Al) but possibly also from garnet.

By their nature, TAPP inclusions plot close to the origin (corresponding to pyrope-almandine composition) in the garnet-perovskite end-member plot of Fig. 6b. In several cases, TAPP is recorded in composite inclusions with mPv; and these TAPP compositions plot a little way along the x-axis (Fig. 6b). Although, the mPv compositions in these composite inclusions are poorly defined, it is clear that the composite TAPP compositions added to a mPv composition (such as those shown in Fig. 6a) would yield an overall composition in the mPv(Al) range. Thus, these composite inclusions appear to represent original aluminous MgSi-perovskite as suggested by Armstrong and Walter (2012).

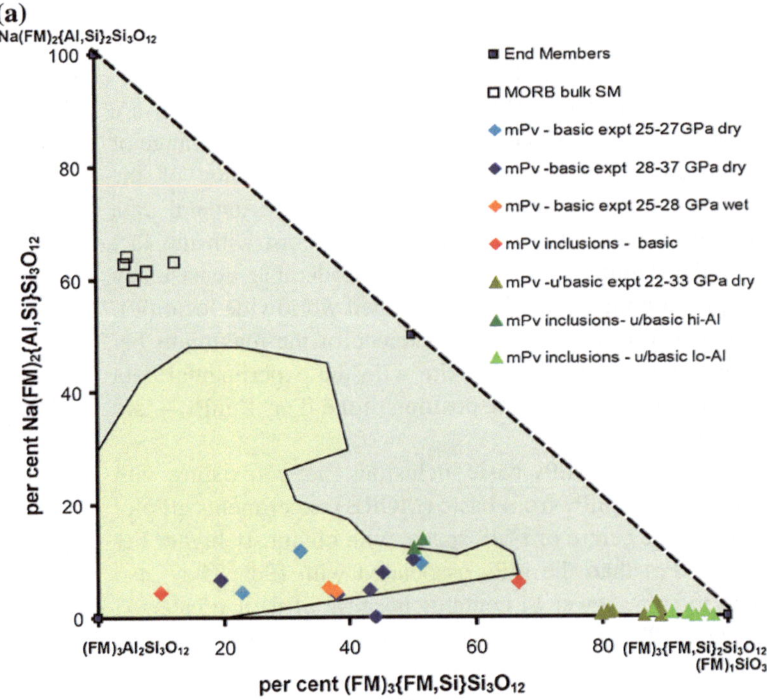

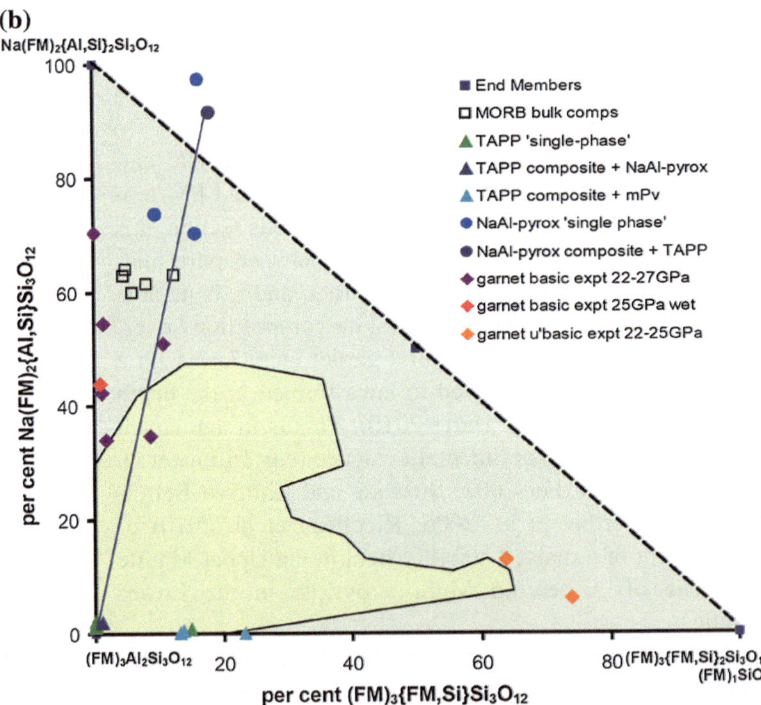

Fig. 6 Plots of MgSi-perovskite (*mPv*), garnet, TAPP and NaAl-pyroxene in garnet-perovskite molecular composition space (see Sect. "Methods", methods for explanation).
a shows the compositions of ultrabasic mPv inclusions that are associated with fPer in the same *diamond* and assumed to indicate ultrabasic bulk compositions (see Fig. 4a for sources) together with two mPv inclusions of basic (MORB-like) composition (Walter et al. 2011). Also shown are the compositions of mPv found in experiments on ultrabasic and basic bulk compositions (see Fig. 4b for sources).
b shows data points for inclusions of TAPP and NaAl-pyroxene (Harte et al. 1999; Stachel et al. 2000; Kaminsky et al. 2001; Hutchison et al. 2001; Hayman et al. 2005; Bulanova et al. 2010); in addition to garnet compositions (Hirose and Fei 2002; Litasov et al. 2004; Litasov and Ohtani 2005) from basic experiments, and the MORB bulk compositions used in such experiments. Garnets found in experiments on ultrabasic and chondritic compositions are from Nishiyama et al. (2004) and Asahara et al. (2004). The *dark blue* line joins the compositions of NaAl-pyroxene and TAPP found in a single-phase inclusion from São Luiz. The pale green shaded field shows the spread of compositions found in majoritic garnet inclusions from Monastery (Moore and Gurney 1989), Jagersfontein (Tappert et al. 2005a, b) and São Luiz (Harte and Cayzer 2007; Harte 2010). Fe calculated as Fe^{total}

In another case, a TAPP occurs in a composite inclusion with NaAl-pyroxene, and the compositions of both phases are shown in Fig. 6b and joined by a tie-line. This tie-line passes through the composition fields of natural maj-grt inclusions and high-pressure experimental basic (MORB) maj-grts, and formation from an original maj-grt is clearly indicated.

However, there remains some enigma in TAPP parageneses. Several occurrences of TAPP are recorded as single-phase inclusions, and these inclusions usually have compositions very close to pure pyrope-almandine with very little additional components such as Ti (Armstrong and Walter 2012). Often these TAPP inclusions are associated with a fPer inclusion in the same diamond, which may be taken as evidence of an ultrabasic bulk composition. Despite their apparent single-phase occurrence, these TAPP compositions cannot be derived one-for-one from a known ultrahigh-pressure phase, and therefore, a reaction relationship is implied. One possibility is that these TAPPs did coexist with other phases in the same inclusion, and that these potentially colourless phases were not detected when the inclusions were investigated. Such a possibility is particularly likely given that early investigations of the inclusions were obtained by cracking the diamonds and hand-picking of the inclusions from the fragmented pieces (Harte et al. 1999; Hutchison et al. 2001).

The NaAl-pyroxene phase listed in Table 2 has a very limited occurrence. It has an enstatite crystal structure, but is believed on experimental grounds to represent a retrogressed Mg- and Na-rich maj-grt, which would have been stable at pressures of 22–25 GPa (Gasparik and Hutchison 2000; Hutchison et al. 2001). In Fig. 6b, the NaAl-pyroxene compositions plot at high Na, near the $Na(FM)_2(Al,Si)_2Si_3O_{12}$ garnet molecule composition and above the bulk compositions of MORB. As noted above, one NaAl-pyrox sample occurs in a composite inclusion with TAPP and appears to have formed from an original maj-grt. Hutchison et al. (2001) also record a composite inclusion of NaAl-pyroxene and garnet (with a small majoritic component) plotting near the TAPP compositions in Fig. 6b. Thus, this inclusion again probably derives from an original garnet with a high-majorite component.

Two of the NaAl-pyroxene inclusions are recorded as single-phase (Table 2), and these pose the same problem as the single-phase TAPP inclusions noted above. Perhaps again, the composite nature of the inclusions was not recognised, and the original inclusions may have been maj-grts. In one case, the NaAl-pyroxene inclusion occurs in association with fPer in the same diamond, and this would commonly be taken to indicate formation in ultrabasic bulk composition.

CaSi-Perovskite

In general, cPv is the expected Ca-bearing phase in the Lower Mantle and as such is expected to occur in both ultrabasic and basic bulk compositions. Inclusions of cPv, including cases where cPv is the only inclusion in a diamond (and therefore not listed in Table 2), are relatively abundant compared to those of mPv (Hayman et al. 2005). Most of the cPv inclusions listed in Table 2, and coexisting with fPer and/or mPv, are characteristically extremely pure, consisting of ca 99 % of the $CaSiO_3$ molecule. Minor impurities are Al, Fe, Mg, Na and Sr, each of which is usually <0.2 wt%, though as much as 0.66 wt% Al_2O_3 and 0.85 wt% SrO are recorded by Stachel et al. (2000). The one cPv (basic) inclusion listed in Table 2 is also very dominantly $CaSiO_3$ in composition, but it is relatively Ti–rich (2.9 wt%), which is believed to indicate a basic rather than an ultrabasic composition (Walter et al. 2011).

Crn, NAL and CF Phases

Inclusions representing the primary original phases of crn (corundum), NAL (new Al–silicate phase) and CF (an aluminous phase with calcium ferrite type structure) have only rarely been reported and so far have largely been found in diamonds containing only one separate inclusion. However, one crn inclusion is recorded in association with mPv(Al) and fPer from Juina-São Juiz (Table 2 and Hutchison et al. 2001). The association with mPv(Al), rather than Al-poor mPv, is in accord with experimental data (e.g. Irifune et al. 1996) showing the occurrence of crn with mPv(Al) as the mPv composition field expands with depth and garnet becomes unstable. The crn inclusion associated with mPv(Al) in the same diamond is approximately 96 % Al_2O_3 in chemical composition with only minor amounts of Si, Mg and Fe (Hutchison et al. 2001).

The NAL and CF phases (Table 2) are represented by inclusions consisting of a composite of retrograde phases, but the bulk compositions of the inclusions fall in the range of experimentally synthesised phases in basic (MORB) rock compositions at upper Lower Mantle P–T conditions (Walter et al. 2011). They are both Mg–Fe–Al–Si–O phases with very high-Al content; but in addition the CF phase carries substantial Na (around 10 wt%), while NAL has moderate Na and K contents (1.0 to 6.0 wt%). They have been widely reported in high-P–T experimental investigations of basic (MORB) bulk compositions (e.g. Ono et al. 2001; Hirose and Fei 2002; Litasov et al. 2004; Litasov and Ohtani 2005; Perillat et al. 2006; Ricolleau et al. 2010) and evidently carry a substantial relative proportion of the Al and Na in basic rock compositions as the garnet stability range contracts near the top of the Lower Mantle (Fig. 1).

The extrapolated NAL and CF inclusion compositions are plotted in Al–Fe–Mg and Na–Fe–Mg space in Fig. 7 and may be seen to be similar to experimental mineral compositions determined in basic (MORB) bulk compositions. Given that the newly found inclusions have undergone extensive retrogression, it seems likely that similar inclusions have been overlooked in the past, and that more NAL and CF inclusions will be discovered in the future.

Inclusion and Experimental Mineral Assemblages in P–T Space

Following the presentation of the range of mineral associations found in the diamonds (Table 2), we have shown that the detailed compositional features of the inclusions show considerable coherence and compatibility, and close similarities to phases found in experimental investigations. In Table 2, we noted the potential subdivision of the putative phase assemblages into several groups with

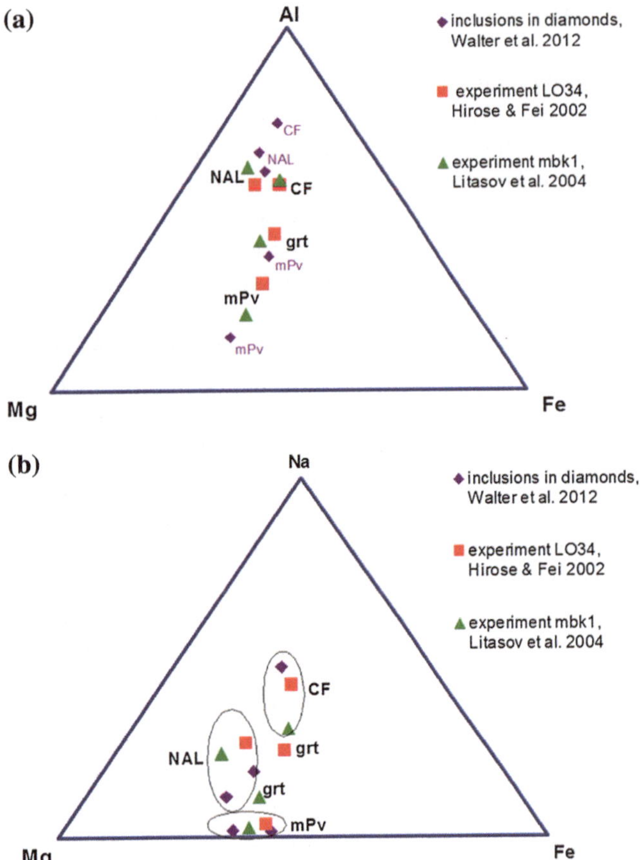

Fig. 7 **a** Al–Fe–Mg and **b** Na–Fe–Mg atomic plots comparing basic natural inclusions compositions with experimental data. Natural compositions from Walter et al. (2011); experimental data from Hirose and Fei (2002), run LO#34 at 27 GPa; Litasov et al. (2004), run mbk-1 at 26.1 GPa. In addition to mPv, garnet, NAL and CF, the experimental run-products also included stv and cPv. All data plotted with Fe as Fe^{total}

fPer + mPv + rw (± cPv) placed at the UM/LM boundary, and a series of other associations involving mPv and mPv(Al) suggesting a range of P–T conditions in the upper Lower Mantle. With respect to the 'uncertain' inclusion associations (group 4, Table 2), we noted that they could not be precisely placed in the Upper or Lower Mantle, but that conditions in the region of the UM/LM boundary were appropriate. Assuming ultrabasic–basic bulk rock compositions, the occurrence of cPv in these group 4 inclusion associations indicates depths of formation in the lower part of the Upper Mantle Transition Zone and deeper, but does not by itself provide tight P–T constraints. However, analysis of the compositional relationships of some TAPP and NaAl-pyrox retrograde inclusions clearly indicates retrograde derivation from maj-grt, and the occurrence of this phase may now be considered in conjunction with the mPv, mPv(Al), crn, NAL and CF phase assemblages.

Extensive experimental work, on both basic and ultrabasic bulk compositions, has shown maj-grt to be the principal Al-bearing phase down to the UM/LM boundary (e.g. see summaries by Stixrude and Lithgow-Bertollini (2007), Perillat et al. (2006), Ricolleau et al. (2010), and Fig. 1). With particular regard to sodic maj-grt compositions, Gasparik and Hutchison (2000) found that garnet compositions, corresponding with the NaAl-pyroxene inclusions, were stable at pressures of 22–25 GPa (ca 620–700 km) depending on temperature. The lowest pressures of stability of the Al–Na-bearing NAL and CF phases are probably close to 23 GPa (Guignot and Andrault 2004), and experimental data show that these phases are established and occur without garnet in basic (MORB) rock compositions by 28 GPa (e.g. Ono et al. 2001; Ricolleau et al. 2010). The coexistence of garnet with NAL and CF at pressures of 26–27 GPa (temperatures 1,700–2,200 °C) was shown by Hirose and Fei (2002) and Litasov et al. (2004). Altogether, these data indicate a change from the stability of maj-grt to NAL and CF in MORB rock compositions between ca 24 and 28 GPa (ca 670–770 km, over a range of temperatures). In addition, in the same depth range, experimental data (e.g. Irifune et al. 1996) show that the mPv composition field widens considerably and this phase becomes capable of carrying considerable Al and may coexist with corundum (crn). Thus, in this narrow range, in the uppermost Lower Mantle, experimental data show a progression from assemblages with maj-grt, to ones with mPv(Al), NAL, CF and crn as the principal Al–Na-bearing phases. The primary natural inclusion assemblages clearly mirror the same sequence of changes, and the whole suite of Perovskite and Periclase Suite mineral associations (Table 2) might be formed within a limited depth range of ca 550–800 km.

The widespread agreement between the natural inclusion associations and the experimental data enables some further exploration of the probable critical mineral reactions and assemblages. Consideration of the tabulated mineral compositions in conjunction with the principal chemical components of peridotitic (ultrabasic) and basaltic/MORB (basic) bulk compositions shows the essential chemical system to be explored is Na_2O–CaO–FeO–MgO–Al_2O_3–SiO_2 (NCFMAS), and if we solely consider potential phases in the Lower Mantle (i.e. at depths below ringwoodite stability), the principal phases found in natural and experimental studies are mPv, garnet, NAL, CF, fPer-mW, cPv, stv and crn. These eight phases in a 6 component system must define an invariant point in P–T space, and if the compositions of the phases at the invariant point may be estimated, then, following Schreinemaker's rules (e.g. Korzhinskii 1959), this chemography will allow definition

Table 3 Univariant reactions in NCFMAS system for Eight Phases[a]

Reaction[b]	Low-P assemblage[c]		High-P assemblage[c]
(fPer) (HF)	124.2 grt + 10 NAL	=	53.3 mPv + 56.2 CF + 17.3 stv + 7.3 cPv + 0.1 crn
(fPer) (Lit)	253.1 grt	=	117.4 mPv + 37.1 CF + 40.9 stv + 41.3 cPv + 10 NAL + 6.3 crn
(NAL) (HF)	24.6 grt	=	10 CF + 7.9 mPv + 4.3 stv + 1.5 cPv + 0.9 fPer + 0.05crn
(NAL) (Lit)	56.5 grt +1.6 fPer	=	10 CF + 29.7 mPv + 7.8 stv + 9.2 cPv + 1.4 crn
(grt) (HF)	13.6 mPv + 5.7 CF	=	10NAL + 4.7fPer + 4.2stv + 0.2cPv + 0.2crn
(grt) (Lit)	15.5 mPv + 7.7 CF + 0.03 cPv	=	10 NAL + 7.1 fPer + 6.1 stv + 0.1 crn
(mPv) (HF)	42.5 grt	=	11.6 stv + 10 NAL + 6.3 fPer + 2.7 cPv + 11.6 CF + 0.2 crn
(mPv) (Lit)	29.6 grt + 2.4 CF	=	10.2 stv + 10 NAL + 6.2 fPer + 4.8 cPv + 0.8 crn
(CF) (HF)	13.9 grt + 9.1 mPv	=	10 NAL + 6.6 stv + 5.2 fPer + 1 cPv + 0.2 crn
(CF) (Lit)	43.2 grt	=	10 NAL + 12 stv + 5.9 fPer + 7 cPv + 7.2 mPv +1.1 crn
(stv) (HF)	24.3 grt + 10 NAL + 3.8 fPer + 0.1 crn	=	21.4 mPv + 15.6 CF + 1.3 cPv
(stv) (Lit)	44.2 grt +10 NAL + 8.3 fPer	=	38.7 mPv + 15.5 CF + 7.2 cPv + 1.0 crn
(cPv) (HF)	10 NAL + 4.6 fPer + 3.7 stv + 2.8 grt + 0.2crn	=	14.5 mPv + 6.8 CF
(cPv) (Lit)	10 NAL + 7.1 fPer + 6.1 stv + 0.1 crn	=	15.4 mPv + 7.6 CF + 0.2 grt
(crn) (HF)	82.8 grt + 10 NAL + 1.6 fPer	=	40.1 mPv + 39.3 CF + 10.1 stv + 4.8 cPv
(crn) (Lit)	2.3 grt + 10 NAL + 7.1 fPer + 5.8 stv	=	16.7 mPv + 8.1 CF + 0.4 cPv

[a] The phases are grt, mPv, fPer, NAL, CF, stv, cpv, crn

[b] The univariant reactions are indicated by the phase absent from each reaction (see text)

[c] In general, garnet (grt) is assumed to be stable on the low-pressure (low-P) side of the reaction; in the case of the (grt) absent reaction, there is probably only a small dependence on pressure

of the eight univariant reactions meeting at the invariant point. The experimental mineral compositions indicated in Fig. 7 potentially provide most of the chemical information needed, and it is noteworthy that similar chemographic sequences are given by both the Hirose and Fei (2002) and Litasov et al. (2004) data sets—with Fe/(Mg + Fe) ratios in the order: CF>grt>mPv>NAL and Na/(Na + Fe + Mg) ratios in the order: CF>NAL>grt>mPv. On the basis of partitioning data for Fe/Mg between mPv and fPer (Fig. 5), fPer coexisting with grt, mPv, NAL and CF will have the most Fe-rich composition, with Fe/(Mg + Fe) near 0.5 (at the boundary of ferropericlase and magnesiowustite) and even into the magnesiowustite field. Such ferropericlase-magnesiowustite compositions have been found as single-phase inclusions in diamonds and in association with cPv (see section on ferropericlase and Figs. 3 and 5). Note that for simplicity, all ferropericlase-magnesiowustite compositions are abbreviated as fPer in Table 3 and Fig. 8. The other phase composition missing from the Hirose and Fei (2002) and the Litasov et al. (2004) data is that of corundum, and for this, we have adopted the composition of inclusion BZ241C from São Luiz (Hutchison et al. (2001).

In Table 3, we show the calculated univariant reactions around the NCFMAS invariant point using the precise mineral compositions given by Hirose and Fei (2002) and Litasov et al. (2004), coupled with the fPer and crn compositions noted above. Figure 8 shows the sequence of reactions around the invariant point using Schreinmakers relationships for the Hirose and Fei (2002) and the Litasov et al. (2004) mineral reaction equations (Table 3), roughly orientated in P–T space on the general assumption that garnet ceases to be stable with increasing pressure as shown by the experimental evidence. For simplicity, we show only five of eight reactions, because we believe cPv, crn and stv may always be present in the full assemblages, and therefore, the reactions in which they are absent will not be relevant. The mineral assemblages are illustrated in Na–Fe–Mg plots as these allow rigorous projection from cPv, crn and stv as ubiquitous phases (though precise phase compositions are somewhat distorted in the figure for clarity of presentation). Experimental data (op. cit.) widely show the occurrence of stv and cPv in equivalent assemblages. The occurrence of crn is much less widely noted, but it is notable that the quantities of crn involved are always very low, irrespective of whether the Hirose and Fei (2002) or Litasov et al. (2004) compositions are used (see Table 3). The precise reaction equations and reaction sequence around the invariant point differ slightly for the two sets of experimental data (Table 3, Fig. 8), but the major features of the assemblage topology are closely similar. The differences principally reflect relative changes in the garnet composition in the experiments. As may be seen in Figs. 7 and 8, the garnet compositions in both experiments are nearly colinear with the mPv-CF and the NAL-fPer tie-lines. Thus, a slight shift in garnet composition moves it across the tie-lines and changes the stoichiometric relations.

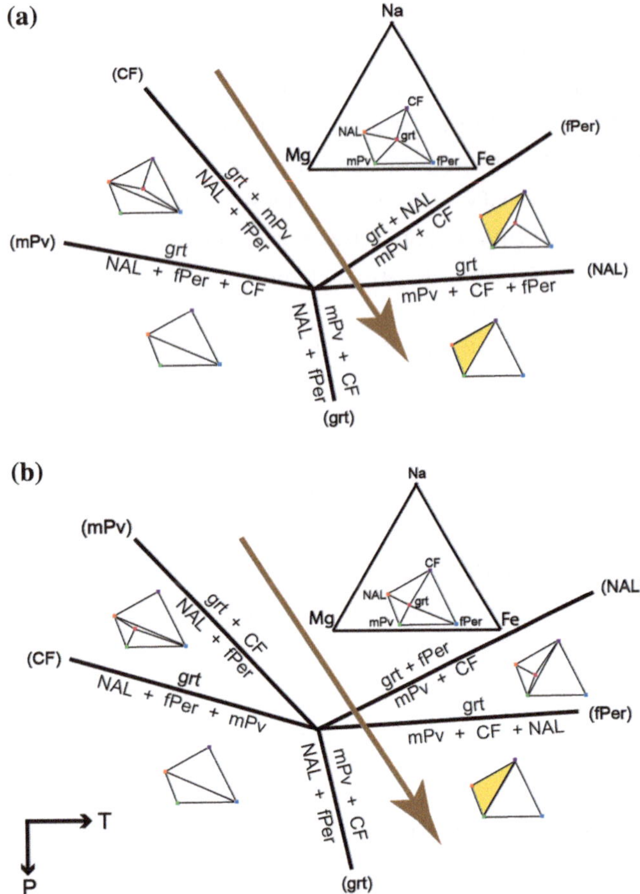

Fig. 8 Schematic illustration of an invariant point for the phases mPv, NAL, CF, garnet (*grt*), ferropericlase-magnesiowustite (*fPer*), cPv, stv and corundum (*crn*) in Na_2O–CaO–FeO–MgO–Al_2O_3–SiO_2 composition space. Reactions are based principally upon mineral compositions from experimental charges of basic (MORB-like) bulk composition: **a** experiment LO34 of Hirose and Fei (2002); **b** experiment mbk1 of Litasov et al. (2004); see also Fig. 7 and text for further details. For clarity of displaying the compatible assemblages in the diagrams, the mineral compositions in compatibility triangles have been portrayed with a widened range of Na–Mg–Fe composition, in particular with fPer being shown with an exaggerated Fe content (see text). Reactions are arranged around the invariant point following Schreinemaker's rules and with the assumption that garnet will be absent at the higher pressure P–T conditions. All assemblages are believed to coexist with stv, cPv and crn, and the (stv), (cPv) and (crn) univariant reactions (not involving these phases) have therefore been omitted. The mPv+CF+NAL compatibility triangles, corresponding to the potential assemblage fields indicated by inclusions in diamonds from Pipe five Juina (Walter et al. 2011), are coloured yellow. The *brown arrow* indicates the potential trajectory of a mantle geotherm giving rise to the mPv+CF+NAL assemblages

It should also be remembered that the single-point compositions used in Table 3 and shown in Fig. 8 only apply to a single pressure and temperature and are based on experimental 6-phase (divariant) assemblages and not the full 8-phase invariant point assemblage. Thus, while the mineral compositions are believed to be close to those of the invariant point, they will not be identical and will change with P–T. With significant change in P–T conditions, there will be marked changes in possible mineral compositions; in particular, the whole composition range for garnet will expand with decreasing pressure, while that for mPv contracts. Prediction of precise changes in mineral compositions and assemblages with P–T and bulk rock composition requires development of an integrated set of thermodynamic parameters (see Holland et al. in preparation); and this will simultaneously allow more accurate estimation of pressures and temperatures of formation for the inclusions in diamonds.

The actual reactions shown in Fig. 8 will most obviously affect basic bulk compositions, since normal ultrabasic compositions do not have enough Na and Al to develop CF, and in some cases, NAL (see for example, Stixrude and Lithgow-Bertelloni 2007; Holland et al. in preparation). From the experimental data, the invariant point of Fig. 8 should be stable in the pressure range ca 24–28 GPa (ca 670–770 km, over a range of temperatures). The disappearance of garnet near the top of the Lower Mantle in mineral abundance diagrams (see Fig. 1) will be strongly controlled by the mineral reactions emanating from the invariant point. Figure 8 thus provides a template for mineral assemblages to be expected in basic rock compositions in both inclusions and experiments. If we assume that the total set of inclusions found in diamonds from Pipe five Juina (Walter et al. 2011) form an equilibrium assemblage of mPv + NAL + CF, then the assemblage would be stable in the relatively high-pressure and high-temperature-phase fields between the (fPer) and (grt) univariant reactions shown in Fig. 8. This suggests that at the time of formation of these assemblages, the mantle geotherm passed to the high-temperature side of the invariant point.

Conclusions

1. The 'Perovskite and Periclase Suite' of inclusions in diamonds includes associations of minerals indicating formation in both ultrabasic (meta-peridotite) and basic (meta-basite) rock compositions. At worldwide localities (South Africa, West Africa, South America, Canada, Australia), it is most particularly represented by the ultrabasic assemblage MgSi-perovskite + ferropericlase. Other primary phases belonging to the suite are the following: CaSi-perovkite (cPv), ringwoodite (rw), corundum (crn), majoritic garnet (maj-grt), new Al–silicate phase (NAL), calcium ferrite structured phase (CF) and stishovite (stv). The MgSi-perovskites show a wide range of composition, and both low-Al and high-Al varieties are recognised and referred to as mPv and mPv(Al), respectively.

2. Many of the primary phases are affected by retrogression to low-pressure phases, but evidence of their original nature can often be determined, except in the case of some inclusions containing the retrograde phase TAPP, which may be derived from maj-grt or mPv(Al). Composite inclusions of TAPP and a NaAl-pyroxene phase clearly point to an origin from maj-grt. Identification of assemblages of primary minerals is believed to be best based upon the occurrence of separate inclusions within the same diamond.
3. The phases largely show consistent and limited ranges of major-minor element composition. This includes Fe–Mg partitioning between mPv and fPer, and between rw and fPer.
4. The Juina province in Brazil shows a much wider range of inclusions than other localities. This includes occurrence of more Fe-rich mPv-fPer associations, and the occurrence of basic assemblages involving the phases maj-grt, NAL and CF.
5. The chemical compositions of the inclusions show many similarities to analyses of phases from experimental investigations. The similarity of experimentally determined mineral compositions over a range of 23–33 GPa (corresponding to a depth range of approaching 300 km) also conforms with the limited variation in inclusion Fe–Mg partition coefficients between worldwide localities. It also appears from experimental data, that mineral compositions may be similar in 'dry' and 'wet' systems, and it is therefore possible for many of the inclusion assemblages to have coexisted with fluid/melt as suggested by Harte (2010).
6. Overall, by correlation with data from experimental investigations, the ultrabasic inclusion assemblages show a progression with depth from mPv + fPer + rw to mPv + fPer to mPv(Al) + fPer; while the basic inclusion assemblages identified indicate a progression with depth from maj-grt to mPv(Al) + NAL + CF. The cPv phase is widely found as inclusions and appears to be the main Ca-rich phase found with all the principal assemblages in the Perovskite and Periclase Suite. SiO_2 (stv) inclusions are also widespread, though assessing their primary or secondary nature is often unclear. Al_2O_3 (crn) has been found as inclusions associated with mPv(Al).
7. From experimental data, these changes in ultrabasic and basic inclusion assemblages may take place in the depth range 670–770 km (ca 24–28 GPa), and it is quite possible for all the inclusion assemblages to have formed within the depth range ca 550–800 km (Harte 2010), though greater depths of formation cannot be excluded. The depth range 550–800 km coincides with that which might be expected if the mineral associations formed in a stagnant subducted slab (Harte 2011; Harte and Richardson 2012).
8. The compositions of the eight principal primary minerals of the Perovskite and Periclase Suite (MgSi-perovskite, cPv, maj-grt, periclase-wustite, NAL, CF, corundum and stishovite) essentially lie in the 6-component system Na_2O–CaO–FeO–MgO–Al_2O_3–SiO_2 (NCFMAS). These eight minerals may, therefore, be linked to a single invariant point in pressure–temperature space. Experimental data suggest this invariant point lies between 24 and 28 GPa (ca 680–780 km depth).

Acknowledgments We wish to thank Michael Walter, Galena Bulanova, Lora Armstrong and their colleagues for giving us access to their articles 'in press'. Thomas Stachel, Jeff Harris and Tim Holland are thanked for helpful reviews and discussions. The 10IKC Organising Committee is thanked for funding BH's attendance at the 10th IKC. Unfortunately, in November 2012, following the acceptance of this chapter for publication, Neil Hudson died of pancreatic cancer; Ben Harte wishes to record Neil's contributions to this manuscript, and his pleasure of collaborating with Neil over more than 40 years.

References

Armstrong LS, Walter MJ (2012) Tetragonal almandine pyrope phase (TAPP): retrograde Mg-perovskite from subducted oceanic crust? Eur J Min 24:587–598

Asahara Y, Kubo T, Kondo T (2004) Phase relations of a carbonaceous chondrite at lower mantle conditions. Earth Planet. Sci Lett 143:421–432

Brenker FE, Vollmer C, Vincze C, Vekemans B, Szmanski A, Janssens K, Szaloki I, Nasdala L, Joswig W, Kaminsky F (2007) Carbonates from the lower part of the transition zone or even the lower mantle. Earth Planet Sci Lett 260:1–9

Brey GP, Bulatov V, Girnis A, Harris JW, Stachel T (2004) Ferropericlase—a lower mantle phase in the upper mantle. Lithos 77:655–663

Bulanova GP, Walter MJ, Smith CB, Kohn SC, Armstrong LS, Blundy J, Gobbo L (2010) Mineral inclusions in sublithospheric diamonds from Collier 4 kimberlite pipe, Juina, Brazil: subducted protoliths, carbonated melts and primary kimberlite magmatism. Contrib Miner Petrol 160:489–510

Cayzer NJ, Odake S, Harte B, Kagi H (2008) Plastic deformation of lower mantle diamonds by inclusion phase transformations. Eur J Miner 20:333–339

Davies RM, Griffin WL, Pearson NJ, Andrew AS, Doyle BJ, O'Reilly SY (1999) Diamonds from the deep: pipe DO-27, Slave craton, Canada. In: Gurney JJ, Gurney JL, Pascoe MD, Richardson SH (eds) Proceedings of the VIIth international kimberlite conference, J.B. Dawson volume, Red Roof Design, Cape Town, pp 148–155

Davies RM, Griffin WL, O'Reilly SY, Doyle BJ (2004) Mineral inclusions and geochemical characteristics of microdiamonds from the DO27, A154, A21, A418, DO18, DD17 and Ranch Lake kimberlites at Lac de Gras, Slave craton, Canada. Lithos 77:39–55

Droop GTR (1987) A general equation for estimating Fe^{3+} concentrations in ferromagnesian silicates and oxides from microprobe analyses, using stoichiometric criteria. Mineral Mag 51:431–435

Dziewonski AM, Anderson DL (1981) Preliminary reference Earth model. Phys Earth Planet Int 25:297–356

Fei Y, Wang Y, Finger LW (1996) Maximum solubility of FeO in (Mg, Fe)SiO_3-perovskite as a function of FeO content in the lower mantle. J Geophys Res 101:11525–11530

Finger LW, Conrad PG (2000) The crystal structure of ''Tetragonal Almandine-Pyrope Phase'' (TAPP): a reexamination. Am Mineral 85:1804–1807

Gasparik T, Hutchison MT (2000) Experimental evidence for the origin of two kinds of inclusions in diamonds from the deep mantle. Earth Planet. Sci Lett 181:103–114

Guignot N, Andrault D (2004) Equations of state of Na-K-Al host phases and implications for MORB density in the lower mantle. Phys Earth Planet Int 142:107–128

Harris JW, Hutchison MT, Hursthouse M, Light M, Harte B (1997) A new tetragonal silicate mineral occurring as inclusions in lower mantle diamonds. Nature 387:486–488

Harte B (2010) Diamond formation in the deep mantle: the record of mineral inclusions and their distribution in relation to mantle dehydration zones. Mineral Mag 74:189–215

Harte B (2011) Diamond window into the lower mantle. Science 334:51–52

Harte B, Cayzer N (2007) Decompression and unmixing of crystals included in diamonds from the mantle transition zone. Phys Chem Miner 34:647–656

Harte B, Richardson S (2012) Mineral inclusions in diamonds track the evolution of a Mesozoic subducted slab beneath West Gondwanaland. Gondwana Res 21:236–245

Harte B, Harris JW, Hutchison MT, Watt GR, Wilding MC (1999) Lower mantle mineral associations in diamonds from Sao Luiz, Brazil. In: Fei Y, Bertka CM, Mysen BO (eds) Mantle petrology: field observations and high pressure experimentation; attribute to Francis R. (Joe) Boyd, Geochim Society Spec Publ No 6 pp 125–153

Hayman PC, Kopylova MG, Kaminsky FV (2005) Lower mantle diamonds from Rio Soriso (Juina area, Mato Grosso, Brazil). Contribs Miner Petrol 149:430–445

Hirose K, Fei Y (2002) Subsolidus and melting relations of basaltic composition in the uppermost lower mantle. Geochim Cosmochim Acta 66:2099–2108

Hirose K, Fei Y, Ono S, Yagi T, Funakoshi K-I (2001) In situ measurements of the phase transition boundary in $Mg_3Al_2Si_3O_{12}$: implications for the nature of the seismic discontinuities in the Earth's mantle. Earth planet. Sci Lett 184:567–573

Holland TJB, Hudson NFC, Powell R, Harte B (2013) New thermodynamic models and calculated phase equilibria in NCF-MAS for basic and ultranbasic compositions through the transition Zone and into the Lower Mantle

Hutchison MT, Hursthouse MB, Light ME (2001) Mineral inclusions in diamonds: associations and chemical distinctions around the 670 km discontinuity. Contribs Miner Petrol 142:119–126

Irifune T, Koizumi T, Ando J-I (1996) An experimental study of the garnet-perovskite transformation in the system $MgSiO_3$-Mg_3Al_2-Si_3O_{12}. Phys Earth Planet Int 96:147–157

Ito E, Takahashi E (1989) Postspinel transformations in the system Mg_2SiO_4-Fe_2SiO_4 and some geophysical implications. J Geophys Res 94:10637–10646

Kaminsky F (2012) Mineralogy of the lower mantle: a review of 'super-deep' mineral inclusions in diamonds. Earth Sci Rev 110:127–147

Kaminsky FV, Zakharchenko OD, Davies R, Griffin WL, Khacatryan-Blinova GK, Shiryaev AA (2001) Superdeep diamonds from the Juina area, Mato Grosso State, Brazil. Contrib Miner Petrol 140:734–753

Kaminsky FV, Sablukov SM, Belousova EA, Andreazza P, Tremblay M, Griffin WL (2009a) Kimberlitic sources of super-deep diamonds in the Juina area, Mato Grosso State, Brazil. Lithos 114:16–29

Kaminsky F, Wirth R, Matsyuk S, Schreiber A, Thomas R (2009b) Nyerereite and nahcolite inclusions in diamond: evidence for lower-mantle carbonatitic magmas. Mineral Mag 73:797–816

Korzhinskii DS (1959) Physicochemical basis of the analysis of the paragenesis of minerals. Translated from the Russian. Consultants Bureau Inc, New York, p 142

Litasov KD, Ohtani E (2005) Phase relations in hydrous MORB at 18–28 GPa: implications for heterogeneity of the lower mantle. Phys Earth Planet Int 150:239–263

Litasov K, Ohtani E, Kawazoe T (2004) Absence of density crossover between basalt and peridotite in the cold slabs passing through 660 km discontinuity. Geophys Res Lett 31:L24607

McCammon CA, Stachel T, Harris JW (2004) Iron oxidation state in lower mantle mineral assemblages II. Inclusions in diamonds from Kankan, Guinea. Earth Planet Sci Lett 222:423–434

Moore RO, Gurney JJ (1989) Mineral inclusions in diamonds from the Monastery kimberlite, South Africa. In: Ross J (ed) Kimberlites and related rocks, Their mantle/crustal setting, diamonds and diamond exploration, vol 2. Geological Society of Australia Special Publication No. 14, pp 1027–1041

Moore RO, Otter ML, Rickard RS, Harris JW, Gurney JJ (1986) The occurrence of moisannite and ferro-periclase as inclusions in diamond. In: Smith CB (ed) Fourth international Kimberlite conference, extended abstracts, Geological Society of Australia Abstracts No. 16, pp 409–411

Nishiyama N, Irifune T, Inoue T, Ando J-I, Funakoshi K-I (2004) Precise determination of phase relations in pyrolite across the 66 km seismic discontinuity by in situ x-ray diffraction and quench experiments. Earth Planet. Sci Lett 143:185–199

Ono S, Ito E, Katsura T (2001) Mineralogy of subducted basaltic crust (MORB) from 25 to 37 GPa, and chemical heterogeneity of the lower mantle. Earth planet. Sci Lett 190:57–63

Perillat J-P, Ricolleau A, Daniel I, Fiquet G, Mezouar M, Guignot N, Cardon H (2006) Phase transformations of subducted basaltic crust in the upmost lower mantle. Phys Earth Planet Int 157:139–149

Ricolleau A, Perillat J-P, Fiquet G, Daniel I, Matas J, Addad A, Menguy N, Cardon H, Mezouar M, Guignot N (2010) Phase relations and equation of state of a natural MORB: Implications for the density profile of subducted oceanic crust in the Earth's lower mantle. J Geophys Res 115:B08202

Ringwood AE, Major A (1971) Synthesis of majorite and other high pressure garnets and perovskites. Earth planet. Sci Lett 12:411–418

Stachel T (2001) Diamonds from the asthenosphere and the transition zone. Eur J Min 13:883–892

Stachel T, Harris JW (2008) The origin of cratonic diamonds—constraints from mineral inclusions. Ore Geol Rev 34:5–32

Stachel T, Harris JW, Brey GP, Joswig W (2000) Kankan diamonds (Guinea) II: lower mantle inclusion parageneses. Contribs Miner Petrol 140:16–27

Stachel T, Brey GP, Harris JW (2005) Inclusions in sublithospheric diamonds: glimpses of deep Earth. Elements 1:73–78

Stixrude L, Lithgow-Bertelloni C (2007) Influence of phase transformations on lateral heterogeneity and dynamics in the Earth's mantle. Earth Planet. Sci Lett 263:45–55

Tappert R, Stachel T, Harris JW, Muelenbachs K, Ludwig T, Brey GP (2005a) Diamonds from Jagersfontein (South Africa): messengers from the sublithospheric mantle. Contribs Miner Petrol 150:505–522

Tappert R, Stachel T, Harris JW, Shimizu N, Brey GP (2005b) Mineral Inclusions in diamonds from the Panda Kimberlite, Slave province, Canada. Eur J Min 17:423–440

Tappert R, Foden J, Stachel T, Muelenbachs K, Tappert M, Wills K (2009) Deep mantle diamonds from South Australia: a record of Pacific subduction at the Gondwanan margin. Geology 37:43–46

Walter MJ, Bulanova GP, Armstrong LS, Keshav S, Blundy JD, Gudfinnson G, Lord OT, Lennie AR, Clark SM, Smith CB, Gobbo

L (2008) Primary carbonatite melt from deeply subducted oceanic crust. Nature 454:622–626

Walter MJ, Kohn SC, Araujo D, Bulanova GP, Smith CB, Gaillou E, Wang J, Steele A, Shirey SB (2011) Deep mantle cycling of oceanic crust: evidence from diamonds and their mineral inclusions. Science 334:54–57

Wirth R, Vollmer C, Brenker F, Matsyuk S, Kaminsky F (2007) Inclusions of nanocrystalline aluminium silicate "Phase Egg" in superdeep diamonds from Juina (Mato Grosso State, Brazil). Earth Planet. Sci Lett 259:384–399

Wirth R, Kaminsky F, Matsyuk S, Schreiber A (2009) Unusual micro- and nano-inclusions in diamonds from the Juina area, Brazil. Earth Planet. Sci Lett 286:292–303

Wirth RL, Dobrzhinetskaya L, Harte B, Schreiber A, Green HW (2013) An Fe-Rich oxide inclusion in diamond; apparently from the lowermost Mantle

Juina Diamonds from Kimberlites and Alluvials: A Comparison of Morphology, Spectral Characteristics and Carbon Isotope Composition

D. P. Araujo, J. C. Gaspar, G. P. Bulanova, C. B. Smith, S. C. Kohn, M. J. Walter, and E. H. Hauri

Abstract

Diamonds from the Juina-5 and Collier-4 kimberlites and alluvials in the Juina area, Brazil (which are important occurrences of ultra-deep diamonds) were characterised and studied using cathodoluminescence, FTIR and SIMS. Resorbed forms are most frequent, followed in abundance by octahedral diamonds. Cathodoluminescence revealed a high abundance of non-luminescent stones with minor occurrence of diamonds with blue luminescence, which is consistent with the high abundance of Type II diamonds (> 69 %). Type I diamonds are IaB or highly aggregated IaAB and most lack platelets, implying degradation of such features due to high temperature annealing after growth. The $\delta^{13}C$ distribution of Juina samples forms two groups: -26.3 to -3 ‰ without any significant mode for Juina-5 and Collier-4 diamonds, and -13.8 to -3.4 ‰ with -5 ‰ mode for alluvial stones, suggesting that these two kimberlites are not the main source of the local alluvial diamonds. Intracrystalline $\delta^{13}C$ and N SIMS measurements showed consistent co-variation in only one of five diamonds, providing possible evidence of carbon isotope fractionation. Resorption horizons and erratic C-N co variation for the other samples suggest episodic growth. The internal growth features and N characteristics of Type I diamonds indicate that some of them were probably formed in the lithospheric mantle, while most Type II diamonds are tentatively related to a sublithospheric "ultradeep" paragenesis, which is yet to be confirmed by mineral inclusions. Previous studies suggest that sublithospheric diamonds were transported from the deep mantle and deposited at the base of the lithosphere prior to exhumation by kimberlite, possibly by a mantle plume. The characteristic carbon isotopic compositions of Juina-5 and Collier-4 diamonds compared with alluvial diamonds suggest that distinct diamond populations exist in the kimberlite source region.

Keywords

Juina • Diamonds • Morphology • CL • FTIR • Carbon isotopes • Ultradeep • Mantle plume

D. P. Araujo (✉) · J. C. Gaspar
Instituto de Geociências, Universidade de Brasília, Brasília, DF 70910-900, Brazil
e-mail: dparaujo@unb.br; deboraparaujo@gmail.com

G. P. Bulanova · C. B. Smith · S. C. Kohn · M. J. Walter
School of Earth Sciences, University of Bristol, Bristol, BS8 1RJ, UK

E. H. Hauri
Department of Terrestrial Magnetism, Carnegie Institution of Washington, Washington, DC 20015, USA

Introduction

Juina has gained significant attention in the last two decades due to the occurrence of diamonds with mineral inclusions formed at sublithospheric depths in the upper mantle, transition zone and lower mantle, referred to here as ultradeep mantle. First reports on Juina were made exclusively on

D. G. Pearson et al. (eds.), *Proceedings of 10th International Kimberlite Conference*, Volume 1, Special Issue of the Journal of the Geological Society of India, DOI: 10.1007/978-81-322-1170-9_16, © Geological Society of India 2013

alluvial diamonds from the São Luiz River (Wilding 1991; Harte and Harris 1994; Watt et al. 1994) and later on from other rivers in the same drainage basin (Hayman et al. 2005; Brenker et al. 2007; Wirth et al. 2007; Kaminsky et al. 2001, 2009) and from Vermelho River (Kaminsky et al. 2009). The first investigations of diamonds from Juina kimberlites were from the Collier-4 pipe (Araujo 2002; Walter et al. 2008; Kaminsky et al. 2009; Bulanova et al. 2010), Juina-5 pipe (Araujo 2002; Walter et al. 2011), Pandrea pipes and Aripuanã-1 pipe (Kaminsky et al. 2009), see Fig. 1.

The collective works about Juina diamonds bring a significant number of results allowing comparison between alluvial and kimberlite sources and among different kimberlite sources. In this report, we present results on internal and external morphology, N characteristics and carbon isotopic composition of selected diamonds with unknown paragenesis from kimberlites and alluvial sources and compare them with literature data in order to make an analysis of the differences among the several populations. We also use our results to discuss diamond growth processes and to tentatively place them into the context of the ultradeep diamond mineral parageneses as determined from inclusions in diamonds from the same sample collection (Bulanova et al. 2010; Walter et al. 2008, 2011).

Geological Setting

The Juina Kimberlite Field is located in the Proterozoic Rio Negro-Juruena Province, formed during 1.8–1.55 Ga accretion onto the Palaeoproterozoic and Archaean terrains at the south-western edge of the Amazon Craton (Tassinari et al. 2000). The Juina pipes (~92–95 Ma; Heaman et al. 1998; Kaminsky et al. 2010) are one of a number of Early to Late Cretaceous alkaline suites occurring along a NW–SE regional lineament (125 deg) in Brazil, see Fig. 1. Dating of the Juina-5 kimberlite (Araujo et al. *in prep.*) places this occurrence in the Upper Cretaceous. Former ages previously reported for Juina-5 and Juina-6 pipes with 80.1 and

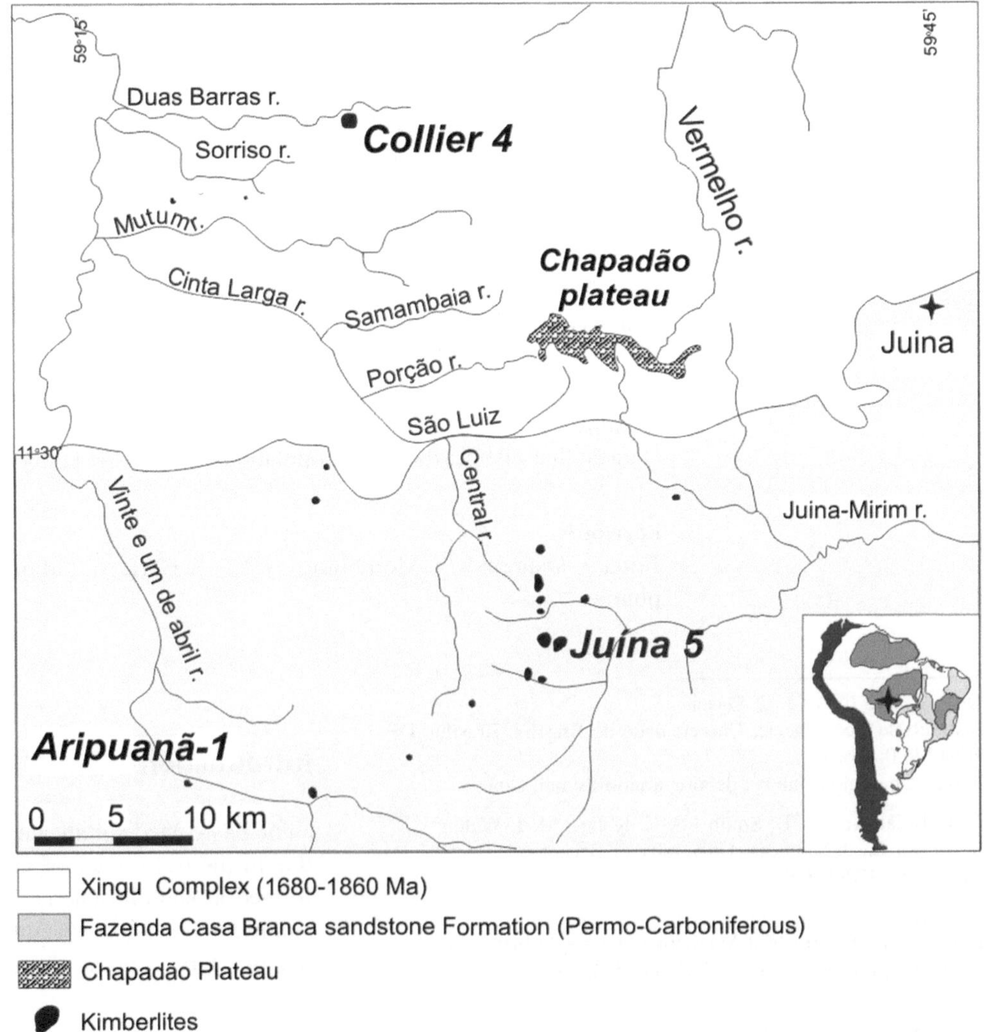

Fig. 1 Location of kimberlites (Haralyi 1991) and diamond-bearing drainages in the Juina Kimberlite Field (Modified from Silva et al. 1980)

79.2 Ma were miscited (Kaminsky et al. 2010) and refer to other occurrences in Brazil (Davis 1977).

Juina kimberlites are intruded into the Proterozoic basement and overlying sediments of the Permo-Carboniferous Parecis Basin. The crystalline basement consists of granite-migmatite terranes and gneisses with tonalitic compositions (Tassinari and Macambira 1999). In the south of the area, the crystalline basement is overlain by Permo-Carboniferous sandstones of the Fazenda Casa Branca Formation, in turn locally overlain by the Cretaceous Parecis Group sandstones (Teixeira et al. 1998) which form the prominent Chapadão plateau outlier (Fig. 1). Diamonds occur both in kimberlites and in Cretaceous palaeoplacers, which are sources for recent alluvial deposits (Haralyi 1991).

Materials and Methods

A total of 193 diamond samples were investigated (Supp. Online Table 1), of which one hundred diamonds are from the Collier-4 kimberlite, 50 from the Juina-5 kimberlite and 43 from alluvial deposits within the *Cinta Larga* Basin (drainages *São Luiz—SL*; *Porção—PO*; *Duas Barras—DB*; and *Cinta Larga—CL*; Fig. 1).

All diamonds were described in detail using an optical microscope following the scheme of Robinson (1980), and 40 stones were imaged using a secondary electron microscope LEO440 at the Mineral Technology Centre (CETEM) in Rio de Janeiro.

A total of 128 diamonds were polished along dodecahedral planes for investigation of internal growth structure by cathodoluminescence (CL) imagery, and among them, twenty were doubly polished to produce plates. CL images were captured with an ELM-3R Nuclide Corporation luminoscope (10–13 kV; 0.5–0.8 mA) coupled to an Olympus BX50 microscope at the Geosciences Institute at the University of São Paulo (São Paulo).

Infrared spectra of 136 diamonds (Supp. Online Table 1) were acquired using a micro-FTIR Bio Radius Excalibur spectrometer, with spatial resolutions of 60, 100 and 400 μm, 128 scans per point and 4 cm^{-1} spectral resolution from 600 to 7,860 cm^{-1} at the Smithsonian National Museum of Natural History, Washington, D.C. All spectra were scaled as absorption coefficients as follows: spectra were baselined using the points at 2,700 cm^{-1} wave number and the lower absorbance between 1,400 and 1,600 cm^{-1}. Absorbance units were converted to absorption coefficients by multiplying the spectra by the factor $11.94/A_{1,994}$, where $A_{1,994}$ is the absorbance at the 1,994 cm^{-1} wave number (dip) in the two-phonon region and 11.94 cm^{-1} is the absorption coefficient for a 1 cm thick diamond at 1,994 cm^{-1}. Nitrogen concentrations were determined from absorption coefficient values at appropriate wave number positions for the various components (Boyd et al. 1994; Kiflawi et al. 1994; Boyd et al. 1995). The data were reduced using the caxbd97n.xls spreadsheet (written by David Fisher, DTC, UK).

Carbon isotopes and N content of sixteen diamond plates imaged by CL and analysed by FTIR, were measured using a Cameca IMS 6F ion microprobe at the Department of Terrestrial Magnetism, Carnegie Institution of Washington. The samples were placed alongside the standard in an indium mount and coated with gold. The C isotope standard was a polished fragment of the Mao diamond that had been previously analysed by conventional gas-source mass spectrometry (Hauri et al. 2002). The measurements were done on three consecutive days, each day representing a single analytical session; standard analyses were interspersed with 6–8 analyses of the Juina diamonds. No drift was observed (within ± 0.3 ‰) on the first and last days; we observed a monotonic drift of −1.4 ‰ during the second day, and this was corrected using a standard linear drift correction. After drift correction, the standard data from the second day had a standard deviation of 0.2 ‰. Over the entire 3 days, the total standard deviation of the standard analyses was ± 0.3 ‰. The uncertainties in the C isotope analyses of the Mao standard combined with the instrumental reproducibility gave an error (1σ) of 0.3 ‰ for each analysis. The spatial isotopic homogeneity of the standard is typically within 0.2 ‰, and the measured values gave a standard error of the mean of 0.20 ‰. After correction for drift and instrument mass fractionation, the $\delta^{13}C$ values were calculated using the $^{13}C/^{12}C$ ratio of 0.0112246 for the Pee Dee belemnite (PDB).

Results

External Diamond Morphology

Diamonds from all groups are mostly light grey and brown, and Juina-5 kimberlite diamonds often have a thin graphite coating. The main morphological features of Juina stones are listed in Supplementary online Table 1, and examples are illustrated in Fig. 2. Diamond morphology spans octahedra, transitional forms (between octahedra and dodecahedra), to dodecahedra and those with undetermined forms due to advanced breakage or resorption, referred to here as irregulars. Table 1 and Fig. 3 summarize the distribution of diamond forms for each population. Some stones show advanced etching. Frequent surface features of the octahedral stones are trigons, frosting (Fig. 2a) and etch channels (Fig. 2b). Resorbed stones have shagreen texture, microhillocks, terraces and lamination lines. The dodecahedral group includes dodecahedron and tetrahexahedral forms,

Fig. 2 Representative diamond morphology of Juina stones. **a** RC1-2—Collier-4 kimberlite—*octahedral* aggregate with coarse frosting. **b** TR2-4—Collier-4 kimberlite—resorbed aggregate with deep etch channels. **c** Ju-5-36–Juina-5 kimberlite—frosted aggregate of macle and octahedral diamond. **d** Ju-5-7—Juina-5 kimberlite—irregular resorbed diamond with shagreen texture and hillocks

the last ones have in their (110) face a medial line usually referred as "c edge" (Robinson 1980), resulting in 24 faces. For simplicity, they were all grouped under the dodecahedra category and their distribution within this group is specified in Table 1.

Diamonds studied from Juina-5 kimberlite are in the size range 0.8–3.2 mm and have a high proportion of octahedra (42 %; Fig. 2a). In addition, 30 % are transitional, 14 % are dodecahedra, and 14 % are so resorbed or broken as to be classified as irregular. Within the octahedral group, 30 % are macles and 6 % are aggregated. In the transitional group, these values are 14 and 2 %, respectively, and for the dodecahedra, 2 and 2 %, also respectively. These figure mean that overall, there are 46 % of macles and 10 % of aggregates (see Fig. 2a and c) in this parcel (also see Supplementary Table 1 and Fig. 3).

The diamond population from Collier-4 kimberlite (0.9–4.5 mm) differs from Juina-5 by the high abundance (54 %) of dodecahedral stones (see Table 1 and Fig. 2d), with only 11 % of octahedra. The remaining proportions are made up of 14 % of transitional stones and 21 % of irregulars (see Fig. 3). A further distinguishing feature is the low abundance of macles (16 %) made up of 4 % from the octahedral, 6 % from the transitional types and 6 % from the dodecahedral forms. The aggregates comprise 11 % of the parcel, making up 4 % from octahedral, 2 % from the transitional forms and 5 % from dodecahedra. In addition, there are no graphite coats in this population.

The alluvial diamonds are in a coarser size fraction range, from 2.3 to 5.6 mm. They are comparable to Collier-4 diamonds regarding their resorption degree and occurrence of macles (15 %) and aggregates (10 %). A total of 58 % of alluvial diamonds are dodecahedra, and 15 % are octahedral diamonds. For the macles, 8, 3 and 5 % are derived from the octahedral, transitional and dodecahedral forms, respectively, and for the aggregates, the percentages

Table 1 Summary of morphology, spectral characteristics and carbon isotope composition of Juina diamond populations in kimberlite and alluvials

	Juina-5 kimberlite (50)[a]	Collier-4 kimberlite (100)[b]	Alluvials (40)[c]
Octahedral	42 %	11 %	15 %
Transition forms	30 %	14 %	5 %
Dodec[d]	14 %	54 %	58 %
Irregular	14 %	21 %	23 %
Macles (of total)	46 %	16 %	15 %
Aggregates (of total)	10 %	11 %	10 %
Size	0.8–3.2 mm	0.9–4.5 mm	2.3–5.6 mm
Type II	89 %	80 %	65 %
N content (of Type I)	67–536 ppm	68–533 ppm	82–919 ppm
% B defect	95–100 %	38–100 %	71–100 %
$\delta^{13}C$	−26.3 to −14.1 ‰	−26.3 to −4.4 ‰	−13.8 to −3.0 ‰

The total number of samples studied is given in parentheses. Carbon isotopic composition and FTIR characteristics were measured for a subset of the samples

[a] Samples (n = 50); FTIR (n = 36); δ^{13} C (n = 3)

[b] Samples (n = 100); FTIR (n = 65); δ^{13} C (n = 8)

[c] Samples (n = 40); FTIR (n = 35); δ^{13} C (n = 6)

[d] Within this category, 43 % are tetrahexahedroid forms for Juina-5, 67 % for Collier-4 and 91 % for alluvials

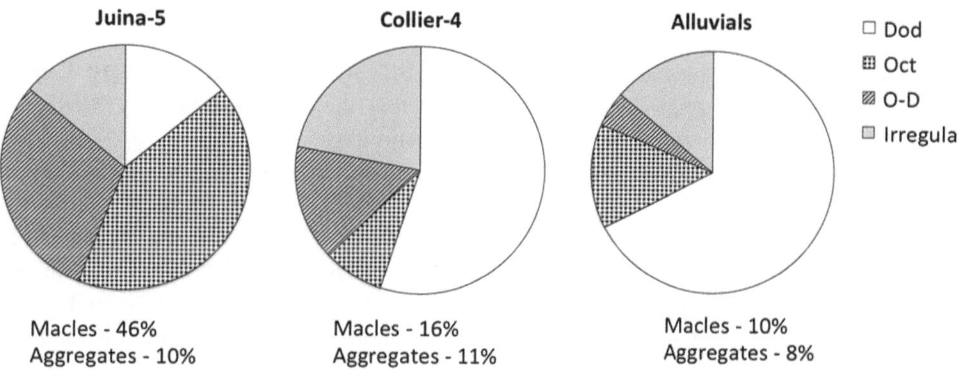

Fig. 3 Pie diagrams summarizing diamond morphology variations between kimberlite pipes and alluvials

are 5 % for both octahedral and dodecahedral shapes. Graphite coating is also absent in alluvial diamonds.

In the limited number of samples we have analysed, there are no unique shape differences in morphology between different drainages. Detailed surface features of all diamonds are reported by Araujo et al. (2001).

Internal Structure

Most of the 128 diamonds examined under CL from Juina kimberlites and alluvial sources gave a low-intensity response, which correspond to the Type II diamonds ascribed for each population (see below). Those specimens with stronger luminescence have features in common to all populations, such as planar octahedral growth and resorption horizons truncating growth zones. Re-filling textures with dark-CL zones and plastic deformation lamination lines are also frequent. Detailed CL interpretation of luminescence diamonds measured by SIMS are further discussed later and illustrated in Fig. 5.

Spectral Characteristics

There is an unusual abundance of Type II (defined as $N < 20$ ppm) samples in all populations, the distribution being as follows: 89 % in Juina-5 kimberlite, 80 % in Collier-4 kimberlite and 65 % among the alluvial diamonds (Table 1). Most Type I diamonds from Juina have N extremely aggregated (>90 % of B defect), except for two diamonds, one from Collier-4 kimberlite (TR2-17) and one from Duas Barras River (DB-5) with 38 and 70 % of B defect, respectively (Supp. Online Table 1 and Fig. 4a). Average N content of Type I diamonds is 281 ppm, reaching 919 ppm (Table 1) for sample SL-2, as measured

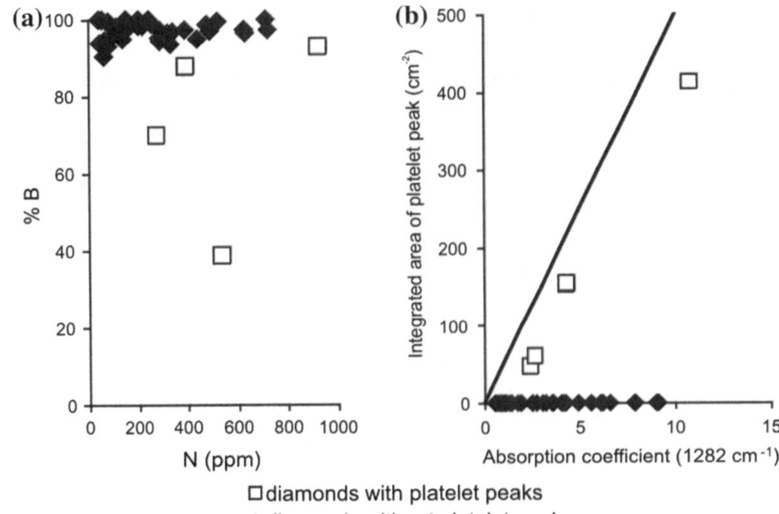

Fig. 4 FTIR characteristics of Type I diamonds from Juina. **a** Nitrogen aggregation (expressed as %1aB) *vs* ppm incorporated nitrogen. **b** Integrated area of the platelet (B′ defect) peak height *vs* the absorption coefficient at 1,282 cm^{-1} resulting from B defects. *The solid line* is the trend for regular diamonds from Woods (1986). In both panes the filled symbols are diamonds with negligible platelet peak intensities and open squares are diamonds with significant platelet peak (TR2-17, DB-5, CL-19 and SL-2). In (**b**) the data show that most samples have undergone severe platelet degradation, but the four samples with measureable platelet peaks form a trend that appears to show a small but constant level of platelet degradation

by FTIR. In situ N content measurements done by SIMS in this diamond give 614 ppm (Table 2), indicating that the infrared analysis crossed zones richer in N not measured by SIMS. This sample, however, also contains a significant component of Type II diamond as measured by SIMS, which will be further discussed in detail.

In Juina Type I diamonds, significant platelet peaks were only detected in four samples: the two with less than 90 % N aggregation mentioned above (38 and 70 %) and another two with 88 and 93 % N aggregation (CL-19 and SL-2; Supp. Online Table 1). These four samples show little or no platelet degradation, as illustrated in Fig. 4b, and are close to the regular diamond trend as defined by Woods (1986), where the intensity of platelet peaks increases with the degree of N aggregation (the correlation line in Fig. 4b). The lack of this positive correlation in aggregated diamonds has been credited to catastrophic degradation of platelets due to high temperature annealing after growth (Woods 1986), which is the status of the rest of Type I diamonds in the population studied—they all lack platelets.

Hydrogen-related peaks at 3,107, 1,405, 2,785 and 3,236 cm^{-1} were detected in all Type I diamonds and correlate positively with the N content. This has been reported before for Juina diamonds (Hutchison 1997; Hayman et al. 2005; Kaminsky et al. 2009) and was attributed to a favourable mechanism for incorporation of both elements (Woods and Collins 1983).

Carbon Isotope Composition and N Content Measured by SIMS

Carbon isotopes were measured at core and rim locations of eleven Type II diamond plates and in traverses across five Type I diamond plates (Table 2).

$\delta^{13}C$ values of Juina diamonds in the present study vary overall from −26.3 to −3.0 ‰, which correspond to the Type II diamonds range. Type I diamonds show a narrower range, from $\delta^{13}C$ −13.8 to −3.4 ‰. The $\delta^{13}C$ range found for Juina-5 kimberlite diamonds is from −26.3 to −14.1 ‰, for Collier-4 kimberlite from −26.3 to −4.9 ‰ and for alluvials, −13.8 to −3 ‰.

Most Type II diamonds show little core to rim internal $\delta^{13}C$ variability, the slight increases or decreases being in the range ≤1.9 ‰. Sample Ju-46, from Juina-5 kimberlite, is one exception, with a 5.3 ‰ internal variation (see Table 2).

Figure 5 brings together CL internal features (highlighted by drawings) and the variation in $\delta^{13}C$ and N content measured in different growth zones along transverses in four Type I diamonds. The N content measured by SIMS varies from 2 to 282 ppm and correlates well with the FTIR data from doubly polished plates, for most diamonds. An exception is diamond SL-2 with high N content measured both by SIMS (614 ppm) and FTIR (919 ppm). The higher N content of the FTIR result relates to growth zones below the surface that were not analysed by SIMS.

Table 2 $\delta^{13}C$ (‰) and N content (SIMS) of Type II (core and rim) and Type I (traverses) Juina diamonds

Type II diamonds							
		$\delta^{13}C$ (‰)				$\delta^{13}C$ (‰)	
		Core	Rim			Core	Rim
Juina-5	Ju-16	−24.6	−23.5	Collier-4	Rc1-2	−9.6	−9.6
	Ju-50	−25.7	−26.3		Rc1-5	−24.7	−25
	Ju-46	−14.1	−19.4		Rc1-8	−4.9	−6
					Rc1-11	−17.7	−18.5
São Luis River	Sl-1	−5.2	−5.3		Tr1-9	−7.4	−6.9
	Sl-7	−3.3	−3		Tr1-13	−26.3	−25.4
Type I diamonds							
		$\delta^{13}C$ (‰)	N (ppm)	São Luis River		$\delta^{13}C$ (‰)	N (ppm)
Cinta Larga River	CL-6p1	−5.2	45		SL2-p1	−12.6	300
	CL-6p2	−5.1	115		SL2-p2	−12.6	26
	CL-6p3	−5.7	38		SL2-p3	−12.6	166
	CL-6p4	−5.5	101		SL2-p4	−13.2	59
	CL-6p5	−5.6	63		SL2-p5	−13.6	405
	CL-6p6	−5.2	10		SL2-p6	−13.1	202
	CL-6p7	−4.6	51		SL2-p7	−12.5	614
	CL-6p8	−4.9	45		SL2-p8	−12.5	25
					SL2-p9	−13.8	9
	CL-11p1	−8.1	71		SL2-p10	−12.7	29
	CL-11p2	−6.8	81		SL2-p11	−13.1	12
	CL-11p3	−6.4	20		SL2-p12	−13.6	333
	CL-11p4	−4.5	2		SL2-p13	−12.8	508
	CL-16p1	−5.9	4	Collier-4	Tr2-3p7	−6.0	65
	CL-16p2	−4.0	282		Tr2-3p6	−6.6	126
	CL-16p3	−3.4	4		Tr2-3p5	−6.3	45
					Tr2-3p4	−5.4	68
					Tr2-3p3	−6.2	90
					Tr2-3p2	−6.0	149
					Tr2-3p1	−5.7	31

N content either decreases towards the rims, see Fig. 5, or shows oscillatory trends. Intra-grain variation in $\delta^{13}C$ in the Type I diamonds is up to 3.6 ‰. $\delta^{13}C$ increases towards the rims (diamonds CL-6, CL-11—Fig. 5; CL-16—Table 2), has an oscillatory trend (diamond TR2-3) or shows fluctuation and constant values along the traverse (SL-2; Fig. 5c). In diamond TR2-3, the change in $\delta^{13}C$ coincides with a resorption horizon as seen in CL imagery (Fig. 5d). There is no consistent correlation between $\delta^{13}C$ and N content for most stones, except for segments along traverses in diamonds TR2-3 (Fig. 5d) and CL-6 (Fig. 5a), which show either positive or negative correlations, and in diamond CL-11 (Fig. 5b), the only diamond that shows a linear negative correlation (Fig. 5).

The internal features of diamond SL-2 are complex (Fig. 5c). Its internal structure suggests that the large area with dark CL (Type II as measured by SIMS) cuts across the growth zones. The $\delta^{13}C$ compositions of Type I and II zones are strikingly similar (−13.6 to −12.6 ‰ for Type I and −13.8 to −12.8 ‰ to Type II; Table 2). $\delta^{13}C$ is constant from intermediate to rim growth zones and fluctuates towards the core.

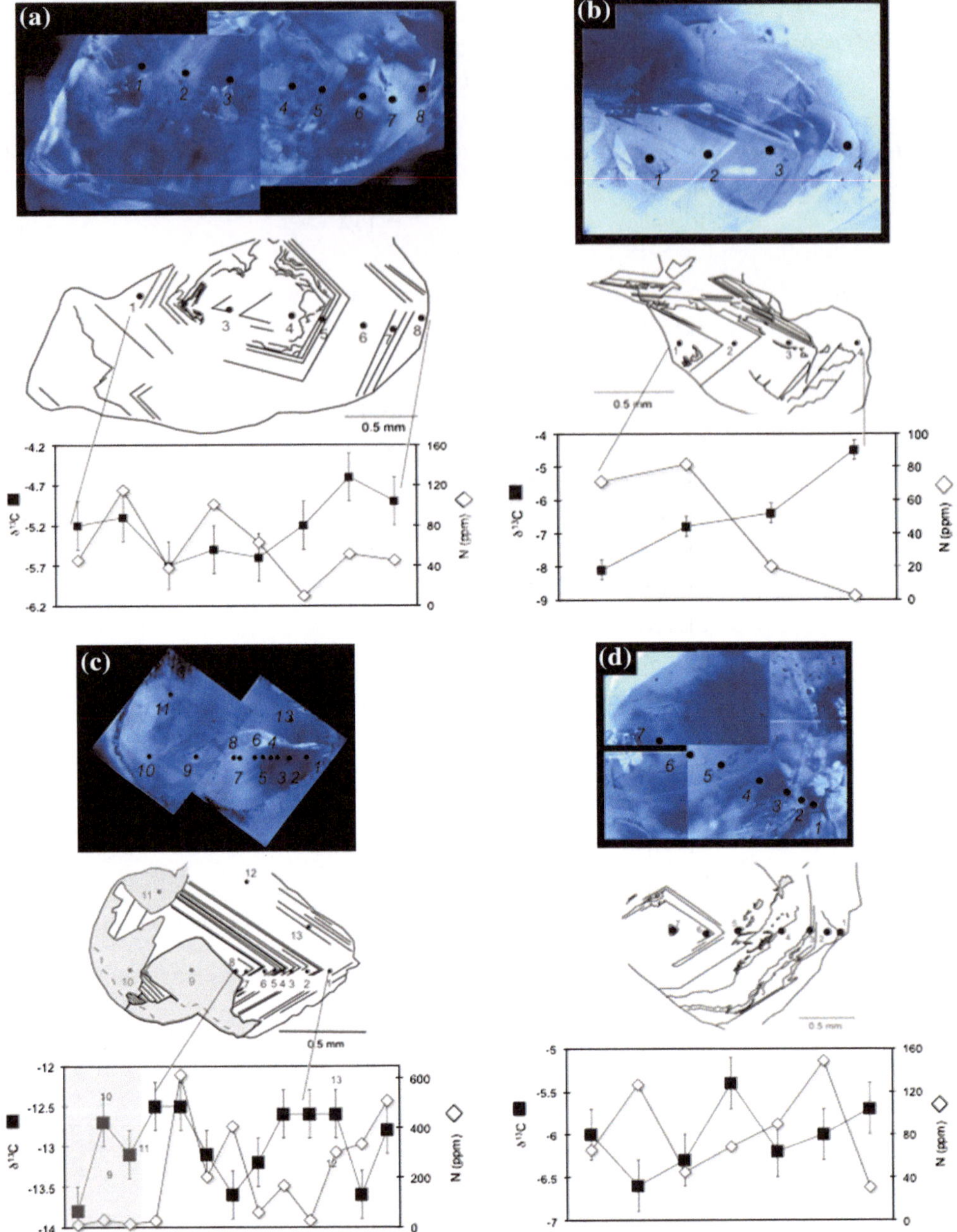

Fig. 5 CL internal features and $\delta^{13}C$ and N variation along traverses in Type I diamonds. **a** CL-6. **b** CL-11 (Cinta Larga river). **c SL-2 (São Luiz River)** and (**d**) TR2-3 (Collier-4 kimberlite). See text for CL description

Discussion

Carbon Isotope Variation and Growth Processes

The carbon isotope composition variation from core to rim zones in the five Type I Juina diamonds studied in detail shows either little increase (up to 3.6 ‰) or oscillatory trends in $\delta^{13}C$ values towards the rims, and for most of them, $\delta^{13}C$ does not correlate with N variation. An exception is sample CL-11, shown in Fig. 6, where the arrows show the $\delta^{13}C$ and N systematically change from core to rim. This diamond has tangential octahedral growth (Fig. 5b) and could possibly have crystallized from a single pulse with increasing uptake of ^{13}C towards the rim due to carbon fractionation. This scenario is in accordance with carbon fractionation from CH_4 fluids in an open system where the loss of CH_4 preferentially depletes the source in ^{12}C (Stachel et al. 2009). However, a similar C–N co-variation could be generated by fractionation of the heavy carbon from CO_2- or CO_3^{2-}-rich fluids in a closed system (Deines 1980). Carbonatite-oxidizing melts derived from subducted slabs were suggested as forming fluids for lower mantle Juina diamonds (Walter et al. 2008, 2011). The lack of other parameters that might constrain the redox state of the forming fluids to our diamonds (such as N partitioning

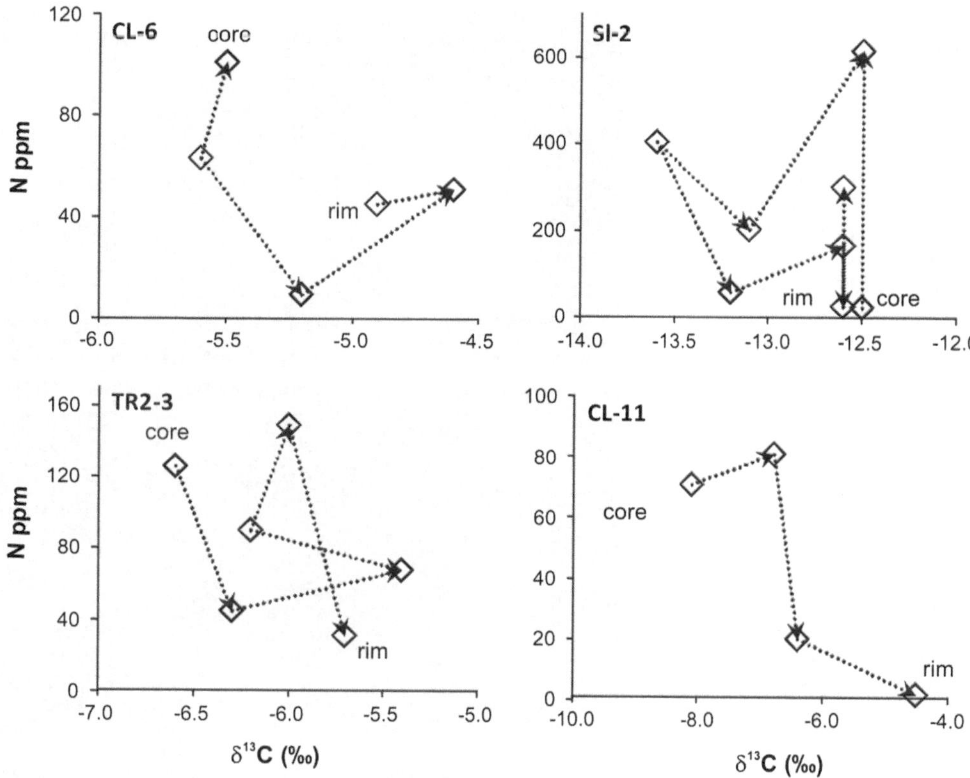

Fig. 6 δ^{13}C-N co-variation of Type I diamonds. The *arrows* indicate the trends from core to rim. **a** Diamond CL-6—point analyses #4 to #8. **b** Diamond SL-2—point analyses #8 to #1. **c** Diamond TR2-3—point analyses #7 to #1. **d** Diamond CL-11—point analyses #1 to #4 (see Fig. 5 for analyses location)

coefficients, temperature and/or mineral inclusions) means that we cannot confirm carbon fractionation as the viable mechanism to produce the C–N variation in sample CL-11, either from reduced or oxidizing fluids. For the remaining samples (CL-6, CL-16 and TR2-3), the erratic C–N co-variations and resorption horizons truncating growth zones (Fig. 5) suggest episodic formation with periods of growth and resorption and fluctuating melt composition.

Diamond SL-2 has an interesting feature regarding its Type II (as measured by SIMS) and Type I zones (Fig. 5c). In the Type I zone, there is intense N fluctuation and the carbon isotope composition is practically constant, with fluctuation of up to 1 ‰ towards lighter composition in the intermediate part of the traverse and back to −12.5 ‰ towards the rim. The intense fluctuation of N is better explained by local shortage of N during diamond growth (Sunagawa 1984). The occurrence of both Type I and Type II zones can be interpreted as an event of strong resorption of Type I diamond and subsequent re-filling with Type II diamond, illustrated in Fig. 5c, where the grey zone (point analyses 9–11) is Type II. The similar carbon isotope composition of both zones suggests re-precipitation of diamond from a fluid/liquid with a similar δ^{13}C, but with low N.

Core and rim carbon isotope compositions of Type II diamonds show variable trends. Most diamonds have little variation (<1.9 ‰), showing either increase or decrease in ^{13}C towards rims. This variation could be explained by carbon isotope fractionation; however, interpretation of the carbon isotope results is difficult due to the little number of analyses and the resolution of their CL images, which did not reveal information on internal growth and resorption features. One diamond, however, shows substantial core to rim variation (JU-46; 5.3 ‰), which is expected in episodic growth from fluids with different carbon isotope compositions. This feature has also been reported in other diamonds from Collier-4 (Bulanova et al. 2010) and Juina-5 (Walter et al. 2011) pipes.

Complexity of Internal Structure and N Characteristics of Ultradeep Diamonds

There are few examples of CL images of ultradeep diamonds. Those published from Juina are from the São Luiz (Hutchison 1997) and Sorriso rivers (Hayman et al. 2005). Invariably, the patterns are highly complex with blocky and brecciated internal structures (see Fig. 7). Although lithospheric diamonds can also sometimes show such complexity, more often in our experience, these diamonds show more continuous and regular growth zones.

Diamonds with non-detectable N and Type I diamonds with strongly elevated N aggregation are frequent in ultradeep populations worldwide (e.g. Deines et al. 1991; Hutchison et al. 1999; Davies et al. 1999; Stachel et al. 2002; Hayman et al. 2005; Bulanova et al. 2010; Walter

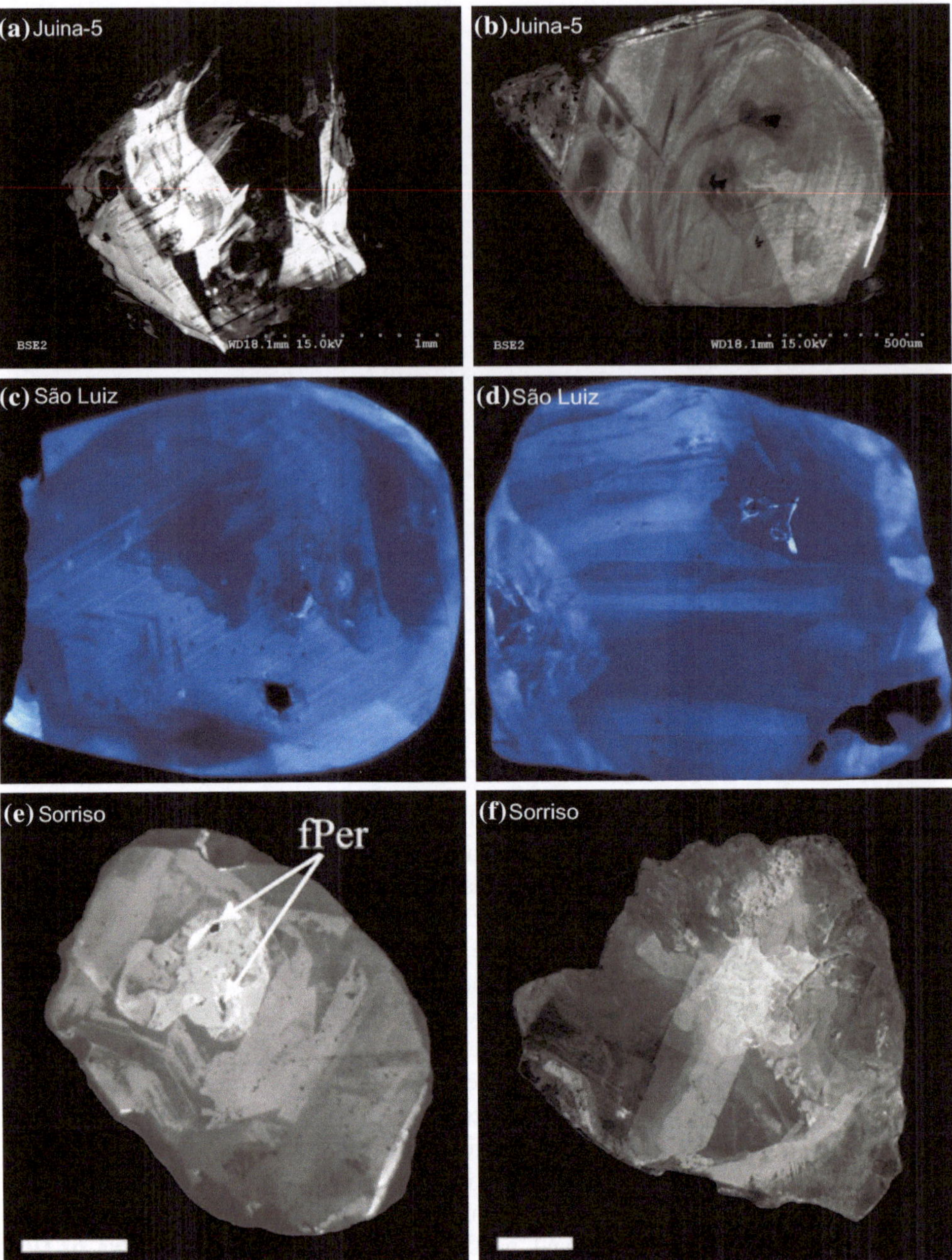

Fig. 7 CL imagery of ultradeep diamonds from Juina. **a** Juina-5 kimberlite: Ju5-20—off-centre, complex internal growth. **b** Juina-5 kimberlite: Ju5-43—off-centre, complex internal features with planar growth and resorption horizons. **c** São Luiz River. BZ251—complex growth and resorption-related zonation (Hutchison 1997) and **d** São Luiz River. BZ254 –off-centre, indistinct concentric zonation (Hutchison 1997). **e** Sorriso River samples 3–5—Strongly resorbed bright CL core with fPer inclusions indicated, episodic octahedral growth and dissolution events and general weakening of CL from core to rim (Hayman et al. 2005). **f** Sorriso River samples 1–4—Bright CL arms radiating from the bright core marking growth on octahedral faces, the *dark zones* between these arms indicating growth on cubic faces. Scale bars are 1 mm (Hayman et al. 2005)

et al. 2011). Although Type II diamonds are reported in significant proportions among lithospheric populations, the accompanying Type I diamonds have, overall, higher N content and lower N aggregation (e.g. Deines et al. 1987, 1989). For example, with respect to nitrogen content, lithospheric Type II diamonds reach 23 % in Koffiefontein mine (Deines et al. 1987), 29 % in Finsch and 25 % in Premier mines (Deines et al. 1989).

Overall, the N characteristics of Juina diamonds with known paragenesis show a correlation with depth. Juina lower mantle diamonds are Type II or Type IaAB with more than 95 % aggregation (Hutchison 1997; Hayman et al. 2005). Type II diamonds are also reported among transition zone and upper mantle diamonds from Juina, but the occurrence of Type I diamonds with a variable degree of aggregation is noteworthy. For example, Hutchison (1999) reports four transition zone diamonds which are Type I with 42–89 % N aggregation and six diamonds with probable transition zone association which are all Type I, with 52–100 % N aggregation. Another two Type IaAB stones are related to the upper mantle. Hayman et al. (2005) reports Type II and Type IaB among lower mantle and transition zone diamonds and Type IaA and IaAB among the upper mantle population, emphasizing the strong relation between the occurrences of A-defects in stones from shallower depths. Both Hutchison et al. (1999) and Hayman et al. (2005)

studies refer to diamonds from alluvial sources. Transition zone diamonds found in the Collier-4 kimberlite by Bulanova et al. (2010) are all Type II; other diamonds in their study with upper mantle inclusions are mostly Type II and two are Type Ia, reinforcing the association of the presence of A-defect with shallower depths.

As shown above, the observation of Type II character does not by itself prove a sublithospheric origin for a specific diamond, but if a population of diamonds shows a high abundance of Type II stones, a consistent pattern of complex internal growth, highly aggregated Type I diamonds and platelet degradation, then an implication may be that these diamonds are derived from the sublithospheric and/or possibly lower mantle sources. However, confirmation of their origin needs to be linked to the identification of the inclusion content.

Our Type I diamonds often show regular octahedral growth (Fig. 5), less internally complex features than those of Type II stones, low to high aggregated N and non-degraded platelet peaks (TR2-17, DB-5, CL-19, SL-2). These diamonds would be candidates for a lithospheric origin. The Type I diamonds with high N aggregation and degraded platelets may be candidates for a sublithospheric origin. Therefore, N characteristics and internal features could possibly be used as a preliminary recognition of shallow and ultradeep diamonds.

Two Different Populations of Juina Kimberlitic and Alluvial Diamonds: Constraints from Morphology, N Characteristics and Carbon Isotope Composition

In order to suggest diamond sources and the existence of different populations, we discuss in the following section the differences in morphology, N characteristics and C isotopic composition between diamonds from Juina kimberlites and alluvial sources, as determined in the present work, and then compare these results with similar information previously published from the same region. Our studies showed that diamond samples from different drainages have similar overall characteristics and are collectively grouped as "alluvials", see below.

Over 1,900 Juina diamonds (both kimberlitic and alluvial) have had their morphological characteristics described in the literature (e.g. Hutchison et al. 1999; Kaminsky et al. 2001, 2009; Hayman et al. 2005), most of them from the Pandrea pipes (1094 diamonds from five pipes; Kaminsky et al. 2009). Juina diamonds are mostly resorbed and octahedral diamonds do not exceed 18 %. For example, Hutchison et al. (1999) reported São Luiz alluvial diamonds distributed as 28 % of dodecahedral stones, 18 % of octahedral, 41 % of irregular diamonds and 13 % of macles. Alluvial diamonds studied by Kaminsky et al. (2001, 2009) are resorbed (61 %, including dodecahedral and transitional forms), octahedral (6 %) and irregulars (33 %). These authors also report high abundance of resorbed stones for Collier-4 kimberlite (57 %), Aripuanã-1 kimberlite (56 %) and Pandrea pipes (48 %), and octahedral forms are up to 7, 5 and 12 %, respectively, for these occurrences. The remainder is reported as irregular (up to 33 %) and fragments (up to 35 %). Hayman et al. (2005) reported the occurrence of 52 % of diamonds in resorption classes (Robinson 1980) 1–3. Comparing our results, the only population that markedly differs from the norm are diamonds from Juina-5 kimberlite, which has almost half of its diamonds with octahedral morphology and a high proportion of macles (Table 1 and Fig. 3). In addition, this is the only population that has graphite coating. The abundance of octahedral diamonds could be directly related to lack of resorption reaction with the transporting kimberlite magma and/or low rate of disaggregation of diamondiferous xenoliths, therefore contributing to the diamond isolation from the kimberlite magma. The etched surface features of the Juina-5 octahedra and their frequent thin graphite coatings recall the morphology of diamonds from Argyle peridotite xenoliths (Hall and Smith 1985; Jaques et al. 1990). Among alluvial Juina diamonds, there are no reports in the literature of abrasive transportation surface features (Hutchison et al. 1999; Kaminsky et al. 2001, 2009; Hayman et al. 2005, and this work), which suggest proximal primary sources for the diamonds or more distance sources if diamond transport was under low energy regimes.

FTIR data reported for Juina in the literature ($n = 1{,}950$, Hutchison et al. 1999; Kaminsky et al. 2001, 2009; Hayman et al. 2005, Bulanova et al. 2010; Walter et al. 2011) showed significant differences for the same kimberlite and alluvial populations as reported by different authors, see Table 3. For example, there is a high proportion (>80 %) of Type II diamonds among Juina-5 (this work, Walter et al. 2011) and Collier-4 kimberlite diamonds (this work, Bulanova et al. 2010), but Kaminsky et al. (2001, 2010) give results showing a low abundance of Type II for Collier-4 (12 %). They report 24 % of Type II diamonds for Aripuanã-1 kimberlite and 42 % for Pandrea pipes. The alluvial diamonds have, overall, a lower proportion of Type II diamonds than kimberlitic ones. For example, we report here 65 % as opposed to 69 % by Hutchison et al. (1999) for Type II stones among São Luiz River diamonds. However, Hayman et al. (2005) found 38 % Type II among Sorriso River diamonds and Kamnisky et al. (2001) reported 21 % for Vermelho River, 18 % for Chicoria and only 17 % for São Luiz (giving the average of 19 % shown in Table 3). Bulanova et al. (2010) noted these differences before and suggested they could be due to sampling bias.

Table 3 Percentage of Type II diamonds for different localities in the Juina area. Number of samples analysed are in parentheses

	% of Type II diamonds in Juina area					
References	This work	1	2	3	4	5
Collier-4 kimberlite (151)	80			12	85	
Juina-5 kimberlite (150)	89					95
Aripuanã-1 kimberlite (120)				24		
Pandrea kimberlites (1,094)				42		
Alluvials (490)	65	69	38	19		

1—Hutchison et al. (1999 v; *2*—Hayman et al. (2005); *3*—Kaminsky et al. (2001, 2009); *4*—Bulanova et al. (2010); *5*—Walter et al. (2011)

When comparing the $\delta^{13}C$ distribution for different kimberlites and alluvial stones, marked differences appear both in the range and modes, and there are no discrepancies among different authors as was found for the FTIR data. Including our results, a total of two hundred and fifty-two diamonds from the Juina area have been analysed for carbon isotopes (Fig. 8). They include diamonds from kimberlites Juina-5, Collier-4, Pandrea and Aripuanã and the drainages of the São Luiz, Cinta Larga, Chicória and Vermelho rivers (see figure caption for references). The range

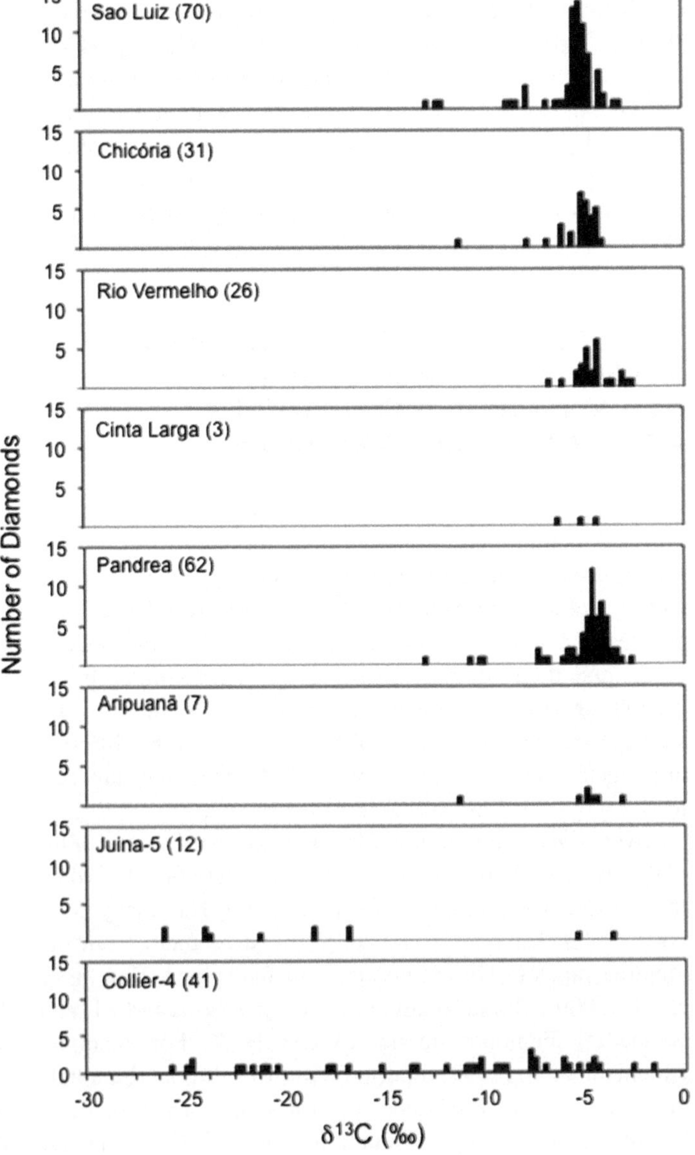

Fig. 8 Histogram of carbon isotope composition of diamonds (with a bin width of 0.25 ‰) from the Juina Kimberlite Field with alluvial deposits in the *upper* four panels and kimberlites in the *lower* panels. Data are shown for: São Luiz (Hutchison 1997; Kaminsky 2001; this work); Chicória, Rio Vermelho, Pandrea, Aripuanã-1 (Kaminsky et al. 2009), Cinta Larga (this work), Juina-5 (this work, Walter et al. 2011), Collier-4 (this work, Kaminsky et al. 2009; Bulanova et al. 2010). Numbers in parenthesis are number of stones analysed

of $\delta^{13}C$ for Collier-4 diamonds found here is similar to literature data (Bulanova et al. 2010), but recent results for Juina-5 diamonds from the same collection studied here (Walter et al. 2011) expanded the interval found in this work. Diamonds with more than one in situ analysis have their composition averaged in the histogram in order to compare with fragments and whole stones analyses. Figure 8 shows that there is a complete overlap in the $\delta^{13}C$ range for diamonds from Juina-5 and Collier-4 pipes, which are evenly spread from −26 to −3.5 ‰ (reaching −0.9 ‰ when in situ analyses are considered), with no "mantle composition" mode near −5 ‰. Diamonds from alluvial sources and those from the Pandrea and Aripuanã kimberlites have a distinct $\delta^{13}C$ range, from −13.8 to −2.7 ‰. Modes are at −5 ‰ for São Luiz and Chicória diamonds and between −4.25 and −4.75 ‰ for Pandrea and Vermelho river diamonds.

Two different groups can be defined by their isotopic compositions and modes: Group 1, with diamonds from Juina-5 and Collier-4 kimberlites (−26 to −3.5 ‰), and Group 2, with diamonds from the other kimberlites and all the alluvial specimens (−13.8 to −2.7 ‰). It is clear from these data that Juina-5 and Collier-4 kimberlites are not the main feeders of diamonds into the local drainages.

The source of diamonds of the Cinta Larga Basin (Fig. 1; including rivers São Luiz, Chicória, Sorriso, Porção, Cinta Larga and Duas Barras) has been related to the Chapadão area, an outlier of the Cretaceous Parecis sediments (Fig. 1; Haralyi 1991). In the Chapadão plateau, two horizons of diamond palaeoplacers separated by 35 metres of sterile sediments were recognized (Haralyi 1991). Kaminsky et al. (2010) assigned these diamond occurrences to several kimberlite pipes, which they termed Pandrea. As previously pointed out by Kaminsky et al. (2010), the Pandrea diamonds are strikingly similar to the alluvial diamonds regarding their carbon isotope composition (see Fig. 8).

The recognition of kimberlites in the Chapadão area, however, is under debate. Units previously described as sediments (Haralyi 1991) have been interpreted by Kaminsky et al. (2010) as air fall kimberlite tuffs, although drilling failed to reveal any primary kimberlite feeders underneath (Kaminsky and Le Noan 2009). Other authorities (e.g. G.M. Gonzaga pers. com.; C.B. Smith pers. com.) interpret these Chapadão sediments as cross-bedded fluviatile units within the Parecis Group carrying diamonds together with weakly abraided picroilmenites. Regardless of their origin, the evidence is that most alluvial diamonds in Juina result from reworking of diamonds from the Chapadão plateau.

The $\delta^{13}C$ distribution in Fig. 8 cannot be used to distinguish between lithospheric upper mantle, sublithospheric upper mantle, transition zone or lower mantle diamonds from Juina, as both kimberlites and alluvials have diamonds from all these depths. Prior to the recent study of Walter et al. (2011), all diamonds that had been shown to originate in the lower mantle (from Juina and worldwide) had carbon isotope compositions close to mantle values (e.g. Hutchison et al. 1999; Kaminsky et al. 2001, 2009; Hayman et al. 2005), while diamonds interpreted as originating from the transition zone and sublithospheric upper mantle tended to span to lighter values. However, Walter et al. (2011) found light $\delta^{13}C$ values down to −26 ‰ for four lower mantle diamonds, which they related to crustal carbon being subducted down to the lower mantle. Similar values were found for the Group 1 diamonds by other studies (this work, Bulanova et al. 2010, Kaminsky et al. 2010), which is likely to be close to the lower limit for carbon isotopic composition in mantle beneath this region.

Heterogeneity in the Mantle Beneath Juina

The $\delta^{13}C$ data indicate that the kimberlites Juina-5 and Collier-4 (Group 1) and the kimberlites and alluvials related to the Chapadão diamonds (Group 2) have sampled different diamond populations in the mantle, though both groups contain ultradeep diamonds. Considering that the ultradeep diamonds were probably brought to the base of the lithosphere by mantle plumes (e.g. Hutchison et al. 1999; Bulanova et al. 2010; Walter et al. 2011), these results suggest that parts of the sublithospheric mantle with different diamond populations were not homogenized at upper mantle levels, where portions of the deep mantle are preserved at the base of the lithosphere.

Conclusions

Juina diamond populations from different kimberlites and alluvial deposits show similar morphological features (except for Juina-5 kimberlite), but differ regarding their N characteristics and carbon isotope compositions. Most Juina diamonds studied here are resorbed with shagreen texture, microhillocks, terraces and lamination lines. Macles are common among Juina-5 pipe diamonds, which also has a higher percentage of octahedral diamonds compared to Collier-4 kimberlite and alluvial stones. Most diamonds show low-intensity cathodoluminescence, which corresponds with the high abundance of Type II diamonds (>69 %). H peaks correlate positively with N content in Type I and are absent in Type II diamonds. The Type I stones are IaB and IaAB with extremely high N aggregation. All but four Juina Type I diamonds show platelet degradation, perhaps reflecting high temperatures present in the mantle. N characteristics, such as high abundance of Type II stones, highly aggregated Type I diamonds and platelet degradation, together with complex internal features, could possibly be used as a preliminary recognition of ultradeep diamonds.

Juina diamonds have low internal C isotopic variability (<3.6 ‰) and often show fluctuation and increase in $^{13}C/^{12}C$ from core to rim. Resorption horizons revealed by CL and the lack of co-variation between C isotope and N content suggest episodic growth with variable melt composition as a common mechanism during formation of the diamonds studied here. Evidence of carbon fractionation during diamond growth was found for one sample (CL-11). The $\delta^{13}C$ results together with literature data revealed two groups with $\delta^{13}C$ ranges from −26.3 to −3.0 ‰ for Juina-5 and Collier-4 kimberlite diamonds (Group 1) and −13.8 to −3.4 ‰ for alluvial diamonds (Group 2). The two carbon isotope groups point to heterogeneity in the mantle source. Such heterogeneity is suggested to reflect parts of the deep mantle that were brought to the base of the lithosphere by a mantle plume without significant homogenization during ascent or in the upper mantle.

Acknowledgments The authors wish to thank Rio Tinto Brasil (RTDM) for providing samples for this study. We thank the analytical support of G. Szabó for CL images at University of São Paulo, Arnaldo Alcover Neto for assistance with SEM images, J. Rougvie for CL images and FTIR work at the Smithsonian Institution and Y. Fei and R. Hemley for the support during the work at Carnegie Institution. D. Fisher (DTC Research Centre) is thanked for providing the CAXBD97.XLS spreadsheet for FTIR data deconvolution. We are greatly thankful to P. Cartigny, W.L. Griffin, O. Navon and T. Stachel, and to the Guest Editor, J. Harris, for constructive reviews made during different periods of building this manuscript. We thank Sami Mikhail for the helpful input in diamond growth modelling discussion. This research was funded by CNPQ (National Counsel of Technological and Scientific Development), FINATEC (Fundação de Empreendimentos Científicos e Tecnológicos) and CAPES (Centro de Aperfeiçoamento de Ensino Superior).

References

Araujo DP, Gaspar JC, Bizzi LA (2001) Morphology and surface features of diamonds from the Juína Kimberlite province, Brazil. Rev Bras Geo 31(4):605–615

Araujo D (2002) Mineralogy of diamonds from the Juina Kimberlite Province, MT (in Portuguese). PhD thesis. University of Brasília. Brasília, p 195

Boyd SR, Kiflawi I, Woods GS (1995) Infrared absorption by the B nitrogen aggregate in diamond. Phil Mag B72(3):351–361

Boyd SR, Kiflawi I, Woods GS (1994) The relationship between infrared absorption and the a defect concentration in diamond. Phil Mag B69(6):1149–1153

Brenker FE, Vollmer C, Vincze L, Vekemans B, Szymanski A, Janssens K, Szaloki I, Nasdala L, Joswig W, Kaminsky FV (2007) Carbonates from the lower part of transition zone or even the lower mantle. Earth Planet Sci Lett 260(1–2):1–9

Bulanova GP, Walter MJ, Smith CB, Kohn SC, Armstrong LS, Blundy J, Gobbo L (2010) Mineral inclusions in sublithospheric diamonds from Collier 4 kimberlite pipe, Juina, Brazil: subducted protoliths, carbonated melts and primary kimberlite magmatism. Contrib Min Petrol 160(4):489–510

Davis GL (1977) The ages and uranium contents of zircons from kimberlites and related rocks, vol 76. Carnegie Institution of Washington Year Book, Washington, pp 631–635

Davies RM, Griffin WL, Pearson NJ, Andrew AS, Doyle BJ, O'Reilly SY (1999) Diamonds from the deep: pipe DO-27, Slave Craton, Canada. In: Gurney JJ, Gurney JL, Pascoe MD, Richardson SH (eds) Proceedings of 7th international Kimberlite conference, vol 1. Red Roof Design, Cape Town, pp 148–155

Deines P (1980) The carbon isotopic composition of diamonds: relationship to diamond shape, color, occurrence and vapor deposition. Geochim Cosmochim Acta 44:943–961

Deines P, Harris JW, Gurney JJ (1987) Carbon isotopic composition, N content and inclusion composition of diamonds from the Roberts Victor Kimberlite, South Africa: evidence for ^{13}C depletion in the mantle. Geochim Cosmochim Acta 51:1227–1243

Deines P, Harris JW, Spear PM, Gurney JJ (1989) Nitrogen and $\delta^{13}C$ content of Finsch and Premier diamonds and their implications. Geochim Cosmochim Acta 53:1367–1378

Deines P, Harris JW, Gurney JJ (1991) The carbon isotopic composition and nitrogen content of lithospheric and asthenospheric diamonds from the Jagersfontein and Koffiefontein kimberlite, South Africa. Geochim Cosmochim Acta 55:2615–2625

Hall AE, Smith CB (1985) Lamproite diamonds—are they different? In: Glover JE, Harris PG (eds) Kimberlite occurrence and origin, vol 8. University of Western Australia, Geology Department Publication, pp 167–212

Haralyi N (1991) Os diamantes de Juina, Mato Grosso. In: Principais depósitos minerais do Brasil, vol IV. (Parte A), pp 155–160

Harte B, Harris JW (1994) Lower mantle mineral associations preserved in diamonds. Min Mag A58:384–385

Hauri EH, Wang J, Pearson DG, Bulanova GP (2002) Microanalysis of δ13C, δ15N, and N abundances in diamonds by secondary ion mass spectrometry. Chem Geol 185(1–2):149–163

Hayman PC, Kopylova MG, Kaminsky FV (2005) Lower mantle diamonds from Rio Soriso (Juina area, Mato Grosso, Brazil). Contrib Min Petrol 149(4):430–445

Heaman LM, Teixeira NA, Gobbo L, Gaspar JCA (1998) U-Pb zircon ages for kimberlites from the Juina and Paranatinga provinces, Brazil. In: 7th international Kimberlite conference, long abstracts, Cape Town, pp 322–324

Hutchison M (1997) Constitution of the deep transition zone and lower mantle shown by diamonds and their inclusions, PhD Thesis. University of Edinburgh

Hutchison MT, Cartigny P, Harris JW (1999) Carbon and nitrogen composition and physical characteristics of transition zone and lower mantle diamonds from São Luiz, Brazil. In: Gurney JJ, Gurney JL, Pascoe MD, Richardson SH (eds) Proceedings of the 7th international Kimberlite conference, Cape Town, pp 372–382

Jaques AL, O'Neill HSC, Smith CB, Moon J, Chappell BW (1990) Diamondiferous peridotite xenoliths from the Argyle (AK1) lamproite pipe, Western Australia. Contrib Min Petrol 104:255–276

Kaminsky F, Zakharchenko O, Davies RM, Griffin W, Khachatryan G, Shiryaev A (2001) Superdeep diamonds form the Juina area, Mato Grosso State, Brazil. Contrib Min Petrol 140(6):734–753

Kaminsky F, Le Noan C (2009) A technical report on the Juina diamond project, Juina, Mato Grosso, Brazil. Technical report 43–101 F1, Diamond Exploration Inc., p 85

Kaminsky FV, Khachatryan GK, Andreazza P, Araujo D, Griffin WL (2009) Super-deep diamonds from kimberlites in the Juina area, Mato Grosso State, Brazil. Lithos 112(Supplement 2):833–842

Kaminsky FV, Sablukov SM, Belousova EA, Andreazza P, Tremblay M, Griffin WL (2010) Kimberlitic sources of super-deep diamonds in the Juina area, Mato Grosso State, Brazil. Lithos 114(1–2):16–29

Kiflawi I, Spear PM, van Wyk JA, Wood GS (1994) Infrared absorption by the single nitrogen and a defect centers in diamond. Phil Mag B69(6):1141–1147
Robinson DN (1980) Surface textures and other features of diamonds. PhD Thesis. University of Cape Town. Natal, South Africa, p 221
Silva GH, Leal J, Montalvão R, Bezerra P, Pimenta O, Tassinari C, Fernandes C (1980) In: RadamBrasil P (ed) Folha SC-21. Juruena. 1—Geologia, p 456
Stachel T, Harris JW, Aulbach S, Deines P (2002) Kankan diamonds (Guinea) III: $\delta^{13}C$ and nitrogen characteristics of deep diamonds. Cont Min Petrol 142:465–475
Stachel T, Harris JW, Muehlenbachs K (2009) Sources of carbon in inclusion bearing diamonds. Lithos 112(Suppl. 2):625–637
Sunagawa I (1984) Morphology of natural and synthetic diamond crystals. In: Sunagawa I (ed) Materials science of the earth's interior, pp 303–330
Tassinari CCG, Macambira MJB (1999) Geochronological provinces of the Amazonian craton. Episodes 22:174–182
Tassinari CCG, Betterncourt JS, Geraldes MC, Macambira MJB, Lafon JM (2000) The Amazonian craton. In: Cordani UG, Milani EJ, Thomaz Filho A, Campos DA (eds) Tectonic evolution of South America, pp 41–96
Teixeira NA, Gaspar JC, Waissel O, Almeida AJ, Belther JA., Gobbo L (1998) Geology in the Juina Diamondiferous province. Extended abstracts. In: Seventh international kimberlite conference, Cape Town, pp 905–907
Walter MJ, Bulanova GP, Armstrong LS, Keshav S, Blundy JD, Gudfinnsson G, Lord OT, Lennie AR, Clark SM, Smith CB, Gobbo L (2008) Primary carbonatite melt from deeply subducted oceanic crust. Nature 454(7204):622–625
Walter MJ, Kohn SC, Araujo D, Bulanova GP, Smith CB, Gaillou E, Wang J, Steele A, Shirey SB (2011) Deep mantle cycling of oceanic crust: evidence from diamonds and their mineral inclusions. Science 334(6052):54–57
Watt GR, Harris JW, Harte B, Boyd SR (1994) A high-chromium corundum (ruby) inclusion in diamond from the Sao Luiz alluvial mine. Brazil. Min Mag 58(3):490–492
Wilding MC, Harte B, Harris JW (1991) Evidence for a deep origin for Sao Luiz diamonds. In: 5th international kimberlite conference, Araxá, Brazil, pp 456–458
Wirth R, Vollmer C, Brenker FE, Matsyuk S, Kaminsky FV (2007) Inclusions of nanocrystalline hydrous aluminium silicate "Phase Egg" in superdeep diamonds from Juina (Mato Grosso State, Brazil). Earth Planet Sci Lett 259(3–4):384–399
Woods G (1986) Platelets and the infrared absorption of type Ia diamonds. Proc Royal Soc London. Series A. Math Phys Sci 407(1832):219–238
Woods GS, Collins AT (1983) Infrared absorption spectra of hydrogen complexes in type I diamonds. J Phys Chem Solids 44(5):471–475

The IR Absorption Spectrum of Water in Microinclusion-Bearing Diamonds

Yakov Weiss, Isaac Kiflawi, and Oded Navon

Abstract

Infrared (IR) spectra of 55 microinclusion-bearing diamonds, combined with the bulk major-element chemistry of the trapped high-density fluids (HDFs) and spectroscopic and mineralogical data on the daughter phases in the microinclusions, were used to investigate the various shapes of IR absorption of water in fibrous diamonds and to decipher their origin. The difference in shape is most prominent for the OH-stretching mode at $\sim$3,400 cm^{-1}. We used the width of this band and the ratio of the absorbance at 3,200, 3,400 and 3,600 cm^{-1} to characterize quantitatively the variation in shape and to correlate it with the chemical composition of the trapped HDFs. Low-Mg carbonatitic to silicic compositions show similar spectrum to that of pure water. Their extra width and high 3,600/3,400 cm^{-1} ratio are attributed to absorption by daughter mica phases. High-Mg carbonatitic HDFs are characterized by a much wider OH-stretching mode, which extends towards lower energies. We attribute this wide band to absorption by hydrated K-bearing carbonates. This is supported by: (1) positive correlation between the K_2O content in these HDF compositions and the width of the OH-stretching band, and (2) correlated reduction in the low-energy absorption of the OH-stretching band and of the absorbance of carbonates with increasing temperatures. In diamonds carrying saline HDFs, the width of the OH-stretching band is narrower than that of pure water. The similarity in width and shape to the spectra of salt-solvated water and the little change in the band shape with increasing temperature suggest that most of the water in saline microinclusions might be solvated in the daughter salts. This allows the identification of saline HDFs in diamonds by IR spectroscopy even though it is insensitive to chlorides. The variation in the shape of the OH-stretching band of water and the quantification of the absorption of the various daughter mineral phases in the microinclusions allow semi-quantitative determination of the composition of fluids trapped in diamonds.

I. Kiflawi—Deceased

Electronic supplementary material The online version of this article (doi:10.1007/978-81-322-1170-9_17) contains supplementary material, which is available to authorized users.

Y. Weiss (✉) · I. Kiflawi · O. Navon
The Fredy and Nadine Herrmann Institute of Earth Sciences, The Hebrew University of Jerusalem, Jerusalem, Israel
e-mail: yakov.weiss@mail.huji.ac.il

D. G. Pearson et al. (eds.), *Proceedings of 10th International Kimberlite Conference*, Volume 1, Special Issue of the Journal of the Geological Society of India,
DOI: 10.1007/978-81-322-1170-9_17, © Geological Society of India 2013

Keywords

IR absorption of water • OH-stretching band • Microinclusion-bearing diamonds • High-density fluid (HDF) • Hydrated carbonates • Salt-solvated water

Introduction

Microinclusions in fibrous and cloudy diamonds trapped high-density fluids (HDFs) of diverse composition: saline, high-Mg carbonatitic and low-Mg carbonatitic to silicic melts/fluids (Navon et al. 1988; Weiss et al. 2009 and references therein). Upon cooling, daughter minerals crystallize and precipitate from the trapped HDFs (Fig. 1); mica, apatite, carbonates and quartz were detected by IR spectroscopy (Chrenko et al. 1967; Navon et al. 1988, 1991; Schrauder and Navon 1994; Zedgenizov et al. 2004; Tomlinson et al. 2007; Klein-BenDavid et al. 2007). Transmission electron microscopy (TEM) analyses confirm the presence of those phases as well as halides in the microinclusions (Guthrie et al. 1991; Walmsley and Lang 1992a, b; Klein-BenDavid et al. 2006). The various compositions of the solids are well separated on an MgO versus $SiO_2 + Al_2O_3$ diagram, best expressing the quantities of carbonates and silicates in the microinclusions, respectively (Fig. 2). Microinclusions of all compositions also carry residual low-density hydrous solutions, which gives rise to the IR absorption bands at ~3,400 and ~1,640 cm^{-1}, corresponding to the OH-stretching and HOH-bending vibration of water, respectively (Chrenko et al. 1967; Navon et al. 1988; Weiss et al. 2009 and references therein).

High-residual internal pressure in the microinclusions, ranging between 1.3 and 2.1 GPa (Navon 1991; Tomlinson et al. 2007), suggests that the trapped water should be in the form of ice VI at room temperature (Bridgman 1935; Datchi et al. 2000). However, the shape of the OH-stretching band measured in many cubic and coated diamonds resembles that of liquid water (Navon 1991; Schrauder and Navon 1994; Klein-BenDavid et al. 2007; Tomlinson et al. 2007; Zedgenizov et al. 2009; Weiss et al. 2009). Navon (1991) attributed the prevention of freezing to the high surface-to-volume ratio in the microinclusions and the high solute content of the residual hydrous fluid. Ice VI coexisting with liquid water was identified in one cubic diamond based on the absorption of the stretching plus bending combination band of molecular water at ~5,200 cm^{-1} (Kagi et al. 2006). Zedgenizov et al. (2006) pointed out that the shape and width of the OH-stretching band is different between cuboid diamonds from Siberia and Congo. They suggested

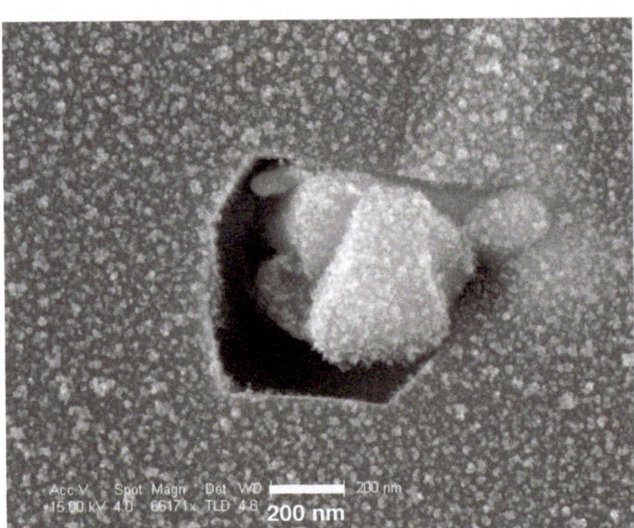

Fig. 1 High-resolution SEM image of an open microinclusion in a fibrous diamond. The microinclusions in such diamonds have a fairly well-defined range of shapes imposed by the diamond preferred growth habit and sizes, most of them are in a few hundred nanometres in size. The inclusion contains a multi-phase assemblage of daughter minerals and a void. IR spectroscopy reveals that prior to the cleaving of the diamond, the void was filled with a hydrous solution and perhaps CO_2. Most of the volume is taken by solids (carbonates, silicates and phosphates) that were initially dissolved in the high-density fluid (HDF) trapped by the diamond during its growth (Weiss et al. 2010)

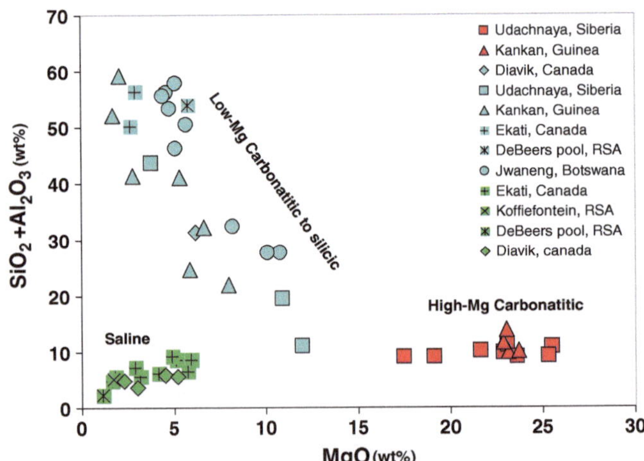

Fig. 2 $SiO_2 + Al_2O_3$ against MgO variation in HDFs trapped in diamonds (water- and CO_2-free basis, normalized composition). The diagram best expresses the three compositional groups: high-Mg carbonatitic HDFs with very high MgO and low silica and alumina, low-Mg carbonatitic to silicic composition with varying amount of carbonates and silicates and saline HDFs which carry mostly K, Na and Cl and are depleted in carbonates and silicates. (See Weiss et al. 2009 for additional variation diagrams with other components)

Table 1 Diamond origin and fluid characteristics

Number of diamonds	Origin	Fluid type
9	Ekati, Canada[a]	Saline
1	DeBeers-Pool, RSA[b]	
2	Koffiefontein, RSA[c]	
4	Diavik, Canada[d]	
10	Udachnaya, Siberia[e]	High-Mg carbonatitic
4	Kankan, Guinea[f]	
9	Jwaneng, Botswana[g]	Low-Mg carbonatitic to silicic
3	Kankna, Guinea[f]	
2	Ekati, Canada[a]	
3	Udachnaya, Siberia[e]	
1	DeBeers-Pool, RSA[b]	
1	Diavik, Canada[d]	
6	Botswana[h]	

Data source [a] McNeill, Pearson and Weiss unpublished data, [b] Weiss et al. (2008); [c] Izraeli et al. (2004); [d] Klein-BenDavid et al. (2007); [e] Klein-BenDavid et al. (2009); [f] Weiss et al. (2009); [g] Schrauder and Navon (1994); [h] Klein-BenDavid et al. (2010)

that the wide band in the spectra of the Siberian diamonds is due to absorption by different hydrous phases: liquid water, ice and/or hydrous minerals.

In the present study, we combine bulk major-element chemistry of the trapped fluids with spectroscopic and mineralogical data on the daughter phases in microinclusions in 55 fibrous diamonds from Africa, Siberia and Canada to investigate the shapes of the IR water bands in the IR spectra of diamonds carrying HDFs with a wide range of compositions.

Samples and Analytical Techniques

Fifty-five fibrous diamonds carrying saline, high-Mg carbonatitic and low-Mg carbonatitic to silicic HDFs were selected for the present study (Table 1). Except for the Ekati diamonds, all other diamonds were described in previous studies. The characteristics of the Ekati set will be published elsewhere.

The diamond's IR spectra were measured on polished slabs using a Bruker IRscope II microscope coupled to a Nicolet 740 FTIR spectrometer (Globar source, KBr beamsplitter, MCT detector, He–Ne laser). Spectra were taken in the range of 550–4,000 cm^{-1} with resolution of 4 cm^{-1}. The diamond bands and the absorption due to nitrogen in the diamond matrix were subtracted from the spectra; both subtractions were accompanied by prior and subsequent baseline correction (spline fit to the lowest points). This procedure yields the residual spectrum due to absorption by the daughter phases in the microinclusions of each diamond. The shape, height and width of the OH stretching band at ~3,400 cm^{-1} and the HOH-bending peak at ~1,650 cm^{-1} were measured directly from this baseline-corrected spectrum.

Four diamonds carrying saline HDFs, six carrying high-Mg carbonatitic HDFs and three carrying low-Mg carbonatitic to silicic compositions were heated using a Linkam THMS600 heating/cooling stage. The stage was clamped to the stage of the IR microscope and connected to a TMS94 temperature programmer with temperature stability of ±0.1 °C. However, as the stage windows are not IR transparent below 2,000 cm^{-1}, they were removed for the IR measurements; hence, the uncertainty is larger and temperatures were probably lower than the nominal reading. Several spectra were recorded between room temperature and ~420 °C for each diamond. After subtracting the diamond absorption and correcting the baseline, the changes in the shape of the water and minerals absorbance due to temperature increase were characterized.

We also prepared pellets of pure NaCl, KCl, $CaCO_3$, Na_2CO_3 and K_2CO_3 and left them to absorb humidity for several days. Subsequently, the IR absorbance of solvated water molecules in these pellets was measured and characterized for comparison with the absorbance of water in the diamonds.

Results

Water Absorbance in Different HDF Compositions

The width and the position of the OH-stretching and HOH-bending modes of water at ~3,400 and ~1,650 cm^{-1}, respectively, vary with the composition of the microinclusions; the change in shape of the OH-stretching mode is prominent (Fig. 3a).

Low-Mg carbonatitic to silicic compositions show similar spectra to that of pure water. However, in most cases, the OH-stretching absorbance is wider than that of pure water; that is, when normalizing the maximum absorbance of the different spectra to the same value, then the diamond spectra reveal higher absorbance in the regions ~3,700–3,500 and ~3,200–2,800 cm^{-1}.

In diamonds carrying saline HDFs, the width of the OH-stretching band is narrower compared to that of pure water (Fig. 3a) and the position of their HOH-bending vibration shifts to lower energies (Fig. 3b). The width of the OH-stretching absorption in diamonds carrying high-Mg carbonatitic HDFs varies significantly from ~440 to ~860 cm^{-1}, and the increase in the band width is always towards lower wavenumbers.

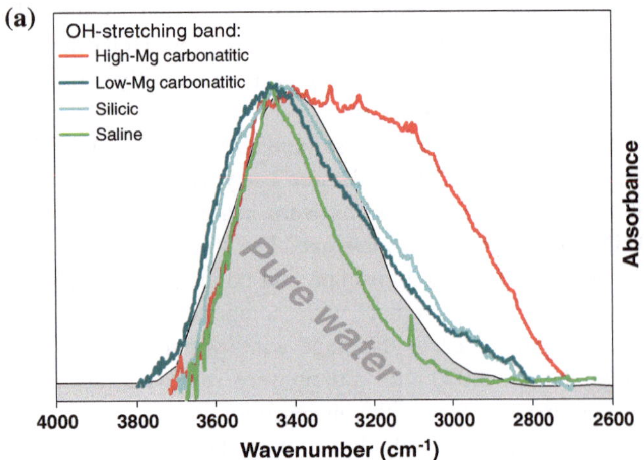

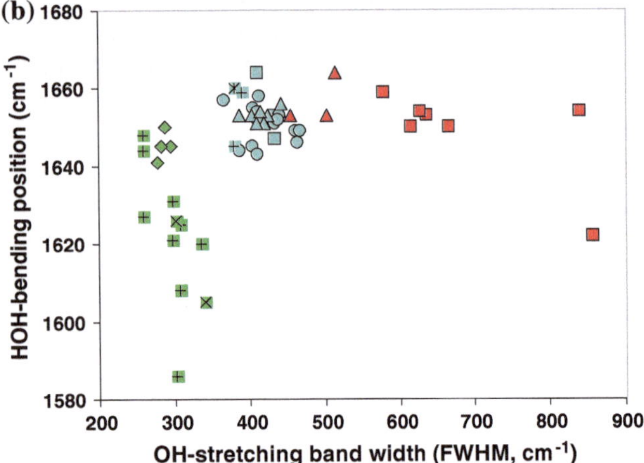

Fig. 3 **a** IR spectrum of the OH-stretching mode of the various HDFs compared with that of pure water (from Venyaminov and Prendergast 1997). **b** The position of the HOH-bending mode versus the width of the OH-stretching band (full width at half-maximum, FWHM). High-Mg carbonatitic (*red symbols*), low-Mg carbonatitic to silicic (*turquoise symbols*) and saline HDFs (*green symbols*). The different symbols mark the provenance of the different HDF (see Fig. 2)

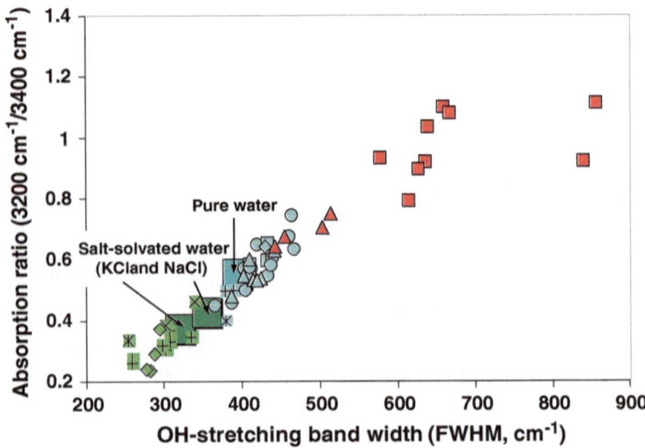

Fig. 4 Absorbance ratio (3,200/3,400 cm^{-1}) versus the width of the OH-stretching band in high-Mg carbonatitic HDFs (*red symbols*), low-Mg carbonatitic to silicic HDFs (*turquoise symbols*) and saline HDFs (*green symbols*). The large squares stand for pure water and salt-solvated water in KCl and NaCl pellets. See Fig. 2 for the provenance of the HDFs

To characterize quantitatively the change in shape of the OH-stretching absorption at low energies and relate it to the composition of the HDFs, we plotted the ratio of the absorption at 3,200 and 3,400 cm^{-1} versus the width of the peak (Fig. 4). The data points form a straight line irrespective of the provenance of the diamonds and the three groups of HDFs are well separated. Similarly, to characterize the change in absorbance in the range ~3,700–3,500 cm^{-1} between HDFs and pure water, we looked at the ratio of the absorption at 3,600 and 3,400 cm^{-1}.

In some of the diamonds, mainly with saline microinclusion compositions, an additional peak situated at 1,580 cm^{-1} has been observed; we could not relate this peak to water or any other daughter phase in the inclusions.

High-Temperature Absorbance of Water in the Various HDFs

Previous studies found that with increasing temperature, the spectrum of pure water in the region of the OH-stretching mode changes considerably: (1) the position of the peak maximum shifts to higher wavenumbers, (2) the absorbance decreases significantly, and (3) the shape of the band changes, that is, at ~200 °C, a shoulder at 3,650 cm^{-1} becomes visible, and at temperatures higher than 250 °C, it appears as a distinct band (Kazarian and Martirosyan 2002; Maréchal 2011).

Figure 5 shows that the absorbance of the OH-stretching mode of water in all diamonds decreases with increasing temperature and that the relative decrease is similar for all HDFs. The peak position and the shape of this band vary as well. However, these variations are compositionally dependent. In the high-Mg carbonatitic and low-Mg carbonatitic to silicic HDFs, the peak position shifts by ~100 cm^{-1} from ~3,420 cm^{-1} at room temperature to ~3,520 cm^{-1} at 420 °C. The shape of the band of these HDFs becomes similar at >300 °C (Fig. 5) and resembles that of pure water at similar temperatures (Kazarian and Martirosyan 2002). In diamonds carrying saline compositions, the shifts due to heating are smaller (~30 cm^{-1}) as is the change in shape.

In most cases, changes in the HOH-bending band are similar to those of the OH-stretching mode. However, it is harder to quantify these changes as the HOH-bending peak partially overlaps the main carbonate band at ~1,450 cm^{-1} and its peak-to-background ratio is low at elevated temperatures.

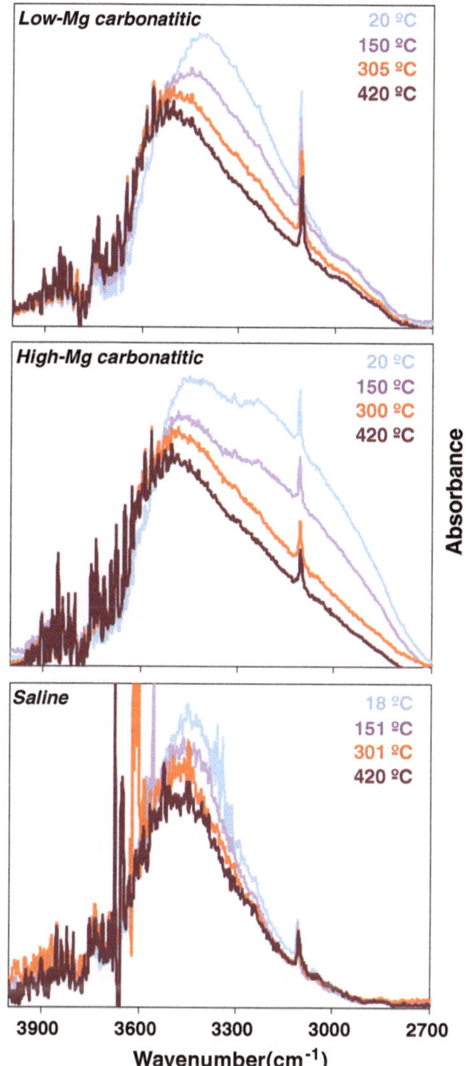

Fig. 5 High-temperature IR spectra of the HO-stretching mode in representative high-Mg carbonatitic, low-Mg carbonatitic and saline HDFs. With increasing temperature, the absorbance of the OH-stretching mode of water decreases irrespective of the HDF composition. Changes in the maximum peak position and in the shape of this band do vary with composition. In high-Mg carbonatitic compositions, the absorbance at lower wavenumbers significantly decreases between 150 and 300 °C (mostly above 200 °C). At >300 °C, the shape of the OH-stretching mode in high-Mg carbonatitic and low-Mg carbonatitic to silicic compositions becomes similar and resembles that of pure water at similar temperatures (Kazarian and Martirosyan 2002)

Associated Changes in Daughter Phases with Increasing Temperature

In addition to the principal changes in the absorbance of water, increasing temperatures also affect the absorbance of carbonate and CO_2. Variations are clearer in diamonds carrying high-Mg carbonatitic HDFs: the absorbance of the main carbonate band at $\sim$1,450 cm^{-1} decreases by $\sim$15 % between room temperature and 420 °C and reached 46 % in diamond ON-UDC-260. At room temperature, the carbonate peak of the high-Mg carbonatitic HDFs at 881 cm^{-1} is accompanied by a second small peak or a shoulder at $\sim$864 cm^{-1}. While the former shows a shift of $\sim$5 cm^{-1} towards lower wavenumbers and its absorbance decreases only slightly, the latter one decreases considerably or disappears with increasing temperature. The absorbance of CO_2 (double peak at 2,344 and 2,361 cm^{-1}) increases with increasing temperature, and the variation in intensity is 25–300 % of the original peak height. Since the spectra were collected with only a limited flushing of the sample by low-CO_2 air, we cannot rule out that the observed variation is external.

In low-Mg carbonatitic to silicic compositions, the absorbance of the main band of carbonate decreases little, the peak at 877 cm^{-1} shifts to 873 cm^{-1}, and the CO_2 absorbance increases by $\sim$50 % with increasing temperature. Saline HDFs exhibit similar change in peak position, but no change in the CO_2 absorbance.

Slight decrease in both phlogopite and quartz absorbance with increasing temperature was observed in a single diamond (ON-DBP-332) that carries silicic HDF. No carbonate absorbance was detected in this diamond.

Solvated Water Molecules in Halides and Carbonates Pellets

Interaction between carbonate ions and solvated water molecules in synthetic $CaCO_3$, Na_2CO_3 and K_2CO_3 pellets that were allowed to absorb humidity leads to the broadening of the OH-stretching band of water and to a significant shift from its $\sim$3,400 cm^{-1} peak position. The largest broadening and the smallest shift appear for $CaCO_3$-solvated water, while the opposite is observed for K_2CO_3-solvated water.

In NaCl and KCl pellets, the OH-stretching band of NaCl- and KCl-solvated water molecules is narrower by 40 and 70 cm^{-1}, respectively, compared with that of pure water. We attribute this narrowing of the OH band to the presence of many isolated water molecules and limited hydrogen bonding between neighbouring water molecules that could cause absorption at lower wavenumbers. It results in a narrower OH band with lower 3,200/3,400 cm^{-1} absorbance ratio (Fig. 3).

The OH spectra of water in the different pellets are presented in appendix A.

Discussion

The rather large difference in the shape of the OH-stretching mode of diamonds carrying microinclusions of different compositions is puzzling; both Figs. 3 and 4 suggest that the differences are compositionally dependent.

Considering the chemical composition of the microinclusions, we would expect the presence of quartz, mica and apatite as well as Ca-, Mg-, Fe- and Na-carbonates. The high abundance of potassium, low chlorine and low mica content in the high-Mg carbonatitic HDFs should lead to the formation of significant, highly soluble K-carbonate (Weiss et al. 2010) in equilibrium with a concentrated solution. In low-Mg carbonatitic to silicic compositions, most of the potassium is bound in mica and the formation of K-carbonate is less likely. Similarly, in saline compositions, Na and K contents are fully compensated by Cl, that is, (Na + K)/Cl < 1, which makes the formation of Na- and K-carbonates unlikely. The residual solution in saline compositions is in equilibrium with chlorides and Mg, Fe and Ca carbonates. We expect it to be mostly saline, with little dissolved carbonate.

Surveying the literature, we found that in water carrying high concentration of dissolved carbonates, the HOH-bending vibration and the OH-stretching mode shift to higher and lower wavenumbers, respectively. Contrarily, high concentration of chloride ions leads to lower frequencies of the HOH-bending vibration and higher ones for the peak of the OH-stretching mode (Venyaminov and Prendergast 1997; Max and Chapados 2001; Zedgenizov et al. 2006; Kagi et al. 2006). The shift in the peak positions in both cases is small and so is the change in their shape.

We now consider the possible contribution of the various daughter phases: ice, sheet silicates, salt-solvated water and hydrated carbonates, as the key to understand the differences in the absorption of water between different HDF compositions.

Solid H_2O: Ice VI and Ice VII

Kagi et al. (2000) argued that the difference in shape between the stretching plus bending combination band of water ($\sim$5,200 cm^{-1}) of a cuboid diamond from Congo and that of liquid water is due to the presence of ice VI together with liquid water in the microinclusions.

Between 3,000 and 4,000 cm^{-1}, both ice VI and ice VII have characteristic IR spectra that are distinct in shape from that of liquid water, their absorption bands are much narrower, and the maximum peak positions are at lower energies (Fig. 6). Compared to the pure water absorption, the absorption band due to the HDFs is broader and extends to both lower and higher wavenumbers. Absorption of ice VI or VII cannot account for the extra width on both sides of the pure water band. As inferred from the general shapes of the OH-stretching band in microinclusion-bearing diamonds of different compositions, it is obvious that ice is not a major part of the secondary hydrous assemblage. Nonetheless, the presence or absence of small amounts of ice is not easy to determine based on the OH-stretching band alone.

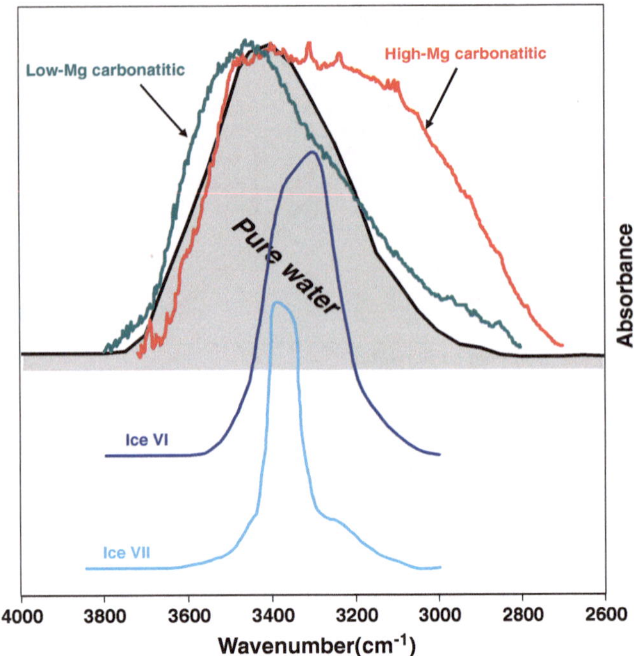

Fig. 6 OH-stretching IR spectra of liquid water, ice VI and ice VII and representative high-Mg carbonatitic and low-Mg carbonatitic HDFs. The spectra of the ices were digitized from Zedgenizov et al. (2006)

Salt Solutions and Salt-Solvated Water

IR spectra of concentrated chloride solutions are only slightly different from that of pure water (Venyaminov and Prendergast 1997; Max and Chapados 2001; Zedgenizov et al. 2006; Kagi et al. 2006). These differences are similar but less significant than those observed in the OH-stretching mode in spectra of saline HDFs (Fig. 3). The effect on the shape of the OH-stretching mode is stronger in the spectra of salt-solvated water absorbed in solid NaCl and KCl (Appendix A; Max and Chapados 2001). In such cases, the narrowing of the bands and the shift to higher wavenumbers are more significant.

Figure 4 shows that the width and the 3,200/3,400 cm^{-1} ratio of the OH-stretching mode of NaCl- and KCl-solvated water are smaller compared to those of pure water and approach the data characterizing saline HDFs. This suggests that most of the water in microinclusions carrying saline compositions might be solvated in the daughter salts; the narrower width of the OH band in such diamonds compared with that of the salt-solvated water molecules in the pellets may be the result of the effect of the residual internal pressure within the inclusions. Furthermore, the small change in shape of the OH-stretching band in saline HDFs at increasing temperature (Fig. 5) suggests restriction of the OH vibration and supports the possibility that water in saline HDFs is present mostly as isolated molecules that are bound in the daughter salts.

Sheet Silicates

Daughter mica is usually detected by IR in diamonds carrying high-Mg carbonatitic and low-Mg carbonatitic to silicic HDFs (Klein-BenDavid et al. 2006, 2009; Weiss et al. 2009, 2010). The hydroxyl groups of micas absorb at ~3,450–3,650 cm^{-1} (Farmer and Russell 1964; Jørgensen 1964; Serratosa and Bradley 1958; Vedder 1964) and may explain the extra absorption at high energies (relative to pure water) characterizing HDFs of low-Mg carbonatitic to silicic compositions (Fig. 7).

Different micas have similar IR spectra and their main band, the Si–O-stretching vibration, is usually situated at ~1,000 cm^{-1} (Farmer 1974). Therefore, it is difficult to determine the exact composition of the mica in the microinclusions by IR spectroscopy alone. EPMA and TEM analyses suggest that the micas found in the microinclusions are of high-Si compositions, midway along the phlogopite–Al-celadonite solid solution (Izraeli et al. 2004; Klein-BenDavid et al. 2006). The spectrum of the OH-stretching vibration of phlogopite, biotite and Fe-celadonite does not provide a close match for the excess absorption at high energies of the OH-stretching band in diamonds relative to pure water (Fig. 7); the spectrum of Al-celadonite is not available. However, as the absorption ratio at 3,600 and 3,400 cm^{-1} shows positive correlation with the Al_2O_3 content of the HDFs (Fig. 8a), we suggest that the presence of mica causes the appearance of the peak/shoulder at ~3,520–3,600 cm^{-1}.

Hydrated Carbonates

Carbonates dissolved in water produce only a small change in the position and shape of the OH-stretching mode

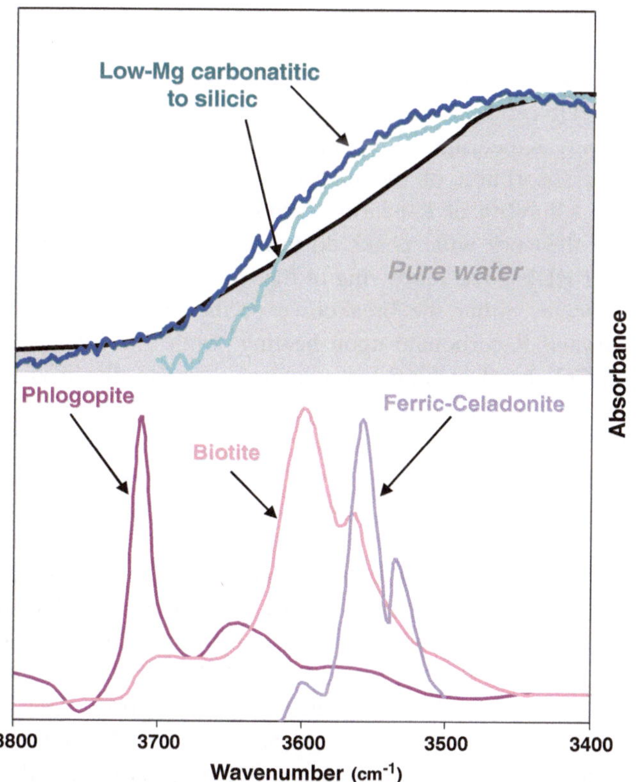

Fig. 7 IR spectrum of the OH-stretching mode in pure water, a low-Mg carbonatitic HDF and a silicic HDF between 3,400 and 3,800 cm^{-1} compared with OH-stretching bands of phlogopite, biotite (Veder 1964) and Fe-celadonite (Farmer 1974)

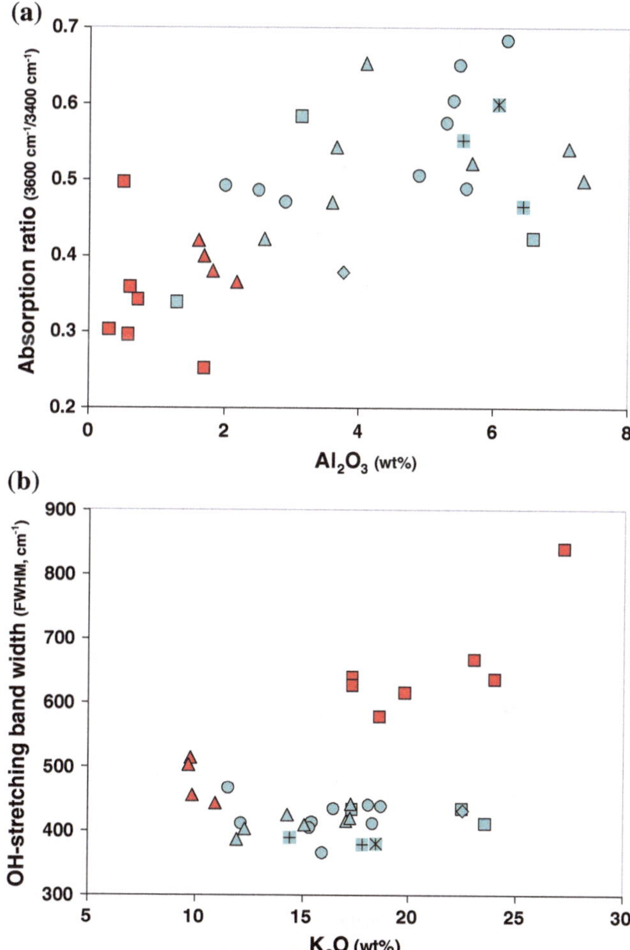

Fig. 8 **a** The absorption ratio at 3,600 and 3,400 cm^{-1} versus the Al_2O_3 content of the HDF (water- and CO_2-free basis). **b** The width of the OH-stretching band versus the K_2O content. High-Mg carbonatitic (*red symbols*) and low-Mg carbonatitic to silicic HDFs (*turquoise symbols*). See Fig. 2 for the provenance of the HDFs

(Zedgenizov et al. 2006) and cannot explain the very wide band observed in the spectra of high-Mg carbonatitic HDFs (Fig. 3). However, the spectrum of the OH-stretching mode of water molecules bound in the structure of hydrous carbonates shows strong absorption between 2,700 and 3,700 cm^{-1} (Moenke 1962; White 1971; Catherine et al. 1977; Frost and Dickfos 2007; Neumann and Epple 2007; http://rruff.info/). A similar range is observed in Ca-, Na- and K-carbonate pellets that were left to absorb atmospheric humidity (carbonate-solvated water, Appendix A).

Combined absorbance of hydrous solution and hydrated carbonates as part of the secondary assemblage in the microinclusions can explain the very wide band in the OH-stretching mode of some of the high-Mg carbonatitic HDFs (Figs. 3, 4).

The width of the OH-stretching band in high-Mg carbonatitic HDFs ranges between ~440 and ~860 cm^{-1}, and the ones with full width at half-maximum (FWHM) <550 cm^{-1} are found in diamonds from Kankan, Guinea, while the ones with FWHM >550 cm^{-1} are all from Udachnaya, Siberia (Figs. 3, 4). Positive correlation is observed between the K_2O content and the width of the OH-stretching band in high-Mg carbonatitic HDFs (Fig. 8b), suggesting the presence of K-carbonate as part of the secondary assemblage and as a key contributor to the width of the OH-stretching band.

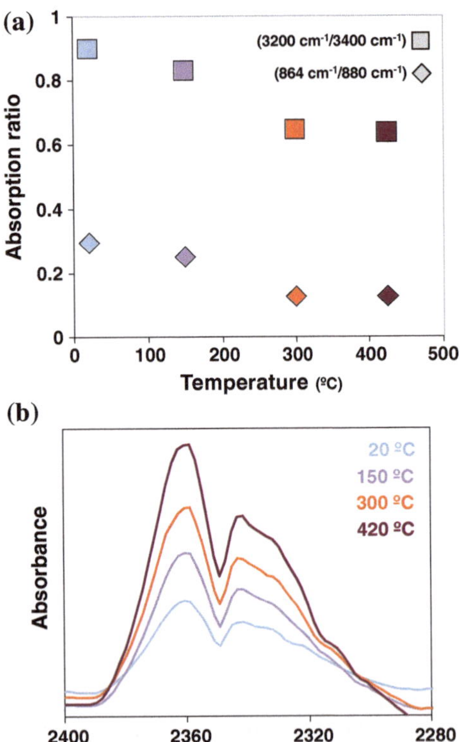

Fig. 9 Variation of spectral characteristics during heating. **a** The absorbance ratio 3,200/3,400 cm^{-1} represents the variation of the OH-stretching band, while the 864/880 cm^{-1} ratio represents the variation in the carbonate phases. **b** The absorbance of CO_2 in the inclusions increases with temperature

The Udachnaya high-Mg carbonatitic diamonds are also characterized by a second small peak or a shoulder at ~864 cm^{-1} that is associated with the sharp carbonate peak at ~880 cm^{-1}. The small peak/shoulder at ~864 cm^{-1} decreases with increasing temperature and its change correlates positively with the reduction in the low-energy absorption of the OH-stretching band and negatively with the height of the CO_2 absorbance in the inclusions (Figs. 5, 9). It is hard to resolve the exact composition for the K-bearing carbonate phase in the microinclusions based on the IR absorption alone. Decomposition of $KHCO_3$ at elevated temperature, $2KHCO_3 \rightarrow K_2CO_3 + H_2O + CO_2$, could have explained these correlations; however, $KHCO_3$ absorbs at ~830 cm^{-1} and not at ~864 cm^{-1}. K_2CO_3 is characterized by a double peak, a sharp peak at ~880 cm^{-1} (like many other carbonates) and a smaller one at 855 cm^{-1}, and two polymorphs of $K_2Ca(CO_3)_2$ absorbs at ~867 cm^{-1} (Schutte and Buijs 1961; Moenke 1962; Farmer 1974; Harrison and Goodyear 1992; Gasc et al. 2009), but bicarbonate is needed to explain the increase in CO_2 absorption. A possible solution may be the presence of a more complex alkali bicarbonate or a mixture of alkali carbonates and bicarbonates.

The low-energy absorption by alkali carbonate-solvated water and the positive correlation between K_2O content of the HDFs and the width of the OH-stretching band strongly support K-bearing carbonates as the reason for the very wide absorption of the high-Mg carbonatitic HDFs. The high solubility of K-bearing carbonates may be the reason why the very wide peaks do not appear in the spectra of other HDFs. The narrowing of the band at high temperature advocates either the breakdown or the dissolution of the hydrated K-carbonate upon heating, while the increase in the CO_2 band at high temperatures suggests that it is not simply dissolution.

Conclusions

The IR absorption of water in microinclusion-bearing diamonds reflects the bulk composition of the trapped HDF. The main hydrous phase is a liquid hydrous solution, which gives rise to the main band at ~3,400 cm^{-1}. Hydrated alkali-bearing carbonates in high-Mg carbonatitic HDFs are responsible for the very wide band of the OH-stretching mode. High mica content in low-Mg carbonatitic to silicic HDFs can explain the shoulder at ~3,600 cm^{-1}. The presence of ice in the microinclusions cannot be ruled out or confirmed based on IR data alone, but its role must be small. Although IR spectroscopy is insensitive to chlorides, a narrow water-stretching band at ~3,400 cm^{-1} may be used to identify saline HDFs based on FTIR alone.

Acknowledgments We thank George Rossman for discussion, Galina Bulanova and an anonymous reviewer for their constructive reviews and Israel Science Foundation (grant #324/07) for funding. Our good friend Dr. Itzhak Kiflawi, who commenced this study and was a major contributor in all aspects of this research, died on the 12th of February 2013 before this paper was published.

Appendix A

OH-stretching band of chloride and carbonate pellets with solvated water absorbed from the humidity in the lab air. The pure water spectrum is from Venyaminov and Prendergast (1997).

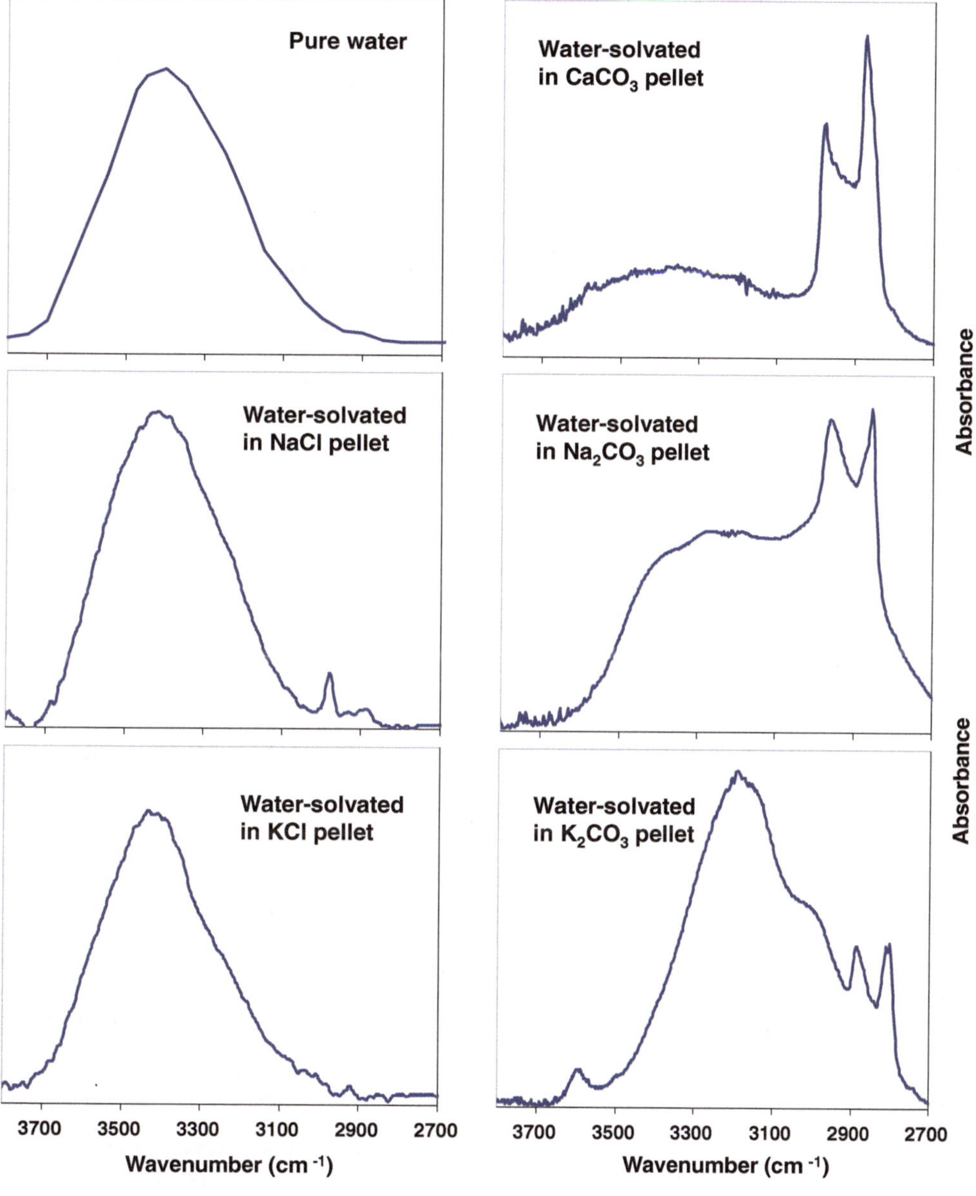

References

Bridgman PW (1935) The phase diagram of water to 45,000 kg/cm^2. J Chem Phys 5:964–966

Catherine H, Skinner K, Osbaldiston GW, Wilner AN (1977) Monohydrocalcite in a guinea pig bladder stone, a novel occurrence. Am Mineral 62:273–277

Chrenko RM, McDonald RS, Darrow KA (1967) Infra-red spectra of diamond coat. Nature 214:474–476

Datchi F, Loubeyre P, LeToullec R (2000) Extended and accurate determination of the melting curves of argon, helium, ice (H2O), and hydrogen (H2). Phys Rev B 61:6535–6546

Farmer VC (1974) The infrared spectra of minerals. Mineralogical Society, London, p 539

Farmer VC, Russell JD (1964) The infrared spectra of layer silicates. Spectrochim Acta 20:1149–1173

Frost RL, Dickfos MJ (2007) Hydrated double carbonates: a Raman and infrared spectroscopic study. Polyhedron 26:4503–4508

Gasc F, Thiebaud-Roux S, Mouloungui Z (2009) Methods for synthesizing diethyl carbonate from ethanol and supercritical carbon dioxide by onepot or two-step reactions in the presence of potassium carbonate. J Supercrit Fluids 50:46–53

Guthrie GD, Veblen DR, Navon O, Rossman GR (1991) Submicrometer fluid inclusions in turbid-diamond coats. Earth Planet Sci Lett 105:1–12

Harrison PG, Goodyear D (1992) Diffuse reflectance FT-IR study of the deactivation and reactivation of potassium carbonate-supported sodium catalysts for propene dimerisation. J Catal 136:76–85

Izraeli ES, Harris JW, Navon O (2004) Fluid and mineral inclusions in cloudy diamonds from Koffiefontein, South Africa. Geoch et Cosmoch Acta 68:2561–2575

Jørgensen P (1964) Infrared absorption of O-H bonds in some micas and other phyllosilicates. Clays Clay Miner 13:263–273

Kagi H, Kiyasu A, Akagi T, Nara M, Sawaki T (2006) Near infrared spectroscopic determination of salinity and internal pressure of fluid inclusions in minerals. Appl Spectrosc 60:430–436

Kagi H, Lu R, Davidson P, Goncharov AF, Mao H-K, Hemley RJ (2000) Evidence for ice VI as an inclusion in cuboid diamonds from high P-T near infrared spectroscopy. Mineral Mag 64:1089–1097

Kazarian SG, Martirosyan GG (2002) ATR-IR spectroscopy of superheated water and in situ study of the hydrothermal decomposition of poly (ethylene terephthalate). Phys Chem Chem Phys 4:3759–3763

Klein-BenDavid O, Izraeli ES, Hauri E, Navon O (2007) Fluid inclusions in diamonds from the Diavik mine, Canada and the evolution of diamond-forming fluids. Geoch et Cosmoch Acta 71:723–744

Klein-BenDavid O, Logvinova AM, Schrauder M, Spetius ZV, Weiss Y, Hauri EH, Kaminsky FV, Sobolev NV, Navon O (2009) High-Mg carbonatitic microinclusions in some Yakutian diamonds: a new type of diamond-forming fluid. Lithos 112:648–659

Klein-BenDavid O, Pearson DG, Nowell GM, Ottley C, McNeill JCR, Cartigny P (2010) Mixed fluid sources involved in diamond growth constrained by Sr–Nd–Pb–C–N isotopes and trace elements. Earth Planet Sci Lett 289:123−133

Klein-BenDavid O, Wirth R, Navon O (2006) TEM imaging and analysis of microinclusions in diamonds: a close look at diamond growing fluids. Am Mineral 91:353–365

Maréchal Y (2011) The molecular structure of liquid water delivered by absorption spectroscopy in the whole IR region completed with thermodynamics data. J Mol Structure 1004:146–155

Max JJ, Chapados C (2001) IR spectroscopy of aqueous alkali halide solutions: pure salt-solvated water spectra and hydration numbers. J Chem Phys 115:2664–2675

Moenke H (1962) Mineralspektren. Akademie-verlag GmbH, Berlin

Navon O (1991) High internal-pressures in diamond fluid inclusions determined by infrared-absorption. Nature 353:746–748

Navon O, Hutcheon ID, Rossman GR, Wasserburg GJ (1988) Mantle derived fluids in diamond micro inclusions. Nature 335:784–789

Neumann M, Epple M (2007) Monohydrocalcite and its relationship to hydrated amorphous calcium carbonate in biominerals. Eur J Inorg Chem 14:1953–1957

Schrauder M, Navon O (1994) Hydrous and carbonatitic mantle fluids in fibrous diamonds from Jwaneng, Botswana. Geoch et Cosmoch Acta 58:761–771

Schutte CJH, Buijs K (1961) The infra-red spectra of K2CO3 and its hydrates. Spectrochim Acta 17:921–926

Serratosa JM, Bradley WF (1958) Determination of the orientation of OH bond axes in layer silicates by infrared absorption. J out Phys Chem 62:1164–1167

Tomlinson EL, McMillan PF, Zhang M, Jones AP, Redfern SAT (2007) Quartzbearing C-O–H fluid inclusions diamond: retracing the pressure–temperature path in the mantle using calibrated high temperature IR spectroscopy. Geoch et Cosmoch Acta 71:6030–6039

Vedder W (1964) Correlations between infrared spectrum and chemical composition of mica. Am Mineral 19:736–767

Venyaminov SY, Prendergast FG (1997) Water (H_2O and D_2O) molar absorptivity in the 1000–4000 cm^{-1} range and quantitative infrared spectroscopy of aqueous solutions. Anal Biochem 248:234–245

Walmsley JC, Lang AR (1992a) On submicrometer inclusions I diamond coat—crystallography and composition of ankerites an related rhombohedral carbonates. Min Mag 56:533–543

Walmsley JC, Lang AR (1992b) Oriented biotite inclusions I diamond coat. Min Mag 56:108–111

Weiss Y, Kessel R, Griffin WL, Kiflawi I, Klein-BenDavid O, Bell DR, Harris JW, Navon O (2009) A new model for the evolution of diamond-forming fluids: evidence from microinclusion-bearing diamonds from Kankan, Guinea. Lithos 112:660–674

Weiss Y, Kiflawi I, Navon O (2010) IR spectroscopy: quantitative determination of the mineralogy and bulk composition of fluid microinclusions in diamonds. Chem Geol 275(1–2):26–34

Weiss Y, Griffin WL, Elhlou S, Navon O (2008) Comparison between LA-ICP-MS and EPMA analysis of trace elements in diamonds. Chem Geol 252(3–4):158–168

White WB (1971) Infrared characterization of water and hydroxyl ion in the basic magnesium carbonate minerals. Am Mineral 56:46–53

Zedgenizov DA, Kagi HK, Shatsky VS, Sobolev NV (2004) Carbonatitic melts in cuboid diamonds from Udachnaya kimberlite pipe (Yakutia): evidence from vibrational spectroscopy. Min Mag 68:61–73

Zedgenizov DA, Ragozin AL, Shatsky VS, Araujo D, Griffin WL, Kagi H (2009) Mg and Fe-rich carbonate–silicate high-density fluids in cuboid diamonds from the Internationalnaya kimberlite pipe (Yakutia). Lithos 112:638–647

Zedgenizov DA, Shiryaev AA, Shatsky VS, Kagi H (2006) Water-related IR characteristics in natural fibrous diamonds. Min Mag 70:219–229

Multiple Growth Episodes or Prolonged Formation of Diamonds? Inferences from Infrared Absorption Data

M. Palot, D. G. Pearson, T. Stachel, J. W. Harris, G. P. Bulanova, and I. Chinn

Abstract

The infrared characteristics of 21 sulphide inclusion-bearing diamonds from Finsch Mine, 1 sulphide inclusion-bearing diamond from Udachnaya and 18 silicate inclusion-bearing diamonds from Premier were examined and modelled to investigate the complexity of diamond genesis. Internal heterogeneities in N-abundance and aggregation state within individual diamonds at Finsch range from 5 to 336 at.ppm and 2–60 % of B-defects, respectively. The Udachnaya diamond 3648 displays a steep decrease from the core to the rim of N-abundance from 482 to 10 at.ppm. Nitrogen aggregation state describes the same trend with value of 86 %B in the core down to 14 %B in the rim. Internal variations in N-abundance and aggregation state within diamonds from Premier are 93–654 at.ppm and of 7–62 %B, respectively. These variations reflect more likely multiple growth episodes of diamond at distinct ages rather than steady changes in temperature conditions during prolonged diamond growth. Modelling of infrared characteristics indicates that some diamonds have experienced distinct growth episodes over extended time periods with estimates up to 2,387 ± 931 Ma. There are implications for dating studies, indicating that isochron ages may be flawed as there appears to be no single formation age for a single diamond. N-abundance and aggregation state mapping by FTIR provide the opportunity to constrain diamond growth history for selecting diamonds for dating.

Keywords

Diamond • Infrared spectroscopy • Cathodoluminescence • Diamond genesis

Introduction

Diamonds are unique samples for studying the secular evolution of the lithosphere and the asthenosphere (e.g. Shirey et al. 2002; Stachel et al. 2005). Inclusions encapsulated in diamonds provide evidence for conditions of diamond growth and represent the most reliable technique of dating diamonds. Dating of silicate inclusions requires generally multiple inclusion composites from individual to several diamonds (e.g. Richardson et al. 1984, 1990). Sulphide inclusions and Re–Os isotopes enhance the possibility of dating individual (e.g. Pearson et al. 1998; Pearson and

M. Palot (✉) · D. G. Pearson · T. Stachel
Department of Earth and Atmospheric Sciences, University of Alberta, Edmonton, Canada
e-mail: palot@ualberta.ca

J. W. Harris
School of Geographical and Earth Sciences, University of Glasgow, Glasgow, UK

G. P. Bulanova
School of Earth Sciences, University of Bristol, Bristol, UK

I. Chinn
De Beers Geoscience Centre, South Africa

D. G. Pearson et al. (eds.), *Proceedings of 10th International Kimberlite Conference*, Volume 1, Special Issue of the Journal of the Geological Society of India,
DOI: 10.1007/978-81-322-1170-9_18,

Shirey 1999a, b; Richardson et al. 2001; Shirey and Richardson 2011). Diamond growth could occur during discrete growth episodes (Rudnick et al. 1993; Taylor et al. 1995), and dating diamonds is not an easy task. Diamond cannot be directly dated using radiometric methods, and the lack of inclusions in diamonds means that only infrared absorption data on the diamond-hosted nitrogen impurities can provide any age constraints (Evans and Qi 1982; Evans and Harris 1989; Boyd et al. 1994, 1995; Mendelssohn and Milledge 1995; Taylor et al. 1996a, b). The aggregation of nitrogen in diamond is sensitive to both temperature and time and could be used as an indication of the diamond residence time in the mantle (e.g. Evans and Harris 1989; Mendelssohn and Milledge 1995).

Silicate inclusion-bearing diamonds at Finsch (Appleyard et al. 2004) and Premier (Chinn et al. 2003) display complex internal growth patterns in cathodoluminescence and distinct infrared absorption characteristics (FTIR) within individual diamonds. These observations suggest that individual diamonds have experienced slight changes in temperature conditions during prolonged diamond formation or multiple diamond growth episodes over time. Complex diamond growth histories indicative of changes in a growth environment within the mantle have been previously stated from cathodoluminescence images, FTIR characteristics and multiple inclusion chemistry in single diamond (e.g. Bulanova 1995; Sobolev and Yefimova 1998; Taylor and Anand 2004; Taylor et al. 1996a, b, 2000; Spetsius and Taylor 2008). Distinct diamond events of growth, resorption and regrowth separated by billions of years have been suggested in Udachnaya diamonds based on Pb and Re–Os isotopes (Rudnick et al. 1993; Bulanova et al. 1998; Pearson et al. 1999b; Wiggers de Vries et al. 2008).

In order to evaluate the complexity of diamond genesis, we have modelled the infrared characteristics of 21 diamonds from Finsch (South Africa), 1 from Udachnaya (Siberia) and 18 from Premier (South Africa).

Geological Settings and Samples

The Finsch kimberlite is located 37 km east of Postmasburg and 160 km north-west of Kimberley, Cape Province, Republic of South Africa. The kimberlite is of Group II variety with an estimated Rb–Sr emplacement age of 118 Ma (Smith et al. 1985). The geology and petrology of the kimberlite have been described in Ruotsala (1975), Clement (1975) and Skinner et al. (1979). The inclusion suite of diamonds at Finsch is predominantly peridotitic (Harris and Gurney 1979). The genesis ages for the peridotitic-type (P) and eclogitic-type (E) diamonds are quite distinct, yielding a CHUR-based model age of 3.19 ± 0.05 Ga (Richardson et al. 1984) for the former and an isochron age of 1.58 ± 0.05 Ga for the latter (Richardson et al. 1990). Diamonds examined in this study all contained sulphide inclusions. All sulphide inclusions in the diamonds except F215 and F218 have high (>44 wt %) Fe, low-to-moderate (<13 wt %) Ni contents and so may be classified as E-type according to the sulphide Ni contents (Yefimova et al. 1983). Diamonds F215 and F218 are therefore classified as P-type which is consistent with a single purple P-type garnet inclusion recovered in the F215. Cathodoluminescence (CL) mapping reveals internal growth structures characterized by different $\delta^{13}C$ and N-abundances (Palot et al. 2013). However, core–rim relationships cannot be derived since the analyses were conducted on diamond fragments. Some diamond fragments exhibit complex growth pattern with non-regular octahedral growth patterns (Fig. 1). Appleyard et al. (2004) also identified CL distinct periods of growth within individual diamonds, resulting in both cubic and octahedral growth layers.

The Udachnaya kimberlite (Siberia) with an age of 350 Ma (Kinny et al. 1997) lies within the central Siberian platform (Sobolev and Nixon 1987). A single diamond examined in this study (diamond plate 3648) has been described in detail by Rudnick et al. (1993), Taylor et al. (1995) and Pearson et al. (1999b). The diamond central, cubo-octahedral zone contains a cluster of sulphide inclusions, a wustite inclusion and an Mg-rich olivine. The intermediate zone is octahedral, truncated by the rim zone indicative of a resorption event. The sulphide inclusions located in the rim zone are surrounded by cracks, one of which is re-healed. The high Ni and Os contents of the sulphide inclusions are consistent with the presence of olivine and therefore the P-type paragenesis of the diamond (Pearson et al. 1998).

The Premier kimberlite is located 30 km east–northeast of Pretoria in Republic of South Africa. The kimberlite is Precambrian in age (Allsopp et al. 1967; Barrett and Allsopp 1973). For a detailed review of its geology and petrology, see for example Scott and Skinner (1979). Samples examined here comprised one peridotitic (diamond AP28) and seventeen eclogitic silicate inclusion-bearing diamonds (Chinn et al. 2003). Most of diamonds exhibit octahedral zonations with indications of slight internal resorption (Chinn et al. 2003). E-type diamond AP28 is much more resorbed than the majority of other E-type diamonds at Premier and presents deformation lamellae features. Chinn et al. (2003) reported the complex diamond growth history of one E-type diamond AP33 studied in double-polished central plate (not analysed in this study). This sample comprises a well-defined core region characterized by high nitrogen aggregation state, which is surrounded by a low aggregation state rim overgrowth.

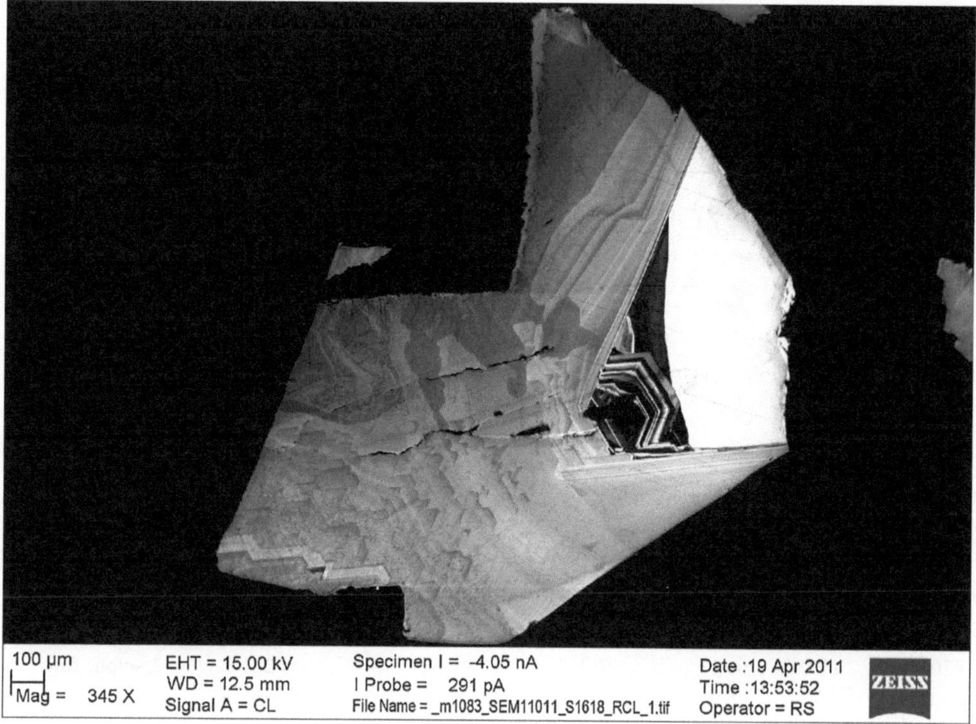

Fig. 1 Cathodoluminescence images of Finsch E-type diamond fragment F213-2 which exhibits non-regular octahedral growth patterns

Analytical Technique

Nitrogen abundances and nitrogen aggregation states, the latter measured as a percentage of nitrogen present in the fully aggregated B-defect, were determined in the De Beers Laboratory of Diamond Research at the University of Alberta, using a Thermo Nicolet Nexus 470 FT-IR spectrometer (bench) fitted with a continuum infrared microscope. Finsch diamond fragments were 1.5–4.5 mm in diameter and mainly were bounded by (111) cleavage planes. Spectra (650–4,000 cm^{-1}) were acquired in transmission mode for 200 s with a resolution of 4 cm^{-1} and an aperture size of 100 μm. Nitrogen concentrations and aggregation states were calculated after spectral decomposition using the Excel program CAXBD97, developed by David Fischer (the Diamond Trading Company, Maidenhead, UK). The absorption coefficients for the A and B centres at 1,282 cm^{-1} were 16.5 and 79.4 at.ppm/cm, respectively (Boyd et al. 1994, 1995). Detection limits and errors typically range from 5 to 20 ppm and about 10 %, respectively. The error of the aggregation state of nitrogen is estimated to be better than ±5 % (2σ).

Udachnaya sample 3648 and the Premier diamonds were diamond plates, polished mainly in (100) and (110) crystallographic directions. The Udachnaya diamond plate is approximately 7 mm long, and Premier plates range in weight between 60 and 200 mg. Data for 3648 were acquired using the equipment and techniques outlined by Mendelssohn and Milledge (1995). For the Premier samples, spectra were acquired with a Nicolet Magna 760 IR spectrometer, over the infrared range from 4,000 to 650 cm^{-1}. Data were processed using the same parameters and method described for the Finsch samples. The Premier data have been described briefly in abstract form by Chinn et al. (2003).

Results: Nitrogen Abundances and Aggregation States

Nitrogen abundances of Finsch samples were determined on 53 diamond fragments from 21 whole diamonds broken to release the inclusions. Overall, the nitrogen content in the studied diamonds ranges from 21 to 1,093 at.ppm (mean 569 ± 271 at.ppm, 1σ) (Table 1). Internal heterogeneity in N-abundance within single diamond ranges from 5 to 336 at.ppm (mean 141 ± 109 at.ppm, 1σ) (Figs. 2 and 3a). Since the analyses were conducted on diamond fragments, core–rim relationships generally cannot be derived. Overall nitrogen aggregation states (expressed as % of B-defects) vary from 0 to 83 % (mean 31 ± 21 %, 1σ) (Table 1). Internal variations of %B are from 2 to 60 % (mean 14 ± 14 %, 1σ) (Figs. 2 and 3a). Diamonds F203, F211, F213 and F215 display the largest variations in N-abundance and aggregation state (Figs. 2 and 3a).

Table 1 Finsch diamond measurements of nitrogen abundance and aggregation state by infrared spectroscopy

Sample	Inclusions	(N) (ppm)	(N) (at.ppm)	% B
F201-1	8 sulph. frag	n.a	n.a.	n.a
F201-2		570	489	6
F201-3		748	641	4
F202-1	15 sulph. frag	425	364	14
F202-2		413	354	8
F203-1	3 sulph. frag	751	644	57
F203-2	+ 1 whole sulph.	770	660	57
F203-3		898	770	56
F203-4		417	357	60
F203-5a		57	49	9
F203-5b		25	21	0
F203-6		225	193	51
F204-1	7 sulph. frag	976	837	36
F204-2		1275	1093	52
F205-1	11 sulph. frag	925	793	27
F205-2		722	619	31
F206-1	1 whole sulph.	767	657	31
F206-2		665	570	40
F206-3		732	627	29
F207-1	1 whole sulph	855	733	79
F207-2		1230	1054	83
F207-3		1223	1048	83
F208-1	1 whole sulph	1261	1081	30
F208-2		1273	1091	33
F208-3		435	373	23
F208-4		645	553	31
F209-1	9 sulph. frag	1160	994	38
F209-2		863	740	25
F210-1	7 sulph. frag	n.a	n.a.	n.a
F210-2		965	827	27
F211-1	19 sulph. frag	779	668	54
F211-2		869	745	31
F212-1	6 sulph. frag	425	364	44
F212-2	from 2 whole sulph	407	349	48
F212-3		526	451	56
F213-1	5 sulph. frag	544	466	58
F213-2		876	751	25
F213-3		620	531	52
F214-1	10 sulph. frag	697	597	28
F214-2	From 3 whole sulph.	679	582	30
F214-3		681	584	28
F214-4		868	744	32
F215-1	4 sulph. frag	359	308	59
F215-2	+ P-type garnet	273	234	41
F215-3		396	339	24
F216-1	5 sulph. frag	71	61	11
F216-2		26	22	2
F216-3		26	22	2
F217-1	19 sulph. frag	562	482	21
F217-2	+ 1 whole sulph.	319	273	19
F217-3		569	488	14
F218-1	3 sulph. frag	315	270	11
F218-2		257	220	2
F218-3		341	292	12
F219-1	16 sulph. frag	1217	1043	17
F219-2		1231	1055	13
F220-1	11 sulph. frag	710	609	23
F220-2		719	616	18
F221-1	2 sulph. frag	595	510	34
F221-2		589	505	23

Errors on nitrogen abundance and aggregation state are 15 % and ± 3 % (2σ), respectively. Abbreviations *sulph* sulphide, *frag* fragment

Nitrogen abundances of Udachnaya diamond 3648 range from 10 to 482 at.ppm (mean 194 ± 143 at.ppm, 1σ) (Table 2). Nitrogen content progressively decreases from the core to the rim. Nitrogen content in the inner core ranges from 482 to 311 at.ppm (mean 376 ± 44 at.ppm), in the outer core from 289 to 135 at.ppm (mean 230 ± 47 at.ppm, 1σ), in the intermediate zone from 113 to 37 at.ppm (mean 57 ± 27 at.ppm, 1σ) and in the rim from 94 to 10 at.ppm (mean 39 ± 25 at.ppm, 1σ) (Fig. 3b). Aggregation states vary from 14 to 86 % (mean of 31 ± 21 %, 1σ). Diamond 3648 also displays a decrease of %B from the core to the rim. The inner core varies from 86 to 69 % (mean 74 ± 4 %, 1σ), the outer core from 69 to 50 % (mean 62 ± 6 %, 1σ), the intermediate zone from 44 to 27 % (mean 35 ± 6 %, 1σ) and the rim from 44 to 14 % (mean 24 ± 8 %, 1σ) (Fig. 3b). The maximum variation in N-abundance and aggregation state between the inner core and the rim in diamond 3648 is 472 at.ppm and 72 %B, respectively (Fig. 2).

Overall, N-abundances in diamonds from Premier range from 21 to 802 at.ppm (mean 408 ± 166 at.ppm, 1σ) (Chinn et al. 2003, Fig. 3c). Internal variation in N-abundance within individual diamonds ranges from 93 to 654 at.ppm (mean of 320 ± 142 at.ppm, 1σ) (Fig. 2). Some diamonds exhibit spatial relationships in N-characteristics (Chinn et al. 2003). Nitrogen aggregation states vary from 2 to 65 % (mean 28 ± 11 %, 1σ) (Chinn et al. 2003, Fig. 3c). Internal variations of %B are from 7 to 62 % (mean of

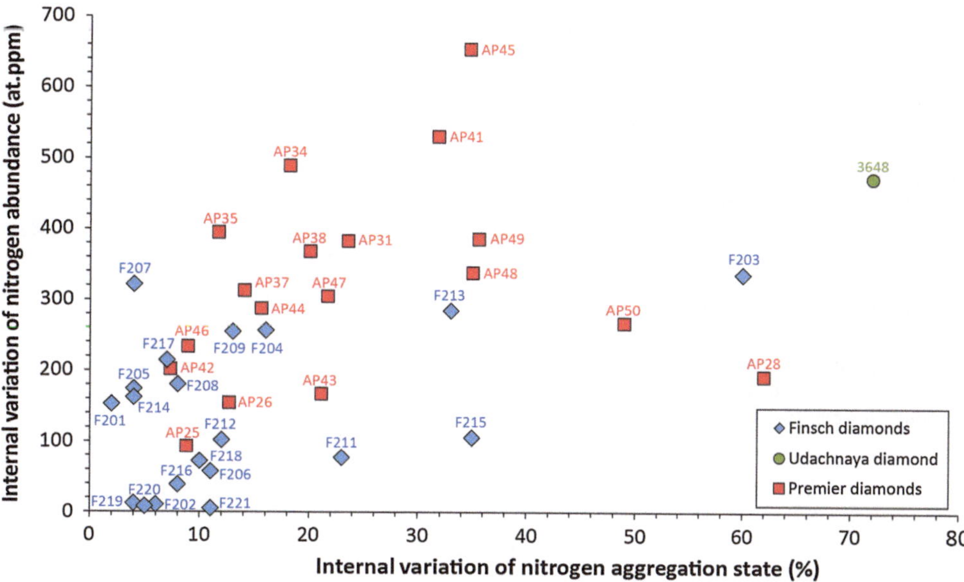

Fig. 2 Internal variation in nitrogen abundance (atomic.ppm) and aggregation state (% of B-defect) within individual diamonds from Premier (*square*), Udachnaya (*circle*) and Finsch (diamond). Internal variation represents the largest measured difference in N-abundance and %B for each diamond

24 ± 15 %, 1σ) (Fig. 2). Diamonds AP28, AP31, AP 41, AP45, AP49 and AP50 display the largest variations in N-abundance and %B (Figs. 2 and 3b, Table 3).

Discussion

Nitrogen Geothermometry

The most common atomic impurity in diamond is nitrogen (Kaiser and Bond 1959), and based on this element, diamonds have been classified as Type I, if nitrogen is present, and Type II, if it is absent (<20 ppm detection level). According to micro-FTIR studies of inclusion-bearing diamonds, 70 % is classified as Type I (Deines et al. 1989; Cartigny 2005; Stachel 2007; Stachel et al. 2009). The different nitrogen-bearing defects in diamonds are linked by a diffusion process. The initially single nitrogen atoms (i.e. the C-defect or diamond Type Ib) diffuse fairly rapidly over millions of years due to a low activation energy (Taylor et al. 1996a, b), to form pairs of nitrogen atoms (i.e. the A-defect or diamond Type IaA). Subsequently, the latter diffuses over billions of years, due to a higher activation energy (Evans and Qi 1982; Evans and Harris 1989), to form aggregates of nitrogen atoms tetrahedrally arranged around a vacancy (i.e. the B-defect or diamond Type IaB). The conversions both from Type Ib to Type IaA and then from Type IaA to Type IaB follow second-order kinetics. In the more important second nitrogen sequence, nitrogen abundance ([N] = [A-defect] + [B-defect]), the average residence time (t) and the storage temperature (T) are all linked by the following formula (Chrenko et al. 1977):

$$\frac{1}{[A-\text{defect}]}-\frac{1}{[N]}=\left(A\times\exp^{-\frac{Ea}{RT}}\right)\times t$$

N-abundance and A-defect are in atomic ppm of nitrogen, t is in seconds and T is in Kelvin, A and R are constants with A = 294,000 $ppm^{-1}\cdot s^{-1}$ and R = 8.317 $J\cdot mol^{-1}\cdot K^{-1}$, Ea = 675,510 $J\cdot mol^{-1}$ is the activation energy (Cooper 1990). Because mantle storage of diamonds is generally over hundreds of Ma, nitrogen aggregation depends chiefly on its time-integrated storage temperature. Nitrogen aggregation is therefore a valuable geothermometer (e.g. Evans and Harris 1989; Palot et al. 2009), which will be used in the next section in order to estimate mantle storage temperatures for the studied diamonds.

Multiple Growth Events for Finsch Sulphide Inclusion-Bearing Diamonds Inferred from Nitrogen Characteristics

In order to investigate the diamond growth histories, we have estimated their probable temperature ranges of formation. To do this calculation, we assume an average mantle residence time of ~1.4 Ga for the Finsch eclogitic sulphide inclusion-bearing diamonds (based on the genesis isochron age of 1.58 Ga for E-type silicate inclusions from Richardson et al. 1990, less than that of the eruption age of 118 Ma from Smith et al. 1985). The estimated mantle storage temperatures for the Finsch diamond fragments range from 1,068 to 1,174 °C with a mean of 1,128 ± 25 °C (1σ) (Fig. 3a). It is noticeable that changes in mantle residence time of the diamonds would affect the absolute storage temperature estimates, but do not alter our conclusions based on relative temperatures between

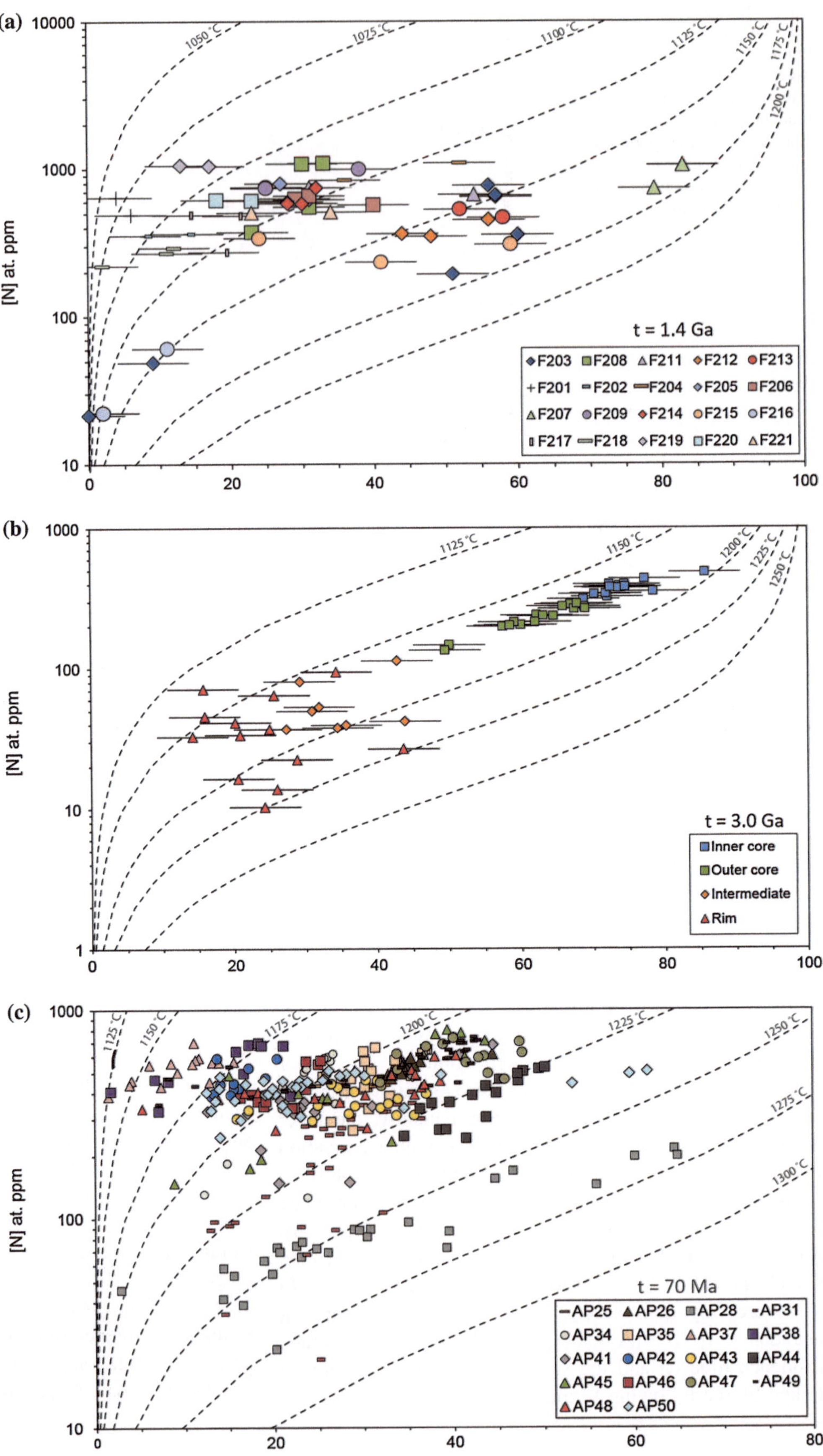

Fig. 3 Nitrogen abundance and aggregation state (%B) of diamonds from **a** Finsch, **b** Udachnaya and **c** Premier. Isotherms are calculated according to the second-order kinetic law for nitrogen diffusion (Chrenko et al. 1977) assuming an averaged mantle residence time of 1.4, 3.0 Ga and 70 Ma, respectively. Total errors of N-abundance and aggregation state are of 15 % and better than ±3 %, respectively. Errors bars are not represented for Premier diamonds for clarity

Table 2 Udachnaya diamond 3648 measurements of nitrogen abundance and aggregation state from core to rim by infrared spectroscopy

Sample	Approximate location	(N) (ppm)	(N) (at.ppm)	%B
B3648TEJ.1	Inner core	461	395	72
B3648TEK.1	Inner core	450	386	74
B3648TEL.1	Inner core	437	375	74
B3648TEW.1	Inner core	462	396	74
B3648TEX.1	Inner core	436	374	74
B3648TEY.1	Inner core	377	323	72
B3648TSJ.1	Inner core	397	340	72
B3648TSK.1	Inner core	444	381	75
B3648TSL.1	Inner core	394	338	70
B3648TSM.1	Inner core	363	311	69
B3648TEH.1	Inner core	412	353	78
B3648TEJ.1	Inner core	506	434	77
B3648TEK.1	Inner core	440	377	72
B3648TEL.1	Inner core	562	482	86
B3648TEF.1	Outer core	231	198	58
B3648TEG.1	Outer core	279	239	62
B3648TEH.1	Outer core	333	285	67
B3648TEM.1	Outer core	321	275	69
B3648TEN.1	Outer core	275	236	65
B3648TEP.1	Outer core	237	203	60
B3648TEZ.1	Outer core	310	266	67
B3648TSE.1	Outer core	171	147	50
B3648TSF.1	Outer core	248	213	59
B3648TSG.1	Outer core	322	276	66
B3648TSH.1	Outer core	337	289	68
B3648TSN.1	Outer core	276	237	63
B3648TSP.1	Outer core	235	201	59
83648TSQ.1	Outer core	157	135	50
B3648TEF.1	Outer core	249	213	62
B3648TEG.1	Outer core	311	267	69
B3648TEA.1	Intermediate	49	42	44
B3648TEB.1	Intermediate	46	39	36
B3648TEC.1	Intermediate	43	37	27
B3648TED.1	Intermediate	62	53	32
B3648TEE.1	Intermediate	94	81	29
B3648TEQ.1	Intermediate	132	113	43
B3648TSR.1	Intermediate	58	50	31
B3648TEA.1	Intermediate	44	38	35
B3648TER.1	Rim	75	64	26
B3648TES.1	Rim	43	37	25
B3648TET.1	Rim	16	14	26
B3648TEU.1	Rim	26	22	29
B3648TEV.1	Rim	110	94	34
B3648TSA.1	Rim	31	27	44
B3648TSB.1	Rim	12	10	24
B3648TSC.1	Rim	19	16	21
B3648TSD.1	Rim	48	41	20
B3648TSS.1	Rim			18
B3648TEB.1	Rim	39	33	21
B3648TEC.1	Rim	38	33	14
B3648TED.1	Rim	53	45	16
B3648TEE.1	Rim	83	71	16

Errors are reported in Table 1

different portions of diamond. Although Finsch diamonds are commonly thought to originate from a relatively homogenous carbon source (Deines et al. 1984), that conclusion was largely based on P-type diamond study. Heterogeneities in N-abundance and aggregation states in some Finsch E-type sulphide inclusion-bearing diamonds suggest that these samples experienced a relatively complex growth and mantle residence history as the data do not plot on a single isotherm (Fig. 3a). For example, diamonds F203, F211, F213, F215 and F218 show internal variations in their integrated mantle residence temperatures of up to ~50 °C (Fig. 3a). These variations reflect either slight changes in temperature conditions during prolonged diamond formation or distinct residence histories for different portions of the diamond, that is, multiple episodes of diamond growth extending over millions to billions of years (e.g. Appleyard et al. 2004; Taylor and Anand 2004). A steady increase in temperatures from the core to the rim is physically unacceptable, since the core must have experienced the same heating event. Therefore, unless these diamonds grew continuously over the course of millions of years while associated with a significant drop in ambient temperature, it is likely that multiple growth episodes occurred at different times. Assuming that the different portions in a single diamond that give the largest heterogeneity in N-abundance and aggregation (Fig. 2) represent distinct growth steps, we modelled the age differences of different portions within the single diamond using the equation of Chrenko et al. (1977) linking temperature, nitrogen abundance and average time residence. We assumed a maximum storage temperature of 1,174 °C based on maximum nitrogen temperature, consistent with the maximum inclusion-based temperature of 1,179 °C reported for E-type silicate-bearing Finsch diamonds (Appleyard et al. 2004). Uncertainties on nitrogen temperatures are of ±12 °C and are based on uncertainties of N-abundance and %B, leading to large uncertainties in time estimates. Our calculations indicate that the time between distinct growth

Table 3 Premier diamond measurements of nitrogen abundance and aggregation state by infrared spectroscopy

Sample	Paragenesis	(N) (at.ppm)	%B	Sample	Paragenesis	N (at.ppm)	%B
AP25	Eclogitic	436	32	AP42	Eclogitic	584	20
AP25	Eclogitic	438	31	AP42	Eclogitic	473	19
AP25	Eclogitic	449	32	AP42	Eclogitic	420	16
AP25	Eclogitic	429	32	AP42	Eclogitic	390	18
AP25	Eclogitic	413	30	AP42	Eclogitic	419	16
AP25	Eclogitic	401	29	AP42	Eclogitic	586	14
AP25	Eclogitic	413	29	AP42	Eclogitic	526	17
AP25	Eclogitic	416	28	AP42	Eclogitic	391	15
AP25	Eclogitic	410	28	AP42	Eclogitic	384	13
AP25	Eclogitic	406	28	AP42	Eclogitic	397	13
AP25	Eclogitic	356	26	AP42	Eclogitic	440	14
AP25	Eclogitic	359	27	AP42	Eclogitic	444	16
AP25	Eclogitic	362	26	AP43	Eclogitic	424	23
AP25	Eclogitic	377	25	AP43	Eclogitic	331	17
AP25	Eclogitic	425	27	AP43	Eclogitic	302	16
AP25	Eclogitic	432	25	AP43	Eclogitic	405	29
AP25	Eclogitic	440	24	AP43	Eclogitic	425	27
AP25	Eclogitic	448	24	AP43	Eclogitic	346	29
AP26	Eclogitic	461	32	AP43	Eclogitic	339	32
AP26	Eclogitic	469	33	AP43	Eclogitic	294	24
AP26	Eclogitic	517	35	AP43	Eclogitic	312	36
AP26	Eclogitic	554	35	AP43	Eclogitic	312	34
AP26	Eclogitic	560	35	AP43	Eclogitic	374	34
AP26	Eclogitic	536	34	AP43	Eclogitic	395	37
AP26	Eclogitic	566	36	AP43	Eclogitic	318	28
AP26	Eclogitic	615	44	AP43	Eclogitic	355	33
AP26	Eclogitic	538	36	AP43	Eclogitic	348	25
AP26	Eclogitic	522	35	AP43	Eclogitic	459	32
AP26	Eclogitic	495	34	AP43	Eclogitic	461	30
AP26	Eclogitic	487	33	AP43	Eclogitic	442	27
AP26	Eclogitic	489	33	AP44	Eclogitic	389	42
AP26	Eclogitic	503	33	AP44	Eclogitic	266	38
AP26	Eclogitic	557	34	AP44	Eclogitic	248	35
AP26	Eclogitic	575	34	AP44	Eclogitic	267	39
AP26	Eclogitic	588	35	AP44	Eclogitic	243	41
AP26	Eclogitic	567	34	AP44	Eclogitic	306	44
AP26	Eclogitic	538	34	AP44	Eclogitic	400	45
AP28	Peridotitic	170	47	AP44	Eclogitic	456	47
AP28	Peridotitic	155	45	AP44	Eclogitic	522	49
AP28	Peridotitic	73	20	AP44	Eclogitic	508	48
AP28	Peridotitic	46	3	AP44	Eclogitic	530	50
AP28	Peridotitic	58	14	AP44	Eclogitic	464	45
AP28	Peridotitic	74	22	AP44	Eclogitic	494	47
AP28	Peridotitic	77	23	AP44	Eclogitic	431	44

(continued)

Table 3 (continued)

Sample	Paragenesis	(N) (at.ppm)	%B	Sample	Paragenesis	N (at.ppm)	%B
AP28	Peridotitic	70	21	AP44	Eclogitic	360	40
AP28	Peridotitic	41	14	AP44	Eclogitic	336	36
AP28	Peridotitic	24	20	AP44	Eclogitic	357	38
AP28	Peridotitic	39	17	AP45	Eclogitic	485	15
AP28	Peridotitic	54	16	AP45	Eclogitic	398	23
AP28	Peridotitic	72	25	AP45	Eclogitic	326	23
AP28	Peridotitic	89	31	AP45	Eclogitic	175	17
AP28	Peridotitic	96	35	AP45	Eclogitic	193	19
AP28	Peridotitic	87	39	AP45	Eclogitic	148	9
AP28	Peridotitic	198	60	AP45	Eclogitic	234	33
AP28	Peridotitic	216	64	AP45	Eclogitic	596	41
AP28	Peridotitic	200	65	AP45	Eclogitic	712	44
AP28	Peridotitic	145	56	AP45	Eclogitic	775	41
AP28	Peridotitic	72	39	AP45	Eclogitic	765	38
AP28	Peridotitic	82	30	AP45	Eclogitic	802	39
AP28	Peridotitic	66	23	AP45	Eclogitic	609	37
AP28	Peridotitic	63	19	AP45	Eclogitic	378	26
AP28	Peridotitic	89	29	AP45	Eclogitic	378	26
AP28	Peridotitic	88	30	AP45	Eclogitic	515	29
AP28	Peridotitic	54	20	AP46	Eclogitic	393	18
AP28	Peridotitic	69	26	AP46	Eclogitic	380	16
AP31	Eclogitic	278	23	AP46	Eclogitic	363	18
AP31	Eclogitic	270	25	AP46	Eclogitic	345	19
AP31	Eclogitic	252	26	AP46	Eclogitic	365	19
AP31	Eclogitic	220	27	AP46	Eclogitic	340	22
AP31	Eclogitic	182	24	AP46	Eclogitic	339	22
AP31	Eclogitic	167	24	AP46	Eclogitic	416	24
AP31	Eclogitic	128	19	AP46	Eclogitic	549	25
AP31	Eclogitic	107	32	AP46	Eclogitic	568	24
AP31	Eclogitic	21	25	AP46	Eclogitic	572	25
AP31	Eclogitic	35	14	AP47	Eclogitic	497	26
AP31	Eclogitic	68	23	AP47	Eclogitic	616	34
AP31	Eclogitic	88	26	AP47	Eclogitic	666	37
AP31	Eclogitic	92	23	AP47	Eclogitic	737	40
AP31	Eclogitic	93	15	AP47	Eclogitic	731	40
AP31	Eclogitic	97	15	AP47	Eclogitic	685	39
AP31	Eclogitic	97	13	AP47	Eclogitic	517	34
AP31	Eclogitic	88	12	AP47	Eclogitic	432	32
AP31	Eclogitic	175	26	AP47	Eclogitic	450	32
AP31	Eclogitic	302	36	AP47	Eclogitic	560	41
AP31	Eclogitic	357	35	AP47	Eclogitic	602	43
AP31	Eclogitic	365	33	AP47	Eclogitic	592	43
AP31	Eclogitic	326	32	AP47	Eclogitic	704	47
AP31	Eclogitic	325	32	AP47	Eclogitic	624	48

(continued)

Table 3 (continued)

Sample	Paragenesis	(N) (at.ppm)	%B	Sample	Paragenesis	N (at.ppm)	%B
AP31	Eclogitic	404	35	AP47	Eclogitic	464	46
AP31	Eclogitic	352	33	AP47	Eclogitic	497	44
AP31	Eclogitic	270	29	AP48	Eclogitic	336	5
AP31	Eclogitic	256	27	AP48	Eclogitic	400	17
AP31	Eclogitic	300	28	AP48	Eclogitic	414	20
AP31	Eclogitic	351	30	AP48	Eclogitic	373	14
AP31	Eclogitic	314	28	AP48	Eclogitic	402	18
AP31	Eclogitic	272	27	AP48	Eclogitic	406	20
AP31	Eclogitic	246	23	AP48	Eclogitic	377	25
AP31	Eclogitic	281	24	AP48	Eclogitic	266	20
AP34	Eclogitic	394	19	AP48	Eclogitic	455	39
AP34	Eclogitic	423	23	AP48	Eclogitic	516	35
AP34	Eclogitic	455	23	AP48	Eclogitic	389	30
AP34	Eclogitic	505	23	AP48	Eclogitic	424	37
AP34	Eclogitic	568	26	AP48	Eclogitic	392	36
AP34	Eclogitic	592	26	AP48	Eclogitic	360	27
AP34	Eclogitic	615	27	AP48	Eclogitic	490	36
AP34	Eclogitic	587	26	AP48	Eclogitic	597	38
AP34	Eclogitic	523	24	AP48	Eclogitic	604	40
AP34	Eclogitic	522	24	AP48	Eclogitic	488	33
AP34	Eclogitic	557	30	AP48	Eclogitic	439	37
AP34	Eclogitic	355	27	AP48	Eclogitic	272	30
AP34	Eclogitic	337	21	AP49	Eclogitic	536	15
AP34	Eclogitic	428	21	AP49	Eclogitic	468	8
AP34	Eclogitic	311	25	AP49	Eclogitic	432	11
AP34	Eclogitic	126	24	AP49	Eclogitic	352	7
AP34	Eclogitic	131	12	AP49	Eclogitic	363	15
AP34	Eclogitic	184	15	AP49	Eclogitic	384	17
AP35	Eclogitic	479	26	AP49	Eclogitic	449	26
AP35	Eclogitic	524	31	AP49	Eclogitic	461	28
AP35	Eclogitic	511	31	AP49	Eclogitic	528	34
AP35	Eclogitic	426	30	AP49	Eclogitic	641	39
AP35	Eclogitic	336	31	AP49	Eclogitic	691	41
AP35	Eclogitic	482	35	AP49	Eclogitic	738	42
AP35	Eclogitic	644	37	AP49	Eclogitic	696	39
AP35	Eclogitic	647	37	AP49	Eclogitic	691	40
AP35	Eclogitic	619	37	AP49	Eclogitic	617	40
AP35	Eclogitic	573	34	AP49	Eclogitic	591	36
AP35	Eclogitic	385	33	AP49	Eclogitic	712	41
AP35	Eclogitic	265	29	AP49	Eclogitic	707	42
AP35	Eclogitic	291	26	AP49	Eclogitic	713	42
AP35	Eclogitic	411	28	AP49	Eclogitic	632	41
AP35	Eclogitic	467	29	AP49	Eclogitic	582	39
AP35	Eclogitic	468	28	AP49	Eclogitic	587	39

(continued)

Table 3 (continued)

Sample	Paragenesis	(N) (at.ppm)	%B	Sample	Paragenesis	N (at.ppm)	%B
AP35	Eclogitic	567	30	AP49	Eclogitic	610	36
AP35	Eclogitic	660	31	AP49	Eclogitic	641	41
AP35	Eclogitic	633	30	AP49	Eclogitic	552	42
AP37	Eclogitic	561	13	AP49	Eclogitic	445	40
AP37	Eclogitic	503	9	AP49	Eclogitic	431	37
AP37	Eclogitic	560	14	AP49	Eclogitic	465	36
AP37	Eclogitic	557	9	AP49	Eclogitic	479	32
AP37	Eclogitic	589	12	AP49	Eclogitic	480	32
AP37	Eclogitic	578	11	AP49	Eclogitic	437	28
AP37	Eclogitic	553	11	AP49	Eclogitic	430	28
AP37	Eclogitic	565	16	AP49	Eclogitic	413	21
AP37	Eclogitic	458	4	AP49	Eclogitic	429	24
AP37	Eclogitic	437	4	AP49	Eclogitic	407	21
AP37	Eclogitic	450	13	AP50	Eclogitic	457	19
AP37	Eclogitic	385	2	AP50	Eclogitic	430	22
AP37	Eclogitic	428	7	AP50	Eclogitic	397	20
AP37	Eclogitic	698	11	AP50	Eclogitic	393	19
AP37	Eclogitic	546	6	AP50	Eclogitic	396	20
AP38	Eclogitic	464	7	AP50	Eclogitic	396	19
AP38	Eclogitic	407	2	AP50	Eclogitic	420	20
AP38	Eclogitic	327	7	AP50	Eclogitic	402	13
AP38	Eclogitic	387	22	AP50	Eclogitic	440	17
AP38	Eclogitic	412	16	AP50	Eclogitic	513	26
AP38	Eclogitic	454	8	AP50	Eclogitic	456	23
AP38	Eclogitic	496	17	AP50	Eclogitic	467	26
AP38	Eclogitic	634	16	AP50	Eclogitic	492	28
AP38	Eclogitic	681	17	AP50	Eclogitic	483	27
AP38	Eclogitic	696	18	AP50	Eclogitic	503	29
AP38	Eclogitic	675	19	AP50	Eclogitic	485	30
AP38	Eclogitic	674	21	AP50	Eclogitic	489	39
AP41	Eclogitic	326	13	AP50	Eclogitic	443	53
AP41	Eclogitic	320	17	AP50	Eclogitic	495	60
AP41	Eclogitic	368	21	AP50	Eclogitic	509	62
AP41	Eclogitic	353	21	AP50	Eclogitic	338	35
AP41	Eclogitic	333	23	AP50	Eclogitic	309	23
AP41	Eclogitic	360	23	AP50	Eclogitic	349	21
AP41	Eclogitic	341	23	AP50	Eclogitic	428	23
AP41	Eclogitic	319	22	AP50	Eclogitic	428	23
AP41	Eclogitic	326	25	AP50	Eclogitic	461	26
AP41	Eclogitic	480	36	AP50	Eclogitic	452	24
AP41	Eclogitic	679	44	AP50	Eclogitic	456	21
AP41	Eclogitic	607	41	AP50	Eclogitic	368	16
AP41	Eclogitic	345	31	AP50	Eclogitic	326	22
AP41	Eclogitic	150	28	AP50	Eclogitic	246	14

(continued)

Table 3 (continued)

Sample	Paragenesis	(N) (at.ppm)	%B	Sample	Paragenesis	N (at.ppm)	%B
AP41	Eclogitic	148	21	AP50	Eclogitic	297	16
AP41	Eclogitic	214	19	AP50	Eclogitic	330	13
				AP50	Eclogitic	366	14
				AP50	Eclogitic	420	14
				AP50	Eclogitic	479	14

Errors are reported in Table 1

steps within individual diamonds varies from minimum estimates (mantle residence at 1,174 °C) of $\sim 3^{+88}_{-3}$ Ma (diamond F208) to 942 ± 368 Ma (diamond F215), to maximum estimates (residence at 1,000–1,140 °C at 4.5 Ga) that are equivalent to the age of the Earth (Fig. 4a). These calculations may be underestimated because the random orientation of fragments may be mixtures of zones with distinct FTIR characteristics. Because of the large uncertainties for some diamonds (F204, F205, F207, F208, F209, F212, F214, F219 and F220, Table 4), it cannot be ruled out that these samples may have experienced a single growth event at relatively constant temperature. The majority of our calculations are, however, compatible with multiple growth events during the formation of sulphide inclusion-bearing diamonds at Finsch. This implies that isochron ages may be flawed as there appears to be no single formation age for a single diamond. This is consistent with distinct carbon isotopic compositions and nitrogen contents observed across some of these diamonds (Palot et al. 2013).

Multiple Growth Events for Udachnaya Peridotitic Sulphide Inclusion-Bearing Diamond Inferred from Nitrogen Characteristics

We assume an average mantle residence time of ~3.0 Ga for the Udachnaya diamond 3648 (based on the model age of 3.1–3.5 Ga on sulphide inclusions in this diamond by Pearson et al. 1999b). The age of inclusions is significantly older than that of the eruption age of 361 Ma (Kinny et al. 1997). The estimated mantle storage temperatures range from 1,137 to 1,200 °C with a mean of 1,166 ± 12 °C (1σ) (Fig. 3b), which are within the range of inclusion-based temperatures of Udachnaya diamonds (e.g. Griffin et al. 1993). The inner and outer cores display a constant estimated storage temperature of 1,163 ± 3 °C (1σ) with data plotting close to the theoretical isotherms at constant time (Fig. 3b). In contrast, the intermediate and rim zones display more scattered time-averaged residence temperatures, with a slightly higher mean of 1,170 ± 17 °C (1σ) (Fig. 3b). A steady increase in integrated temperature from the core to the rim within mantle diamonds is physically implausible, since the former and the latter must have experienced the same temperature during any heating event, and diamonds conduct heat very efficiently. We therefore explore the possibility of multiple growth episodes at different times suggested by Rudnick et al. (1993) based on large variations in Pb isotopic compositions of sulphides included in this diamond and by Bulanova et al. (1996) and Pearson et al. (1999b) based on enriched Os, Re, Pb and Zn and more radiogenic Os and Pb isotopes in one sulphide inclusion in the rim of diamond 3648. Calculations indicate that the maximum relative time between the core and the rim varies from a minimum estimate (mantle residence at 1,200 °C) of ~2,387 ± 931 Ma to maximum estimate (residence at 1,186 °C) that is equivalent to the age of the Earth (Fig. 4b). The minimum estimate equals the estimate of 2 Ga as the age difference between different growth zones of this diamond determined by Pb-isotope analysis of sulphide inclusions by Rudnick et al. (1993).

Multiple Growth Events for Premier Silicate Inclusion-Bearing Diamonds Inferred from Nitrogen Characteristics

Eclogitic silicate inclusion-bearing diamonds at Premier display ages from 1,150 ± 60 to 1,254 Ma (Richardson 1986; Burgess et al. 1989; Phillips et al. 1989) compared to a pipe eruption age of 1,180 ± 30 Ma (Allsopp et al. 1989). We assume a maximum average mantle residence time of ~70 Ma for these diamonds. The estimated mantle storage temperatures range from 1,127 to 1,286 °C with a mean of 1,209 ± 23 °C (1σ) (Fig. 3c) and are slightly lower than estimates from inclusions in Premier diamonds (between 1,199 and 1,363 °C, Gurney et al. 1985). The large variations in N-abundance and aggregation states in some Premier diamonds suggest that these samples also have experienced complex growth histories. Variation in time-integrated storage temperature within individual diamonds ranges from 10 °C (diamond AP26) to 79 °C (diamond AP28) (Fig. 3c). These variations, which are similar to Finsch and Udachnaya diamonds, are interpreted as reflecting multiple episodes of diamond growth separated in time, rather than steady increases or decreases in storage temperature. Calculations indicate that the maximum time between distinct growth steps varies from minimum estimates (mantle residence at 1,286 °C) of 1.5 ± 0.5 Ma (diamond AP37) to 76 ± 23 Ma (diamond AP31), to

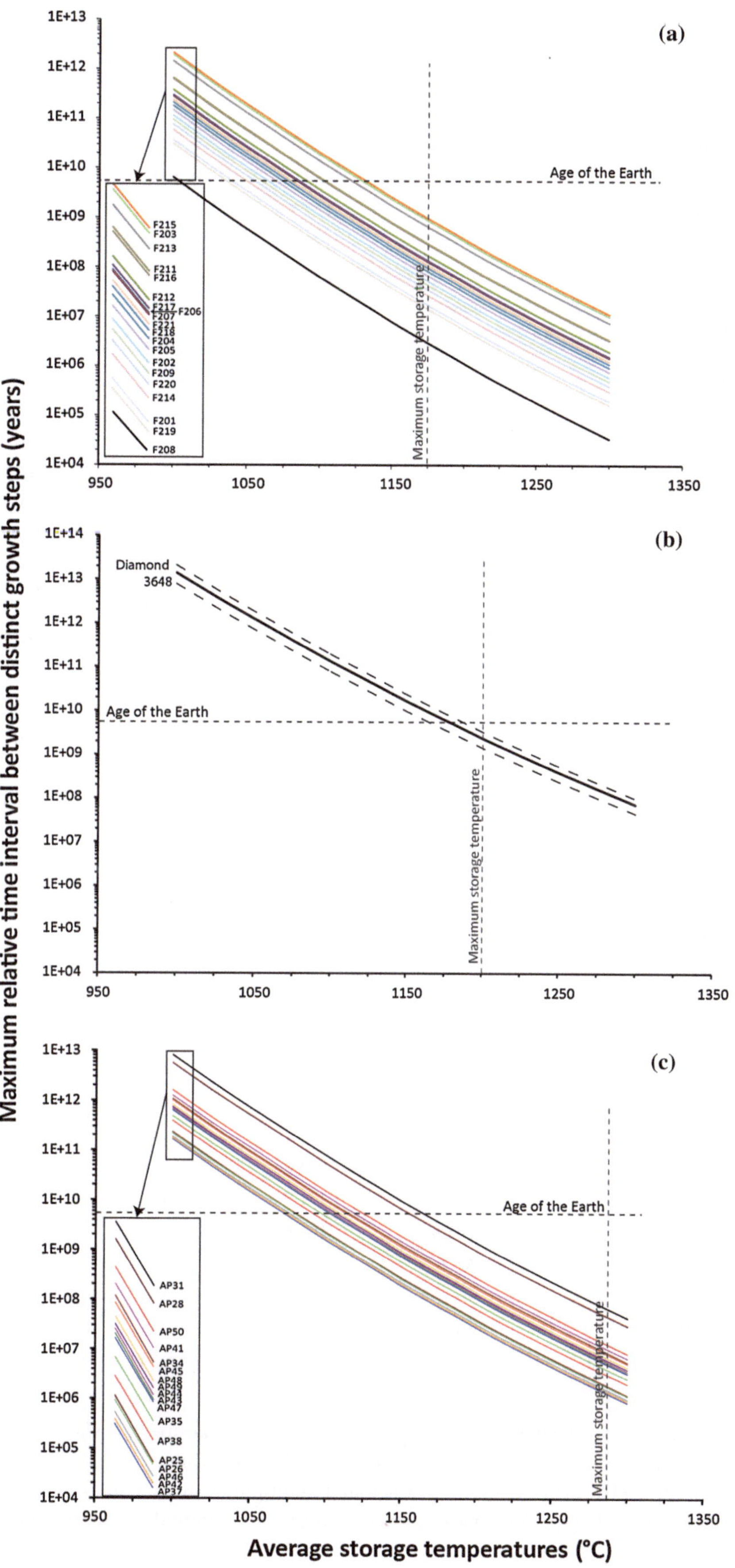

Fig. 4 Maximum relative times between distinct growth steps against averaged storage temperatures within individual diamonds from **a** Finsch, **b** Udachnaya and **c** Premier. The maximum averaged storage temperature for a given locality is based on the highest nitrogen (and inclusion)-based temperature estimate reported for each population. The age of the Earth is reported as the upper limit for time difference between distinct diamond portions. Errors bars (*dotted lines*) are only reported for Udachnaya diamond for clarity

Table 4 Relative errors (± %, 2σ) on temperature estimates for Finsch, Udachnaya and Premier diamonds at temperatures from 1,000 to 1,300 °C (50 °C increments)

Samples/Temperature	1,000 °C	1,050 °C	1,100 °C	1,150 °C	1,200 °C	1,250 °C	1,300 °C
F201	78.6	75.5	72.6	69.7	66.9	64.2	61.6
F202	89.6	86.1	82.7	79.4	76.2	73.2	70.2
F203	60.5	58.1	55.8	53.6	51.4	49.3	38.3
F204	132.9	127.7	122.6	117.7	113.0	108.5	104.1
F205	117.6	113.0	108.6	104.2	100.0	96.0	92.2
F206	91.3	87.7	84.3	80.9	77.6	74.5	72.4
F207	486.9	467.9	449.4	431.4	414.1	397.5	381.5
F208	3501.5	3364.7	3231.5	3102.4	2977.9	2858.1	2743.2
F209	162.9	156.5	150.3	144.3	138.5	133.0	127.6
F211	56.9	54.7	52.5	50.4	48.4	46.5	44.6
F212	188.7	181.3	174.1	167.2	160.5	154.0	142.0
F213	42.1	40.4	38.8	37.3	35.8	34.3	33.0
F214	330.6	317.7	305.1	292.9	281.2	269.9	262.2
F215	45.0	43.3	41.6	39.9	38.3	36.8	41.0
F216	70.7	67.9	65.2	62.6	60.1	57.7	55.4
F217	62.1	59.7	57.3	55.0	52.8	50.7	53.9
F218	45.0	43.2	41.5	39.9	38.3	36.7	34.6
F219	156.5	150.4	144.4	138.6	133.1	127.7	122.6
F220	159.3	153.1	147.0	141.2	135.5	130.1	124.8
F221	99.0	95.1	91.3	87.7	84.2	80.8	77.5
3648	22.9	22.0	21.2	20.3	19.5	18.7	18.0
AP25	112.4	108.0	103.8	99.6	95.6	91.8	88.1
AP26	144.6	139.0	133.5	128.1	123.0	118.1	113.3
AP28	37.7	36.2	34.8	33.4	32.1	30.8	29.5
AP31	38.1	36.6	35.2	33.8	32.4	31.1	29.9
AP34	47.8	45.9	44.1	42.3	40.6	39.0	37.4
AP35	69.5	66.8	64.1	61.6	59.1	56.7	54.4
AP37	40.8	39.2	37.6	36.1	34.7	33.3	32.0
AP38	37.8	36.3	34.9	33.5	32.2	30.9	29.6
AP41	43.0	41.4	39.7	38.1	36.6	35.1	33.7
AP42	73.9	71.0	68.2	65.5	62.8	60.3	57.9
AP43	58.0	55.7	53.5	51.4	49.3	47.3	45.4
AP44	97.6	93.8	90.1	86.5	83.0	79.7	76.5
AP45	43.4	41.7	40.0	38.4	36.9	35.4	34.0
AP46	105.8	101.6	97.6	93.7	90.0	86.3	82.9
AP47	62.2	59.7	57.4	55.1	52.9	50.7	48.7
AP48	39.7	38.1	36.6	35.1	33.7	32.4	31.1
AP49	40.6	39.0	37.4	35.9	34.5	33.1	31.8
AP50	39.9	38.4	36.9	35.4	34.0	32.6	31.3

maximum estimates (residence at 1,076–1,170 °C) that are equivalent to the age of the Earth (Fig. 4c). Marked variation in FTIR data suggests distinct and multiple growth episodes for E-type diamonds at Premier with ages minimal 1 Ma apart, in agreement with some conclusions of Chinn et al. (2003). A single diamond growth event cannot be ruled out for diamonds AP25, AP26 and AP46 due to large uncertainties (Table 4).

Overall, this study therefore suggests multiple episodes of diamond growth based on distinct infrared spectroscopy characteristics within individual macro-diamonds from different localities and parageneses. Similar evidence was found previously for eclogitic diamonds from Siberian kimberlites (Bulanova et al. 1998 and 2012) and by Wiggers de Vries et al. (2008, 2011) in a detailed study of sulphide-bearing diamond plates from Udachnaya, Mir and Mwadui kimberlites. The study by Wiggers de Vries et al. (2008) also revealed complex Re–Os isotope systematics in sulphides derived from different diamond growth zones, clearly indicating the complexity present in single diamond. FTIR study is therefore a valuable diagnostic tool in characterizing samples for radiometric dating. In addition, the technique is an excellent complement to classical CL images in selecting diamonds for in situ carbon isotopic profiles, allowing the ability to determine the likelihood of multiple growth events. Measurements for such work are best taken in situ on polished diamond plates, as for the Udachnaya and Premier diamonds, to preserve the relationships between the data points, using the CL image as a map of the growth history. In this way, any changes in crystal growth or periods of resorption can be integrated with the data measurements for interpretation.

Conclusions

The examination of 21 sulphide inclusion-bearing diamonds from Finsch, 1 sulphide inclusion-bearing diamond from Udachnaya and 18 silicate inclusion-bearing diamonds from Premier revealed distinct N-abundances and aggregation states of nitrogen within individual samples. These variations more likely reflect multiple growth episodes of diamond at different times rather than steady changes in temperature conditions during prolonged diamond formation. Modelling of these heterogeneities indicates that some diamonds have experienced distinct growth episodes over extended time periods, implying that great care is needed in the evaluation of isochron and model ages in radiometric dating studies, as there may be multiple phases of diamond growth in single diamond.

The diamonds investigated in this study were derived from different localities and parageneses, suggesting that such complex growth histories are not rare. Infrared characteristics are a critical tool for constraining diamond growth histories and for selecting diamonds for dating.

Acknowledgments The Finsch diamond samples used in this study were gifted to JWH by the Diamond Trading Company, a member of the DeBeers Group of Companies. The authors sincerely thank DeBeers for this material. The Udachnaya diamond plate was generously provided by the Diamond and Precious Metal Geology Institute, Siberian Branch, Russian Academy of Sciences. The Premier diamonds were made available for study by De Beers. We thank Judith Milledge for her FTIR measurements. The study was funded by a CERC award to D.G. Pearson. The authors express their appreciation to L.A. Taylor and D. Wiggers de Vries for their reviews, which has resulted in improvement to the manuscript and helped improve the ideas presented here.

References

Allsopp HL, Bristow JW, Smith CB, Brown R, Gleadow AJ, Kramers JD, Garvie O (1989) A summary of radiometric dating methods applicable to kimberlites and related rocks. In: Ross JL (ed) Kimberlite and related rocks: their composition, occurrence, origin and emplacement. Blackwell, Oxford, p 349

Allsopp HL, Burger AJ, van Zyl C (1967) A minimum age for the Premier kimberlite pipe yielded by Rb-Sr measurements with related galena isotopic data. Earth Planet Sci Lett 3:161–166

Appleyard CM, Viljoen KS, Dobbe R (2004) A study of eclogitic diamonds and their inclusions from the Finsch kimberlite pipe, South Africa. Lithos 77:317–332

Barrett DR, Alsopp HL (1973) Rubidium-strontium age determinations on South African kimberlite pipes. Extended abstract 1st international kimberlite conference, Cape Town, pp 23–25

Boyd SR, Kiflawi I, Woods GS (1994) The relationship between infrared absorption and A-defect concentration in diamond. Phil Mag B69:1149–1153

Boyd SR, Kiflawi I, Woods GS (1995) Infrared absorption by the B nitrogen aggregate in diamond. Phil Mag 72:351–361

Bulanova GP (1995) The formation of diamond. J Geochem Explor 53:1–23

Bulanova GP, Griffin WL, Ryan CG, Shestakova OY, Barnes S-J (1996) Trace elements in sulphide inclusions from Yakutian diamonds. Contrib Miner Petrol 124:111–125

Bulanova GP, Shelkov D, Milledge HJ, Hauri EH, Smith BC, Chris B (1998) Nature of eclogitic diamonds from Yakutian kimberlites: evidence from isotopic composition and chemistry of inclusions. In: Proceedings of 7th international kimberlite conference, vol 1. Red Roof Design, Cape Town, pp 57–65

Bulanova GP, Wiggers de Vries DF, Beard A, Pearson DG, Mikhail S, Smelov AP, Davies GR (2012). Two-stage origin of eclogitic diamonds recorded by a single crystal from Mir pipe (Yakutia). 10th international kimberlite conference extended abstract no. 10IKC-220

Burgess R, Turner G, Laurenzi M, Harris JW (1989) $^{40}Ar/^{39}Ar$ laser probe dating of individual clinopyroxène inclusions in Premier eclogitic diamonds. Earth Planet Sci Lett 94:22–28

Cartigny P (2005) Stables isotopes and the origin of diamond. Elements 1:79–84

Chinn I, Pienaar C, Kelly C (2003) Diamond growth histories at premier mine. Extended abstract 8 h international kimberlite conference victoria (British Columbia), pp 1–4

Chrenko RM, Tuft RE, Strong HM (1977) Transformation of the state of nitrogen in diamond. Nature 270:141–144

Clement CR (1975) The emplacement of some diatreme-facies kimberlite. In: Ahrens LH, Dawson JB, Duncan AR, Erlank AJ (eds) Physics and chemistry of the earth, vol 9. Pergamon Press, New York, pp 51–60

Cooper GI (1990) Infrared spectroscopy of diamond in relation to mantle processes. Ph.D. thesis, University of London

Deines P, Gurney JJ, Harris JW (1984) Associated chemical and carbon isotopic composition variations in diamonds from Finsch and Premier kimberlite, South Africa. Geochimica and Cosmochimica Acta 48:325–342

Deines P, Harris JW, Spear PM (1989) Nitrogen and ^{13}C content of Finsch and Premier diamonds and their implications. Geochimica and Cosmochimica Acta 53:1367–1378

Evans T, Harris JW (1989) Nitrogen aggregation, inclusion equilibration temperatures and the age of diamonds. In: Ross N (ed) Kimberlites and related rocks, vol 2. Geological Society Special Publication No. 14, pp 1001–1006

Evans T, Qi Z (1982) The kinetics of the aggregation of nitrogen atoms in diamond. Proceedings of the Royal Society of London A381, pp 159–178

Griffin WL, Sobolev NV, Ryan CG, Pokhilenko NP, Win TT, Yefimova ES (1993) Trace elements in garnets and chromites: diamond formation in the Siberian lithosphere. Lithos 29:235–256

Gurney JJ, Harris JW, Rickard RS, Moore RO (1985) Inclusions in premier mine diamonds. Trans Geol Soc S Afr 88:301–310

Harris JW, Gurney JJ (1979) Inclusions in diamond. In: Field JE (ed) Properties of diamond. Academic Press, London, pp 555–594

Kaiser W, Bond WL (1959) Nitrogen—a major impurity in common type I diamond. Phys Rev 115:857–863

Kinny PD, Griffin BJ, Heaman LM, Brakhfogel FF, Spetsius ZV (1997) SHRIMP U/Pb ages of perovskite and zircon from Yakutian kimberlites. Russ Geol Geophys 38:97–105

Mendelssohn M, Milledge HJ (1995) Geologically significant information from routine analysis of the mid-infrared spectra of diamonds. Int Geol Rev 37:95–110

Palot M, Cartigny P, Viljoen KS (2009) Diamond origin and genesis: A C and N stable isotope study on diamonds from a single eclogitic xenolith (Kaalvallei, South Africa). Lithos 112S:758–766

Palot M, Pearson DG, Stern T, Stachel T, Harris JW (2013) Multiple growth events, processes and fluid sources involved in the growth of sulphide-bearing diamonds from Finsch mine, RSA: a micro-analytical study. Geochim Cosmochim Acta 106:51–70

Pearson DG, Shirey SB (1999a) Isotopic dating of diamonds. In: Lambert DD, Ruiz J (eds) Application of radiogenic isotopes to ore deposit research and exploration, vol 12. Society of Economic Geologists, Boulder, pp 143–172

Pearson DG, Shirey SB, Bulanova GP, Carlson RW, Milledge HJ (1999b) Re-Os isotope measurements of single sulfide inclusions in a Siberian diamond and its nitrogen aggregation systematics. Geochim Cosmochim Acta 63:703–711

Pearson DG, Shirey SB, Harris JW, Carlson RW (1998) Sulfide inclusions in diamonds from the Koffiefontein kimberlite, S. Africa: constraints on diamond ages and mantle Re-Os systematics. Earth Planet Sci Lett 160:311–326

Phillips D, Onstott TC, Harris JW (1989) $^{40}Ar/^{39}Ar$ laser-probe dating of diamond inclusions from Premier kimberlite. Nature 340:460–462

Richardson SH (1986) Latter-day origin of diamonds of eclogitic paragenesis. Nature 322:623–626

Richardson SH, Erlank AJ, Harris JW, Hart SR (1990) Eclogitic diamonds of Proterozoic age from cretaceous kimberlites. Nature 346:54–56

Richardson SH, Gurney JJ, Erlank AJ, Harris JW (1984) Origin of diamonds in old enriched mantle. Nature 310:198–202

Richardson SH, Shirey SB, Harris JW, Carlson RW (2001) Archean subduction recorded by Re-Os isotopes in eclogitic sulfide inclusions in Kimberley diamonds. Earth Planet Sci Lett 191:257–266

Rudnick RL, Eldridge CS, Bulanova GP (1993) Diamond growth history from in situ measurement of Pb and S isotopic compositions of sulfide inclusions. Geology 21:13–16

Ruotsala AP (1975) Alteration of the Finsch kimberlite pipe, South Africa. Econ Geol 700:582–590

Scott BH, Skinner EMW (1979) The premier kimberlite pipe, Transvaal, South Africa. Extended abstract kimberlite symposium II

Shirey SB, Harris JW, Richardson SH, Fouch MJ, James DE, Cartigny P, Deines P, Viljoen SK (2002) Diamond genesis, seismic structure, and evolution of the Kaapvaal-Zimbabwe craton. Science 297:1683–1686

Shirey SB, Richardson SH (2011) Start of the Wilson cycle at 3 Ga shown by diamonds from subcontinental mantle. Science 333:434–436

Skinner EM, Clement CR (1979) Mineralogical classification of southern African kimberlites. In: Boyd FR, Meyer HOA (eds) Kimberlites, diatremes and diamonds: their geology, petrology, and geochemistry. Proceedings of the 2nd international kimberlite conference, vol 1. American Geophysical Union, pp 129–139

Smith CB, Allsopp HL, Kramers JD, Hutchinson G, Roddick JC (1985) Emplacement ages of Jurassic—cretaceous South African kimberlites by the Rb–Sr method on phlogopite and whole rock samples. Trans Geol Soc S Afr 88:249–266

Sobolev NV, Nixon PH (1987) Xenoliths from the USSR and Mongolia: a selective and brief review. In: Nixon PH (ed) Mantle xenoliths. Wiley, Chichester, pp 159–166

Sobolev NV, Yefimova ES (1998) The variation of chromite inclusions composition as indicator of zonation of diamonds. Dokladi Russian Akademii nauk 358(5):649–652

Spetsius ZV, Taylor LA (2008) Diamonds of Siberia photographic evidence for their origin. Tranquility Base Press, Lenoir City

Stachel T (2007) Diamond. Mineral Assoc Can Short Course Ser 37:1–22

Stachel T, Brey GP, Harris JW (2005) Inclusions in sublithospheric diamonds: glimpses of deep Earth. Elements 1:73–78

Stachel T, Harris JW, Muehlenbachs K (2009) Sources of carbon in inclusion bearing diamonds. Lithos 112S:625–637

Taylor LA, Anand M (2004) Diamonds: time capsules from the Siberian Mantle. Chem Erde 64:1–74

Taylor LA, Keller RA, Snyder GA, Wang W, Carlson WD, Hauri E, Kim KR, Sobolev NV, Bezborodov SM (2000) Diamonds and their mineral inclusions, what they tell us: a detailed "pull-apart" of a diamondiferous eclogite. Int Geol Rev 42:959–983

Taylor LA, Snyder GA, Crozaz G, Sobolev VN, Yefimova ES, Sobolev NV (1996a) Eclogitic inclusions in diamonds: evidence of complex mantle processes over time. Earth Planet Earth Sci Lett 142:535–551

Taylor WR, Bulanova GP, Milledge HJ (1995) Quantitative nitrogen aggregation study of some Yakutian diamonds: constraints on the growth, thermal and deformation history of peridotitic and eclogitic diamonds. Extended abstract 6th international kimberlite conference, Novosibirsk, pp 608–610

Taylor WR, Canil D, Milledge HJ (1996b) Kinetics of Ib to IaA nitrogen aggregation in diamond. Geochimica and Cosmochimica Acta 60:4725–4733

Wiggers de Vries DF, Harris JW, Pearson GP, Davis GR (2011) Re-Os isotope constraints on the ages of diamonds from Mwadui, Tanzania. Extended abstracts 10th international kimberlite conference, Bangalore 10IKC-203

Wiggers de Vries DF, Pearson DG, Bulanova GP, Pavlushin AD, Molotkov AE, Davies GR (2008). Comprehensive petrological and geochemical study of Yakutian diamonds focussing on Re-Os dating of multiple sulphides in single diamonds. Extended abstracts 9th international kimberlite conference, Frankfurt, p A309

Yefimova ES, Sobolev NV, Pospelova LN (1983) Sulfide inclusions in diamond and specific features of their paragenesis. Zapiski Vseoyuznogo Mineralogicheskogo Obshcestva 112:300–310

Surface Dissolution Features on Kimberlitic Chromites as Indicators of Magmatic Fluid and Diamond Quality

Yana Fedortchouk and Elizabeth McIsaac

Abstract

Mantle minerals undergo dissolution while being transported by kimberlite magma from the depth to the surface of the Earth. In this study, we experimentally investigate morphology of dissolution features developed on chromite crystals in the presence of H_2O-rich and CO_2-rich fluid and compare them to the features on natural kimberlite-borne chromite macrocrysts from Misery, Beartooth, and Grizzly kimberlites, Ekati Mine, Canada. The experiments were done in a piston-cylinder apparatus at 1,350–1,400 °C and 1 GPa in a synthetic CaO–MgO–$SiO_2 \pm H_2O$–CO_2 system with 0, 5, 13, and 15 wt % H_2O, and 5 and 27 wt % CO_2. The surface of chromite crystals with experimentally and naturally induced resorption was studied using a field-emission scanning electron microscope (FE-SEM). It showed angular or smooth steplike dissolution features developed in the presence of H_2O-rich fluid and highly irregular surfaces with cavities, knobs, and polyhedrons developed in the presence of CO_2 fluid. Naturally resorbed chromite grains from volcaniclastic Ekati kimberlites (Misery and Beartooth) show steplike features that are like the products of H_2O-bearing experiments, whereas grains from hypabyssal (coherent) Grizzly kimberlite have irregular surfaces with flakes and spikes that do not resemble any of our experimental products. We propose a high H_2O:CO_2 ratio in kimberlitic fluid and a potential for using chromite morphology for investigating the volatile system within kimberlite magma and its effect on diamond preservation.

Keywords

Kimberlite • Chromian spinel • Dissolution features • Kimberlitic fluid • Piston-cylinder apparatus • Ekati Mine • Diamond preservation

Introduction

During the ascent to the Earth's surface, kimberlite magmas entrain mantle minerals—chromian spinel, ilmenite, garnet, and the most desirable—diamond. Partial dissolution of these minerals during the ascent leaves the imprint of dissolution conditions on the surface of these minerals. Experiments showed that surface dissolution features on diamond are sensitive to the H_2O:CO_2 ratio and the presence of magmatic fluid in kimberlite magma (Fedortchouk et al. 2010; Khokhryakov and Pal'yanov 2010). However, the scarcity of diamond and its limitation to only the

Electronic supplementary material The online version of this article (doi:10.1007/978-81-322-1170-9_19) contains supplementary material, which is available to authorized users.

Y. Fedortchouk (✉) · E. McIsaac
Department of Earth Sciences, Dalhousie University, 1459 Oxford Street, 15000Halifax, NS B3H 4R2, Canada
e-mail: yana@dal.ca

D. G. Pearson et al. (eds.), *Proceedings of 10th International Kimberlite Conference*, Volume 1, Special Issue of the Journal of the Geological Society of India,
DOI: 10.1007/978-81-322-1170-9_19, © Geological Society of India 2013

deepest rare rocks, kimberlites and lamproites, restrict the widespread application of diamond dissolution features for fluid studies. In addition, diamond recovery is a very expensive process that requires industry-funded exploration and mining. At the same time, other mantle minerals transported by kimberlites, namely chromian spinel, ilmenite, and garnet, similarly undergo dissolution during interaction with the hot volatile-rich magma and perhaps can also serve as indicators of the magmatic fluid. These kimberlite indicator minerals (KIMs) are widely used in diamond exploration as the companions of diamond in the mantle, in magmatic, and in post-eruption history. Composition of KIMs helps to evaluate the diamond potential of the mantle source sampled by kimberlites, whereas their dissolution surface features can be equally important for constraining the history of volatiles in the kimberlite magmas. KIMs are more abundant than diamond and not restricted to kimberlites and lamproites but common for mantle-derived volcanic rocks. Most importantly, their chemical composition can be used to determine the depth of entrainment in kimberlite magma and the host mantle lithology allowing examination of the evolution of the fluid phase during the magma ascent.

Volatiles play an important role in any magmatic system and especially in kimberlite magmas (Skinner and Marsh 2004) for which geological evidences of extremely explosive eruptions suggest high content of volatiles. However, petrographic and geochemical methods cannot provide robust estimates of the extent of volatile loss during the eruption or firmly distinguish between the secondary phases formed due to magmatic or meteoric fluid. Fluid plays an important role in diamond preservation and quality and influences the ascent rate of kimberlite magma (Fedortchouk et al. 2010; Hilchie 2011). If other mantle minerals develop distinct type of dissolution morphologies in the presence of different volatiles, they could provide an easy tool to assess the composition of volatiles and the presence of free fluid not only in diamond-bearing kimberlites but also in a larger range of mantle magmas.

The existing data on the dissolution morphology of chromite, ilmenite, and garnet are very limited. Leblanc (1980) described different types of dissolution surface features on chromite from stratiform and podiform chromite deposits but did not provide any conclusive explanation for the nature of these variations. Lee et al. (2004) proposed using chromite morphology for diamond prospecting and exploration to distinguish between kimberlitic chromites and those from other common rock types in heavy-mineral samples. They examined over 6,000 grains and found that chromite grains from kimberlite and lamproite sources are typically elongate, whereas common chromites more often have undistorted octahedral shape. They explained the high proportion of distorted and elongate grains in kimberlites by plastic deformation happening in the asthenosphere. Although surface features on chromites were not a focus of their study, the authors mention that fine stepping and remnants of crystal faces may still be visible on grains transported in streams for distances over 20 km (Lee et al. 2004, and references therein). Most of the morphological studies of KIMs used in diamond exploration are focused on the surficial processes in the post-eruption history of kimberlite-derived material to determine the distances and the environment of diamond transportation from the source rock. However, a comprehensive study of the morphology of KIMs by Afanasiev et al. (2001) described at least two different types of kimberlite-induced surface dissolution features produced in the magmatic stage on kimberlite-borne chromite and ilmenite macrocrysts. For example, in Siberian kimberlites, the majority of ilmenite grains have "pyramidal" surface features and only small proportion of kimberlites has ilmenite with the less common irregular "rough" type of surface. The most common features on chromite are "regular steps" developing along the edges of the {111} octahedral crystal faces, whereas some pipes contain chromite with "irregular" resorption that covers the whole grain (Afanasiev et al. 2001). Lee et al. (2004) also described the "regular steps" type as a common chromite resorption style in kimberlites and lamproites based on their study of twelve kimberlite bodies from South Africa, Canada, India, Russia, and Australia. Yet the reasons for the development of any of these dissolution styles are not known. In addition, garnet and ilmenite often develop a reaction rim during interaction with the kimberlite melt, which may complicate interpretation of their surface features (Garvie and Robinson 1982; Nikolenko et al. 2012; Robles-Cruz et al. 2009). The existing experimental studies focused mainly on the kinetics of chromite or garnet dissolution (Brearley and Scarfe 1986; Canil and Fedortchouk 1999) or the development of reaction rims around ilmenite (Nikolenko et al. 2012). Here, we report the initial results of our experimental study on dissolution morphology of chromite by exploring the effect of magmatic fluid. We also compare the experimental results to the natural chromite grains recovered from three Canadian kimberlites and demonstrate a correlation between the chromite dissolution features and kimberlite geology. Our study confirms that surface features on xenocrystal minerals can help to examine volatiles in mantle-derived magmas.

Sample Selection, Experimental, and Analytical Methods

The study uses natural spinel grains from heavy-mineral concentrates of exploration samples from the Misery, Beartooth, and Grizzly kimberlites, Ekati Mine Property,

Table 1 Starting Compositions

	Chr PC-32	Chr PC-35	Chr PC-37	SD	SDw	SDc	SW
SiO_2	0.07	0.40	0.00	62.62	52.71	52.71	37.51
TiO_2	0.04	1.38	0.04				
Al_2O_3	6.43	19.80	11.30				
Cr_2O_3	65.13	38.24	58.35				
FeO	13.61	23.10	18.11				
MnO	0.29	0.31	0.42				
MgO	13.73	16.19	11.83	21.01	17.68	17.68	0.00
CaO	0.07	0.03	0.10	16.37	24.61	24.60	34.99
V_2O_3	0.22	0.32	0.33				
H_2O[a]			–		5.00	–	–
CO_2[a]			–	–		5.00	27.49
Total	99.59	99.77	100.47	100.00	100.00	100.00	100.00

[a] H_2O added as brucite, CO_2 added as calcium carbonate

Northwest Territories, Canada. Spinel grains were selected under a binocular microscope and analysed using energy dispersion spectrometry mode (EDS) to confirm their composition. For this study, we used only chromites with Cr_2O_3 content above 40 wt % in order to eliminate the effect of composition on the dissolution process (Table 1). We recorded naturally occurring surface features of the selected grains using a scanning electron microscope (SEM) and then ground them into spheres following the method used in Canil and Fedortchouk (1999). After abrasion, the spheres were washed ultrasonically for 5 min in dilute HCl and examined optically. One grain was examined by SEM to confirm that the abrasion technique removed all the pre-existing surface features. The EDS spectra and SEM images were obtained using the field-emission scanning electron microscope (FE-SEM) Hitachi S-4700 FEG at Institute for Research in Materials, Dalhousie University, at 15 keV and 10 uA on carbon-coated grains.

The chromite dissolution experiments were conducted using a synthetic $CaO–MgO–SiO_2–H_2O–CO_2$ system. This composition was chosen to approximate the composition of kimberlite melt and to minimize interaction between the melt and chromite grain to observe the effect of volatiles on dissolution. In order to be confident that the observed surface features arose from interaction between the chromite and volatiles, we ran one experiment in volatile-free diopside melt. The starting compositions were prepared by mixing reagent-grade SiO_2, MgO, $Mg(OH)_2$, and $CaCO_3$ in a mortar under alcohol and confirmed by electron microprobe analyses (EMP) of the run products (Table 1). $CaCO_3$ was decarbonated at 850 °C prior to the addition of brucite or additional $CaCO_3$ as a source of CO2. The runs were conducted in "dry" diopside composition, in diopside melt with 5, 13, and 15 wt % H_2O, with 5 wt % CO_2, and in wollastonite melt with 27 wt % CO_2 (Table 1). Water was added as $Mg(OH)_2$ (diopside stoichiometry allows only 5 wt % of H_2O by this method) and as distilled water using a microsyringe (in 13 and 15 wt % H_2O runs). CO_2 was added as $CaCO_2$ in stoichiometry of diopside ($CaMgSi_2O_6$) and wollastonite ($CaSiO_3$) and run at temperatures above the reaction $CaCO_3 + SiO_2 = CaSiO_3 + CO_2$. For each experiment, about one-third of the starting mixture was packed in a 3-mm-diameter platinum capsule. A chromite sphere was placed in the middle of the capsule and packed around with the remaining material (Fig. 1a). The capsule was then sealed by arc welding and weighed before and after welding to ensure no fluid was lost. Dissolution experiments were conducted at 1,350–1,400 °C and 1 GPa in a piston-cylinder apparatus with 12.5-mm $BaCO_3$ assemblies. Temperature was measured with a $W_{95}Re_5$–$W_{74}Re_{26}$ thermocouple, separated from the capsule by an Al_2O_3 disc. The accuracy of the temperature measurements and the position of the hot zone were confirmed from the thickness of spinel growth layer (Watson et al. 2002) at 1,350–1,550 °C and 1 GPa. Pressure was calibrated at 1,090–1,140 °C and 1–1.5 GPa using the melting point of NaCl (Bohlen 1984) and a 5 % friction correction was applied. For each run, after initial pressure increases to 0.3 GPa, the temperature was brought to 600 °C at a rate of 100°/min and held for 6 min, during which the pressure was brought up to the final value. During the following heating to the final run temperature, pressure was adjusted at 1,000 °C and once again at the run temperature, and no more pressure adjustments were made. Quenching was achieved by terminating power to the graphite furnace. After the experiments, the capsules appeared expended and produced a smell or a sound of escaping gas during the opening, confirming retention of fluid during the run.

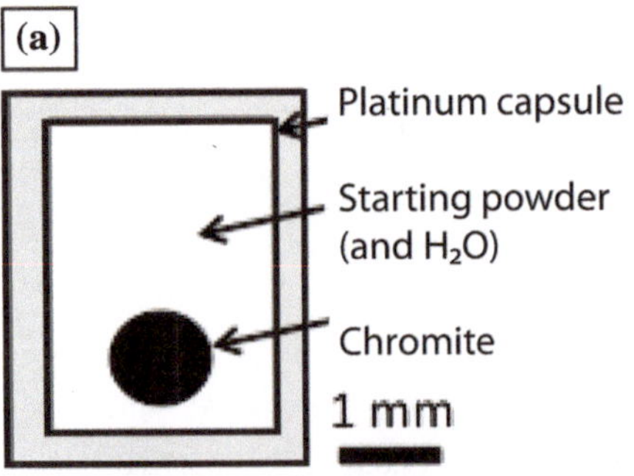

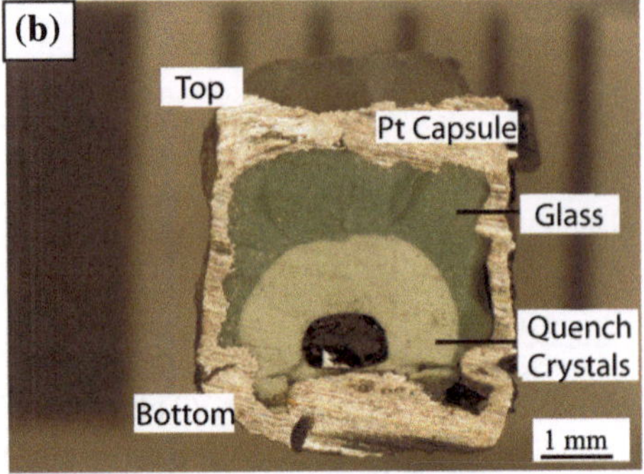

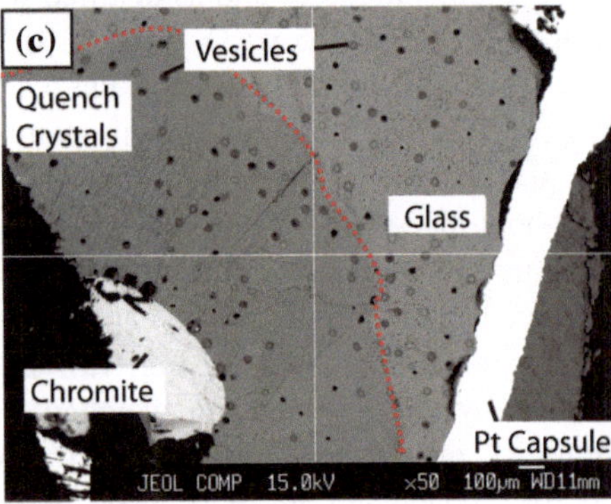

Fig. 1 **a** The set-up of chromite dissolution experiments; **b** Photograph of a typical product (PC-32) showing a chromite grain sitting at the *bottom* of the platinum capsule and surrounded by a *pale* quenched rim in a *green glass* matrix; **c** BSE image of run product showing abundant large vesicles that confirm presence of fluid during the run

The capsules were cut vertically, ground through about half of the chromite grain, and then polished. The matrix was examined under the optical microscope and SEM. Presence of free aqueous fluid in the CaO–MgO–SiO_2–H_2O system was predicted from the water solubility in diopside melts (Eggler and Burnham 1984) and confirmed petrographically for each run product. Large vesicles, which were not associated with quench crystals in the matrix, indicated the presence of a free fluid phase during the experiment. The polished mounts of the experimental products were examined under SEM in backscatter electrons (BSE) and used for EMP analyses to determine the composition of the chromite grains, the matrix quench crystals and glass, and the secondary phases in the reaction zone from PC-35 and PC-37 runs. The analyses of the silicates (Table 3 and Table ES) were done in wavelength-dispersion spectrometry mode (WDS) with a JEOL 8200 Electron Microprobe, Dalhousie University, at 15 kV and 20 nA, using a 2 μm beam width and no online correction for V–Ti signal using a set of well-characterized standards. The analyses of oxides (Table 1) were done using a 3 μm beam width and online corrections for the V–Ti and Cr–V signal overlaps. For three runs PC-32, PC-35, and PC-37, we examined the chemical changes on both sides of the chromite—matrix interface by composition profiles varying from 20–25 μm (Table and Fig. ES). Probably due to insufficient polishing of the samples, these analyses have totals diverging from 100 % by several percent and while cannot be used in a quantitative sense, they clearly show the nature of the chemical exchange during the experiments. After the EMP analyses and BSE imaging, chromite grains were recovered from the matrix using HF acid and their surface features were examined under SEM. Although we cannot completely rule out a possible minor effect of HF on the chromite surface, much higher resistance of chromite compared to silicates (diopside and wollastonite) and consistent diversity of the surface features in experimental runs at different conditions insure that the experiment-induced dissolution features are well-preserved after the liberation of the grains.

We chose Misery, Beartooth, and Grizzly kimberlites for comparison between experimentally and naturally developed dissolution features on kimberlitic chromian spinel owing to the different history of volatile behaviour in these kimberlites determined from our previous work on diamond resorption features and water content in olivine macrocrysts (Fedortchouk et al. 2010). Both Misery and Beartooth magma were proposed to have plenty of aqueous fluid on arrival at the surface, whereas Grizzly probably experienced significant fluid loss during the ascent. Diamond grade is highest in Misery (4 cts/t), intermediate in Beartooth (1.2 cts/t), and lowest in Grizzly (0.5 cts/t) (Fedortchouk et al. 2005, and references therein). Grizzly also has poor quality diamonds perhaps due to resorption in fluid-poor magma. Beartooth has many octahedral stones, whereas Misery diamond population has mostly rounded tetrahexa-hedron forms (Gurney et al. 2004). After the grains were selected under the optical microscope, they were carbon-coated and studied by SEM to confirm their composition and examine the surface features.

Results

Experimentally Induced Dissolution Features on Chromite

All dissolution experiments in CaO–MgO–SiO_2–H_2O and in CaO–MgO–SiO_2–CO_2 systems (Table 2) contain chromite grains with newly developed dissolution features set in

Table 2 Experimental conditions of chromite dissolution experiments at 1 GPa (all runs are 1 h)

Run no.	Starting mixture	Composition (wt %)				Temp (°C)	Phases present[a]
		$CaMgSi_2O_6$	$Ca_2Si_2O_6$	H_2O	CO_2		
PC-43	SD	100				1,400	chr + di
PC-30	SDw	95		5		1,400	chr + gl + q(di)
PC-32	SDw	87		13		1,400	chr + gl + v + q(di)
PC-29	SDw	85		15		1,350	chr + gl + v + q(di)
PC-35	SDc	95			5	1400	chr + gl + v + q(di) + r/r
Pc-37	SW		73		27	1,400	chr + gl + v + q(wo) + r/r

[a] *chr* chromite, *di* diopside, *wo* wollastonite, *gl* glass, *v* vapour bubbles, *q* quench crystals with type of the quench crystals in parenthesis, *r/r* reaction rim with secondary phases

a matrix consisting of dark green glass, light green long blades of quench diopside (or wollastonite in PC-37) and bubbles (Fig. 1b, c). Quench crystals form a halo around chromite grains and mark the top and bottom of the capsules. The difference in colour between the glass and the areas with quench diopside or wollastonite is due to the concentration of chromium serving as chromophore. Chromium content is higher in glass (dark green) than in the areas with quench blades of pyroxenes (pale green) that can become pale green above 1,160 °C when chromium is high-spin and occupies octahedral sites (Ikeda and Yagi 1982). The run in volatile-free diopside composition contained poorly consolidated matrix indicating that it did not melt. The chromite grain from this run had no dissolution features but a smooth, recrystallized surface with apparent linear fractures. This confirms that the angular surface features observed in this study are a product of interaction between the chromite grain and the fluid or volatile component of the melt. In 5 wt % H_2O run, water-undersaturated conditions of diopside melt are confirmed by the absence of vesicles in the matrix consisting of glass and quench crystals. In 13 % H_2O run, the chromite grain surrounded by light green quench crystals sits in a dark green glass loaded with bubbles. In 15 % H_2O run, the matrix consists of the light green quench crystals with interstitial glass and large bubbles showing no association with the quench crystals. This texture (Fig. 1c) confirms the presence of a free aqueous fluid in the two later runs. Both CO_2-bearing runs produced matrix with large and abundant bubbles in the glass portion of the experimental product that showed no association with the quench crystals confirming the presence of CO_2 fluid under these T-P conditions. The slightly rounded shape of the capsules and a hissing sound during the puncture of the capsule after the experiment, caused by the escape of free gas, also confirm presence of a free fluid phase during the run.

Dissolution features induced by interaction of chromite grains with aqueous fluid take the form of trigonal steps that became more angular with an increase in water content. The experiment in H_2O-undersaturated melt (5 wt % H_2O) developed a smooth surface with dissolution features only along the top edge of the grain (Fig. 2a). The trigonal steps have rounded edges and tips occasionally developing into wavy steps (Fig. 2b). The chromite grain from the 13 wt % H_2O run is mostly covered with trigonal steps that become more angular, regular, and distinct (Fig. 2d–f). There are also some trigonal etch pits similar to the ones commonly described on diamond. At 15 wt % H_2O, chromite developed similar angular regular trigonal steps that merge into smooth areas on the grain (Fig. 2h). The higher content of aqueous fluid in this run resulted in the development of rounded depressions (Fig. 2i) similar to the "disc-like" dissolution features known on naturally and experimentally resorbed diamond. There are also wavy ruts (Fig. 2g) and rounded trigonal step features. All angular surface features show consistent orientation, whereas orientation of the rounded features is more random. It is possible that smoothness of some features in this run is due to a partial fluid loss.

Chromite dissolution in the presence of CO_2- rich fluid produced drastically different features that are less regular and more complex than those from the H_2O runs. At 5 wt % CO_2, chromite developed a smooth surface (Fig. 3b) with small nodular features visible at higher magnification (Fig. 3c), which were well rounded with no angularity common for H_2O runs. In addition to the nodules, very fine linear features were formed across the entire chromite grain. At 27 wt % CO_2, chromite developed a very rough surface and some areas covered with rounded, protruding features (Fig. 3e). Another type of surface feature, smaller but ubiquitous on the grain, at high magnification appeared as well-formed polyhedrons (Fig. 3f) interspersed with angular depressions. In some places, the top faces of all polyhedrons corresponded with the plane marking the surface of the host grain, with the side faces outlining the boundaries between the individual polyhedrons. In other areas, each polyhedron forms its own individually oriented faces. The individual polyhedrons consist of four-, five-, and six-edged polygonal faces, with three meeting in vertices. They most closely resemble pyritohedra and trapezohedra.

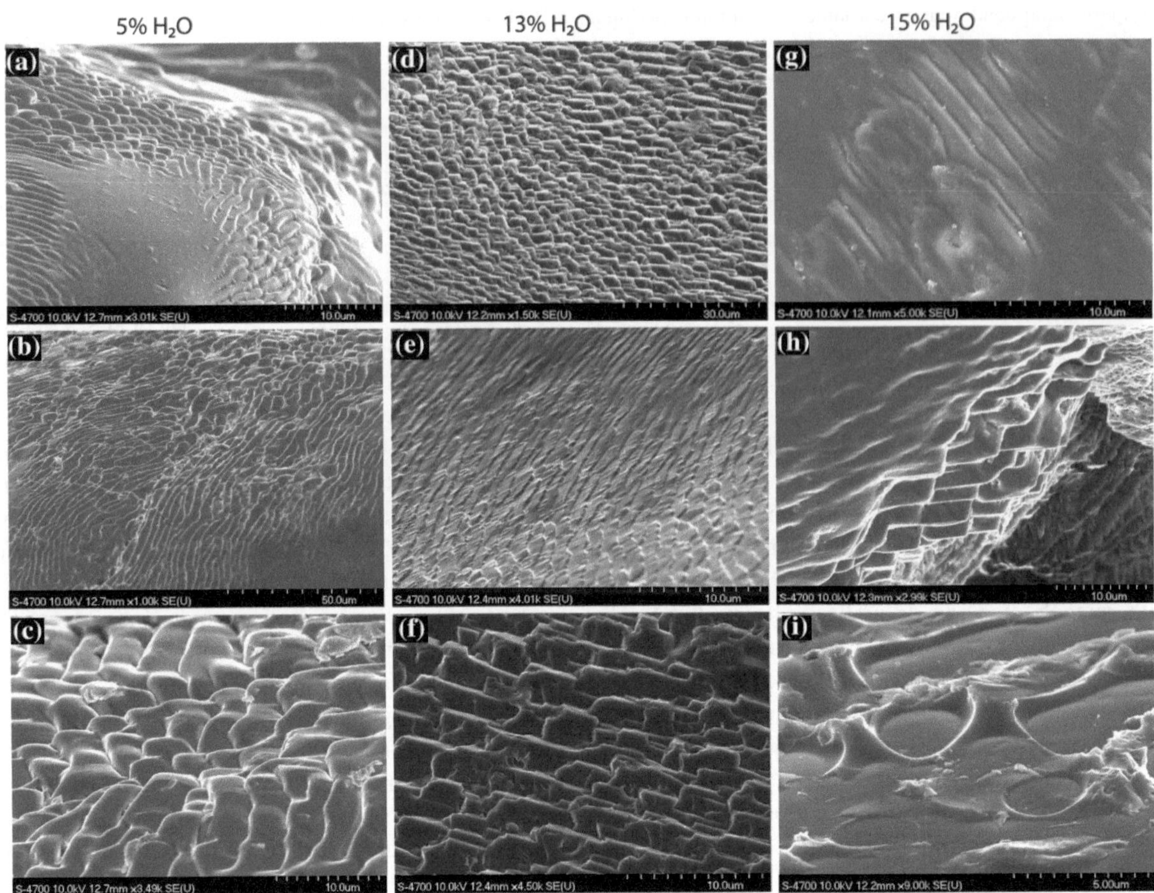

Fig. 2 Surface dissolution features produced on chromite in H_2O-bearing experiments. Note smooth wavy steps in 5 wt % H_2O run (**a–c**), regular well-defined steps and corner pyramids in 13 and 15 wt % H_2O run (**d–f**, **h**), ruts (G), and circular and trigonal depressions (I) in 15 wt % H_2O run. All images are in secondary electrons

In the presence of CO_2-rich fluid, rims of secondary phases develop around chromite at the expense of the initial grain (Fig. 3a, d). The run in the diopside melt with 5 wt % CO_2 contained chromite with a few patches up to 10 μm thick consisting of two phases (Fig. 3a): bright Cr-rich blobs are surrounded by melt with composition close to that of the diopside matrix. The size of the blobs and the areas of interstitial liquid is much less than 1 μm and cannot be analysed quantitatively. EMP analysis gives Cr content of the blobs higher than that of the host chromite (Table 3, an. 1) in spite of the significant contamination of the EMP analyses by Si, Ca, and Mg from the surrounding glass due to the secondary fluorescence of X-rays in the adjacent area. Wollastonite melt with 27 wt % CO_2 entirely rimmed the chromite grain with embayments and melt pockets as deep as 50 μm into the grain (Fig. 3d). They consist of glass and hopper crystals of what is possibly high-chromium chromite (Table 3, an. 2) with 88 wt % of Cr_2O_3 compared to the 58 wt % of the host chromite grain. The interstitial melt (Table 3, an. 8) has CaO and SiO_2 in the proportion of wollastonite stoichiometry but also high contents of Cr_2O_3 and Al_2O_3 (25 and 3 wt %, respectively). The size of the interstitial melt areas is up to 5 μm so that the high Cr and Al concentrations must be real and not an artefact of the secondary fluorescence of X-rays in the surrounding crystals of the secondary chromite. High Cr and Al content of the melt is also supported by its formation at the expense of the initial chromite grain. The Cr/Al ratio is higher in the melt than in the initial chromite grain that suggests slower diffusion of Cr compared to Al from the reaction zone into the surrounding melt. The margins of the chromite grains in both H_2O and CO_2-bearing systems show compositional changes within a 20-μm-wide zone (Fig. ES) where the outer 10-μm zone is slightly enriched in Ca, Mg, Si (in diopside melt) or in Ca and Si (in wollastonite melt) and depleted in Fe, Cr, and Al showing no diffusion loss. The enrichment of the matrix adjacent to the chromite in Al and Cr is probably due to the dissolution of the chromite grain rather than diffusion of Cr and Al into the melt.

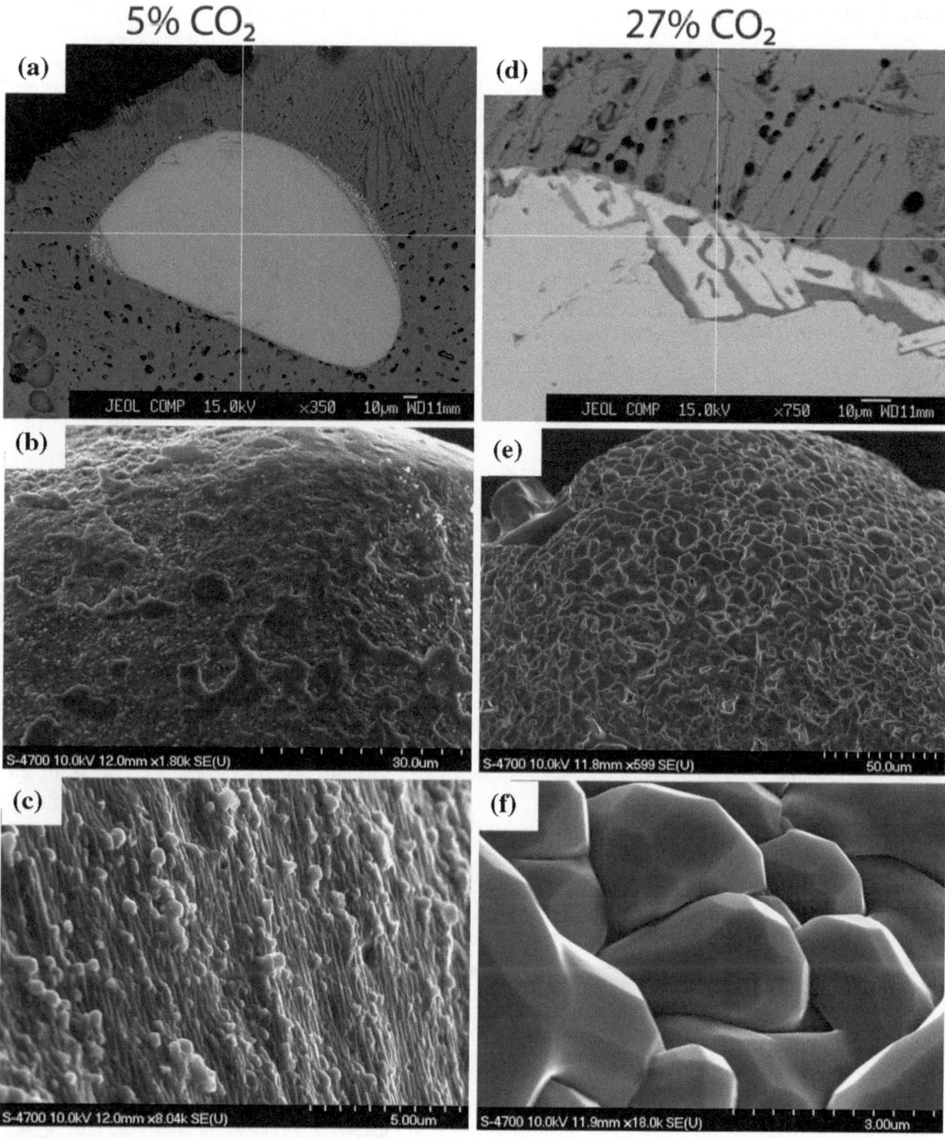

Fig. 3 Surface dissolution features produced on chromite in CO_2-bearing experiments. BSE images (**a** and **d**) show development of a reaction rim around chromite grains. SE images (**b** and **e**) show the irregular complex surface of chromite in both runs with linear and nodular features at 5 wt % CO_2 (**c**) and polygonal microfeatures in (**f**)

Natural Features on Kimberlite-Hosted Chromites

Many chromian spinel grains from the Misery, Beartooth, and Grizzly kimberlites show natural surface features resulted from dissolution in kimberlite magma (Figs. 4, 5). Picking through similar amounts of heavy-mineral concentrate returned only five chromite grains from Grizzly sample, 24 grains from Misery sample, and 53 grains from Beartooth sample. The grain size varies between 0.5 and 2 mm. Chromite population has 60 % of grain fragments with magmatic dissolution features on broken surfaces, 23 % of elongate grains (using terminology from Lee et al. 2004), 13 % with high elongation, only 4 % rounded grains with equant shape, and no undistorted octahedral grains. Studied in details 15 chromite grains from Misery and 14 grains from Beartooth have very similar dissolution features. The {111} face can be flat or slightly rounded (Fig. 4a–e). Often, it shows intensive development of trigonal flat-bottomed etch pits similar to those observed on diamond crystals (Fig. 5i), but combined with trigonal flat-topped prisms. Occasionally, coalescence of trigonal pits develops more complex polygonal shapes, but still very regular. The edges of the pits and the prisms can be rounded to various degrees. The grain edges consist of very regular steps that evolve into pointy "corners" or pyramids by approaching [100] vertices (Fig. 5d–h). These pyramids typically four-sided can also develop a more complex polyhedral shape. The steps and pyramids can have sharp or rounded edges (Fig. 5 d and g, respectively). Few chromite grains from Beartooth show irregular ruts on {111} faces. Some grains from both Misery and Beartooth have all the steps and pyramids along the edges and vertices actively etched by the development of small (~1–2 μm in diameter) pointed-bottomed trigonal or

Table 3 Microprobe analyses of run products (glass, areas with quench phases, secondary phases in reaction rims)

Run no	PC-35	PC-37	PC-30	PC-30	PC-32	PC-29	PC-35	PC-37	PC-37
Anal. no	1	2	3	4	5	6	7	8	9
Phase	r/r: oxide	r/r: oxide	Matrix: gl	Matrix: q(di)	Matrix: q(di)	Matrix: q(di)	Matrix: q(di)	r/r: gl	Matrix: q(wo)
SiO_2	25.14	0.06	52.28	53.74	54.48	52.21	55.44	35.63	51.73
TiO_2	0.04	0.05	0.04	0.03	0.00	0.04	0.00	0.04	0.00
Al_2O_3	0.82	7.04	0.47	0.47	0.73	0.43	0.43	2.67	0.01
Cr_2O_3	44.11	88.32	0.46	0.70	1.24	1.20	0.97	25.62	0.02
FeO	1.06	3.85	0.25	0.27	0.35	0.54	0.25	1.51	0.00
MnO	0.01	0.25	0.04	0.09	0.00	0.05	0.00	0.11	0.07
MgO	20.04	0.09	16.79	17.65	15.33	16.40	18.95	0.66	0.13
CaO	10.85	0.49	25.78	26.07	23.40	25.99	25.75	33.44	49.39
Total	102.06	100.14	96.11	99.02	95.51	96.87	101.79	99.68	101.35

Abbreviations as in Table 2

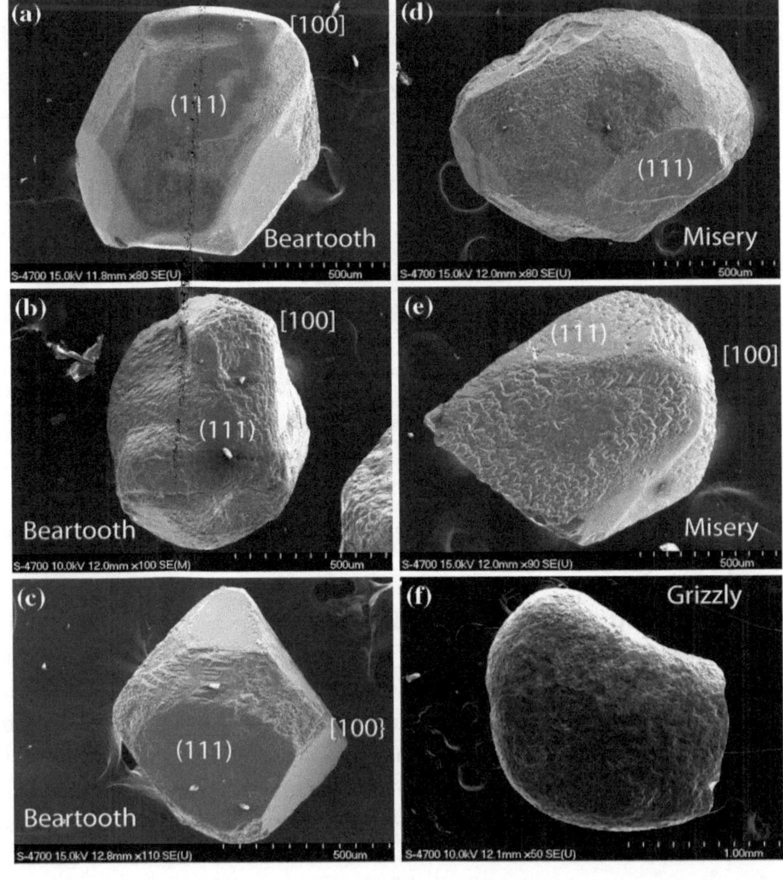

Fig. 4 BSE images showing differences in the dissolution morphology of kimberlite-born chromian spinel from Beartooth (**a–c**), Misery (**d**, **e**), and Grizzly (**f**) kimberlites. Note smooth fine features on what we described in the text as type 1b (**a**), and sharp coarser features on four grains of different morphologies described in the text as type 1a (**b–e**). The figure shows that given a larger dataset type 1a may need to be subdivided further according to the general morphology of the chromite grains

square pits. Variations of the dissolution features in the form of steps and pyramids observed on all the studied 29 grains from Misery and Beartooth samples allow them to be subdivided into three types (1, 2, 3) with two subtypes (a, b). Type 1 is the prevailing morphology with features slightly sharper (type 1a—Fig. 5d–f) or more rounded, sometimes almost droplike (type 1b—Fig. 5g). Few crystals from both samples develop uneven surfaces with flake-like features (type 2) or show sharp and less regular steps and corners (type 3—Fig. 5h). Overall, the dissolution morphology of chromian spinel from Misery and Beartooth kimberlites is similar and most closely resembles the product of chromite dissolution in the presence of aqueous fluid (runs with 13 and 15 wt % H_2O).

Grizzly has chromian spinel with highly irregular surface. The pattern of pyramidal corners pointed in [111]

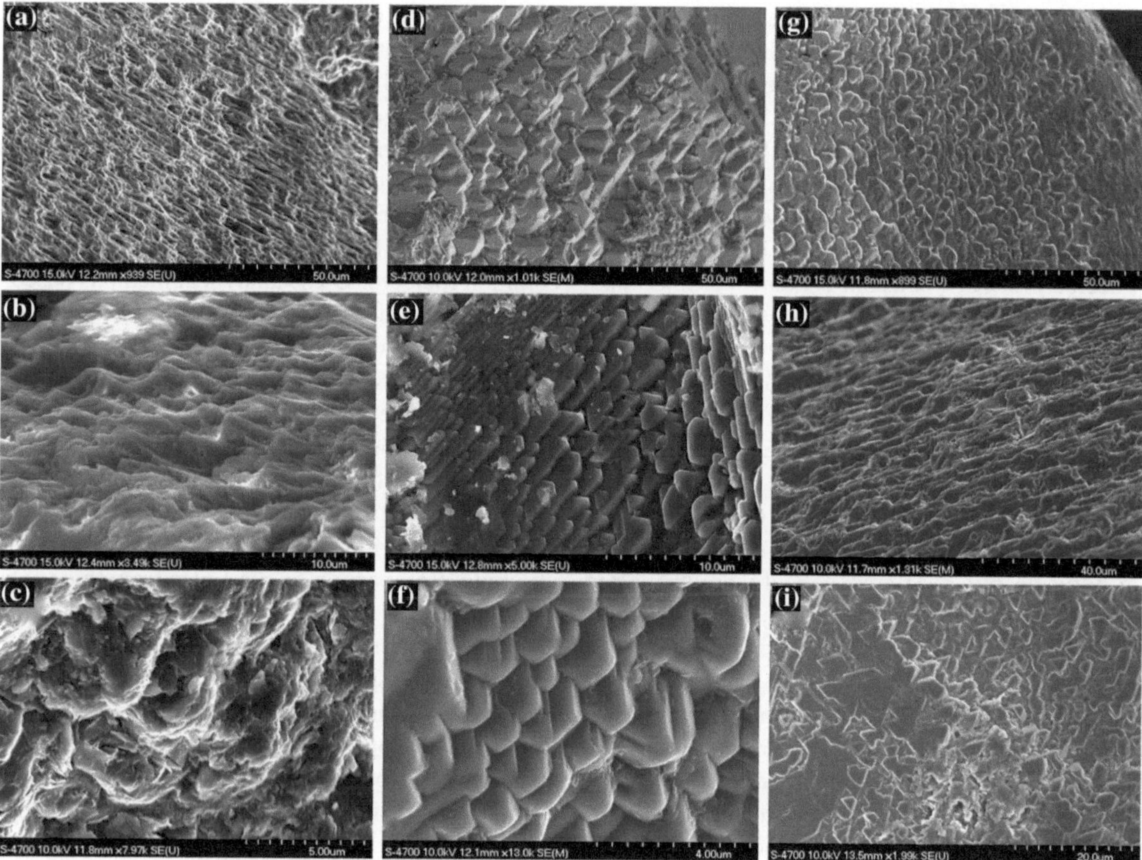

Fig. 5 BSE images of surface features on natural chromian spinel from Grizzly, Beartooth, and Misery kimberlites. Note irregular flaky or spiky surfaces of Grizzly grains (**a–c**) compared to the regular well-defined steps on Misery and Beartooth grains (**d–I**). **a–c** show undetermined crystallographic direction on rounded Grizzly grain; **d** [111] vertices of type 1a (Beartooth); **e** steps along the edge of {111} face on type 1a (Misery); **f** [111] vertices of type 1a (Beartooth); **g** rounded features in [111] direction of type 1b (Beartooth); **h** sharp steps along the edge along [110] direction on type 3 (Beartooth); **I** combination of trigonal flat-bottomed pits and flat-topped prisms on {111} face (Misery)

direction can typically be seen underneath the irregularities, but it is very different from the well-defined features on Misery and Beartooth grains. Surface features on Grizzly chromian spinel often consist of flakes or even shows "layering" (Fig. 5a–c). The pyramidal corners are either rounded or covered with a complex net of micro-protrusions or pointy with many irregular cavities (Fig. 5b). We observed no flat well-defined microfaces on these crystals. There are also some features that are unique to Grizzly samples. An example is small rounded protrusions. Surface features on Grizzly chromian spinel vary and a larger population could be subdivided into several types. The studied Grizzly grains show no resemblance to those of Misery and Beartooth chromite grains for which regular steps, pyramidal corners, and combination of trigonal pits and prisms are common features. They are also very different from any of our experimental products. At the same time, one of the Misery grains developed irregular features somewhere similar to the Grizzly specimens.

Discussion

Compositional Changes Along the Crystal/Melt Interface

H_2O-bearing experiments show no reaction texture around chromite grains and CO_2-fluid runs produced reaction pockets around chromite crystals at the original crystal/melt interface. 5 wt % CO_2 runs developed reaction texture in only two areas on the crystal surface (Fig. 3a) and 27 wt % CO_2 runs completely surrounded the chromite grain with a reaction rim (Fig. 3d) consisting of high Cr oxide and a glass (Table 3). Brearley and Scarfe (1986) proposed that development of reaction rims during dissolution are related to mineral stability in the melt at the experimental conditions where minerals stable near the liquidus exhibit no reaction (H_2O-bearing runs) and dissolution of minerals that are relatively unstable in the melt is accompanied by complex reaction textures (CO_2-bearing runs).

The diffusion profiles through crystal/melt interface were obtained for one experiment with 13wt % H_2O and both CO_2 runs (El. Sup. Table and Fig.) and allow evaluation of the relative dissolution rate in both media by comparing it to the rate of the element diffusion from the melt into the chromite grain. Only the H_2O-fluid run showed apparent diffusion profiles for CaO, SiO_2, and FeO within the 10 μm marginal zone of the chromite grain with increase in CaO and SiO_2, decrease in FeO, constant contents of Cr_2O_3 and Al_2O_3, and insignificant variations of MgO. Both experiments with CO_2 fluid have disturbed concentration patterns for CaO, SiO_2, and FeO and no apparent steady increase towards the grain boundary governed by diffusion. This may indicate that the dissolution rate in a H_2O system is slower than diffusion of Ca, Si, and Fe, but faster than diffusion of Al and Cr at these conditions. On the contrary, in a CO_2 system, the chromite dissolution rate might be faster than diffusion of Ca, Si, and Al and therefore faster than dissolution in the H_2O system. Presence of the reaction zone in CO_2 runs would agree with the lesser stability of chromite in the CO_2 system.

The glass adjacent to the chromite crystal from the H_2O run shows a flat pattern of elevated concentrations of Cr and Fe within the 15-μm zone probably marking the initial location of the crystal/melt boundary and then decreases in the concentration of both elements. On the contrary, both CO_2-fluid runs show constant decrease in Cr and Fe outward from the present interface and no change in Al content. If the dissolution rate is governed by the rate of diffusion, then a possible explanation for the lesser chromite stability in the presence of CO_2 fluid could be the faster diffusion of Cr and Fe in this system. At the same time, formation of Cr-rich oxide in both runs with CO_2 fluid may indicate that Cr diffusion from the dissolving interface is not fast enough and creates local oversaturation in Cr that results in the formation of the new Cr-rich phase. Another possibility is perhaps a different effect of H_2O and CO_2 on the structure of the melt that could make chromite less stable in CO_2 systems.

Chromite Morphology and Diamond Exploration

Discriminating Kimberlitic Chromite

Chromite is a stable mineral resistant to chemical and mechanical weathering and common in heavy-mineral concentrates from stream sediments. It can be transported to significant distances from the source rock and along with pyrope garnet, and picro-ilmenite can help to trace back the kimberlite or lamproite body. It is also an abundant mineral in kimberlites, and a relatively small sample of weathered kimberlite material can provide enough chromite grains for study. Presence of kimberlitic chromite in stream sediments can help to identify presence of kimberlite-derived material (Lee et al. 2004) and dissolution features on chromite surface can help to constrain the fluid and emplacement history of kimberlite magma, to evaluate the characteristics and preservation of diamond population.

Chemistry of chromite cannot provide a clear discrimination between kimberlitic chromites and those from other common rock types. The analytical methods based on major elements are not rigorous enough, and more accurate methods based on minor and trace elements are expensive for studying large chromite populations. Lee et al. (2004) proposed to use elongate and distorted shape of chromite grains as an indicator of kimberlite-derived chromites. High proportion of elongate crystals in our study supports this conclusion. Surface dissolution features can offer additional criteria for discriminating between kimberlitic and common chromite. Comparison of chromites from this study to those from stratiform and podiform chromite deposits hosted in continental-layered mafic intrusions and in alpine-type peridotites, respectively (Leblanc 1980), shows that resorption features on kimberlitic chromite are different. The well-developed steps in [110] and pyramids in [100] on kimberlitic chromites from Misery and Beartooth and on chromite from other kimberlites worldwide (Afanasiev et al. 2001; Lee et al.2004) were not described on grains from mafic–ultramafic complexes (Leblanc 1980). Irregular vermiculated pits on cleavage and {111} planes on grains from chromitite ore are notably different from the flat-bottomed trigonal pits and flat-topped prisms on Misery and Beartooth or irregular flakes on Grizzly chromites. Thus, "step-type" resorption seems to be a characteristic feature of kimberlitic chromite, while some kimberlites may also have chromite grains with other resorption styles.

Predicting Characteristics of Diamond Population

The main focus of this study was to test whether surface features on kimberlite-hosted chromite can provide information about crystallization conditions and H_2O/CO_2 ratio of the magmatic fluid and can help in predicting characteristics of diamond population. Our experiments have shown that similarly to diamond, chromite also develops very different dissolution features in the presence of H_2O and CO_2 fluids and in the absence of a free fluid. In H_2O-bearing system, regular steps develop along the edges of octahedral crystal and evolve into corner pyramids towards the vertices. These features are smooth and rounded in water-undersaturated melt and become sharp and well defined in the presence of free aqueous fluid. A CO_2-bearing system produces irregular features with cavities, knobs, and polyhedrons, which morphology changes significantly as fluid content increases. This makes it possible not only to use chromite grains for estimating H_2O/CO_2 ratio in the fluid but also to examine saturation of the system in volatiles and formation of a free fluid phase.

Application of our experimental results to the three kimberlites from Ekati Mine have shown a good agreement between the surface features on chromite grains, kimberlite geology, diamond characteristics, and fluid history determined earlier from diamond resorption features and water content in olivine macrocrysts (Fedortchouk et al. 2010). Chromite grains from volcaniclastic facies kimberlites (Misery and Beartooth) have surfaces covered with regular steps evolving towards the vertices into corner pyramids pointed in [100] and combination of flat-bottomed trigonal pits and flat-topped prisms on {111} faces. They most resemble the products of our experiments in H_2O-fluid that agrees with the abundant aqueous fluid proposed for both magmas in our earlier study (Fedortchouk et al. 2010). Chromite grains from hypabyssal facies (coherent) Grizzly kimberlite have irregular surfaces with spiky, pointy features and cavities that differ greatly from those observed on Misery and Beartooth grains and from the experimentally induced in H_2O and CO_2 fluids confirming a different behaviour of volatiles in this magma. In addition, the low abundance of chromite macrocrysts in the Grizzly concentrate may indicate its poor preservation in Grizzly kimberlite magma. This would agree with the previously proposed early loss of aqueous fluid by the Grizzly kimberlite magma (Fedortchouk et al. 2010) that slowed down the ascent rate and increased the duration of the reaction between the mantle minerals and kimberlite magma. Thus, the "step-type" resorption features on chromite correlate with good-quality diamond populations, which regardless of the degree of crystal rounding (octahedral or hexoctahedral) have clear smooth surfaces with minor etching.

Emplacement and Crystallization Conditions of Kimberlite Magma

Surface features observed on chromite grains in H_2O-bearing experiments and in two volcaniclastic Ekati kimberlites known for their high-grade and good-quality diamonds (Misery and Beartooth) are similar to the type 1 "step" morphology of chromite grains from Siberian kimberlites described by Afanasiev et al. (2001) as the most common type of magma-induced surface features on kimberlite-derived chromite and to the description of Lee et al. (2004) for kimberlite-born chromite from twelve different kimberlite bodies worldwide. At the same time, our CO_2-bearing experiments produced features that do not resemble any of those described on kimberlitic chromites in this study, in works by Afanasiev et al. (2001) and Lee et al. (2004). This confirms the high H_2O:CO_2 ratio of kimberlitic fluid proposed earlier from diamond dissolution morphology (Fedortchouk et al. 2010), composition of kimberlites (Kjarsgaard et al. 2009), and solubility of both volatiles in kimberlite melt (Brooker et al. 2011). Furthermore, our study shows morphological diversity of the most common "step type" of chromite resorption, which includes variation in step size, degree of rounding, presence of micropits, and other features that could be a basis for further subdivision of this common type and might help further refinement of crystallization conditions of the magma. As very common mineral in mafic and ultramafic igneous rocks chromite can help to investigate fluid composition in a large variety of rocks. For example, dissolution features on some chromite grains from podiform deposits (Leblanc 1980) form polygons that are very similar to those produced in our CO_2-bearing experiments.

Summary

Our study shows that similarly to diamond, dissolution features on chromite differ greatly in the presence of H_2O and CO_2 fluids. In H_2O-bearing systems, chromite develops regular steps along the crystal edges that are smooth and rounded in water-undersaturated melt and become sharp in the presence of free aqueous fluid. In CO_2-bearing systems, chromite develops irregular features with cavities, knobs, and polyhedrons. Dissolution features on natural kimberlite-born chromite from Ekati Mine kimberlites differ between volcaniclastic and hypabyssal (coherent) facies kimberlites. The former (Misery and Beartooth pipes) host chromite macrocrysts with steplike features on {110} and trigonal pits and flat-topped prisms on {111} that most resemble the products of experiments in aqueous fluid. The later (Grizzly pipe) hosts chromites with irregular spiky and pointy features and cavities that are different from any experimentally induced features in this study. None of the studied kimberlites host chromite grains with features resembling those from CO_2-bearing experiments, which confirms the high H_2O:CO_2 ratio of kimberlitic fluid. Comparison of naturally and experimentally developed dissolution features on kimberlite-hosted chromite can help to constrain magmatic fluid in kimberlite and in other mafic magmas making it possible to use chromite in diamond exploration. Preliminary results suggest that kimberlites carry chromite grains with angular steplike surface features had abundant aqueous fluid during the emplacement and may associate with good-quality diamonds. The observed variation in the size and morphology of the "step" features on Misery and Beartooth chromite grains could provide additional constrains of the crystallization conditions of kimberlite magma such as pressure and/or H_2O content, which would require further experimental investigation.

Acknowledgments We thank Jon Carlson and BHPBilliton Diamonds Inc. for providing kimberlite concentrate for this study. Dan MacDonald is thanked for the help with EMP analyses, Patricia Scallion for the help with SEM work, and Institute for Research in Materials for the access to the FE-SEM funded by Canada Foundation for Innovation. We thank Chris Smith for very helpful review of the earlier version of this manuscript. This research was supported by NSERC of Canada Discovery and RTI grants to YF.

References

Afanasiev VP, Zinchuk NN, Pokhilenko NP (2001) Morphology and morphogenesis of kimberlite indicator minerals. Publishing house of SO RAN, Russia, Novosibirsk, p 276

Bohlen SR (1984) Equilibria for precise pressure calibration and a frictionless furnace assembly for the piston-cylinder apparatus. Neues Jahrbuch fuer Mineralogie 9:404–412

Brearley M, Scarfe CM (1986) Dissolution rates of upper mantle minerals in an alkali basalt melt at high pressure: an experimental study and implications for ultramafic xenolith survival. J Petrol 27:1157–1182

Brooker RA, Sparks RSJ, Kavanagh JL, Field M (2011) The volatile content of hypabyssal kimberlite magmas: some constraints from experiments on natural rock compositions. Bull Volcanol 73(8):959–981

Canil D, Fedortchouk Y (1999) Garnet dissolution and the emplacement of kimberlites. Earth Planet Sci Lett 167:227–237

Eggler DH, Burnham CW (1984) Solution of H_2O in diopside melts; a thermodynamic model. Contrib Miner Petrol 85(1):58–66

Fedortchouk Y, Canil D, Carlson JA (2005) Dissolution forms in Lac de Gras diamonds and their relationship to the temperature and redox state of kimberlite magma. Contrib Miner Petrol 150:54–69

Fedortchouk Y, Matveev S, Carlson JA (2010) H_2O and CO_2 in kimberlitic fluid as recorded by diamonds and olivines in several Ekati Diamond Mine kimberlites, Northwest territories, Canada. Earth Planet Sci Lett 289:549–559

Garvie OG, Robinson DN (1982) The formation of kelyphite and associated sub kelyphitic and sculptured surfaces on pyrope from kimberlite. In: Kornprobst J (ed) Third international kimberlite conference, vol 11A. Elsevier Sci. Publ., Amsterdam, pp 371–382

Gurney JJ, Hildebrand PR, Carlson JA, Fedortchouk Y, Dyck DR (2004) The morphological characteristics of diamonds from the Ekati property, Northwest territories, Canada. Lithos 77:21–38

Hilchie L (2011) Zonation of hydrogen in kimberlitic and mantle olivines: a possible proxy for the water content of kimberlite magmas. Department of Earth Sciences, MSc, p 85. Dalhousie, Halifax

Ikeda K, Yagi K (1982) Crystal-field spectra for blue and green diopsides synthesized in the join $CaMgSi_2O_6$-$CaCrAlSiO_6$. Contrib Miner Petrol 81:113–118

Khokhryakov AF, Pal'Yanov YN (2010) Influence of the fluid composition on diamond dissolution forms in carbonate melts. Am Mineral 95(10):1508–1514

Kjarsgaard BA, Pearson DG, Tappe S, Nowell GM, Dowall DP (2009) Geochemistry of hypabyssal kimberlites from Lac de Gras, Canada: comparisons to a global database and applications to the parent magma problem. Lithos 112S:236–248

Leblanc M (1980) Chromite growth, dissolution and deformation from a morphological view point: SEM investigations. Mineral Deposita 15:201–210

Lee DC, Maddren J, Griffin BJ (2004) The importance of chromite morphology in diamond exploration. In: 8th international kimberlite conference long abstract, Victoria, p 5

Nikolenko EI, Afanasiev VP, Chepurov AI (2012) Fe-rich ilmenite and kimberlite melt interaction, experimental researches, long abstract: 10th international kimberlite conference, Bangalore

Robles-Cruz SE, Watangua M, Isidoro L, Melgarejo JC, Galí S, Olimpio A (2009) Contrasting compositions and textures of ilmenite in the Catoca kimberlite, Angola, and implications in exploration for diamond. Lithos 112 S:966–975

Skinner EMW, Marsh JS (2004) Distinct kimberlite pipe classes with contrasting eruption processes. Lithos 76(1–4):183–200

Watson EB, Wark DA, Price JD, Vanorman JA (2002) Mapping the thermal structure of solid-media pressure assemblies. Contrib Mineral Petrol 142:640–652

Diamonds from the Behradih Kimberlite Pipe, Bastar Craton, India: A Reconnaissance Study

D. Mainkar, T. Gupta, S. C. Patel, B. Lehmann, P. Diwan, F. V. Kaminsky, and G. K. Khachatryan

Abstract

The Behradih kimberlite is a diamondiferous, diatreme facies kimberlite intrusion with surface dimension of ~300 m × 160 m and is the largest among six intrusions known in the end-Cretaceous Mainpur Kimberlite Field of the Bastar Craton in east-central India. From drillcore material, 34 macrodiamonds with sizes in the range 0.6–2.2 mm and 86 microdiamonds of <0.6 mm size were recovered. The majority of the Behradih diamonds are colourless, while a small population consists of light brown and pale yellow diamonds. The microdiamonds have experienced significant resorption after crystallisation as evidenced by the development of dodecahedral faces in 50 % of the studied population. The macrodiamonds are mostly broken fragments, but several grains exhibit identifiable octahedral or dodecahedral faces. Sixteen macrodiamonds from the Behradih kimberlite were analysed for structurally bound nitrogen aggregation using infrared absorption spectroscopy. All grains belong to the IaAB diamond Type with a predominance of A-centre over B-centre. Their total nitrogen content (N_{total}) falls in the range of 358–950 at.ppm. The proportion of aggregated nitrogen [$\%N_B = 100N_B/(N_A + N_B)$] in the studied diamonds varies from 22 to 45 %. Based on this parameter, the Behradih diamonds are interpreted to have formed at c. 1100–1125 °C, assuming a diamond age of c. 3 Ga. Based on N_{total} and $\%N_B$, the Behradih diamonds are similar to those from major primary diamond deposits worldwide.

Keywords

Kimberlite • Diamond • Nitrogen • Infrared • Bastar Craton • Behradih • India

D. Mainkar (✉)
Directorate of Geology and Mining, Sonakhan Bhavan, Ring Road # 1, Raipur, 492006, India
e-mail: dmainkar@gmail.com

T. Gupta · S. C. Patel
Department of Earth Sciences, Indian Institute of Technology, Powai, Mumbai 400076, India

B. Lehmann
Mineral Resources, Technical University of Clausthal, 38678, Clausthal Zellerfeld, Germany

P. Diwan
Department of Applied Geology, National Institute of Technology, Raipur, 492010, India

F. V. Kaminsky · G. K. Khachatryan
KM Diamond Exploration Ltd, 815 Evelyn Drive, West Vancouver, BC V7T 1J1, Canada

Introduction

The earliest find of diamond was recorded from India between 2500 and 1700 BC (Balfour 2000). India had the monopoly in diamond production for over 3,500 years until the discovery of diamonds in Brazil in the early eighteenth century. Presently, India is a negligible contributor (<0.05 %) to the world diamond production, though a large number of 'globally most celebrated' diamonds including the Akbar Shah (cut diamond wt 116 ct), Darya-i-Noor (cut diamond wt 190.9 ct), Great Mugal (rough diamond wt 787.5 ct), Hope (cut diamond wt 110.5 ct), Koh-i-noor (cut diamond wt 105.60 ct), Nizam (rough diamond wt 277

D. G. Pearson et al. (eds.), *Proceedings of 10th International Kimberlite Conference*, Volume 1, Special Issue of the Journal of the Geological Society of India,
DOI: 10.1007/978-81-322-1170-9_20,

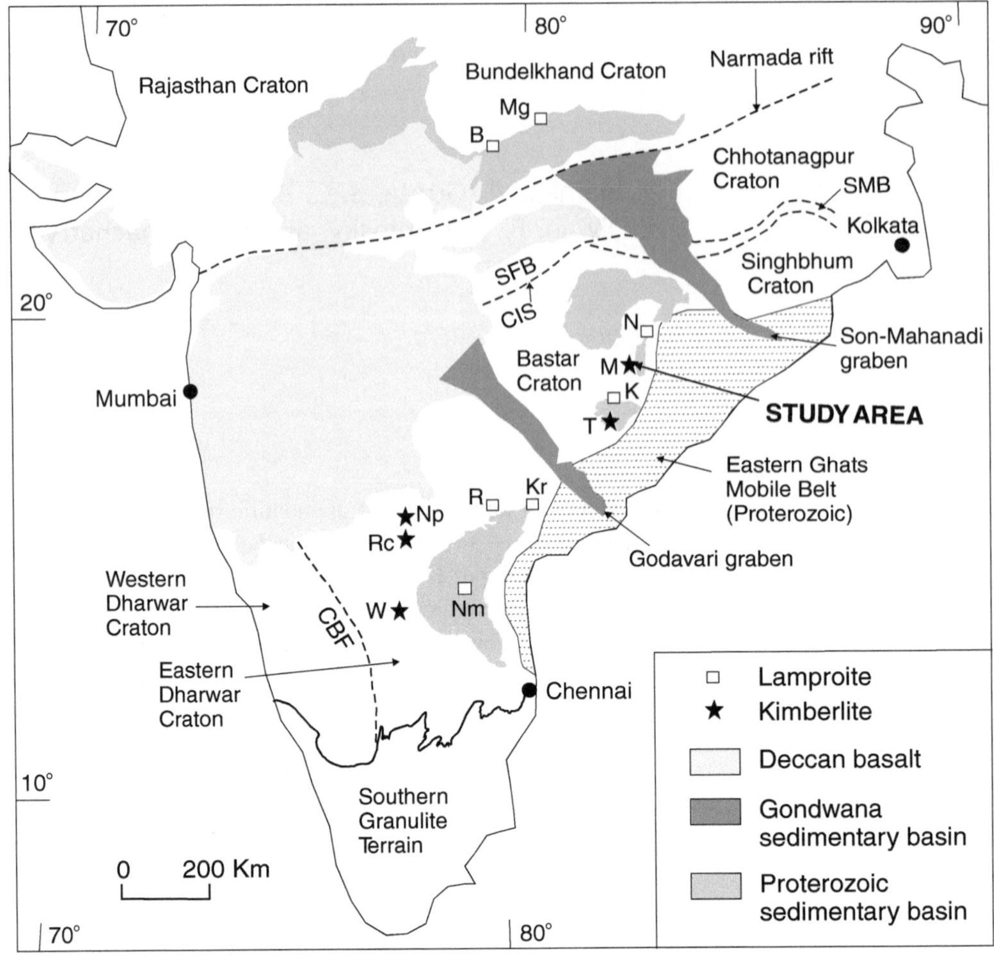

Fig. 1 Schematic geological map of India showing occurrences of kimberlites and lamproites. *CBF* Chitradurga Boundary Fault; *CIS* Central India Shear Zone; *SFB* Satpura Fold Belt; *SMB* Singhbhum Mobile Belt; *B* Bunder; *K* Khadka; *Kr* Krishna; *M* Mainpur; *Mg* Majhgawan; *N* Nuapada; *Nm* Nallamalai; *Np* Narayanpet; *R* Ramadugu; *Rc* Raichur; *T* Tokapal; *W* Wajrakarur

ct), Orlov (cut diamond wt 189.6 ct), Queen of Holland (cut diamond wt 135.92 ct) and Regent (cut diamond wt 140.5 ct) were found in India in ancient alluvial gravel workings (Balfour 2000). Diamond-bearing primary and secondary sources are located in three cratonic blocks in India, viz. Dharwar Craton, Bundelkhand Craton and Bastar Craton (see compilation by Fareeduddin and Mitchell 2012) (Fig. 1).

Diamonds recovered from primary kimberlitic or lamproitic sources exhibit varied physical properties in terms of morphology, size, colour, inclusions, impurities and internal structure, which are functions of the source rock, pressure–temperature condition, activity of chemical components, mechanism of diamond growth and age of formation (Orlov 1973; Thomassot et al. 2007; Stachel and Harris 2008, 2009). Natural diamonds most commonly contain nitrogen as a substitutional impurity and are categorised by the dominant form of N-bearing defect in the lattice. One of the most effective and widely used methods of investigation of structurally bound nitrogen aggregation state in diamonds is Fourier transform infrared spectroscopy (FTIR) (Mendelssohn and Milledge 1995). Diamonds are broadly divided into two 'Types' (I and II) on the basis of the presence or absence of nitrogen impurities, and further subdivided according to the arrangement of nitrogen atoms (isolated or aggregated) and the occurrence of boron impurities. Knowledge of the Type is important in differentiating natural diamonds from synthetic and colour-treated ones (Thongnopkun and Ekgasit 2005; Breeding and Shigley 2011).

Hitherto, the infrared characteristics of Indian diamonds are poorly known. Jha et al. (1995) studied infrared characteristics of 7 diamonds from the Behradih kimberlite and reported IaA–IaB spectral characteristics with low N-aggregation. The present work was undertaken with the aim of morphological and infrared characterisation of a larger sample size of both micro- and macrodiamonds obtained from the bulk drillcore material of the Behradih kimberlite.

Kimberlites of Bastar Craton

The Bastar Craton covers an area of ~1,80,000 km^2 in east-central India (Fig. 1). Most part of the craton lies in the states of Chhattisgarh and Maharashtra, while a small part in the east extends into the state of Odisha. The craton is dominantly occupied by an Archaean quartzofeldspathic

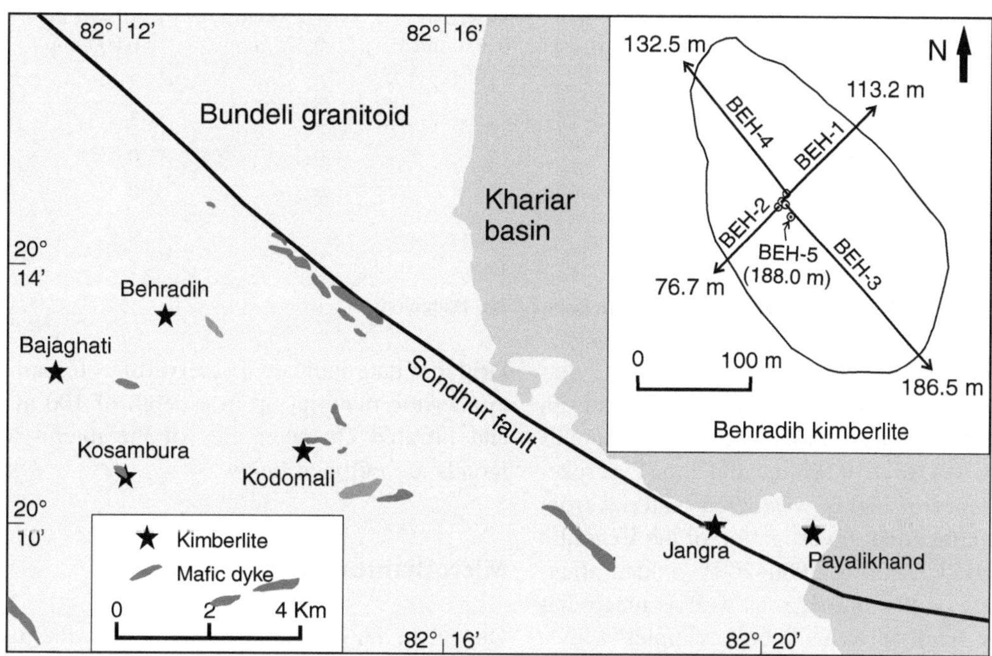

Fig. 2 Generalised geological map of Mainpur Kimberlite Field. Inset shows surface outline of Behradih kimberlite with location, direction and depth of exploratory boreholes. Borehole angles are 45° for BEH-1 to BEH-4 and 90° for BEH-5. Granite contacts in the inclined boreholes were encountered at depths of 108.0 m (BEH-1), 73.3 m (BEH-2), 186 m (BEH-3) and 126.8 m (BEH-4)

gneiss–granitoid complex, within which a number of metamorphosed supracrustal belts of different ages occur (Ramachandra et al. 2001; Ramakrishnan and Vaidyanadhan 2008). Together, they constitute the basement for several Proterozoic sedimentary basins, which from north to south include the Chhattisgarh basin, Khariar basin, Ampani basin, Indravati basin and Sukma basin (Das et al. 2001). Crustal xenoliths in some of the kimberlites of the craton include fragments of granitoids, basic intrusives and basinal sedimentary rocks.

Diamonds were first reported from the Bastar Craton in the early 1990s in colluvium near Payalikhand village in the Raipur district of Chhattisgarh state. This initiated detailed survey for the search of kimberlitic rocks in the area by the Directorate of Geology and Mining (DGM), Madhya Pradesh (now Chhattisgarh), and the Geological Survey of India which led to the discovery of four kimberlites, namely Payalikhand, Behradih, Jangara and Kodomali, respectively (Newlay and Pashine 1993; Chatterjee et al. 1995). These kimberlites were grouped under the name Southeastern Raipur Kimberlite Field (SRKF) by Chatterjee et al. (1995) which was renamed as the Mainpur Kimberlite Field (MKF) by Mainkar et al. (1997). Further investigation for diamond in the MKF was carried out by M/s B. VijayKumar Chhattisgarh Exploration Pvt. Ltd. (BVCE), Raipur, in technical collaboration with M/s Oropa Ltd. (now M/s Sihayo Gold Ltd.) during 2000–2001 under the Large Area Prospecting License. This work added two more intrusions located at Bajaghati (Temple) and Kosambura, respectively. The six kimberlites of the MKF discovered so far are located in a WNW-trending, ~20-km-long and ~5-km-wide corridor in the Palaeoproterozoic Bundeli granitoid, which forms the basement for the Neoproterozoic Khariar basin (Fig. 2). The Behradih, Payalikhand, Kodomali and Bajaghati kimberlites are reported to be diamondiferous (Newley and Pashine 1993; Jha et al. 1995; Oropa 2000; Mainkar 2011).

The Behradih kimberlite (82° 12′ 06″: 20° 12′ 54″) is a diatreme facies, oval-shaped intrusion with surface dimension of ~300 m × 160 m and is the largest among all kimberlites in the MKF (Mainkar and Lehmann 2007). The intrusion is covered by a thick blanket of smectite-rich yellow ground, which locally preserves the texture of the parent rock. The DGM explored the Behradih kimberlite by four angular, and one vertical drill hole down to a depth of 188 m (inset of Fig. 2).

Diamond in Behradih Kimberlite

Jha et al. (1995) reported physical characteristics of 7 macrodiamonds from Behradih kimberlite in the weight range of 0.16–4.58 ct and found them to be brown

Table 1 Microdiamond populations in different size fractions obtained from caustic soda dissolution of Behradih kimberlite samples. Borehole numbers correspond to those shown in the inset of Fig. 2

Borehole no.	Sample weight (in kg)	+0.105 to − 0.150 mm	+0.150 to − 0.212 mm	+0.212 to − 0.300 mm	+0.300 to − 0.425 mm	+0.425 to − 0.600 mm	No. of stones
BEH-1	24.96	2	4	1	0	2	9
BEH-2	24.98	10	12	4	2	3	31
BEH-3	24.92	1	7	7	0	0	15
BEH-4	24.96	1	8	3	2	0	14
BEH-5	25.12	6	8	2	1	0	17
Total	124.94	20	39	17	5	5	86

Data source Belmont laboratory, WA, through Rio Tinto India Exploration, Bangalore

dodecahedra. Verma and Saxena (1997) recovered 2 diamonds weighing 0.12 ct and 0.08 ct, respectively, from surface samples of the Behradih kimberlite, and 4 diamonds weighing 0.284 ct, 0.118 ct, 0.031 ct and 0.031, respectively, from 2.7 tonnes of highly weathered material from shallow pits in the kimberlite. Investigation of the Behradih kimberlite by the BVCE during 2000–2001 yielded abundant microdiamonds (< 0.4 mm size) as well as macrodiamonds (> 0.4 mm size). In total, 9 bulk samples with a combined weight of ~ 318 kg yielded 486 microdiamonds (262 of 0.10 mm, 141 of 0.15 mm, 55 of 0.21 mm and 28 of 0.30 mm mesh size) and 11 macrodiamonds (5 of 0.43 mm, 3 of 0.60 mm, 2 of 0.85 mm and 1 of 1.18 mm mesh size) (Oropa 2000). Broad physical characteristics of these microdiamonds are outlined in Oropa (2000), which indicate that the majority of the stones are colourless and inclusion-free, while a small population comprises milky white stones. Brown and yellow stones occur rarely. Identifiable crystal forms of the microdiamonds are mostly octahedra and octahedral aggregates, while dodecahedra occur occasionally.

For the present microdiamond testing, 5 bulk samples including one each from the 5 boreholes (inset of Fig. 2) were processed comprising minus 5 mm fraction of drillcore material. The samples weighing ~25 kg each were tested by the caustic soda dissolution (CSD) process at the Belmonte laboratory in Western Australia (courtesy: M/s Rio Tinto India Exploration). This yielded a total of 86 microdiamonds of < 0.6 mm size in different size fractions (Table 1). Another ~ 300 kg drillcore bulk material of minus 5 mm fraction was treated by dense media separator (DMS) for kimberlite indicator minerals and diamond incidence in the De Beers India laboratory at Bangalore (courtesy: M/s De Beers India). The non-magnetic fraction recovered from the DMS yielded 34 hand-picked macrodiamonds including 20 stones between 1 mm and 2.2 mm in size and 14 stones in the size range 0.6–1.0 mm. Physical observation of the recovered macrodiamonds suggests the possibility of breakage of larger-sized stones into smaller fragments. Total weight of the macrodiamonds is 0.29 ct. Drillcore data indicate a reserve of ~6.7 million tonnes of kimberlitic material up to a depth of 100 m. The physical and infrared characteristics of the micro- and macrodiamonds are outlined below.

Microdiamonds

Out of the 86 studied microdiamonds, the majority (88 %) are colourless, while the rest includes light brown (11 %) and yellow (1 %) stones. Crystal forms and surface features of a set of 29 colourless microdiamonds have been studied using scanning electron microscope. The forms observed in this sample set include single octahedra (8), octahedral aggregates (6), dodecahedra (5), combined octahedra-dodecahedra (9) and an irregular fraction (1). The surfaces of octahedral crystals range from flat and finely pitted (Fig. 3a) to frosted. Some microdiamonds have sharp and bevelled edges, while others show stepped growth (Fig. 3b). Intergrowth of octahedra with stacked lamellae (Fig. 3c) is seen in 10 % of the microdiamonds. Octahedral crystals are sometimes distorted along the L4 axis (both flattened and elongated) and often show polycentric growth. Nearly 50 % of the microdiamonds have been affected by resorption, the initial stage of which is expressed in negatively oriented trigonal etch pits. Further resorption has resulted in rounding of edges and apices that have lead to the formation of combined octahedral–dodecahedral crystals (Fig. 3d) (e.g. Robinson et al. 1989).

Macrodiamonds

The majority (80 %) of Behradih macrodiamonds are colourless crystals, while the rest includes light brown (17 %) and pale yellow (3 %) stones. The crystals are mostly broken fragments, but five grains exhibit rhombical dodecahedral faces (B1, B2, B3, B6 and B14), while four grains show octahedral faces (B5, B9, B21 and B22) (Fig. 4). One of the macrodiamonds is a triangular macle

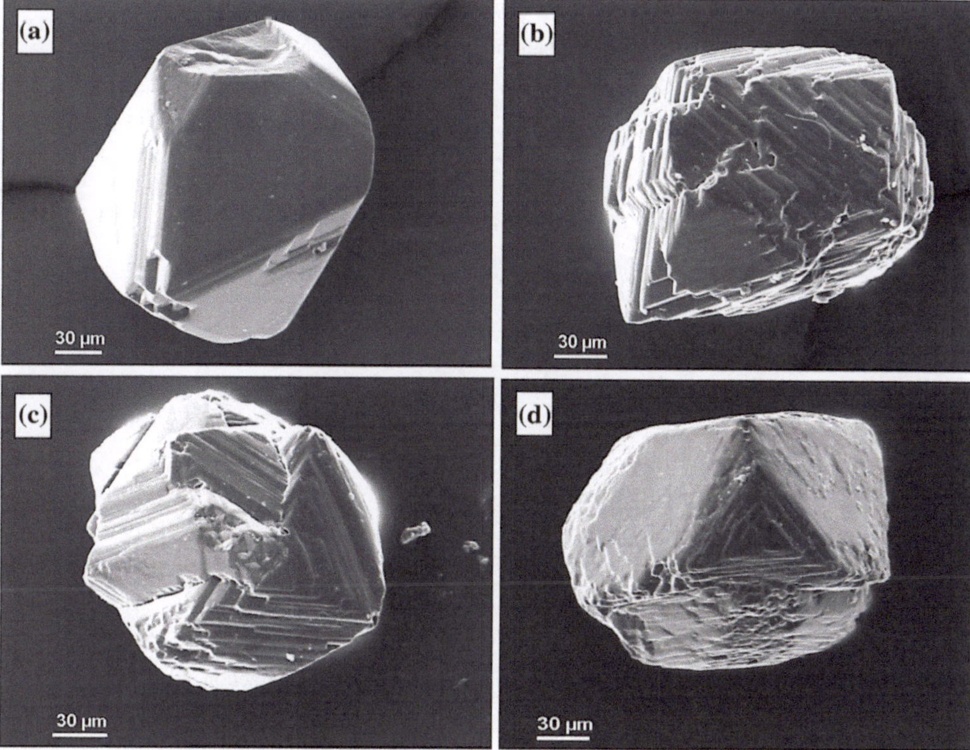

Fig. 3 SEM images of Behradih microdiamonds. Length of bar scale is 30 micron. See text for description

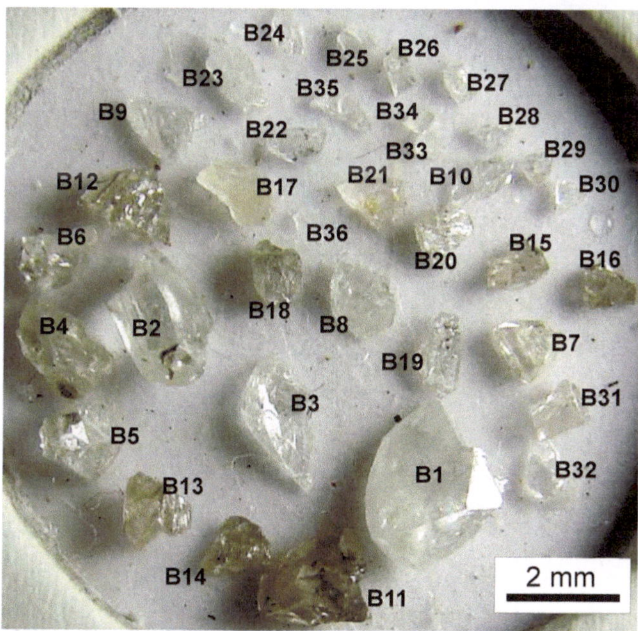

Fig. 4 Photograph of macrodiamonds from the Behradih kimberlite

(B7), whose surface bears negatively oriented trigonal etch marks. An irregular fraction (B8) shows dissolution pits shaped like negative quadrangular pyramids, implying the cubic habit of the crystal. The macrodiamonds are mostly clear crystals except for a few stones in which black inclusions, probably graphite, are noticed.

Infrared Characteristics

Sixteen macrodiamonds from the Behradih kimberlite were analysed by FTIR for structurally bound nitrogen aggregation state, the small number of diamonds studied being a consequence of machine-time availability. The absorption spectra were obtained using a Bruker Vertex 80 Infrared spectrometer (fitted with Hyperion 3,000 microscope) at the Indian Institute of Technology, Bombay. The diamonds were placed on an IR-transparent disc of zinc selenide and analysed integrally. No compositional zoning was identified during the course of the analyses. The spectral resolution was 4 cm^{-1}, and each diamond was scanned 100 times. In all, 33 spectra were obtained and detailed information is shown in Table 2. Concentrations of N_A and N_B nitrogen impurities were determined from the absorption coefficients measured for the corresponding bands in IR spectra using the following equations:

$$N_A = K_A \times \mu_{1,282(A)};\ N_B = K_B \times \mu_{1,282(B)} = K'_B \times \mu_{1,175(B)};$$

where N_A and N_B are the concentrations of A- and B-centres in at.ppm, respectively; $K_A = 16.5 \pm 1$ at.ppm (Boyd et al. 1994), $K_B = 79.4 \pm 8$ at.ppm (Boyd et al. 1995) and $K_B' = 26.6$ at.ppm (Khachatryan 2003) are the coefficients for A- and B-centres, respectively; μ_{1282} and μ_{1175} are the values of absorption peaks at frequencies of 1282 cm^{-1} and

Table 2 Results of IR spectral decomposition. Sample numbers correspond to those in Fig. 4

Sample no.	Shape	Color	Number of spectra	N_A, ppm	N_B, ppm	N_{tot}, ppm	% N_B	P, cm^{-1}
B-1	D–F	CL	3	471	228	699	33	11.8
B-2	D–F	CL	3	573	313	886	35	8.0
B-3	D–F	CL	2	640	310	950	33	10.1
B-4	F	CL	2	312	118	430	27	7.2
B-5	O–F	CL	2	511	417	929	45	10.7
B-6	D–F	CL	2	431	292	725	40	9.1
B-7	M	CL	2	362	409	771	53	28.3
B-8	Fr	CL	2	523	253	776	33	14.2
B-9	O–F	CL	2	456	270	726	37	15.1
B-11	I	BR	2	347	130	477	27	8.8
B-12	Fr	BR	2	307	145	457	32	9.0
B-13	Fr	BR	3	343	96	439	22	8.8
B-14	D–F	BR	1	279	79	358	22	8.3
B-15	Fr	BR	1	282	246	528	47	15.1
B-18	I	CL	2	550	341	891	38	18.3
B-23	Fr	CL	2	415	241	656	37	13.9

Note D rhombical dodecahedron, *O* octahedron, *M* macle, *F* fragment, *Fr* fraction, *I* irregular, *CL* colourless, *BR* light brown

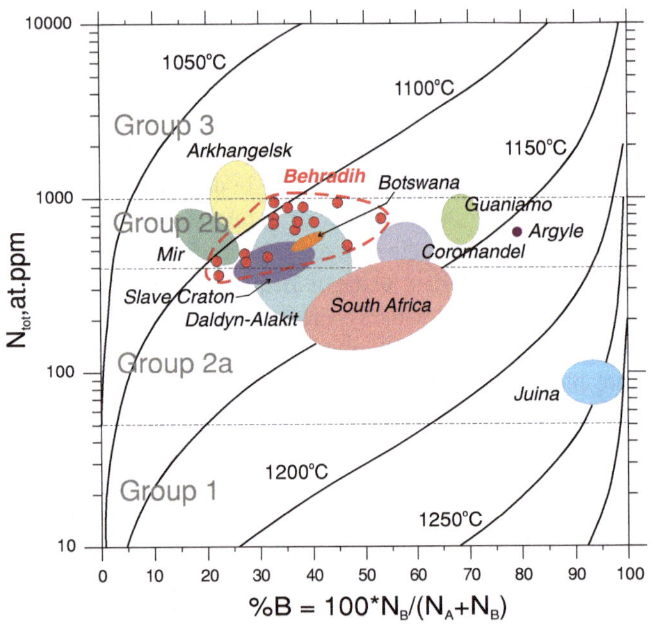

Fig. 5 Temperature estimates for diamonds from different areas (modified after Kaminsky and Khachatryan 2001). Data on diamonds from South Africa (Premier, Finsch, Jagersfontein, Koffiefontein and Venetia), Botswana (Jwaneng and Orapa), Slave Craton (DO-27, Panda and Point Lake) and Argyle after Cartigny et al. (1998), Deines et al. (1984, 1989, 1991, 1993), Iakoubovskii and Adriaenssens (2002), Stachel et al. (2003), Tappert et al. (2005), Taylor et al. (1990, 1995), Viljoen (2002) and Westerlund et al. (2003). Isotherm curves for 3 Ga after Taylor and Milledge (1995)

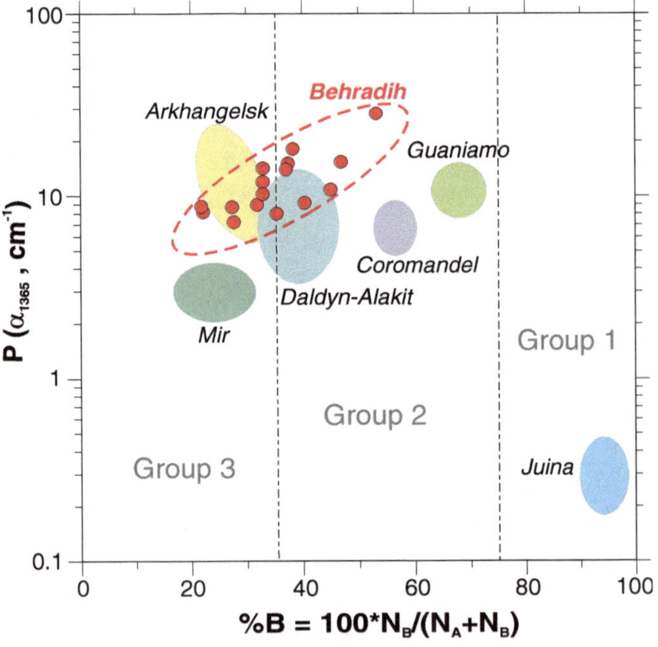

Fig. 6 Distribution of 'platelets' (P) and aggregated nitrogen proportion [$\%N_B = 100*N_B/(N_A + N_B)$] in diamond from different regions (after Kaminsky and Khachatryan 2001). P is an average 'platelet' content in diamond in arbitrary units as a function of the absorption coefficient value measured at 1365 cm^{-1}

1175 cm^{-1}, respectively; $\mu_{1282(A)}$ and $\mu_{1282(B)}$ for A- and B-centres in each analysis were calculated according to Bokii et al. (1986). The results of these calculations are in full agreement with those determined by the method of Mendelssohn and Milledge (1995). Concentrations of *platelet* (planar defects several atoms thick; Woods 1986) in the diamonds were calculated in arbitrary units (cm^{-1}) as a function of the absorption coefficient value measured at 1365 cm^{-1}.

All the Behradih macrodiamonds belong to the IaAB diamond Type with a predominance of A-centres (a pair of nitrogen atoms) over B-centres (an aggregate of nitrogen atoms tetrahedrally arranged around a vacancy). The total nitrogen content (N_{total}) of the diamonds falls in the range of 358–950 at.ppm, based on which they are classified as Group 2b (medium nitrogen) diamond, to which the majority of natural diamonds belong (Kaminsky and Khachatryan 2001) (Fig. 5). Platelet concentration in the diamonds falls in the range of 8.0–28.3 cm^{-1}. The proportion of aggregated nitrogen [$\%N_B = 100N_B/(N_A + N_B)$] in the diamonds varies from 22 to 45 %, based on which the Behradih diamonds belong to Groups 2 ('intermediate') and 3 ('low-aggregated diamond') of Kaminsky and Khachatryan (2001) (Fig. 6).

The proportion of aggregated nitrogen in diamonds depends on the mantle residence temperature and age of diamond and initial concentration of nitrogen in diamond (Evans 1992). These parameters indicate that the Behradih diamonds formed at c. 1100–1125 °C, assuming a diamond age of c. 3 Ga, like for most peridotitic diamonds in different parts of the world (e.g. Shirey et al. 2004). Among mineral inclusions in Behradih and Payalikhand macrodiamonds, both peridotitic-type (two samples) and eclogitic-type inclusions (one sample) were identified (Jha et al. 1995), but the number of the studied samples remains too small for any statistical conclusion. Without additional data on this subject, we assume the peridotitic nature of the Behradih diamonds because the vast majority of inclusions in diamonds worldwide belongs to the peridotitic association (65–75 % for macrodiamonds and 90–95 % for microdiamonds; e.g. Hawthorne et al. 1978; Stachel and Harris 2008). However, the temperature estimate needs to be corrected when the real age of diamonds from the Bastar Craton is known.

In terms of N_{total} and $\%N_B$, the Behradih diamonds are similar to those from major primary diamond deposits in Siberia (Mir, Daldyn-Alakit), Archangelsk (Kaminsky and Khachatryan 2001) and South Africa (Deines et al. 1984, 1989, 1991; Viljoen 2002). Also, diamonds from the Slave Craton (DO-27, Panda, Point Lake) (Taylor et al. 1995; Kaminsky and Khachatryan 2001; Stachel et al. 2003; Westerlund et al. 2003; Tappert et al. 2005) and Botswana (Jwaneng, Orapa) (Deines et al. 1993; Cartigny et al. 1998) belong to the same group, while diamonds from Argyle (Iakoubovskii and Adriaenssens 2002) and placer deposits in Venezuela (Guaniamo area) and Brazil (Coromandel, and particularly Juina area) (Kaminsky and Khachatryan 2001) have more aggregated nitrogen. Based on platelet concentration, the Behradih diamonds are also similar to the ones from the Arkhangelsk and Daldyn-Alakit deposits as well as from the Slave Craton (Kaminsky and Khachatryan 2001).

Acknowledgements Mr. Vijayendra, Secretary, Government of Chhattisgarh, Mineral Resources Department, is thanked for permission to present and publish the paper. Dr. M. N. Gandhi and Mrs. P. Nikam of the SAIF, IIT Bombay, are thanked for help in FTIR analysis. The paper is a contribution to the IGCP-557. Detailed and constructive reviews by Jeff Harris and an anonymous reviewer resulted in significant improvement of the paper.

References

Balfour I (2000) Famous diamonds. Christie Manson & Woods Ltd, London, p 320

Bokii GB, Bezrukov GN, Klyuev YA, Naletov AM, Nepsha VI (1986) Natural and synthetic diamonds. Nauka Press, Moscow (in Russian)

Boyd SR, Kiflawi I, Woods GS (1994) The relationship between infrared absorption and A-defect concentration in diamond. Phil Mag B 69:1149–1153

Boyd SR, Kiflawi I, Woods GS (1995) Infrared absorption by the B nitrogen aggregate in diamond. Phil Mag B 72:351–361

Breeding CM, Shigley JE (2011) The "Type" classification system of diamonds and its importance in gemology. Gems Gemology 45:96–111

Cartigny P, Harris JW, Javoy M (1998) Eclogitic diamond formation of Jwaneng: no room for a recycled component. Science 280:1421–1424

Chatterjee B, Smith CB, Jha N, Khan MWY (1995) Kimberlites of Southeastern Raipur kimberlite Field, Raipur district, Madhya Pradesh, Central India. In: 6th International kimberlite conference extended abstracts. Novosibirsk, pp 106–108

Das N, Dutta DR, Das DP (2001) Proterozoic cover sediments of southeastern Chhattisgarh state and adjoining parts of Orissa. Geol Surv India Spl Pub II(55):237–262

Deines P, Gurney JJ, Harris JW (1984) Associated chemical and carbon isotopic composition variations in diamonds from Finsch and Premier kimberlite, South Africa. Geochim Cosmochim Acta 48:325–342

Deines P, Harris JW, Spear PM, Gurney JJ (1989) Nitrogen and ^{13}C content of Finsch and Premier diamonds and their implications. Geochim Cosmochim Acta 53:1367–1378

Deines P, Harris JW, Gurney JJ (1991) The carbon isotopic composition and nitrogen content of lithospheric and asthenospheric diamonds from Jagersfontein and Koffiefontein kimberlite, South Africa. Geochim Cosmochim Acta 55:2615–2625

Deines P, Harris JW, Gurney JJ (1993) Depth-related carbon isotope and nitrogen concentration variability in the mantle below the Orapa kimberlite, Botswana, Africa. Geochim Cosmochim Acta 57:2781–2796

Evans T (1992) Aggregation of nitrogen in diamond. In: Field J (ed) The Properties of Natural and Synthetic Diamond. Academic Press, London, pp 259–290

Fareeduddin, Mitchell RH (2012) Diamonds and their Source Rocks in India. Geological Society of India, Bangalore, p 434

Hawthorne JB, Gurney JJ, Harris JW (1978) Inclusions in diamonds from South Africa. International Mineralogical Association, Eleventh General Meeting, September 1978, Novosibirsk (USSR), Abstracts

Iakoubovskii K, Adriaenssens GJ (2002) Optical characterization of natural Argyle diamonds. Diamonds Relat Mater 11:125–131

Jha N, Smith CB, Griffin BJ, Chatterjee B, Pooley GD (1995) Diamonds from the kimberlites of southeastern Raipur kimberlite Field, Raipur district, Madhya Pradesh, central India. In: 6th International kimberlite conference extended abstracts, Novosibirsk, pp 266–268

Kaminsky FV, Khachatryan GK (2001) Characteristics of nitrogen and other impurities in diamond, as revealed by infrared absorption data. Can Mineralogist 39:1733–1745

Khachatryan GK (2003) Improved methods of evaluating nitrogen concentration in diamond and their practical application. Geological Material Base ALROSA. Mirny, pp 319–321 (in Russian)

Mainkar D (2011) Petrological and geochemical investigation of the Behradih kimberlite from the Bastar craton, central India, with special reference to its diamond potential. Unpublished Ph.D thesis, Pt. R.S. University, Raipur (Chhattisgarh), p 175

Mainkar D, Lehmann B (2007) The diamondiferous Behradih kimberlite pipe, Mainpur Kimberlite Field, Chhattisgarh, India. Reconnaissance petrography and geochemistry. J Geol Soc India 69:547–552

Mainkar D, Chandrakar NK, Khatedia MS (1997) A report on the photogeological studies in the southeastern part of Raipur district, M.P., F.S. 1992–93, Unpublished Report of DGM, M.P

Mendelssohn MJ, Milledge HJ (1995) Geologically significant information from routine analysis of the mid-infrared spectra of diamonds. International Geol. Review 37:95–110

Newlay SK, Pashine J (1993) New find of diamond-bearing kimberlite in Raipur district, Madhya Pradesh. India Curr Sci 65:292–293

Orlov YL (1973) The mineralogy of the diamond. John Wiley & Sons, New York 235p

Oropa (2000) Office announcement on Raipur diamond project, Block D–7, Madhya Pradesh, India, November 23, 2000, www.sihayogold.com

Ramachandra HM, Roy A, Mishra VP, Dutta NK (2001) A critical review of the tectonothermal evolution of the Bastar craton. Geol Surv India Spl Pub II(55):161–180

Ramakrishnan M, Vaidyanadhan R (2008) Geology of India. vol I. Geological Society of India, Bangalore, p 556

Robinson DN, Scott JA, Niekerk AV, Anderson VG (1989) The sequence of events reflected in the diamonds of some southern African kimberlites. In: Ross J, Ferguson J, Green DH, O'Reilly SY, Danchin RV, Janse AJA (eds) Kimberlites and related rocks, vol 2. Geological Society of Australia Special Publication No. 14, pp 990–1000<Query ID="Q3" Text="Please provide publisher location for reference 'Robinson et al. (1989)'. –>

Shirey SB, Richardson SH, Harris JW (2004) Age, paragenesis and composition of diamonds and evolution of the Precambrian mantle lithosphere of southern Africa. S Afr J Geol 107:91–106

Stachel T, Harris JW (2008) The origin of cratonic diamonds: constraints from mineral inclusions. Ore Geol Rev 34:5–32

Stachel T, Harris JW (2009) Formation of diamond in the Earth's mantle. J Phys: Condens Matter 21(364206):10

Stachel T, Harris JW, Tapperts R, Brey G (2003) Peridotitic diamonds from the Slave and the Kaapvaal cratons: similarities and differences based on a preliminary data set. Lithos 71:489–503

Tappert R, Stachel T, Harris JW, Shimizu N, Brey GP (2005) Mineral inclusions in diamonds from the Panda kimberlite, Slave Province. Canada Eur J Mineral 17:423–440

Taylor WR, Gurney JJ, Milledge HJ (1995) Nitrogen aggregation and cathodoluminescence characterisics of diamond from the Point Lake kimberlite pipe, slave province, NWT Canada In: 6th International kimberlite conference extended abstracts, Novosibirsk, pp 614–616

Taylor WR, Jaques LA, Ridd M (1990) Nitrogen-defect aggregation characteristics Australian diamonds: Time-temperature constraints on the source regions of pipe and alluvial diamonds. Am Mineralogist 75:1290–1310

Taylor WR, Milledge HJ (1995) Nitrogen aggregation character, thermal history and stable isotope composition of some xenolith-derived diamonds from Roberts Victor and Finch. In: Sixth international kimberlite conference extended abstracts, Novosibirsk, pp 620–622

Thomassot E, Cartigny P, Viljoen KS, Harris JW (2007) Methane-related diamond crystallization in the Earth's mantle. Earth Planet Sci Lett 257:362–371

Thongnopkun P, Ekgasit S (2005) FTIR spectra of faceted diamonds and diamond simulants. Diam Relat Mater 14:1592–1599

Verma D, Saxena VK (1997) A report on investigation for diamond and other precious–semiprecious stones in southeastern parts of district Raipur, M.P., F.S. 1992–94, Unpublished Report of DGM, M.P

Viljoen KS (2002) An infrared investigation of inclusion-bearing diamonds from the Venetia kimberlite, Northern Province, South Africa: implications for diamonds from craton-margin setting. Contrib Mineral Petrol 144:98–108

Westerlund K, Hauri EH, Gurney JJ (2003) FTIR absorption and stable nitrogen and carbon isotope microanalysis of Mid-Archaean diamonds from the Panda kimberlite. In: 8th International kimberlite conference extended abstracts, Victoria, No. 8IKC-137, p 5

Woods GS (1986) "Platelets" and the infrared absorption of Type Ia diamonds. In: Proceedings of the royal society, vol A407. London, pp 219–238

Wear of Diamond: An Experimental Study and Field Evidence

V. P. Afanasiev and N. P. Pokhilenko

Abstract

Wear of diamond and kimberlite indicator minerals (pyrope, Mg-ilmenite, and chromite) has been investigated in experiments. The minerals were subjected to mechanical action using an ultrasonic dispersant. Over the 132 h of the total duration of experiment, magnesian ilmenite and chromite were nearly eliminated, while pyrope became well rounded and lost about 65 % of its initial weight. In contrast, diamond acquired a very weak signature of wear restricted to sharp edges and tips, and no weight loss was recorded. In the nature, minerals, including diamond, are most strongly abraded in coastal environments being exposed to the action of waves in the swash zone. Therefore, slightly abraded placer diamond may coexist with strongly rounded pyrope, but the assemblages may lack Mg-ilmenite, which becomes worn almost to zero. As one may infer by extrapolation, all indicator minerals except diamond must be eliminated from placers where the wear of diamond is of a medium degree.

Keywords

Diamond • Pyrope • Mg-ilmenite • Wear • Placer • Coastal placer formation • Abrasion indicator

Introduction

Although being exceptionally hard, diamond shows signature of wear in some placers. The degree of wear ranges from very low, expressed as defects on crystal edges and tips, to very high, in which crystals become round to oval and their edges are rounded off (Afanasiev et al. 2010) (Fig. 1).

There are placers of three types according to diamond wear style. (1) Placers in India, some placers in Brazil, Africa, Kalimantan, southeastern Australia, Burma, and Siberia (East Sayan Mountains) consisting of subrounded and rounded diamonds not accompanied by other indicator minerals. (2) Many placers in the Siberian craton (that of the Kyutyungde graben in northern Yakutia, the Tarydak placer in the Krasnoyarsk region, the D'yukunnakh placer along the Alamdzha River in Yakutia, etc.), all with signature of coastal origin, that include slightly worn diamonds, with polished edges, in association with indicator minerals, most often only pyropes. (3) Rich placers in the eastern Siberian craton, the Urals, Africa, the Hunan Province of China, etc., in which subrounded and rounded diamonds coexist with slightly worn or subrounded indicator minerals (pyrope, Mg ilmenite, and chromite). Diamonds and indicator minerals in these placers originate from sources of different ages, the sources of diamonds being older than those of other minerals.

Thus, two questions arise that concern (1) the behavior of indicator minerals while they experience wear together with diamonds and (2) the wear conditions. These issues were investigated through laboratory testing the wear

V. P. Afanasiev (✉) · N. P. Pokhilenko
V.S. Sobolev Institute of Geology and Mineralogy, Siberian Branch of the Russian Academy of Sciences, 3 Koptyuga Avenue, Novosibirsk, 630090, Russia
e-mail: avp-diamond@mail.ru

D. G. Pearson et al. (eds.), *Proceedings of 10th International Kimberlite Conference*, Volume 1, Special Issue of the Journal of the Geological Society of India,
DOI: 10.1007/978-81-322-1170-9_21, © Geological Society of India 2013

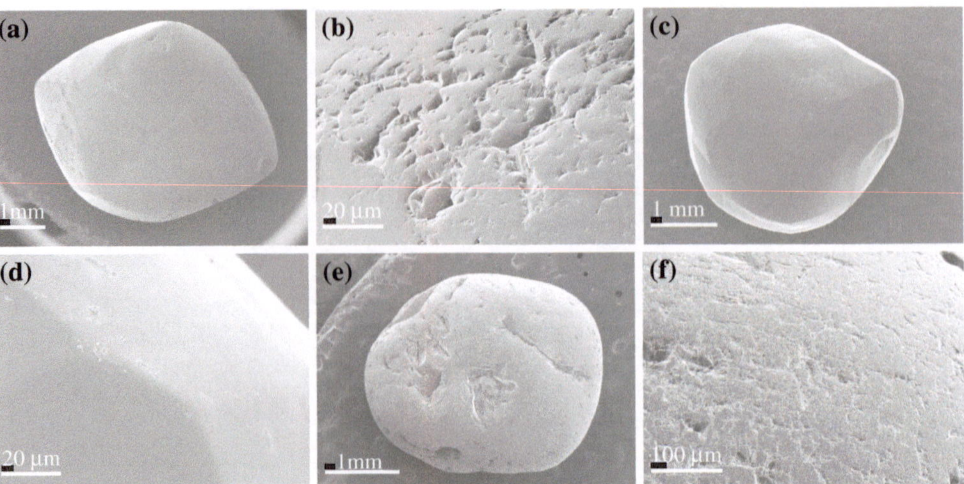

Fig. 1 Diamonds with signatures of wear. **a**, **b** Hunan Province (China): crystals with rough abraded surfaces, subsequently polished; **c**, **d** Kalimantan Island: polished edges of a dodecahedral crystal; **e**, **f** northeastern Siberian craton: most strongly rounded variety V diamond, with multiple defects and a rough abraded surface

resistance of indicator minerals and diamond complemented with modeling the formation conditions of placers that contain rounded diamonds.

Experimental Methods and Results

The tests were performed with an ultrasonic dispersant at 22 kHz. Grains of pyrope (3.796 g), Mg-ilmenite (3.752 g), and chromite (0.781 g), from 1 to 2 mm in size, were placed in a cylindrical steel container together with eight diamonds (from 0.008 to 0.014 g and 1 to 2 mm). The diamonds comprised four colorless crystals or their fragments, two yellow cuboids, a gray cuboid and a crystal of variety VII according to Orlov's (1977) classification, which was a high-defect aggregate of dodecahedra with numerous fluid inclusions. The minerals were immersed in water at room temperature and ultrasonically forced to move whereby they experienced wear as a result of high-energy collisions with one another and with the cylinder walls. Altogether twenty-nine tests were performed over 132 h of the total duration. The minerals were weighed and photographed after each test.

After 8 h of processing, Mg-ilmenites and chromites were worn so heavily as to become hard to discriminate visually and for this reason were weighed together. Chromites were observed to be more stable than ilmenites: the chromite-to-ilmenite ratio changed from 1:5 in the beginning of the test to 5:4 in the end. Over the time of the experiment, the ilmenite + chromite grains became very small (less than 0.25 mm) and were not extracted for weighing, while pyropes lost 65 % of their original weight and acquired spherical or oval shapes (Figs. 2 and 3). The diamond crystals were abraded only slightly on sharp edges and tips, which was noticeable on close examination only;

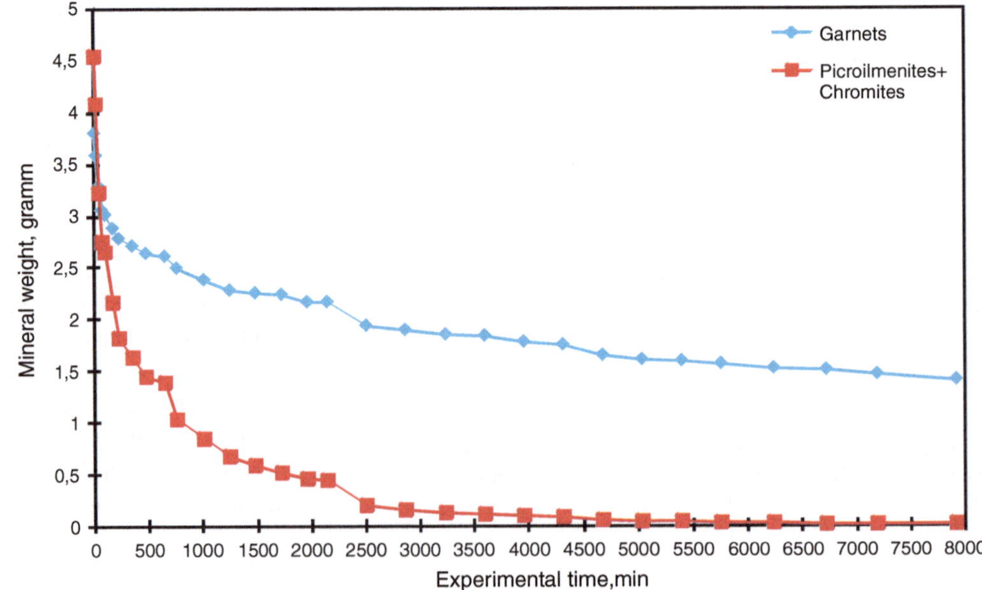

Fig. 2 Weight loss in kimberlite indicator minerals (pyropes, magnesian ilmenites + chromites)

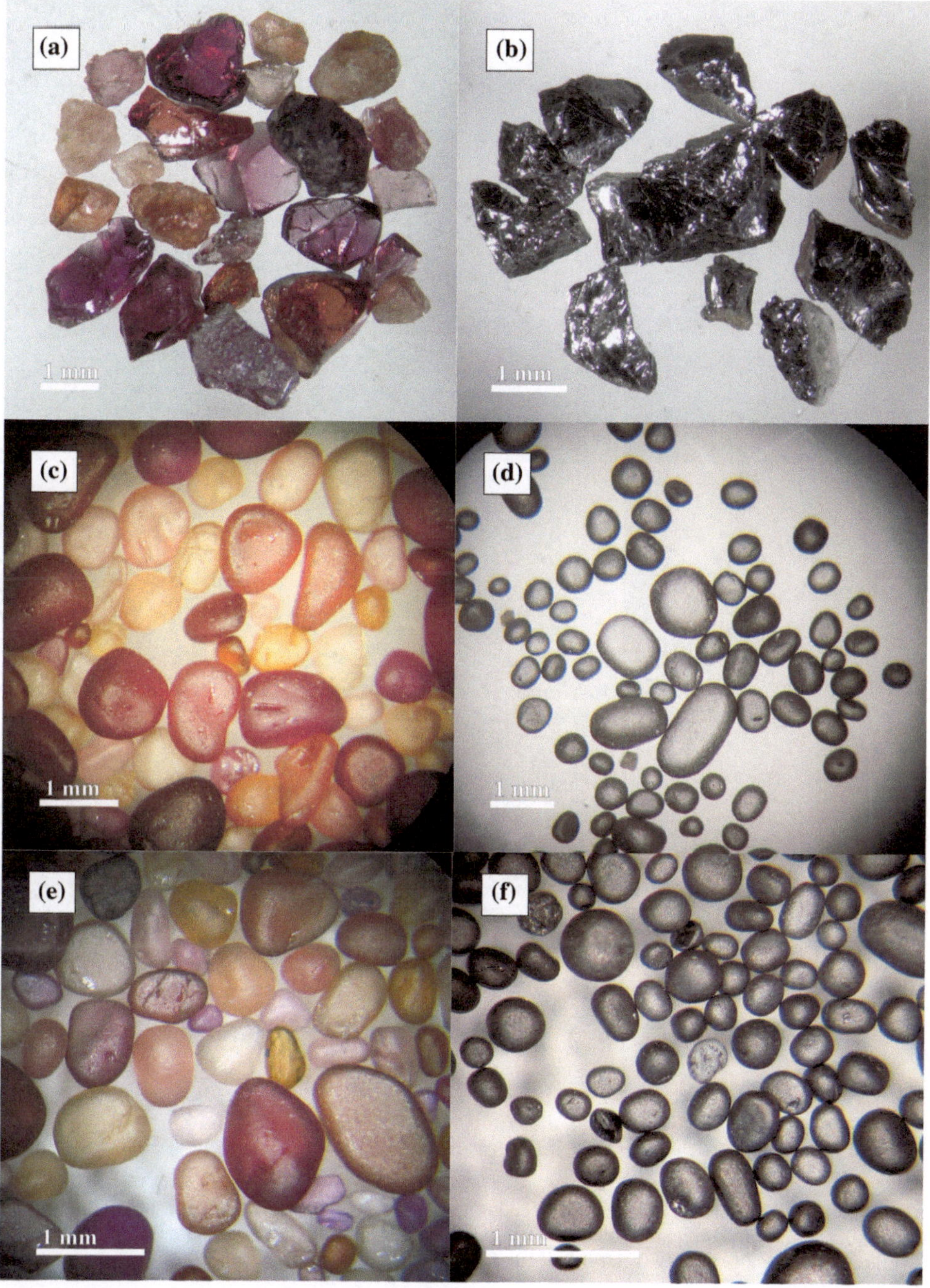

Fig. 3 Pyropes and ilmenites at the beginning of the experiment (**a**, **b**), after 430 min of ultrasonic treatment (**c**, **d**), in the end of experiment, after 7920 min (**e**, **f**). **a**, **c**, and **e** are pyropes; **b**, **d**, and **f** are ilmenites

no weight loss was recorded at an instrument precision of ±0.01 mg. Wear was more evident in octahedra and their fragments, whereas the cuboids and the variety VII crystal demonstrated high resistance to wear, possibly owing to their fibrous and thus more "viscous" structure and lacking simple cleavage directions (Fig. 4).

Note that the intent of the experiment was not to simulate the natural wear conditions for diamond and indicator minerals, this being almost impossible because the natural conditions are too diverse. Rather, the objective was to estimate the relative stability of these minerals to wear, in order to gain insights into natural phenomena.

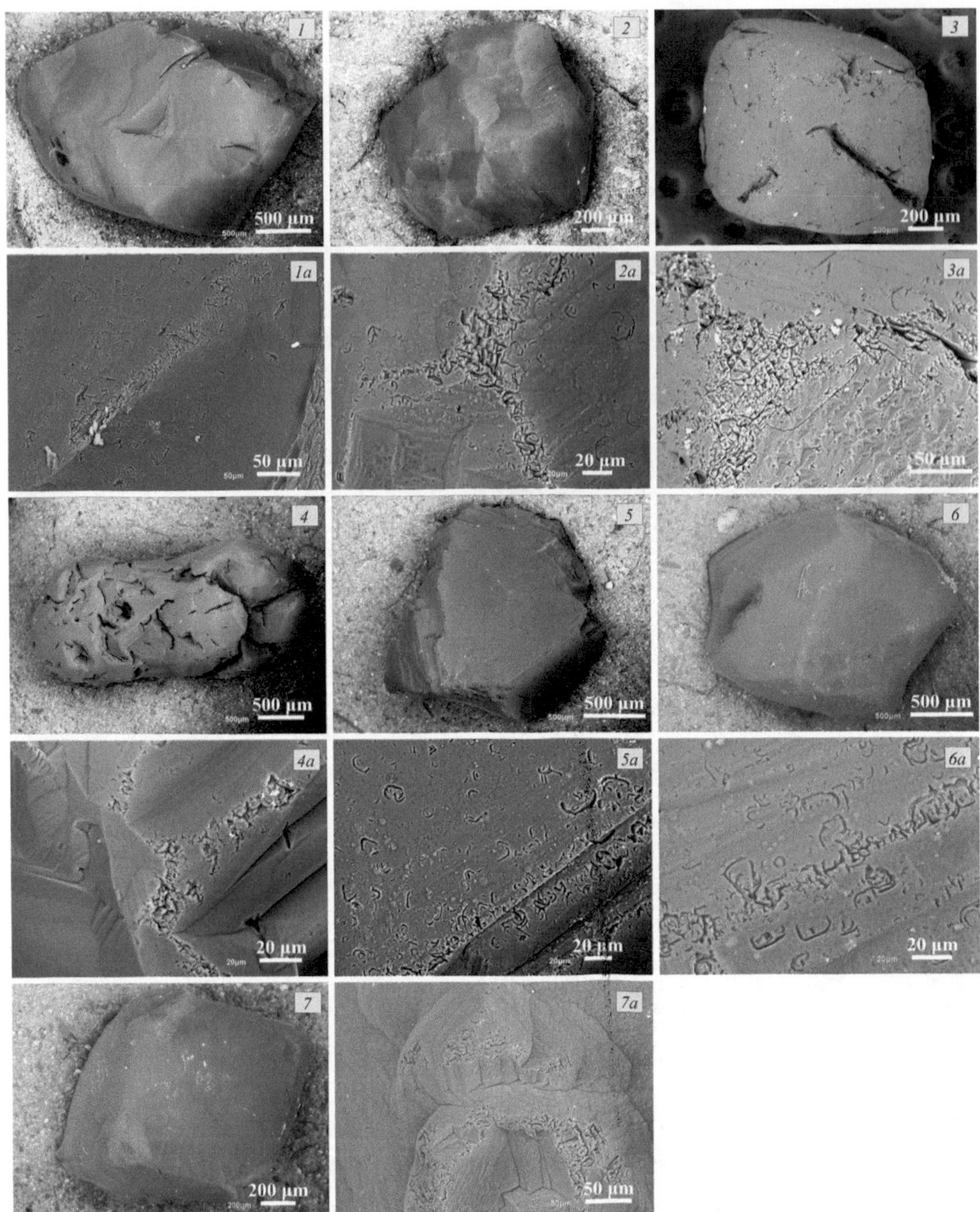

Fig. 4 Diamonds in the end of experiment. (*1*) a piece of pseudo-dodecahedron, colorless, 0.012 g; (*2*) an aggregate (boart), colorless, 0.008 g; (*3*) cuboid, yellow, 0.008 g; (*4*) variety VII (Orlov 1977), colorless, 0.014 g; (*5*) a piece of octahedron, colorless, 0.008 g; (*6*) cuboid, yellow, 0.013 g; (*7*) cuboid, gray, 0.009 g

Discussion

The results of the experiments suggest that wear of diamond is controlled by the conditions of dispersal during placer formation. Diamond in continental placers that originate from commercial primary (kimberlite) deposits and contain a complete assemblage of kimberlite indicator minerals bears no wear signature, while kimberlite indicator minerals are abraded to low or medium degrees. Some examples are the placers Vodorazdelnye Galechniki (Russian for *Watershed Pebbles*) or Novinka (*Novelty*) near Mirny city, and placers near the Nakyn field in Yakutia. In marine (coastal) placers, pyropes are well rounded, diamonds show minor wear, and ilmenite is fully eliminated (e.g., placers Tarydak in Krasnoyarsk region, Kyutyungde and Alamdzha in Yakutia). There are placers in which diamonds are abraded to different degrees, from weak but well evident to medium or strong (Ural, Kalimantan, Burma, southeastern Australia, placers of Brazil, Africa, etc.), while indicator minerals are absent (Kukharenko 1955; Metelkina et al. 1976; Trofimov 1980).

Based on the results of our experiments, one may infer that all indicator minerals which formerly coexisted with diamond in these types of placers have been worn to zero (reduced to an irrecoverable size) during dispersion from their primary sources.

According to our study of dispersal conditions, diamond can reach high roundness degrees only in a hard abrasive environment provided by high hydrodynamic activity. This occurs in coastal swash zones where the bedload rests upon metamorphic or igneous bedrock with clastic material of the same composition, which is a setting more typical of Precambrian rather than Phanerozoic time.

Rounded diamonds can become incorporated into young placers either when old placers in exposed inliers of Precambrian basement become eroded or via intermediate Phanerozoic placers.

The latter is, for instance, the case of Kalimantan (Sobolev 1951; Erlich and Hausel 2002) where the intermediate placers are Mesozoic. On redeposition, old diamonds may mix with indicator minerals from Phanerozoic kimberlites, including those from barren ones. As a result, there arises a paradox: the indicator mineral compositions evidence that kimberlites are poor or barren, while diamonds are abundant. For example, there are commercial placers under development in northeastern Siberia where all discovered kimberlites show very low or zero diamond contents. Indicator minerals from the area do not show the chemical signature of economic kimberlite occurrences, while up to 50 % of diamonds are heavily worn, which attests to Precambrian ages of the primary deposits (Afanasiev et al. 2011).

Concluding Remarks

Experiments on wear of diamond and kimberlite indicator minerals have shown that at the point when pyropes are well rounded, magnesian ilmenites become almost fully eliminated, whereas diamonds bear only a slight wear signature. Assemblages in which diamonds exhibit weak (though evident) or heavier wear degrees may be monomineralic and lack indicator minerals, because the latter have been worn to irrecoverable sizes. Thus, heavily worn diamonds may very likely come from Precambrian sources, but become incorporated into younger placers as the Precambrian ones are being eroded.

References

Afanasiev VP, Zinchuk NN, Pokhilenko NP (2010) Mineralogy for diamond exploration (in Russian). GEO Academic Publishers, Novosibirsk, p 650

Afanasiev VP, Lobanov SS, Pokhilenko NP, Koptil VI, Mityukhin SI, Gerasimchuk AV, Pomazanskii BS, Gorev NI (2011) Polygenesis of diamonds in the Siberian platform. Russ Geol Geophys 52(3):259–274

Erlich E, Hausel W (2002) Diamond deposits: origin, exploration, and history of discovery. SME, Littleton, p 343

Kukharenko AA (1955) Diamonds of the Urals (in Russian). Gosgeoltehizdat, Moscow, p 514

Metelkina MP, Prokopchuk BI, Sukhodolskaya OV, Frantsesson EV (1976) World Precambrian diamondiferous formations (in Russian). Nedra, Moscow, p 134

Orlov YL (1977) The mineralogy of diamond. Wiley, New York, p 234

Sobolev VS (1951) Geology of diamond deposits in Africa, Australia, Borneo, and North America (in Russian). Gosgeoltekhizdat, Moscow, p 126

Trofimov VS (1980) Geology of natural diamond deposits (in Russian). Nedra, Moscow, p 304

Internal Structure and Color of the Natural Plastically Deformed Diamonds from the Internatsionalnaya Kimberlite Pipe (Yakutia)

E. N. Fedorova, A. M. Logvinova, R. I. Mashkovtsev, and N. V. Sobolev

Abstract

Plastically deformed differently colored diamond crystals from the Internatsionalnaya pipe (Yakutia) have been studied using synchrotron radiation in Laue geometry (Laue-SR method) and UV-Vis-IR absorption spectroscopy. According to the Laue-SR data, brown, smoky gray and gray with a light purple-pink hue diamonds have suffered a high degree of deformation during the post-crystallization period. Strong asterism of Laue spots, intense continuous optical absorption throughout the visible spectrum, monotonically increasing toward shorter wavelengths, and "amber centers" in the IR region from 4,000 to 6,000 cm^{-1} are typical of these diamond crystals. Light purplish pink diamonds with gray or brown hue have experienced lesser degree of deformation. This group of diamonds has two absorption bands with maximums at 480 and 550 nm and "amber centers" in the IR region. A distinct polygonization is typical of the inner structure of light purplish pink diamonds. Light purplish pink diamonds have an intensive absorption band with maximum at 550 nm and are characterized by the lack of "amber centers." The relationship between the deformation levels, diamond color characteristics and manifestation of the "amber centers" in the IR region has been traced based on the obtained results.

Keywords

Diamond • Plastic deformation • Polygonization • Synchrotron radiation • Spectroscopy

Introduction

The abundance of colored, plastically deformed diamonds in different kimberlite and lamproite pipes widely varies and is one of the characteristics of a particular deposit. Brown and pink diamonds from the Argyle deposit in Australia (Clackson and Moore 1992; King et al. 2002; Nailer et al. 2007), rare violet diamonds from the kimberlites of South Africa (Raal 1958) are known. Diamond colors range from pink to purple color occur in Mirny kimberlite field (the Internatsionalnaya and Dachnaya pipes, Yakutia) (Titkov et al. 2008, 2012). Much progress has been made in recent years toward understanding the defects responsible for the yellow and green color of diamonds (Collins 2001). The origin of brown, pink, rare violet, or purple color of diamonds is still under discussion (Gaillou et al. 2010). The theoretical work of Hounsome et al. (2006) suggested vacancy disks in {111} planes as possible causes of brown color, but the positron annihilation results in Fisher et al. (2009) indicate three-dimensional vacancy clusters as the most likely defect. It is assumed that these vacancy clusters lead to monotonically increasing

E. N. Fedorova (✉) · A. M. Logvinova · R. I. Mashkovtsev
N. V. Sobolev
VS Sobolev Institute of Geology and Mineralogy, Siberian Branch of Russian Academy of Sciences, Koptyug Avenue 3, Novosibirsk 90, 630090, Russia
e-mail: fedorova@igm.nsc.ru

D. G. Pearson et al. (eds.), *Proceedings of 10th International Kimberlite Conference*, Volume 1, Special Issue of the Journal of the Geological Society of India,
DOI: 10.1007/978-81-322-1170-9_22,

absorption in the range from 1 to 5.5 eV (1,240–225 nm). The role of nitrogen in the stiffening of the diamond structure against distortion by natural plastic deformation and relations between the color of the stone and its total nitrogen content and nitrogen aggregation state have been studied by Nailer et al. (2007). Brown diamonds are known to have a series of absorption bands in the IR range from 4,000 to 6,000 cm^{-1} known in literature as "amber centers" (Massi et al. 2005). "Amber centers" were observed in brown and brown with red or yellow hue diamonds (Reinitz et al. 2000). Large clusters of vacancies dissociate during HPHT treatment and brown color and "amber centers" disappear (Reinitz et al. 2000). The study of pink diamonds of the Type Ia has shown that color is distributed in the parallel lamellae oriented along the octahedral planes {111} named "pink graining" (King et al. 2002). The optical absorption band with a maximum at 550 nm, which is responsible for the pink color, is considered to be related to plastic deformation of the diamond crystalline lattice, but the defect model has not been established (Gaillou et al. 2010).

The aim of the present investigation is to study colored, plastically deformed diamonds from the Internatsionalnaya pipe (Yakutia), to determine the relations between the deformation level and color that is of interest not only for diamond genesis, but also for understanding the physics of optical absorption leading to the appearance of diamond color.

Fig. 1 Plastically deformed diamonds from the Internatsionalnaya kimberlite pipe (Yakutia): **a** Brown diamond In-10 (group I); **b** Light purplish pink diamond In-3 (group IV)

Samples and Experimental Techniques

Forty-five diamond crystals from the Internatsionalnaya kimberlite pipe have been studied by optical spectroscopy and synchrotron Laue-SR method: 35 colored diamonds (Fig. 1) of brown and gray (group I), gray smoky with purplish pink hue (group II), light purplish pink with gray hue (group III), and light purplish pink (group IV) color and 10 colorless diamonds for comparison (Table 1). The crystal dimensions were about 1–2 mm. Colored diamond crystals were octahedral fragments or octahedra with different plastic deformations, as confirmed by microscopic examination. The surface of some samples was corrugated, while others demonstrated parallel striation, which reflected plastic deformation bands. All crystals demonstrated characteristic anomalous birefringence pattern in polarized light.

X-ray synchrotron radiation in Laue geometry (Laue-SR method) has been used to study the inner structure of plastically deformed crystals. It allows quick and effective study of imperfect crystals and determination of deformation degree (Rylov et al. 2001). Laue patterns have been obtained in the Institute of Nuclear Physics, Novosibirsk, using VEPP-3 synchrotron radiation source with X-ray wavelength in the range of 0.5–1.5 Å. Diamond is almost transparent in this wavelength range (Tzeng 1991). High intensity in the wide wavelength range and small beam divergence make it possible to obtain single step diffraction images of strongly deformed crystals (Rylov et al. 2001; Nailer et al. 2007). Contrary to the traditional X-ray method, Laue-SR method makes it possible to increase resolution by approximately two orders of magnitude (Rylov et al. 2001).

The Laue patterns were recorded on flat photographic film placed behind the crystal perpendicular to the incident white radiation (transmission Laue method). The diamonds with octahedral morphology were oriented with one octahedral face {111} perpendicular to the beam of synchrotron radiation X-rays using the goniometric head. The position of the diffraction sports is determined by the Bragg's law. For the Laue patterns recorded on the film in the direction of the incident beam, the reflection angle can be determined by the equation $tg2\theta = l/D$, where l is the distance from the center of Laue pattern to the spot, and D is the distance from crystal to flat photographic film (Bokii and Poray-Koshits 1964, pp. 420–428). The distance from the crystal to the flat photographic film is 50 mm.

Optical absorption spectra in the wavelength range from 300 to 900 nm were recorded using a Specord M40 spectrophotometer. The absorption coefficient is expressed in cm^{-1}.

Table 22.1 Characteristics of plastically deformed diamonds from the Internatsionalnaya pipe (Yakutia)

				Spectroscopy, α cm^{-1}					
Group	Sample	Laue-SR		UV-Vis		IR			
	In-	PD	P-G	480nm	550nm	4065cm^{-1}	4170cm^{-1}	N, ppm	B,%
I	12	+	-	-	-	0.9	0.7	818	13
	10	+	-	-	-	0.6	0.6	588	16
	21	+	-	-	-	0.6	0.5	693	9
	15	+	-	-	-	1.2	0.5	822	9
	25	+	-	-	-	0.7	0.5	487	10
II	11	+	-	1.2	1.4	0.0	0.2	418	15
	9	+	-	-	1.2	0.7	sh	620	14
	22	+	-	1.5	2.3	0.7	sh	803	14
	33	+	-	1.5	1.6	0.6	sh	573	21
	34	+	-	1.2	2.2	0.6	sh	930	19
	24	+		0.6	0.9	0.4	0.3	697	10
	16	+	-	0.6	0.9	0.1	sh	312	7
III	14	+	+	0.6	1.0	0.4	0.3	670	8
	19	+	+	0.5	1.0	0.3	0.2	778	12
	35	+	+	0.5	0.8	0.4	0.3	570	11
	18	+	+	0.4	0.8	0.2	0.1	667	8
	17	+	+	0.4	0.6	0.1	0.1	880	15
	36	+	+	-	1.8	0.1	sh	808	9
	13	+	+	0.5	1.2	0.8	sh	776	13
	37	+	+	1.4	1.9	0.4	sh	676	21
	23	+	+	0.4	0.5	0.2	sh	707	10
	30	+	+	0.4	0.7	0.3	sh	819	12
	28	+	+	0.4	0.5	0.1	sh	676	10
	31	+	+	-	1.8	0.1	sh	776	11
IV	1	+	+	-	2.9	-	-	774	17
	3	+	+	-	3.0	-	-	589	11
	2	+	+	-	3.2	-	-	836	20
	4	+	+	-	0.6	-	-	531	18
	5	+	+	-	2.2	-	-	640	12
	7	+	+	-	2.3	-	-	1,151	12
	8	+	+	-	3.5	-	-	550	13
	26	+	+	-	3.0	-	-	678	12
	27	+	+	-	2.3	-	-	943	23
	29	+	+	-	3.2	-	-	1,004	7
	32	+	+	-	2.1	-	-	525	8

Note: PD - plastic deformation; PG - polygonization; α - absorption coefficient; sh - shoulder; (+) - well expressed phenomenon; (-) – no phenomenon observed. N - the total nitrogen concentration in ppm; B,% - nitrogen aggregation states

Infrared (IR) spectra were recorded on a Bruker Vertex 70 FTIR spectrometer equipped with a Hyperion 2000 microscope with aperture value of 50 μm in the wavenumber range from 500 to 5,000 cm^{-1} with resolution of 1 cm^{-1}. The absorption coefficients of *A*- and *B*-defects at 1,282 cm^{-1} have been determined using the OPUS program, which allows the mixed IaAB diamond spectrum to be resolved into the component corresponding to the absorption spectra of A- and B-defects. Relative nitrogen content in A- and B-centers was determined according to the absorption coefficients of the corresponding systems in IR spectra by the standard procedure (Boyd et al. 1994, 1995).

Results and Discussion

Laue-SR Methods

The characteristics of plastically deformed diamonds studied by Laue-SR method are presented in Table 1 and Figs. 2, 3, 4. The Laue patterns were obtained in [111] direction for the octahedral diamonds In-22, In-10, In-18, and In-27 and in a random direction for the diamond In-11 that has a highly distorted shape of uncertain morphology. The reflection angles for each spot are shown in the figure captions.

The Laue-SR results make it possible to reveal the deformation levels (Panin et al. 1990, p. 16). It is known that the dislocation density of nondeformed crystal is usually 10^4–10^6 lines/cm^2, which increases to 10^{10}–10^{12} lines/cm^2 during deformation (Friedel 1967, p. 251). The dislocation density observed in TEM studies of brown diamonds are typically 2×10^9 lines/cm^2 (Fisher 2009). The increase in density of dislocations leads to the formation of various defects, and as a result, the structural networks of crystal atoms strongly change and the local zones with bending of crystal atom networks appear (Friedel 1967, Chap. I, II). As a consequence, Laue spots of the strongly deformed diamonds are elongated in the radial direction together with continuous blackened spots (Figs. 2 and 3). This phenomenon is known as asterism of Laue spots (Friedel 1967, pp. 489–495; Nailer et al. 2007). The study of plastically

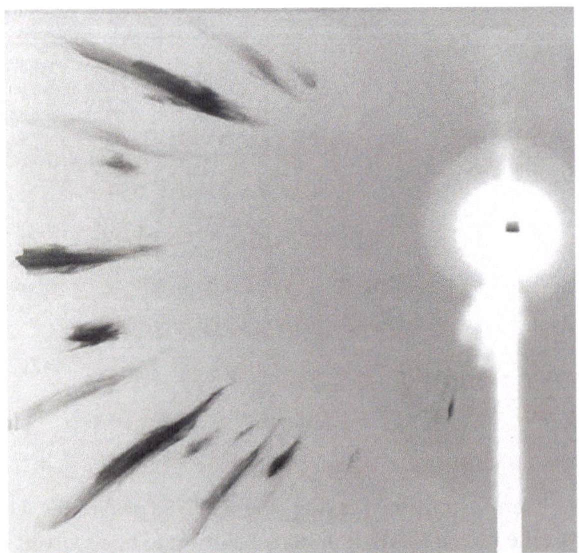

Fig. 2 Synchrotron Laue pattern of the strongly deformed gray diamond (sample In-22) taken in a [111] direction. Veiled-like Laue spots are elongated in a radial direction

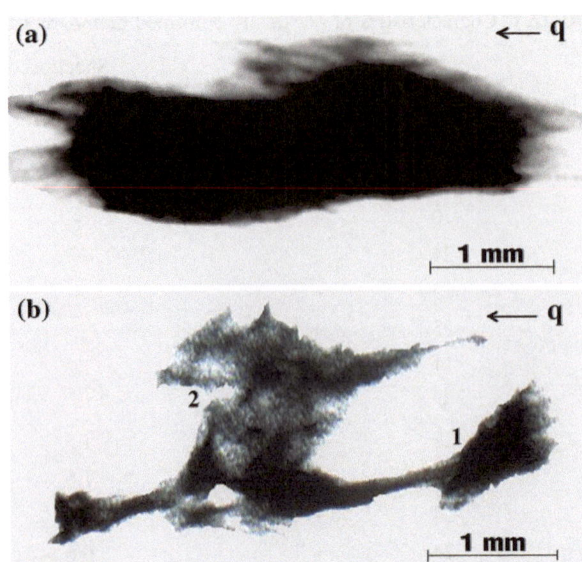

Fig. 3 Laue-SR spots show the different degree of deformation of brown and gray diamonds (groups I and II): **a** Strong deformation of brown diamond (sample In-10) leads to the strong asterism of the spot (2θ is about of 35.3°); **b** Irregular reorientation of individual local zones of the crystalline lattice leads to the fragmentation of the spot, and reflected angles 2θ are about 29° and 31° for the fragments 1 and 2, respectively (gray diamond In-11). Vector q is the radial direction

deformed diamonds showed that spot extension corresponds to lattice rotation around the <112> direction lying in the slip plane and perpendicular to the slip direction, with the (111)[110] slip system known for the diamond (Urusovskaya and Orlov 1964).

The higher the degree of crystal deformation, the more elongated and distorted the Laue spots (Guinier 1961, pp. 283–286; Friedel 1967, pp. 489–495). In the brown and gray diamonds of groups I and II, respectively, the strong asterism of the Laue spots is observed (Table 1, Figs. 2, 3). It is caused by the local bending of the slip planes and reorientation of individual sections with respect to each other in the deformed crystal (Urusovskaya and Orlov 1964; Klassen-Nekludova 1960, p. 194). In the crystals displaying high deformation levels, it is impossible to identify the levels of deformation as slip, climb, deformation bands, kinking, fragmentation, and Laue spots appear completely darkened (Fig. 3a). Reoriented sections can be regularly arranged relative to the basic part of the crystal contributing to the formation of mechanical twins. Irregular reorientation of individual local zones of the crystalline lattice can occur during plastic deformation along with mechanical twinning, in which case a Laue spot is split into several reflections mutually misoriented at small angles (Fig. 3b). It has been found (Mokievsky et al. 1962) that splitting of the Laue spot directly depends on the reorientation angles of microtwin layers, which diverge from the basic part of the crystal from 2° to 8°. Splitting corresponds to the lattice reorientation through the angle of about 2° for the diamond In-11 (group II) (Fig. 3b). Such misorientated microtwin interlayers are not twins, but fault regions which differ by sets of shear systems. An inhomogeneous slip distribution in the crystal can cause misorientation of its individual local zones of the crystalline lattice relative to the host crystal (Klassen-Nekludova 1960, p. 194; Mokievsky et al. 1962; Rylov et al. 2001).

Numerous point defects formed during plastic deformation are thermodynamically unstable and annealed due to atom diffusion (Collins 2001). However, the defects related to dislocations can be restored at higher annealing temperature than the temperature of plastic deformation, for example, the temperature which can be reached in laboratory conditions during the annealing of brown diamonds or natural long annealing (Fisher et al. 2009). During annealing, the dislocations move so that elastic energy decreases, and the dislocations are redistributed forming low-angle boundaries (dislocation walls of the same sign) between domains of different orientation, thereby resolving the diffraction streaks into the Laue spots (Friedel 1967, p. 327; Hirth and Lothe 1972, pp. 473–489; Sumida and Lang 1981). This phenomenon is known as polygonization. Laue spots of crystals, which underwent polygonization, have a striation that reflects the distribution of dislocations in the crystal (Urusovskaya and Orlov 1964; pp. 489–495; Rylov et al. 2001; Smith et al. 2010).

In some cases, deformation lamellae are mechanical microtwins as suggested by Titkov et al. (2008). The presence of microtwins was established by TEM in some pink diamonds from South Africa, Venezuela, and Australia (Gaillou et al. 2010) and by X-ray diffraction in pink-purple diamond from the Internatsionalnaya kimberlite pipe (Titkov et al. 2012). Plastic deformation by spinel-law slip or mechanical twining was observed in some brown diamonds with insignificant nitrogen concentrations from Snap Lake deposit of the Slave province (Canada) (Rylov et al. 2006).

According to Laue-SR data, brown and gray diamonds (groups I and II) have not been subjected to natural annealing and do not show polygonization (Table 1). Veiled tails and needles beyond the main diffraction spots, observed for the diamonds of groups I and II (Fig. 3), characterize the presence of elastic residual stresses of crystalline lattice (Kuznetsov 2003).

Diamonds of groups III and IV underwent polygonization (Table 1, Fig. 4). The Laue spots of purplish pink with brown or gray hue diamonds (group III) and purplish pink diamonds (group IV) from the Internatsionalnaya pipe have a striation as a result of polygonization (Rylov et al. 2007), but the veils extending beyond the main diffraction spots of some samples characterize the presence of residual elastic stresses of the crystalline lattice for the diamonds of group III (Fig. 4a). The diffraction streak forms of some samples are insignificantly curved and that may indicate preserved elastic strain (Rylov et al. 2007). The shape of the Laue spots of group IV diamonds becomes more isometric (Fig. 4b), which is typical of the crystals with low levels of elastic stresses (Guinier 1961; Friedel 1967, pp. 489–495).

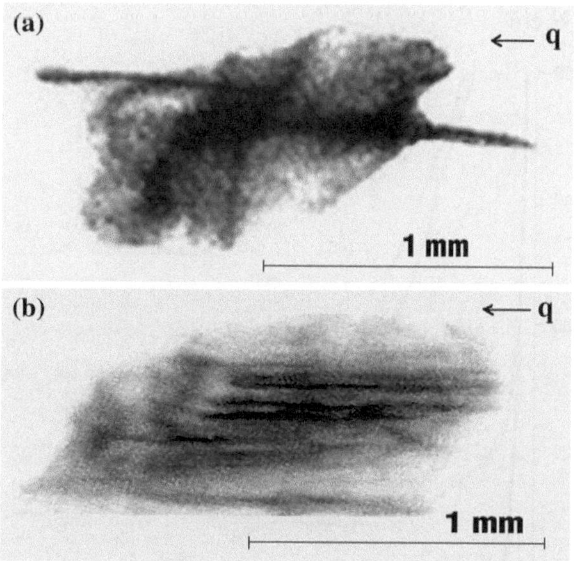

Fig. 4 Laue spots of light purplish pink with gray hue (group III) and light purplish pink (group IV): **a** The veiled tails extending beyond the main diffraction spot (2θ is about 35.4°), and the deformation bands are observed almost perpendicular to the radial direction inside the spot (sample In-18); **b** Slip system deformation bands extending throughout the crystal in the radial direction (2θ is 35.1°, sample In-27). Vector q is the radial direction

UV-Vis Spectroscopy

A continuum of absorption rising progressively from the near IR to UV region of the spectrum has been observed for brown, gray diamonds, and purplish pink diamonds with gray or brown hue and that is different from the optical spectrum of purplish pink diamonds of group IV and non-deformed diamonds of good optical quality (Fig. 5).

As shown above, the diamonds of groups I and II are characterized by the highest level of plastic deformation, which is followed by continuous and discrete misorientations throughout the crystal and local bending of the slip planes. The dislocation density is high at this deformation level. The interaction of dislocations causes the vacancies and vacancy disks, which can be the sources of new dislocations (Hirth and Lothe 1972, pp. 509–545). It is considered that the continuous absorption from the near IR to UV region in diamonds, which is responsible for the brown color, can be caused by the extended defects such as vacancy clusters (Hounsome et al. 2006; Fisher et al. 2009).

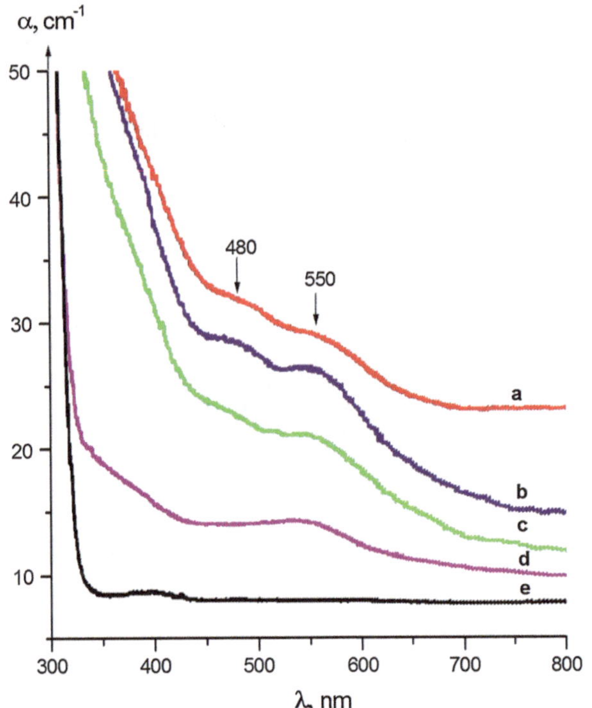

Fig. 5 Optical absorption spectra of plastically deformed diamonds (*curves a–d*) and stainless diamond (curve *e*): (*a*) sample In-10 (group I), (*b*) sample In-22 (group II), (*c*) sample In-13 (group III), (*d*) sample In-3 (group IV), (*e*) sample In-52 (gem-quality diamond)

For the nondeformed diamond, the UV absorption edge is conditioned by the presence of nitrogen in A-form and begins from ~320 nm (Bokii et al. 1986, p. 11), and in the IR range, band at 1,282 cm^{-1} is observed (Figs. 5e, 6e). For purplish pink diamonds, the band at 550 nm is observed along with the continuum absorption which increases from the wave length below ~400 nm (Fig. 5d). Two bands with absorption maxima at 480 and 550 nm have been registered for smoky-gray, brown with purplish pink hue (group II) and purplish pink with brown or gray hue diamonds (group III) against the background of intense continuous absorption (Fig. 5b, c). According to the literature, the band at 480 nm was observed for the amber and orange Argyle diamonds (Collins 1999). It is suggested that the absorption band with maximum at 480 nm is strictly vibronic and zero-phonon line is not observed due to its low intensity (Collins and Mohammed 1982). It seems possible that the broad absorption feature centered at 480 nm could simply be the vibronic band of the H3 center associated with a nitrogen-vacancy-nitrogen defect observed in the plastically deformed Type IaAB diamonds, but the zero-phonon line (ZPL) at 503.2 nm was not observed in this study. This is probably due to the strong absorption in this wavelength region observed for the I and II diamond groups. It is known that HPHT treatment of natural brown Type Ia diamonds causes the appearance of H3 center with ZPL at 503.2 nm (Collins et al. 2000, 2005; Fisher 2009; Vins and Yelisseyev 2010). The observed broad absorption band with maximum about 550 nm (Fig. 5) has been found earlier in the plastically deformed brown, pink, and violet diamond crystals by other authors (Collins 2001; Deljanin et al. 2008).

Purplish pink diamonds with gray hue have less intense continuous absorption background in UV-Vis spectral region and a more pronounced absorption band at 550 nm as compared to the diamonds of group I (Fig. 5c). Purplish pink diamonds of group IV are characterized by the absorption band at 550 nm (Fig. 5d, Table 1). It should be noted that some crystals of group I have a broad absorption band in the form of a shoulder at 550 nm (Fig. 5a). As regards the broad absorption band in the visible spectral region with maximum at 550 nm, this band should be responsible for the pink color in plastically deformed diamonds.

It is found that the color of the plastically deformed pink and brown Argyle diamonds, which is defined by a broad absorption band with maximum at 550 nm, is concentrated along the parallel lamellae consisting of the grains of different color intensity (King et al. 2002; Deljanin et al. 2008). Colored granular lamellae are directed along the octahedral planes and form parallel bands crossed by other similar bands in several places. These bands appear along another set of octahedral planes (King et al. 2002). It was suggested that the absorption band at 550 nm was of vibronic origin, and a zero-phonon line did not appear in the absorption spectra due to the strong electron–phonon interaction (Collins and Mohammed 1982; Fritsch et al. 2007). However, the investigations at liquid-nitrogen temperature (77 °K) did not reveal any fine vibronic structure of the defect (Massi et al. 2005). According to our data, the absorption band with maximum at 550 nm has been revealed for all studied diamonds in groups II–IV from the Internatsionalnaya pipe (Table 1), including the diamonds with irregularly disoriented microblocks (Fig. 5). The absorption band at 550 nm is the most intensive for the purplish pink diamonds of group IV whose inner structure is characterized by the lamellae along the octahedral planes (Table 1, Fig. 5d).

Thus, the nature of the absorption bands in the visible spectral region, which defines a diamond color, is related to a wide variety of optically active defects, which form during plastic deformation and depends on the deformation level.

FTIR Spectroscopy

According to IR spectroscopic data, plastically deformed diamonds are classified as Type IaA (Table 1). The IR spectra of groups I-III diamonds are characterized by two "amber centers" absorption bands (Massi et al. 2005) with maximum at 4,065 and 4,170 cm^{-1} (inset in Fig. 6). Purplish pink and pink diamonds (group IV) have no absorption bands related to "amber centers" (Fig. 6d). The absence of "amber centers" in the IR spectra of pink natural diamonds has been shown by De Weerdt and Van Royen (2001). It is shown that "amber centers" are annealed during HPHT treatment just like a brown component (Reinitz et al. 2000).

The "amber centers" of IaA diamond crystals have been studied in more detail in the work of Massi et al. (2005) who showed that the registered absorption bands in IR region from 4,000 to 9,000 cm^{-1} are a superposition of independent zero-phonon lines at 4,067, 4,098, 4,170 cm^{-1} (for the Type IaA diamonds) together with vibronic bands at higher energies. The authors also point out that the "amber centers" have been found only for the crystals with "brown grain" (lamellae) and nitrogen, which basically enters into A-defect for the Type IaA and IaAB diamonds but is never observed in nitrogen-free Type II diamonds further suggesting the relationship between the "amber centers" and nitrogen. However, there is no direct correlation between the "amber centers" intensity and total nitrogen content and the nitrogen aggregation state (Table 1). Nevertheless, different "amber centers" are formed during plastic deformation depending on whether the diamond is Type Ia or Type Ib (Massi et al. 2005).

There is no correlation between the total nitrogen content and the intensity of the "amber centers" for the collection of plastically deformed diamonds studied in this work. According to our data, the intensity of the "amber centers" absorption bands depends on the deformation level. Diamonds with strongly deformed crystalline lattice due to plastic deformation have more intense "amber centers," in this case, both absorption bands at 4,065 and 4,170 cm^{-1} are intense for brown and gray diamonds (Table 1, Fig. 6a). More intense absorption bands at 4,065 and 4,175 cm^{-1} are registered mostly like a shoulder for diamonds of groups II and III (Fig. 6b, c).

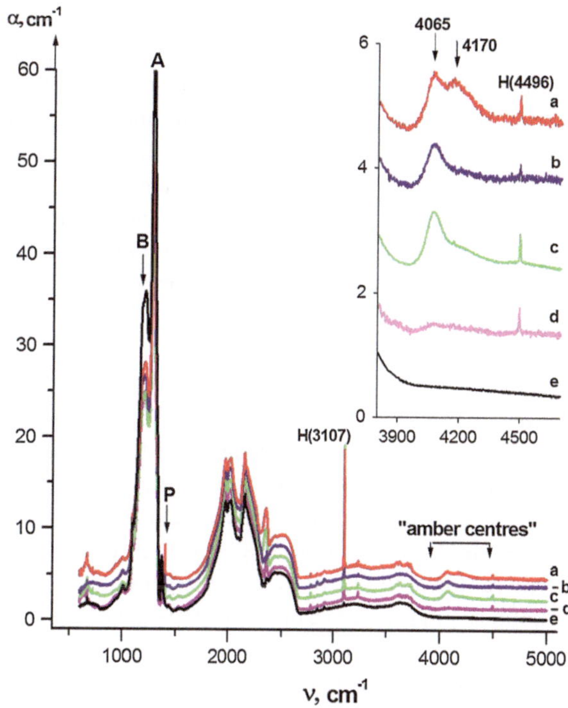

Fig. 6 IR spectra of plastically deformed diamonds (*curves a–d*) and stainless diamond (*curve e*): (*a*) sample In-10 (group I), (*b*) sample In-22 (group II), (*c*) sample In-13 (group III), (*d*) sample In-3 (group IV), (*e*) sample In-52 (gem-quality). Inset: the "amber centers" in low energy part of the spectra

Plastic Deformation of Natural Diamonds

It is widely believed that diamond plastic deformation can occur in nature under the effect of dynamic processes followed by the disintegration of the deep-seated rocks and transportation of diamonds to the Earth's surface by kimberlite and lamproite magmas (Orlov 1977, pp. 128–131). In addition, the plastic deformation may occur either during any stage of diamonds mantle residence or during the proto-phase of transportation of the diamonds by kimberlite and lamproite to the surface (Fisher 2009). It is known that the centers responsible for the brown color of diamonds are unstable thermally and annealed at HPHT treatment (T > 1,800 °C and 6 GPa). These laboratory conditions contrast with the more likely diamond forming conditions within the mantle of 1,150 °C and 5GPa (Stachel and Harris 2008). Also, De Vries (1975) showed experimentally that the plastic deformation of diamond was essentially linear after 2GPa (1,050 °C) up to 7 GPa (900 °C). Thus, brown centers could be formed in the extreme conditions of the Earth's upper mantle, but would be subsequently annealed out due to the elevated temperatures and, additionally because of long time diamonds reside in the mantle, with determined genesis ages ranging from 3.5 to 1.0 Ga (Fisher 2009). On the other hand, the defects responsible for pink color are more stable to HPHT treatments and would be expected to survive in the mantle conditions for a long time (Collins et al. 2000). It is therefore possible to explain the rare pink color compared to brown because pink color could be associated with the defects created by plastic deformation in the upper mantle or at a very early stage of proto-magma ascent while the brown color is acquired later, during the proto-magmatic phase. The brown and pink colors of the natural diamonds are associated with plastic deformation. It is suggested that the possibility of both defects responsible for brown and pink colors could be interconnected and the absorption in the same centers may be observed simultaneously (Fisher et al. 2009).

Brownish-pink and pinkish-brown diamonds are well known (Deljanin et al. 2008). It should be noted that the same diamonds of groups I and II have a small broad absorption band centered around 550 nm underlying on the continuously rising absorption (Fig. 5a, b). Thus, taking into account the conditions of the diamond formation (Stachel and Harris 2008) and the plastic deformation (De Vries 1975), the possibility of the plastic deformation followed by prolonged annealing with the modification of the defects responsible for the color of the diamonds in the upper mantle cannot be excluded (Fisher et al. 2009; Smith et al. 2010).

Diamonds of groups I and II of the collection under study did not experience natural annealing and therefore were probably deformed immediately before the extraction to the Earth's surface by kimberlite rocks. During the magma ascent to the Earth's surface, a sharp decrease in both temperature (as compared to the temperature of plastic deformation) and pressure can occur (Hirsch et al. 1986). In this case, the diffusion processes are slowed down or stopped altogether, and lattice distortions caused by previous deformation would remain in the crystal, that is, the cause of the residual thermal and dynamic elastic strains (Gaillou et al. 2010). Uniformly blacking of the Laue spots due to diffuse scattering of X-rays on local lattice misorientations, emissions outside the contrast of the main diffraction image observed for the diamonds of groups I and II of the collection characterize the availability of crystalline lattice elastic residual stresses.

Using theoretical and experimental values of activation energy associated with the dislocation motion 7.7 ± 0.4 eV (Davies et al. 1992) and 8.0 ± 0.3 eV (Fisher et al. 2009), Smith et al. (2010) extrapolated the data on HPHT treatment to subcontinental lithospheric mantle (SCLM) conditions to evaluate the treatment times for brown diamonds. The time required to remove the brown color during mantle storage at or below 1,000 °C in the SCLM is significant on a geological time scale (10^8 years or more). At the higher temperatures, 1,000–1,300 °C, brown color may reduce over a reasonable geological time scale (millions years) and at 1,300–1,400 °C (toward the base of lithosphere) may eliminate within thousands of years (Smith et al. 2010). Nevertheless, the question whether the brown color is created and retained for millions of years or formed during the events closely preceding magma ascent is still an open question.

The pink color of plastically deformed diamonds is concentrated in the so-called pink graining associated with the lamellae parallel to different sets of {111} planes of deformation (King et al. 2002). The origin of lamellae might be a result of the deformation of these diamonds by twining (Gaillou et al. 2010; Titkov et al. 2012) or polygonization of deformed diamonds in the event of prolonged annealing in nature (Fisher 2009; Smith et al. 2010).

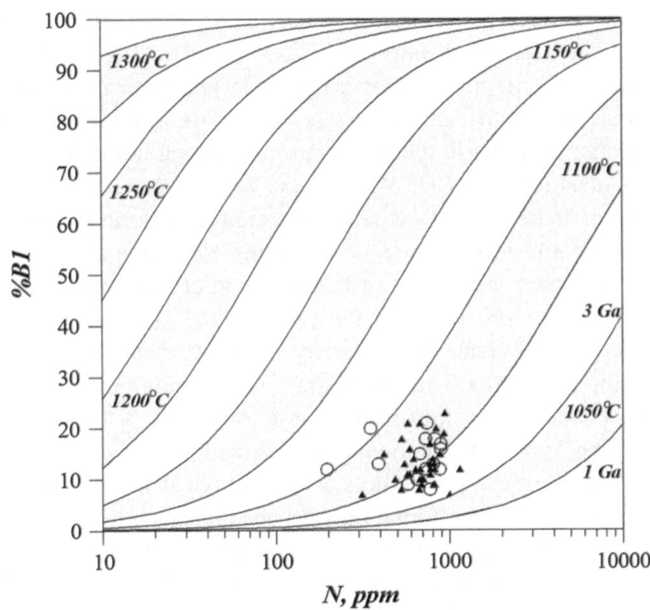

Fig. 7 Nitrogen aggregation state (%B) plotted as a function of content in ppm for diamonds from the Internatsionalnaya pipe. Isotherms are based on the second-order kinetic equation for an assumed mantle residence of 1 and 3 billion years (Ga) for the temperature from 1,050 to 1,300 °C (Taylor et al. 1996). Thus, the lines marked 1 Ga and 3 Ga at the bottom right are for the temperature of 1,050 °C

It should be noted that the formation of dislocation walls, known as "graining," may be formed both by polygonization as a result of annealing after drastic deformation (static polygonization) and during plastic deformation, when sliding along just one system of parallel planes take place (dynamic polygonization) (Friedel 1967, p. 328).

In diamond, the plastic deformation by dislocation slipping starts at 1,300 °C (Davey et al. 1984). According to our estimates, the mantle residence temperatures determined by the second-order kinetic equation are 1,075–1,125 °C or 1,050–1,100 °C for an assumed mantle residence of 1 or 3 Ga, respectively (Fig. 7). Low nitrogen aggregation state (%B) observed for the studied deformed diamonds (Table 1) is typical for diamonds from the Internatsionalnaya pipe (Logvinova et al. 2008; Bogush and Vasilyev 2008; Skuzovatov et al. 2011). In practice, however, the nitrogen aggregation state has a very high sensitivity to temperature and an equally weak dependence on mantle residence time (Stachel and Harris 2008).

Relatively low temperature and short duration of stress application are the main contribution factors to dynamic polygonization of the high-symmetry crystals such as diamonds (Friedel 1967, pp. 326–357). Also, taking into account this factor and a low mantle residence temperature, it can be assumed that the samples under study suffered dynamic polygonization by short duration of the stress during eruption of magma to the Earth's surface or by a short impact of shock loading due to the events causing lithospheric instability.

On the other hand, the study of the effect of HPHT treatment on the change in the set and concentration of the optically active defects in the natural brown diamonds showed that the annealing of brown color in natural diamonds started at the temperature of 1,800 °C (Fisher et al. 2009), but aggregation of the *A*-to *B*-center was observed at the temperatures above 2,200 °C (Vins and Yelisseyev 2010). The activation energy associated with *A*- to *B*-center conversion is of 7.0 ± 0.3 eV as defined by Cooper et al. (1989) and Taylor et al. (1990), and the activation energy of dislocations is of 8.0 ± 0.3 eV (Fisher et al. 2009). These results imply that dislocation rearrangement is preceded by aggregation of *A*-to *B*-center. In this case, it is possible to assume that it could be a static polygonization during prolonged annealing which was not accompanied by a simultaneous aggregation of *A*-to *B*-centers. However, nitrogen aggregation depends not only on residence temperature but also on the concentration of the nitrogen and mantle residence time, and plastic deformation accelerates nitrogen diffusion and consequently aggregation (Evans and Harris 1989; Evans 1992; Stachel and Harris 2008). Thus, a high degree of aggregation of the pink Argyle diamonds (Type Ia $A < B$) might be interpreted as a evidence for long residence time of the diamonds in the upper mantle after a plastic deformation (Mendelssohn and Milledge 1995; Gaillou et al. 2010). Furthermore, it is necessary to take into account that a dislocation rearrangement is accompanied by dissociation of A-centers and by formation of H3, H4, N3 defects (Fisher et al. 2009; Vins and Yelisseyev 2010). Thus, a clear answer to the question of what kind of polygonization (static or dynamic) takes place in nature is problematic. The complex study by spectroscopy and luminescence is required to obtain information about the annealing of diamond crystals (Wang et al. 2005; Fisher 2009). It should be noted that the mechanism of plastic deformation of diamonds in nature is rather complicated and defined not only by temperature, time, and applied stress, but also by the inherent properties of the diamond crystal (initial growth dislocations, point defects, inclusions), mechanical properties, and degree of homogeneity of the medium in which the crystal is deformed (Shiryaev et al. 2007).

Conclusions

According to Laue-SR data, the brown and gray (group I) and smoky-gray with purplish pink hue diamonds (group II) have not been subjected to annealing after deformation. Laue spots are demonstrated as strong asterism and fragmentation, indicating a high deformation level of these diamonds. This deformation level is followed by the extended defects in the form of vacancy clusters, which cause a continuous absorption throughout the visible region of the spectrum responsible for the brown color of diamonds. The "amber centers" in IR region with the absorption at 4,065 and 4,170 cm^{-1} are typical of these diamonds.

The inner structure of purplish pink with gray hue diamond crystals (group III) is characterized by the deformation lamellae with negligible local discrete disorientations in some directions. Broad absorption bands with maxima at 480 and 550 nm appear on the background of complete absorption in UV-Vis area with weaker absorption intensity as compared to the diamonds of groups I and II. "Amber centers" of lower intensity than in brown and gray diamonds of the studied collection are manifested in the IR region.

The inner structure of the purplish pink diamonds (group IV) is characterized by polygonization lamellae along the octahedral planes. The lack of asterism of Laue spots is indicative of the absence of disoriented regions in the sliding traces. No "amber centers" were found in the IR region. These crystals are characterized by the intense absorption band with maximum at 550 nm related to the pink color. In contrast to nondeformed diamonds, the absorption is observed at 400 nm instead of the typical band at 320 nm for type IaA diamonds. This absorption is, probably, responsible for the purplish hue of the studied diamonds.

Acknowledgments The authors are grateful to G.M. Rylov for developing the technique for studying the internal structure of naturally faceted diamonds using synchrotron radiation and measurements of Laue diffraction patterns of diamonds. We thank David Fisher and Mederic Palot for their useful reviews of our manuscript and helpful comments. Finally, we thank Jeff Harris for his editorial handling and suggestions for revision.

References

Bokii GB, Poray-Koshits MA (1964) X-ray analysis, vol 1. Publishing House of Moscow University, Moscow (in Russian)

Bokii GB, Bezrukov GN, Klyuev Yu A, Naletov AM, Nepsha VI (1986) Natural and synthetic diamonds. Nauka, Moscow (in Russian)

Bogush IN, Vasilyev TA (2008) Features of diamond from deposits of Siberia revealed by FTIR. In: 9th international kimberlite conference. Extended abstract, No. 9IKC-A-00215

Boyd SR, Kiflawi I, Woods GS (1994) The relationship between infrared absorption and the A defect concentration in diamond. Phil Mag 69:1149–1153

Boyd SR, Kiflawi I, Woods GS (1995) Infrared absorption by B nitrogen aggregate in diamond. Phil Mag B 72:351–361

Clackson SG, Moore M (1992) An X-ray study of some Argyle diamonds. Ind Diamond Rev 52:192–194

Collins AT (1999) Things we still don't know about optical centres in diamond. Diamond Relat Mater 8:1455–1462

Collins AT (2001) The colour of diamond and how it may be changed. J Gemmol 27(6):341–359

Collins AT, Mohammed K (1982) Optical studies of vibronic bands in yellow luminescing natural diamonds. J Phys C Solid State Phys 15:147–158

Collins AT, Kanda H, Kitawaki H (2000) Colour changes produced in natural brown diamonds by high pressure, high-temperature treatment. Diamond Relat Mater 9:113–122

Collins AT, Connor A, Ly C-H, Shareef A (2005) High-temperature annealing of optical centers in type-I diamond. J Appl Phys 97:083517(1-10)

Cooper GI, Mandelssohn MJ, Milledge JH (1989) High pressure/temperature experiments with natural diamonds. In: 28th international geological congress, pp 14–17

Davies G, Lawson SC, Collins AT, Mainwood A, Sharp SJ (1992) Vacancy-related centres in diamond. Phys Rev B 46:13157–13170

Davey ST, Evans T, Robertson SH (1984) An investigation of plastic deformation in sintered diamond compacts using photoluminescence spectroscopy. J Mat Sci Lett 3:1090–1092

De Vries RC (1975) Plastic deformation and the work-hardening of diamond. Mater Res Bull 10:1193–1199

De Weerdt F, Van Royen J (2001) Defects in coloured natural diamonds. Diamond Relat Mater 10:474–479

Deljanin B, Simic D, Zaitsev A, Chapmen J, Dobrinets I, Wideman A, Del Re N, Middleton T, Deljanin E, De Stefano A (2008) Characterization of pink diamonds of different origin: natural (Argyle, non- Argyle), irradiated and annealed, treated with multi-process, coated and synthetic. Diamond Relat Mater 17:1169–1178

Evans T (1992) Aggregation of nitrogen in diamond. In: Field JE (ed) The properties of natural and synthetic diamond. Academic Press, London, pp 259–290

Evans T, Harris JW (1989) Nitrogen aggregation, inclusion equilibration temperatures and the age of diamonds. In: Ross J, Jaques AL, Ferguson J, Green DH, O'Reilly SY, Danchin RV, Janse AJA (eds) Kimberlites and related rocks, geological society of Australia special publication, vol 14. Blackwell, Carlton, pp 1001–1006

Fisher D (2009) Brown diamonds and high pressure high temperature treatment. Lithos 112S:619–624

Fisher D, Sibley SL, Kelly CJ (2009) Brown colour in natural diamond and interaction between the brown related and other colour-inducing defects. J Phys: Condens Matter 21:364213 (10)

Friedel G (1967) Dislocations [Russian translation], Mir, Moscow

Fritsch E, Rondeau B, Hainschwang T, Quellier M-H (2007) A contribution to the understanding of pink color in diamond: the unique, historical «Grand Condé». Diamond Related Mater 16:1471–1474

Gaillou E, Post JE, Bassim ND, Zaitsev AM, Rose T, Fries MD, Stroud RM, Steele A, Butler JE (2010) Spectroscopic and microscopic characterizations of color lamellae in natural pink diamonds. Diamond Relat Mater 19:1207–1220

Guinier A (1961) Radiography of crystals. Theory and practice. Nauka, Moscow [in Russian]

Hirth JP, Lothe J (1972) Theory of dislocation [Russian translation]. Atomizdat, Moscow, (McGraw-Hill, New York 1968)

Hirsch PB, Pirouz PP, Barry JC (1986) Platelets, dislocation loops and voidites in diamond. Proc Roy Soc Lond A 407:239–258

Hounsome LS, Jones R, Martineau PM, Fisher D, Shaw MJ, Briddon PR, Oberg S (2006) Origin of brown coloration in diamond. Phys Rev B 73:125203

King JM, Shigley JE, Guhin SS, Gelb TH, Hall M (2002) Characterization and grading of natural-color pink diamonds. Gems Gemol 38:128–147

Klassen-Nekludova MV (1960) Mechanical twinning of crystals. Izd. Acad. Nauk SSSR, Moscow (in Russian)

Kuznetsov GF (2003) X-ray identification and measurement of plastic deformation and elastic strain in individual crystallites of polycrystalline diamond layers. J Thech Phys RAH 73:46–53 Moscow [in Russian]

Logvinova AM, Wirth R, Fedorova EN, Sobolev NV (2008) Nanometre-sized minerals and fluid inclusions in cloudy Siberian diamonds: new insights on diamond formation. Eur J Mineral 20:317–331

Massi L, Fritsch E, Collins AT, Hainschwang T, Notari F (2005) The "amber centers" find their relation to the brown colour in diamond. Diamond Relat Mater 14:1623–1629

Mendelssohn MJ, Milledge HJ (1995) Geologically significant information from routine analysis of the mid-infrared spectra of diamonds. Int Geol Rev 37:95–110

Mokievsky VP, Titova VM, Bartoshinsky ZV (1962) Manifestation of plastic deformation in diamonds and some problems related to crystal plasticity. Zap VMO 91(4):381–393 (in Russian)

Nailer SG, Moore M, Chapman J, Kowalski G (2007) On the role of nitrogen in stiffening the diamond structure. J Appl Cryst 40:1146–1152

Orlov YuL (1977) Mineralogy of the diamond. Wiley Interscience, New York

Panin VE, Grinyaev YV, Danilov VI (1990) Structural levels of plastic deformation and fracture. Nauka, Novosibirsk (in Russian)

Raal FA (1958) A new absorption bands in diamond and its likely cause. Proc Phys Soc Lon 71:846–847

Reinitz IM, Buerki PR, Shigley JE, McClure SF, Moses TM (2000) Identification of HPHT-treated yellow to green diamonds. Gems Gemol 34:128–137

Rylov GM, Yefimova ES, Sobolev NV, Kulipanov GN, Kondratyev VI, Tolochko BP, Sharafutdinov MR (2001) Study of imperfect natural diamonds with the application of the X-ray synchrotron radiation (the "Laue-SR" method). Nucl Instr Meth A 470:182–188

Rylov GM, Fedorova EN, Pokhilenko NP, Kulipanov GN, Sobolev NV (2006) Digest reports of the XVI international synchrotron radiation conference. 10–14 July 2006. Novosibirsk, Russia, p 56

Rylov GM, Fedorova EN, Logvinova AM, Pokhilenko NP, Kulipanov GN, Sobolev NV (2007) The peculiarities of natural plastically deformed diamond crystals from Internatsionalnaya pipe (Yakutia). Nucl Instr Meth A 575:152–154

Shiryaev AA, Frost DJ, Langenhorst F (2007) Impurity diffusion and microstructure in diamond deformed at high pressures and temperatures. Diamond Relat Mater 16:503–511

Skuzovatov SYu, Zedgenizov DA, Shatsky VS, Ragozin AL, Kuper KE (2011) Composition of cloudy microinclusions in octahedral diamonds from the Internatsionalnaya kimberlite pipe (Yakutia). Russ Geol Geophys 52:85–96

Smith EM, Helmstaed HY, Flemming RL (2010) Survival of brown color in diamond during storage in the subcontinental lithospheric mantle. Can Mineral 48:571–582

Stachel T, Harris JW (2008) The origin of cratonic diamonds-constraints from mineral inclusions. Ore Geol Rev 34:5–32

Sumida N, Lang AR (1981) Cathodoluminescence evidence of dislocation interactions in diamond. Phil Mag A 43:1277–1287

Taylor WR, Jaques AL, Ridd M (1990) Nitrogen-defect aggregation characteristics of some Australasian diamonds: time-temperature constraints on the source regions of pipe and alluvial diamonds. Am Mineral 75:1290–1310

Taylor WR, Canil D, Milledge HJ (1996) Kinetics of Ib to IaA nitrogen aggregation in diamond. Geochim Cosmochim Acta 60(23):4725–4733

Titkov SV, Shigley JE, Breeding CM, Minceva RM, Zudin NG, Sergeev AM (2008) Natural-color purple diamonds from Siberia. Gems Gemol 44(1):56–64

Titkov SV, Krivovichev SV, Organova NI (2012) Plastic deformation of natural diamonds by twinning: evidence from X-ray diffraction studies. Mineral Mag 76:143–149

Tzeng Y (1991) Diamond transmissive windows. Diamond Films Technol 1(1):31–54

Urusovskaya AA, Orlov YA (1964) On the features of plastic deformation of diamond crystals. Doklady Academii Nauk SSSR 154(5):1099–1102

Vins VG, Yelisseyev AP (2010) Effect of annealing at high pressures and temperatures on the defect-admixture structure of natural diamonds. Inorg Mater: Appl Res 1(4):303–310

Wang W, Smith CP, Hall MS (2005) Treated-color pink-to-red diamonds from Lucent Diamonds Inc. Gems Gemol 41:6–19

Diamond Potential of the Eastern Dharwar Craton, Southern India, and a Reconnaissance Study of Physical and Infrared Characteristics of the Diamonds

S. Ravi, M. V. Sufija, S. C. Patel, J. M. Sheikh, M. Sridhar, F. V. Kaminsky, G. K. Khachatryan, S. S. Nayak, and K. S. Bhaskara Rao

Abstract

Diamond potential of three sources in the Eastern Dharwar Craton (EDC) in southern India, namely kimberlites of the Wajrakarur Kimberlite Field (WKF), conglomerates of the Banganapalle Quartzite Formation and the famed alluvial gravels of the Krishna and Penner Rivers in southern India, has been investigated. The majority of the 24 studied WKF kimberlite pipes are poorly diamondiferous (<2 cpht). Diamond grade of the conglomerates is generally low (average 3 cpht or less) and erratic, though high grades of up to 35 cpht are locally recorded. Diamond incidence in the Krishna gravels is also erratic with spot values varying from 0.07 to 19.46 cpht. A set of 89 diamonds, including 65 from WKF kimberlites, 7 from conglomerates and 17 from gravels, have been characterised on the basis of their physical features. The WKF diamond population is dominated by colourless (51 %) and brown (35 %), while the remaining stones are yellow (8 %), transparent green coated (3 %), grey (1.5 %) and black (1.5 %). The diamonds from the conglomerates are colourless (3), brown (2), pink (1) and transparent green coated (1). The gravel diamonds include 4 each of brown, transparent green and cloudy milky white varieties, and 3 yellow and 2 colourless stones. In the WKF diamond population, octahedral and dodecahedral forms occur in subequal proportions, comprising 48 and 52 %, respectively. The conglomerate diamonds include 2 octahedra and 5 dodecahedra, while the gravel diamonds include 2 octahedra and 15 dodecahedra. The diamonds from the three sources exhibit a variety of surface resorption features including shield laminae, trigonal and hexagonal etch pits, and elongated and pyramidal hillocks. Infrared spectroscopic study was carried out on 51 diamonds (36 from the WKF, 4 from the conglomerates and 11 from the gravels). The total nitrogen content (N_{total}) for WKF diamonds falls in the ranges of 0–586 at. ppm (except for one stone with 1,718 at. ppm), 41–496 at. ppm for conglomerate diamonds and 30–750 at. ppm for gravel

S. Ravi (✉) · M. V. Sufija · M. Sridhar · K. S. Bhaskara Rao
Geological Survey of India, Bandlaguda Complex, Hyderabad, 500068, India
e-mail: sraviiyers@gmail.com

S. C. Patel · J. M. Sheikh
Department of Earth Sciences, Indian Institute of Technology, Powai, Mumbai, 400076, India

F. V. Kaminsky · G. K. Khachatryan
KM Diamond Exploration Ltd, 2446 Shadbolt Lane, West Vancouver, British Columbia, V7S 3J1, Canada

S. S. Nayak
Geological Survey of India, Kumaraswamy Layout, Bangalore, 560078, India

D. G. Pearson et al. (eds.), *Proceedings of 10th International Kimberlite Conference*, Volume 1, Special Issue of the Journal of the Geological Society of India,
DOI: 10.1007/978-81-322-1170-9_23,

diamonds. Type IaAB diamonds constitute the major proportion (58 %) of the WKF diamonds, while the remainder of the population comprises Type IaA (25 %) and Type II (17 %) diamonds. The conglomerate diamonds include equal proportions of Type IaAB and Type II, while the gravel diamonds are mostly of Type IaAB. The concentration of platelets varies widely in the ranges of 0–9.2, 0.4–21.4 and 1.3–11.2 cm^{-1} for diamonds from the WKF kimberlites, conglomerates and gravels, respectively.

Keywords
Kimberlite • Diamond • Nitrogen • Infrared • Dharwar Craton • Wajrakarur • India

Introduction

Occurrences of kimberlites, lamproites and old workings of diamond in India can be broadly grouped into three diamond provinces, namely Southern Indian Diamond Province (SIDP), Central Indian Diamond Province (CIDP) and Eastern Indian Diamond Province (EIDP) (Fig. 1) (Babu 1998; Satyanarayana 2000). The SIDP is marked by both primary and secondary source rocks of diamond, with the latter including Neoproterozoic conglomerates of the Banganapalle Quartzite Formation and the famed quaternary gravels of the Krishna and Penner Rivers (Neelakantam 2001). Many of the globally celebrated diamonds such as the Koh-i-Noor (186 ct), Great Mogul (787 ct), Pitt or Regent (410 ct), Orloff (189 ct) and Hope (45 ct) were recovered from the Krishna gravel in the historic past. The link between the secondary and primary source rocks of diamonds in the SIDP remains unconstrained. A comparative study of physical properties and infrared characteristics of diamonds from both types of source rocks could provide important clues to the provenance of diamonds in the secondary source rocks.

Infrared characteristics of diamonds are a function of the concentration and form of substitutional nitrogen in the diamond lattice and are used to classify diamonds into spectral types (Harris et al. 1975; Evans et al. 1981).

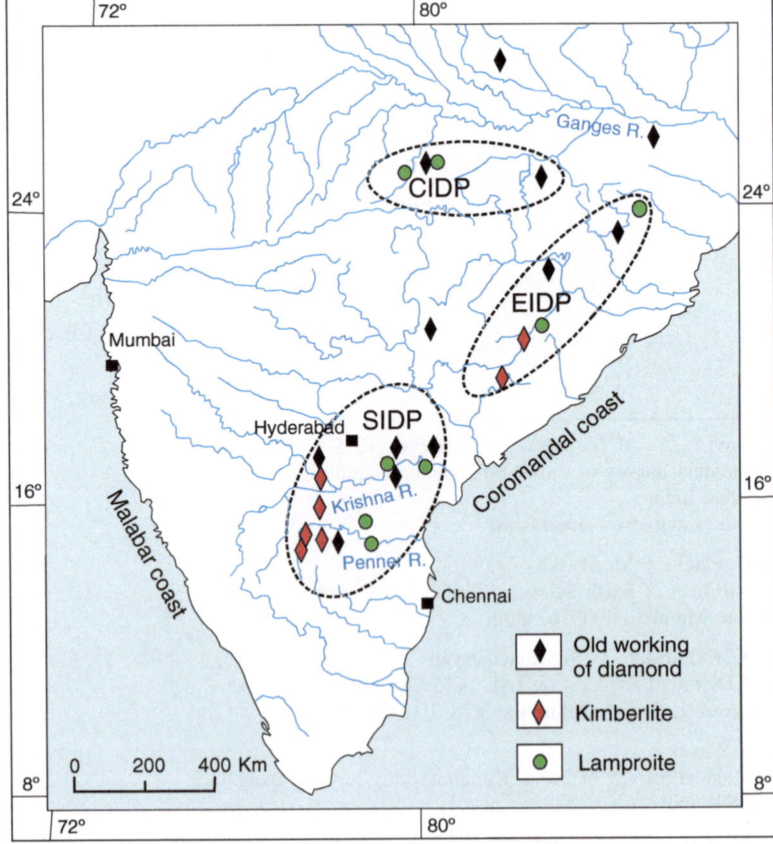

Fig. 1 Map of peninsular India showing major river systems, diamond provinces and old workings of diamond. *EIDP* Eastern Indian Diamond Province, *CIDP* Central Indian Diamond Province, *SIDP* Southern Indian Diamond Province

Available infrared study on southern Indian diamonds is limited to Wiens et al. (1990), who analysed a total of 8 diamonds including 3 from P1 kimberlite and 1 from P4 kimberlite of the WKF (Fig. 2) and another 4 from the Krishna gravels. These diamonds were supplied to them by the Geological Survey of India (GSI). The reported diamond types include IaA and II for WKF diamonds, and IaA, IaAB and IaB for gravel diamonds, but no data on nitrogen concentration of the diamonds are given in the paper.

In the present study, 89 diamonds, mostly from the WKF (65) and others from the Banganapalle conglomerates (7) and Krishna gravels (17), have been examined for their physical characteristics. Fifty-one diamonds were then further characterised by infrared spectroscopy for their nitrogen contents and aggregation state.

Southern Indian Diamond Source Rocks

The SIDP is confined to the Eastern Dharwar Craton (EDC), which is an Archaean gneiss–granite–greenstone terrane with a major Proterozoic platformal basin, known as the Cuddapah Basin (inset of Fig. 2). The WKF is located west of this basin and is the largest diamondiferous kimberlite field (~ 80 km × 70 km) in the Indian peninsula. So far, a total of 45 kimberlite bodies are known from the WKF, 31 of which are distributed in four clusters: Wajrakarur–Lattavaram (14 intrusions, P1–P14), Chigicherla (5 intrusions, CC1–CC5), Kalyandurg (6 intrusions, KL1–KL6) and Timmasamudram (6 intrusions, TK1–TK6) (Neelakantam 2001; Srinivas Choudary et al. 2007). Most of these kimberlites were discovered by the GSI in the last three decades. The rest 14 intrusions of the WKF (WK1–WK14) have recently been discovered by M/s Rio Tinto Exploration (Chatterjee et al. 2008). The WKF kimberlites are generally small intrusions (surface area of <10 ha) except for a few large ones such as P10 (63 ha) and P1 (19 ha) (Neelakantam 2001) (Table 1). Apart from the WKF, the Narayanpet Kimberlite Field (NKF), the Raichur Kimberlite Field (RKF) and the Tungabhadra Kimberlite Field (TKF) are three other potentially diamondiferous areas within the EDC (see inset of Fig. 2). Available radiometric ages indicate that the southern Indian kimberlites erupted episodically in the Mesoproterozoic with a peak at 1,100 Ma (Chalapathi Rao et al. 2012).

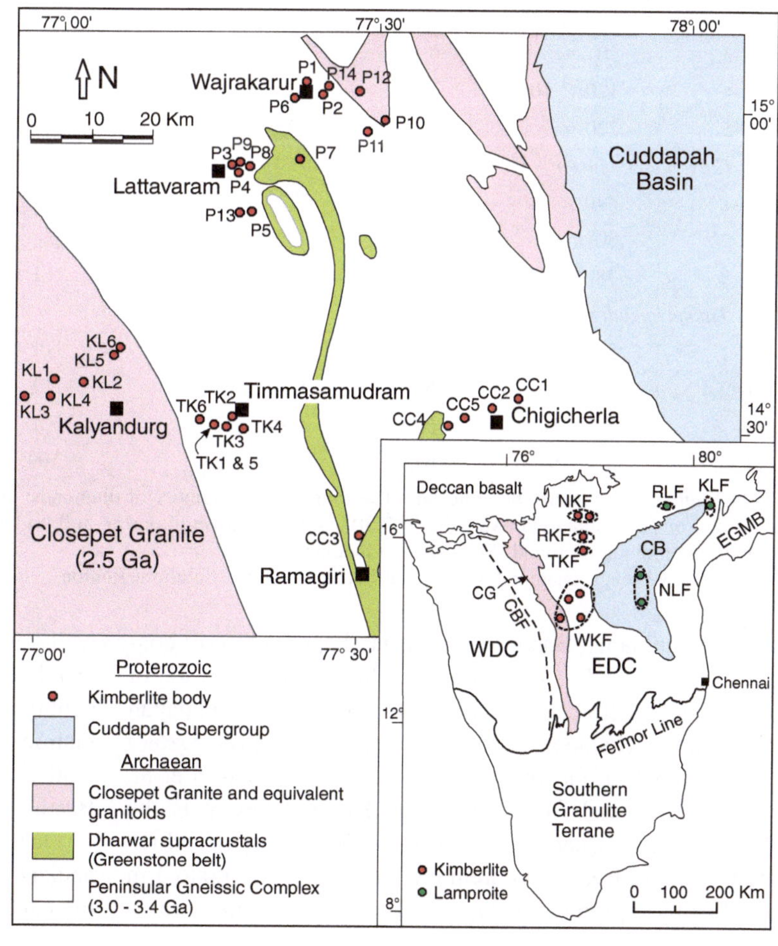

Fig. 2 Generalised geological map of Wajrakarur Kimberlite Field (*WKF*). Inset shows sketch map of southern India with location of kimberlite and lamproite fields in the Eastern Dharwar Craton (*EDC*). *CB* Cuddapah Basin, *CBF* Chitradurga Boundary Fault, *CG* Closepet Granite, *KLF* Krishna Lamproite Field, *NKF* Narayanpet Kimberlite Field, *NLF* Nallamalai Lamproite Field, *RLF* Ramadugu Lamproite Field, *TKF* Tungabhadra Kimberlite Field, *WDC* Western Dharwar Craton

Table 1 Diamond resources in the Wajrakarur Kimberlite Field

Pipe no.	Dimension (m)	Area (ha)	Material processed (t)	Diamond recovery (no.; wt.ct.)[a]	Diamond incidence (ct/100t)
(A) Wajrakarur–Lattavaram cluster					
P1	1080 × 180	19	13780	384; 80.88	0.56
P2	380 × 70	2.1	788	Nil	Nil
Extension-1	580 × 160	9.28	278	3; 3.94	1.56
Extension-2	500 × 120	6.12			
P3	120 × 40	0.48	907	12; 2.55	0.28
P4	265 × 130	3.45	2400	30; 6.02	0.25
P5	240 × 45	1	482	Nil	Nil
P6	260 × 240	6	8471	371; 56.81	0.67
P7	900 × 25	1.8	1515	1,142; 119.52	7.89
			Overburden		
			1,221	2,838; 543.26	44.49
P8	110 × 25	0.50	1773	40; 5.95	0.33
P9	37 × 21	0.07	229	4; 1.17	0.55
P10	1200 × 1000	63	1434	48; 14.71	1.00
P11	143 × 82	1.1	275	2; 2.15	0.78
P12	130 × 50	0.48	69	Nil	Nil
P13	125 × 100	1.7	Not tested		
(B) Chigicherla cluster					
CC-1	315 × 185	5.8	436	5; 1.31	0.30
CC-2	200 × 175	3.5	302	2; 1.05	0.35
CC-3	70 × 15	0.1	20	Nil	Nil
CC-4	120 × 105	1.3	138	15; 2.17	3.55
CC-5	220 × 70	1.5	143	6; 1.38	1.28
(C) Kalyandurg cluster					
KL-1	440 × 240	8.7	273	2; 0.22	0.08
KL-2	80 × 80	1.0	202	ψ	
KL-3	380 × 130	11.0	216	Nil	Nil
(D) Timmasamudram cluster					
TK-1	50 × 30	N/A	120	Nil	Nil
TK-2	50 × 40	N/A	126	Nil	Nil
TK-3	50 × 30	N/A	100	Nil	Nil
TK-4	14 × 6	N/A	228	509; 134.21	65

Abbreviations m = metre; ha = hectare; t = tonne; no. = number of diamonds; wt.ct. = total weight in carat; ct/100t = carat/100 tonnes
[a] A total of 5,383 diamonds weighing 977.30 cts were recovered during bulk sampling
ψ One microdiamond found. *N/A* Data not available
Data source Project Diamond, Geological Survey of India, Southern Region

The Cuddapah Basin is a crescent shaped, easterly concave basin, which is infilled by >10 km thick succession of clastic and chemical sediments belonging to the Cuddapah Supergroup (Palaeoproterozoic to early Mesoproterozoic) and Kurnool Group (Neoproterozoic) (Nagaraja Rao et al. 1987). The basal formation of the Kurnool Group is the Banganapalle Quartzite (maximum 50 m thick), which mainly comprises quartz arenite with local lenses (up to 3 m thick) of conglomerate (Fig. 3). This conglomerate was a well-known secondary source of diamonds in the seventeenth century, and other old workings are found within pebbly sandstones of the Mesoproterozoic Cumbum Formation of the Cuddapah Supergroup that occur in the Kolluru area in the north-eastern part of the Cuddapah Basin (Fig. 4) (Munn 1929; Dutt 1953).

Quaternary gravels, which constitute the host rock of the ancient diamond workings, occur in alluvial terraces along the banks of the Krishna River for over 300 km upstream of

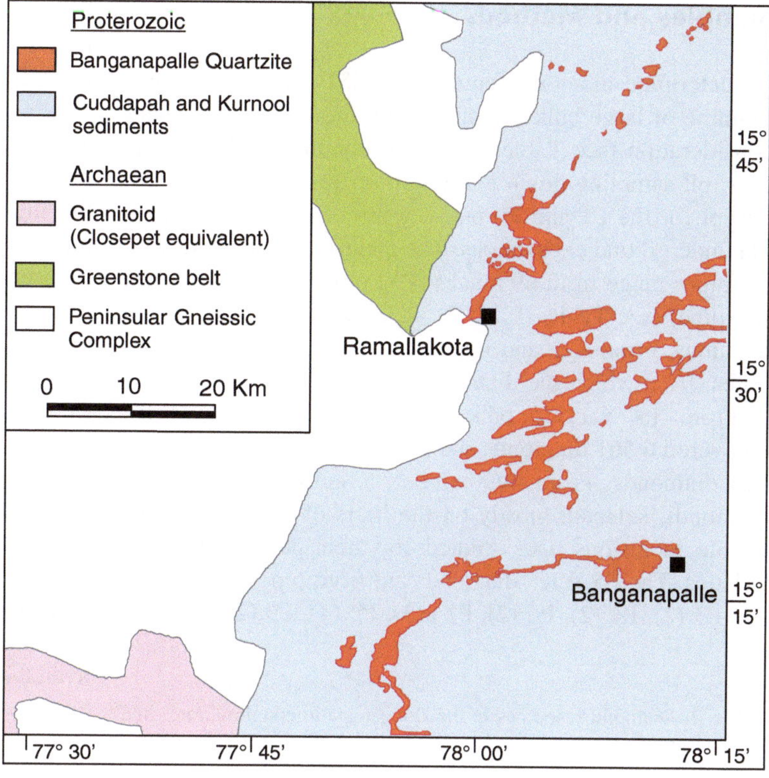

Fig. 3 Generalised geological map showing outcrops of Banganapalle Quartzite around Banganapalle

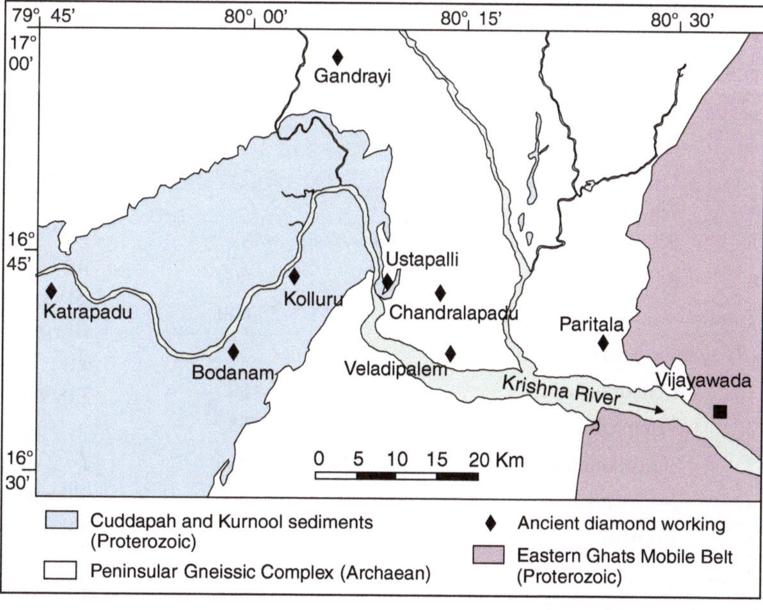

Fig. 4 Generalised geological map around Krishna River showing ancient diamond workings. The area falls on the north-eastern tip of the Cuddapah Basin

Ustapalle (Fig. 4). Downstream of Ustapalle is a vast flood plain, and the gravels in palaeochannels are now buried under thick alluvial cover. Most of the ancient diamond workings in the Krishna gravels are concentrated over a stretch of ~60 km between Kolluru and Paritala. The quaternary gravels consist of well-rounded pebbles, cobbles and boulders of quartzite, sandstone, jasper and agate in a matrix of sand. Other old workings are recorded in quaternary gravels along the banks of the Penner River (Fig. 1) (Krishnan 1950; Iyer 1961).

Samples and Methods

To determine diamond content, the GSI completed a programme of large bulk sampling from the three areas under consideration (see Tables 1 and 2). For the WKF kimberlites, pit sampling down to a depth of 6 m was completed, except for the P1 and P6 pipes, where sampling was done via underground excavations. The materials processed from the pipes range from 69 tonnes (P12 pipe) to 13,780 tonnes (P1 pipe) (see Tables 1 and 2). Rock material was crushed to minus 5 mm size and jigged for heavy mineral concentrates, from which the diamonds were hand-picked.

From its various bulk sampling programmes, GSI recovered 6,561 diamonds and presently holds only a set of 725 diamonds. From this number, a subset of 89 macrodiamonds, selected mostly on the basis of the presence of visible inclusions, was studied for their physical characteristics. The 65 WKF diamonds are from pipes: P1 (4), P2 (2), P3 (2), P4 (2), P6 (2), P7 (13), P8 (4), P9 (2), P10 (7), P11 (2), TK4 (14), CC1 (2), CC2 (1), CC4 (3), CC5 (3) and KL1 (2). In addition, 7 came from conglomerates and 17 from gravels.

Crystal forms and surface features of the diamonds were examined using optical microscope and scanning electron microscope. A few diamonds were cut by laser to expose the inclusions, which were then analysed by an electron microprobe at the GSI, Hyderabad. Fourier transform infrared spectroscopy (FTIR) was performed on 51 diamonds: 36 from the WKF, 4 from the conglomerates and 11 from the gravels (Table 3). FTIR absorption spectra were obtained using a Bruker Vertex 80 infrared spectrometer (fitted with Hyperion 3,000 microscope) at the Indian Institute of Technology, Bombay. Each diamond was represented by 1–4 spectra. In total, 87 spectra were recorded by averaging the signals of 100 scans for each spectrum. The spectral resolution was 4–8 cm^{-1}. Nitrogen concentrations were calculated from the IR spectra following the method outlined in Mainkar et al. (2013). The number of diamonds studied by FTIR was constrained because of the availability of machine time.

Table 2 Diamond resources in the Banganapalle conglomerates and Krishna gravels

Sl. No.	Block	Rock resource (mt)	Diamond resource (ct)
(A) Banganapalle conglomerates			
1.	Munimadugu	0.0607	2,000
2.	Allahabad West and West extension	1.044	15,600
3.	Allahabad SW and SW Extension	0.394	11,400
4.	Racherla–Munimadugu	1.523	N/A
5.	Banganapalle	0.067	N/A
6.	Rallakotturu–Lingambadi	2.137	26,600
Total =		5.2257[a]	55,600
(B) Krishna gravels			
1.	Chandralapadu	0.4	1,160
2.	Patempadu	0.224	3,763
3.	Veladikottapalem	10.809	35,676
4.	Kodavatikallu	0.74	6,290
5.	Sagileru	0.0055	ψ
Total =		12.1785[b]	46,889

Abbreviations mt = million tonne; ct = carat

[a] A total of 1,060 diamonds weighing 381 cts were recovered during bulk sampling

[b] A total of 120 diamonds weighing 100.13 cts were recovered during bulk sampling

N/A Not available. ψ Diamondiferous

Data source Project Diamond, Geological Survey of India, Southern Region

Results

Diamond Potential

The diamond grades are presented in Table 1 for the kimberlites and Table 2 for the conglomerates and gravels. The results are based on unpublished reports of the GSI by Rajaraman (1969), Govinda Rao and Misra (1977), Subba Rao (1988), Burhanuddin and Rao (1991), Rao et al. (1997), Nayak et al. (1999), Nayak and Dhakate (2002) and Sridhar and Sinha (2005) for the kimberlites, by Deshpande (1968), Sivaji et al. (1989) and Sivaji and Satyanarayana (1990) for the conglomerates and by Deshpande (1970), Deshpande and Sagar (1971, 1972), Nayak (1992) and Sivaji et al. (1992) for the gravels.

Overall, the WKF kimberlites are poorly diamondiferous (<2 cpht), but the diamonds are mostly gem quality. The largest diamond recovered is 16.30 cts from the P7 pipe. Diamonds from the Banganapalle conglomerates also have low grades (average 3 cpht or less) with an erratic distribution, although occasional relatively high grade up to 35 cpht has been recorded (Sivaji and Satyanarayana 1990). The diamonds are gem quality, but <0.5 ct in weight, the largest being 6.2 cts. Krishna gravel diamonds are also erratic in distribution and vary widely in caratage from 0.07 to 19.5 cpht. Most of these diamonds are gem quality with an average weight of ~0.83 cts (Nayak 1992).

Table 3 Physical characteristics and results of infrared spectral decomposition of southern Indian diamonds

Pipe	Sample no.	Weight (ct)	Colour	Morphology	Surface features	No. of spectra	N_A	N_B	N_{tot}	$\%N_B$	H (cm^{-1})	P	Type
(A) Wajrakarur Kimberlites													
P1	P1-149	0.15	Colourless	Octahedron	Flat-bottomed etch trigons	2	32	10	42	24	0	0	Near IIa
	P1-274	0.23	Colourless	Macle	Coarse frosting	1	<10	0	0	0	N/A	N/A	IIa
	P1-292	0.32	Colourless	Dodecahedron	Elongated hillocks; fine frosting	1	77	80	156	51	0	5.2	IaAB
P3	P3-6	0.07	Light brown	Irregular octahedron	Small areas of zigzag texture; striations	2	80	88	168	52	2.2	4.7	IaAB
	P3-8	0.50	Brown	Irregular octahedron	Stepped growth; deformation lines (1 set)	2	50	36	86	42	0.7	1.3	IaAB
P4	P4-4	0.25	Light brown	Irregular octahedron	Shield laminae; pointed and flat-bottomed etch trigons; edges replaced by curved surfaces with striations	2	62	187	249	75	N/A	N/A	IaAB
P6	P6-339	0.18	Brown	Flattened dodecahedron	Deformation lines (1 set); fine frosting; ruts	2	13	0	13	0	0.7	0	IIa
P7	P7-1303	0.08	Colourless	Octahedron	Stepped growth; enhanced lustre	2	483	78	561	14	N/A	5.3	IaAB
	P7-1381	0.05	Colourless	Dodecahedron	Striations; enhanced lustre	2	296	204	500	41	0	6.5	IaAB
	P7-1820	0.12	Colourless	Flattened dodecahedron	Striations; enhanced lustre	2	300	199	499	40	0.1	7.0	IaAB
	P7-2290	0.27	Colourless	Octahedron	Edges replaced by curved surfaces with striations	2	181	13	194	7	0	2.4	Near IaA
	P7-2563	0.07	Green	Dodecahedron	Striations	1	283	226	509	44	0	7.2	IaAB
	P7-3037	0.05	Smoky brown	Irregular dodecahedron (stellate twin)	Deformation lines (1 set); striations; elongated hillocks; ellipsoidal cavity	1	90	0	90	0	N/A	N/A	IaA
	P7-3238	0.72	Light brown	Octahedral aggregate	Stepped growth	2	188	93	281	33	0	4.0	IaAB
	P7-3244	0.05	Colourless	Flattened dodecahedron	Deformation lines (1 set); striations; enhanced lustre	1	1245	473	1718	28	N/A	5.5	IaAB
	P7-3840	0.17	Deep brown	Octahedron	Deformation lines (1 set); ruts	2	263	164	427	38	0	4.7	IaAB
	P7-3906	0.14	Colourless	Dodecahedron	Striations; fine frosting	2	265	188	453	42	0	5.1	IaAB
	P7-3920	0.14	Colourless	Dodecahedron	Striations; enhanced lustre	1	294	0	294	0	N/A	N/A	IaA

(continued)

Table 3 (continued)

Pipe	Sample no.	Weight (ct)	Colour	Morphology	Surface features	No. of spectra	N_A	N_B	N_{tot}	$\%N_B$	H (cm^{-1})	P	Type
P8	P8-R4	0.12	Light yellow	Dodecahedron	Elongated hillocks; ruts; fine frosting	1	308	39	347	11	0	2.6	IaAB
	P8-R11	0.05	Honey brown	Flattened dodecahedron	Deformation lines (2 sets); fine frosting	2	274	0	274	0	1.1	0	IaA
	P8-R17	0.74	Light brown	Dodecahedron	Deformation lines (1 set); striations; fine frosting; ruts	2	324	69	393	18	N/A	N/A	IaAB
P9	P9-S1	0.18	Colourless	Dodecahedron	Pyramidal hillocks; fine frosting	2	55	90	145	62	3.9	5.0	IaAB
P10	P10-4049	0.09	Colourless	Irregular dodecahedron	Deformation lines (3 sets); fine frosting	1	26	146	171	85	1.5	7.5	IaAB
	P10-4080	0.08	Green	Dodecahedron	Elongated hillocks; fine frosting	1	35	83	118	70	N/A	N/A	IaAB
	P10-4112	0.30	Pale yellow	Irregular dodecahedron	Coarse frosting; dark green radiation spots	1	0	0	0	N/A	0	0	IIa
TK4	TK4-4146	0.14	Colourless	Irregular octahedron	Shield laminae	2	76	0	76	0	0	0	IaA
	TK4-4147	0.19	Colourless	Elongated dodecahedron	Striations	2	399	213	612	35	0.7	9.2	IaAB
	TK4-4149	0.27	Colourless	Irregular octahedron	Serrate laminae	1	204	0	204	0	N/A	N/A	IaA
	TK4-4206	0.11	Colourless	Octahedron	Edges replaced by curved surfaces with striations; enhanced lustre	4	335	7	342	2	0.1	0.6	IaA
	TK4-4232	0.24	Yellow	Octahedron	Stepped growth; flat-bottomed etch trigons	2	492	94	586	16	0	4.5	IaAB
	TK4-4372	0.46	Colourless	Macle	Striations	2	155	0.4	155.4	0.3	0.1	1.1	IaA
	TK4-4430	0.63	Brown	Irregular dodecahedron (twinned)	Deformation lines (2 sets); striations; fine frosting	1	234	240	474	51	N/A	4.0	IaAB
	TK4-4579	0.23	Colourless	Octahedron	Edges replaced by curved surfaces with striations	2	105	12	117	10	0	0.2	Near IaA
CC1	CC1-5-10	0.94	Yellow	Irregular octahedron	Incomplete octahedral growth; coarse frosting	1	0	0	0	N/A	0	0	IIa
CC2	CC2-2	0.38	Colourless	Irregular octahedron	Pointed etch trigons; coarse frosting	1	0	0	0	N/A	0	0	IIa
CC5	CC5-4100	0.26	Colourless	Irregular octahedron	Fine frosting	4	151	177	328	54	0.1	13.4	IaAB

(continued)

Table 3 (continued)

Pipe	Sample no.	Weight (ct)	Colour	Morphology	Surface features	No. of spectra	N_A	N_B	N_{tot}	$\%N_B$	H (cm^{-1})	P	Type
(B) Banganapalle Conglomerates													
	RAC-110	0.13	Pale green	Deltoid dodecahedron	Corrosion sculptures; fine frosting	2	259	237	496	48	0.4	21.4	IaAB
	RAC-203	0.37	Colourless	Dodecahedron	Deformation lines (1 set); elongated hillocks; fine frosting	2	23	25	48	52	0.6	1.8	Near IIa
	RAC-242	0.62	Light brown	Dodecahedron	Deformation lines (1 set); ruts; fine frosting	2	22	19	41	46	0.4	1.2	Near IIa
	RAC-717	0.30	Pink	Dodecahedron	Deformation lines (1 set); elongated hillocks; fine frosting	2	39	22	61	36	0	0.4	IaAB
(C) Krishna Gravels													
	CP-24	0.35	Pale green	Dodecahedron	Enhanced lustre	2	~300	~100	~400	~25	0.3	1.8	IaAB
	CP-45	0.46	Pale green	Dodecahedron	Fine frosting	2	~600	~150	~750	~20	0	4.0	IaAB
	CP-48	0.32	Dark green	Irregular octahedron	Edges replaced by curved surfaces with striations	2	23	31	52	60	2.6	1.3	IaAB
	CP-87	0.08	Colourless	Deltoid dodecahedron	Striations; dark green radiation spots	2	442	117	559	21	0	6.0	IaAB
	SL-29	0.31	Light brown	Irregular dodecahedron	Deformation lines (2 sets); corrosion sculpture; fine frosting	2	231	36	267	13	0	2.3	IaAB
	VK-9	0.30	Colourless	Dodecahedron	Striations; fine frosting	2	282	160	442	36	0.5	11.2	IaAB
	CP-50	0.47	Yellowish white	Dodecahedron	Fine frosting	1	25	61	86	71	N/A	N/A	IaAB
	CP-60	0.38	Yellow	Irregular dodecahedron	Fine frosting	1	247	253	500	51	N/A	N/A	IaAB
	VK-2	2.26	Green	Irregular dodecahedron	Coarse frosting	1	27	36	63	57	N/A	N/A	IaAB
	KLR-4	0.35	White	Irregular dodecahedron	Fine frosting	1	30	0	30	0	N/A	N/A	Near IIa
	KLR-8	0.31	Honey yellow	Irregular dodecahedron	Dark green radiation spots	1	220	0	220	0	N/A	N/A	IaA

Physical Characteristics of Diamonds

The WKF diamonds are mostly colourless (51 %) and brown (35 %), while the remainder of the population comprises yellow (8 %), green (3 %), grey (1.5 %) and black (1.5 %) stones. The conglomerate diamonds are colourless (3), brown (2), pink (1) and green (1). The gravel diamond population comprises 4 each of brown, green and cloudy milky white varieties, and 3 yellow and 2 colourless stones. The green colour of some of the stones is a transparent green coat resulting from alpha-particle irradiation (Vance et al. 1973). Radiation damage in the form of dark green spots is observed in a few stones (P10-4112 from Wajarakarur kimberlites and CP87 and KLR8 from Krishna gravels, Table 3).

The WKF diamond morphologies include subequal proportions of octahedra (48 %) and dodecahedra (52 %). The octahedral stones are mostly single crystals (81 %) (Figs. 5a, b), with the remaining 19 % being aggregates or twins. The shape of single octahedra varies from regular (56 %) to irregular (44 %). The dodecahedra vary in shape from equant (50 %) (Figs. 5c, d) to flattened (12 %) (Fig. 5e), elongated (3 %), deltoid (9 %) (Fig. 5f) and irregular (26 %). Diamonds from different kimberlite clusters of the WKF (P, TK and CC—see Table 3) show broadly similar morphologies. Diamonds from the conglomerates include 2 octahedra and 5 dodecahedra (4 equant and 1 deltoid shapes), while the gravel diamonds include 2 octahedra and 15 dodecahedra (8 equant, 1 flattened, 1 deltoid and 5 irregular shaped).

Fig. 5 Representative crystal forms of southern Indian diamonds. **a** Octahedron; **b** resorbed octahedron; **c** equant dodecahedron with rhombic faces; **d** equant dodecahedron with triangular faces; **e** flattened dodecahedron; **f** deltoid dodecahedron. Stones (**a**) and (**e**) are from kimberlites; stones (**b**) and (**c**) from conglomerates; and stone (**d**) is from gravels. Bar scale is 1 mm long

Some octahedral diamonds show incomplete development of crystal faces (Fig. 6a). Deformation lines are observed in a small population of diamonds, especially the brown and pink diamonds (Fig. 6b). Such lines are a manifestation of glide planes in the crystal and indicate plastic deformation of the diamonds in the mantle source (Mokievsky et al. 1962; Robinson 1979). The southern Indian diamonds exhibit a host of surface resorption features, the initial stage being expressed by shield-shaped laminae that consist of superimposed laminae of progressively diminishing areal extent in octahedral crystals (e.g. Robinson 1979; Laiginhas 2008). Further resorption results in etch pits of trigonal (Fig. 6e) and hexagonal (Fig. 6f) geometry and serrate laminae. Etch trigons vary in form from pointed (i.e. triangular pits in the shape of negative pyramids) to flat-bottomed (i.e. triangular pits with flat bottom on {111} plane). The hexagonal etch pits are flat-bottomed, and the pits of larger size (up to 0.6 mm long) contain smaller ones. Continued resorption leads to the rounding of edges and apices of octahedra (Fig. 5b) and then to the conversion of the octahedra to dodecahedral forms (Fig. 5c–f) (e.g. Robinson et al. 1989). Pyramidal and elongated hillocks (Fig. 6c) are seen on several dodecahedral stones. Ruts (Fig. 6g) are occasionally present which have resulted from resorption or etching. Fine to coarse frosting (Fig. 6d) is exhibited by 51 % of the diamonds, whereas 14 % of the stones show enhanced lustre. A comparison of diamonds from the TK4 (Timmasamudram cluster) and P7 (Wajrakarur cluster) kimberlites, from which larger numbers of diamonds have been examined in the present study compared to the other pipes, shows that the TK4 diamonds, with 9 octahedral and 5 dodecahedral forms, have undergone lower degree of resorption than the P7 diamonds, with 3 octahedral and 10 dodecahedral forms. The gravel diamonds commonly display physical evidences of transportation such as percussion marks (Fig. 6h) and network patterns.

Most of the diamonds selected for this study contain visible mineral inclusions. Preliminary investigation shows that black flakes, probably graphite, are the most common inclusions (Fig. 7a), while other suspected inclusions include colourless olivine (Fig. 7b) and orange, eclogitic garnet (Fig. 7c) in diamonds from the TK4 kimberlite. One of the diamonds from the P9 kimberlite contains an inclusion of Cr-spinel (Fig. 7d), which has the composition (in wt percentage): SiO_2-0.14; TiO_2-0.16; Al_2O_3-7.48; Cr_2O_3-62.86; FeO-12.19; MgO-16.23; MnO-0.33; total-99.39. The chromite mineral chemistry is indicative of its primary origin (Deines et al. 2009). A magnetite inclusion, possibly of epigenetic origin, occurs in a diamond from the P4 kimberlite and has a total iron content expressed as FeO = 90.45 wt %. One of the conglomerate diamonds contains an inclusion of silica (98.65 wt % SiO_2), which

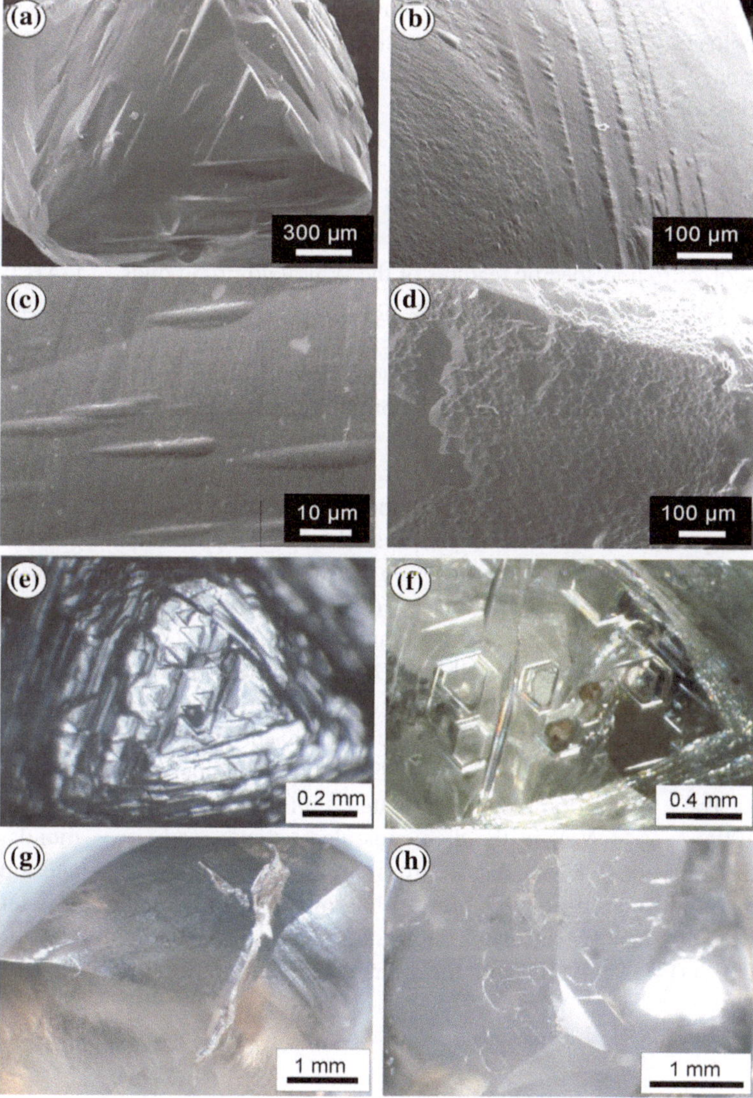

Fig. 6 SEM images (**a–d**) and optical photographs (**e–h**) showing surface features of southern Indian diamonds. **a** Stepped octahedral growth; **b** deformation lines; **c** elongated hillocks; **d** frosting; **e** trigonal etch pits; **f** hexagonal etch pits; a larger one in the *left side* contains two smaller ones; **g** ruts or etch channels; **h** percussion marks. Stones **a** and **e** are from kimberlites, stones **b**, **c**, **d**, **f** and **g** are from conglomerates, and stone **h** is from a gravel

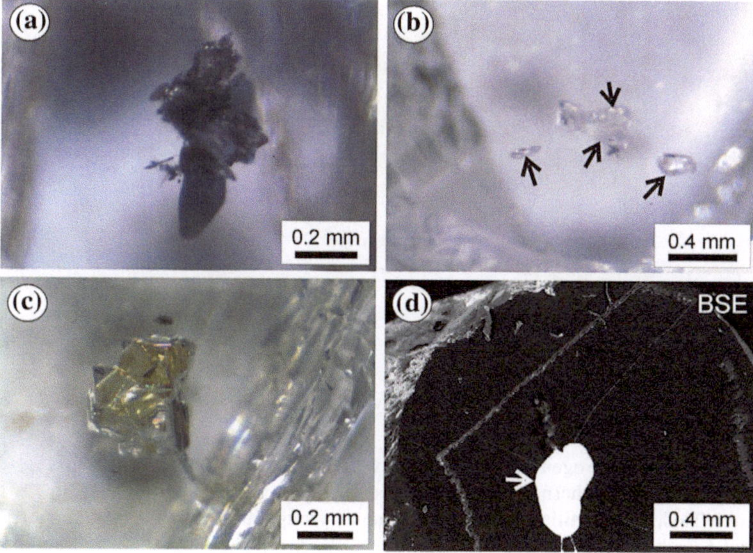

Fig. 7 Optical photographs (**a–c**) and back scattered electron image (**d**) showing mineral inclusions in southern Indian diamonds. **a** Black, flaky inclusions of graphite; **b** colourless inclusions of olivine; two euhedral crystals occur together at the centre; **c** orange, octahedral inclusions of garnet; **d** inclusion of Cr-spinel; note multiple fractures in the diamond. Stones **a**, **b** and **c** are from the TK4 kimberlite, and stone **d** is from P9 kimberlite

may be coesite. The reconnaissance study of the mineral inclusions suggests to the presence of both the main parageneses of diamonds (peridotitic and eclogitic) in all the three diamond sources.

Infrared Characteristics

In order to find specific features of the Wajrakarur diamonds, we performed a study of infrared spectra and the subsequent calculation of the nitrogen characteristics. A prominent feature of the southern Indian diamonds is the great variety of stones in respect of their nitrogen concentrations and nitrogen aggregation (Table 3). Diamonds from different sources (kimberlites, conglomerates and gravels) have broadly similar nitrogen contents. The diamonds belong to four major groups: Group 1 (low nitrogen, less than 50 at. ppm N_{tot}), Group 2a and Group 2b (medium nitrogen, 50–800 at. ppm N_{tot}), and Group 3 (high nitrogen, more than 800 at. ppm N_{tot}) of Kaminsky and Khachatryan (2001) (Fig. 8). Most diamonds fall into Groups 2a and 2b and are similar, in this respect, to diamonds from major Siberian, Arkhangelsk, Canadian (Kaminsky and Khachatryan 2001) and South African (Deines et al. 1991, 1997, 2009) deposits.

The proportion of aggregated nitrogen [$\%N_B = 100N_B/(N_A + N_B)$] in the diamonds also varies greatly, in the range of 0–85 %, that is, from Group 1 (highly aggregated nitrogen) through Group 2 (intermediate) to Group 3 (low-aggregated nitrogen) (Kaminsky and Khachatryan 2001) (Fig. 9). As a result, the diamonds belong to different types of nitrogen-based classification: IaAB, IaA and II. Table 3 shows that Type IaAB diamonds constitute the major proportion (58 %) of the WKF diamonds with N_{total} = 86–612 at. ppm (except one with 1,718 at. ppm), while the rest of the population comprises Type IaA (N_{total} = 76–342 at. ppm) diamonds (25 %) and Type II (N_{total} < 20 at. ppm) or near Type II (N_{total} <50 at. ppm) diamonds (17 %). The gravel diamonds are mostly of Type IaAB (9 out of 11) with N_{total} = 63–750 at. ppm, while the rest are one each of Type IaA (N_{total} = 220 at. ppm) and near Type II (N_{total} = 30 at. ppm). The conglomerate diamonds include two of Type IaAB (N_{total} = 61 and 496 at. ppm) and two of near Type II (N_{total} = 41 and 48 at. ppm).

The 'platelet' (planer defects several atoms thick; Woods 1986) concentrations in arbitrary units (cm^{-1}) were calculated as a function of the absorption coefficient value measured at 1,365 cm^{-1}. Their values, like the nitrogen characteristics, vary in a wide range of 0–21.4 cm^{-1} (Fig. 9). In this respect, they are similar to diamonds from other localities in the world, only the presence of low-platelet–low-nitrogen-aggregated stones is unusual. In earlier studied diamonds from different areas, low-platelets diamonds are usually high nitrogen aggregated (Kaminsky and Khachatryan 2001). In contrast, 16 % of the WKF diamonds are low-platelet–low-nitrogen-aggregated (Fig. 9), and three of them (P6-339, P8-R11 and TK4-4146) have no platelet and zero nitrogen aggregation that was never observed in diamonds from other regions (Kaminsky and Khachatryan 2001).

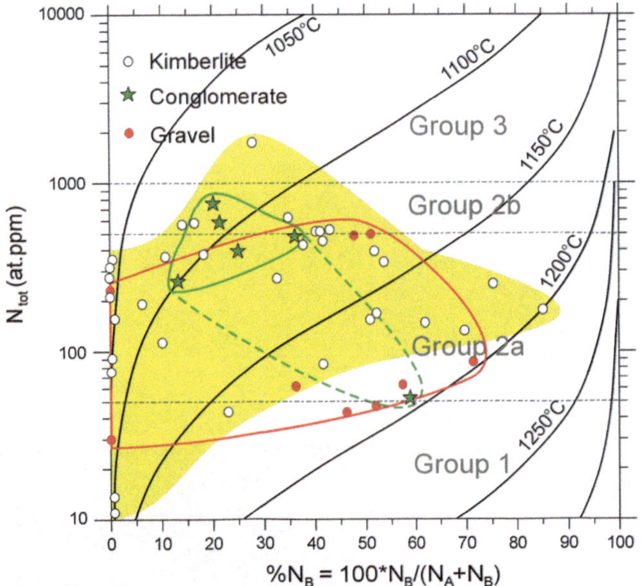

Fig. 8 Total nitrogen concentration versus aggregated nitrogen proportion in southern Indian diamonds. Isotherm curves for 3 Ga after Taylor and Milledge (1995). Diamond groups after Kaminsky and Khachatryan (2001)

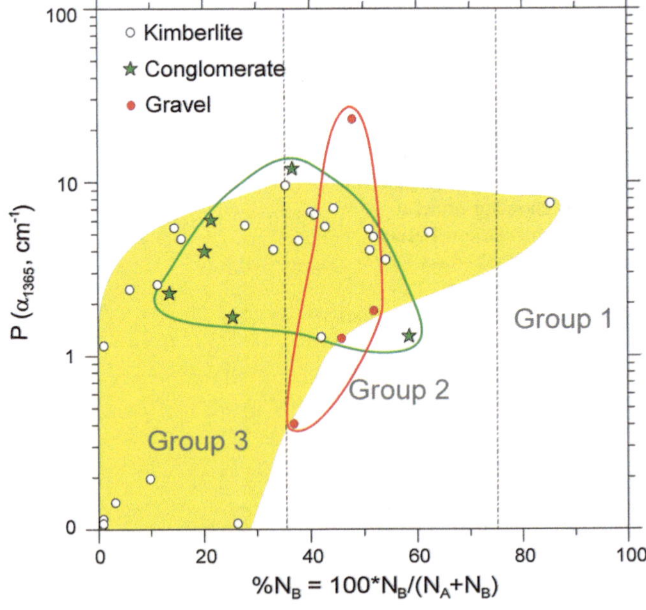

Fig. 9 Nitrogen platelet content (*P*) versus aggregated nitrogen proportion in southern Indian diamonds. Diamond groups after Kaminsky and Khachatryan (2001)

Discussion

The great variations in nitrogen as well as platelet characteristics indicate different thermal regimes for different stones, assuming the same duration of thermal action to all the studied diamonds. If the conventional mantle residence time of 3 Ga is considered, the temperature range of formation of these diamonds is 1,050–1,200 °C from the isotherms of Taylor and Milledge (1995) (Fig. 8). Different thermal regimes recorded by the diamonds may signify derivation of the diamonds from different depths in the mantle source region.

Of particular interest and practical importance is the distribution of nitrogen characteristics in diamonds from kimberlites, conglomerates and gravels. In both the plots of Figs. 8 and 9, they form overlapping halos. This may be considered as an indication of the common source for all the three groups. However, there are some diamonds from conglomerates and gravels which fall outside the kimberlite field in these plots. It is possible that these secondary diamonds are derived from hitherto undiscovered kimberlite pipes in the region. This conclusion is only a suggestion based on a limited number of the stones analysed. A final conclusion can be made only on a basis of statistically representative number of analyses (50–100 stones from each geological unit).

Acknowledgments This paper is a contribution to the IGCP-557. Thanks are due to Mrs. S. V. Netravali of the Department of Earth Sciences, IIT Bombay for help in SEM imaging. Dr. M. N. Gandhi and Mrs. P. Nikam of the SAIF, IIT Bombay, are thanked for help in FTIR analysis. Detailed and constructive reviews by Jeff Harris and an anonymous reviewer greatly improved the quality of the manuscript.

References

Babu TM (1998) Diamonds in India. Geological Society of India, Economic Geology Series Publication, Bangalore 331

Burhanuddin Md, Rao KRP (1991) Testing of Chigicherla kimberlite pipe, Anantapur district, Andhra Pradesh. Unpublished progress report of the Geological Survey of India for field season 1990–1991

Chalapathi Rao NV, Wu F, Mitchell RH, Li Q, Lehmann B (2012) Mesoproterozoic U–Pb ages, trace element and Sr–Nd isotopic composition of perovskite from kimberlites of the Eastern Dharwar craton, southern India: Distinct mantle sources and a widespread 1.1 Ga tectonomagmatic event. Chem Geol. doi:10.1016/j.chemgeo.2012.04.023

Chatterjee B, Haggerty SE, Beards AD, Smith CB, Townend R (2008) Kimberlite-carbonatite relationships revisited: evidence from Khaderpet pipe, Andhra Pradesh, India. In 9th international Kimberlite conference extended abstract 9IKC-A-00070

Deines P, Harris JW, Gurney JJ (1991) The carbon isotopic composition and nitrogen content of lithospheric and asthenospheric diamonds from Jagersfontein and Koffiefontein kimberlite, South Africa. Geochim Cosmochim Acta 55:2615–2625

Deines P, Harris JW, Gurney JJ (1997) Carbon isotope ratios, nitrogen content and aggregation state, and inclusion chemistry of diamonds from Jwaneng, Botswana. Geochim Cosmochim Acta 61:3993–4005

Deines P, Stachel T, Harris JW (2009) Systematic regional variations in diamond carbon isotopic composition and inclusion chemistry beneath the Orapa kimberlite cluster, in Botswana. Lithos 112S:776–784

Deshpande ML (1968) Report on the preliminary investigation for diamonds in the Banganapalli conglomerate and Tungabhadra–Krishna gravels in parts of Kurnool and Mahabubnagar districts, Andhra Pradesh. Unpublished progress report of the Geological Survey of India for field season 1967–1968

Deshpande ML (1970) A report on the diamond investigation in the Krishna gravels and reconnaissance work in Srisailam and Pennar river area, Krishna, Kurnool and Cuddapah districts, Andhra Pradesh. Unpublished progress report of the Geological Survey of India for field season 1969–1970

Deshpande ML, Sagar AK (1971) A report on the assessment of diamond resources, Krishna river gravels and reconnaissance work of Ustapalli and Malavalli areas of Krishna district, Andhra Pradesh. Unpublished progress report of the Geological Survey of India for field season 1970–1971

Deshpande ML, Sagar AK (1972) A report on the assessment of diamond resources, Krishna river gravels in Partiala block, Krishna district, Andhra Pradesh. Unpublished progress report of the Geological Survey of India for field season 1971–1972

Dutt NVBS (1953) Ancient diamond mining in Andhra Pradesh and its future. Indian Miner 7:138–150

Evans T, Qi Z, Maguire J (1981) The stages of nitrogen aggregation in diamond. J Phys C Solid State Phys 14:L379–L384

Govinda Rao M, Misra RC (1977) Investigation for diamond in Wajrakarur and Lattavaram areas, Anantapur district, Andhra Pradesh. Unpublished progress report of the Geological Survey of India for field seasons 1974–1975; 1975–1976 and 1976–1977

Harris JW, Hawthorne JB, Oosterveld MM, Wehmeyer E (1975) A classification scheme for diamond and a comparative study of South African diamond characteristics. Phys Chem Earth 9:165–783

Iyer LAN (1961) Indian precious stones. Geol Surv India Bull 18:105p

Kaminsky FV, Khachatryan GK (2001) Characteristics of nitrogen and other impurities in diamond, as revealed by infrared absorption data. Can Mineral 39:1733–1745

Krishnan MS (1950) Mineral resources of Madras. Geol Surv India Mem 80:99–105

Laiginhas FATP (2008) Diamonds from the Ural Mountains: their characteristics and the mineralogy and geochemistry of their inclusions. Ph.D. Thesis, University of Glasgow, United Kingdom

Mainkar D, Gupta T, Patel SC, Lehmann B, Diwan P, Kaminsky FV, Khachatryan GK (2013) Diamonds from the Behradih kimberlite pipe, Bastar Craton, India: a reconnaissance study. J Geol Soc India (this volume)

Mokievsky VA, Titova VM, Bartoshinsky ZV (1962) Manifestation of plastic deformation in diamond and some problems relating to the plasticity of crystals. Zap Vses Mineralog. Obshch Part 91, No. 4 (in Russian)

Munn L (1929) The Golconda diamond mines. J Hyderabad Geol Surv I(I):21–62

Nagaraja Rao BK, Rajurkar ST, Ramalingaswamy G, Ravindra Babu B (1987) Stratigraphy and structure and evolution of the Cuddapah basin. Geol Soc India Mem 6:33–86

Nayak SS (1992) Report on exploration for Diamond in Kolluru area, Guntur district, Andhra Pradesh. Unpublished progress report of the Geological Survey of India for field seasons 1986–1987, 1987–1988, 1988–1989, 1989–1990, 1990–1991 and 1991–1992

Nayak SS, Dhakate MV (2002) Report on delineation and assessment of the diamondiferous nature of the pipe P-2, Wajrakarur, Anantapur district, Andhra Pradesh. Unpublished progress report of the Geological Survey of India for field seasons 2000–2001 and 2001–2002

Nayak SS, Rao KRP, Suresh G, Dhakate MV (1999) Report on delineation and assessment of the diamondiferous nature of Chintalampalle (P-12) and Gollapalle (CC-4 and CC-5) kimberlite pipes, Anantapur district, Andhra Pradesh. Unpublished progress report of the Geological Survey of India for field seasons 1997–1998 and 1998–1999

Neelakantam S (2001) Exploration for diamond in southern India. Geol Surv India Spl Pub 58:521–555

Rajaraman S (1969) A report on detailed investigation for diamonds in pipe P-1, Wajrakarur, Anantapur district, Andhra Pradesh. Unpublished progress report of the Geological Survey of India for field season 1968–1969

Rao KRP, Dhakate MV, Srinivasa Chowdary V, Bhaskara Rao KS, Satyanarayana G, Nayak SS, Reddy TAK (1997) Assessment of the diamond potential of Anumpalle and Dibbsanipalli kimberlites, Anantapur district, Andhra Pradesh. Unpublished progress report of the Geological Survey of India for field season 1996–1997

Robinson DN (1979) Surface textures and other features of diamonds. Ph.D. Thesis, University of Cape Town, South Africa

Robinson DN, Scott JA, Van Nieker A, Anderson VG (1989) The sequence of events reflected in the diamonds of some Southern African kimberlites. Geol Soc Aust Spl Pub 14(2):990–1000

Satyanarayana SV (2000) Diamonds in the Deccan: an overview. In: Gupta HK, Parasher-Sen A, Balasubramanian D (eds) Deccan heritage. Universities Press (India) Ltd., Hyderabad, pp 135–156

Sivaji K, Satyanarayana SV (1990) Preliminary surveys for diamond in the Banganapalle conglomerates of Balapuram–Rangapuram area, Kurnool district, Andhra Pradesh. Unpublished progress report of the Geological Survey of India for field season 1989–1990

Sivaji K, Rao AK, Satyanarayana SV (1989) Assessment of diamond resources in the Banganapalle conglomerates of Tammarajupalle–Undutla area, Kurnool district, Andhra Pradesh. Unpublished progress report of the Geological Survey of India for field season 1988–1989

Sivaji K, Singa Raju V, Rao AK (1992) Preliminary assessment of gravels for diamonds in the Sagileru river basin in Prakasam and Cuddapah districts, Andhra Pradesh. Unpublished progress report of the Geological Survey of India for field seasons 1988–1989, 1989–1990 and 1991–1992

Sridhar M, Sinha KK (2005) Testing the diamondiferous nature of the newly discovered Timmasamudram kimberlites and Ramadugu lamproites. Unpublished progress report of the Geological Survey of India for field season 2004–2005

Srinivas Choudary V, Rau TK, Bhaskara Rao KS, Sridhar M, Sinha KK (2007) Timmasamudram kimberlite cluster, Wajrakarur Kimberlite Field, Anantapur district, Andhra Pradesh. J Geol Soc India 69:597–609

Subba Rao MV (1988) Assessment of diamond resources in kimberlite Pipe 8, Lattavaram area, Anantapur district, Andhra Pradesh. Unpublished progress report of the Geological Survey of India for field season 1987–1988

Taylor WR, Milledge HJ (1995) Nitrogen aggregation character, thermal history and stable isotope composition of some xenolith-derived diamonds from Roberts Victor and Finch. In: 6th international kimberlite conference extended abstract, Novosibirsk, 1995, 620–622

Vance ER, Harris JW, Milledge HJ (1973) Possible origins of alpha-particle damage in diamonds from kimberlite and alluvial sources. Mineral Mag 39:349–360

Wiens R, Lal D, Craig H (1990) Helium and carbon isotopes in Indian diamonds. Geochim Cosmochim Acta 54:2587–2591

Woods GS (1986) "Platelets" and the infrared absorption of Type Ia diamonds. Proc R Soc, Lond, A 407:219–238

Author Index

D. G. Pearson et al. (eds.), *Proceedings of 10th International Kimberlite Conference*, Volume 1, Special Issue of the Journal of the Geological Society of India, DOI: 10.1007/978-81-322-1170-9, © Geological Society of India 2013